BIOLOGY AND CONSERVATION OF THE MONARCH BUTTERFLY

On 15 February 1987 the smoke of a nearby fire elicited this panic response from perhaps a million aggregated monarchs overwintering on Sierra Chincua in central Mexico. The response began suddenly and lasted for about 30 minutes, with approximately one-half the colony taking flight above the forest canopy. The remaining butterflies plummeted to the forest floor. Photograph by John I. Glendinning.

BIOLOGY AND CONSERVATION OF THE MONARCH BUTTERFLY

EDITED BY
STEPHEN B. MALCOLM
AND
MYRON P. ZALUCKI

NO. 38
SCIENCE SERIES
NATURAL HISTORY MUSEUM
OF LOS ANGELES COUNTY

Natural History Museum of Los Angeles County
Los Angeles, California 90007
ISSN 0079-0943
Published on 18 February 1993
Printed in the United States of America

CONTENTS

PREFACE ix

ACKNOWLEDGMENTS xi

1: INTRODUCTION 1

The Monarch Butterfly: Research and Conservation, *by Stephen B. Malcolm and Myron P. Zalucki* 3

2: SYSTEMATICS 9

Systematic Perspectives on the Evolution of the Monarch Butterfly, *by Ian J. Kitching, Phillip R. Ackery, and Richard I. Vane-Wright* 11

3: CHEMICAL COMMUNICATION 17

Danaine Butterflies: A Didactic Story about Chemical Ecology, *by Dietrich Schneider* 19

The American Monarch: Courtship and Chemical Communication of a Peculiar Danaine Butterfly, *by Michael Boppré* 29

Pheromone-Sensitive Receptor Neurons in *Danaus gilippus, by Rolf Kittmann and Dietrich Schneider* 43

4: MATING BEHAVIOR 47

Non-random Mating in Monarch Butterflies Overwintering in Mexico, *by Tonya Van Hook* 49

Is Multiple Mating an Adaptive Feature of Monarch Butterfly Winter Aggregation?, *by Harrington Wells, Patrick H. Wells, and Steffen H. Rogers* 61

Sex around the Milkweed Patch—The Significance of Patches of Host Plants in Monarch Reproduction, *by Myron P. Zalucki* 69

5: HOST PLANT USE, CARDENOLIDE SEQUESTRATION, AND DEFENSE AGAINST NATURAL ENEMIES 77

Host Plant Exploitation by Aposematic Nymphalid Butterflies: Pre-adaptation versus Stepwise Sequestration, *by Phillip R. Ackery* 79

A Model for Cardenolide and Cardenolide Glycoside Storage by the Monarch Butterfly, *by Carolyn J. Nelson* 83

Sequestration and Storage of Cardenolides and Cardenolide Glycosides by *Danaus plexippus plexippus* and *D. chrysippus petilia* When Reared on *Asclepias fruticosa*: With a Review of Some Factors that Influence Sequestration, *by Carolyn J. Nelson* 91

Milkweed Host Plant Utilization and Cardenolide Sequestration by Monarch Butterflies in Louisiana and Texas, *by Steven P. Lynch and Ronald A. Martin* 107

Rejection of *Apocynum androsaemifolium* and *A. sibiricum* (Apocynaceae) as Food Plants by Larvae of *Danaus plexippus*: Refutation of Early Accounts, *by Susan Sullivan Borkin* 125

A Reassessment of the Mimicry Relationship among Viceroys, Queens, and Monarchs in Florida, *by David B. Ritland and Lincoln P. Brower* 129

6: PHYSIOLOGICAL ECOLOGY AND THE ANNUAL CYCLE 141

Endocrinology of the Monarch Butterfly, *by William S. Herman* 143

Temperature and Thermoregulation in the Monarch Butterfly, *by Alan R. Masters* 147

Cold-Hardiness in the Annual Cycle of the Monarch Butterfly, *by James B. Anderson and Lincoln P. Brower* 157

The Comparative Biology and Ecology of the Australian Danaines, *by R. L. Kitching and E. Scheermeyer* 165

7: MIGRATION 177

The Columbus Hypothesis: An Explanation for the Dramatic 19th Century Range Expansion of the Monarch Butterfly, *by Richard I. Vane-Wright* 179

Migration Biology of the Monarch Butterfly in Australia, *by David G. James* 189

Seasonal Migration of Monarchs and Other Butterflies in Costa Rica, *by William A. Haber* 201

Do California Monarchs Undergo Long-Distance Directed Migration?, *by Adrian M. Wenner and Ann M. Harris* 209

Spring Migration of Monarch Butterflies in California, *by Christopher D. Nagano, Walter H. Sakai, Stephen B. Malcolm, Barbara J. Cockrell, Julian P. Donahue, and Lincoln P. Brower* 219

Time, Temperature, and Latitudinal Constraints on the Annual Recolonization of Eastern North America by the Monarch Butterfly, *by Barbara J. Cockrell, Stephen B. Malcolm, and Lincoln P. Brower* 233

Spring Recolonization of Eastern North America by the Monarch Butterfly: Successive Brood or Single Sweep Migration?, *by Stephen B. Malcolm, Barbara J. Cockrell, and Lincoln P. Brower* 253

Spring Migration and Oviposition of the Monarch Butterfly in Louisiana, *by Thomas J. Riley* 269

Orientation of Autumn Migration in the Monarch Butterfly, *by Klaus Schmidt-Koenig* 275

8: OVERWINTERING BIOLOGY 285

Vegetation Types of Monarch Butterfly Overwintering Habitat in Mexico, *by José Carmen Soto Núñez and Leonila Vázquez García* 287

Investigations of Possible Monarch Butterfly Overwintering Areas in Central and Southeastern Mexico, *by Javier de la Maza E. and William H. Calvert* 295

Comparative Phenology of Variation in Size, Weight, and Water Content of Eastern North American Monarch Butterflies at Five Overwintering Sites in Mexico, *by William H. Calvert and Robert O. Lawton* 299

The Influence of Temperature on Crawling, Shivering, and Flying in Overwintering Monarch Butterflies in Mexico, *by Alfonso Alonso-M., John I. Glendinning, and Lincoln P. Brower* 309

Interspecific Comparisons of the Foraging Dynamics of Black-Backed Orioles and Black-Headed Grosbeaks on Overwintering Monarch Butterflies in Mexico, *by Alfredo Arellano G., John I. Glendinning, James B. Anderson, and Lincoln P. Brower* 315

Comparative Feeding Responses of the Mice *Peromyscus melanotis, P. aztecus, Reithrodontomys sumichrasti,* and *Microtus mexicanus* to Overwintering Monarch Butterflies in Mexico, *by John I. Glendinning* 323

Overwintering Monarch Butterflies in California: Past and Present, *by John Lane* 335

Overwintering of Three Australian Danaines: *Tirumala hamata hamata, Euploea tulliolus tulliolus,* and *E. core corinna, by Elly Scheermeyer* 345

9: CONSERVATION 355
Conservation of Monarch Butterfly Migration in North America: An Endangered Phenomenon, *by Stephen B. Malcolm* 357
Conservation of the Monarch Butterfly Reserves in Mexico: Focus on the Forest, *by Laura C. Snook* 363
Conservation Actions Taken by Monarca, A.C., to Protect the Overwintering Sites of the Monarch Butterfly in Mexico, *by Rodolfo Ogarrio* 377
Monarch Conservation in Mexico: The Challenge of Membership and Fund-raising for Monarca, A.C., *by Carlos F. Gottfried Joy* 379
Education to Promote Protection of the Monarch Butterfly in Mexico: A Discussion of the Conservation Education Program Initiated by Monarca, A.C., *by Fernando Ortíz Monasterio* 383
Operative Programs in the Monarca, A.C., Project, *by María Elena Camus de Castro* 385
The Importance of Alternative Sources of Income to "Ejidatarios" (Local Residents) for Conservation of Overwintering Areas of the Monarch Butterfly, *by Susana Rojas González de Castilla* 389
The Monarch Project: A Program of Practical Conservation in California, *by Melody Mackey Allen and Katrin B. Snow* 393
10: CONCLUSIONS 395
Biology and Conservation of the Monarch Butterfly: Concluding Remarks, *by Stephen B. Malcolm and Myron P. Zalucki* 397
CONTRIBUTORS 403
INDEX 407

PREFACE

This volume brings together a diverse array of papers on many aspects of the biology and conservation of the monarch butterfly, *Danaus plexippus* (L.). The biology of monarchs has attracted much interest almost everywhere that this large and conspicuous butterfly occurs: from both North and South America and the Caribbean to Australia, New Zealand, the oceanic islands of the Pacific, Mauritius, the Canary Islands of the Atlantic, and, most recently, western Europe. Interest has not simply focused on the most remarkable aspects of the monarch's life history, such as migration and defenses against natural enemies, but runs to the full gamut of research into the biochemistry, molecular biology, genetics, physiology, behavior, ecology, and evolution of this mobile butterfly.

The title of this volume, *Biology and Conservation of the Monarch Butterfly,* reflects a strongly multidisciplinary approach to understanding the many concepts that this remarkable butterfly exemplifies. We feel that the full breadth of the word "biology" is needed to describe the multifaceted contents of this volume and the use of much of this painstakingly gathered information to help solve problems of species and habitat conservation.

Such diversity of interest in the monarch prompted two recent international meetings to compare and develop understanding of the biology and conservation of the monarch. The first, *Symposium on the Biology and Conservation of Monarch Butterflies,* was dubbed "Moncon-1" and occurred at Cocoyoc, Morelos, Mexico, in August 1981. The dominant theme of this select meeting was the urgent need for objective information to provide a sound footing for conservation efforts in Mexico. Almost the entire North American population of monarchs east of the Rocky Mountains coalesces each autumn to spend the winter in a few tiny areas west of Mexico City. Inevitably, such a marked coalescence makes overwintering monarchs extremely vulnerable to habitat destruction, and so this meeting helped develop a scientific rationale to protect monarch overwintering sites in the high-altitude oyamel fir forests of central Mexico.

The *Second International Conference on the Monarch Butterfly* was, not surprisingly, called "Moncon-2" and met at the Natural History Museum of Los Angeles County, in Los Angeles, California, U.S.A., during 2–5 September 1986. Moncon-2 honored the half-century of research on the monarch by Fred A. Urquhart, of the University of Toronto in Canada, which culminated with the comparatively recent discovery in 1975 of the millions of monarchs that spend each winter in central Mexico. In contrast to Moncon-1, Moncon-2 was located near the heart of the overwintering range of the second, but smaller, population of monarchs in North America, that which occurs west of the Rocky Mountains and overwinters at many sites along the coast of California. This conference was considerably larger than the first and enjoyed a strong research presentation, participants from seven nations and four continents, and the announcement of significant conservation initiatives by the government of Mexico. Julian P. Donahue of the Natural History Museum of Los Angeles County was the host, prime motivator, and program chairman of the meeting. Without his help, enthusiasm, and encouragement this volume would not exist.

The collection of refereed papers that makes up this volume reflects the diversity of the second monarch meeting and is an up-to-date product of the interest and synergism generated by the meeting. We think that the papers demonstrate the enthusiasm, skill, and concern with which biologists face the problems of understanding monarch biology and the use of such understanding to ensure that this remarkable butterfly does not fly, like the passenger pigeon, into oblivion.

Stephen B. Malcolm
Kalamazoo, Michigan

Myron P. Zalucki
St. Lucia, Queensland

July 1992

ACKNOWLEDGMENTS

The editors are very grateful to Julian P. Donahue of the Natural History Museum of Los Angeles County for initiating this collection of papers on the monarch butterfly through his organization of the Moncon-2 meeting in Los Angeles and his wish that such a collection be published. Lincoln Brower has provided constant encouragement throughout the process of cajoling authors into production and revision, and much of the research described in this collection is a direct product of his enthusiasm for monarch research and conservation. Grace Kiltie word-processed revised manuscripts with great efficiency and good humor and Daryl Harrison redrew and photographed many of the figures with skill and patience. Many others helped with references, Spanish–English translations, systematic names, chemistry, facilities, the index, and various conceptual questions, including: Alfredo Arellano, Mick Crawley, Gary Dodson, Julian Donahue, Clementina Equiha, Kathy Ewel, Jim Fordyce, Carolyn Nelson, Jack Putz, Jonathan Pratt, and James Seiber. We are especially grateful to the following persons for their reviews and comments on the papers in this volume: Alfonso Alonso-M., Jim Anderson, Paul Bailey, Lincoln Brower, C. M. Bull, Bill Calvert, Barbara Cockrell, Ian Common, Steve Courtney, Rex Dadd, W. Danthaṇaraya, John Edgar, Thomas Emmel, David Evans, Luke George, Peter Gregg, Dinah Hales, Larry Harris, Jon Hastings, Chris Hill, Jane Hughes, David James, R. E. Jones, Roger Kitching, David Lambert, Larry Marshall, Rachel McFadyen, Dave Merritt, Andy Moore, Carolyn Nelson, Austin Platt, James Rawls, Martin Rice, Miriam Rothschild, Ronald Rutowski, D. P. A. Sands, Elly Scheermeyer, Frank Slansky, C. N. Smithers, Bea Van Horne, Dick Vane-Wright, Thomas Walker, G. H. Walter, and Ward Watt.

Finally, the financial sponsorship of the Moncon-2 conference by The David and Lucile Packard Foundation, World Wildlife Fund–US, Chevron U.S.A. Inc., BioQuip Products, Inc., the Emilio Guglielmo Winery, and the Foundation of the Natural History Museum of Los Angeles County is gratefully acknowledged.

1

INTRODUCTION

The bright orange and black warning coloration of the monarch butterfly and its habit of feeding as larvae only on poisonous milkweed plants prompted E. B. Poulton to call in 1914 for chemists to show whether monarchs use toxic host plant chemicals in their defenses against predators. More than 50 years later, Lincoln Brower described such collaboration with chemists and ecologists in Europe and North America in his 1969 article, "Ecological Chemistry." This illustration from that article shows the blue jay (*Cyanocitta cristata*) as a predator confronted by toxic and conspicuously colored queen and monarch butterflies. The butterflies are defended against the predator by both mimicry and sequestered host plant toxins. Painting by Rudolf Freund.

THE MONARCH BUTTERFLY: RESEARCH AND CONSERVATION

Stephen B. Malcolm and Myron P. Zalucki

Many biologists seek model systems with which to test their hypotheses and develop their understanding of evolution, ecology, behavior, physiology, genetics, and cellular and molecular biology. Some organisms are now so well known that they have become almost synonymous with a particular research activity. For example, the fruit fly, *Drosophila melanogaster,* is an almost ubiquitous model that has facilitated understanding of evolutionary genetics and recent multidisciplinary approaches to developmental biology and neurobiology (Rubin, 1988). Likewise, some plants, retroviruses, bacteria, yeasts, nematodes, frogs, rats, and primates have been used as models to develop important and influential concepts, especially in mechanistic or reductionist approaches to understanding biology—especially molecular biology (Botstein and Fink, 1988; Dawid and Sargent, 1988; Goldberg, 1988; Jaenisch, 1988; Kenyon, 1988; King et al., 1988; Magasanik, 1988; Varmus, 1988; White and Caskey, 1988). Although the advances that these species have promoted are significant and important, the organisms described have been used to develop highly reductionist approaches to biology.

In contrast, cockroaches, locusts, rats, and pigeons are also well-known models that have been used extensively to develop more holistic ideas in behavior and physiology that owe much to evolutionary and ecological interpretations. Although the search for model organisms may have been determined subjectively by personal preference, or objectively by the fulfillment of rational criteria, very few organisms are used more universally as models to investigate both mechanistic and interactive, holistic approaches to biological science.

Insects, particularly the Lepidoptera, are good candidates for wide use in research because they usually have short and tractable life histories that are often very well known. Immature and adult stages are easily sampled, and all aspects of their biology are usually readily observable and can be manipulated in both the laboratory and field. These factors combine to facilitate the task of understanding the variability of an insect's responses to its biotic and abiotic environment, which in turn makes it comparatively simple to place reductionist, mechanistic science into broader ecological or evolutionary perspectives.

Among insects, the monarch butterfly, *Danaus plexippus* (L.), is a large and conspicuous species with a comparatively short life history, ranging from weeks to months. As described in this volume, the monarch has been used as a model in many aspects of biology, from systematics through molecular biology, biochemistry, physiology, and behavior to ecology and evolution. As yet, the monarch has attracted little interest from molecular biologists, but, although this volume does not consider the genetics of monarchs, the population genetics of the butterfly have been considered as a model of genetic structure in highly dispersive species by Eanes (1978, 1979) and Eanes and Koehn (1978) in North America and by Hughes and Zalucki (1984) and Zalucki et al. (1987) in Australia. In addition, the monarch was used as a model for investigations into the interaction among weather, activity times, and genotype (Carter et al., 1989).

We think this volume shows that the monarch is an effective and tractable model for testing and developing a wide range of ideas in all aspects of biology and helps us to integrate a strongly multidisciplinary approach to understanding biology in general.

The monarch is a specialist insect herbivore that has expanded its range from the American tropics to include both tropical and temperate habitats worldwide (Young, 1982; Ackery and Vane-Wright, 1984; Strong et al., 1984). The papers by Ackery, Kitching et al., and Vane-Wright, in this volume, each expand on this theme. Such an evolutionary shift in range and ecology provides us with excellent opportunities to test ideas about large- and small-scale interactions between plants and animals in space and time, interactions between prey defenses and predator foraging, interactions among environmental variation, behavior, and physiology, and the evolutionary implications of intraspecific interactions. The evolutionary shift of the monarch from the primarily tropical range of all other species in the nymphalid family Danaidae to a temperate range provides many opportunities for controlled tropical–temperate comparisons.

Two prominent aspects of monarch biology attracted early interest in this species. The first is the conspicuous warning, or aposematic coloration of both larvae and adults, that advertises their distastefulness derived from the toxic steroids that the larvae sequester from their milkweed host plants. The second is the equally conspicuous seasonal migration of adult monarchs in relation to the spatial and temporal distribution of the North American milkweed flora.

The use of plant-derived toxins by this brightly colored insect as a defense against predators was suspected by early naturalists and observers. Back in 1914, Poulton called for North American chemists to apply their skills to determine whether or not plant-derived chemicals made the monarch toxic to birds. Unfortunately his plea was ignored, although in 1932 Jones showed that extracts of the milkweed host plants of monarch larvae repelled ants. Despite much circumstantial evidence, the rejection of monarchs by bird predators was not clearly demonstrated until the elegant series of experiments by J.V.Z. Brower in 1958.

It was not until the 1960s that a Swiss chemist, the Nobel laureate Tadeus Reichstein, answered Poulton's plea to show conclusively that monarch larvae sequestered cardiac-active steroids, or cardenolides, from their milkweed host plants (Reichstein, 1967). Reichstein worked collaboratively with both J.V.Z. and L.P. Brower in the United States and Parsons and Rothschild in England, and together they showed that plant-derived cardenolides in monarchs are a very effective defense against bird predators (Parsons, 1965; Brower et al., 1967, 1968; Reichstein, 1967; Reichstein et al., 1968). Monarch cardenolides are bitter tasting and emetic if ingested by birds, or other vertebrate predators, and provide protection to monarchs and other similarly colored, mimetic butterflies (Ritland and Brower, this volume). Lincoln Brower's research went on to give a major impetus to the developing field of chemical ecology (Brower, 1969, 1984; Brower and Glazier, 1975; Malcolm, 1987a; Harborne, 1988; Boppré, this volume; Schneider, this volume). The monarch, and closely related species in the family Danaidae, have now been studied extensively by researchers around the world interested in interactions among two and three trophic levels. For example, chapter 5 of this volume comprises six papers on such interactions, including host plant exploitation (Ackery; Borkin), cardenolide sequestration (Lynch and Martin; two papers by Nelson), and a reassessment of the classical mimicry example among monarchs, queens, and viceroys (Ritland and Brower). In addition, Zalucki considers the influence of host plant patches on monarch reproduction in chapter 4.

Of course, it is highly unlikely that cardenolides alone determine the nature of interactions among monarchs, milkweeds, and natural enemies. Other secondary chemicals have recently been implicated in such interactions. For example, in chapter 3, both Schneider and Boppré discuss the role of plant-derived pyrrolizidine alkaloids (PAs) in the biology of monarchs and other danaids. They describe the role of nonmilkweed PAs as pheromones and compare the pheromonal and possible defensive functions of PAs with the use of cardenolides. In the same chapter, Kittmann and Schneider describe the pheromone-sensitive receptors of another danaid butterfly. Both Schneider and Boppré also mention the recent discovery of Rothschild et al. (1984)—that monarchs contain pyrazines, which appear to be intrinsically synthesized and act as a warning odor to repel predators (Guilford et al., 1987).

Much of the understanding of monarch migration has come from the pioneering research of Fred Urquhart, working from the University of Toronto in Canada. Following Williams (1930), he recognized early that the monarch is an ideal model for studies of insect migration, and since 1937, he and his wife Norah, along with several thousand enthusiastic collaborators spread across North America, have regularly tagged and released monarchs to try and describe their patterns of migration (Urquhart, 1960, 1966). Urquhart's work reached its apogee in the winter of 1974–1975 when his associates, Ken and Cathy Brugger, found the extraordinary overwintering aggregations of monarch butterflies in the high-altitude oyamel fir forests, west of Mexico City, in Mexico's transvolcanic range of mountains (Urquhart, 1976, 1987; Urquhart and Urquhart, 1976a, 1977, 1978, 1979). Each winter, hundreds of millions of monarchs coalesce from a continent-wide distribution to overwinter at approximately 10 small sites that range in area from 0.03 to 3.34 hectares (Calvert and Brower, 1986; Calvert et al., 1989). The discovery of these sites has led to a steady stream of research on monarch butterflies as biologists realize their value as testable ecological, evolutionary, behavioral, and physiological models (Urquhart and Urquhart, 1976a,b, 1977, 1978, 1979; Brower, 1977, 1985; Brower et al., 1977, 1985; Calvert et al., 1979, 1982, 1983, 1984, 1986; Calvert and Brower, 1981, 1986; Fink and Brower, 1981; Calvert and Cohen, 1983; Fink et al., 1983; Brower and Calvert, 1985; Brower and Fink, 1985; Herman, 1985; Seiber et al., 1986; Malcolm, 1987b; Glendinning et al., 1988;

Masters et al., 1988; Herman et al., 1989; Malcolm and Brower, 1989).

This volume contributes further to our understanding of monarch overwintering biology with eight papers on the subject, including the vegetation and distribution of overwintering habitat in Mexico (de la Maza E. and Calvert; Soto Núñez and Vázquez García), overwintering physiology (Alonso-M. et al.; Calvert and Lawton), predation by birds and mice in overwintering colonies (Arellano G. et al.; Glendinning), the history and natural history of overwintering in California (Lane), and overwintering by three additional danaine butterflies in Australia (Scheermeyer). In addition, the four papers in chapter 6, on the physiological ecology of monarchs, each include information on overwintering biology, and both Van Hook and Wells et al. consider the mating behavior and reproductive physiology of overwintering monarchs in chapter 4.

Equally, the discovery of the overwintering sites focused efforts to describe both the autumn and spring migrations to and from the high-altitude locations in Mexico. Thus in this volume, Schmidt-Koenig describes the flight orientation of autumn migrant monarchs to Mexican overwintering sites, and Nagano et al. describe the orientation away from overwintering sites in California. Three papers discuss patterns of spring migration in the eastern United States relative to host plant distribution and exploitation (Cockrell et al.; Malcolm et al.; Riley), and Wenner and Harris use similar methods to ask the provocative question of whether or not monarchs migrate at all in California. Researchers of monarch movement in relation to milkweed distribution are fortunate to have available Woodson's (1954) exhaustive monograph on the genus *Asclepias.* This genus includes all of the primary larval host plant species of monarchs in North America. Chapter 7 on migration is completed by Haber, who describes seasonal butterfly migration in Costa Rica, and James, who gives an overview of monarch migration in Australia.

Four papers in this volume by Lynch and Martin, Malcolm et al., and two by Nelson, develop a chemical fingerprinting technique (Roeske et al., 1976; Brower et al., 1982, 1984a,b; Lynch and Martin, 1987; Martin and Lynch, 1988; Malcolm et al., 1989) that links the two main themes of interest in monarchs—namely, host-plant-derived cardenolides and migration. In particular, Malcolm et al. show how specific, milkweed-derived cardenolide fingerprints in monarchs can be used to describe patterns of migration in relation to host plant distribution.

Although the monarch butterfly is a common insect in North and South America, Australia, New Zealand, and islands of the Pacific and Caribbean (Vane-Wright, this volume), its extraordinary migration in temperate environments makes it vulnerable to habitat perturbations. The annual autumn migration of monarchs in North America, both east and west of the Rocky Mountains, in South America, and in southeastern Australia, to tightly aggregated overwintering sites, makes this phase of the monarch's life history especially vulnerable. Consequently, conservation of these highly vulnerable overwintering aggregations is crucial to ensure the survival of the spectacular, birdlike migrations (Wells et al., 1983; Calvert et al., 1989; Brower and Malcolm, 1991). Thus, the International Union for Conservation of Nature and Natural Resources has given monarch migration in North America the designation of a "threatened phenomenon," in which "aggregates or populations of organisms that together constitute major biological phenomena, [are] endangered as phenomena but not as taxa" (Wells et al., 1983; Brower and Malcolm, 1991).

This theme is developed in chapter 9, with eight papers that describe many of the ecological, political, economic, and sociological problems associated with conserving monarch overwintering habitat in Mexico and California. Chapter 9 is prefaced by an overview of monarch conservation (Malcolm) that offers a framework for the more specific accounts of the following seven descriptions of the various ways that are being implemented to protect monarchs.

Because the milkweed food plants of monarch larvae are no longer grown commercially, monarchs do not suffer from the status of being a "pest insect." Thus, although the monarch has little direct relevance to applied agricultural and forestry problems, the papers published in this volume show that the monarch is a valuable model for research in biology, including conservation biology. Like any good model, its value is determined primarily by the generation of both understanding and new ideas. As discussed in the concluding chapter of this volume, the papers herein show that the monarch is indeed advancing our understanding of concepts in biology and that many new questions are developing hand-in-hand with current advances.

LITERATURE CITED

Ackery, P.R., and R.I. Vane-Wright. 1984. *Milkweed butterflies: Their cladistics and biology.* London: British Museum (Natural History); and Ithaca, NY: Cornell University Press, 425 pp.

Botstein, D., and G.R. Fink. 1988. Yeast: An experimental organism for modern biology. *Science* 240: 1439–1443.

Brower, J.V.Z. 1958. Experimental studies of mimicry in some North American butterflies. Part I. The monarch, *Danaus plexippus,* and viceroy, *Limenitis archippus archippus. Evolution* 12(1):32–47.

Brower, L.P. 1969. Ecological chemistry. *Scientific American* 220(2):22–29.

Brower, L.P. 1977. Monarch migration. *Natural History* 86:40–53.

Brower, L.P. 1984. Chemical defence in butterflies. In *The Biology of Butterflies,* eds. R. I. Vane-Wright and P. R. Ackery, *Symposia of the Royal Ento-*

mological Society of London 11:109–134. London and New York: Academic Press (1989, Princeton, NJ: Princeton University Press).

Brower, L.P. 1985. New perspectives on the migration biology of the monarch butterfly, *Danaus plexippus* L. In *Migration: Mechanisms and adaptive significance,* ed. M.A. Rankin, 748–785. Austin: University of Texas, *Contributions in Marine Science,* Supplement, 27.

Brower, L.P., and W.H. Calvert. 1985. Foraging dynamics of bird predators on overwintering monarch butterflies in Mexico. *Evolution* 39:852–868.

Brower, L.P., and L.S. Fink. 1985. A natural toxic defense system: Cardenolides in butterflies versus birds. *Annals of the New York Academy of Sciences* 443: 171–188.

Brower, L.P., and S.C. Glazier. 1975. Localization of heart poisons in the monarch butterfly. *Science* 188: 19–25.

Brower, L.P., and S.B. Malcolm. 1991. Animal migrations: Endangered phenomena. *American Zoologist* 31:265–276.

Brower, L.P., J.V.Z. Brower, and J.M. Corvino. 1967. Plant poisons in a terrestrial food chain. *Proceedings of the National Academy of Sciences* 57:893–898.

Brower, L.P., W.N. Ryerson, L.L. Coppinger, and S.C. Glazier. 1968. Ecological chemistry and the palatability spectrum. *Science* 161:1349–1351.

Brower, L.P., W.H. Calvert, L.E. Hedrick, and J. Christian. 1977. Biological observations on an overwintering colony of monarch butterflies (*Danaus plexippus,* Danaidae) in Mexico. *Journal of the Lepidopterists' Society* 31:232–242.

Brower, L.P., J.N. Seiber, C.J. Nelson, S.P. Lynch, and P.M. Tuskes. 1982. Plant-determined variation in the cardenolide content, thin-layer chromatography profiles, and emetic potency of monarch butterflies, *Danaus plexippus* reared on the milkweed, *Asclepias eriocarpa* in California. *Journal of Chemical Ecology* 8:579–633.

Brower, L.P., J.N. Seiber, C.J. Nelson, S.P. Lynch, and M.M. Holland. 1984a. Plant-determined variation in the cardenolide content, thin-layer chromatography profiles, and emetic potency of monarch butterflies, *Danaus plexippus* L. reared on milkweed plants in California: 2. *Asclepias speciosa. Journal of Chemical Ecology* 10:601–639.

Brower, L.P., J.N. Seiber, C.J. Nelson, S.P. Lynch, M.P. Hoggard, and J.A. Cohen. 1984b. Plant-determined variation in cardenolide content and thin-layer chromatography profiles of monarch butterflies, *Danaus plexippus* reared on milkweed plants in California 3. *Asclepias californica. Journal of Chemical Ecology* 10:1823–1857.

Brower, L.P., B.E. Horner, M.A. Marty, C.M. Moffitt, and B. Villa-R. 1985. Mice (*Peromyscus maniculatus, P. spicilegus,* and *Microtus mexicanus*) as predators of overwintering monarch butterflies (*Danaus plexippus*) in Mexico. *Biotropica* 17:89–99.

Calvert, W.H., and L.P. Brower. 1981. The importance of forest cover for the survival of overwintering monarch butterflies (*Danaus plexippus,* Danaidae). *Journal of the Lepidopterists' Society* 35:216–225.

Calvert, W.H., and L.P. Brower. 1986. The location of monarch butterfly (*Danaus plexippus* L.) overwintering colonies in Mexico in relation to topography and climate. *Journal of the Lepidopterists' Society* 40:164–187.

Calvert, W.H., and J.A. Cohen. 1983. The adaptive significance of crawling up onto foliage for the survival of grounded overwintering monarch butterflies (*Danaus plexippus*) in Mexico. *Ecological Entomology* 8:471–474.

Calvert, W.H., L.E. Hedrick, and L.P. Brower. 1979. Mortality of the monarch butterfly (*Danaus plexippus* L.): Avian predation at five overwintering sites in Mexico. *Science* 204:847–851.

Calvert, W.H., W. Zuchowski, and L. P. Brower. 1982. The impact of forest thinning on microclimate in monarch butterfly (*Danaus plexippus* L.) overwintering areas of Mexico. *Boletín de la Sociedad Botánica de México* 42:11–18.

Calvert, W.H., W. Zuchowski, and L.P. Brower. 1983. The effect of rain, snow and freezing temperatures on overwintering monarch butterflies in Mexico. *Biotropica* 15(1):42–47.

Calvert, W.H., W. Zuchowski, and L.P. Brower. 1984. Monarch butterfly conservation: Interactions of cold weather, forest thinning and storms on the survival of overwintering monarch butterflies (*Danaus plexippus* L.) in Mexico. *Atala* 9(1&2):2–6.

Calvert, W.H., M.B. Hyatt, and N.P. Mendoza Villaseñor. 1986. The effects of understory vegetation on the survival of overwintering monarch butterflies (*Danaus plexippus* L.) in Mexico. *Acta Zoológica Mexicana* 18:1–17.

Calvert, W.H., S.B. Malcolm, J.I. Glendinning, L.P. Brower, M.P. Zalucki, T. Van Hook, J.B. Anderson, and L.C. Snook. 1989. Conservation biology of monarch butterfly overwintering sites in Mexico. *Vida Silvestre Neotropical* 2(1):38–48.

Carter, P.A., J.M. Hughes, and M.P. Zalucki. 1989. Genetic variation in a continuously breeding population of *Danaus plexippus* L.: An examination of heterozygosity at four loci in relation to activity times. *Heredity* 63:191–194.

Dawid, I.B., and T.D. Sargent. 1988. *Xenopus laevis* in developmental and molecular biology. *Science* 240: 1443–1448.

Eanes, W.F. 1978. Morphological variance and enzyme heterozygosity in the monarch butterfly. *Nature* 276: 263–264.

Eanes, W.F. 1979. The monarch butterfly as a paradigm of genetic structure in a highly dispersive species. In *Movement of highly mobile insects: Concepts and methodology in research,* eds. R.L. Rabb and G.G. Kennedy, 88–102. Raleigh: North Carolina State University.

Eanes, W.F., and R.K. Koehn. 1978. An analysis of genetic structure in the monarch butterfly, *Danaus plexippus* L. *Evolution* 32(4):784–797.

Fink, L.S., and L.P. Brower. 1981. Birds can overcome

the cardenolide defence of monarch butterflies in Mexico. *Nature* 291:67–70.

Fink, L.S., L.P. Brower, R.B. Waide, and P.R. Spitzer. 1983. Overwintering monarch butterflies as food for insectivorous birds in Mexico. *Biotropica* 15: 151–153.

Glendinning, J.I., A. Alonso Mejía, and L.P. Brower. 1988. Behavioral and ecological interactions of foraging mice (*Peromyscus melanotis*) with overwintering monarch butterflies (*Danaus plexippus*) in Mexico. *Oecologia* 75:222–227.

Goldberg, R.B. 1988. Plants: Novel developmental processes. *Science* 240:1460–1467.

Guilford, T., C. Nicol, M. Rothschild, and B.P. Moore. 1987. The biological role of pyrazines: Evidence for a warning odour function. *Biological Journal of the Linnean Society* 31:113–128.

Harborne, J.B. 1988. *Introduction to ecological chemistry.* London: Academic Press, 356 pp.

Herman, W.S. 1985. Hormonally mediated events in adult monarch butterflies. In *Migration: Mechanisms and adaptive significance,* ed. M.A. Rankin, 799–815. Austin: University of Texas, *Contributions in Marine Science,* Supplement, 27.

Herman, W.S., L.P. Brower, and W.H. Calvert. 1989. Reproductive tract development in monarch butterflies overwintering in California and Mexico. *Journal of the Lepidopterists' Society* 43:50–58.

Hughes, J.M., and M.P. Zalucki. 1984. Genetic variation in a continuously breeding population of *Danaus plexippus* (L.) (Lepidoptera: Nymphalidae). *Heredity* 59:213–221.

Jaenisch, R. 1988. Transgenic animals. *Science* 240:1468–1474.

Jones, F.M. 1932. Insect coloration and the relative acceptability of insects to birds. *Transactions of the Entomological Society of London* 80:345–385 (+ 28 pl.).

Kenyon, C. 1988. The nematode *Caenorhabditis elegans. Science* 240:1448–1453.

King, F.A., C.J. Yarbrough, D.C. Anderson, T.P. Gordon, and K.G. Gould. 1988. Primates. *Science* 240:1475–1482.

Lynch, S.P., and R.A. Martin. 1987. Cardenolide content and thin-layer chromatography profiles of monarch butterflies, *Danaus plexippus* L., and their larval host-plant milkweed, *Asclepias viridis* Walt., in northwestern Louisiana. *Journal of Chemical Ecology* 13:47–69.

Magasanik, B. 1988. Research on bacteria in the mainstream of biology. *Science* 240:1435–1439.

Malcolm, S.B. 1987a. The fourth chemical ecology meeting. *Trends in Ecology and Evolution* 2:292–293.

Malcolm, S.B. 1987b. Monarch butterfly migration in North America: Controversy and conservation. *Trends in Ecology and Evolution* 2:135–138.

Malcolm, S.B., and L.P. Brower. 1989. Evolutionary and ecological implications of cardenolide sequestration in the monarch butterfly. *Experientia* 45:284–295.

Malcolm, S.B., B.J. Cockrell, and L.P. Brower. 1989. The cardenolide fingerprint of monarch butterflies reared on the common milkweed, *Asclepias syriaca. Journal of Chemical Ecology* 15:819–853.

Martin, R.A., and S.P. Lynch. 1988. Cardenolide content and thin-layer chromatography profiles of monarch butterflies, *Danaus plexippus* L., and their larval host-plant milkweed, *Asclepias asperula* subsp. *capricornu* (Woods.) Woods., in north central Texas. *Journal of Chemical Ecology* 14:295–318.

Masters, A.R., S.B. Malcolm, and L.P. Brower. 1988. Monarch butterfly (*Danaus plexippus*) thermoregulatory behavior and adaptations for overwintering in Mexico. *Ecology* 69:458–467.

Parsons, J.A. 1965. A digitalis-like toxin in the monarch butterfly, *Danaus plexippus* L. *Journal of Physiology* 178:290–304.

Poulton, E.B. 1914. Mimicry in North American butterflies: A reply. *Proceedings of the Academy of Natural Sciences of Philadelphia* January 1914:161–195.

Reichstein, T. 1967. Cardenolide (herzwirksame Glykoside) als Abwehrstoffe bei Insekten. *Naturwissenschaftliche Rundschau* 20:499–511.

Reichstein, T., J. von Euw, J.A. Parsons, and M. Rothschild. 1968. Heart poisons in the monarch butterfly. *Science* 161:861–866.

Roeske, C.N., J.N. Seiber, L.P. Brower, and C.M. Moffitt. 1976. Milkweed cardenolides and their comparative processing by monarch butterflies (*Danaus plexippus* L.). *Recent Advances in Phytochemistry* 10: 93–167.

Rothschild, M., B.P. Moore, and W.V. Brown. 1984. Pyrazines as warning odour components in the monarch butterfly, *Danaus plexippus,* and in moths of the genera *Zygaena* and *Amata* (Lepidoptera). *Biological Journal of the Linnean Society* 23:375–380.

Rubin, G.M. 1988. *Drosophila melanogaster* as an experimental organism. *Science* 240:1453–1459.

Seiber, J.N., L.P. Brower, S.M. Lee, M.M. McChesney, H.T.A. Cheung, C.J. Nelson, and T.R. Watson. 1986. Cardenolide connection between overwintering monarch butterflies from Mexico and their larval food plant, *Asclepias syriaca. Journal of Chemical Ecology* 12:1157–1170.

Strong, D.R., J.H. Lawton, and Sir Richard Southwood. 1984. *Insects on plants: Community patterns and mechanisms.* Oxford: Blackwell Scientific Publications. 313 pp.

Urquhart, F.A. 1960. *The monarch butterfly.* Toronto: University of Toronto Press, 361 pp.

Urquhart, F.A. 1966. A study of the migrations of the Gulf Coast population of the monarch butterfly (*Danaus plexippus* L.) in North America. *Annales Zoologici Fennici* 3:82–87.

Urquhart, F.A. 1976. Found at last: The monarch's winter home. *National Geographic* 150:160–173.

Urquhart, F.A. 1987. *The monarch butterfly: International traveler.* Chicago: Nelson-Hall, 232 pp.

Urquhart, F.A., and N.R. Urquhart. 1976a. The overwintering site of the eastern population of the monarch butterfly (*Danaus p. plexippus*: Danaidae) in southern Mexico. *Journal of the Lepidopterists' Society* 30:153–158.

Urquhart, F.A., and N.R. Urquhart. 1976b. A study of

the peninsular Florida populations of the monarch butterfly (*Danaus p. plexippus*: Danaidae). *Journal of the Lepidopterists' Society* 30:73–87.

Urquhart, F.A., and N.R. Urquhart. 1977. Overwintering areas and migratory routes of the eastern population of the monarch butterfly (*Danaus p. plexippus*, Lepidoptera: Danaidae) in North America with special reference to the western population. *Canadian Entomologist* 109:1583–1589.

Urquhart, F.A., and N.R. Urquhart. 1978. Autumnal migration routes of the eastern population of the monarch butterfly (*Danaus p. plexippus* L.; Danaidae; Lepidoptera) in North America to the overwintering site in the neovolcanic plateau of Mexico. *Canadian Journal of Zoology* 56:1759–1764.

Urquhart, F.A., and N.R. Urquhart. 1979. Vernal migration of the monarch butterfly (*Danaus p. plexippus*, Lepidoptera: Danaidae) in North America from the overwintering site in the neo-volcanic plateau of Mexico. *Canadian Entomologist* 111:15–18.

Varmus, H. 1988. Retroviruses. *Science* 240:1427–1435.

Wells, S.M., R.M. Pyle, and N.M. Collins. 1983. *The IUCN invertebrate red data book*. Gland, Switzerland: IUCN, 632 pp.

White, R., and C.T. Caskey. 1988. The humans as an experimental system in molecular genetics. *Science* 240:1483–1488.

Williams, C.B. 1930. *The migration of butterflies*. Edinburgh: Oliver and Boyd, 473 pp.

Woodson, R.E., Jr. 1954. The North American species of *Asclepias* L. *Annals of the Missouri Botanical Garden* 41:1–211.

Young, A.M. 1982. An evolutionary-ecological model of the evolution of migratory behavior in the monarch butterfly, and its absence in the queen butterfly. *Acta Biotheoretica* 31:219–237.

Zalucki, M.P., J.M. Hughes, and P.A. Carter. 1987. Genetic variation in *Danaus plexippus* L.: Habitat selection or differences in activity times? *Heredity* 59: 213–221.

2

SYSTEMATICS

Part of a research collection of several subspecies of *Danaus plexippus.* Clockwise form upper left: *plexippus plexippus* (Linnaeus, 1758), *plexippus nigrippus* (Haensch, 1909), *plexippus erippus* (Cramer, 1776), *plexippus plexippus* (L., 1758) (Papua New Guinea), *plexippus* (undescribed subspecies, Trinidad), and *plexippus portoricensis* (Clark, 1941). Photograph by Dick Meier.

SYSTEMATIC PERSPECTIVES ON THE EVOLUTION OF THE MONARCH BUTTERFLY

Ian J. Kitching, Phillip R. Ackery, and Richard I. Vane-Wright

ABSTRACT. To answer comparative biological problems posed by the migratory butterfly, *Danaus (Danaus) plexippus plexippus* (Linnaeus), it is essential that its systematic relationships be understood. The most recent reclassification of the nymphalid subfamily Danainae, to which the monarch belongs, is discussed in light of additional data obtained from the immature stages and allozyme analyses. Overall, the hypothesis that the subgenus *Danaus (Danaus)*, which includes only the monarch and two very closely related species, is the sister-group of the two other subgenera of *Danaus, D. (Salatura) + D. (Anosia)*, is corroborated. Evidence is also presented indicating that the polytypic *D. (D.) plexippus* is a monophyletic taxon. The allozyme data also permit a tentative time of origin to be calculated for subgenus *Danaus*. This figure, of approximately 250,000 yr, is discussed in relation to Poulton's suggestion that the monarch invaded the New World by way of the Aleutian Islands. However, it is concluded that the systematic relationships of the genus as a whole favor a New World origin for subgenus *Danaus*, including the migratory monarch itself.

INTRODUCTION

Where and when did the monarch evolve? These questions are important if we seek to understand the remarkable biology of the monarch, including its long-distance migration, distribution, chemical ecology, and mimicry. Because such questions are essentially historical, it is necessary to elucidate the phylogenetic relationships of *Danaus plexippus*.

The monarch is a member of the Danainae—the milkweed butterflies. This subfamily (as conceived by Ackery and Vane-Wright, 1984) comprises 157 mainly tropical, medium- to large-sized species, belonging to the Nymphalidae—one of the major divisions of the true butterflies (Papilionoidea). Within the Nymphalidae, the Danainae are most closely related to the neotropical glasswings (Ithomiinae) and the monobasic Papuasian Tellervinae (Ackery and Vane-Wright, 1984; Ackery, 1987).

The Danainae are generally rather tough, conspicuous, and somewhat evil-smelling creatures—characteristics shared with a number of other aposematic insects. Among the Nymphalidae, members of the Danainae can be recognized unequivocally: The males have paired eversible abdominal hair-pencils (Ackery and Vane-Wright, 1984; Boppré and Vane-Wright, 1989), and the females have clubbed, "ankylosed" foretarsi (Ehrlich, 1958). The cylindrical, glabrous larvae are also usually distinctive, having two or more pairs of fleshy "filaments." The only exceptions are members of the two small genera *Lycorea* (including *Ituna*) and *Anetia*, which have only a single pair on postcephalic segment 2, a feature shared with larvae of a few genera of Ithomiinae. The food plants of the Danainae are also largely characteristic: Asclepiadaceae, Apocynaceae, and Moraceae predominate as the normal hosts (Ackery, 1988), although there are a number of significant exceptions (Ackery and Vane-Wright, 1984).

CURRENT CLASSIFICATION OF THE DANAINAE

The most recent reclassification of the Danainae (Ackery and Vane-Wright, 1984) recognized two constituent tribes, the Euploeini and the Danaini (Fig. 1). The Euploeini were further divided into two subtribes, the Euploeina (*Euploea, Idea,* and *Protoploea,* in an unresolved trichotomy) and the Itunina (*Anetia* and *Lycorea*). The Danaini also comprised two subtribes. In the Amaurina, *Parantica* and *Ideopsis* (including subgenus *Radena*) were interpreted as being more closely related to each

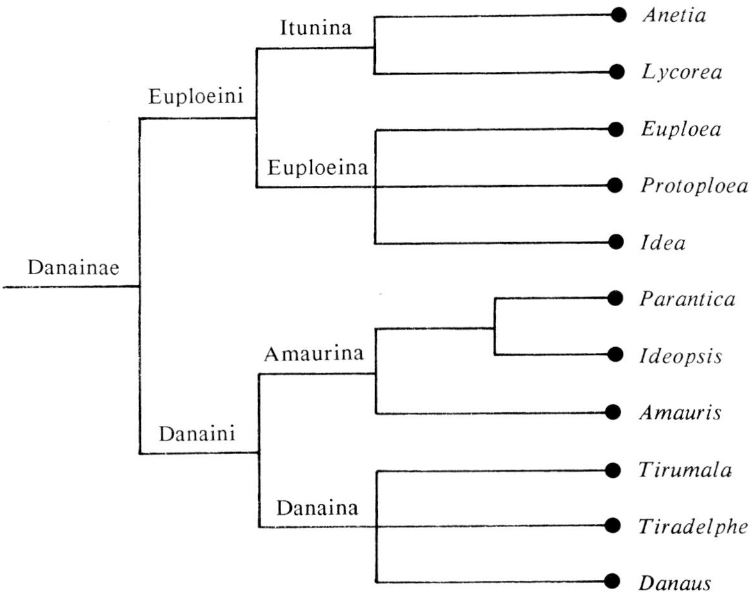

Figure 1. Cladogram depicting the generic interrelationships of the Danainae (after Ackery and Vane-Wright, 1984, and Kitching, 1985).

other than to *Amauris*; i.e., *Parantica* and *Ideopsis* shared a sister-group relationship. The three genera of the Danaina, *Tirumala, Tiradelphe,* and *Danaus,* were placed in an unresolved trichotomy. The classical concept of the genus *Danaus,* as enshrined by "Seitz" (Ackery and Vane-Wright, 1984), included all those orange, yellow, brown, or black and white species that remained, following removal of the obviously divergent taxa *Idea, Euploea, Lycorea, Anetia, Amauris,* and *Ideopsis* (s.s.). This concept of *Danaus,* as employed by many specialists until very recently, was extremely unnatural as members of other genera of the Danaini (*Tirumala, Parantica, Ideopsis* (*Radena*)) and even one species of the Euploeina (*Protoploea apatela*) were included within it. Removal of these heterogeneous elements rendered the genus *Danaus* monophyletic.

Danaus now contains just 11 species, distributed among three subgenera (Fig. 2). Subgenus *Danaus* comprises three of the accepted species: *plexippus, erippus,* and *cleophile*; subgenus *Salatura* includes the Asian *ismare, genutia, melanippus,* and *affinis*; while subgenus *Anosia* comprises the American species *eresimus, plexaure,* and *gilippus,* together with the Old World *chrysippus.* Ackery and Vane-Wright (1984) concluded that (*Anosia*) and (*Salatura*) shared a sister-group relationship.

THE MONARCH'S CLOSEST RELATIVES

These systematic results are important for proper perspectives on the monarch's historical origins and comparative biology. First, the monarchs of North, Central, and tropical South America, the Atlantic islands, the Pacific, and Southeast Asia are seen to share a sister-species relationship with the southern South American *erippus,* and these two taxa together have a sister-group relationship with the Caribbean endemic *cleophile.* The subgenus *Danaus,* comprising just these three species, is thus seen as originally endemic to the Americas, and it is most parsimonious to assume that *plexippus* arose through a speciation event in the New World. It should be noted here that difficulties over the status of *erippus* (Ackery and Vane-Wright, 1984; Kitching, 1985; see also below) and different views as to its relationship to the various subspecies of *plexippus* do not materially affect this conclusion.

In turn, the subgenus *Danaus* is considered to share a sister-group relationship with (*Anosia*) + (*Salatura*), a group of three New World and five Old World species. To date, only two of these have been studied in depth, and thus they have a special significance in the history and appreciation of comparative work on the monarch: the queen and plain tiger butterflies, *Danaus* (*Anosia*) *gilippus* (Cramer) and *Danaus* (*Anosia*) *chrysippus* (Linnaeus). It should be noted that (*Anosia*), like *Danaus* sensu stricto, is an essentially American group. Furthermore, *chrysippus* and *gilippus* are sister-species, spanning the Old and New Worlds, and could be classified as a single pantropical species if one chose to ignore the autapomorphies of *chrysippus.*

Contrast these systematic perspectives with E.B. Poulton's very influential views, first published in 1909, that *plexippus* is a member of "an Old World genus that has invaded the New," and that *Danaus*

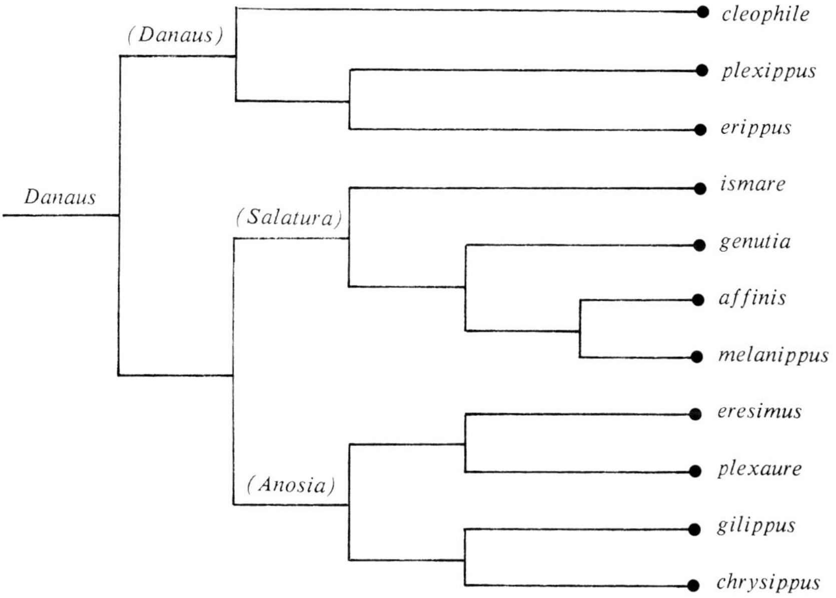

Figure 2. Cladogram depicting the subgeneric and specific interrelationships of the genus *Danaus* (after Ackery and Vane-Wright, 1984).

[*Salatura*] *genutia* is "the probable representative of its Old World ancestor." While it is plausible that the tribe Danaini, and even the subtribe Danaina, could be correctly viewed as Old World groups, the immediate ancestors of both *plexippus* and *chrysippus* appear, on the Ackery and Vane-Wright analysis, to have been New World species. Such a departure from established wisdom makes it desirable to check the accuracy of the Ackery and Vane-Wright cladograms, to ensure that any new perspective is soundly based.

ACKERY AND VANE-WRIGHT CORROBORATED—AND QUESTIONED

Recently, new data were obtained from the early stages (Kitching, 1985) and from the electrophoresis of allozymes (Kitching, 1986) to test the classification of Ackery and Vane-Wright (1984). While most of the danaine relationships proposed by Ackery and Vane-Wright were corroborated, there were two areas of disagreement.

The first concerned the relationships of the three genera of the Amaurina. As noted above, Ackery and Vane-Wright interpreted *Amauris* as being the sister-group of *Parantica* + *Ideopsis*. Evidence from the morphology of the pupae suggested that, instead, *Amauris* might represent the sister-group of the remainder of the Danaini (i.e., of *Parantica, Ideopsis,* and the Danaina). If so, the Old World origin of the Danaini would appear more certain. However, very few species were examined and the conclusion remains tentative.

The second area of uncertainty concerns the interrelationships of the three subgenera of *Danaus*. As already noted, all danaine larvae bear fleshy filaments on the body, the number of pairs and distribution of which are often useful taxonomically. Taken individually by segment, the filaments are highly homoplasious, i.e., subject to loss and convergent gain. Homoplasy is markedly reduced when the filaments are coded as "formulae." Thus the monarch has a filament formula of 2-11, indicating pairs of filaments on body segments 2 (mesothorax) and 11 (abdominal segment 8). It should be noted that this formula also occurs in *Tirumala, Parantica, Ideopsis,* and *Euploea.* All known members of (*Salatura*) and (*Anosia*), however, have a formula of 2-5-11, which is unique to them. It was solely on this basis that Ackery and Vane-Wright hypothesized a sister-group relationship between the two subgenera, although they acknowledged this conclusion as tentative.

New evidence drawn from the immature stages was equivocal. No characters were found in the eggs that would distinguish the various taxa of *Danaus.* The widely utilized features of transverse and longitudinal rib number were found to be useless. The number of ribs varied not only around the egg but was also dependent upon the angle at which the egg had been laid (Kitching, 1985). It has also been found in other nymphalids that rib number can be dependent upon the age of the ovipositing female: As a female gets older, the eggs laid have fewer and fewer longitudinals (Shirôzu and Setoya, 1973; Dennis, 1985).

The larvae also provided no additional characters, but the pupae did. Ackery and Vane-Wright's arrangement was corroborated by the discovery that

(*Salatura*) and (*Anosia*) lack the lateral ocular gold spot that is found in *Tirumala* and *Danaus* (s.s.). However, a contradictory character state was also found that suggested that (*Anosia*) shared a sister-group relationship with *Danaus* (s.s.) instead of with (*Salatura*). In the former two subgenera, the transverse gold abdominal band remains after adult eclosion as a yellow stripe, whereas in (*Salatura*) and *Tirumala* no trace of the gold band persists. The form of cremaster sclerotization may yield further evidence to help resolve this conflict, but the lack of pupae of critical species of all three subgenera is impeding progress on this aspect.

Electrophoresis of adult enzymes (Kitching, 1984, 1986) demonstrated that danaines are as variable in this respect as numerous other taxa. The three *Danaus* subgenera proved to be highly distinctive and each has several unique enzyme bands or combinations of bands. However, there was very little in the enzyme data that was of help in resolving the subgeneric relationships. In only one enzyme was there a band that occurred in only two of the three subgenera. Band E of esterase-D (terminology of Kitching, 1986) was found in (*Salatura*) and (*Anosia*) but not in *Danaus* (s.s.) (which possessed bands A and B). However, band E is also present in *Euploea, Amauris,* and *Tirumala.* It therefore may be a plesiomorphic feature and thus provide no evidence for treating (*Salatura*) and (*Anosia*) as sister-groups.

Overall, although there is a great deal of uncertainty, the current consensus still favors Ackery and Vane-Wright's hypothesis of subgeneric relationships within *Danaus,* implying a New World origin for the monarch. The most likely alternative, having (*Anosia*) + (*Danaus*) as sister-groups, suggests a New World origin for the monarch even more strongly. In particular, Poulton's implication that the Asian *Danaus* (*Salatura*) *genutia* could be treated as the sister-species of *plexippus* is rejected by all modern evidence.

THE SEPARATION OF *PLEXIPPUS* AND *ERIPPUS*

Within the subgenus *Danaus,* it has long been held that the southern *erippus* represents a good species—but this has often been questioned. Indeed, although they maintained the two as separate, Ackery and Vane-Wright (1984) could find no convincing apomorphic feature to distinguish *plexippus* from *erippus.* The two species are classically separated by the color of the posterior margin of the forewing upperside (black in *plexippus*; orange, brown, or gray in *erippus*).

Three characters that provided apomorphies for *plexippus* were found on the pupae (Kitching, 1985). The best is the presence of a median dorsal spot on abdominal segment 9. This feature proved constant for all examined specimens of the two species and showed that, judging from the limited available material, the species *plexippus* extends into South America at least as far as Ecuador. A less convincing transformation series, because it involves character reversal, concerns the dorsal surface of abdominal segments 4–7. In *Tirumala,* (*Salatura*), and (*Anosia*) these segments are rugose, a feature that is unique within the Danaini. In *plexippus,* however, these segments are smooth (as they are in the rest of the Danainae), with *erippus* showing an intermediate state. The third character state is the presence of a dorsal spot at the junction of abdominal segments 9 and 10. This feature was also recorded in *chrysippus* and only in some samples of each of these two species. Thus, it is of doubtful value. Nevertheless, there is now at least one pupal character to suggest that *plexippus* is a monophyletic entity, although *erippus* remains uncharacterized. This character could prove useful in future biogeographic studies, to help ascertain the geographic limits of the two taxa.

GENETIC DISTANCES WITHIN *DANAUS*

A second aspect of the electrophoretic study (Kitching, 1986) compared the allozymes of the eastern United States (EUSA) and Australian (AUS) populations of the monarch. The band frequencies found for the enzymes phosphoglucomutase (PGM), phosphohexose isomerase (PHI), nonspecific dehydrogenase-1 (NDH-1), malate dehydrogenase (MDH), aspartate aminotransferase (AAT), and glucose-6-phosphate dehydrogenase (G6PD) for the EUSA population were extremely similar to those found by Eanes and Koehn (1979) 7 years earlier. This temporal stability in band frequencies is very remarkable (it plausibly relates to the annual migration cycle leading to true panmixia of the entire eastern population each spring in Mexico). The AUS population, in comparison, was found to be markedly divergent. The common S bands for MDH and AAT, parts of dimorphisms in the EUSA sample, were fixed in the AUS sample. Bands F_1, M, and S_1 in NDH-1, PHI, and G6PD, respectively, which were very rare in the EUSA sample, were much more common in Australia. Finally, while the frequency of PGM band B was more or less the same in the two populations, the frequencies of bands A and C in the EUSA population (0.375 and 0.167, respectively) were virtually reversed in Australia (0.151 and 0.368, respectively).

These differences give rise to a Nei's standard genetic distance between the two populations of 0.0168. Using figures derived from studies of other organisms (Nei, 1971; Nixon and Taylor, 1977) for the fraction of detectable amino acid substitutions ($c = 0.4$), amino acid substitution rate per year ($n = 2.1 \times 10^{-9}$), and average number of codons per protein ($\lambda = 800$), this converts to a time since divergence (t) of about 12,000 yr. This should be compared to a similarly estimated time for the three *Danaus* subgenera of 250,000 yr. Although methods exist to place confidence limits on genetic distance estimates, none is yet available for the derivative divergence times. This is primarily due to there being no methods available for adequately estimating errors on *c*, *n*, and λ. However, to give some notion of confidence limits, however imprecise, by arbitrarily selecting ranges for *c* (0.25–0.45), *n* ($1 \times 10^{-9} - 3 \times 10^{-9}$), and λ (400–1,200), and by choosing the appropriate limits, it is possible to calculate minimum and maximum values for *t*. Using this procedure, the time of divergence for the AUS and EUSA populations of *D. plexippus* is estimated at 5,000–84,000 yr, while that of the three *Danaus* subgenera lies between 100,000 and 1,750,000 yr.

Dates such as these must be treated with great caution. It has long been accepted that *plexippus* spread across the Pacific during the middle of the nineteenth century and has been established in Australia for <150 yr (Walker, 1914; Vane-Wright, 1986, this volume). However, Malcolm and Brower (1987) have recently questioned the evidence for this as anecdotal, and so the figure of 12,000 (5,000–84,000) yr should not be rejected. Furthermore, there is no evidence as to which aboriginal monarchs gave rise to the Australian colony: they could have come from the eastern United States, western United States, or central America—or even some population now extinct. Perhaps most plausibly they would have come from the western United States; regrettably, enzyme data for the California population are not available to us. The only clue is that all the Pacific (and Atlantic) monarchs do appear to be derived from the *plexippus plexippus* phenotype, which extends from the United States south to Costa Rica and Jamaica.

THE TIME OF ORIGIN OF SUBGENUS *DANAUS* IN THE AMERICAS

Accepting the divergence time of 250,000 (100,000–1,750,000) yr for *Danaus* sensu stricto from (*Salatura*) also leads to some interesting ideas. According to Poulton (1909), "there can be little doubt that *D. plexippus* invaded America by way of the north, probably following the line of the Aleutian islands." Even if Poulton was wrong in the narrowest sense, if we substitute subgenus *Danaus* for *plexippus*, the origin of the *plexippus* group in the New World appears to have occurred during the last 1.75 million yr. Most of the last 1.6 million yr have seen the northern land masses icebound (Sutcliffe, 1985), making Poulton's favored Aleutian route seem a little unlikely. But even if the North American origin occurred at about 1.75–2 million yr ago, this date might still fit comfortably with Poulton's passionate desire to establish a relatively recent Old World origin for the monarch, the evidence for which was the comparative lack or even poverty of Batesian mimics in the Americas: "we are made to realize the recent date of the invasion of South America when we remember that nowhere else in the world do Danaine butterflies of equal abundance 'range through a crowded area without producing any effect on any member of the Lepidopterous fauna, or without themselves being affected thereby'" (Poulton, 1909).

Have we vindicated Poulton's bold speculation? Certainly not, if the present cladograms are correct. If we place the origin of subgenus *Danaus* in the Americas as recently as 1.75 million yr ago, to explain the absence of mimicry, then we would have to accept the origin of (*Anosia*) *chrysippus* in the *Old World* from the New at something <1.75 million yr—and *it* has more putative mimics than any other species of *Danaus*. Furthermore, in addition to *plexippus*, the two more restricted North American *Danaus* species, *eresimus* and *gilippus*, are also very well represented in the Caribbean and Central and South America—all three species fly as far south as Peru and Brazil. Is it not more likely that the presence of these essentially tropical insects in North America during an interglacial period, such as the present, has been brought about by crossing the Darien Gap in eastern Panama rather than crossing the Bering Sea from Siberia to Alaska? Could it be significant that the Pliocene Darien connection between North and South America formed about 2 million yr ago?

IN CONCLUSION

Although there are too few data at present for some of these ideas to be more than speculation, they merit further investigation. The problem of the subgeneric relationships needs more work, and the separation times based on genetic distances need radical reinvestigation before any estimate can be taken seriously. Nevertheless, if either of the best-supported subgeneric relationships is correct, then it seems almost certain that the subgenus *Danaus* is a New World group, and that the monarch evolved in the Americas—most probably South or Central America, north of the Amazon. Furthermore, it would seem likely that subgenus *Danaus* is a much older American group than either Poulton thought—or our preliminary enzyme data suggest.

ACKNOWLEDGMENTS

We would like to thank the organizers of Moncon-2 for the opportunity to attend the meeting and to present this paper.

LITERATURE CITED

Ackery, P.R. 1987. The danaid genus *Tellervo* (Lepidoptera, Nymphalidae)—A cladistic approach. *Zoological Journal of the Linnean Society* 89:203–274.

Ackery, P.R. 1988. Hostplants and classification: A review of nymphalid butterflies. *Biological Journal of the Linnean Society* 33:95–203.

Ackery, P.R., and R.I. Vane-Wright. 1984. *Milkweed butterflies: Their cladistics and biology*. London: British Museum (Natural History); and Ithaca, NY: Cornell University Press, 425 pp.

Boppré, M., and R.I. Vane-Wright. 1989. Androconial systems in Danainae (Lepidoptera): Functional morphology of *Amauris, Danaus, Tirumala* and *Euplo-*

ea. Zoological Journal of the Linnean Society 97: 101–133.

Dennis, R.L.H. 1985. Egg-keel number in the small tortoiseshell butterfly. *Entomologist's Record and Journal of Variation* 97:162–164.

Eanes, W.F., and R.K. Koehn. 1979. An analysis of genetic structure in the monarch butterfly, *Danaus plexippus* L. *Evolution* 32:784–797.

Ehrlich, P.R. 1958. The comparative morphology, phylogeny and higher classification of the butterflies (Lepidoptera: Papilionoidea). *Kansas University Science Bulletin* 39:305–370.

Kitching, I.J. 1984. Enzyme variation in the Danainae. In *The biology of butterflies,* eds. R.I. Vane-Wright and P.R. Ackery, 191–192. *Symposia of the Royal Entomological Society of London* 11.

Kitching, I.J. 1985. Early stages and the classification of the milkweed butterflies (Lepidoptera: Danainae). *Zoological Journal of the Linnean Society* 85:1–97.

Kitching, I.J. 1986. Allozyme variation in the milkweed butterflies (Lepidoptera: Danainae). *Zoological Journal of the Linnean Society* 86:367–389.

Malcolm, S.B., and L.P. Brower. 1987. White monarchs. *Antenna* 11:2–3.

Nei, M. 1971. Interspecific gene differences and evolutionary time estimated from electrophoretic data on protein identity. *American Naturalist* 105:385–398.

Nixon, S.E., and R.J. Taylor. 1977. Large genetic distances associated with little morphological variation in *Polycelis coronata* and *Dugesia tigrina* (Planaria). *Systematic Zoology* 26:152–164.

Poulton, E.B. 1909. *Charles Darwin and the Origin of Species. Addresses, etc., in America and England in the year of the two anniversaries.* London: Longmans, Green & Co., 302 pp.

Shirôzu, T., and K. Setoya. 1973. Variation of longitudinal ridge number of eggs laid by a female of *Hypolimnas bolina* Linnaeus. *Butterflies and Moths* 23:51–53.

Sutcliffe, A.J. 1985. *On the track of the Ice Age mammals.* London: British Museum (Natural History), 244 pp.

Vane-Wright, R.I. 1986. White monarchs. *Antenna* 10: 117–118.

Walker, J.J. 1914. The geographical distribution of *Danaida plexippus.* L. (*Danais archippus,* F.) with especial reference to its recent migrations. *Entomologist's Monthly Magazine* 50:181–193, 233–237.

3

CHEMICAL COMMUNICATION

Everted abdominal hairpencils of two daniane butterflies. Left, *Tirumala petiverana*; right, *Euploea core*. Photographs by Michael Boppré.

DANAINE BUTTERFLIES: A DIDACTIC STORY ABOUT CHEMICAL ECOLOGY

Dietrich Schneider

ABSTRACT. Many Danaine species feed as larvae on milkweeds and sequester their bitter-tasting and toxic cardiac glycosides, to render them unpalatable for predators (the Brower–Reichstein–Rothschild paradigm of "ecological chemistry").

Danaine body odors have been found recently to be characterized by the presence of pyrazines, which may also function as warning signals to predators.

A third group of allelochemicals, the bitter-tasting and toxic pyrrolizidine alkaloids (PAs), occur in many plant groups and are actively searched for and ingested by adult butterflies, especially males. All closely observed danaine species eagerly search for PAs, which may serve as (a) additional, or the only, protective chemicals; (b) essential precursors for the biosynthesis of one or more of the male dihydropyrrolizine pheromones: danaidone, danaidal, and hydroxydanaidal. The only known exception to this rule of pheromone production is *Danaus plexippus* (and *Danaus erippus*); (c) pheromones are dissipated by abdominal hairbrushes as an odorous signal to induce female receptivity.

The striking odor of many hairbrush signals is due to a wide spectrum of substances and is to some extent species-specific. The role of these odors is unclear, but they may facilitate species and sex discrimination. This would be important in those genera, or groups of genera (African *Amauris,* Asian *Euploea,* Asian *Tirumala/Ideopsis/Parantica*), that look similar and live in close sympatry.

The androconial (male scent-producing) organs of the danaines show an extraordinary diversity of structure and function, for which the adaptive value is only partly understood. In the genera *Danaus, Amauris,* and *Tirumala,* hairbrushes and wing glands must interact to ensure pheromone production.

Odor-sensitive cells on the antennae respond specifically to the principal pheromones and their precursors. Other odor receptors respond to the less predominant chemicals of the hairbrush pheromone spectrum.

INTRODUCTION

Chemical factors in plant–animal interactions have puzzled biologists for a long time, and the so-called "secondary plant substances" play a major role in these considerations. Because insects are significant consumers of plants, it is not surprising that an insect physiologist, Gottfried Fraenkel, in 1959, presented the first strong evidence that such compounds protect plants against herbivores. This producer–consumer interaction must have taken place throughout evolutionary time, and so Brower and Brower (1964) and Ehrlich and Raven (1965) suggested that host plants and butterfly herbivores were coevolved. Detailed studies now describe some of the chemical complexity of the ecological and evolutionary interactions between plants and insects, based on the formalization of allelochemical interactions by Whittaker and Feeny (1971).

Following the demonstrations by Parsons (1965), Reichstein (1967), and Reichstein et al. (1968) that adult danaines stored cardiac glycosides (CGs) from their larval milkweed host plants, Brower et al. (1968) and Brower (1969) found that CGs in monarchs butterflies are an effective defense against some bird predators (see also Rothschild et al., 1970). CGs are toxic, bitter, and emetic and restrict the number of consumers of these plants to some adapted, specialist species such as the larvae of milkweed butterflies—the Danainae. CGs are stored by the larvae of *Danaus plexippus, Danaus gilippus,* and *Danaus chrysippus* and are used by the aposematic adults as a defense against predation by some birds (Brower, 1969; Brower et al., 1975; Cohen, 1985). Such protection is also extended to the viceroy (*Limenitis archippus*), a mimic of the monarch, which is also avoided by birds, presumably after they learned the visual pattern of the bitter-tasting monarch (but see Brower, 1958, 1984; and

Ritland and Brower, this volume, for the question of the unpalatability of the viceroy in its own right). The Browers called the interaction among birds, butterflies, and plant poisons "ecological chemistry" (Brower and Brower, 1964; Brower, 1969).

Stimulated by the seminal papers of Fraenkel (1959) and Ehrlich and Raven (1965), more and more relevant research was undertaken (Pasteels et al., 1983; Rosenthal, 1986) and the subsequent proceedings of meetings and monographs carry similar titles that reflect the new developments—for example, "molecular ecology" (Florkin and Schoffeniels, 1969), "chemical ecology" (Sondheimer and Simeone, 1970), "phytochemical ecology" (Harborne, 1972), "biochemical adaptations" (Hochachka and Somero, 1973), "ecological chemistry" (Harborne, 1977), and "chemical ecology" (Barbier, 1979). In 1974, Silverstein and Simeone founded the *Journal of Chemical Ecology*, and since 1983 the International Society of Chemical Ecology has united many interested scientists. Ecology as a discipline deals with interactions between organisms and their environment, and chemical ecology is now a firmly established subdiscipline with particular emphasis on chemical interactions.

The case of the danaine butterflies and their host-plant allelochemicals is a particularly didactic example of the role plant compounds play in the life histories of animals. Danaines are thought to exploit their host plants, but is this a single-sided affair, or do the plants also benefit (Rothschild, 1985)? As Rothschild (1985) suggests, danaines may search more often on their host plants for nectar than on other plants and thus act as beneficial pollinators. What other benefits might danaids offer the plants? Only strong insects such as butterflies, beetles, and bees are able to pollinate milkweeds with their unique floral morphology (Good, 1974; Morse, 1986). Thus do the milkweed herbivores among these insects benefit the plants more than nonherbivores do?

The phenomena described here are, of course, part of a general plant–herbivore theme within chemical ecology. It is interesting to note that some ecologists have begun to pay special attention to herbivore foraging and feeding behaviors, but it is also important to consider the sensory and nervous systems that control these behaviors. Specific animal preferences may thus channel the flux of energy in an ecosystem. In a thoughtful article, Remmert (1982) considers "animal amplifiers" in the context of "physiological ecology" and in his textbook (Remmert, 1980) speaks of "neuro-ecology." In an overview, the problems of plant recognition by insects are treated by Schneider (1987), who calls for a "neuro-ethological" approach, by the study of the neural bases of respective animal behaviors.

In this paper, I describe some aspects of the chemical ecology of several danaine butterfly species in which other secondary plant substances, in addition to CGs, were found to be not only protective, but also essential in connection with male pheromone production. Such chemical communication studies are influenced by modern sociobiological thinking, and the term "eco-ethology" is sometimes used to describe this field, which of course overlaps with "chemical ecology."

DEFENSIVE CHEMISTRY OF THE DANAINAE—THE MILKWEED BUTTERFLIES

Chemical defense of butterflies is a complex subject, which has recently been treated extensively by Brower (1984). The defensive role of CGs in *D. plexippus*, *D. gilippus*, and *D. chrysippus* has been clearly established by Brower and Rothschild and their colleagues. Larvae of these danaids sequester CGs from their milkweed host plants; however, some populations of these species do not always contain CGs, perhaps because their milkweed host plants contain insufficient CG for storage (Fink and Brower, 1981; Brower, 1984).

Do other danaines sequester cardenolides as effectively as these *Danaus* species? The subfamily comprises some 150 species in nearly all tropical and some subtropical countries (Ackery and Vane-Wright, 1984). Among them, species of the genus *Danaus* feed as larvae on many plant species of the families Asclepiadaceae and Apocynaceae containing variable amounts and different kinds of CGs. However, larvae of *Amauris, Parantica, Ideopsis,* and even most *Euploea* do not feed on CG-containing plants. As long as the majority of the members of a population of a species contains enough CG to render them unpalatable, Brower's claim (Brower, 1969)—that Batesian mimicry operates—is substantiated, and thus automimetic, CG-deficient individuals are also protected. But if, for example, species of the African genus *Amauris* never ingest any CGs but do prove to be protected from predation because they are thought to be models in mimetic systems, one needs to look for protective chemicals other than CGs (Ackery and Vane-Wright, 1984).

Could the strong odor of danaines also be a repellent? Even before the importance of CGs for *Danaus* species became known, various field workers reported that many danaines have an unpleasant, probably protective odor. These observations were sometimes related to the "obvious" aposematism and membership of these butterflies in mimicry rings. The earlier literature on such butterfly odors was recently discussed in connection with new findings of pyrazines in danaine bodies (Rothschild et al., 1984), although pyrazines have yet to be found in hairpencil odors (S. Schulz and W. Francke, pers. comm.). Because pyrazines may be general defensive agents among aposematic insects (e.g., coccinellids, zygaenid and ctenuchid moths, as well as monarch butterflies), it is supposed that they may also play such a general role in all danaines (Rothschild et al., 1984; Guilford et al., 1987). To date, we know little about the biosynthesis of these odors, which interestingly also occur in some plants, such as the *Asclepias* milkweed hosts of danaines (Rothschild et al., 1984), but could possibly also be produced by the insects themselves. In this respect, one wonders about Brower's (1969) CG-free raised monarchs that were eaten by blue jays. Did these monarchs still have the strong pyrazine odor? And if so, did the jays ignore the odor, while other more sensitive predators would respond?

The adults of many danaine species have been seen visiting specific plants where they fed on the plant surface (Boppré, 1981, 1986). Butterflies have also been seen feeding on seedpods of *Crotalaria* (Fabaceae) or drying or wilting leaves and seeds of *Heliotropium* (Boraginaceae). Usually only male butterflies show this behavior, but this sex preference may be partly erroneous, because females of many danaines are often less frequently observed than males.

The secondary substances, which all these plants share, are the bitter and toxic pyrrolizidine alkaloids (PAs), which have now been detected in many of the danaines, in more or less unchanged form. This argues strongly that butterfly PAs are derived from the plants visited by adult butterflies (Edgar and Culvenor, 1974; Edgar, 1975; Edgar et al., 1971, 1973, 1976a, 1979; Rothschild and Marsh, 1978; Rothschild et al., 1979; Brown, 1984). In other groups of Lepidoptera, the arctiid and ctenuchid moths, and in the "sister" subfamily of the danaine butterflies, the Ithomiinae, PA uptake is either performed through the proboscis, as described, or by larval feeding on PA plants, as in some of the Arctiidae (Pliske, 1975; Edgar et al., 1976b; Goss, 1979; Boppré, 1978, 1981, 1984a, 1986; Brower, 1984; Schneider, 1986). The attraction of these insects to PA-containing plants is impressive. Danaines assemble in numbers in traps that are baited with dry *Heliotropium. Tirumala* males in Africa have been seen tearing fresh leaves of these plants open with their legs after beetles started biting holes in the leaves (Boppré, 1983). Such "addicted" search and uptake of a substance that has no nutritional value, but adds to the animal's fitness, was called "pharmacophagy" by Boppré (1984b).

Evidence for protection of insects by PAs is strong in the Arctiidae (Rothschild et al., 1979; Conner et al., 1981; Eisner and Meinwald, 1987) and the Ithomiinae (Brown, 1984) but weak in the Danainae (Kelley et al., 1987). When added to the food, PAs are repellent to locusts (Bernays and Chapman, 1977), cockroaches, and vertebrates (M. Boppré, pers. comm.).

Edgar and Culvenor in Australia and our group in Seewiesen (influenced by T. Eisner and J. Meinwald and their Cornell University groups) found that PA uptake by danaines is (with few exceptions) necessary because these substances are essential precursors for male pheromones. Following Rothschild and Marsh (1978), we consider that PAs are more general plant toxins than CGs, and PAs are stored by most, or perhaps all, danaines for both communication and adult defense. Because PA-containing spermatophores in the Arctiidae and Ithomiinae are transferred from males to females and eventually to their eggs, the potential protection by PAs is not restricted to males (Brown, 1984; Dussourd et al., 1984; Eisner and Meinwald, 1987). So far, Brower's (1984) reluctance to accept the notion of the general protective action of PAs in the danaines might still be justified in principle, because large-scale palatability tests have yet to be performed. However, the evidence is strongly indicative of a defensive role for PAs in adult danaine butterflies.

PHEROMONES OF THE DANAINAE

The comparatively recent research discipline of "intraspecific chemical communication," or "pheromone biology" until a few years ago, dealt largely with female attractants of moth and beetle pests and combined basic biological and chemical research with the pragmatism of applied entomologists. While female moths possess powerful chemical attractants, little is known of such odors in female butterflies. As diurnal insects, they rely mainly on vision for sexual encounters. However, odor-producing epidermal organs, of often bizarre structure (brushes, fans, etc., composed of hairs or scales), have been seen in male Lepidoptera of most groups (Illig, 1902; Barth, 1960; Varley, 1962; Birch, 1979). Such androconial organs are found on different body parts, such as scattered on wings or arranged as denser packages on structures that allow the controlled exposure of the odor-dissipating hairs.

All male Danainae have a pair of abdominal tubular brush-organs. The tubes can be everted, like the fingers of a glove, by hemolymph pressure and retracted by a muscle. Either only the tip or the large parts of the tube are covered with hairs, which have in some species different shapes and functions. Some, but probably not all, of the trichogen cells at the base of these hairs are also the glandular cells, which are responsible for the production of the pheromone or, at least, are involved in the biosynthetic process. During eversion, the brushes open from the compact, paintbrush shape to a spherical or a bottlebrush structure. Such brushes (which are also called hairpencils) have been observed in danaines for more than a century (Müller, 1877a,b) and have been described and reviewed by a number of subsequent authors (e.g., Illig, 1902; Eltringham, 1913, 1915; Brower et al., 1965; Pliske and Salpeter, 1971; Ackery and Vane-Wright, 1984; Boppré 1984a).

Hairpencils were always suspected of having a function in the sexual behavior of danaines but the final proof did not come until Pliske and Eisner (1969) reported on *D. gilippus*, the queen butterfly. This research followed the earlier ethological and histological work of Brower et al. (1965) with the same species. These observers noted that in the final courtship phase, the male butterfly "hairpencils" the front end of the female in flight and again while she is sitting on vegetation. This leads to mating if the female is receptive. In this and several other species, the chemical signal is not only exposed in the gaseous state, but also as a cuticular hairpencil-dust (later called pheromone-transfer particles (PTPs; Boppré and Fecher, 1977). The odor-laden PTPs are released when the male expands the pencils, and they then stick to the female antennal surface, where the pheromone receptor sensilla are located (Brower et al., 1965; Pliske and Eisner, 1969; Schneider and Seibt, 1969; Boppré, 1979, 1984a, this volume; Kittmann and Schneider, this volume).

The major volatile substance found in the hairpencils of two American danaines (*Lycorea ceres* = *L. cleobaea*) and *D. gilippus* (Meinwald and Meinwald, 1966; Meinwald et al., 1966, 1969) is a dihydropyrrolizine, which was later named danaidone (Boppré et al., 1978). The evidence that danaidone is indeed a pheromone came from the work of Pliske and Eisner (1969) with *D. gilippus*. They found that laboratory-raised males lacked the danaidone and were unsuccessful suitors; however, females accepted these males after danaidone was applied to their hairpencils.

Two related dihydropyrrolizines, danaidal and hydroxydanaidal, were later found in other danaines, in some cases all three forms together. Interestingly, hydroxydanaidal is also a common pheromone of several male arctiid moths (Ackery and Vane-Wright, 1984; Boppré, 1984a).

Recently, a fourth dihydropyrrolizine (methylhydroxydanaidoate) was discovered in wing hairbrushes of three male ithomiine butterfly species (Schulz et al., 1988). This volatile is very likely also a PA derivative and a pheromone. In earlier analytical studies, such hairbrushes were only found to contain a lactone, which was suspected to derive from the acidic moiety of the PA (Edgar et al., 1976b), while the danaidoate probably derives from the heterocyclic, alcoholic moiety. Ithomiine males are also attracted by PA plants and often visit such plants for nectaring.

Why are laboratory-raised *D. gilippus*, and several other danaine species, unable to produce danaidone or any of the other two dihydropyrrolizines? The solution to this puzzle was suggested by Meinwald and Meinwald (1966) and Meinwald et al. (1966) because they doubted that these insects might be able to biosynthesize the pyrrolizines *de novo*, and they called attention to the similarly composed heterocyclic moiety of the PAs known from some plants. Observations of the Old World *D. chrysippus* on wilting PA plants (*Heliotropium*) eventually led to the respective experiments in Australia (Edgar et al., 1973) and independently by our group, with specimens from Africa (Schneider et al., 1975). In both studies, it was found that adult males needed to ingest PAs to synthesize their danaidone. We also found that the behavior of the males, when searching for the PA plant, was initiated by antennal odor-receptor cells, which respond to PA or PA plants. In a subsequent study with this species, we reported on another unexpected prerequisite for the pheromone production, namely the behavioral (and probably biochemical) interaction of the hairpencils with the hindwing androconial glands. The function of these glands had been a matter of speculation since Müller (1877a; see also Brower et al., 1965) but was now clear in principle (Boppré et al., 1978). This finding was a surprise because such a behavioral interaction of two integumental glands has not been known for the production of any biologically active substance. Male reproductive behavior occurs during the early-morning daylight hours and was verified functionally for other danaines with alar glands, including *Amauris* and *Tirumala* (Boppré, 1977, 1979, 1984a, 1986). However, because some Danainae like *Lycorea* lack alar androconia, their danaidone biosynthesis must be fully performed in the hairpencils without such contacts.

The described interaction between hairpencils and wing pockets is easily seen in the males of *Amauris ochlea* from East Africa (Boppré, 1977, 1979), a species that produces danaidone. All the species of the genus *Amauris* have open (glandular) alar patches, which allows direct observation of the mechanical contact between the hairpencil and wing patch. During their early-morning reproductive activity, *A. ochlea* males rub an outer fan of black hairs over the scales and hairs of their alar patches, which partly break off during this procedure. We suspect in *Amauris* that danaidone is only produced when this behavior occurs, provided the males have previously imbibed PAs. Critical tests of this assumption have yet to be done. The *A. ochlea* courtship differs from *D. gilippus* and *D. chrysippus* in that the male, after following his female, hovers approximately 50 cm above her, for up to 15 min after she settles. Then suddenly he dives directly down to the female and fully opens his hairpencils immediately before touching her. The black hair-fan is then completely folded back and several types of white hairs come into play. One type fractionates to PTPs (dust), which can later be found on the female antennae. Other hair types function as bordering fringes or as seals for the innermost hair cone, which seem to contain some of the other odorous substances of unknown function (Petty et al., 1977). This male behavior does not always lead to copulation. The danaidone content of the hairpencil probably plays a role, but the physiological state of the female may also be important.

Dihydropyrrolizine pheromones have been found in most danaines examined, except the American monarch, *D. plexippus* (Meinwald et al., 1968; Bellas et al., 1974; Boppré, 1984a); however, Kelley et al. (1987) have recently found that PA source compounds, both diesters and monoesters, are present in *D. plexippus* that overwinter in Mexico. All three species of the *Danaus* (*Danaus*) subgenus, *D. plexippus*, *Danaus cleophile* in Jamaica and His-

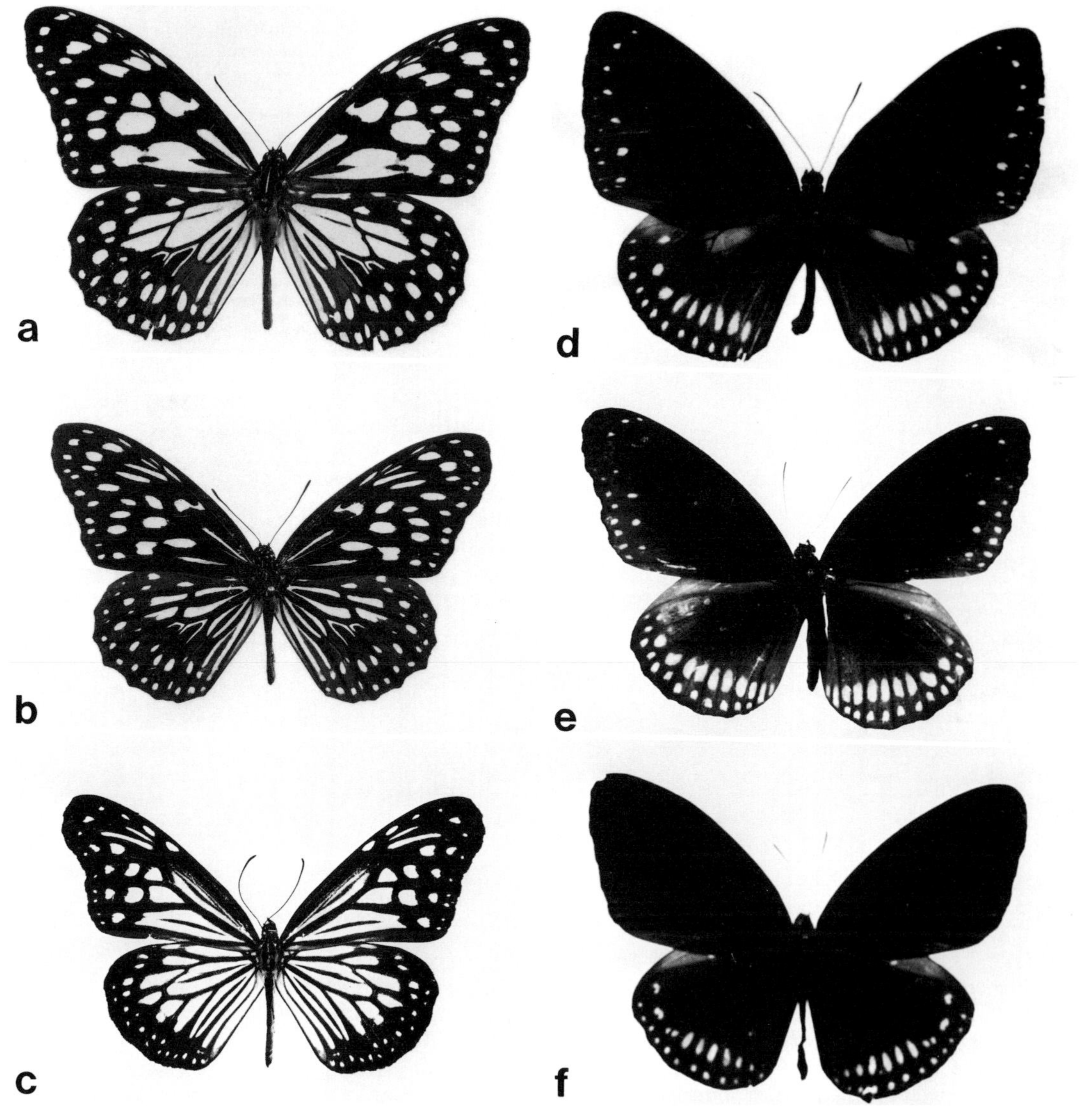

Figure 1. Wing patterns of the male danaines described in Table 1 (females are similar). Photographs of the *Tirumala* species are courtesy of R.I. Vane-Wright from the collection at The Natural History Museum, London. All others are from the Seewiesen collection. **a,** *Tirumala limniace* (Cramer); **b,** *T. septentrionis*; **c,** *Parantica aglea* (Stoll); **d,** *Euploea klugii* Moore; **e,** *E. sylvester* (Fabricius); **f,** *E. core* (Cramer).

paniola, and *D. erippus* in South America, are probably unable to metabolize PAs to dihydropyrrolizines, although they visit PA plants and contain PAs. We have recently seen *D. erippus* males feeding at *Eupatorium* flowers (Schneider, unpubl.), which are known to contain PA-rich nectar (Brown, 1984). However, the preliminary chemical analysis of the hairpencils shows no dihydropyrrolizine (S. Schulz, W. Francke, and D. Schneider, unpubl.).

The induction of mating by danaidone is quite clear in *D. gilippus* and appears similar in *D. chrysippus* and in some *Amauris* and *Tirumala* species. However, we wonder if we can generalize for all three dihydropyrrolizines and include all the danaines, except the *Danaus* (*Danaus*) group.

Because of its "seductive" effect, danaidone was often called an aphrodisiac pheromone; however, the behavior of the Asian species *Euploea core* (Boppré, 1979) indicates that additional or more complex functions of the dihydropyrrolizines are possible, because hydroxydanaidal is active in this species. The *E. core* males not only use hairpencils during the mating prelude (similar to *A. ochlea* as described above), but also during apparently territorial patrol flights. Such behavior has been seen in other *Euploea* species by field observers. This be-

Table 1. Volatile substances identified in extracts from the hairpencils of six different species of danaine butterflies from Sri Lanka (Schulz, 1987; Schulz et al., 1988; W. Francke, S. Schulz and D. Schneider, unpubl.). Each of the two sets of three columns unites species with very similar wing patterns (see Fig. 1) that live in similar habitats (all six are sometimes found together). MC = major component, + = the species-specific set of substances is similar to the scent of the respective hairpencil.

	Euploea			*Parantica/Tirumala*		
Volatile	*core*	*sylvester*	*klugii*	*P. aglea*	*T. septentrionis*	*T. limniace*
Dihydropyrrolizines[a]						
Hydroxydanaidal	MC	+	+	MC		+
Danaidone		+		+	MC	MC
Other volatiles						
Maltol	+					
Methyl salicylate	+	MC	+			
Linalool		MC				
Hydroxyisophorone		MC				
2-Hexenal		+				
Vanillin		+				
Methylanthranilate		+				
o-methoxyphenol		+				
Molecule W.210			MC			
2-Phenylethanol			+			+
2-Phenylacetate			+			
β-ionone		+				
2-Dodecenal						+
2,4-Dodecadienal						+
Unknown				MC		
Nonanal			+			
Decanal			+			
Hexadecylacetate				+		
3,4-Dimethoxyacetophenone				+		

[a] Dihydropyrrolizines are odorless to the human nose.

havior suggests that dihydropyrrolizines, or additional odorous components, are signaling the presence of the emitter to either its female, as an attractant, or another male, as a warning, or simply as a species-specific recognition signal. Recognition is likely for the species-specific odor bouquet, but not for the ubiquitous dihydropyrrolizines, because very similar-looking danaines, such as *Euploea* species, live together in the same locality, chase one another, roost together, and must solve their "who-is-who" problem. This is obvious in *Amauris* species, and the *Tirumala–Parantica–Ideopsis* groups of Asia or even the queen and monarch in America (e.g., see Fig. 1 and Table 1).

Thus, in summary, PAs have both specific and general sociobiological functions. They may function as both direct sexual signals and as more general signals, with an odor bouquet (Table 1), which may be related to intra- and interspecific competition. According to Conner et al. (1981), Eisner (1982), and Eisner and Meinwald (1987), female choice could be related directly to the amount of PA stored by a male, which is indirectly signaled to her by the PA-derived pheromone during courtship. Based on these authors' evidence that male danaines and arctiids transfer PAs to their females in spermatophores, females may select males with a strong dihydropyrrolizine odor, because the nuptial PA gift would enhance female and egg protection against natural enemies (Conner et al., 1981; Dussourd et al., 1984).

Future experiments must show whether the females of these butterflies (and arctiid moths; see Conner et al., 1981) respond to the male dihydropyrrolizines with simple accept or reject reactions, dependent upon a certain threshold concentration of pheromone odor, or they select males in a graded manner. Danaine mating behavior, as described here and also observed by myself in several species, does not conform to a typical lek behavior, in which males present themselves in a group and compete for selection by a female. The problem of male advertisement by a briefly displayed chemical (pheromonal) signal is the difficulty for a female to compare competitors, in contrast to classical visual leks. In this respect, the function of pheromone-laden PTP particles, when put close to the pheromone receptors on the female antenna, remains obscure. This permanent odor signal leads to sensory adaptation of the female danaidone receptors (Schneider and Seibt, 1969), which only an imaginary, but not likely existing, competing super-male (with 5 mg instead of the normal 0.05–0.5 mg of danaidone on his hairpencil) could perhaps overcome.

The sensory receptors for dihydropyrrolizines were examined by Myers and Brower (1969) and Schneider and Seibt (1969). Recently, by means of single sensillum recordings from female antennae of *D. gilippus*, we have found receptor cell types specific for either danaidone or hydroxydanaidal (Kittmann and Schneider, this volume). The known "side spectra" of the hairpencils are like those produced by general flower odors and are probably received by antennal odor-receptor cells, which are thought to recognize specific odor images (Schneider, 1987). But this also requires detailed further study, as do the receptors for PAs or their derivatives. In some arctiids, receptors for these latter compounds have been found (Bogner and Boppré, 1985). Danaines must also have such receptors, as indicated by earlier electroantennogram studies (Schneider and Seibt, 1969; Schneider, 1975). Only closely integrated and comparative studies of the behavior, chemistry, biochemistry, and physiology of the danaines and their host plants will eventually reveal all the details of their chemical ecology.

SYNOPSIS

The danaines are a particularly didactic example for chemical ecology by virtue of their complex interactions with plants and within their own species. The allelochemicals within these systems that deter herbivores were initially found to be CGs, which are typical components of the milkweed and apocynale larval host plants of danaines. CGs function to (1) defend plants against many herbivores, (2) defend *Danaus* species that are adapted to feed as larvae on these plants against some of their bird predators, and (3) defend Batesian mimics of *Danaus* species, such as *L. archippus*, the viceroy butterfly, against birds.

Danaines are also members of an even larger system of insect–plant interactions, in which other secondary plant substances—PAs—are significant. PAs also protect plants of several families as feeding deterrents against herbivores. Some adapted insects (e.g., larvae of several arctiid moths) eat these plants, and some even use the PAs for their own protection. Adult danaines, like some ctenuchids, arctiids, and ithomiines, search for plants with PAs on their surface, after drying or when wilting. Lepidoptera either ingest a PA solution from moist plant surfaces or wet the plant surfaces themselves with "saliva" before imbibing the solution. Evidence is accumulating for the postulate that PAs acquired by this behavior function as (1) defensive allomones in addition to CGs sequestered by larval feeding; (2) an important chemical defense in those cases where CG storage is absent, as in the majority of danaines; (3) a transferable chemical defense since males can contribute PAs to females in their spermatophore, and the female may then supply the eggs with PAs; and (4) indispensable precursors of pheromone components in male hairpencils.

ADDENDUM

After the completion of this paper, some important findings became known which relate to the theme of danaines and secondary plant compounds. (1) The role of the PAs in the Lepidoptera was critically reviewed by Boppré (1990; see also Boppré, this volume). (2) Details of danaine androconial structures which relate to the taxonomy of three danaine genera were published by Boppré and Vane-Wright (1989). (3) Another paper on the antennal receptor cells which initiate insect attraction to PA plants was published by Bogner and Boppré (1989). (4) S.B. Malcolm answered my question whether female monarch butterflies sense milkweed CGs during their ovipositing flight and said: . . . "the evidence for female perception of plant cardenolides is accumulating" because (a) Malcolm and Brower (1986) found that monarchs laid more eggs on *Asclepias curassavica* (with high CG content) than on *A. incarnata* (with lower CG content); and (b) Zalucki et al. (1990) reported that female monarchs preferred plants with intermediate CG content over those with either high or low content. (5) In two moth species, the uptake of PAs and CGs—which play such a crucial role in the danaines—was studied in some detail. The presumption that these compounds do not pass passively through the cell wall of the gut (because of their chemical nature) was found to be true. Furthermore, evidence for preferential PA storage in the integument was published by Wink and Schneider (1988), Egelhaaf et al. (1990) and Wink et al. (1990).

ACKNOWLEDGMENTS

I acknowledge with sincere gratitude the constructive criticism of this text by Drs. S.B. Malcolm and L.P. Brower. The former did not spare any effort to improve on the English and eventually had the paper retyped while I was ill. Thanks also to Dr. M. Boppré, who kindly criticized an earlier version. Dr. R. Kittmann (Konstanz) collected some of the danaines of Figure 1 in Sri Lanka and Dr. Sunil Ranasinghe (Peradenyia, Sri Lanka) provided information on their habitats.

LITERATURE CITED

Ackery, P.R., and R.I. Vane-Wright. 1984. *Milkweed butterflies: Their cladistics and biology.* London: British Museum (Natural History); and Ithaca, NY: Cornell University Press, 425 pp.

Barbier, M. 1979. *Introduction to chemical ecology.* London and New York: Longman, 128 pp.

Barth, R. 1960. Órgãos odoríferos dos Lepidópteros. *Boletim Parque Nacional do Itatiaia* 7:1–159.

Bellas, T.E., R.G. Brownlee, and R.M. Silverstein. 1974. Isolation, tentative identification, and synthesis studies of the volatile components of the hairpencil secretion of the monarch butterfly. *Tetrahedron Letters* 1974(30):2267–2271.

Bernays, E.A., and R.F. Chapman. 1977. Deterrent chemicals as a basis of oligophagy in *Locusta migratoria* (L.). *Ecological Entomology* 2:1–18.

Birch, M.C. 1979. Eversible structures. In *Moths and butterflies of Great Britain and Ireland,* Vol. 9, *Sphingidae-Noctuidae,* eds. J. Heath and A.M. Emmet, 9–18. London: Curwen Books.

Bogner, F., and M. Boppré. 1985. Ein spezieller Riechrezeptor zur Wahrnehmung von Alkaloiden bei Lepidopteren. *Verhandlungen der Deutschen Zoologischen Gesellschaft* 78:281.

Bogner, F., and M. Boppré. 1989. Single cell recordings reveal hydroxydanaidal as the volatile compound attracting insects to pyrrolizidine alkaloids. *Entomologia experimentalis et applicata* 50:171–184.

Boppré, M. 1977. Pheromonbiologie am Beispiel des Monarchfalter. *Biologie in Unserer Zeit (Verlag Chemie, Weinheim)* 7(6):161–169.

Boppré, M. 1978. Chemical communication, plant relationships, and mimicry in the evolution of danaid butterflies. *Entomologia Experimentalis et Applicata* 24:264–277.

Boppré, M. 1979. Untersuchungen zur Pheromonbiologie bei Monarch-faltern (Danaidae). Ph.D. thesis. Munich: Ludwigs-Maximilians-University.

Boppré, M. 1981. Adult Lepidoptera 'feeding' at withered *Heliotropium* plants (Boraginaceae) in East Africa. *Ecological Entomology* 6:449–452.

Boppré, M. 1983. Leaf-scratching—A specialized behaviour of danaine butterflies (Lepidoptera) for gathering secondary plant substances. *Oecologia (Berlin)* 59:414–416.

Boppré, M. 1984a. Chemically mediated interactions between butterflies. In *The biology of butterflies,* eds. R.I. Vane-Wright and P.R. Ackery, *Symposia of the Royal Entomological Society of London* 11: 259–275. London and New York: Academic Press (1989, Princeton, NJ: Princeton University Press).

Boppré, M. 1984b. Redefining "pharmacophagy." *Journal of Chemical Ecology* 10:1151–1154.

Boppré, M. 1986. Insects pharmacophagously utilizing defensive plant chemicals (pyrrolizidine alkaloids). *Naturwissenschaften* 73:17–26.

Boppré, M. 1990. Lepidoptera and pyrrolizidine alkaloids: Exemplification of complexity in chemical ecology. *Journal of Chemical Ecology* 16:165–185.

Boppré, M., and K. Fecher. 1977. Structural complexity of scent organs of male danaid butterflies. *Scanning Electron Microscopy* 2:639–644.

Boppré, M., and R.I. Vane-Wright. 1989. Androconial systems in Danainae (Lepidoptera): Functional morphology of *Amauris, Danaus, Tirumala* and *Euploea. Zoological Journal of the Linnean Society* 97: 101–133.

Boppré, M., R.L. Petty, D. Schneider, and J. Meinwald. 1978. Behaviorally mediated contacts between scent organs: Another prerequisite for pheropmone production in *Danaus chrysippus* males (Lepidoptera). *Journal of Comparative Physiology* 126:97–103.

Brower, J.V.Z. 1958. Experimental studies of mimicry in some North American butterflies. I. The monarch, *Danaus plexippus,* and viceroy, *Limenitis archippus archippus. Evolution* 12:32–47.

Brower, L.P. 1969. Ecological chemistry. *Scientific American* 220(2):22–29.

Brower, L.P. 1984. Chemical defence in butterflies. In *The biology of butterflies,* eds. R.I. Vane-Wright and P.R. Ackery, 109–134. London and New York: Academic Press.

Brower, L.P., and J.V.Z. Brower. 1964. Birds, butterflies and plant poisons. A study in ecological chemistry. *Zoologica (New York)* 49:137–159.

Brower, L.P., J.V.Z. Brower, and F.P. Cranston. 1965. Courtship behavior of the queen butterfly, *Danaus gilippus berenice. Zoologica (New York)* 50:1–39.

Brower, L.P., W.N. Ryerson, L.L. Coppinger, and S.C. Glazier. 1968. Ecological chemistry and the palatability spectrum. *Science* 161:1349–1351.

Brower , L.P., M. Edmunds, and C.M. Moffitt. 1975. Cardenolide content and palatability of a population of *Danaus chrysippus* butterflies from West Africa. *Journal of Entomology (A)* 42:183–196.

Brown, K.S., Jr. 1984. Adult-obtained pyrrolizidine alkaloids defend ithomiine butterflies against a spider predator. *Nature* 309:707–709.

Cohen, J.A. 1985. Differences and similarities in cardenolide contents of queen and monarch butterflies in Florida and their ecological and evolutionary implications. *Journal of Chemical Ecology* 11:85–103.

Conner, W.E., T. Eisner, R.K. Vander Meer, A. Guerrero, and J. Meinwald. 1981. Precopulatory sexual interaction in an arctiid moth (*Utetheisa ornatrix*): Role of a pheromone derived from dietary alkaloids. *Behavioral Ecology and Sociobiology* 9:227–235.

Dussourd, D., K. Ubik, J.F. Resch, J. Meinwald, and T. Eisner. 1984. Egg protection by parental investment of plant alkaloids in Lepidoptera. Abstract. *XVII International Congress of Entomology, Hamburg,* p. 840.

Edgar, J.A. 1975. Danainae (Lep.) and 1,2-dehydropyrrolizidine alkaloid-containing plants—With reference to observations made in the New Hebrides. *Philosophical Transactions of the Royal Society of London B* 272:467–476.

Edgar, J.A., and C.C.J. Culvenor. 1974. Pyrrolizidine ester alkaloid in danaid butterflies. *Nature* 248:614–616.

Edgar, J.A., C.C.J. Culvenor, and L.W. Smith. 1971. Dihydro-pyrrolizine derivatives in the "hairpencil" secretions of danaid butterflies. *Experientia* 27:761–762.

Edgar, J.A., C.C.J. Culvenor, and G.S. Robinson. 1973. Hairpencil dihydropyrrolizines of Danainae from the New Hebrides. *Journal of the Australian Entomological Society* 12:144–150.

Edgar, J.A., P.A. Cockruum, and J.L. Frahn. 1976a. Pyrrolizidine alkaloids in *Danaus plexippus* L. and *Danaus chrysippus* L. *Experientia* 32:1535–1537.

Edgar, J.A., C.C.J. Culvenor, and T.E. Pliske. 1976b. Pyrrolizidine alkaloid-derived pheromone on the costal fringes of male Ithomiinae. *Journal of Chemical Ecology* 2:263–270.

Edgar, J.A., M. Boppré, and D. Schneider. 1979. Pyrrolizidine alkaloid storage in African and Australian danaid butterflies. *Experientia* 35:1447–1448.

Egelhaaf, A., K. Cölln, B. Schmitz, M. Buck, M. Wink, and D. Schneider. 1990. Organ specific storage of dietary pyrrolizidine alkaloids in the arctiid moth *Creatonotos transiens. Zeitschrift für Naturforschung* 45c:172–177.

Ehrlich, P.R., and P.H. Raven. 1965. Butterflies and plants: A study in coevolution. *Evolution* 18:586–608.

Eisner, T. 1982. For love of nature: Explorations and discovery at biological field stations. *BioScience* 32: 321–326.

Eisner, T., and J. Meinwald. 1987. Alkaloid-derived pheromones and sexual selection in Lepidoptera. In *Biochemistry of insect pheromones,* eds. G. Prestwich and G. Blomquist. New York: Academic Press.

Eltringham, H. 1913. On the scent apparatus in the male of *Amauris niavius* Lin. *Transactions of the Entomological Society of London* 1913:399–406.

Eltringham, H. 1915. Further observations on the structure of the scent organs of certain male danaine butterflies. *Transactions of the Entomological Society of London* 1915:152–176.

Fink, L.S., and L.P. Brower. 1981. Birds can overcome the cardenolide defence of monarch butterflies in Mexico. *Nature* 291:67–70.

Florkin, M., and E. Schoffeniels. 1969. *Molecular approaches to ecology.* New York and London: Academic Press.

Fraenkel, G. 1959. The raison d'être of secondary plant substances. *Science* 129:1466–1470.

Good, R. 1974. *Features of evolution in the flowering plants.* New York: Dover.

Goss, G.J. 1979. The interaction between moths and plants containing pyrrolizidine alkaloids. *Environmental Entomology* 8:487–493.

Guilford, T., C. Nicol, M. Rothschild, and B.P. Moore. 1987. The biological roles of pyrazines: Evidence for a warning odour function. *Biological Journal of the Linnean Society* 31:113–128.

Harborne, J.B. (ed.). 1972. *Phytochemical ecology.* New York and London: Academic Press.

Harborne, J.B. 1977. *Introduction to ecological biochemistry.* London: Academic Press.

Hochachka, P.W., and G.N. Somero. 1973. *Strategies of biochemical adaptation.* Philadelphia: Saunders.

Illig, K.G. 1902. Duftorgane der männlichen Schmetterlinge. *Zoologica (Stuttgart)* 38:1–34.

Kelley, R.B., J.N. Seiber, A.D. Jones, H.J. Segall, and L.P. Brower. 1987. Pyrrolizidine alkaloids in overwintering monarch butterflies (*Danaus plexippus*) from Mexico. *Experientia* 43:943–946.

Malcolm, S.B., and L.P. Brower. 1986. Selective oviposition by monarch butterflies (*Danaus plexippus* L.) in a mixed stand of *Asclepias curassavica* L. and *A. incarnata* L. in south Florida. *Journal of the Lepidopterists' Society* 40(4):255–263.

Meinwald, J., and Y.C. Meinwald. 1966. Structure and synthesis of the major component in the hairpencil secretion of a male butterfly, *Lycorea ceres ceres* (Cramer). *Journal of the American Chemical Society* 88:1305–1310.

Meinwald, J., Y.C. Meinwald, J.W. Wheeler, T. Eisner, and L.P. Brower. 1966. Major components in the exocrine secretion of a male butterfly (*Lycorea*). *Science* 151:583–585.

Meinwald, J., A.M. Chalmers, T.E. Pliske, and T. Eisner. 1968. Pheromones. III. Identification of trans, trans 10-hydroxy-3,7-dimethyl-2,6-decadienoic acid as a major component in "hairpencil" secretions of the male monarch butterfly. *Tetrahedron Letters* 47: 4893–4896.

Meinwald, J., Y.C. Meinwald, and P.H. Mazzochi. 1969. Sex pheromone of the queen butterfly: Chemistry. *Science* 164:1174–1175.

Morse, D.H. 1986. Inflorescence choice and time allocation by insects foraging on milkweed. *Oikos* 46: 229–236.

Müller, F. 1877a. As maculas sexuaes do individuos masculinos das especies *Danaus erippus* e *D. gilippus. Archivos do Museo Nacional, Rio de Janeiro* 2:25–29. (English translation in Longstaff, G.B. 1912. *Butterfly hunting in many lands,* 604–615. London: Longmans, Green & Co.).

Müller, F. 1877b. Uber Haarpinsel, Filzflecke und ähnliche Gebilde auf den Flügeln männlicher Schmetterlinge. *Jenaische Zeitschrift für Naturwissenschaften* 5:25–29.

Myers, J., and L.P. Brower. 1969. A behavioural analysis of the courtship pheromone receptors of the queen butterfly, *Danaus gilippus berenice. Journal of Insect Physiology* 15:2117–2130.

Parsons, J.A. 1965. A digitalis-like toxin in the monarch butterfly *Danaus plexippus. Journal of Physiology, London* 178:290–304.

Pasteels, J.M., J.-C. Grégoire, and M. Rowell-Rahier. 1983. The chemical ecology of defense in arthropods. *Annual Review of Entomology* 28:263–289.

Petty, R.L., M. Boppré, D. Schneider, and J. Meinwald. 1977. Identification and localization of volatile hairpencil components in male *Amauris ochlea* butterflies (Danaidae). *Experientia* 33:1324–1326.

Pliske, T.E. 1975. Attraction of Lepidoptera to plants containing pyrrolizidine alkaloids. *Environmental Entomology* 4:455–473.

Pliske, T., and T. Eisner. 1969. Sex pheromones of the queen butterfly: Biology. *Science* 164:1170–1172.

Pliske, T.E., and M.M. Salpeter. 1971. The structure and development of the hairpencil glands in males of the queen butterfly, *Danaus gilippus berenice. Journal of Morphology* 134:215–240.

Reichstein, T. 1967. Cardenolide (herzwirksame Glykoside) als Abwehrstoffe bei Insekten. *Naturwissenschaftliche Rundschau* 20:499–511.

Reichstein, T., J.V. Euw, J.A. Parsons, and M. Rothschild. 1968. Heart poisons in the monarch butterfly. *Science* 161:861–866.

Remmert, H. 1980. *Ecology.* Berlin: Springer-Verlag.

Remmert, H. 1982. And now? Ecosystem research! In *Trends in ecological research in the 1980's,* eds. J.H. Cooley and F.B. Golley, 179–191. New York and London: Plenum Press.

Rosenthal, G.A. 1986. The chemical defenses of higher plants. *Scientific American* 254(1):76–81.

Rothschild, M. 1985. British aposematic Lepidoptera. In *The moths and butterflies of Great Britain and Ireland,* Vol. 2, 9–62. Essex, Colcester: Harley Books.

Rothschild, M., and N. Marsh. 1978. Some peculiar aspects of danaid/plant relationships. *Entomologia Experimentalis et Applicata* 24:437–450.

Rothschild, M., T. Reichstein, J.V. Euw, R. Aplin, and R.R.M. Harman. 1970. Toxic lepidoptera. *Toxicon* 8:293–299.

Rothschild, M., R.T. Aplin, P.A. Cockrum, J.A. Edgar, P. Fairweather, and R. Lees. 1979. Pyrrolizidine alkaloids in arctiid moths (Lep.) with a discussion on host plant relationships and the role of these secondary plant substances in the Arctiidae. *Biological Journal of the Linnean Society* 12:305–326.

Rothschild, M., B.P. Moore, and W.V. Brown. 1984. Pyrazines as warning odour components in the monarch butterfly, *Danaus plexippus,* and in moths of the genera *Zygaena* and *Amata* (Lepidoptera). *Biological Journal of the Linnean Society* 23:375–380.

Schneider, D. 1975. Pheromone communication in moths and butterflies. In *Sensory physiology and behavior,* eds. R. Galun, P. Hillman, I. Parnas, and R. Werman, 173–193. New York: Plenum Press.

Schneider, D. 1986. The strange fate of the pyrrolizidine alkaloids. In *Perspectives in chemoreception and behavior,* eds. R.F. Chapman, E.A. Bernays, and J.G. Stoffolano, 123–142. New York: Springer-Verlag.

Schneider, D. 1987. Plant recognition by insects: A challenge for neuroethological research. In *Insect plants. Proceedings of the 6th International Symposium on Insect-Plant Relationships,* eds. V.G. Labeyrie, G. Fabres, and D. Lachaise. Amsterdam: W. Junk.

Schneider, D., and U. Seibt. 1969. Sex pheromone of the queen butterfly: Electroantennogram responses. *Science* 164:1173–1174.

Schneider, D., M. Boppré, H. Schneider, W.R. Thompson, C.J. Boriack, R.L. Petty, and J. Meinwald. 1975. A pheromone precursor and its uptake in male *Danaus* butterflies. *Journal of Comparative Physiology* 97:245–256.

Schulz, S. 1987. *Die Chemie der Duftorgane männlicher Lepidopteren.* Ph.D. thesis. Hamburg: University of Hamburg.

Schulz, S., W. Francke, J. Edgar, and D. Schneider. 1988. Volatile compounds from androconial organs of danaine and ithomiine butterflies. *Zeitschrift für Naturforschung* 43c:99–104.

Sondheimer, E., and J.B. Simeone. 1970. *Chemical ecology.* New York: Academic Press.

Varley, G.C. 1962. A plea for a new look at Lepidoptera with special reference to the scent distributing organs of male moths. *Transactions of the Society for British Entomology* 15:29–40.

Whittaker, R.H., and P.P. Feeny. 1971. Allelochemics: Chemical interactions between species. *Science* 171: 757–770.

Wink, M., and D. Schneider. 1988. Carrier-mediated uptake of pyrrolizidine alkaloids in larvae of the aposematic and alkaloid-exploiting moth *Creatonotos. Naturwissenschaften* 75:524–525.

Wink, M., E. von Nickisch-Rosenegh, and D. Schneider. 1990. Processing of pyrrolizidine alkaloids in three moths, *Syntomis mogadorensis, Syntomeida epilais* and *Creatonotos transiens. Symposia Biologica Hungarica* 39:53–61.

Zalucki, M.P., L.P. Brower, and S.B. Malcolm. 1990. Oviposition by *Danaus plexippus* in relation to cardenolide content of three *Asclepias* species in the southeastern USA. *Ecological Entomology* 15(2):231–240.

THE AMERICAN MONARCH: COURTSHIP AND CHEMICAL COMMUNICATION OF A PECULIAR DANAINE BUTTERFLY

Michael Boppré

ABSTRACT. The androconial organs, courtship behavior, and the roles of chemical signals and allelochemicals in *Danaus* (*Danaus*) *plexippus plexippus*, the American monarch, are discussed in comparison with other danaine butterflies. In the monarch, and in other members of *Danaus* sensu stricto, a change in mating strategy has evolved so that chemical communication plays a minor role in comparison with butterflies in other danaine genera. To date, there is no plausible explanation for this curious evolutionary change because many basic questions on the behavioral and chemical ecology of the monarch remain unanswered.

The monarch butterfly could well be designated nature's prime example of the male chauvinistic pig.
Miriam Rothschild (1978)

INTRODUCTION

Considering how much is known about the biology of the monarch butterfly, *Danaus* (*Danaus*) *plexippus plexippus* (L.), it is curious how poorly understood courtship and chemical communication is in this widely distributed species. The literature on these aspects of monarch biology is strikingly scant, and one can summarize the available information almost in one sentence: Compared to most of its danaine relatives, *D. plexippus* has abolished elaborate androconial organs and complex courtship rituals, as well as complex bouquets of male pheromones; instead, it employs comparatively simple mating behavior where male pheromones play a minor role and in which rape is common.

This paper reviews evidence for the simplicity of monarch reproductive behavior and compares the morphology, histology, and chemistry of the androconial organs of *D. plexippus* with some other danaines. Because little information on the behavioral and chemical ecology of reproduction in the American monarch is available, neither its sexual peculiarities can be interpreted from an ecological perspective, nor is an evolutionary explanation yet possible. The discussion and the new data establish the questions that require investigation, and my aim is to stimulate further work on the monarch's sex life.

The chemical ecology of the subfamily Danainae, the milkweed butterflies, as well as their reproductive biology, has recently been reviewed extensively by Ackery and Vane-Wright (1984) and Boppré (1978, 1984a, 1986).

MALE SECONDARY SEXUAL CHARACTERS—RUDIMENTARY FEATURES?

MORPHOLOGY AND HISTOLOGY OF MONARCH ANDROCONIAL ORGANS

Binate androconial organs, used to secrete and disseminate pheromones during courtship, are one of the basic characters of the Danainae. The males possess extrudable abdominal "hairpencils," and in most species they are also equipped with alar glands. Male *D. plexippus* possess hairpencils as well as wing pockets, and their basic morphology has long been known (Müller, 1877 [see Fig. 1]); Illig, 1902). More recent investigations (Hausman, 1951; Urquhart, 1958, 1960, 1976) have added little information to the widely "forgotten" early papers. Instead, our understanding of the "scent organs" in the American monarch is confused and even misleading in part. Therefore, I summarize and illustrate the basic morphology and histology of monarch androconial organs (cf. Boppré and Vane-Wright, 1989, for basic features of danaine androconial organs and terminology).

The alar androconial organs of *D. plexippus* are

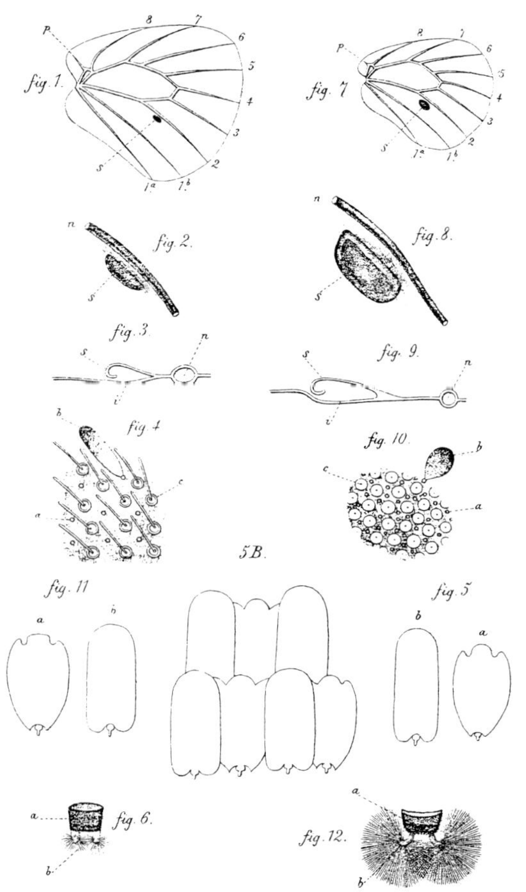

Figure 1. Figure reproduction from Müller (1877) comparing the androconia of *D. erippus* (**1–6**) with that of *D. gilippus* (**7–12**). (*Danaus* (*Danaus*) *erippus* and *D. plexippus* are similar; *D.* (*Anosia*) *gilippus* resembles *D.* (*A.*) *chrysippus*.) **1** and **7**, position of pockets on hindwings; **2** and **8**, magnified pocket (3.2×); **3** and **9**, transverse section through pocket (9.7×); **4** and **10**, pocket scale, cushions, and cushion scales inside pockets (120×) (cf. Fig. 2A and 2B); **5** and **11**, ordinary scales from the upper surface of the hindwings (120×); **6** and **12**, expanded hairpencils (1.3×) (cf. Fig. 2C and 2D).

located on the hindwings adjacent to vein Cu_{1b} and show up as pocketlike cavities (folds of the upper wing membranes) with orifices toward the inner margin of the hindwings (cf. Fig. 1.1–1.3). The inside of these alar pockets (Figs. 1.4 and 2A) is lined by "pocket scales," which (1) are larger and less densely arranged than the ordinary covering scales (Fig. 1.5) and (2) have a double lattice structure (cf. Boppré and Vane-Wright, 1989). Intermittently, roundish projections ("cushions") occur, each with a long hair ("cushion scale") arising from the center (Figs. 1.4 and 2A). While the alar organs of *D. plexippus* and *Danaus chrysippus* are located on corresponding hindwing areas and are made of the same structural elements (cf. Fig. 2A and 2B), the pockets of the two species differ significantly in their histology: In *D. chrysippus* the cells under the pocket scales are the most prominent structures (Fig. 3B), but in *D. plexippus* large cells occur under both pocket scales and cushions (Fig. 3A; see also Figs. 4 and 5, and respective legends). In addition, vein Cu_{1b} of the monarch contains specialized cells (like other danaines) and is, thus, strikingly different from all other veins, as well as from all female wing veins. Hence, vein Cu_{1b} must be considered as part of the androconial organs.

The hairpencils are usually hidden inside the abdomen, but they can be everted by hemolymph pressure from the end of the abdomen to appear as two spheres of hair at the tips of stalks (Figs. 1.6 and 1.12 and 2C and 2D). In *D. plexippus,* each hairpencil consists of about 400 hairs, which are 2.1 mm long and 10 μm in diameter (Table 1), and their surface is sculptured (Fig. 2E). Viewed macroscopically, hairpencils of the monarch appear miniaturized compared to those of other danaines such as *D. chrysippus* (Fig. 2D), which has about 1,300 hairs, each 4.5 mm long (Table 1). Unlike *D. chrysippus, D. plexippus* also possesses only one type of hair and lacks pheromone-transfer particles (PTPs) (cf. below and Fig. 2E and 2F) entirely. Histologically, the hairs are very similar to *D. chrysippus,* and in freshly hatched males (<24 hr) prominent cells (80 × 25 μm) are found beneath the sockets of the hairs, which contain large nuclei and microvilli. These cells appear very similar but are slightly larger in *D. chrysippus,* at approximately 115 × 30 μm (Table 1), and in both species they are probably glandular and release a secretion into the hair shafts. In contrast to the glandular cells of the alar pockets, the hairpencil cells atrophy very early in adult life. *Danaus plexippus* also differs in that the glandular cells contain round to oval inclusions of 1–2 μm, and their number increases with age. At present, these inclusions cannot be properly interpreted, and their possession of a capsule and a membrane contradicts the assumption that they would represent secretions. Nevertheless, they appear to be typical for the monarch and were found in preparations of both indoor-raised and field-caught males (O.W. Fischer, pers. comm.).

In all other Danaini, except *Danaus* sensu stricto, both the hairpencils and the alar organs are much larger compared to those of *D. plexippus,* and often they are very elaborate (Boppré and Fecher, 1977; Boppré and Vane-Wright, 1989). For example, the hairpencils of *Amauris* comprise up to five structurally different hair types, which unfold sequentially. The Amaurina and the genera *Danaus* (*Salatura*), *Danaus* (*Anosia*), and *Tirumala* produce (in different ways) cuticular, pheromone-bearing particles, or PTPs (Boppré, 1976), which are disseminated during courtship behavior. In comparison with such specialized and complex structures, the androconial organs of *D. plexippus* appear considerably less complex and unspecialized. Nevertheless, the abdominal and alar secondary char-

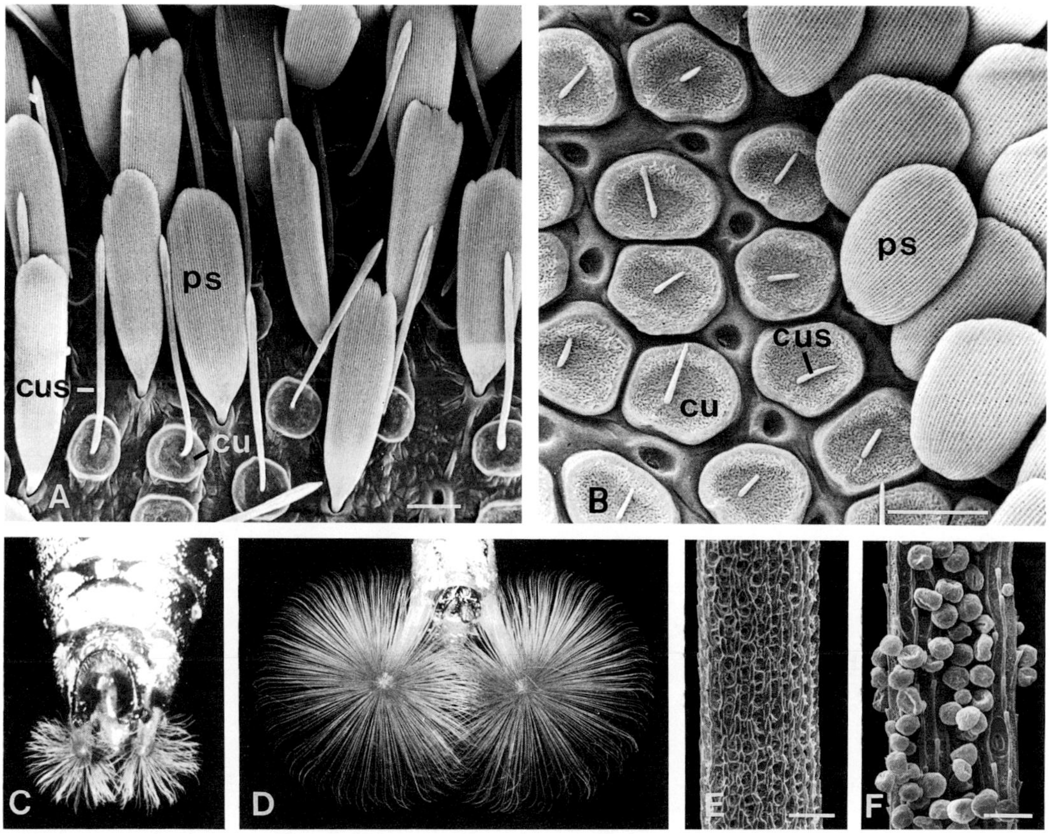

Figure 2. Scanning electron micrographs (**A**, **B**, **E**, **F**) and macrophotographs (C, D) of androconial organs in *D. plexippus* (**A**, **C**, **E**) and *D.* (*A.*) *chrysippus* (**B**, **D**, **F**) to contrast respective structures. **A** and **B**, inside alar pockets (cu = cushion; cus = cushion scales; ps = pocket scales); C and **D**, artificially expanded hairpencil; **E** and **F**, hairpencil hair (Fischer and Boppré, unpubl. originals). Scale bars: **A** and **B** = 20 μm; C and **D** = ca. 3×; **E** and **F** = 5 μm.

acters in male monarchs consist of all the microscopical and histological elements, with the exception of PTPs, which are typical of androconial organs in all Danainae, and there are no signs of an atrophied secretory function in these structures. The organs are simply smaller in the monarch, and also in the doubtfully distinct species *Danaus erippus* and the *plexippus/erippus* sister-species *Danaus* (*D.*) *cleophile.* As can be seen in Boppré and Vane-Wright (1989), the androconial organs of *Danaus* sensu stricto are secondarily modified and do not represent an early stage of phylogenetic development.

CHEMISTRY OF MONARCH ANDROCONIAL ORGANS

With respect to the chemistry of the androconial secretions, prior to the recent study by Schulz (1987) there was little information available. Some authors reported that the hairpencils emit a "flower-like aroma" (Urquhart, 1958), or a "faint, fragrant odor" (Hausman, 1951). By chemical analyses, Meinwald et al. (1968, 1969) identified two bishomoterpenes ((*E,E*)-10-hydroxy-3,7-dimethyl-2,6-decadienoic acid and (*E,E*)-3,7-dimethyl-2,6-decadiendioic acid) in extracts of hairpencils. Bellas et al. (1974) found benzyl caproate, and a ketone with unknown configuration, in extracts of abdominal tips. Schulz (1987) reinvestigated the hairpencils and found, in addition to the above-mentioned chemicals, numerous acetogenins, aromatic esters, and terpenoids in extracts of hairpencils. In extracts of alar pockets, he detected only traces of the major hairpencil compounds but larger amounts of ubiquitous fatty acids and hydrocarbons, which are typical cuticular components. Table 2 lists the compounds and Figure 6 shows representative gas chromatograms (cf. Schulz, 1987).

Compared with other danaine butterflies (Schulz, 1987; Schulz et al., 1988; see also Ackery and Vane-Wright, 1984), the American monarch thus seems similarly equipped with chemicals that could serve as male pheromones. However, male *D. plexippus* lack plant-derived dihydropyrrolizines, which are the characteristic, major hairpencil components in

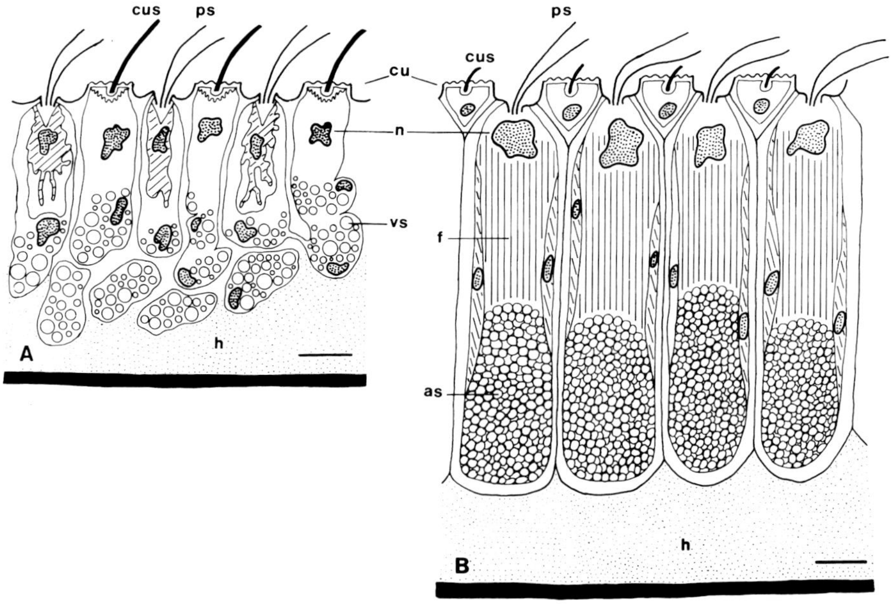

Figure 3. Schematic sketches of the cellular situation in the alar pocket of *D. plexippus* (**A**) and *D. chrysippus* (**B**), reconstructed from light and electron micrographs (O.W. Fischer, unpubl.). Within the area of a pocket, the space between the wing lamellae is widened and filled with hemolymph (h) as well as with prominent cells, aligned toward the lumen of the pocket and containing large nuclei (n). Two major differences between the species are evident: (1) the cells underneath the cushions (cu) are very small in *D. chrysippus* but prominent in *D. plexippus*; and (2) the intermittently located cells underneath the pocket scales (ps) in *D. chrysippus* contain an amorphous substance (as) at their bases, and fibrous structures (f) medially, while in *D. plexippus,* all cells are filled with vesicles (vs). In addition, the arrangement of the glandular cells in *D. chrysippus* is quite regular, while in *D. plexippus* the basal parts of the cells form interleaved processes. Because the cellular structure of the alar organs undergo marked differentiation during the first days of adult life, 14-day-old males were used for comparison. The thick black line marks the cuticle of the outer wing membrane. Scale bar = 20 μm.

other Danainae (see below). Apart from dihydropyrrolizines, the bouquets of volatiles released by male danaine hairpencils are not understood functionally. They may serve a role in species recognition, but proof is most difficult to establish, and comparison with other danaine butterflies does not help interpret monarch behavior.

Thus, in contrast to previous understanding, the histology and chemistry of androconial organs in the monarch do not indicate that chemical communication should be less elaborate than in other species of the subfamily. However, more detailed comparisons among danaine androconia are required for a fuller interpretation.

BEHAVIORAL USE OF ANDROCONIAL ORGANS— NO FUNCTION?

In most species of the tribe Danaini, male hairpencils in general are used in two different contexts. First, independent of courtship, hairpencils are protruded and introduced into the alar pockets, or fanned out on the alar patches. For *D. chrysippus,* it has been demonstrated that this contact behavior is required for biosynthesis of the important pheromone component "danaidone" (see below) in physiologically normal amounts. For *Tirumala* species, similar mechanical contacts between the two organs load the hairpencils with PTPs produced in the alar pouches. Second, during the last phases of courtship flight, a male briefly expands its hairpencils close to a female's antennae and disseminates the volatile secretions and PTPs.

Other danaine groups (e.g., *Lycorea, Euploea*) either have no alar organs or contact behavior has never been observed. However, after visual courtship initiation, the use of hairpencils in close-range stimulation of the female with pheromones seems generally to be required by the female to accept a mate. The most detailed proof for the importance of hairpencil secretion comes from studies on *Danaus gilippus berenice* by Brower et al. (1965) and Pliske and Eisner (1969). These authors clearly demonstrated that hairpencilling by the male is a necessity for successful courtship and, moreover, that danaidone is a key stimulus. Although the behavioral role of hairpencils may be partially modified in other danaine species, the findings obtained with *D. gilippus* basically seem to hold true for

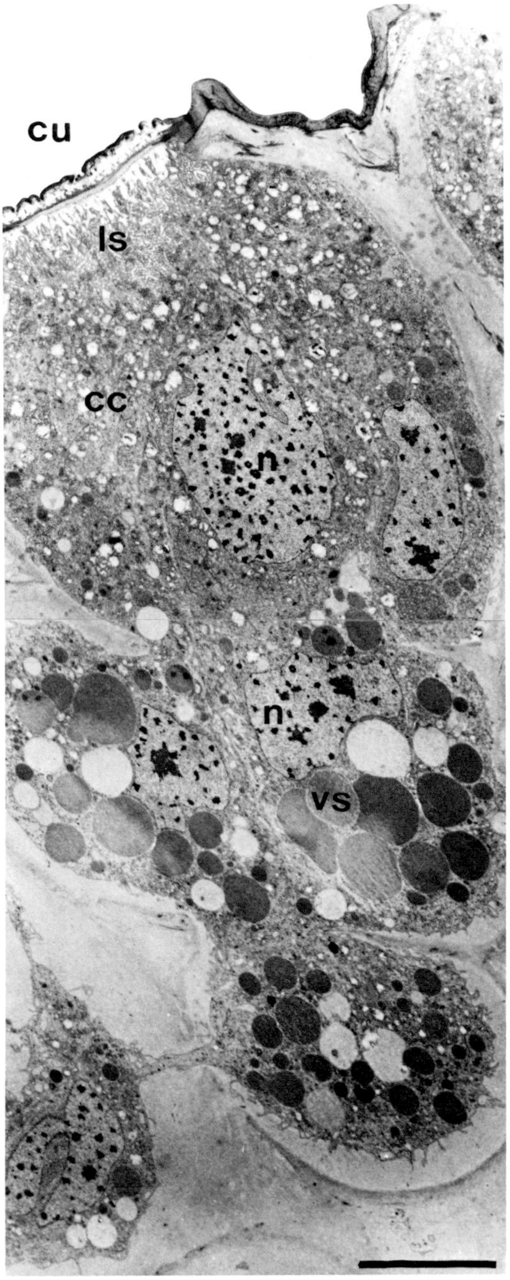

Figure 4. Transmission electron micrograph of an alar pocket of a 14-day-old *D. plexippus* male showing the functional unit of glandular cells underneath the cushions (cu) (cf. Fig. 3A). A central cell (cc) is bordered with a heavily lamellated structure (ls) at the thin, perforated cuticle of the cushion and shows many empty vacuoles. Basal to the central cell, further interlaced cells occur, characterized by numerous vesicles (vs), which appear gradually during adult life (O.W. Fischer, unpubl. original). n = nucleus. Scale bar = 10 μm.

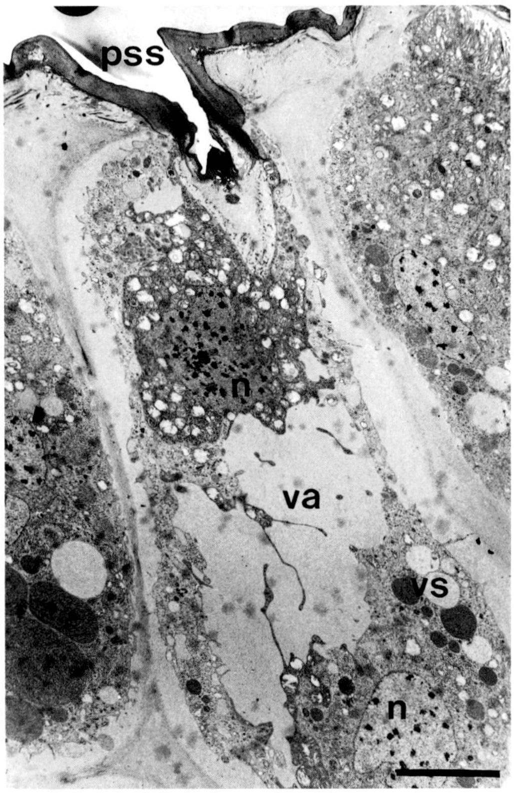

Figure 5. Transmission electron micrograph of an alar pocket of a 14-day-old *D. plexippus* male showing glandular cells underneath a pocket scale (cf. Fig. 3A). The cell close to the socket of the pocket scale (pss) is characterized by dense cytoplasm and a prominent nucleus (n) and is surrounded by the glandular cell type shown in Figure 4, with which it is also connected (O.W. Fischer, unpubl. original). va = vacuole; vs = vesicles. Scale bar = 10 μm.

most danaines (Seibt et al., 1972; Boppré, 1977; Boppré and Schneider, unpubl.). However, thorough analyses have not been performed.

In contrast to other Danaini, androconial contact behavior has never been observed in the American monarch, and Urquhart (1960) believes that it does not occur. Despite speculation by Urquhart on the role of the androconial organs, no evidence indicates a function. Male "hairpencilling" during courtship is not required, because when hairpencils are removed artificially there is no reduction in sexual vigor or "courtship" success (Pliske, 1975b); nevertheless, the abdominal organs are occasionally expanded prior to copulation (see the next section).

RAPE—AN ALTERNATIVE TO SEDUCTION?

The mating behavior of *D. plexippus* was analyzed in detail by Pliske (1975b; Fig. 7). According to this study, monarch courtship has an aerial phase and

Table 1. Comparison of morphometric data of alar and abdominal androconial organs in *D. plexippus* and *D. chrysippus* (Fischer and Boppré, unpubl.).

	D. plexippus	*D. chrysippus*
Alar organs		
Size of pocket scales (μm)	70 × 15	50 × 30
Cushion diameter (μm)	20	23
Cushion density (mm^{-2})	≈500	≈1,000
Size of cushion scales (μm)	55 × 2	7 × 1
Abdominal organs		
Hair diameter (μm)	10	10
Hair length (mm)	2.1	4.5
Number of hair types	1	2
Pheromone-transfer particles	no	yes
Number of hairs	≈400	≈1,300
Size of glandular cells (μm)	≈80 × 25	≈115 × 30

a ground phase. The aerial phase includes nudging, pursuit, hairpencilling, and takedown behaviors by the male, and the ground phase concludes courtship with landing and antennal palpation by the male, followed by copulation and postnuptial flight. Unlike other, closely related danaines, hairpencilling is not necessary for male reproductive success and occurs only in some courtships. Instead, aerial takedowns predominate.

The literature also contains several observations on parts of *D. plexippus* mating behavior. Urquhart (1960) describes male aerial pursuit of females to which females either respond with an ascending, pre-nuptial spiral flight or a fast, zig-zag escape flight. Like Urquhart (1960), Miller and Clench (1968), Pliske (1975b), Hill et al. (1976), Rothschild (1978), and Shull (1979) recorded male monarchs as the active partner in courtship. Hill et al. (1976) also described male aerial pursuit of other monarchs, followed by capture, rapid descent to the ground, and contact discrimination between male and female. Attempts at copulation on the ground were vigorous, and males occasionally extruded hairpencils during struggles with females.

In her account of male monarch mating behavior, Rothschild (1978) emphasizes the apparently indiscriminate and violent chauvinism often observed in both intra- and interspecific interactions with reproductively active males. Like Rothschild's (1978) observations, there are several reports of faulty discrimination by males (see also Van Hook, this volume). Tilden (1979) noted numerous attempts at mating between male monarchs at an overwintering site in California during dispersal from the site when sex ratios are strongly male-biased (see also Van Hook, this volume). Winter (1985) illustrated a male monarch pinning a *Vanessa virginiensis* to the ground after the monarch had attacked the nectaring butterfly.

Although Tuskes and Brower (1978) and Brower (1985) assumed that monarchs mate randomly at overwintering colonies in Mexico and California as the butterflies disperse in spring, Van Hook (this volume) shows that they do not mate randomly. Instead she found that the aggressive mating behavior of small, worn males resulted in their nonrandom copulation with large females in good condition. Such active aerial pursuit and takedown of females by males precluded female mate choice. Furthermore, Zalucki (this volume) shows that male presence at patches of host plants also influences mating patterns.

All these reports are basically consistent with Pliske's study, and there is little doubt that the American monarch has drastically changed its mating strategy from the typical danaine pattern. Although its courtship contains elements that are found in other Danainae, the monarch is unusual in the absence of both female choice and use of pheromones during courtship.

PLANT CHEMICALS—REQUIRED TO INCREASE FITNESS!?

It is another basic feature of danaine butterflies that males, and in a few species occasionally females, are strongly attracted to withering and decaying parts of plants containing pyrrolizidine alkaloids (PAs) and that they ingest these secondary plant chemicals (for review, see Boppré, 1986). PAs are required by the males for biosynthesis of dihydropyrrolizine pheromone components such as "danaidone," "danaidal," and/or "hydroxydanaidal," which make up the quantitatively major components of the hairpencil secretion (up to 500 μg) in the majority of Danainae, and appear as key chemicals in female mating decisions. Not only is male courtship success thus dependent on the plant chemicals, but the insects also store unconverted PAs in their body as a distasteful protective device against predation (see below).

Danaus plexippus can also be seen visiting dead and decaying PA plants, particularly *Senecio, Heliotropium, Crotalaria,* and *Parsonsia* (Edgar et al., 1973, 1974; Pliske, 1975a), and they gather PAs with nectar from plants such as *Eupatorium* (Edgar et al., 1976). Rothschild and Edgar (1978) have also shown that the larvae of both sexes of *D. plexippus* can sequester and store PAs sprayed onto a milkweed host plant, but that males store more PA than females. However, despite such abilities to handle PAs in *D. plexippus,* numerous investigations have clearly demonstrated that hairpencils of monarchs do not secrete PA derivatives (Meinwald et al., 1968; Edgar et al., 1971; Schulz, 1987).

Edgar et al. (1976) found unmodified PAs in extracts of the bodies of field-caught *D. plexippus* and first suggested that PAs might contribute to the

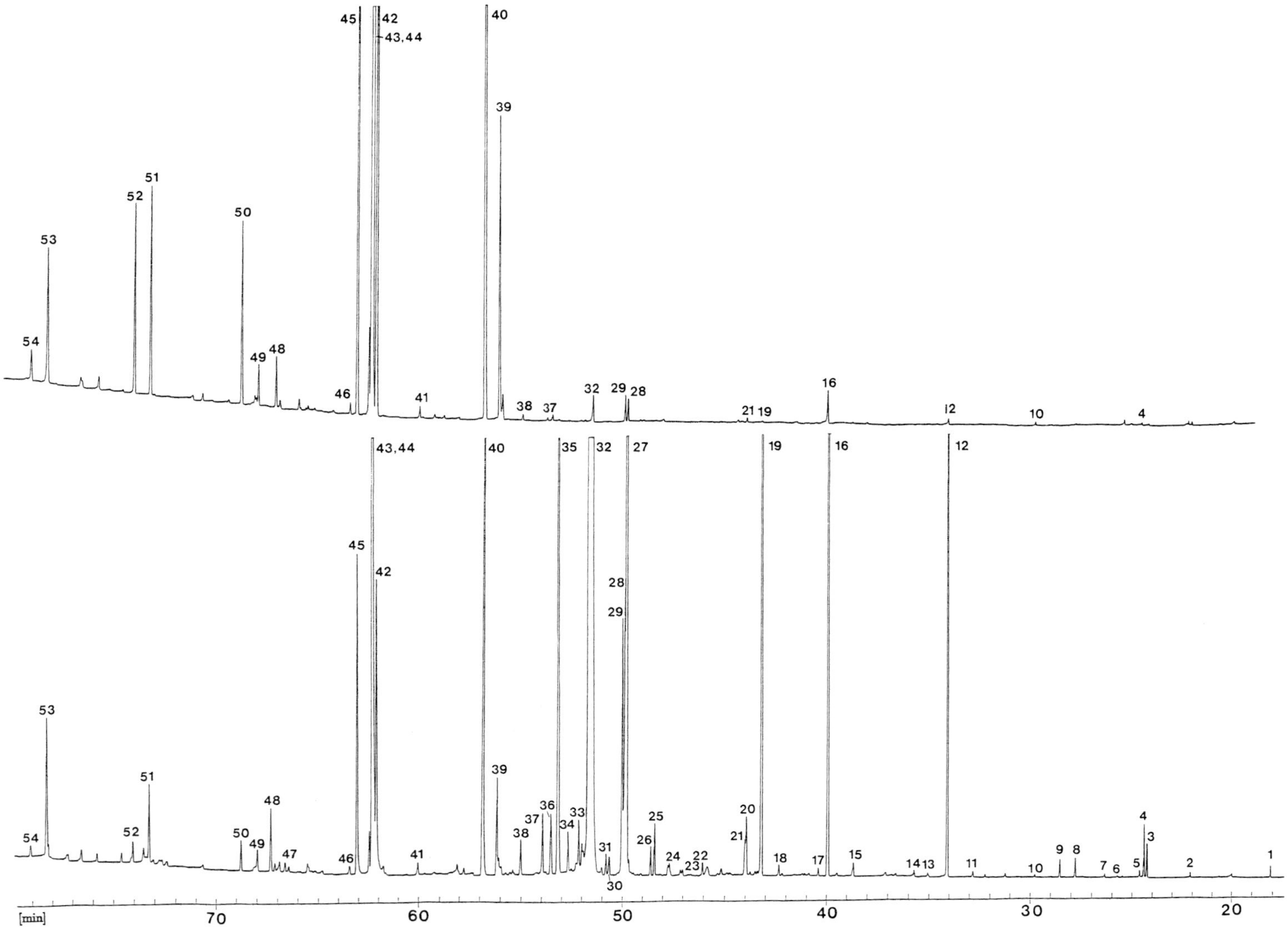

Figure 6. Gas chromatograms of methylated extracts of pockets (upper trace) and hairpencils (lower trace) of *D. plexippus* (50m CP-sil-8-CB, programmed from 60–300°C with 3°C/min) (original, courtesy of S. Schulz).

Table 2. Compounds identified from extracts of androconia (hairpencils and alar pockets) of *D. plexippus* (courtesy of S. Schulz, Hamburg; cf. Schulz, 1987) (cf. Fig. 6). * = tentative identification; +++ = major component; ++ = minor component; + = trace component.

	Hair-pencils	Alar pockets
1 Heptanoic acid	+	
2 Contamination	+	
3 Benzoic acid	+	
4 Linalool	+	+
5 Nonanal	+	
6 Oxoisophorone	+	
7 Nonanoic acid	+	
8 Ethyl benzoate	+	
9 Methyl salicylate	+	
10 Decanoic acid	+	+
11 Nerolic acid	+	
12 Geranic acid	++	
13 Undecanoic acid	+	
14 Benzyl butanoate	+	
15 Benzyl pentanoate	+	
16 (1R*, 3R*, 6S*)-1,3,7,7-Tetramethyl-2-oxabicyclo[4.4.0]-9-decen-7-one	+++	+
17 2-Phenylethyl pentanoate	+	
18 Pentadecane	+	
19 Benzyl caproate	+++	+
20 Lactone	+	
21 (*E*)-Nerolidol	+	
22 Unknown terpene	+	
23 2-Phenylethyl caproate	+	
24 Unknown compound	+	
25 3,7-Dimethyl-10-hydroxy-2,6-decadienoic acid	+	
26 (*E,Z*)- and/or (*Z,E*)-α-Farnesol	+	
27 3,7-Dimethyl-10-hydroxy-2,6-decadienoic acid	++	
28 (*E,E*)-α-Farnesol	++	+
29 Myristic acid	++	+
30 Benzyloctanoate	+	
31 Contamination	+	
32 (*E,E*)-3,7-Dimethyl-10-hydroxy-2,6-decadienoic acid	+++	+
33 (*E,E*)-Methyl farnesenoate	+	
34 Unknown compound	+	
35 (*E,E*)-3,7-Dimethyl-2,6-decadien-1,10-dioic acid	+++	
36 14-Hydroxy-3,7,11-trimethyl-2,6,10-tetradecatrienoic acid*	+	
37 Pentadecanoic acid	+	+
38 Contamination	+	+
39 Hexadecenoic acid	+	++
40 Palmitic acid	+++	+++
41 Heptadecanoic acid	+	+
42 Octadecadienoic acid	++	+++
43 Octadecatrienoic acid	+++	+++
44 Octadecenoic acid	+++	+++
45 Stearic acid	++	+++
46 Tricosane	+	
47 Eicosanoic acid	+	++
48 Pentacosane	+	++
49 Docosanoic acid	+	++
50 Heptacosane	+	++
51 Tetracosanoic acid	+	+

unpalatability of the species. Kelley et al. (1987) found that female monarchs overwintering in Mexico contained more PA than males (means of 119 and 59 μg, respectively), and that monoesters comprised 90% of the PAs identified. Of the monoesters, echinatine was the most abundant in males and lycopsamine was the most abundant in females. Two macrocyclic diesters were also found in most monarch samples. Recently, Stelljes and Seiber (1990) also reported details of PA sequestration by monarchs overwintering in California and suggested that monarchs alter sequestered flower PAs to other PAs, like retronecine, that are more toxic to potential predators.

Since Edgar et al. (1976) suggested an additional defensive role for PAs, storage of PAs gathered by larvae and adult Lepidoptera of several families has been well documented (Edgar et al., 1979; Eisner, 1982; Brown, 1984a,b; Dussourd et al., 1984, 1988; for overviews, see Boppré, 1986, 1990). According to studies with various vertebrates, as well as invertebrates, PAs are taste-rejected, and it appears possible that PAs are the primary chemical defense of the entire danaine subfamily (see Ackery and Vane-Wright, 1984; Boppré, 1986, 1990; Glendinning, this volume; Schneider, this volume). Thus, in the Danainae in general, PAs establish a link between chemical communication and chemical defense. Individual differences in the amounts of hairpencil pheromones among males may reflect amounts of defensive chemicals gathered by males during adult life. An attempt to interpret this link was advanced by Eisner (1980) and Conner et al. (1981), who suggested that the amount of dihydropyrrolizine pheromone could be an indication of the defensive vigor of the male and, thus, decisive for the female in sexual selection.

Large amounts of PAs can be transferred from the male to the female via the spermatophores, as has been demonstrated for ithomiine butterflies, *Danaus gilippus*, and the moth, *Utetheisa ornatrix* (Brown, 1984a,b: Dussourd et al., 1984, 1988; Eisner and Meinwald, 1987). Thus, the male pheromone content might not only indicate to the female the defensive vigor of a pursuer, but also how much PA she can expect to receive from him during mating. However, it is yet to be proven that females actually discriminate among males on the amount of dihydropyrrolizine pheromones.

A lot more could be said about PAs in the context of sexual selection, but for *D. plexippus* these plant-derived chemicals have become less important than for other danaines; in common with reduced androconia and reluctant use of hairpencils in courtship. Despite the retention of PA-gathering behavior in adult monarchs, PAs are not needed for male courtship success, and so male monarchs have become emancipated from the danaine dependency upon plant chemicals for reproductive success. Nevertheless, *D. plexippus* may utilize PAs as defensive chemicals (see also Glendinning, this volume, and Schneider, this volume); perhaps, male monarchs donate PAs to females with their spermatophores during mating to enhance protection of females and eggs (see also Wells et al., this volume).

A discussion of chemical defense of the American monarch is not possible without considering cardenolides (cardiac glycosides) (reviewed by Brower, 1984). The predation experiments of J.V.Z. Brower (1958), the pharmacological research of Parsons (1965), and the chemical characterization of Reichstein (1967) established the American monarch as the first proven case of secondary plant chemical use by insects as a defense against predators and heralded the discipline of "ecological chemistry" (Brower and Brower, 1964; Brower, 1969)—now more often referred to as "chemical ecology" or "chemoecology." In summary, milkweed host plants (Asclepiadaceae) of *D. plexippus* contain varying quantities and kinds of cardenolides. These are ingested by monarch larvae and retained and stored by the adults. The cardenolide content of a butterfly is thus a characteristic fingerprint of its larval host plant, although it is modified with respect to the kind and amount of cardenolide stored (see Lynch and Martin, this volume; Malcolm et al., this volume; Nelson, this volume). Cardenolides have been clearly recognized to protect *D. plexippus* from several potential predatory bird species. However, because the cardenolide content of monarchs depends on available host-plant cardenolides, there is a "palatability spectrum," with individuals ranging from completely palatable to highly unpalatable. Furthermore, because no protection is perfect, some predators can circumvent or tolerate the cardenolide-based defenses of monarchs (Arellano G. et al., this volume; Glendinning, this volume).

As important as cardenolide sequestration must be for *D. plexippus,* I am certain that PAs significantly enhance monarch defense against predation—especially for individuals with little, or no, cardenolide content and against cardenolide-tolerant predators. Apart from the chemical analyses by Edgar et al. (1976), Kelley et al. (1987), and Stelljes and Seiber (1990), no experiments have been performed on the defensive potency of PAs in *D. plexippus.* However, studies with other "PA insects" clearly demonstrate the defensive use of PAs (Eisner, 1982; Brown, 1984a,b; Vasconcellos-Neto and Lewinsohn, 1984; Dussourd et al., 1988). Therefore, these allelochemics must not be neglected when considering the ecology of chemical defense and communication in the American monarch.

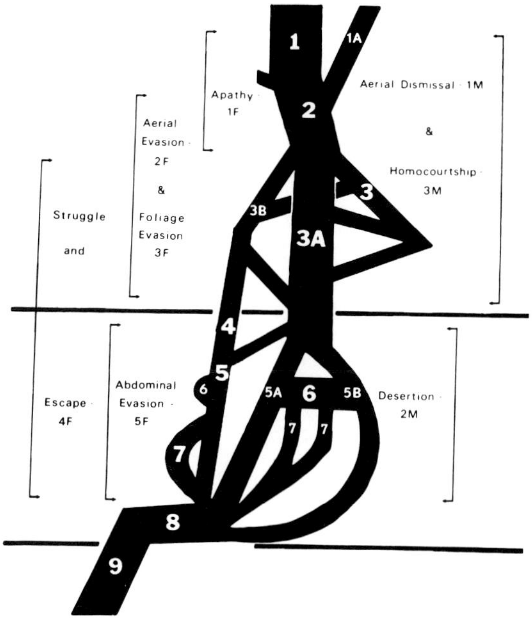

Figure 7. Sequence of mating behavior in *D. plexippus* (from Pliske, 1975b). The flow diagram illustrates monarch mating behavior: (1) male nudging, (1A) female nudging, (2) aerial pursuit, (3) aerial pounce, (3A) aerial takedown, (3B) aerial hairpencilling, (4) pair alighting, (5) lateral copulation attempt, (5A) dorsal copulation attempt, (5B) ventral copulation attempt, (6) antennal palpation (occurs with all copulation attempts), (7) female proboscis extension, (8) copulation, (9) postnuptial flight. 1–3 = aerial phase I, 4–8 = ground phase, 9 = aerial phase II. Brackets indicate female (left) and male (right) terminations.

Bird predation studies at the Mexican overwintering sites of the American monarch revealed that orioles and grosbeaks prey heavily on the butterflies (Calvert et al., 1979) and kill them randomly with respect to cardenolide content (Fink and Brower, 1981). Nevertheless, although most overwintering monarchs contain <1 emetic unit of cardenolide, observations of 98 oriole and 92 grosbeak attacks showed that 75 and 65, respectively, were released without apparent harm (Fink and Brower, 1981). Do such contradictory findings not indicate that a noncardenolide element is involved in the chemical defense of American monarchs? Most of the accepted butterflies may have been females since they seem to be reluctant PA-feeders and are likely to contain little or no PAs to enhance their chemical defense, particularly while they are unmated. Thus it is a pity that the sex of the eaten and rejected butterflies was not determined. However, it is interesting that Arellano G. et al. (this volume) found that grosbeaks killed more male than female monarchs at a Mexican overwintering colony, particularly in view of the fact that Kelley et al. (1987) found more PA in female than in male monarchs. This suggests that PAs do have a role in enhancing the chemical defense of overwintering monarchs, especially at a time when their cardenolide defenses are diminished (Malcolm and Brower, 1989).

It is also unfortunate that the series of force-feeding experiments (in which birds are fed powdered monarchs in capsules) were not designed to determine aversion to PAs, because, like cardenolides, their defensive potency is mediated by taste-rejection. However, Glendinning et al. (1988) and Glendinning (this volume) showed that one mouse species, *Peromyscus melanotis,* fed preferentially on male monarchs overwintering in Mexico (much like the grosbeaks described by Arellano G. et al., this volume). Presumably PAs in males had little influence on the foraging mice. Glendinning (this volume) also suggests that monarch consumption by *P. melanotis* was probably not influenced strongly by PAs, but that cardenolides were important in excluding three additional mouse species as predators of overwintering monarchs.

Despite different published conclusions as to the roles of PAs and cardenolides, research that takes both of these groups of chemicals into consideration, and includes possible synergism, is likely to make the story of sequestered plant chemicals that protect adult insects even more fascinating and instructive. In this context, it needs to be stressed that there are basic differences in the source and utilization between cardenolides and PAs and in their modes of action. Whereas cardenolides are ingested erratically with food by the larvae, PAs are actively searched for by adults and taken up independently of feeding behavior, i.e. pharmacophagously (cf. Boppré, 1984b). It also needs to be emphasized that the pharmacologically well-known noxious effects of PAs on vertebrate metabolism (e.g., hepatotoxicity) cannot be directly responsible for their protective potency in insects. Instead, it is their "bad" taste that deters predation. Although both PAs and cardenolides are thought to be bitter tasting to predators, there are basic differences between their activities in vertebrates. Cardenolides often produce an immediate emetic response, whereas PAs are nonemetic and do not harm an organism instantly. However, because mixed-function oxidases in the liver of vertebrates produce toxic metabolites of PAs, and some are carcinogenic, PAs can be severe vertebrate poisons (Robins, 1982).

The chemical defense of the American monarch seems to be still more complicated yet since Rothschild et al. (1978) suggested that *D. plexippus* may also contain self-produced cardio-active chemicals. Furthermore, Rothschild et al. (1984) suggested that 2-methoxy-3-alkylpyrazines in monarchs function as warning odors to deter predators (Guilford et al., 1987). In any case, while all Danainae utilize PAs, cardenolide-containing Asclepiadaceae are exclusively exploited by *Danaus* species (cf. Ackery and Vane-Wright, 1985), including those with typical danaine courtship behavior. Thus, a consideration of chemical defenses does not help explain the monarch's unusual mating habits.

PECULIAR SEX LIFE—ECOLOGICALLY MEDIATED?

Although the American monarch's chemical communication should be reconsidered, it is beyond doubt that *D. plexippus* is a peculiar danaine butterfly with respect to its mating habits (although not necessarily a peculiar lepidopteran). Its evolutionary shift in mating behavior from the general danaine pattern must have an ecological basis.

The American monarch is the only danaine that inhabits cool-temperate northern climates (apart from the Asian *Parantica sita*), and its extraordinary annual migration is thought to relate to the exploitation of the asclepiad (milkweed) flora of North America (Brower, 1985; Vane-Wright, 1986; Malcolm and Brower, 1989; Malcolm et al., 1989; Cockrell et al., this volume; Malcolm et al., this volume; Vane-Wright, this volume). By extending its range a long way north, did the monarch leave behind the rich tropical resources of PAs? This, and perhaps less selective pressure from tropical Batesian mimics (see Ritland and Brower, this volume), might well have caused adoption of a novel mating strategy (cf. Boppré, 1978). However, such thinking appears to be too simple. On one hand, *D. plexippus* is partly sympatric with other danaine species in which male courtship success depends on gathering PAs. On the other hand, other species of the genus *Danaus* sensu stricto, such as *D. erippus* and *D. cleophile,* together with the tropical American races of *D. plexippus,* all have androconial organs that are morphologically similar to those of the American monarch in that they are also secondarily reduced in size. Unfortunately, we know virtually nothing about the courtship behaviors and the androconial secretions of the closest relatives of *D. plexippus plexippus.* Thus, we have as yet no meaningful hypothesis as to what might have caused the American monarch to adopt such a modified mating strategy.

CONCLUSIONS—LEADING TO CONSEQUENCES?

In conclusion, courtship and chemical communication in the American monarch is a fascinating subject, although as yet it is rather incompletely investigated. Questions of importance for further research on the monarch include the following. (1) Are contacts between hairpencils and alar pockets definitely not established? Doubleday (1846) noted in some males of *D. erippus* (which is doubtfully recognized as a distinct species by Ackery and Vane-Wright, 1984) that a small portion of the wing, close to the orifice of the pocket, was denuded of scales, as if they had been rubbed away by introducing something into the slit. This observation would indicate that contact behavior may occur, but the observation has never been investigated further. (2) What are the androconial secretions and courtship behaviors of other members of the genus *Danaus* sensu stricto? Only comparative studies can shed light on why the American monarch became such a peculiar danaine. (3) Is the search for PAs, and their uptake and sequestration, a significant behavior in *D. plexippus*? At present, seemingly contradictory statements are found in the literature. While Pliske (1975a) provided some evidence for the monarch's reduced visitation to PA sources (compared to *D. gilippus*), Edgar et al. (1973) and Rothschild and Marsh (1978) state that male *D. plexippus* are avid feeders on PAs. (4) Are there subspecific or regional differences in PA use and PA occurrence? (5) Last, but by no means least, further work should be carried out on the likely supplementary roles that cardenolides and PAs play in the chemical defense of the American monarch. As outlined above, both cardenolides and PAs need to be considered to understand fully antipredator mechanisms in the American monarch. Combined efforts at both behavioral and chemical levels would be more than useful. Unlike most other insects sequestering secondary plant chemicals, danaines can be reared on various host plants and the amount of cardenolides a specimen contains can be determined, and adult danaines can be fed PAs in varying amounts. Thus, feeding experiments with an array of predators can be easily manipulated without unnatural treatment of the insects.

It is likely that more detailed knowledge of the chemical ecology of *D. plexippus* will not only aid in understanding monarch courtship, but will also lead to understanding the dual relationship that danaines have with both cardenolide- and PA-containing plants. The debate on automimicry (Brower et al., 1967; Gibson, 1984) could also receive a helpful interjection.

I hope to have shown here that we need to know much more about the reproductive and defensive biology of *D. plexippus*, and that there are challenging and rewarding avenues to be followed in the future. By virtue of our poor understanding, it seems to me that it is too early to designate the American monarch as nature's prime example of the male chauvinistic pig.

ACKNOWLEDGMENTS

Support of my studies by the Deutsche Forschungsgemeinschaft is acknowledged with thanks. I am greatly indebted to R.I. Vane-Wright and, in particular, to S.B. Malcolm for critical comments and valuable suggestions on the manuscript and its linguistic revision. I am also most grateful to O.W. Fischer and S. Schulz for permission to incorporate their unpublished data into this paper.

LITERATURE CITED

Ackery, P.R., and R.I. Vane-Wright. 1984. *Milkweed butterflies: Their cladistics and biology.* London: British Museum (Natural History), 425 pp.

Ackery, P.R., and R.I. Vane-Wright. 1985. Patterns of plant utilization by danaine butterflies. *Proceedings of the 3rd Congress of European Lepidopterology 1982,* 3–6. Karlsruhe: Societas Europaea Lepidopterologica.

Bellas, T.E., R.G. Brownlee, and R.M. Silverstein. 1974. Isolation, tentative identification, and synthesis studies of the volatile components of the hairpencil secretion of the monarch butterfly. *Tetrahedron Letters* 30:2267–2271.

Boppré, M. 1976. Pheromon-Transfer-Partikel auf einem Duftpinselhaar eines Monarchfalters (*Danaus formosa*). *Naturwissenschaftliche Rundschau* 29(9): cover photograph.

Boppré, M. 1977. Pheromonbiologie am Beispiel der Monarchfalter (Danaidae). *Biologie in unserer Zeit* 7(6):120–124.

Boppré, M. 1978. Chemical communication, plant relationships, and mimicry in the evolution of danaid butterflies. *Entomologia Experimentalis et Applicata* 24:264–277.

Boppré, M. 1984a. Chemically mediated interactions between butterflies. In *The biology of butterflies,* eds. R.I. Vane-Wright and P.R. Ackery, *Symposia of the Royal Entomological Society of London* 11: 259–275, London and New York: Academic Press (1989, Princeton, NJ: Princeton University Press).

Boppré, M. 1984b. Redefining "pharmacophagy." *Journal of Chemical Ecology* 10:1151–1154.

Boppré, M. 1986. Insects pharmacophagously utilizing defensive plant chemicals (pyrrolizidine alkaloids). *Naturwissenschaften* 73:17–26.

Boppré, M. 1990. Lepidoptera and pyrrolizidine alkaloids: Exemplification of complexity in chemical ecology. *Journal of Chemical Ecology* 16:165–185.

Boppré, M., and K. Fecher. 1977. Structural complexity of scent organs of male danaid butterflies. *Scanning Electron Microscopy,* Vol. II, 639–644, Chicago: IIT Research Institute.

Boppré, M., and R.I. Vane-Wright. 1989. Androconial systems in Danainae (Lepidoptera): Functional morphology of *Amauris, Danaus, Tirumala* and *Euploea. Zoological Journal of the Linnean Society* 97: 101–133.

Brower, J.V.Z. 1958. Experimental studies of mimicry in some North American butterflies. Part I. The monarch, *Danaus plexippus,* and viceroy, *Limenitis archippus archippus. Evolution* 12(1):32–47.

Brower, L.P. 1969. Ecological chemistry. *Scientific American* 220(2):22–29.

Brower, L.P. 1984. Chemical defence in butterflies. In *The biology of butterflies,* eds. R.I. Vane-Wright and P.R. Ackery, *Symposia of the Royal Entomological Society of London* 11:109–134, London and New York: Academic Press (1989, Princeton, NJ: Princeton University Press).

Brower, L.P. 1985. New perspectives on the migration biology of the monarch butterfly, *Danaus plexippus* L. In *Migration: Mechanisms and adaptive significance,* ed. M.A. Rankin, 748–785. Austin: University of Texas. *Contributions in Marine Science,* Supplement, 27.

Brower, L.P., and J.V.Z. Brower. 1964. Birds, butterflies,

and plant poisons: A study in ecological chemistry. *Zoologica, New York* 49:137–159.

Brower, L.P., J.V.Z. Brower, and F.P. Cranston. 1965. Courtship behavior of the queen butterfly, *Danaus gilippus berenice* (Cramer). *Zoologica, New York* 50:1–39.

Brower, L.P., J.V.Z. Brower, and J.M. Corvino. 1967. Plant poisons in a terrestrial food chain. *Proceedings of the National Academy of Sciences* 57:893–898.

Brown, K.S. 1984a. Adult-obtained pyrrolizidine alkaloids defend ithomiine butterflies against a spider predator. *Nature* 309:707–709.

Brown, K.S. 1984b. Chemical ecology of dehydropyrrolizidine alkaloids in adult Ithomiinae (Lepidoptera: Nymphalidae). *Revista Brasileira de Biologia* 44(4):435–460.

Calvert, W.H., L.E. Hedrick, and L.P. Brower. 1979. Mortality of the monarch butterfly (*Danaus plexippus* L.): Avian predation at five overwintering sites in Mexico. *Science* 204:847–851.

Conner, W.E., T. Eisner, R.K. Vander Meer, A. Guerrero, and J. Meinwald. 1981. Precopulatory sexual interaction in an arctiid moth (*Utetheisa ornatrix*): Role of a pheromone derived from dietary alkaloids. *Behavioral Ecology and Sociobiology* 9:227–235.

Doubleday, E. 1846. In *The genera of diurnal Lepidoptera 1*, by E. Doubleday and J.O. Westwood. London, 250 pp.

Dussourd, D.E., K. Ubik, J.F. Resch, J. Meinwald, and T. Eisner. 1984. Egg protection by parental investment of plant alkaloids in Lepidoptera. *XVII International Congress of Entomology, Hamburg*, 840.

Dussourd, D.E., K. Ubik, C. Harvis, J. Resch, J. Meinwald, and T. Eisner. 1988. Biparental defensive endowment of eggs with acquired plant alkaloid in the moth *Utetheisa ornatrix*. *Proceedings of the National Academy of Sciences USA* 85:5992–5996.

Edgar, J.A., C.C.J. Culvenor, and L.W. Smith. 1971. Dihydropyrrolizine derivatives in hairpencil secretion of danaid butterflies. *Experientia* 27:761–762.

Edgar, J.A., C.C.J. Culvenor, and G.S. Robinson. 1973. Hairpencil dihydropyrrolizines of Danainae from the New Hebrides. *Journal of the Australian Entomological Society, Brisbane* 12:144–150.

Edgar, J.A., C.C.J. Culvenor, and T.E. Pliske. 1974. Coevolution of danaid butterflies with their host plants. *Nature* 250:646–648.

Edgar, J.A., P.A. Cockrum, and J.L. Frahn. 1976. Pyrrolizidine alkaloids in *Danaus plexippus* L. and *Danaus chrysippus* L. *Experientia* 32:1535–1537.

Edgar, J.A., M. Boppré, and D. Schneider. 1979. Pyrrolizidine alkaloid storage in African and Australian danaid butterflies. *Experientia* 35:1447–1448.

Eisner, T. 1980. Chemistry, defense, and survival: Case studies and selected topics. In *Insect biology in the future*, eds. M. Locke and D.S. Smith, 847–878. New York: Academic Press.

Eisner, T. 1982. For love of nature: Exploration and discovery at biological field stations. *Bioscience* 32: 321–326.

Eisner, T., and J. Meinwald. 1987. Alkaloid-derived pheromones and sexual selection in Lepidoptera. In *Pheromone biochemistry*, eds. G.D. Prestwich and G.J. Blomquist, 251–269. Orlando, Florida: Academic Press.

Fink, L.S., and L.P. Brower. 1981. Birds can overcome the cardenolide defence of monarch butterflies in Mexico. *Nature* 291:67–70.

Gibson, D.O. 1984. How is automimicry maintained. In *The biology of butterflies*, eds. R.I. Vane-Wright and P.R. Ackery, *Symposia of the Royal Entomological Society of London* 11:163–165, London and New York: Academic Press (1989, Princeton, NJ: Princeton University Press).

Glendinning, J.I., A. Alonso Mejía, and L.P. Brower. 1988. Behavioral and ecological interactions of foraging mice (*Peromyscus melanotis*) with overwintering monarch butterflies (*Danaus plexippus*) in México. *Oecologia* 75:222–227.

Guilford, T., C. Nicol, M. Rothschild, and B.P. Moore. 1987. The biological roles of pyrazines: Evidence for a warning odour function. *Biological Journal of the Linnean Society* 31:113–128.

Hausman, S.A. 1951. The scent-producing organ of the male monarch butterfly. *The American Naturalist* 85:389–391.

Hill, H.F., Jr., A.M. Wenner, and P.H. Wells. 1976. Reproductive behavior in an overwintering aggregation of monarch butterflies. *The American Midland Naturalist* 95(1):10–19.

Illig, K.G. 1902. Duftorgane der männlichen Schmetterlinge. *Zoologica*, Stuttgart 38:1–34.

Kelley, R.B., J.N. Seiber, A.D. Jones, H.J. Segall, and L.P. Brower. 1987. Pyrrolizidine alkaloids in overwintering monarch butterflies (*Danaus plexippus*) from Mexico. *Experientia* 43:943–946.

Malcolm, S.B., and L.P. Brower. 1989. Evolutionary and ecological implications of cardenolide sequestration in the monarch butterfly. *Experientia* 45:284–295

Malcolm, S.B., B.J. Cockrell, and L.P. Brower. 1989. Cardenolide fingerprint of monarch butterflies reared on common milkweed, *Asclepias syriaca* L. *Journal of Chemical Ecology* 15(3):819–853.

Meinwald, J., A.M. Chalmers, T.E. Pliske, and T. Eisner. 1968. Pheromones. III. Identification of *trans,trans*-10-hydroxy-3,7-dimethyl-2,6-decadienoic acid as a major component in "hairpencil" secretion of the male monarch butterfly. *Tetrahedon Letters* 47:4893–4896.

Meinwald, J., A.M. Chalmers, T.E. Pliske, and T. Eisner. 1969. Identification and synthesis of *trans,trans*-3, 7-dimethyl-2,6-decadien-1,10-dioic acid, a component of the pheromonal secretion of the male monarch butterfly. *Chemical Communications* 1969D: 86–87.

Miller, L.D., and H.K. Clench. 1968. Some aspects of mating behavior in butterflies. *Journal of the Lepidopterists' Society* 22(3):125–132.

Müller, F. 1877. On the sexual spots of the males of *Danais erippus* and *D. gilippus*. *Archivos do Museu Nacional do Rio de Janeiro* II:25–29. (Reprinted in Longstaff, G.B. 1912. *Butterfly hunting in many lands: Notes of a field naturalist*, 616–620. London: Longmans, Green, and Co.)

Parsons, J.A. 1965. A digitalis-like toxin in the monarch butterfly, *Danaus plexippus* L. *Journal of Physiology, London* 178:290–304.
Pliske, T.E. 1975a. Attraction of Lepidoptera to plants containing pyrrolizidine alkaloids. *Environmental Entomology* 4(3):455–473.
Pliske, T.E. 1975b. Courtship behavior of the monarch butterfly, *Danaus plexippus* L. *Annals of the Entomological Society of America* 68(1):143–151.
Pliske, T.E., and T. Eisner. 1969. Sex pheromone of the queen butterfly: Biology. *Science* 164:1170–1172.
Reichstein, T. 1967. Cardenolide (herzwirksame Glykoside) als Abwehrstoffe bei Insekten. *Naturwissenschaftliche Rundschau* 20:499–511.
Robins, D.J. 1982. The pyrrolizidine alkaloids. In *Progress in the chemistry of organic natural products 41*, eds. W. Herz, H. Grisebach, and G.W. Kirby, 115–203. Wien: Springer-Verlag.
Rothschild, M. 1978. Hell's angels. *Antenna* 2:38–39.
Rothschild, M., and J.A. Edgar. 1978. Pyrrolizidine alkaloids from *Senecio vulgaris* sequestered and stored by *Danaus plexippus. Journal of Zoology, London* 186:347–349.
Rothschild, M., and N. Marsh. 1978. Some peculiar aspects of danaid/plant relationships. *Entomologia Experimentalis et Applicata* 24:637–650.
Rothschild, M., N. Marsh, and B.O.C. Gardiner. 1978. Cardio-active substances in the monarch butterfly and *Euploea core* reared on leaf-free artificial diet. *Nature* 275:649–650.
Rothschild, M., B.P. Moore, and W.V. Brown. 1984. Pyrazines as warning odour components in the monarch butterfly, *Danaus plexippus,* and in moths of the genera *Zygaena* and *Amata* (Lepidoptera). *Biological Journal of the Linnean Society* 23:375–380.
Schulz, S. 1987. *Die Chemie der Duftorgane männlicher Lepidopteren.* Ph.D. thesis. Hamburg: University of Hamburg.
Schulz, S., W. Francke, and M. Boppré. 1988. Carboxylic acids from hairpencils of *Amauris* butterflies (Lep.: Danainae). *Biological Chemistry Hoppe-Seyler* 369:633–638.
Seibt, U., D. Schneider, and T. Eisner. 1972. Duftpinsel, Flügeltaschen und Balz des Tagfalters *Danaus chrysippus* (Lepidoptera: Danaidae). *Zeitschrift für Tierpsychologie* 31:513–530.
Shull, E.M. 1979. Mating behavior of butterflies (Papilionoidea) and skippers (Hesperioidea) in Indiana. *Proceedings of the Indiana Academy of Science* 88: 200–208.
Stelljes, M.E., and J.N. Seiber. 1990. Pyrrolizidine alkaloids in an overwintering population of monarch butterflies (*Danaus plexippus*) in California. *Journal of Chemical Ecology* 16:1459–1470.
Tilden, J.W. 1979. Attempted mating between male monarchs. *Journal of Research on the Lepidoptera* 18(1):2.
Tuskes, P.M., and L.P. Brower. 1978. Overwintering ecology of the monarch butterfly, *Danaus plexippus* L., in California. *Ecological Entomology* 3:141–153.
Urquhart, F.A. 1958. Scent receptors in Lepidoptera. *Contributions of the Royal Ontario Museum* 49:3–16.
Urquhart, F.A. 1960. *The monarch butterfly.* Toronto: University of Toronto Press, 361 pp.
Urquhart, F.A. 1976. Alar pocket of the male monarch butterfly (*Danaus p. plexippus*) (Danaidae: Lepidoptera). *Canadian Entomologist* 108:777–782.
Vane-Wright, R.I. 1986. White monarchs. *Antenna* 10: 117–118.
Vasconcellos-Neto, J., and T.M. Lewinsohn. 1984. Discrimination and release of unpalatable butterflies by *Nephila clavipes,* a neotropical orb-weaving spider. *Ecological Entomology* 9:337–344.
Winter, D. 1985. Misdirected monarch mating behavior (Danaidae: *Danaus plexippus*) or *noblesse oblige*? Journal of the Lepidopterists' Society 39(4):334.

PHEROMONE-SENSITIVE RECEPTOR NEURONS IN *DANAUS GILIPPUS*

Rolf Kittmann and Dietrich Schneider

ABSTRACT. Pheromone-sensitive receptor neurons on the antennae of male and female *Danaus gilippus* were investigated electrophysiologically. All stimuli (synthetic pheromones) were applied with the cartridge method. By single-cell recordings with glass microelectrodes from the base of short sensilla basiconica, we found two types of receptor cells. Both cells showed a phasic response to the synthetic pheromone stimuli and a threshold of <1 μg of pheromone on the odor source. One cell type reacted to danaidone, the other to danaidal and hydroxydanaidal.

INTRODUCTION

Male danaine butterflies possess abdominal pheromone-releasing organs (hairpencils), which are exposed in the final courtship phase close to the female, who receives the pheromone with antennal receptor cells (Schneider and Seibt, 1969; Myers and Brower, 1969; Schneider, 1975). In many danaines, the major chemical components of the odor are heterocyclic dihydropyrrolizines: a ketone (danaidone) and two aldehydes (danaidal and hydroxydanaidal). These pheromone molecules promote mating in some (probably in most) danaines (Pliske and Eisner, 1969). Analysis from hairpencils of *Danaus gilippus* (Cramer) showed only danaidone, and not the two aldehydes (Meinwald et al., 1969). Pyrrolizine biosynthesis in male butterflies essentially depends on the search for, and active uptake

Figure 1. Antennal surface with sensilla basiconica (s) surrounded by microtrichiae (SEM).

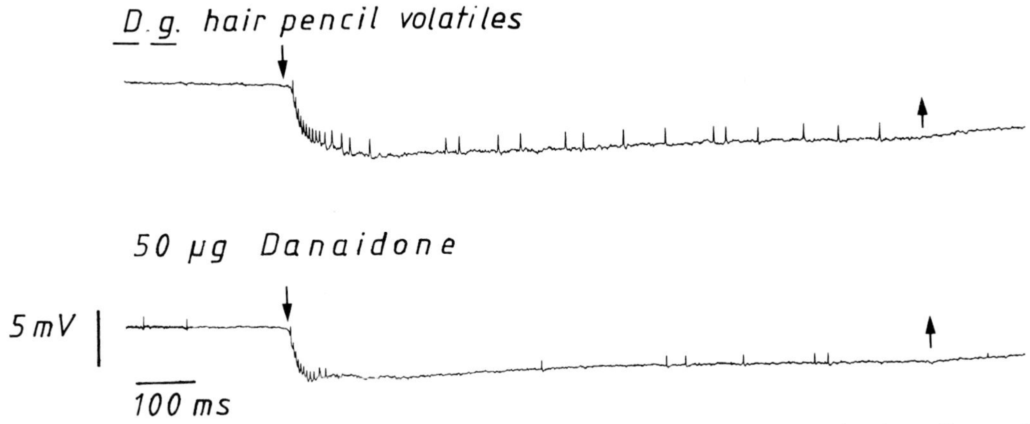

Figure 2. Ketone cell: DC-responses with receptor potentials and superimposed nerve impulses (recordings are from two different sensilla). Upper trace is the response to *Danaus gilippus* hairpencil volatiles; lower trace is the response to 50 mg of Danaidone. ↓ stimulus on, and ↑ off.

of, pyrrolizidine alkaloids (PAs) from the surface of plants such as *Heliotropium* (Boraginaceae), *Eupatorium* (Asteraceae), and *Crotalaria* (Fabaceae). The heterocyclic moiety of the PA is used as the precursor for the pheromone. Antennae of both sexes were found to respond to volatiles from alkaloid-producing plants, for which the males (and in some species also the females) search in an upwind flight (overviews in Ackery and Vane-Wright, 1984; Boppré, 1984; Schneider, this volume).

This review is a preliminary report on electrophysiological recordings from single sensilla on the antennae of 12 females and 3 males of *D. gilippus* from Florida.

METHODS

We used standard equipment and penetrated the bases of the sensilla with glass microelectrodes. Volatiles were applied with the cartridge method at 3-min intervals. Between stimuli, the antennae were flushed with air. Stimulus duration was 1 sec (air speed 1 m/sec). Stimuli were preceded by a control airpuff through an empty cartridge. We tested the three pheromones and two PAs: lycopsamine and heliotrine. During the stimulus program, cartridges loaded with 0.1, 1, 10, and 50 µg of each substance were tested. The response frequency of the receptor cells to stimuli was evaluated for an initial "peak" period of 100 msec and also for 1 sec.

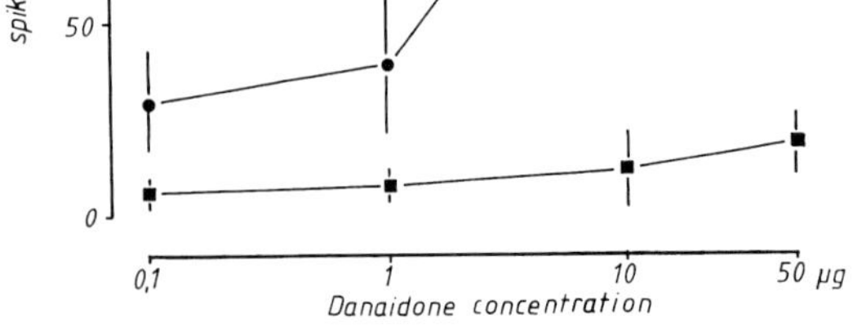

Figure 3. Ketone cells: Dose–response curves to danaidone stimuli. Circles show the frequency of the first 100 msec, and squares show the frequency of the first second. Mean values with standard deviation; n > 8.

RESULTS AND DISCUSSION

Receptor cell responses were, in principle, equal in males and females, but only the female recordings allow statistical treatment.

In the short sensilla basiconica (Fig. 1; cf. Myers, 1968), on the female antennae, we found two physiological types of pheromone-receptor cells (although never together in the same sensillum). Both types showed a low spontaneous activity (<10 Hz) and very phasic responses to the pheromones. They did not respond to the PA stimuli.

The first receptor cell type (*ketone cell*) responded to the hairpencil volatiles and to danaidone (Fig. 2) but not to danaidal or hydroxydanaidal. In the recordings from these sensilla, up to 3 units with different impulse amplitudes appeared, but only the cell with the largest impulses responded to danaidone and the hairpencil volatiles. The electrophysiologically detectable threshold of this cell type was <1 µg of danaidone at the odor source (Fig. 3) and the response approached a plateau >50 µg. Hairpencil volatiles elicited a higher response than a 50-µg stimulus, which corresponds to the known contents of up to several hundred micrograms of danaidone per hairpencil (Meinwald et al., 1969; Boppré et al., 1978). In 8 out of 13 recordings we could test the whole stimulus program.

The second receptor cell type (*aldehyde cell*) responded equally to danaidal and hydroxydanaidal

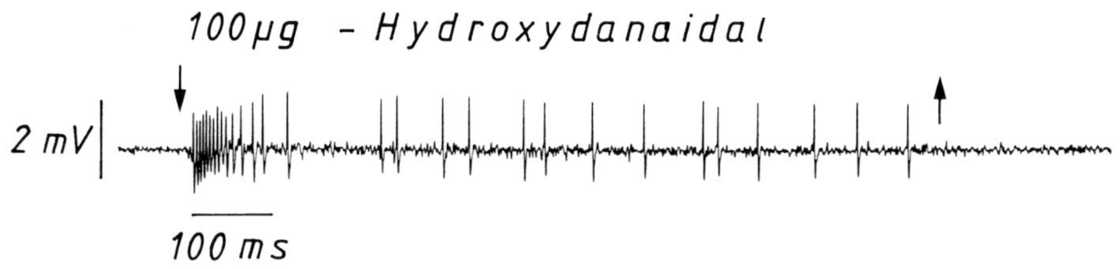

Figure 4. Aldehyde cell: AC-response to 100 µg (−)-hydroxydanaidal (receptor potential electronically cancelled). ↓ stimulus on, and ↑ off.

(Fig. 4) but not to danaidone. The response of this cell type to hairpencil volatiles was weak or even doubtful. Hydroxydanaidal exists in two enantiomeric forms, (−) and (+). We tested both substances and found that all cells of this type responded equally well to danaidal and (−)-hydroxydanaidal (Fig. 5). Interestingly, some cells gave good, and others only weak, reactions to the (+) enantiomer.

As a pheromone receptor, the ketone cell is likely to be important in courtship behavior. The biological role of the aldehyde cell is not clear because *D. gilippus* does not produce these substances. Further experiments with this and other danaines are needed to clarify the specificity of these pheromone receptor cells and also to locate the receptor cells for the PA and its derivatives (see Bogner and Boppré, 1984).

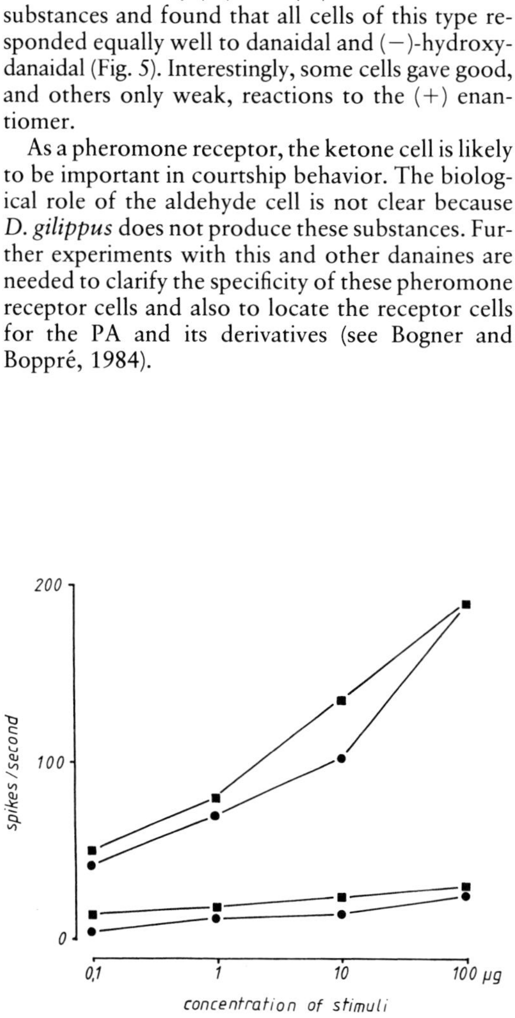

Figure 5. Aldehyde cells: Dose–response curves for danaidal (■) and (−)-hydroxydanaidal (●). The upper curve pair shows the frequency of the first 100 msec, and the lower curve pair shows the frequency of the first second.

ACKNOWLEDGMENT

We thank W. Francke and S. Schulz, University of Hamburg, for the gifts of hydroxydanaidals in high enantiomeric purity.

LITERATURE CITED

Ackery, P.R., and R.I. Vane-Wright. 1984. *Milkweed butterflies: Their cladistics and biology.* London: British Museum (Natural History); and Ithaca, NY: Cornell University Press, 425 pp.

Bogner, F., and M. Boppré. 1984. Ein spezieller Riechrezeptor zur Wahrnehmung von Alkaloiden bei Lepidopteren. *Verhandlungen der Deutschen Zoologischen Gesellschaft* 78:281 pp.

Boppré, M. 1984. Chemically mediated interactions between butterflies. In *The biology of butterflies,* ed. R.I. Vane-Wright and P.R. Ackery, *Symposia of the Royal Entomological Society of London* 11:259–275. London and New York: Academic Press (1989, Princeton, NJ: Princeton University Press).

Boppré, M., R.L. Petty, D. Schneider, and J. Meinwald. 1978. Behaviorally mediated contacts between scent organs: Another prerequisite for pheromone production in *Danaus chrysippus* males (Lepidoptera). *Journal of Comparative Physiology A* 126:97–103.

Meinwald, J., Y.C. Meinwald, and P.H. Mazzocchi. 1969. Sex pheromone of the queen butterfly: Chemistry. *Science* 164:1174–1175.

Myers, J. 1968. The structure of the antennae of the Florida queen butterfly, *Danaus gilippus berenice* (Cramer). *Journal of Morphology* 125:315–328.

Myers, J., and L.P. Brower. 1969. A behavioural analysis of the courtship pheromone receptors of the queen butterfly, *Danaus gilippus berenice. Journal of Insect Physiology* 15:2117–2130.

Pliske, T., and T. Eisner. 1969. Sex pheromones of the queen butterfly: Biology. *Science* 164:1170–1172.

Schneider, D. 1975. Pheromone communication in moths and butterflies. In *Sensory physiology and behavior,* eds. R. Galun, P. Hillman, I. Parnas, and R. Werman, 173–193. New York: Plenum Press.

Schneider, D., and U. Seibt. 1969. Sex pheromone of the queen butterfly: Electroantennogram responses. *Science* 164:1173–1174.

4

MATING BEHAVIOR

Monarchs mating in early March at an overwintering site in Mexico, just before the northward spring migration. Van Hook (this volume) describes the aggressive non-random selection of large females in good condition by small, battered males. Photograph by Tonya Van Hook.

NON-RANDOM MATING IN MONARCH BUTTERFLIES OVERWINTERING IN MEXICO

Tonya Van Hook

ABSTRACT. Each autumn an extensive migration brings nearly the entire North American population of monarch butterflies, *Danaus plexippus* (L.), into dense overwintering aggregations in central Mexico for several months. The following spring, a brief mating period occurs just before remigration from these sites. Males make large copulatory investments and females have substantial nutrient needs for vitellogenesis and remigration. I hypothesized that male investments and female nutrient needs should together lead to positive assortative mating based on phenotypic characters that indicate the quantity or quality of material benefits to be gained. Relevant characters include body size, body weight, age, and mating history.

Contrary to my prediction of positive assortative mating, mated males were light weight, had small wings, and were in poor condition; conversely, mated females were heavy, had large wings, and were in good condition. Further analyses within this mating subgroup indicated no negative assortative pairing with respect to body weight or wing size. Wing condition showed a slight but significant negative association within mated pairs. The non-random mating pattern appears to result from two factors: (1) smaller, worn males begin mating earlier and account for most of the mating at overwintering sites in Mexico; and (2) these males capture females that are large and in good condition during aerial pursuit. This takedown strategy apparently constrains female choice during precopulatory interactions. After takedown, males may then reject previously mated females. These findings are discussed in terms of sexual selection theory.

INTRODUCTION

Monarch butterflies, *Danaus plexippus* (L.), from eastern North America migrate each autumn to a few tightly packed aggregation sites in the volcanic highlands of central Mexico (Calvert and Brower, 1986; Calvert and Lawton, this volume). A brief mating period occurs just before colony breakup and remigration from these sites in early spring. Migration and aggregation behaviors bring nearly every adult in the entire population into close proximity during the time of mating. Such a concentration of adults, at up to 10 million per hectare (Brower et al., 1977), presents a unique opportunity for mate discrimination before colony breakup and the onset of remigration. Such mate selection may benefit mating partners because females produce and carry eggs extensive distances from overwintering colonies to oviposition sites (Cockrell et al., this volume), and males provide nutrients through spermatophore transfer (Boggs and Gilbert, 1979; Wells et al., this volume). These benefits are likely to influence reproductive fitness and select for premigratory mate selection. Characteristics available for mate discrimination, which should reflect the capacity to provide these benefits, include body size, age, and mating history (Rutowski, 1985).

Males are predicted to be selective in their choice of females for four reasons: (1) they make costly prezygotic nutrient investments (Boggs and Gilbert, 1979; Oberhauser, 1988; Wells et al., this volume); (2) mating occurs over a brief time period, which may limit spermatophore production, thereby limiting opportunity for polygyny (Knowlton, 1979); (3) both egg production and migration ability are related to female size and condition (Brower and Glazier, 1975; Suzuki, 1978; Gibo and Pallett, 1979); and (4) the mating behavior involves male capture and takedown of females in an aerial pursuit (Hill et al., 1976; Pliske, 1975; Urquhart, 1987).

Female monarchs use nutrients provided by males through spermatophore transfer for both egg production and somatic maintenance (Boggs and Gilbert, 1979, Wells et al., this volume). Acquisition of such material benefits is thought to be especially advantageous in a long-lived migrant species (Trivers, 1972; Byers, 1978). Spermatophore size, and therefore amount of material benefits available to females, is positively correlated with male size and negatively correlated with male age and mating history (Oberhauser, 1988). This suggests that females should mate preferentially with large, young males (Rutowski, 1985). Reproductive advantages of mating with males that produce larger spermatophores have been demonstrated in *Colias eurytheme* by Rutowski et al. (1987), who showed that larger male contributions lead to increased female longevity, egg-laying rate, and lifetime egg production.

Brower et al. (1977) (also see Eanes and Koehn, 1978) described the mass mating activity at Mexican overwintering sites as panmictic with respect to geographic origin but did not address non-random mating based on phenotype. In this paper, I test the hypothesis that extremely high population densities and the brief mating period at Mexican overwintering sites, together with male investments and material benefits available to both sexes, could lead to positive assortative mating via both male and female choice. Non-mating and mating butterflies were compared with respect to body weight, wing length, and wing condition during the mass mating period at a Mexican overwintering site in 1985. In 1986, I tested whether or not the mated pair sample as collected in 1985 was representative of the mating population, and asked when in the precopulatory sequence non-random pairing might originate. Taken together, these data provide strong evidence for non-random mating, but in a direction that was not anticipated.

METHODS

The site of the overwintering colony was similar in 1985 and 1986 and was located at approximately 3,000-m elevation on the Sierra Chincua, Michoacán, Mexico, at 19°41′N, 100°18′W (Calvert and Brower, 1986). In both years the monarch overwintering colony covered approximately 1 ha and was nestled between two small ravines. In 1985, data were collected from 11 February through 25 March, concurrent with the timing of mating activity. In 1986, data were collected from 1 March through 24 March after mating began in mid-February. In both years nearly all the butterflies had left the overwintering colony by the end of the data collection period.

GENERAL POPULATION: NON-MATING BUTTERFLIES, 1985

Three phenotypic characters (wet weight, wing length, and wing condition) were measured in four categories of non-mating monarchs (i.e., butterflies in or near the colony that were not mating when collected). The four categories were (1) *actively flying* within 3 m of the ground, (2) *drinking* at water or mud sources (proboscis contact was not determined), (3) *clustered* at up to 6 m above the ground (hanging immobile on tree trunks or leaves), and (4) *nectaring* at flowers.

Butterflies were netted and immediately placed in glassine envelopes. Within 6 hr they were stored in plastic bags on ice until measured the following day. Wet weight was measured to the nearest milligram, and forewing length was measured to the nearest 0.5 mm along the costal margin from base to apex. Wing condition was rated subjectively as good, medium, or poor (1, 2, and 3, respectively) based on wing scale loss, fading, scratches, and degree of tattering along the margin. To ascertain that there were no significant differences for either males or females among the four categories of non-mating monarchs, I performed separate Kruskal-Wallis tests for each phenotypic character. To control for changes with time, the data were divided into two subgroups: early (collected before 5 March) and late (collected on or after 5 March).

MATED BUTTERFLIES, 1985

In copula mating pairs were collected from 11 February to 25 March, usually in the afternoon when they were most abundant. Because they were not found outside the colony and were generally restricted to scrubby vegetation, collections were made only inside the cluster area and within 3 m of the ground. Each mated pair was stored together overnight in a glassine envelope on ice. Once uncoupled the next day, they were individually weighed, measured, and rated for wing condition. Mated females were palpated to determine presence or absence of a freshly transferred spermatophore. This was done by gently applying pressure between the thumb and index finger while feeling along the abdomen from anterior to posterior for a swollen bursa copulatrix. Phenotypic characters of mated males and females were compared with those from the general overwintering population. Using female abdominal palpations as the criterion for distinguishing mating status, males and females that successfully transferred spermatophores were compared with those that were unsuccessful.

TESTS FOR BIAS OF MATED PAIRS, 1986

In 1985, comparisons of mating and non-mating monarchs involved only comparisons of *quiescent mated pairs* collected within the colony boundaries with the general overwintering population. Because the normal behavior of mating pairs is to fly off in tandem after coupling and then settle on vegetation, it was possible that the mated pair sample of 1985 involved only those pairs that did not or could not fly out of the colony or those that crawled up nearby vegetation and became quiescent without a tandem flight. To test this possibility, I made two indirect control samples in 1986.

First, I compared quiescent mated males collected in the same manner as in 1985 to males that successfully coupled during ground-phase interactions, i.e. those that had just attained genital coupling and were caught before postnuptial flight. This latter sample would not be biased toward weak fliers. Second, I tested whether or not just-coupled males could takeoff and carry the female in tandem flight. Each pair was placed immediately upon capture into a glassine envelope and transported to an open area near the colony. This area was approximately 50 m in diameter with only grass and sparse scrubby vegetation and surrounded by dense stands of trees. Pairs that uncoupled during transport were not tested. I tested 13 pairs on 5 and 16 March between 1415 and 1615 hours. Both days were clear and sunny with mild to strong breezes and ambient temperatures $\geq$16°C (i.e., above flight threshold; Masters et al., 1988; Alonso M. et al., this volume). Males that were able to carry females out of the open area within 2 min were scored as successful, regardless of whether flight was directly from the ground or followed a climb up nearby vegetation. Unsuccessful males either failed to take flight directly or climbed up vegetation and then became quiescent for at least 2 min. Males in the unsuccessful group were compared with those that successfully flew away with respect to wing size and wing condition.

COMPARISONS OF BUTTERFLIES COLLECTED AT VARIOUS POINTS DURING GROUND-PHASE PRECOPULATORY INTERACTIONS, 1986

For purposes of this study, the mating event can be summarized as follows: Males pursue females in aerial flight. If successful at capturing the female, the male folds his wings and together the pair falls to the ground. On the ground, males attempt to couple with the assistance of genital claspers. Females appear to struggle throughout the ground-phase precopulatory interaction but they rarely escape. A male may leave a female at any point during the ground-phase interaction. If successful genital coupling occurs, the female becomes quiescent and the male carries her away in a postnuptial flight. Males also sometimes pursue and takedown other males. Ground-phase male–male interactions appear indistinguishable from male–female interactions except the pursuing male always eventually abandons the male it has taken down.

To determine when in the mating sequence phenotypic non-random pairing might originate, monarchs collected from four sequential ground-phase precopulatory categories were collected. These samples were handled in the same fashion as mated pairs. Collections were made from 1 March through 25 March between 1100 and 1600 hours and included (1) *attempted copulations,* pairs in which a male held a female while probing her genital area with his abdomen (these were collected upon sighting without allowing the precopulatory interaction to continue and, therefore, do not distinguish successful from unsuccessful mating attempts); (2) *just-coupled* butterfly pairs, as described above; (3) *abandoned individuals,* either male or female butterflies left by their potential mates during ground-phase interactions; and (4) *homocourtships,* male–male pairs in which one male was attempting copulation with a second male. The homocourtship category is identical to the attempted copulation category except the pursued individual was a male.

Males and females from the attempted copulation group were compared to the general population to determine if all male phenotypes successfully takedown females and if they do so randomly. Attempted copulation pairs were also compared to just-coupled pairs to determine if all pairs that fall in aerial takedown are equally likely to couple successfully. To assess whether or not females abandoned during precopulatory interactions were distinct from those that successfully coupled, females that were left by their potential partners during ground-phase courtship were compared with those that had just coupled successfully. Finally, sexually aggressive males and males with which they were attempting to copulate were compared to each other, to the general population, and to *in copula* mated males to determine if these males were of distinct phenotypes. Females in all categories were palpated to determine recent mating status using the method described above for mated females.

STATISTICAL ANALYSES

For data analyses I used SAS (1985). Data were tested for normality with the Kolmogorov *D* test. Because the data in several cases were not normally distributed, I used nonparametric statistics: Wilcoxon two-sample rank-sum (two-tailed tests), Kruskal-Wallis multiple-sample (chi-square approximation), and Spearman's rank-order correlation coefficient.

RESULTS

GENERAL POPULATION, 1985

Mating was first observed on 11 February, gradually increased in frequency through 25 February, and remained high through the end of the overwintering period, except during unusually cool and overcast days. Almost all the butterflies had left the overwintering site by 19 March.

The four general population sampling categories of actively flying, drinking, clustered, and nectaring varied widely in sample sizes and sex ratios across collection dates (Table 1). In early season collections, both sexes differed in wet weight among sampling categories, but neither sex differed in forewing length or wing condition (Table 2). In late season samples, females differed among sampling categories in wet weight and wing condition, while males showed no differences in any of the three characters measured. Phenotypic differences among sampling categories were not consistently different for either males or females in early- and late-season samples. For further statistical comparisons, I chose to combine all four sampling categories listed in Table 1 to represent the general non-mating overwintering population.

In the general population, male and female mean wet weight and mean wing size differed among good, medium, and poor wing condition groups. Larger mean wet weights and wing sizes were associated with better wing conditions (Table 3). These trends were consistent in early and late samples (only male forewing length was not significantly positively associated with wing condition in the early-season sample; $\chi^2 = 3.95$, $p = 0.14$, df = 2, n = 233). Mated individuals showed the same trend with heavier wet weights and longer wing lengths associated with better wing conditions (Table 3). Early- and late-season samples showed the same trend

Table 1. Sex ratios (percent males) and sample sizes of the general non-mating population, 1985, listed by sampling category and date. * = sex ratio not taken.

Active Date	Active (n)	Active % male	Drinking Date	Drinking (n)	Drinking % male	Clustered Date	Clustered (n)	Clustered % male	Nectaring Date	Nectaring (n)	Nectaring % male
19 Feb	(107)	39	20 Feb	(24)	37	13 Feb	(103)	50	16 Feb	(40)	53
26 Feb	(24)	35	4 Mar	(36)	19	19 Feb	(40)	*	24 Feb	(100)	35
3 Mar	(50)	28	9 Mar	(15)	43	2 Mar	(25)	28	4 Mar	(103)	32
5 Mar	(15)	*	11 Mar	(10)	58	10 Mar	(25)	36	15 Mar	(26)	18
11 Mar	(109)	37	17 Mar	(20)	10	10 Mar	(25)	27	16 Mar	(15)	7
19 Mar	(20)	41	25 Mar	(14)	*	11 Mar	(25)	32	23 Mar	(15)	28
25 Mar	(7)	*				16 Mar	(20)	38			
						23 Mar	(15)	30			
Total	(332)			(119)			(278)			(299)	

except that in the early season, female wet weights and wing sizes did not differ among the three wing condition categories ($\chi^2 = 3.39$, $p = 0.18$, df = 2, n = 122; $\chi^2 = 1.83$, $p = 0.40$, df = 2, n = 125, respectively).

As determined by abdominal palpation, 90% of the 450 *in copula* mated pairs collected in 1985 successfully transferred spermatophores. Males that were unsuccessful in transferring spermatophores weighed less and were in worse condition than successful males (weight: $Z = -2.22$, $p = 0.03$, n = 434; condition: $Z = -3.53$, $p = 0.0004$, n = 447), while they did not differ in wing length ($Z = -1.05$, $p = 0.30$, n = 444). The test for weight difference

Table 2. Comparisons of wet weight (mg), forewing length (mm), and wing condition (no units; see text) of males and females among general population sampling categories in 1985 (Kruskal-Wallis test, χ^2 approximation, df = 3) for early (before 5 March) and late (on or after 5 March) samples.

Measure	Active (n)	Active $\bar{X}$	Active SD	Drinking (n)	Drinking $\bar{X}$	Drinking SD	Clustered (n)	Clustered $\bar{X}$	Clustered SD	Nectaring (n)	Nectaring $\bar{X}$	Nectaring SD	χ^2	*p*
Early (before 5 March)														
Wet weight														
Males	(51)	436	65	(16)	425	65	(81)	471	58	(89)	445	48	13.01	0.005
Females	(80)	402	76	(44)	402	61	(86)	449	70	(154)	382	54	45.69	0.0001
Wing length														
Males	(48)	52.2	2.3	(16)	51.5	2.2	(81)	52.3	2.0	(88)	52.4	1.9	3.77	0.29
Females	(80)	51.5	2.3	(44)	51.9	1.6	(85)	52.0	1.9	(154)	51.3	2.5	5.43	0.14
Wing condition														
Males	(51)	1.6	0.7	(16)	1.6	0.5	(82)	1.6	0.6	(89)	1.7	0.8	2.50	0.48
Females	(80)	1.4	0.7	(44)	1.6	0.7	(86)	1.6	0.7	(154)	1.6	0.7	5.37	0.15
Late (after 5 March)														
Wet weight														
Males	(15)	471	49	(15)	466	54	(25)	452	58	(18)	430	56	5.72	0.13
Females	(57)	427	60	(49)	413	71	(65)	393	62	(58)	362	61	26.97	0.0001
Wing length														
Males	(15)	53.2	2.3	(15)	52.4	2.2	(25)	52.6	1.7	(18)	52.3	1.7	2.14	0.54
Females	(57)	52.3	1.5	(49)	51.7	2.3	(65)	51.8	1.9	(58)	51.5	2.2	3.47	0.32
Wing condition														
Males	(15)	1.8	0.7	(15)	1.7	0.7	(25)	1.5	0.7	(18)	1.7	0.7	2.11	0.55
Females	(57)	1.5	0.6	(49)	1.9	0.6	(65)	1.6	0.6	(58)	1.8	0.6	11.61	0.009

Table 3. **Butterflies in each of the wing condition categories (good, medium, and poor) compared with respect to wet weights (mg) and forewing lengths (mm) for both the general population and mated pair samples in 1985 using Kruskal-Wallis tests (χ^2 approximation, df = 2). $p < 0.005$, except * $p < 0.006$.**

	Wet weight						Wing size					
	General population			Mated pairs			General population			Mated pairs		
Wing condition	(n)	$\bar{X}$	SD	(n)	$\bar{X}$	SD	(n)	$\bar{X}$	SD	(n)	$\bar{X}$	SD
Males												
Good	(149)	468	53	(147)	465	59	(149)	52.7	1.8	(152)	52.2	2.0
Medium	(126)	443	56	(194)	432	69	(124)	52.1	2.1	(201)	51.2	2.5
Poor	(35)	411	58	(94)	385	78	(33)	51.5	2.0	(93)	50.3	3.1
χ^2		30.69			66.74			12.19			26.60	
Females												
Good	(294)	414	68	(332)	478	65	(294)	51.9	2.1	(339)	52.5	1.8
Medium	(240)	397	66	(103)	452	62	(240)	51.6	2.0	(105)	51.8	1.9
Poor	(59)	364	65	(7)	448	35	(58)	50.6	2.7	(7)	51.7	2.1
χ^2		27.66			13.87			13.91			10.29*	

is probably conservative because males that successfully transferred spermatophores had just lost the weight of a fresh spermatophore, yet they still weighed significantly more than unsuccessful males. In contrast, successful spermatophore transfer did not reflect female wet weight ($Z = -0.81$, $p = 0.42$, n = 442), wing length ($Z = 1.56$, $p = 0.12$, n = 450), or wing condition ($Z = -0.81$, $p = 0.42$, n = 450).

COMPARISONS OF *IN COPULA* MATED PAIRS VERSUS NON-MATED BUTTERFLIES OF THE GENERAL POPULATION, 1985

Mated males weighed less, had shorter wings, and were in worse condition than non-mating males in the general population (Table 4). Mated females showed opposite trends relative to non-mating females. They were heavier, larger, and in better condition. These differences between mating and non-mating individuals were of such magnitude that mated monarchs showed opposite male–female weight size, and wing condition trends relative to non-mating monarchs. Males in the general population weighed more and had longer wings than females, while they did not differ in wing condition. In contrast, mating males weighed less, were smaller, and were in poorer condition than mating females (Table 4).

Within the mating population, there was no evidence of negative assortative pairing by wet weight or wing length. Spearman rank-order analyses showed no significant correlation in wet weight

Table 4. **Wet weight (mg) and forewing length (mm) comparisons for males and females in the general population and mated pairs (11 February–25 March 1985) using Wilcoxon two-sample rank-sum tests (normal approximation; continuity correction of 0.05). $p < 0.001$, except * = not significant at $p = 0.05$.**

	General population			Mated pairs			
	(n)	$\bar{X}$	SD	(n)	$\bar{X}$	SD	Z
Wet weight							
Males	(310)	451	58	(435)	433	74	3.13
Females	(593)	402	68	(442)	472	64	−15.41
Z		10.90			−7.56		
Forewing length							
Males	(306)	52.3	2.0	(446)	51.4	2.6	5.14
Females	(592)	51.7	2.1	(451)	52.3	1.8	−4.49
Z		4.64			−5.05		
Wing condition							
Males	(311)	1.6	0.7	(449)	1.9	0.7	4.48
Females	(593)	1.6	0.7	(451)	1.3	0.5	−8.78
Z		0.59*			−13.25		

Table 5. Wet weight (mg) and forewing length (mm) comparisons for males and females in the general population and mated pairs for early- (before 5 March) and late-season (on or after 5 March) 1985 samples using Wilcoxon two-sample rank-sum tests (normal approximation; continuity correction of 0.5).* $p < 0.0005$; NS = not significant at $p = 0.05$.

	Early								Late		
	General population			Mated pairs					General population		
	(n)	$\bar{X}$	SD	(n)	$\bar{X}$	SD	Z	p	(n)	$\bar{X}$	SD
Wet weight											
Males	(237)	451	58	(122)	416	71	−4.20	*	(73)	453	56
Females	(364)	405	69	(122)	490	61	11.00	*	(229)	398	67
Wing length											
Males	(233)	52.3	2.0	(124)	51.1	2.7	−3.94	*	(73)	52.6	1.9
Females	(363)	51.6	2.2	(125)	52.7	1.6	5.24	*	(229)	51.8	2.0
Wing condition											
Males	(238)	1.6	0.7	(125)	2.1	0.7	6.04	*	(73)	1.7	0.7
Females	(364)	1.6	0.7	(125)	1.3	0.5	−3.99	*	(229)	1.7	0.6

($r_s = 0.06$, $p = 0.23$) or wing length ($r_s = -0.01$, $p = 0.78$) between mated pairs. Male and female wing condition showed a slight but significant negative correlation ($r_s = -0.17$, $p = 0.0004$).

Phenotypic characteristics of mated monarchs changed through the mating period. Males mating early weighed less and were in worse condition than those mating after 5 March (Z = 3.11, $p = 0.002$; Z = 4.42, $p = 0.0001$, respectively). However, male wing size did not differ between early and late mating samples (Z = −1.39, $p = 0.17$). Mated females showed different trends. Those that mated early weighed more and had longer wings than those mating late (Z = 3.50, $p = 0.0005$; Z = 3.07, $p = 0.002$, respectively). Unlike mated males, there were no differences in female wing condition with time (Z = −0.17, $p = 0.86$).

In contrast, the general population changed little with time. Non-mating males collected early versus late in the breeding period did not differ in wet weight (Z = 0.57, $p = 0.57$), wing size (Z = 1.40, $p = 0.16$), or wing condition (Z = 0.42, $p = 0.68$). Similarly, wet weight and wing size of early and late collections of non-mating females did not differ (Z = −0.85, $p = 0.40$; Z = 1.24, $p = 0.22$, respectively). Only female wing condition changed significantly through the breeding period (Z = 3.19, $p = 0.001$). Mating monarchs more closely resembled the general non-mating overwintering population as the breeding season progressed (Table 5).

TESTS FOR BIAS OF MATED PAIRS: 1986 SAMPLE

Wet weights, wing sizes, and wing conditions of *in copula* males and females did not differ from those that had just coupled. This suggests that the 1985 mated pair sample was not biased due to differences in male behavior after genital coupling (Table 6). Consistent with the results of 1985, just-coupled butterflies also showed opposite male–female phenotypic trends from those of the general population; males weighed less, were smaller, and were in worse condition than females (Table 7).

The results of the tandem flight experiment further support the use of *in copula* mated pairs as an unbiased sample of the mating population. Of the 13 males tested in the tandem flight experiment, 10 (77%) were able to carry females out of the testing area. There were no differences in wing size (Z = −0.65, $p = 0.52$, n = 12) or wing condition (Z = 0.81, $p = 0.42$, n = 13). between successful and unsuccessful males. Furthermore, the male : female wing size ratio did not influence flight capacity (Z = 0.9, $p = 0.36$, n = 12). The small sample size may have prevented detection of differences in male tandem flight capacity.

COMPARISONS OF BUTTERFLIES COLLECTED AT VARIOUS POINTS DURING GROUND-PHASE PRECOPULATORY INTERACTIONS: 1986

Although mated females did not differ from females attempting copulation or just-coupled females in wing size or wing condition, they did weigh significantly more. This difference could be a result of the weight of the just-transferred spermatophore. However, mated males, weighed after transferring the weight of a fresh spermatophore, did not weigh less than those attempting copulation or those that had just coupled (Table 6).

Males and females that had just coupled did not differ from those in the attempted copulation sample (Table 6). Because males and females in attempted copulation and just-coupled categories also did not differ from mated pairs of that season, these sample categories were combined to form a sexually active sample and compared to the general pop-

Table 5. Extended.

Late				
Mated Pairs				
(n)	X̄	SD	Z	p
(313)	440	74	1.36	NS
(320)	465	65	−10.77	*
(322)	51.5	2.5	3.63	*
(326)	52.2	1.9	−1.53	NS
(324)	1.8	0.7	−1.27	NS
(326)	1.3	0.5	8.75	*

ulation of 1986 (Table 6). Mated females were excluded from the sexually active sample for analysis of wet weight to avoid biasing the sample by the weight of just-transferred spermatophores. As in 1985, sexually active females were heavier, had longer wings, and were in better condition than females in the general population (Table 6). Again as in 1985, sexually active males weighed less, were smaller, and were in worse condition than males in the general population (Table 6).

Females left by males during ground-phase precopulatory interactions did not differ from those that successfully coupled with males in wet weight ($Z = 0.80$, $p = 0.43$), wing length ($Z = -0.77$, $p = 0.44$), or wing condition ($Z = 1.59$, $p = 0.11$) (Table 6). However, as determined by abdominal palpation, 90% (n = 53) of females left by males had mated while only 43% (n = 21) of females that had just coupled had detectable spermatophores in the bursa copulatrix. These differences were significant ($\chi^2 = 13.58$, $p = 0.0001$, df = 1, n = 74). Due to the small sample size, males left by females during ground-phase interactions were not compared to males that successfully coupled (Table 7).

Males attempting copulation with other males were smaller and in poorer condition than males being pursued by other males, but they did not differ in wet weight (Table 7). Males that took down other males did not differ from mated males in wing length or wing condition ($Z = 1.56$, $p = 0.12$; $Z = -1.60$, $p = 0.11$, respectively), but they did weigh more ($Z = 2.78$, $p = 0.005$).

DISCUSSION

Monarch butterflies that mate at overwintering sites in Mexico are a non-random subset of the overwintering population. Small, light-weight males in poor condition were found mating with large, heavy females in good condition. Large males and small females were not represented in mated pairs in proportions expected from the general population if mating were random. This is contrary to my prediction of positive assortative mating, i.e., that large butterflies should pair preferentially. Although there was a slight negative association between mated male and female wing condition late in the breeding period, I found no evidence of negative assortative pairing with respect to wet weight or wing length; i.e., these characters were not correlated within mated pairs.

Table 6. Male and female wet weight (mg), forewing length (mm), and wing condition (no units; see text) are compared among various sampling categories for 1986 data using Wilcoxon two-sample rank-sum tests (normal approximation; continuity correction of 0.05). * $p < 0.05$.

	Compared sample categories							
	Just-coupled versus mated pairs		Attempted copulation versus mated pairs		Attempted copulation versus just-coupled		Combined sexually active sample[a] versus general population	
Measure	Z	p	Z	p	Z	p	Z	p
Wet weight								
Males	−0.13	0.90	−0.90	0.37	−0.54	0.59	−4.8	0.0001*
Females	−2.11	0.03*	−3.22	0.001*	1.02	0.31	10.5	0.0001*
Forewing length								
Males	−1.38	0.17	0.34	0.74	−1.28	0.20	−6.3	0.0001*
Females	−1.42	0.15	0.08	0.93	−1.23	0.22	3.5	0.0004*
Wing condition								
Males	−1.57	0.12	−1.88	0.06	−0.15	0.88	−3.2	0.002*
Females	0.42	0.68	0.60	0.55	−0.20	0.84	8.3	0.0001*

[a] Combined sexually active sample = combined just-coupled, attempted copulation, and mated pair samples from 1986 field season. The mated pair sample was excluded from the combined sample for comparison of female wet weight; see text.

Table 7. **Comparison of male versus female wet weights (mg), forewing lengths (mm), and wing conditions (no units; see text) from five sampling categories from 1 March to 25 March 1986 using Wilcoxon two-sample rank-sum tests (normal approximation). * $p < 0.05$; ** $p < 0.005$; NS = not significant at $p = 0.05$.**

	Sample											
	Mated pairs			Just-coupled			Attempted copulation			Homocourtship[a]		
Measure	(n)	$\bar{X}$	SD	(n)	$\bar{X}$	SD	(n)	$\bar{X}$	SD	(n)	$\bar{X}$	SD
Wet weight												
Males	(65)	427	73	(51)	423	107	(66)	436	75	(47)	464	64
Females	(67)	506	77	(63)	478	67	(66)	464	62	(50)	485	66
Z		−5.45**			−2.81**			−1.92 (NS)			−1.35 (NS)	
Forewing length												
Males	(57)	51.6	2.6	(55)	50.8	3.3	(59)	51.6	2.4	(40)	52.6	1.7
Females	(67)	52.5	1.6	(75)	53.0	1.8	(64)	52.5	2.0	(50)	53.5	1.6
Z		−1.74 (NS)			−3.87**			2.20*			−2.47*	
Wing condition												
Males	(66)	1.9	0.7	(65)	2.1	0.8	(66)	2.2	0.7	(48)	2.1	0.6
Females	(67)	1.4	0.6	(77)	1.3	0.5	(66)	1.3	0.5	(51)	1.7	0.6
Z		−4.85**			6.32**			−6.83**			−3.69**	

[a] In the homocourtship category, the heading "Females" represents males pursued by other males during ground-phase courtship.

The magnitude of wet weight difference between mated and non-mated females is greater than that accounted for by spermatophore weight. Females in attempted copulation and just-coupled samples were significantly heavier than those in the general population, even though they had not yet received a spermatophore.

Wet weight and wing length were positively correlated with wing condition in both the general overwintering and mating populations. Because both wet weight and wing length reflect body size, it is not surprising that these two measures covary. That superior wing condition was associated with heavier wet weight and larger wing size was not expected. This finding suggests that either smaller monarchs age faster under overwintering conditions or physiologically older individuals reaching the overwintering site are more likely to be small.

The phenotypic character of the mating population changed through the breeding season in contrast to that of the general overwintering population. Early breeding males weighed less and were in worse condition than those breeding later, but they did not differ in wing length, suggesting that early breeders may be a physiologically distinct group of small males. Early-breeding females were larger and weighed more than those breeding later in the breeding period, while they did not differ in wing condition. Whether or not a decrease in the mean wet weight and wing size of breeding females was due to larger females leaving the overwintering site after mating is not known (see Calvert and Lawton, this volume). However, these characters did not change in females sampled from the general overwintering population with progression of the breeding period, as would be expected if large, heavy females left the area after mating. It is also unlikely that this change in the phenotypic character of breeding females was because large, heavy females were more receptive to mating attempts earlier in the season because males also took down large, heavy males. Males may resort to mating with smaller, less weighty females later in the breeding season to avoid sperm competition (see below).

An absence of differences in size or wing condition among males or females in attempted copulation, just-coupled, and mated pair samples suggests four points. First, only small males capture females in aerial takedown. Whether large males in good condition do not attempt to mate or are less competent at aerial takedown awaits further study. Possible advantages of small size in male Dipterans that mate on the wing are suggested by McLachlan and Allen (1987). Marshall (1988) found that lighter-weight (but not smaller-winged) males had an advantage in aerial intraspecific competition in a Lepidopteran and attributed the advantage to lower wing loading. Second, males takedown potential mates differentially, capturing large females in good condition. Third, females do not, or cannot, reject ground-phase mating attempts by males based on male size or wing condition. Fourth, although males apparently abandon some females during ground-phase courtship, this abandonment is not based on female size or wing condition.

Abandoned females were more likely to have recently mated than those successfully paired by males, suggesting that males may discriminate against recently mated females during ground-phase precopulatory interactions. Oberhauser (1988) found that virgin females were more likely to mate in caged conditions than non-virgin females. Al-

Table 7. Extended.

Sample					
Abandoned			General population		
(n)	$\bar{X}$	SD	(n)	$\bar{X}$	SD
(5)	433	54	(429)	463	61
(23)	466	67	(452)	418	59
	0.95 (NS)			10.61**	
(5)	52.3	2.8	(391)	52.9	1.8
(23)	52.8	1.7	(435)	52.1	1.9
	0.09 (NS)			6.07**	
(5)	1.4	0.6	(429)	1.9	0.7
(23)	1.5	0.5	(452)	1.8	0.7
	0.28 (NS)			1.91 (NS)	

though Oberhauser attributes this difference to female reluctance, this study suggests that males may reject non-virgin females at overwintering sites in Mexico.

Assuming fresh spermatophores originated from *in copula* males, monarchs mating at Mexican overwintering sites showed a high rate (90%) of successful spermatophore transfer. In agreement with Drummond (1984), these data suggest that success in spermatophore transfer is generally high in monarchs under field conditions.

FEMALE PERSPECTIVES

Pliske (1975), studying summer breeding monarch populations, suggested that female monarch butterflies exercise indirect mate choice by evading copulation attempts, thereby permitting only physically strong males to mate. Although overwintering females did apparently resist copulation attempts by males during ground-phase courtship, evasion seems limited as a means of female mate choice because (1) males that took down females in aerial pursuit were small and worn, and presumably should not be favored by females seeking material benefits; (2) females only rarely successfully evaded copulation attempts by these males during ground-phase interactions (pers. obs.); and (3) males that successfully coupled with females did not differ in the phenotypic characters measured from those attempting copulation. Even though males were more likely to abandon mated than unmated females during ground-phase precopulatory interactions, the consistency of female behavior and the general inability of males to evade mating attempts by other males (pers. obs.) suggest that males, and not females, were in control.

Aerial takedown also appears more likely to be under male than female control since female needs for sperm and male-provided nutrients cannot account for the absence of small females in poor condition in ground-phase precopulatory samples. Evidence that males captured by other males in aerial takedown were, like females, large and in good condition also suggests that differences in female behavioral receptivity do not account for the outcome of male aerial takedown. Although female monarchs begin gonad development later in the overwintering season than males (Herman, 1985), there were no obvious differences in female precopulatory behavior through the breeding period. Actively flying females available for mating were of similar size, weight, and condition as clustered females (see Table 2). Furthermore, because gonad development and aging are associated, possible female sexual receptivity differences seem especially unlikely to account for the bias toward large females in good condition in mated pairs.

A lack of female control in mate pair formation at Mexican overwintering sites and apparent male abandonment of recently mated females support the hypothesis (Gwynne, 1984) that male investments in large nutritious spermatophores are likely to have developed through direct male reproductive benefits (as suggested by Herman, 1981) and/or sperm competition avoidance rather than precopulatory female choice. Oberhauser (1989) found that females remated sooner in the laboratory when they received smaller spermatophores, but her study did not address whether this resulted from female refusal to mate or male avoidance of sperm competition. The present study supports the second of these alternative hypotheses.

Although precopulatory female choice was not supported in this study, females may still influence which male fathers the offspring during copulation or postcopulation (Thornhill and Alcock, 1983; Drummond, 1984). In this study, males that successfully transferred spermatophores weighed more and were in better condition (shown to produce larger spermatophores; Oberhauser, 1988) than unsuccessful males, but they did not differ in wing length. Success in obtaining a spermatophore did not reflect female phenotype. Females terminating copulation before the complete spermatophore is transferred may gain nutrients passed into the spermatophore before the sperm (see Drummond, 1984). Even if sperm is transferred, females may be able to fertilize all or most of their eggs, via last male reproductive advantage, by males that successfully remigrate and maintain a position at oviposition sites (see Zalucki, this volume).

POSTULATED ALTERNATE MALE REPRODUCTIVE BEHAVIORS

Possible small male advantages in aerial takedown could account for the bias toward small males in mated pairs and cannot be ruled out in this study.

But the bias toward light-weight, worn males only in early-breeding samples merits further explanation. Juvenile hormone increases associated with increasing temperature and day length influence the onset of mating behavior (Herman, 1973) and senescence (Herman, 1985) in monarchs at overwintering sites (Herman, this volume). Differences in the rate of gonad development could account for why lighter-weight, more worn males predominate early-mating pair samples. Because in this study wing condition was positively correlated with wing size (assuming wing condition reflects physiological age rather than distance traveled [see Herman, 1985]), it is not clear whether initiation of sexual activity by males is size- and/or age-dependent. Variation in the initiation of sexual activity associated with male age or physiological condition could account for the observed bias toward small, light-weight, worn males in mating samples. This variation in the initiation of sexual activity, nested within the timing of colony breakup and remigration, could result in two functionally discontinuous conditional male reproductive behaviors; smaller (less weighty, more worn) males mate primarily before migration, whereas larger (heavier, less worn) males mate primarily during or after remigration.

Whether or not possible alternate male reproductive behaviors are the result of sexual selection cannot be determined from this study, but it is likely that males showing these alternate mating behaviors incur different costs and reproductive benefits. Small males are at a disadvantage in remigration because larger wings are advantageous in soaring, which is thought to be the primary means of migration in monarchs (Gibo and Pallett, 1979). If lower wet weight reflects lower fat content, light-weight males may have little chance of successful remigration, especially if followed by intraspecific competition at oviposition sites. Beall (1946) found that monarchs that were washed ashore on Lake Erie during the autumn migration had both lower lean weight (the weight of water and fat-free material) and significantly shorter wings than a group of clustered monarchs from beside the lake. Wing wear is also likely to hinder worn butterflies during the long remigration. Wing scale loss is an important contributor to wing wear. Because monarch wing scales contain significant concentrations of defensive cardiac glycosides (Brower and Glazier, 1975; Nishio, 1980; Brower et al., 1988) and probably also pyrrolizidine alkaloids (Brown, 1984; Kelly et al., 1987), males in good condition may be better protected from potential predators during remigration (Brower, 1985).

Males that mate at overwintering sites are ensured of at least a possibility of fathering some offspring without the need to incur costs of remigration, mate-searching, or possible male competition at oviposition sites. They may increase their reproductive success directly through nutrient contributions used for egg production and, indirectly, by providing their mates with resources (Boggs and Gilbert, 1979; Wells et al., this volume) for speedy remigration and early oviposition. Such early oviposition may give larvae competitive advantages, increasing the reproductive value of each offspring produced (Waltz and Wolf, 1984).

In contrast, assuming last male sperm precedence, males that migrate and afford the possible rigors of intraspecific competition at oviposition sites will be the likely winners in sperm competition. Males that mate at oviposition sites after migration may even fertilize eggs produced from nutrients supplied by the males that mated at overwintering colonies. Thus, males that migrate and mate at oviposition sites have a safer paternity assurance bet that possibly only large males in superior condition can make.

MALE MATE DISCRIMINATION AT MEXICAN OVERWINTERING SITES

Males that mate at overwintering sites are probably limited in their immediate opportunity for polygyny since time is needed to produce successive resource-rich spermatophores (Oberhauser, 1988), especially if spermatophore size is positively related to reproductive success. Male resource investments may also lower future mating opportunities by decreasing the male's longevity (e.g., Boggs, 1981; Shapiro, 1982) or reducing his chances of remigration. Such immediate and future constraints on male promiscuity increase the benefits of pairing with large, healthy females at overwintering sites (Drummond, 1984), as was found in this study. Whether or not this pairing represents active male mate choice (sensu Parker, 1983) will be difficult to determine in the field.

Males mating at overwintering sites should also show mechanisms of sperm competition avoidance even if last male reproductive advantage occurs, if the degree of sperm mixing increases with time between mating and oviposition (Boggs, 1981). Although male abandonment of recently mated females was supported in this study, a possible mechanism for this discrimination was not investigated. Male monarchs have been shown to transfer substances that could further reduce sperm competition by stimulating gonad development (Herman and Barker, 1977; Herman, 1985; Herman et al., 1989), speeding vitellogenesis (Herman and Barker, 1977), and possibly stimulating female movement from the area (Zalucki, this volume).

SUMMARY

In this paper I have shown that small, worn males mate at Mexican overwintering sites and these males pair with large females in good condition. The male takedown strategy apparently precludes precopulatory female choice. Such a non-random mating pattern does not necessarily imply that differential reproductive advantages are in effect, or that these mating patterns are the result of sexual selection.

The evolutionary implications of such a non-random mating system await future work.

ACKNOWLEDGMENTS

I give special thanks to Lincoln P. Brower for continual support and especially for his genuine and tireless enthusiasm for monarch biology, which has made studying this animal a truly inviting endeavor. Thanks to Bruce Ferguson for technical assistance and evolutionary perspective. I also thank John Sivinski, James Lloyd, and Donald Dewsbury, Tom Walker, Steve Malcolm, and again Lincoln P. Brower for suggestions on versions of the manuscript. This research was partially funded by a Sigma Xi Grant-in-Aid of Research.

LITERATURE CITED

Beall, G. 1946. Seasonal variation in sex proportion and wing length in the migrant butterfly, *Danaus plexippus* L. *Transactions of the Royal Entomological Society of London* 97:337–353.

Boggs, C.L. 1981. Selection pressures affecting male nutrient investment at mating in heliconiine butterflies. *Evolution* 35(5):931–940.

Boggs, C.L., and L.E. Gilbert. 1979. Male contribution to egg production in butterflies: Evidence for transfer of nutrients at mating. *Science* 206:83–84.

Brower, L.P. 1985. New perspectives on the migration biology of the monarch butterfly, *Danaus plexippus* L. In *Migration: Mechanisms and adaptive significance,* ed. M.A. Rankin, 748–785. Austin: University of Texas, *Contributions in Marine Science,* Supplement, 27.

Brower, L.P., and S.C. Glazier. 1975. Localization of heart poisons in the monarch butterfly. *Science* 188: 19–25.

Brower, L.P., W.H. Calvert, L.E. Hedrick, and J. Christian. 1977. Biological observations on an overwintering colony of monarch butterflies (*Danaus plexippus* L., Danaidae) in Mexico. *Journal of the Lepidopterists' Society* 31(4):232–242.

Brower, L.P., C.J. Nelson, J.N. Seiber, L.S. Fink, and C. Bond. 1988. Exaptation as an alternative to coevolution in the cardenolide-based chemical defense of monarch butterflies (*Danaus plexippus* L.) against avian predators. In *Chemical mediation of coevolution,* ed. K.C. Spencer, 447–475. San Diego: Academic Press.

Brown, K.S., Jr. 1984. Chemical ecology of dehydropyrrolizidine alkaloids in adult Ithomiinae (Lepidoptera: Nymphalidae). *Revista Braziliera de Biologia* 44(4):435–460.

Byers, J.R. 1978. Biosystematics of the genus *Euxoa* (Lepidoptera: Noctuidae). X. Incidence and level of multiple mating in natural and laboratory populations. *Canadian Entomologist* 110:193–200.

Calvert, W.H., and L.P. Brower. 1986. The location of monarch butterfly (*Danaus plexippus* L.) overwintering colonies in Mexico in relation to topography and climate. *Journal of the Lepidopterists' Society* 40(3):164–187.

Drummond, B.A., III. 1984. Multiple mating and sperm competition in the Lepidoptera. In *Sperm competition and the evolution of animal mating systems,* ed. R.L. Smith, 291–370. Orlando, FL: Academic Press.

Eanes, W.F., and R.K. Koehn. 1978. An analysis of genetic structure in the monarch butterfly *Danaus plexippus* L. *Evolution* 32:784–797.

Gibo, D.L., and M.J. Pallett. 1979. Soaring flight of monarch butterflies, *Danaus plexippus* (Lepidoptera: Danaidae), during the late summer migration in southern Ontario. *Canadian Journal of Zoology* 57:1393–1401.

Gwynne, D.T. 1984. Male mating effort, confidence of paternity, and insect sperm competition. In *Sperm competition and the evolution of animal mating systems,* ed. R.L. Smith, 117–149. Orlando, FL: Academic Press.

Herman, W.S. 1973. The endocrine basis of reproductive inactivity in monarch butterflies overwintering in central California. *Journal of Insect Physiology* 19:1883–1887.

Herman, W.S. 1981. Studies on the adult reproductive diapause of the monarch butterfly, *Danaus plexippus. Biological Bulletin* 160:89–106.

Herman, W.S. 1985. Hormonally mediated events in adult monarch butterflies. In *Migration: Mechanisms and adaptive significance,* ed. M.A. Rankin, 799–815. Austin: University of Texas, *Contributions in Marine Science,* Supplement, 27.

Herman, W.S., and J.F. Barker. 1977. Effect of mating on monarch butterfly oogenesis. *Experientia* 33:688–689.

Herman, W.S., L.P. Brower, and W.H. Calvert. 1989. Reproductive tract development in monarch butterflies overwintering in California and Mexico. *Journal of the Lepidopterists' Society* 43(1):50–58.

Hill, H.R., Jr., A.M. Wenner, and P.H. Wells. 1976. Reproductive behavior in an overwintering aggregation of monarch butterflies. *American Midland Naturalist* 95(1):10–19.

Kelly, R.B., J.N. Seiber, A.D. Jones, H.J. Segall, and L.P. Brower. 1987. Pyrrolizidine alkaloids in overwintering monarch butterflies (*Danaus plexippus*) from Mexico. *Experientia* 43:943–946.

Knowlton, N. 1979. Reproductive synchrony, parental investment and evolutionary dynamics of sexual selection. *Animal Behaviour* 27:1022–1033.

Marshall, L.D. 1988. Small male advantage in mating in *Parapediasia teterrella* and *Agriphila plumbifimbriella* (Lepidoptera: Pyralidae). *American Midland Naturalist* 119(2):412–419.

Masters, A.R., S.B. Malcolm, and L.P. Brower. 1988. Monarch butterfly (*Danaus plexippus*) thermoregulatory behavior and adaptations for overwintering in Mexico. *Ecology* 69(2):458–467.

McLachlan, A.J., and D.F. Allen. 1987. Male mating success in Diptera: Advantages of small size. *Oikos* 48:11–14.

Nishio, S. 1980. The fates and adaptive significance of cardenolides sequestered by larvae of *Danaus plexippus* (L.) and *Cycnia inopinatus* (Hy. Edwards). Ph.D. dissertation. Athens: University of Georgia.

Oberhauser, K.S. 1988. Male monarch butterfly spermatophore mass and mating strategies. *Animal Behaviour* 36:1384–1388.

Oberhauser, K.S. 1989. Effects of spermatophores on male and female monarch butterfly reproductive success. *Behavioral Ecology and Sociobiology* 25: 237–246.

Parker, G.A. 1983. Mate quality and mating decisions. In *Mate choice,* ed. P. Bateson, 141–166. Cambridge: Cambridge University Press.

Pliske, T.E. 1975. Courtship behavior of the monarch butterfly, *Danaus plexippus* L. *Annals of the Entomological Society of America* 68(1):143–151.

Rutowski, R.L. 1985. Evidence for mate choice in a sulfur butterfly (*Colias eurytheme*). *Zeitschrift für Tierpsychologie* 70:103–114.

Rutowski, R.L., G.W. Gilchrist, and B. Terkanian. 1987. Female butterflies mated with recently mated males show reduced reproductive output. *Behavioral Ecology and Sociobiology* 20:319–322.

SAS. 1985. *SAS user's guide: Statistics,* Version 5 ed. Cary, NC: SAS Institute, 956 pp.

Shapiro, A.M. 1982. Survival of refrigerated *Tatochila* butterflies (Lepidoptera: Pieridae) as an indicator of male nutrient investment in reproduction. *Oecologia (Berlin)* 53:139–140.

Suzuki, Y. 1978. Adult longevity and reproductive potential of the small cabbage white, *Pieris rapae crucivora* Boisduval (Lepidoptera: Pieridae). *Applied Entomology and Zoology* 13(4):312–313.

Thornhill, R., and J. Alcock. 1983. *Sexual selection and the evolution of insect mating systems.* Harvard: Harvard University Press, 547 pp.

Trivers, R.L. 1972. Parental investment and sexual selection. In *Sexual selection and the descent of man, 1871–1971,* ed. B. Campbell, 136–179. Chicago: Aldine-Atherton.

Urquhart, F.A. 1987. *The monarch butterfly: International traveler.* Chicago: Nelson-Hall, 232 pp.

Waltz, E.C., and L.L. Wolf. 1984. By Jove!! Why do alternative mating tactics assume so many different forms? *American Zoologist* 24:333–343.

IS MULTIPLE MATING AN ADAPTIVE FEATURE OF MONARCH BUTTERFLY WINTER AGGREGATION?

Harrington Wells, Patrick H. Wells, and Steffen H. Rogers

ABSTRACT. During multiple mating, in late-winter aggregations of monarch butterflies, male nutrient contributions to females via spermatophore transfer may occur. We have confirmed this possibility by radioautographic techniques. Females inseminated by C^{14}-primed males became radioactive. First, bursa copulatrix tissues became lightly labeled, followed within 8 hr by dense labeling of all abdominal organs.

Stored lipid energy reserves regressed linearly in both male and female premating monarchs during the winter. When multiple mating occurred in late winter, male lipids continued to deplete while those of females significantly increased. A male energetic contribution by spermatophore transfer is inferred.

A mathematical model was developed to explore the effect of multiple mating on monarch butterfly populations. The model predicts that multiple mating is a specific adaptive feature of the monarch butterfly aggregations within overwintering areas. Male energetic contribution through spermatophore transfer appears to enhance female survival and reproductive success. In the model system, increased survival and reproductive success of females resulting from male energy transfer enhances the long-term persistence of monarch populations under adverse conditions.

INTRODUCTION

Monarch butterfly (*Danaus plexippus*) aggregation within overwintering areas has been thought to result from long-distance mass movements by which butterflies escape the rigors of northern winters, as do some migratory birds (Urquhart, 1960). Alternatively, at least in parts of Australia and California, the winter aggregations may simply reflect seasonal contraction of the environmentally suitable monarch butterfly range, with clusters populated by locally produced butterflies (Smithers, 1977; James, this volume; Wenner and Harris, this volume). Neither the migration nor the seasonal range contraction hypothesis explains winter monarch aggregation; dispersed roosting in these overwintering regions would be as beneficial under either of these theories.

The wintering areas that we studied in California are frost-free, support vegetation for roosting, and offer moisture sources. They are also cool throughout the night and during most hours of the day. Consequently, lipid energy reserves are not rapidly depleted by basal metabolic processes of inactive butterflies within the winter clusters (Chaplin and Wells, 1982). Adaptive as these features may be, they are independent of aggregation per se within the wintering areas. Dispersed roosting would serve as well.

Mexican winter aggregation locations are similar environmentally to sites in California, except that occasional freezing periods occur during some years (Anderson and Brower, this volume). Although clustering may increase survival during an occasional freeze, three lines of evidence suggest that aggregations are not the result of selection by this environmental factor. First, aggregations in California have remained coastal, and Mexican populations inland. Selection would be expected to have caused either Californian overwintering aggregations to move to cooler elevations, where metabolic rates would be lower, or Mexican overwintering clusters to move to lower elevations, where the likelihood of death due to freezing would be minimized. Second, new aggregation site formation can occur, as evidenced by monarch clusters in *Eucalyptus* forests in California (*Eucalyptus* species having been introduced within the last 150 yr). Thus, elevation of Californian and Mexican aggregation sites cannot be ascribed to innate behavior, which

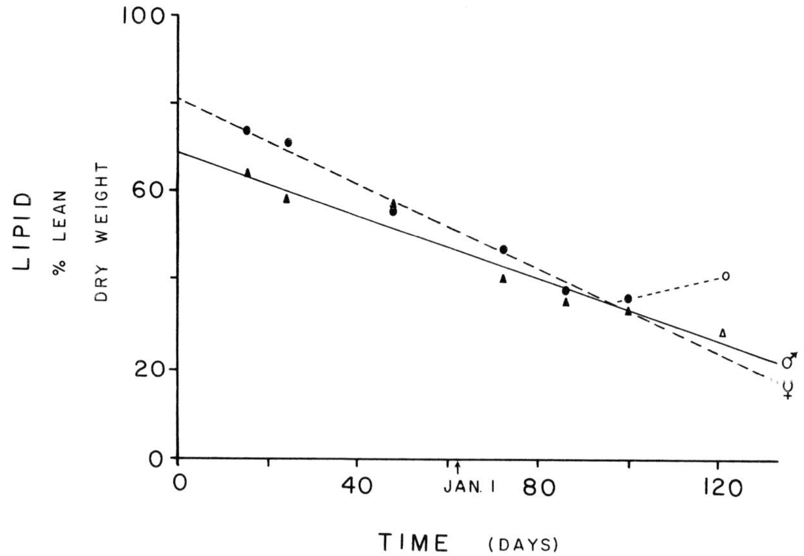

Figure 1. Least-squares linear regression of data for lipid content (% lean dry weight) versus time performed separately on male and female data. Each point represents 25 pooled individuals. Regression analyses were performed on premating samples (solid points). Male and female regressions are significantly different. Only the female postmating sample (0, male = Δ) was significantly different than predicted by the 95% prediction limits based on the regression analyses (see text).

inhibits selection from causing new cluster localities. Finally, if clustering were simply the result of selection to protect butterflies from occasional freeze periods, then Californian populations would not be expected to aggregate.

One behavioral feature that has not been observed in dispersed populations is the "frenzied" mating activity, which occurs in late winter just prior to and during dispersal of the aggregations. This activity has been described in detail (Hill et al., 1976; Van Hook, this volume), and spermatophore counts from dissected bursa copulatrices of February females show that mating is universal and multiple (Hill et al., 1976). We hypothesize that this mating frenzy is an important aspect of the monarch butterfly's life history that markedly increases fitness, not simply in terms of classical aspects of reproduction, but rather for energetic reasons that improve the chances of living to successfully reproduce.

Prior studies of overwintering energetics have focused on the months of passive roosting, which precede mating and dispersal. During the premating period, large initial lipid reserves of a population at Santa Barbara, California, were depleted linearly at a rate predicted by the oxygen consumption of inactive butterflies under prevailing temperatures (Chaplin and Wells, 1982). Similar winter depletion of large autumn lipid reserves has been recorded in a population at Muir Beach, California (Tuskes and Brower, 1978), in New South Wales, Australia (James, 1984), and in Mexico (Masters et al., 1988; Brower et al., this volume).

When the mating frenzy starts, monarch butterfly behavior and energetics begin to change. Butterflies actively fly, chase, and engage in reproductive activity during parts of warmer days. Subsequently, monarch butterflies disperse from cluster locations and females visit oviposition sites. Active butterflies have elevated body temperatures and a higher metabolism than the basal metabolic rates of passively roosting butterflies. Reproductive success requires that females in particular must expend energy in egg production as well as in the search for suitable larval host plants. In this context, it has been reported that butterfly spermatophores may actually contain nutrient reserves that can be transferred to recipient females (Boggs and Gilbert, 1979). In Lepidoptera, approximately 6% of male body mass may be passed to the female during copulation (Rutowski et al., 1983). The male secretions that form the spermatophore are rich in proteins and a variety of lipid compounds (Marshall, 1985). Thus, predispersal copulatory nutrient transfer is a potentially significant male energetic investment in egg production and in female survival to oviposition in the monarch butterfly.

This study asks: Is multiple mating in late winter a specific adaptive feature of the aggregation phenomenon within monarch butterfly overwintering areas?

METHODS

Samples of 25 male and 25 female monarch butterflies were collected on each of seven dates from November through February (see Chaplin and Wells, 1982, table 1) at the Ranchos Dos Pueblos aggregation 10 km west of Goleta, California. Fat content of male and female butterflies separately were determined for each collection date using an ether reflux lipid extraction technique (Morton and Tung, 1971). Females collected at each date (N = 25) were pooled for lipid extraction. Males collected at each date (N = 25) were also pooled for lipid extraction. Data were divided into premating samples (14 November through 7 February; N = 6) and a postmating sample (29 February). Data analysis was performed using least-squares linear regression (Sokal and Rohlf, 1969; Zar, 1984) separately on male and female data. Premating data were used to calculate each regression line.

An additional 80 butterflies (40 males, 40 females) collected in January were brought into the laboratory. Males and females were kept in separate cages. Males were fed 1 *M* sucrose:2.5 m*M* glucose solution having 4 μCiC[14]/μ*M* glucose; females were fed 1 *M* sucrose:2.5 m*M* nonradioactive glucose. Males and females were then put together into four cages 0.5 m × 0.5 m × 1.0 m tall (10 males and 10 females per cage). Time at which copulation occurred was noted for each pair; specimens were killed and their abdomens fixed in 2% glutaraldehyde 0.1 *M* sodium cacodylate buffer 0.5, 2, 4, 8, 12, 24, and 48 hr after copulation was completed. Controls consisted of precopulatory females and females that copulated with males fed only on nonradioactive glucose solution (sampled at 24 hr). Five specimens were killed at each time period and for each control. Abdomens were embedded in paraffin and serially sectioned at 10 μm on a rotary microtome. Autoradiographs were made of sectioned material coated with Ilfard Nuclear Research Emulsion type L/4 and incubated 3 wk before development.

Isotope label was quantified utilizing a Nikon microscope with a Photometrics charged coupling device (CCD): liquid-cooled CH 220 camera head containing a Thomson CSF TH 7882DA CCD and a CE 200 camera electronics unit, integrated with a Photometrics 3000 Computer Digital Image Acquisition and Processing System.

Light microscope images were exposed to the CCD system for 2 dasec. Next, the entire CCD image was computer flat-fielded to equalize light exposure across the specimen. The digitized image was then displayed on an Image Display Module and specific organs were selected for quantification by electronically eliminating on the computer screen display those CCD pixels that were not of interest. A second "blank," or "background," area of equal size was selected, analyzed, and compared to the "experimental" area. By electronically subtracting background from experimental pictures, the total light absorbed (blocked) by the silver grains in the autoradiographic emulsion can be quantified within 0.1% error. The accuracy of the system permits exact quantification and comparison of isotope uptake from area to area within the same tissue section or between different sections. Section thickness can affect total light absorbed; however, this error was corrected by analyzing multiple exposures of the same tissues in different sections.

A theoretical model, based on observed data on monarch life history and energetics, was used to examine the effects of spermatophore energy transfer and the aggregation phenomenon on monarch butterfly population dynamics.

RESULTS

LIPID LEVELS

The linear regression performed on premating male lipid levels was significant ($Y = -0.362X + 69.372$; ANOVA test: $F = 53.7$, df = 1,4, $p < 0.005$), and the slope of the regression line was significantly different from zero ($t = 7.33$, df = 4, $p < 0.002$). Similarly, the linear regression performed on premating female lipid data was significant ($Y = -0.470X + 80.864$; ANOVA test: $F = 273.0$, df = 1,4, $p < 0.005$), and the slope of the regression line was significantly different from zero ($t = 16.32$, df = 4, $p < 0.001$). Male and female regressions are significantly different ($t = 16.9$, df = 8, $p < 0.05$ [Note: $F = \sigma_1^2/\sigma_2^2 = 3.0$, df = 4,4, $p > 0.05$ so test is legitimate]). Therefore, lipid reserves declined throughout the winter in both male and female butterflies (Fig. 1).

The 95% prediction limits (95% PL) based on the regression analyses were calculated next for the postmating sample, 29 February, and compared to observed values for that date. The observed male lipid percent lean dry weight (28.9) did not differ from that predicted by the regression analysis using premating male data (95% PL: 39.86, 11.28). However, female lipid percent lean dry weight (40.5) was significantly higher than predicted by the female premating regression analysis (95% PL: 32.24, 15.78). In fact, female lipid percent lean dry weight had increased and was greater than it was for males or females on either of the two previous sampling dates. Also, female lipid percent lean dry weight was greater than male lipid percent lean dry weight on 29 February. Male lipid reserves had continued to decline over this same period of time.

NUTRIENT TRANSFER

Female butterfly assimilation of nutrients transferred via copulation is depicted in Figure 2. The bursa copulatrix was always labeled lightly first (Figs. 2 and 3a), followed by other parts of the reproductive tract and tissues immediately surrounding the bursa (Fig. 3b). Organs not associated with reproduction were the last to become radioactive. All tissues in all female abdomens sectioned showed uniform radioactivity (light to heavy) 8 hr after copulation (Fig. 3b). Thus, assimilation, and the potential for using transferred energy reserves, occurs quite rapidly. Unmated control females and those mated with unlabeled males showed no radioactivity (Fig. 3c).

THEORETICAL ANALYSIS

Lipid energy reserves are critical for butterfly overwinter survival. Consequently, once a butterfly's lipid reserves are depleted, its chances for survival or successful reproduction would be substantially di-

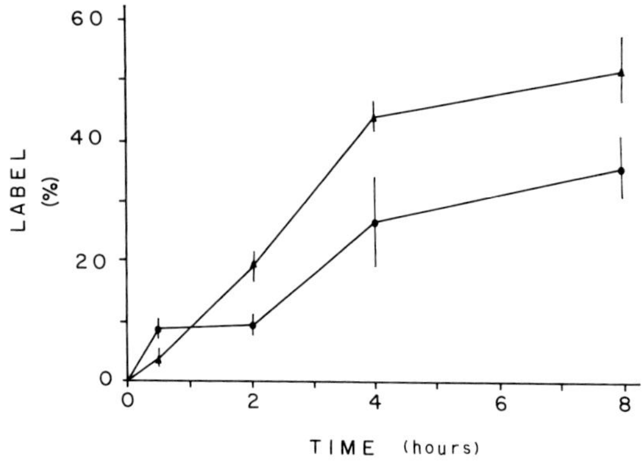

Figure 2. Density of label in female butterfly abdomen tissues versus time postcopulation. Label density was measured using a CCD (see Materials and Methods) from autoradiograph slides. Values recorded did not differ appreciably between serial sections of the same butterfly. Mean and standard deviation relative label (%) using the spermatophore as 100% are shown for the bursa (●) and upper reproductive tract tubes (▲).

minished. The following model examines the effect energy transfer from male to female butterflies has on the population dynamics of monarchs. Since aggregating facilitates multiple male–female contacts, the model considers the potential effect of aggregation versus dispersed overwinter roosting on butterfly population dynamics.

The model is based on Californian populations from which the above data were collected, and that we have observed for the past 20 yr, but may also be applicable to Mexican populations. Female butterfly energetics, survival, and population size are followed in the model. Assume that (1) metabolism is temperature- and activity-dependent, (2) metabolic rate is constant through winter, and (3) no environmental energy source (nectar) is harvested during winter. Then:

$$\frac{dR}{dt} = -\lambda \qquad (1)$$

where R = lipid reserve = energy reserve, t = time, and λ = metabolic rate. Integration of equation (1) gives:

$$R_t = -\lambda t + R_0 \qquad (2)$$

where R_0 = lipid reserve at time of cluster formation and R_t = lipid reserve at time t. Let $R_{0,i}$ = lipid reserve at t = 0 for individual i. The lipid reserve for the population of butterflies is: $\{R_{0,i} \mid i = 1, \ldots, n\}$. Assume R_0 is normally distributed with mean μ_{R0} and variance $\sigma_{R0}{}^2$.

Death occurs from either of two independent events: (1) random misfortune (chance) or (2) starvation. Assume that death due to random misfortune in the cluster is density-independent. Therefore, $dN/dt = -mN$, which upon integration yields $N_t = N_0\, e^{-mt}$ and can be expressed as:

$$C_t = N_t/N_0 = e^{-mt} \qquad (3)$$

where N_t = population size at time t, N_0 = initial population size, $-m$ = instantaneous rate of population change, and C_t = probability of surviving death from random misfortune at least to time t.

Environmental energy sources (nectar) throughout winter are essentially nonexistent. Death due to starvation occurs when $R_{t,i} \le R_{min}$; we assume death occurs at the point of energy depletion $R_{min} = 0$ (i.e. death occurs when $R_{t,i} = 0$). Since $R_{t,i} = -\lambda t + R_{0,i}$ then depletion of energy reserves occurs when $0 = -\lambda t + R_{0,i}$ and therefore time to energy reserve depletion is $t = R_{0,i}/\lambda$. Denote time to death by starvation for individual i as $T_i = R_{0,i}/\lambda$. Since R_0 is normally distributed, then T is normally distributed with mean $\mu_T = \mu_{R0}/\lambda$ and variance $\sigma_T{}^2 = \sigma_{R0}{}^2/\lambda^2$. Let Φ_T = probability of death due to starvation by time t = T:

$$\Phi_T = \int_{-\infty}^{T} \frac{e^{(t-\mu_T)^2/2\sigma_T}}{\sqrt{2\pi}\,\sigma_T}\, dt \qquad (4)$$

Therefore, the proportion of the female population still alive at time t = T is S_T, where $S_T = C_T(1 - \Phi_T) = e^{-mT}(1 - \Phi_T)$.

Mating in natural population occurs early in February. Oviposition occurs after winter cluster dispersal in late February, until death. Dispersal occurs in a "wave" from the winter cluster, proceeding to new areas coinciding with the geographic appearance of milkweed plants. Dispersal of the cluster, but not later monarch generations, precedes abundant food resources (nectar). Survival of cluster butterflies is largely dependent on lipid reserves, even after winter aggregation dispersal. Assume that the number of eggs oviposited occurs at an average rate, ϵ. Then:

$$\frac{dE}{dT} = \frac{\epsilon}{2} N_0\, e^{-mT}(1 - \Phi_T) \qquad (6)$$

where E = total number of eggs oviposited by females from the overwinter aggregation ($\epsilon/2$ is used because only half of the eggs oviposited are female). Integration of equation (6) gives the reproductive success, E, of overwintering females (equation (7)), where τ = time of first milkweed appearance:

$$E = \int_{\tau}^{\infty} \frac{\epsilon}{2} N_0\, e^{-mT}(1 - \Phi_T)\, dT \qquad (7)$$

Following oviposition by females from the aggregation, and their subsequent death, population growth occurs through a series of generations over summer. Population growth over summer is modeled as density-dependent (dependent on larvae and adult food supply) following Pielou (1977): $dN/df = rN(1 - N/K)$. Integration yields:

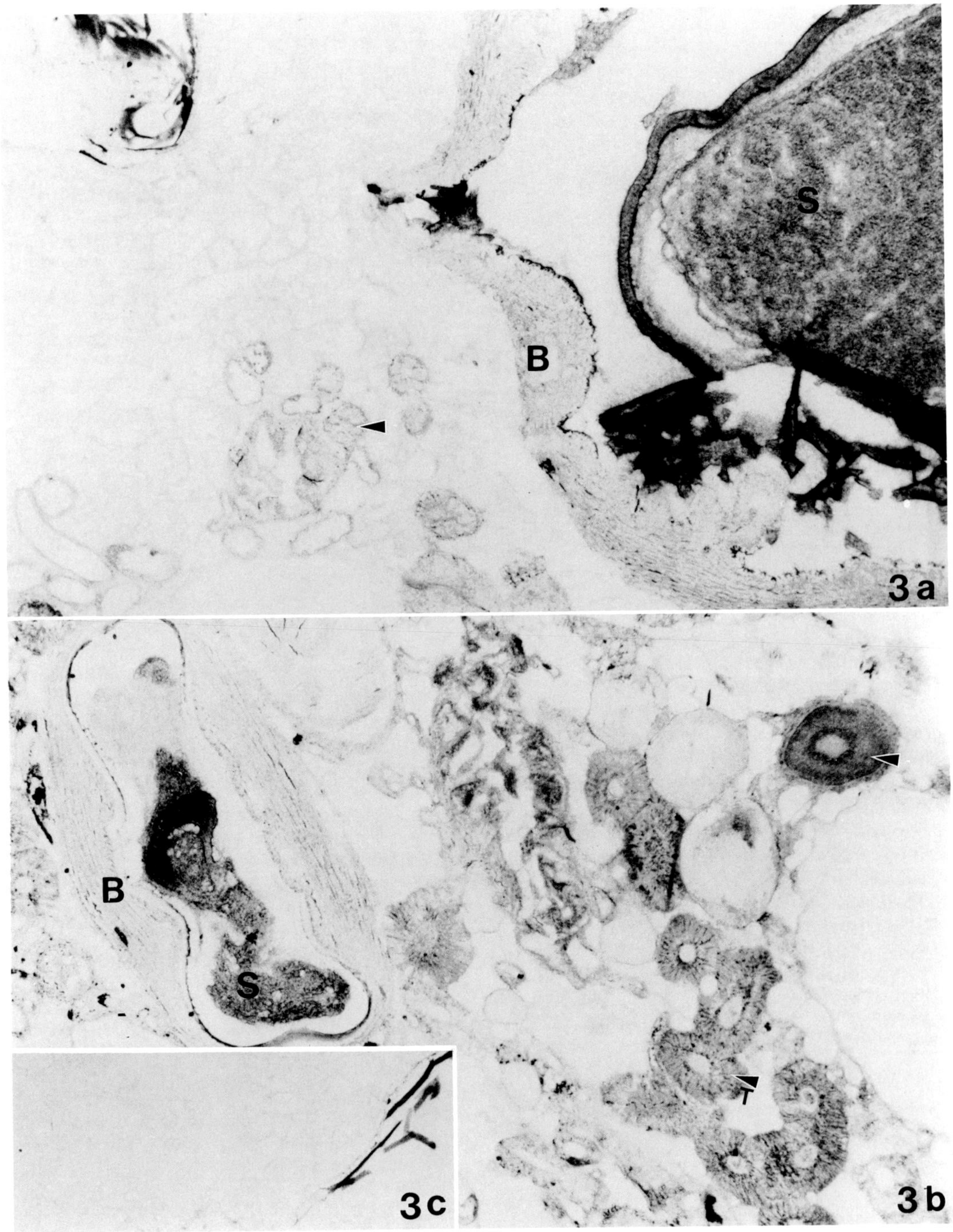

Figure 3. **a.** Unstained section of unlabeled female abdomen 2 hr postcopulation with labeled male. Note intense labeling in spermatophore (S), light labeling in bursa copulatrix (B), light labeling in some ducts (arrows), and lack of label in remaining tissue. **b.** Unstained section of unlabeled female abdomen 4 hr postcopulation with labeled male. Note intense labeling in spermatophore (S), complete labeling of bursa copulatrix (B), and intense labeling of most surrounding tissue, especially ducts (arrows). **c.** Control: Unstained section of unlabeled female abdomen 24 hr postcopulation with unlabeled male.

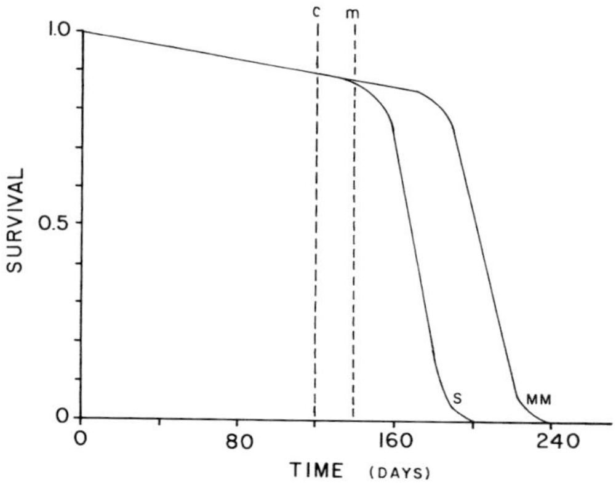

Figure 4. Mathematical model of female butterfly survival versus time with and without the multiple mating resulting from February mating frenzy. Survival is based on chance death (m = −0.0009) and on lipid energy reserves, which are depleted while overwintering (death due to starvation). Lipid depletion rates and fat content used were those observed in Figure 1. C = mean time of cluster dispersal, m = mean time of first milkweed appearance, MM = multiple mating, S = single mating.

$$N_f = \frac{K}{1 + \left(\frac{K - \epsilon}{\epsilon}\right) e^{-rf}} \qquad (8)$$

where K = carrying capacity, ϵ = number of eggs oviposited by the cluster population that survived to produce female adults (α percent of E), r = instantaneous rate of population change, f = time to fall population reclustering, and N_f = fall population size before condensation into overwintering sites. A portion of the fall female population, p, survives to reach the winter cluster that year ($\hat{N}_0$): $\hat{N}_0 = pN_f$.

The preceding discussion describes monarch life history without multiple mating. Spermatophores, however, not only contain sperm but also are an energy source for female butterflies. Spermatophore counts in females dissected in mid-February varied from 1 to 7 ($\bar{X}$ = 2.6, SD = 1.5, N = 25; see Hill et al., 1976, table 2). Since spermatophores are degraded by female butterflies, the actual number of predispersal copulations probably is several times greater. In fact, the idea that the Hill et al. (1976) data represent a minimum estimate is supported by more refined methods of calculating number of matings (Suzuki and Zalucki, 1986; Zalucki, this volume). Let $R_{M,i}$ be the total energy reserve gained by female i after the mating frenzy period. Assume that $\{R_{M,i} \mid i = 1, \ldots, n\}$ is normally distributed with mean μ_{R_M} and variance $\sigma_{R_M}^2$. Female survival, now with multiple mating, is based on the energy depletion equation:

$$R_{t,i} = -\lambda t + R_{0,i} + R_{M,i} \qquad (9)$$

Energy reserve to be depleted by a female before it will starve to death now is $(R_{0,i} + R_{M,i})$. This energy reserve per female is normally distributed with mean $\mu_R = \mu_{R_0} + \mu_{R_M}$ and variance $\sigma_R^2 = \sigma_{R_0}^2 + \sigma_{R_M}^2$. As a result, the time to death due to starvation is now: $\hat{T}_i = (R_{0,i} + R_{M,i})/\lambda$. $\hat{T}$ is normally distributed with mean $\mu_{\hat{T}} = (\mu_{R_0} + \mu_{R_M})/\lambda$ and variance $\sigma_{\hat{T}}^2 = (\sigma_{R_0}^2 + \sigma_{R_M}^2)/\lambda^2$. Figure 4 displays survival of female butterflies over time, with and without multiple mating. Finally, the number of eggs oviposited, given that aggregated females have multiply mated, is:

$$E = \frac{\epsilon}{2} N_0 \int_{\tau}^{\infty} e^{-m\hat{T}} (1 - \Phi_{\hat{T}})\, d\hat{T} \qquad (10)$$

where $\Phi_{\hat{T}}$ = probability of death due to starvation by time t = $\hat{T}$ (see equation (4); only mean and variance have been changed).

To examine the effect of the multiple mating frenzy versus nonenergy transfer from male to female butterflies, the availability of oviposition sites must be considered. In a typical year, milkweed is first available in approximately mid-March (California) but may first appear considerably earlier or later in response to fluctuating patterns of seasonal rainfall and temperature (Wenner and Harris, this volume). Thus, first appearance of milkweed in spring is a variable that will affect reproductive success from year to year. Assume that the first appearance of milkweed in spring is normally distributed.

Monte Carlo computer simulations (Spain, 1982) varying first appearance of milkweed in spring were performed. They track butterfly population size, based on the mathematics previously described. The assumptions that energy reserves continue to decline at the rate observed throughout winter and that time of milkweed first appearance in spring is variable were made in each of two situations: (1) dispersed winter roosting, where multiple mating is relatively rare, and (2) winter aggregation, which leads to a spring mating frenzy and extensive multiple mating. Typical results (based upon μ_τ = 140, σ_τ = 10) are displayed in Figure 5. Without multiple mating, monarch populations declined to the point where extinction was inevitable in poor years. The aggregation model, however, resulted in population persistence, at approximately the levels observed in nature.

If butterflies actively fly for 1 hr/day after dispersing from the overwintering areas, energy use is predictably increased (Chaplin and Wells, 1982). Monte Carlo simulations, which assume increased metabolic rates consistent with 1 hr of active flight per day beyond what occurs in overwintering clusters, were performed next (Fig. 6). The results show that extinction occurs within 60 yr when multiple mating does not occur, but that the population stabilizes at numbers within the range of those observed in nature with multiple mating.

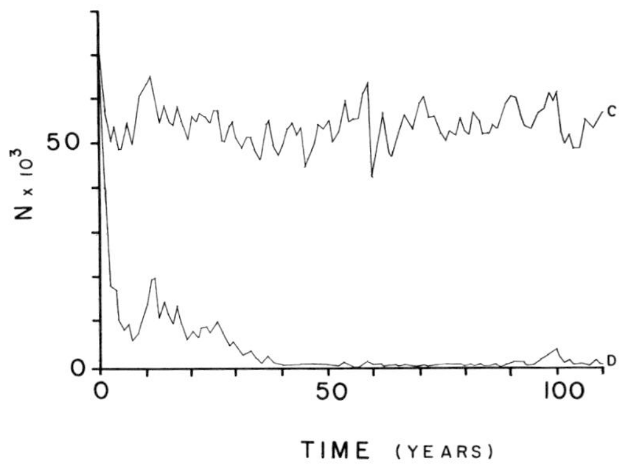

Figure 5. Typical result of monarch butterfly population dynamics for Monte Carlo simulation censused at the time of autumn aggregation. Time of first milkweed appearance in spring was the normally distributed random variable. Mating frenzy at end of winter produces a stable population, whereas nonmultiple mating results in populations that eventually result in extinction (energy use and mean lipid reserve taken from Figure 1: $\mu_T = 172$, $\sigma_T = 10.5$, $\mu_{\hat{T}} = 206$, $\sigma_{\hat{T}} = 12.6$; K = 100,000,000; $p = 0.001$; r = 1.0; $\alpha = 0.1$; $\epsilon = 20$/day; f = 6 generations, m = −0.0009; $\mu_\tau = 140$, $\sigma_\tau = 10$). C = cluster, D = dispersed overwinter roosting.

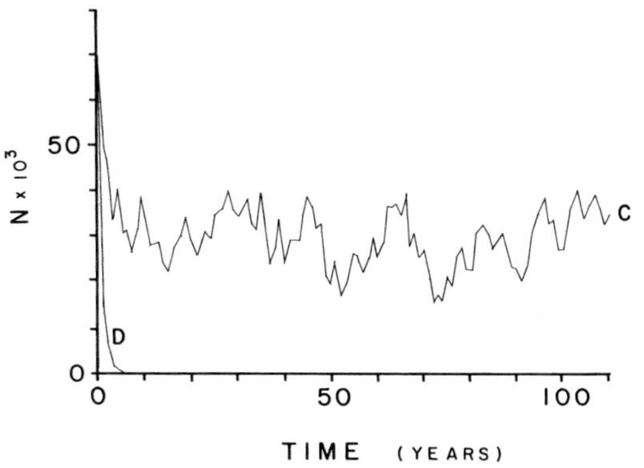

Figure 6. Typical result of monarch butterfly population dynamics from Monte Carlo simulation censused at the time of autumn aggregation. Butterfly energy use is assumed to be that of overwintering clusters plus 1 hr active flight per day (parameters same as Fig. 5 except for energy use rate). C = cluster, D = dispersed overwinter roosting.

DISCUSSION

In summary, the data show that material transferred from male to female monarch butterflies during copulation significantly slows female butterfly energy reserve depletion in natural populations; it apparently even increases the female's energy reserves. This occurs when nectar sources are few and not abundant (Tuskes and Brower, 1978). The continued energy impoverishment of males during this period confirms that nectar foraging is not a primary source of increased female energy reserves.

After spermatophores are deposited in the female bursa copulatrix, they are disrupted mechanically (Rogers and Wells, 1984) and nutrient materials are rapidly assimilated by the female. Rapid assimilation suggests that a specialized mechanism for nutrient absorption may have evolved in the female reproductive tract. The physiological mechanisms allowing energy reserve transfer between sexes thus appear to be of high selective value to both male and female butterflies because it assures successful reproduction. Successful reproduction requires that females disperse to locations with proper species of plants for oviposition. The male butterflies need not survive since they have already reproduced.

This system of significant nutrient transfer from males to females in late winter requires multiple matings, which are facilitated by aggregation (Wells et al., 1990; Wells and Wells, 1991). Thus, aggregation of monarch butterflies is inherently advantageous over dispersed roosting in the overwintering areas. Theoretical analyses suggest that overwintering aggregations contribute to population persistence in monarch butterflies. Increased reproductive success of multiply mating individuals would select for individuals with the clustering behavior. Winter aggregations are predicted to minimize population crashes in poor years by maximizing the potential for energy transfer from male to female butterflies just prior to dispersal in early spring. In fact, multiple mating may be energetically important year round to female butterflies since monarchs tend to coalesce around patches of host plants where female multiple mating with nutritional transfer occurs, followed by oviposition (Suzuki and Zalucki, 1986; Zalucki and Suzuki, 1987; Zalucki, this volume). Reproductively active winter populations, as observed by James (1981, this volume) in the Sydney area of New South Wales, may represent a transitional state and, in climates with seasonal energy sources and oviposition sites, the multiple mating effect on fitness appears to be an important factor selecting for aggregated rather than dispersed overwintering.

ACKNOWLEDGMENTS

We thank Rancho Dos Pueblos for courtesies extended during this study; Adrian M. Wenner and Myron P. Zalucki for reviewing draft versions of this manuscript; H. Frederick Hill, Andrew Senyei, and students in our entomology classes for assisting us in field and laboratory studies; and the Natural History Museum of Los Angeles County and the Organizing Committee (Julian P. Donahue, Chairman) for convening the Second International Conference on the monarch butterfly.

LITERATURE CITED

Boggs, C.L., and L.E. Gilbert. 1979. Male contribution to egg production in butterflies: Evidence for transfer of nutrients at mating. *Science* 206:83–84.

Chaplin, S.B., and P.H. Wells. 1982. Energy reserves and metabolic expenditures of monarch butterflies overwintering in southern California. *Ecological Entomology* 7:249–256.

Hill, F.H., A.M. Wenner, and P.H. Wells. 1976. Reproductive behavior in an overwintering aggregation of monarch butterflies. *The American Midland Naturalist* 95:10–19.

James, D.G. 1981. Studies on the winter breeding population of *Danaus plexippus* (L.) (Lepidoptera: Nymphalidae) at Spencer, New South Wales. *General Applied Entomology* 13:47–53.

James, D.G. 1984. Phenology of weight, moisture and energy reserves of Australian monarch butterflies, *Danaus plexippus*. *Ecological Entomology* 9:421–428.

Marshall, H.D. 1985. Protein and lipid composition of *Colias philodice* and *C. eurytheme* spermatophores and their changes over time (Pieridae). *Journal of Research on the Lepidoptera* 24:21–30.

Masters, A.R., S.B. Malcolm, and L.P. Brower. 1988. Monarch butterfly (*Danaus plexippus*) thermoregulatory behavior and adaptations for overwintering in Mexico. *Ecology* 69:458–467.

Morton, M.L., and H.L. Tung. 1971. The relationship of total body lipid to fat depot weight in the belding ground squirrel. *Journal of Mammalogy* 52:839–842.

Pielou, E.C. 1977. *Mathematical ecology*. New York: John Wiley & Sons, 385 pp.

Rogers, S.H., and H. Wells. 1984. The structure and function of the bursa copulatrix of the monarch butterfly (*Danaus plexippus*). *Journal of Morphology* 180:213–221.

Rutowski, R.H., M. Newton, and J. Schaefer. 1983. Interspecific variation in the size of the nutrient investment made by male butterflies during copulation. *Evolution* 37:708–713.

Smithers, C.N. 1977. Seasonal distribution and breeding status of *Danaus plexippus* (L.) (Lepidoptera: Nymphalidae) in Australia. *Journal of the Australian Entomological Society* 16:175–184.

Sokal, R.R., and F.J. Rohlf. 1969. *Biometry*. San Francisco: W.H. Freeman, 776 pp.

Spain, J.D. 1982. *Basic microcomputer models in biology*. London: Addison-Wesley, 368 pp.

Suzuki, Y., and M.P. Zalucki. 1986. The influence of sex ratio on female dispersal in *Danaus plexippus* (L.) (Lepidoptera: Danaidae). *Journal of the Austrian Entomological Society* 25:31–35.

Tuskes, P.M., and L.P. Brower. 1978. Overwintering ecology of the monarch butterfly, *Danaus plexippus* (L.) in California. *Ecological Entomology* 3:141–153.

Urquhart, F.A. 1960. *The monarch butterfly*. University of Toronto Press: Toronto.

Wells, H., and P.H. Wells. 1991. The monarch butterfly: A review. *Southern California Academy of Sciences Bulletin*. In press.

Wells, H., P.H. Wells, and P. Cook. 1990. The importance of overwinter aggregations for reproductive success of monarch butterflies (*Danaus plexippus*). *Journal of Theoretical Biology* 147:115–131.

Zalucki, M.P., and Y. Suzuki. 1987. Milkweed patch quality, adult population structure, and egg laying in the monarch butterfly. *Journal of the Lepidopterists' Society* 41:13–22.

Zar, J.E. 1984. *Biostatistical analysis*. Englewood Cliffs, NJ: Prentice-Hall, 718 pp.

SEX AROUND THE MILKWEED PATCH—THE SIGNIFICANCE OF PATCHES OF HOST PLANTS IN MONARCH REPRODUCTION

Myron P. Zalucki

ABSTRACT. A simulation model based on observations of female monarchs predicts egg distribution patterns in small, man-made patches but fails for field patches. The number of eggs per plant is much higher on single plants around the margins of a field patch than the simulation model would predict. Previous evidence suggested that these findings might be explained by frequent male–female encounters in field patches, which could promote female dispersal. This would result in male-biased sex ratios in patches, female-biased sex ratios around the margins of such patches, and subsequently higher numbers of eggs per plant in such situations.

Sex ratio and adult density within patches were manipulated in both summer and winter: Contrary to expectations, removing males *decreased* female recapture rates in a patch. Males may be providing females with the limiting resources of sperm and, more likely, nutrients for oogenesis. Other circumstantial evidence supporting this interpretation is presented.

Current ideas of adult monarch population structure around host-plant patches and its relationship to egg dispersion are discussed in light of these unexpected results.

INTRODUCTION

The strategy used by an animal to locate and utilize a resource is one of the central issues of population ecology. Butterflies are appropriate organisms for investigating such strategies. The distribution of resources such as host plants, nectar, and mates can be quantified, and associated movement patterns can be observed. Thus, movement and resource location can be related.

The monarch butterfly, *Danaus plexippus* (L.), was accidentally introduced to Australia around the late 1860s (Vane-Wright, this volume). Here it utilizes a number of introduced milkweed species, and in southeastern Queensland where the butterfly breeds year-round, the major food plants are *Asclepias curassavica* L. and *Asclepias fruticosa* L. (=*Gomphocarpus fruticosus*). The latter species predominates and grows in patches (dense stands of host plants) that vary in distribution, size, shape, and age (Zalucki et al., 1981). Monarchs and milkweeds provide a model system with which to investigate how an animal utilizes unevenly distributed resource patches.

In this paper, I summarize observations of egg-laying behavior in host-plant patches of different sizes (fully reported in Zalucki, 1983). New observations (presented for the first time) indicate a discrepancy between simulated and observed egg distributions. Many more eggs were found on single isolated plants around the margins of a field patch (see Table 1) than the simulation model would predict.

A number of authors have observed higher numbers of eggs on single isolated plants than on plants in a patch (Cromartie, 1975; Jones, 1977; Mackay and Singer, 1982; Shapiro, 1981; Thompson and Price, 1977; Wiklund and Ahrberg, 1978; Zalucki and Kitching, 1982b). There are a number of ways in which such a pattern could arise (see Mackay and Singer, 1982) including (1) differences in plant quality between the two situations, (2) insect preference for isolated plants, (3) insects treat patches and isolated plants as equivalent resources, (4) searching is biased to areas outside patches for some other reason (e.g., shade, which affects plant density), and (5) initiation of search is at random with respect to local plant densities. The effect of male behavior on female dispersal (a special case of (4) above) might account for the discrepancy and, thus, field experiments were conducted to test this hypothesis (Bull et al., 1985; Suzuki and Zalucki, 1986). Our results have necessitated a revision of our un-

Table 1. A comparison of egg-laying patterns observed on a man-made and a natural milkweed patch. Predicted patterns using a simulation model are also shown.

	Eggs/plant											
	Man-made patches[a]				Simulated[b]				Natural patches[c]			
Patch size	N[d]	X̄	SE	Range	N[e]	X̄	SE	Range	N[d]	X̄	SE	Range
Single plants[a]	40	0.72	0.17	0–3	24	1.27	0.60	0–4	185	2.50	0.19	0–15
Ca. 100[a]	200	0.44	0.06	0–3		0.42		(0–3)				
"Large"[f]					24	0.13			305	0.15	0.02	0–3

[a] Based on plants used to collect adult movement rules by Zalucki and Kitching (1982c).

[b] Based on Zalucki's (1983) model.

[c] Based on a patch and nearby single plants sampled from 10 August to 10 November 1982 (see text for details).

[d] N is the number plant × sampling occasions in this case. For "man-made" single plants, this was 20 plants sampled twice. For ca. 100, it is a patch of 90 and one of 110 sampled once each. For the natural patch, details of plant numbers are in the text.

[e] Number of females flown in the simulation.

[f] The simulated large patch was of the same size as the natural patch (200 m^2). The egg density is calculated assuming a plant occurs each time the egg-lay decision is made in the model and a plant density of 15 plants/m^2 (see nearest neighbor distances in text).

derstanding of male and female monarch interactions around patches of host plants.

OBSERVATIONS OF FEMALE EGG LAYING AND DISPERSION

Female monarchs were observed ovipositing on a man-made patch of *A. fruticosa* and *A. curassavica*. Potted plants were arranged outdoors in a regular grid of either 90 (9 × 10 m) or 110 plants (10 × 11 m), which included single outlying plants (ca. 3–10 m from a patch boundary). Details of plant arrangement can be found in Zalucki and Kitching (1982c). Butterfly movements in relation to plants were recorded using modified theodolites (Zalucki et al., 1980) and information characterizing the animal's flight patch (=track) was calculated (Zalucki and Kitching, 1982c). By using the criteria developed by Kitching and Zalucki (1982) and Zalucki and Kitching (1982a), the development of an animal's path may be simulated as a series of (x, y) horizontal locations (Zalucki, 1983). At each iteration, the position of the animal in relation to some prespecified host-plant locations can be checked, and if a plant is within perceptive distance a decision to land and/or oviposit on the plant can be made. In this fashion, the model can be used to generate egg-dispersion patterns on an array of plants. The predicted pattern of egg laying can be compared with the observed pattern on a similar plant array. This provides a check on how well we think we understand the process generating egg dispersion (or "patch use") in monarchs. Movement and egg laying in *Pieris rapae* has been similarly described by Jones (1977) and Jones et al. (1980).

Observed and predicted egg-dispersion patterns are compared in Table 1. As claimed by Zalucki (1983), there is reasonable agreement between the two, particularly for egg dispersions on small man-made milkweed patches (Table 1). These types of patches were also used when collecting the basic movement information. Agreement is not so good when we look at dispersions around large natural milkweed areas and the simulated results.

To determine how well the simulation model could predict egg dispersion, I measured egg laying in a natural situation analogous to the man-made plant dispersion used to gather movement data. The area chosen was on the University of Queensland Experimental Farm at Moggill in southeast Queensland, Australia. There was a central patch measuring roughly 20 × 10 m and surrounding this were scattered milkweeds. Fifteen of the latter, i.e., plants on the margin of the patch, and 25 plants within the patch were randomly selected and the number of eggs on each plant was checked each week from 10 August to 10 November 1982. I recorded number of eggs, plant height, distance to nearest neighbor, and distance of each single plant to the central host-plant patch.

The single plants were on average 65 m (range 15–120 m) from the main patch and were generally larger (X̄ height = 0.89 m ± 0.38 SD, range 0.5–1.6 m as measured half way through the experiment) than the patch plants (X̄ height = 0.53 ± 0.18 m, range 0.3–0.9 m). Patch plants were also (as might be expected!) much closer together (nearest neighbor distance, X̄ = 0.14 ± 0.10, range 0.02–0.05 m) than single plants (X̄ = 4.1 ± 2.72, range 0.6–10.4 m). The natural situation differed from the man-made setup and the simulation. Plants were younger within the patch than the surrounding plants (this *a priori* should lead to a higher number of eggs within the patch; see Zalucki and Kitching, 1982b). However, egg counts on single plants were much higher than the model would predict (Table 1). Statistical comparisons are difficult because I do not know how many females generated the ob-

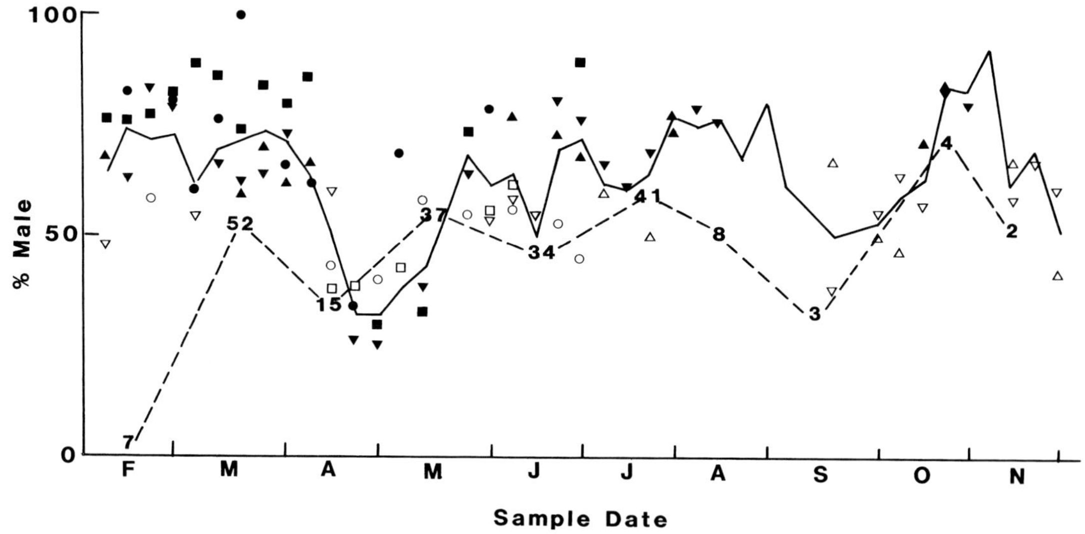

Figure 1. Percentage of males for each of four milkweed patches (△, ▽, □, ○) plotted against sample time. Solid symbols denote significant deviation from 50% ($p < 0.05$); open symbols indicate nonsignificance. Samples with <10 individuals are not shown. Solid line indicates average percentage of males across all sites using all sample data; dashed line indicates percentage of males in newly emerged category caught in all sites for each month; numbers indicate the total (♀ + ♂) of newly emerged butterflies caught.

served egg dispersions (in Table 1). One way around this is to run the simulation until eggs per plant in one situation (e.g., patch) are similar to the observed value and then to make comparisons for single plants. The important point here is that the maximum number of eggs per plant in the simulation was 4 compared with 15 in the field. Why the discrepancy?

THE INFLUENCE OF MALES ON FEMALE DISPERSAL: AN HYPOTHESIS

For monarchs, the simulation model partially takes into account two of the mechanisms commonly proposed to account for the higher number of eggs on single isolated plants, namely factors (3) and (5) (see Introduction). A number of the model components may be varied (e.g., plant attractiveness) so as to "improve" predictions (euphemistically called calibration). However, the most striking difference between small, man-made patches and large, natural ones is the presence of resident males in the latter. Males did encounter the small patches but were not resident (Zalucki and Kitching, 1982a). A number of lines of evidence suggest that the presence of males could explain the observations of egg distribution and provide a special case of mechanism (4) above (i.e., searching is biased to areas outside the patch).

I have found that the percentage of males in samples of adults taken regularly from four patches

Table 2. Mean residence times (days) for four milkweed patches by sex and age (based on Zalucki and Kitching, 1984).

	Males		Females	
Patch size (m²)[a]	Fresh[b]	Other	Fresh[b]	Other
435	8.8 (4)[c]	11.5 (14)	—	8.2 (6)
350	9.4 (3)	14.9 (29)	7.0 (3)	10.1 (20)
300	9.5 (2)	10.0 (20)	—	11.4 (8)
1,200	15.9 (11)	13.7 (99)	7.0 (6)	10.3 (51)
Overall[d]	13.00 (±11.4)	13.3 (±8.87)	7.0 (±2.5)	10.2 (±6.18)

[a] Estimated for areas of dense milkweed cover in which the mark–recapture study was undertaken.
[b] Newly emerged; wings soft.
[c] Values in parentheses are the number of recaptures recorded in each category.
[d] Calculated from the overall distribution of recapture times (±SD), except for fresh females, where it is ±½ the intercapture period.

Table 3. Percentage of males in samples of monarch butterflies from patches and nonpatches (Zalucki and Suzuki, 1987).

	Nonpatches	Patches
Number sampled	8	24
Mean % males[a]	40	60
Standard error	2.8	2.0
Range	24–50	35–72

[a] The mean % males in patches versus non-patches are significantly different (*t* test on arcsine square root transformed values).

over a 10-mo-period was generally >50% (Fig. 1). This male bias is not due to bias in the sex ratio at birth (Fig. 1), nor is it due to differential mortality between the sexes, since survival patterns and mean longevity (16–20 days, 40 days maximum) in flight cages were the same for both sexes (Zalucki, 1981). Under field conditions, "mean-minimum life spans" based on all recapture data were also similar (13.5 days ♂; 10 days ♀; Zalucki and Kitching, 1984). The longer male "longevity" in the field reflects differential male–female residence times in patches (Table 2) and not differences in survival per se. Males tend to be resident in patches and more likely to be recaptured than females, which are recaptured at random with respect to marked status (Zalucki and Kitching, 1984). So where do females go?

Extensive surveys of adults from patches and nonpatches (i.e. areas in which milkweed was scattered at low density among other weed species and where there was no central area with plants at high density) indicated that butterflies around the latter have a female-biased sex ratio. The difference in sex ratio between patches and non-patches is significant ($p < 0.05$; Table 3; Zalucki and Suzuki, 1987).

As part of a continuing study of genetic variation in monarch populations in southeast Queensland (Hughes and Zalucki, 1984; Zalucki et al., 1987), we have repeatedly sampled the same patch at different times throughout 1 day (Table 4). In 15 out of 17 samples, the sex ratios are male-biased (significantly so in 7 of these; Table 4). If the population structure around patches as depicted in Figure 2 is correct, then repeated intensive sampling from a patch with removal of the captured males should lead to a shift in the sex ratio to a female basis, as did occur on some occasions (Table 4).

These results imply that females are differentially dispersed away from patches (Fig. 2). The disproportionate number of eggs on single plants on the margins of patches is due to the bias of ovipositing females searching in such areas. How is such a dispersion pattern of females maintained? And why shouldn't females remain within patches (the large dense areas of milkweed) since they have an array of host plants at minimum search times and abundant nectar sources when milkweeds (and other weeds within the patch) are in flower?

Male–female interaction suggested a possible explanation for the pattern of patch-related sex ratios and egg dispersion illustrated in Figure 2. I have observed many aspects of courtship behavior (as described by Pliske, 1974) in monarchs during visits to patches over many years. Males tend to patrol and perch in and around the margins of a patch. A male will investigate anything that moves—other males, other butterfly species, falling leaves, and even small birds. Male–male encounters usually lead to a short spiral flight followed by a return to the original activity. Occasionally a male will pursue another male, or after an encounter, one or both males fly off erratically at high speed (similar to the flight pattern after an unsuccessful netting attempt or encounter with a predatory bird). If a female is encountered, a high-speed chase usually ensues. Most such chases do not result in copulation. Unsuccessful courtships result from females apparently outdistancing the male, loosing him via a power dive, or by flying through foliage. I have seen females leave a patch and fly across country at high speed, following an unsuccessful mating attempt by a male.

These observations suggest that interactions between males and females could provide an explanation for why females are differentially dispersed from a local population, such as exists within a patch, and the subsequent fewer eggs per plant in large patches than on a single plant ("patches").

Table 4. Proportion of males in samples of monarchs (sample size in parentheses) taken at single patches at different times during the day (Hughes and Zalucki, unpubl.). — = no collections at these times. Proportions within a sampling day followed by the same letter are not significantly different ($p > 0.05$). Proportions followed by * are significantly different from 0.50 ($p < 0.05$).

		Time period			
Site	Date	Early morning	Mid-morning	Mid-day	Afternoon
Wessels	19/11/84	0.60[a] (47)	0.64[a]* (47)	0.63[a] (48)	0.73[a]* (37)
	17/1/85	0.78[a]* (55)	0.54[b] (39)	—	—
Mt. Crosby	17/4/85	—	0.53[a] (49)	0.59[a] (49)	0.59[a] (41)
	24/6/85	—	0.60[a] (43)	0.58[ab] (79)	0.41[b] (44)
	29/11/85	0.68[a]* (104)	0.62[a]* (123)	0.48[b] (102)	—
	5/12/85	0.77[a]* (162)	0.61[b]* (144)	—	—

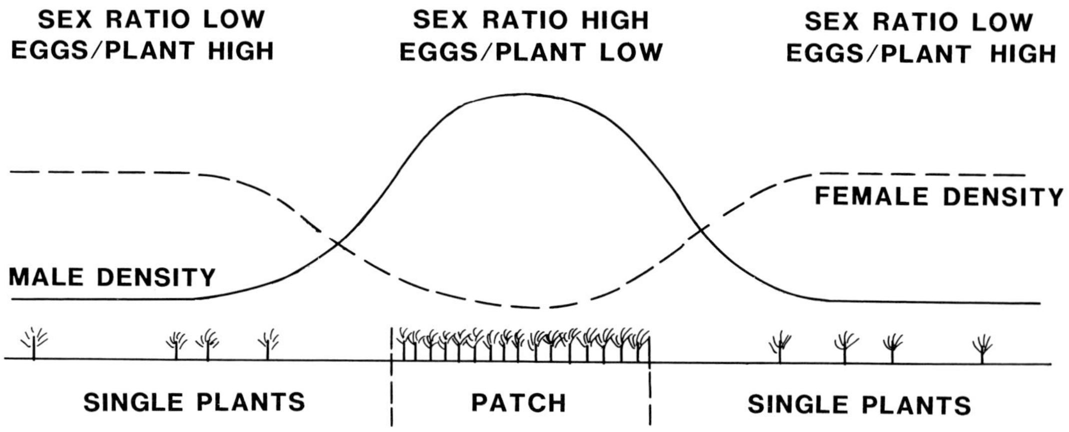

Figure 2. Schematic representation of host-plant dispersion, adult population density, sex ratio, and egg dispersion.

Such a reason for dispersal has also been proposed for pierids by Shapiro (1970) and Yamamoto (1983). Implicit in such an explanation is that a female has already mated successfully, and further mating theoretically would be costly in terms of time and energy, particularly for any short-lived species, and a single mating will fertilize most eggs. For instance, David and Gardiner (1962) showed that in *Pieris brassicae* one mating provides sufficient sperm to fertilize eggs for 7 days; fertility declines to ca. 95% after 14 days and to 50% after 18 days. It is possible to test experimentally some of the predictions of the female dispersal model proposed here. In particular, if males were to be removed from a patch, then female residence times should increase.

FIELD EXPERIMENTS MANIPULATING PATCH SEX RATIOS

Experiments to test the effect of removing males from patches were conducted at four milkweed patches in the winter of 1983 (Bull et al., 1985) and at four different sites in the summer of 1984 (Suzuki and Zalucki, 1986). I will summarize those studies briefly.

The methods were similar in both experiments. Butterflies were captured (for 1 person-hr), tagged, and released at each site during an initial period (15 days in winter, 5 days in summer) to establish a base recapture rate. Subsequently, two experimental treatments were used. On each sampling day, either all the males in the sample were removed and only females tagged and released, or every alternate individual of each sex was removed, such that half the sample was marked and released. The treatments led to an equivalent density reduction but the sex ratio of released animals was altered only in the male-removal treatment. In any treatment period, two sites had male-removal and two sites equal-sex removal treatments. During the winter experiment, the treatment of each site was reversed twice in consecutive periods (7 days between each). In summer, treatments were applied once (2 days between control treatment and manipulation).

A recapture at a site was interpreted as implying that conditions in that patch were sufficient to induce that individual to remain in or return to the patch. The assumption is that butterflies will remain in (or return to) a patch they assess as having high quality.

The experiments were designed to test the response of female monarchs to reduced presence of males. Females marked and recaptured at least once during a treatment period were recorded, and the percentage of all females released over the first part of a treatment period, which were recaptured at the same site in the same sampling period was calculated (the % recapture referred to below). The short sampling period could lead to an error because individuals marked on the first sampling date of a period were more likely to be recaptured than those marked on later sampling dates, even if they were equally likely to stay in the patch. These errors should be similar between sites and allow unbiased comparisons within a period. A full treatment of the results is given in Bull et al. (1985) and Suzuki and Zalucki (1986). Here I will concentrate on female recaptures only (Table 5). Male recaptures, where applicable, were generally higher than for females, mirroring the result of Zalucki and Kitching (1984).

Contrary to expectation, the removal of males from a patch *reduced* female recaptures in both summer and winter experiments (Table 5). In the summer experiment, male removal reduced female recaptures at both patches (Table 5) although the reduction was significant only at patch 2 ($\chi^2 = 5.84$, $p < 0.05$; Table 5). Equal-sex removals in summer had no effect on female recaptures (Table 5). During the winter, female recaptures rose during the second period relative to the control. The rise was not significant in the patches with the male-removal

Table 5. Summary of female recaptures (%) during experimental manipulation of monarch populations around four patches during winter 1983 and summer 1984. Treatments in parentheses: C = control period, E = equal-sex removal, M = male-removal. N = number of females marked and available for recapture in each experimental period.

Winter

1st period		2nd period		3rd period		4th period	
10–25 June	N	27 June–11 July	N	18 July–1 August	N	8–22 August	N
11.1 (C)	90	11.8 (M)	76	20.0 (E)	40	13.6 (M)	118
2.8 (C)	72	8.2 (M)	61	18.2 (E)	44	3.9 (M)	130
11.9 (C)	101	25.9 (E)	58	16.3 (M)	196	8.8 (E)	137
14.0 (C)	43	22.7 (E)	22	16.5 (M)	85	5.6 (E)	54

Summer

16–20 January	N	23–27 January	N
33.7 (C)	69	22.0 (M)	54
43.3 (C)	65	22.0 (M)	59
32.6 (C)	85	38.6 (E)	64
20.0 (C)	108	17.3 (E)	77

treatment (1 and 2) but was significant in the patches with equal-sex removal (3 and 4). In the third period, female recaptures rose in patches 1 and 2 with equal-sex removal but fell with male removal in patches 3 and 4. The female recapture data were homogeneous over the second and third periods for equal-sex and for male-removal treatments. Combining the data for all sites over the second and third periods, there were significantly fewer female recaptures in male-removal than in equal-sex removal treatments (14.4% of 418 vs. 22.0% of 164; $\chi^2 = 4.94$, $p < 0.05$). During the fourth period, female recaptures fell in both male-removal and equal-sex removal treatments (Table 5). The fourth period was characterized by a decline in the sex ratio at all sites independent of treatments. Female recaptures were unrelated to the average sample size during a sampling period, the average number of males captured, the average number of females captured, or the percentage of young butterflies in the samples (see Bull et al., 1985). However, there was a significant relationship between sex ratio and female recaptures across all sampling periods and sites ($r = 0.601$, $p < 0.05$). The chances of recapturing a marked female at a site decreased as the proportion of males at the site decreased.

During the summer and winter experiments, both male-removal and equal-sex removal treatments did not noticeably influence sex ratios nor population sizes on subsequent sampling days, suggesting the presence of a large, mobile population of adults, which move around and replace removed individuals. However, sex ratios must have been reduced, albeit for a short time after each treatment period. Marked females presumably responded to patch quality during that time and made the decision to stay or leave accordingly. Although we did not record the effect of our treatment on egg dispersion, Zalucki and Suzuki (1987) subsequently surveyed adult population parameters and their relationship to measures of egg dispersion in a diverse array of field patches. Neither nectar availability nor sex ratio influenced measures of egg laying. Total number of eggs in a patch was unrelated to the number of males in a patch (Zalucki and Suzuki, 1987), in part corroborating the earlier results.

INTERPRETATION

Two hypotheses may account for the unexpected changes in female recaptures: (1) female recapture rate is inversely related to female density, and (2) the recapture rate is positively related to sex ratio (see Bull et al., 1985; Suzuki and Zalucki, 1986).

During the winter experiments, equal-sex removal, which would have lowered female density relative to male removal, led to an increased recapture rate in periods 2 and 3. This could be explained by a reduced encounter rate with other females. A possible explanation for such a response is that females assess the quality of a patch as an oviposition site by the number of other females they encounter. If there are many encounters, then perhaps most oviposition sites have been used and the female leaves the patch to seek other sites. This explanation can be partly discounted on two grounds: (1) there was no correlation between recapture rates and female density, and (2) in the summer experiments, equal-sex removal had no effect on female recapture rates. However, this hypothesis will need to be tested further.

The second hypothesis relates female recapture to sex ratio. An explanation for a direct causal link between sex ratio and female recapture is that males that aggregate around the oviposition sites to find females are an important resource for females. Shields (1967) found that unmated female butterflies of some species seek males by aggregating at hilltops. Males provide not only gametes necessary for fertilization but also nutritive materials that can be utilized in egg production (Boggs and Gilbert, 1979; Marshall, 1982; Rutowski et al., 1983; Rutowski, 1984). Further circumstantial evidence supporting this view is that monarch females mate many times. Suzuki and Zalucki (1986) found in a dissection of 97 females that number of matings increased with female age up to a maximum of 10 times. This mating frequency is very high compared with other butterfly species (Shields, 1967; Burns, 1968; Pliske, 1973; Ehrlich and Ehrlich, 1978). Multiple matings in other butterfly species have been shown to increase egg fertility (Lederhouse, 1981). In monarchs mating stimulates oogenesis (Herman and Barker, 1977) and male investment is incorporated into egg production (Boggs and Gilbert, 1979) (see also Wells et al., this volume).

DISCUSSION

The very high egg densities I recorded on single plants around the margins of dense field patches led me to investigate the role of male–female interactions on monarch population structure. My basic conceptual model is summarized in Figure 2, and my initial hypothesis was that males promote female dispersal away from patches, which leads to the egg dispersion I observed. My field experiments testing this hypothesis indicate that this is not the whole story. Males may well promote female dispersal initially after mating, but females treat such patches with resident males as resources. As such, these results represent a special case of female searching bias to areas outside patches and is, to my knowledge, the first time such an explanation has been proposed for the high numbers of eggs observed on single isolated plants. Such egg dispersions are not uncommon among the Lepidoptera (see references in Mackay and Singer, 1982). Male-biased sex ratios around patches of host plants are also common (pers. obs.). The more general nature of the phenomenon I have proposed here is well worth investigating.

In summary, my current view of the population structure of monarchs in relation to milkweed patches is that patches act as foci of adult activity—particularly mating, feeding, and oviposition. Good-quality patches, which are usually large with a high density of milkweed plants, have a large population of monarchs with a male-biased sex ratio. Females appear to assess patch quality, in part, by the relative rate of encounters with males and females. If this is high, they will tend to "hang around" such a patch site, leading to the high number of eggs on the surrounding low-density milkweeds. Such females return to the patch to oviposit and remate. Having mated, a female would disperse from a patch proper, because further male encounters will reduce oviposition time and immediate remating is presumably impossible. Oviposition can continue relatively unhindered on the surrounding single plants (Fig. 2) and leads to the high number of eggs on these plants. The biased sex ratio presumably reflects a balance between attraction to a site for oviposition and mating (in males and females) and "repulsion" immediately after mating in females. If the encounter rate with females is low, presumably males will also leave. There are many questions raised by these results. Are males and females simply responding to patch chemical quality? What is male investment per mating? Do females "need" to remate to lay a complement of fertile eggs? What fraction of a female's eggs end up in a patch versus outside the patch? The monarch–milkweed system is a good system in which to pursue the answers.

ACKNOWLEDGMENTS

I wish to thank all those who have joined in and contributed to this work, especially Jane Hughes, Mike Bull, Duncan Mackay, Yoshito Suzuki, and Roger Kitching. Martin Rice, Gimme Walter, Gary Dodson, and Hugh Paterson made constructive comments on an earlier draft of this manuscript. The review comments of Ronald Rutowski and Steven Courtney were also much appreciated. The Ian Potter Foundation, the University of Queensland, and Moncon-2 provided support for the presentation of this paper.

LITERATURE CITED

Boggs, C., and L. Gilbert. 1979. Male contribution to egg production in butterflies: Evidence for transfer of nutrients at mating. *Science* 206:83–84.

Bull, C.M., M.P. Zalucki, Y. Suzuki, D.A. Mackay, and R.L. Kitching. 1985. An experimental investigation of resource use by female monarch butterflies, *Danaus plexippus* (L.). *Australian Journal of Ecology* 10:391–398.

Burns, J.M. 1968. Mating frequency in natural populations of skippers and spermatophore butterflies as determined by spermatophore counts. *Proceedings of the National Academy of Sciences USA* 61:852–859.

Cromartie, W.J. 1975. The effect of stand size and vegetational background on the colonization of cruciferous plants by herbivorous insects. *Journal of Applied Ecology* 12:517–533.

David, W.A.L., and B.O.C. Gardiner. 1962. Oviposition and the hatching of the eggs of *Pieris brassicae* (L.) in a laboratory culture. *Bulletin of Entomological Research* 53:91–109.

Ehrlich, A.H., and P.R. Ehrlich. 1978. Reproduction strategies in the butterflies: Mating frequency, plugging and egg number. *Journal of the Kansas Entomological Society* 51:666–697.

Herman, W.S., and J.F. Barker. 1977. Effect of mating on monarch butterfly oogenesis. *Experientia* 33:688–689.

Hughes, J.M., and M.P. Zalucki. 1984. Genetic variation in a continuously breeding population of *Danaus plexippus* L. (Lepidoptera: Nymphalidae). *Heredity* 52:1–7.

Jones, R.E. 1977. Movement patterns and the egg distributions of cabbage butterflies. *Journal of Animal Ecology* 46:195–212.

Jones, R.E., N. Gilbert, M. Guppy, and V. Nealis. 1980. Long distance movement in *Pieris rapae*. *Journal of Animal Ecology* 49:629–642.

Kitching, R.L., and M.P. Zalucki. 1982. Component analysis and modelling of the movement process: Analysis of simple tracks. *Researches on Population Ecology* 24:224–238.

Lederhouse, R.C. 1981. The effect of female frequency on egg fertility in the black swallowtail, *Papilio polyxenes asterius* (Papilionidae). *Journal of the Lepidopterist's Society* 35:266–277.

Mackay, D.A., and M.C. Singer. 1982. The basis of an apparent preference for isolated host plants by ovipositing *Euptychia libye* butterflies. *Ecological Entomology* 7:299–303.

Marshall, L.D. 1982. Male nutrient investment in the Lepidoptera: What nutrients should males invest? *American Naturalist* 120:273–279.

Pliske, T.E. 1973. Factors determining mating frequencies in some New World butterflies and skippers. *Annals of the Entomological Society of America* 66: 164–169.

Pliske, T.E. 1974. Courtship behavior of the monarch butterfly, *Danaus plexippus* L. *Annals of the Entomological Society of America* 68:143–151.

Rutowski, R.L. 1984 [1985]. Sexual selection and the evolution of butterfly mating behavior. *Journal of Research on the Lepidoptera* 23:125–142.

Rutowski, R.L., M. Newton, and J. Schaefer. 1983. Interspecific variation in the size of the nutrient investment made by male butterflies during copulation. *Evolution* 37:708–713.

Shapiro, A.M. 1970. The role of sexual behavior in density-related dispersal of pierid butterflies. *American Naturalist* 104:367–372.

Shapiro, A.M. 1981. The pierid red-egg syndrome. *American Naturalist* 117:276–294.

Shields, O. 1967 [1968]. Hilltopping: An ecological study of summit congregation behavior of butterflies on a southern California hill. *Journal of Research on the Lepidoptera* 6:69–178.

Suzuki, Y., and M.P. Zalucki. 1986. The influence of sex ratio on dispersal in *Danaus plexippus* (L.) (Lepidoptera: Danaidae). *Journal of the Australian Entomological Society* 25:31–35.

Thompson, J.N., and P.W. Price. 1977. Plant plasticity, phenology, and herbivore dispersion: Wild parsnip and the parsnip webworm. *Ecology* 58:1112–1119.

Wiklund, C., and C. Ahrberg. 1978. Host plants, nectar source plants, and habitat selection of males and females of *Anthocharis cardamines* (Lepidoptera). *Oikos* 31:169–183.

Yamamoto, M. 1983. Relationship between sex ratio and flight activity on two related butterflies, *Pieris rapae crucivora* and *P. napi nesis* (Lepidoptera, Pieridae). *Kontyû, Tokyo* 51:528–533.

Zalucki, M.P. 1981. The effects of age and weather on egg laying in *Danaus plexippus* L. (Lepidoptera: Danainae). *Research on Population Ecology* 23:318–327.

Zalucki, M.P. 1983. Simulation of movement and egg laying in *Danaus plexippus* (Lepidoptera: Nymphalidae). *Researches on Population Ecology* 25: 353–365.

Zalucki, M.P., and R.L. Kitching. 1982a. Component analysis and modelling of the movement process: The simulation of simple tracks. *Researches on Population Ecology* 24:239–249.

Zalucki, M.P., and R.L. Kitching. 1982b. The dynamics of oviposition of *Danaus plexippus* L. on *Asclepias* spp. *Journal of Zoology, London* 198:103–116.

Zalucki, M.P., and R.L. Kitching. 1982c. Movement patterns in *Danaus plexippus* L. *Behaviour* 80:174–198.

Zalucki, M.P., and R.L. Kitching. 1984. The dynamics of adult *Danaus plexippus* L. within patches of its food plant, *Asclepias* spp. *Journal of the Lepidopterist's Society* 38:209–219.

Zalucki, M.P., and Y. Suzuki. 1987. Milkweed patch quality, adult population structure, and egg laying in the monarch butterfly. *Journal of the Lepidopterist's Society* 41:13–22.

Zalucki, M.P., R.L. Kitching, D. Abel, and J. Pearson. 1980. A novel device for tracking butterflies in the field. *Annals of the Entomological Society of America* 73:262–265.

Zalucki, M.P., A. Chandica, and R.L. Kitching. 1981. Quantifying the distribution and abundance of an animal's resource using aerial photography. *Oecologia* 50:176–183.

Zalucki, M.P., J.M. Hughes, and P. Carter. 1987. Genetic variation in *Danaus plexippus* L.: Habitat selection or differences in activity times? *Heredity* 59: 213–221.

5

HOST PLANT USE, CARDENOLIDE SEQUESTRATION, AND DEFENSE AGAINST NATURAL ENEMIES

Fifth instar monarch larva feeding on the abundant milkweed *Asclepias speciosa* in northern California. Monarch larvae are milkweed specialists, sequestering toxic cardenolides from these plants for use in their chemical defenses against predators. Photograph by Lincoln P. Brower.

HOST PLANT EXPLOITATION BY APOSEMATIC NYMPHALID BUTTERFLIES: PRE-ADAPTATION VERSUS STEPWISE SEQUESTRATION

Phillip R. Ackery

ABSTRACT. Recent investigations into unpalatability in various nymphalid butterfly groups question some long-held assumptions concerning the fundamental role of larval host-plant chemistry. Intrinsic unpalatability and pre-adaptation, instead of evolved, stepwise sequestration, could account for the host preferences shown by members of the Acraeinae, Heliconiinae, Ithomiinae, and even, perhaps, Danainae.

INTRODUCTION

Monarch butterflies, *Danaus plexippus* (L.), are well known for their oligophagous preference for milkweeds and their ability to sequester host-derived cardenolides as the basis of an aposematic defense against predators (Brower, 1984). However, to what degree does the monarch–host plant–cardenolide defense system provide a general model for understanding host plant exploitation by all aposematic butterflies? While the *Aristolochia*-feeding papilionids (Rothschild et al., 1970, 1972), *Chelone*-feeding *Euphydryas* (Bowers, 1980, 1981, 1983), and cycad-feeding *Eumaeus* (Rothschild et al., 1986) may parallel the monarch–Asclepiadaceae system, some recent studies (Nahrstedt and Davis, 1981, 1983; Brown, 1984, 1985) have raised doubts concerning the host plant-based unpalatability in three important aposematic nymphalid groups: the Ithomiinae, Heliconiinae, and Acraeinae. In these butterflies, pre-adaptation might provide an equally plausible alternative hypothesis to account for their larval host preferences (see Rothschild, 1972; Davis and Nahrstedt, 1985; Ackery, 1988). Rather than host-plant chemistry universally accounting for unpalatability, might intrinsic unpalatability sometimes provide the basis for host plant preferences?

HOST PLANTS AND UNPALATABILITY

Since Bates published his original ideas on mimetic complexes, based on the butterflies of the Amazon, members of several nymphalid groups, most particularly the Danainae, Ithomiinae, Acraeinae, and Heliconiinae, have been regarded as primary models in a variety of mimicry complexes (Bates, 1862; Müller, 1879; Eltringham, 1910). This overall generalization is still supported by workers familiar with these butterflies in the field (e.g., Owen and Owen, 1972; Brown, 1973). Although members of all four groups exploit larval hosts that contain an array of secondary compounds, some evidence now suggests that this may not always be the primary source of their unpalatability and, hence, the chemical basis of their aposematism (Nahrstedt and Davis, 1981, 1983; Brown, 1984, 1985). Is it possible, then, that the "chicken came before the egg"; might chemically protected plants be vulnerable to exploitation by intrinsically unpalatable pre-adapted butterflies? Futuyma (1983) suggests that many insects are pre-adapted to deal with a great variety of chemical compounds. This pre-adaptation has been brought about through their adaptation to by-products of their own metabolism or other harmful compounds with which they naturally have to contend. How might the Danainae, Ithomiinae, Heliconiinae, and Acraeinae fit within such a system?

Consider as an example a hypothetical proto-danaine producing self-synthesized cardio-active substances; among extant species, Rothschild et al. (1978) report such compounds in both *D. plexippus* (L.) and *Euploea core* (Cramer). These compounds might then pass from the adult female to her eggs, in a comparable manner to the known passage of plant-derived cardenolides to the eggs of the succeeding generation (Reichstein et al., 1968).

In this context, the first-instar larval habit of eating the eggshell assumes particular importance; it could condition the newly emerged larva to accept potential hosts that also contain cardio-active substances.

ITHOMIINAE

Such a system seems particularly plausible for the Ithomiinae/Solanaceae association. Adult sequestration of pyrrolizidine alkaloids (PAs), a special feature of the Ithomiinae–Danainae–Tellervinae group, almost certainly contributes toward adult unpalatability in much of this infrafamily (Edgar et al., 1976, 1979; Brown, 1984, 1985). Could adult tolerance of PAs have facilitated a host plant shift by many ithomiine species to the alkaloid-containing Solanaceae? Both chemical groups share certain affinities, being derived from the same evolutionary nitrogen pathway (see Harborne, 1982). The presence of adult-derived PAs in ithomiine eggs (they are absent from both larvae and pupae; Brown, 1985) matches the hypothetical model. Generally, Solanaceae-feeding ithomiine larvae are cryptic, suggesting that, while they can tolerate the pharmacologically active secondary compounds of their hosts, none is stored to permit an aposematic larval life-style. Brown (1985) concluded that "contrary to general wisdom . . . the Ithomiinae derive few or no predator-defense compounds from their larval hostplants." But species of *Melinaea* perhaps provide a striking exception. Despite clearly aposematic, danaoid larvae, most recent records (Drummond, 1976) indicate solanaceous hosts. Members of this genus, at least, may thus have evolved a further way of exploiting the Solanaceae, either by actively storing toxins that permit an aposematic larval life-style or by synthesizing new toxins.

Based on host plant chemistry, Edgar's attractive hypothesis (see Edgar et al., 1974) suggests that PAs were present in the ancestral larval hosts of the Ithomiinae–Danainae–Tellervinae. An ancestor of *Parsonsia,* or a similar apocynaceous plant, probably represents the larval host of the entire infrafamily (Edgar, 1984). As members of various scattered ithomiine genera (*Aeria, Tithorea,* and *Elzunia*; Ackery, 1988) and *Tellervo* (Ackery, 1987a) exploit the Apocynaceae, this seems to be supported phylogenetically (Ackery and Vane-Wright, 1985). But the assumption of ancestral larval tolerance of PAs again provides a pre-adaptation that might facilitate a host switch to the Solanaceae. In danaines, at least, limited evidence suggests that the caterpillars "remain" PA-tolerant although these compounds are only rarely encountered in the larval hosts (Rothschild and Edgar, 1978; Edgar, 1982).

HELICONIINAE AND ACRAEINAE

Virtually all heliconiines feed on members of the Passifloraceae (Benson et al., 1976), which together with the Flacourtiaceae and Turneraceae are well known as sources of cyanogenic glycosides (cyclopentenoid cyanogens; see Seigler, 1981). Passifloraceous hosts are also widely exploited by acraeine butterflies, with members of the Urticaceae, Malvaceae, and Tiliaceae additionally providing significant contributions (Pierre, 1984; Ackery, 1987b, 1988). Analyses by Nahrstedt and Davis (1981, 1983) of adult heliconiines and acraeines revealed high concentrations of cyanoglucosides. Their conclusion that these were entirely synthesized from amino acids during both the larval and adult stages, and were not dependent on sequestration during larval feeding, is challenged by Raubenheimer's (1987) recent findings. Investigating unpalatability in the highly derived acraeine species *Acraea horta* (Pierre, 1984), he confirmed the presence of cyanogenic glycosides originating from the larval hosts. This could be interpreted as reinforcing the established view that host plant chemistry plays the paramount role in the origin of unpalatability. Alternatively, given intrinsic unpalatability, it is perhaps what we should expect: Aposematic insects with their own well-developed defensive system will tend to add to their repertoire by sequestering and storing plant secondary compounds (Rothschild, 1972). The possible role of pollen-derived amino acids in relation to adult unpalatability remains obscure. Grimshaw (1983) suggests that *Heliconius melpomene* derives its toxicity from the pollen that adults ingest, particularly the pollen of *Gurania* species (Cucurbitaceae). However, as Brown (1981) treats pollen utilization as a character of a subset within *Heliconius* (i.e., *Heliconius* with the exception of *godmani, metharme, aoede,* and *doris*), any protective function cannot be universal to both the Heliconiinae and Acraeinae.

Is it likely, then, that in both the Heliconiinae and Acraeinae exploitation of the Passifloraceae has been facilitated through their tolerance of self-synthesized cyanogenic compounds? If the Heliconiinae and Acraeinae are indeed closely related, this need not have arisen independently in both subfamilies. Pierre's (1984) cladistic analysis of the Acraeinae perhaps surprisingly suggests larval host plant shifts in the evolutionary sequence Violales to Urticales to Malvales—members of the Violaceous Passifloraceae could represent the larval hosts of a common ancestor.

DANAINAE

All known danaine larvae appear to be aposematic (see Ackery and Vane-Wright, 1984). This suggests host plant-derived unpalatability, although *de novo* synthesis cannot be ruled out. While adult-sequestered PAs (Edgar et al., 1974; Boppré, 1984) could perhaps pre-adapt the ithomiines to exploit the Solanaceae, a similar hypothesis seems less likely for danaines and their hosts. But to what degree do cardenolides function as a generalized source of danaine unpalatability (see Brower, 1984, for a review)?

In the popular literature, statements such as "adult [danaine] butterflies . . . store heart-poisons synthesized from foodplant poisons eaten by the caterpillar" (e.g., Watson and Whalley, 1975) might suggest that cardenolides have a widespread protective role. Even the "Edgar hypothesis" (Edgar et al., 1974; see above) presupposes the presence of both cardenolides and PAs in the ancestral danaine larval host. However, cardenolides are not universal features of danaine food plants; in fact, it seems they only make a significant contribution toward unpalatability in *Danaus* itself (Ackery and Vane-Wright, 1984, 1985). So, despite the obviously aposematic nature of all known danaine larvae, cardenolides can be only one of perhaps several factors influencing unpalatability within the subfamily. Possibly, self-secreted cardio-active substances (Rothschild et al., 1978) may have played a role in pre-adapting danaines toward their host plants, but it seems likely that many as-yet unknown factors remain to be elucidated.

CONCLUSION

In their seminal review of "Birds, butterflies and plant poisons," Brower and Brower (1964) suggested that the origins of unpalatability in the Ithominae, Heliconiinae, and Acraeinae lay with their larval hosts—"phytophagous insects . . . have developed the ability to incorporate poisonous and repellent plant chemicals [from their larval hosts] into their tissues *with the result that* [my italics] they are unpalatable to vertebrate predators." Brown's (1984, 1985) findings seriously question this for the Ithomiinae. In the Acraeinae and Heliconiinae, Nahrstedt and Davis (1981, 1983) confirmed cyanogenesis in both larvae and adults but favored independent *de novo* synthesis. It seems that members of these three nymphalid subfamilies may be pre-adapted to feed on toxic plants, rather than the larval hosts contributing significantly toward overall unpalatability. With host plant-derived cardenolides only having a significant function as predator deterrents in the genus *Danaus,* the chemical basis of larval aposematism in much of the Danainae remains unknown.

ACKNOWLEDGMENTS

I wish to thank Campbell Smith, Dick Vane-Wright, and two anonymous reviewers for their helpful and constructive comments. Dr. David Raubenheimer kindly allowed me access to his unpublished thesis.

LITERATURE CITED

Ackery, P.R. 1987a. The danaid genus *Tellervo* (Lepidoptera, Nymphalidae)—A cladistic approach. *Zoological Journal of the Linnean Society* 89:203–274.

Ackery, P.R. 1987b. Diversity and phantom competition in African acraeine butterflies. *Biological Journal of the Linnean Society* 30:291-297.

Ackery, P.R. 1988. Hostplants and classification: A review of nymphalid butterflies. *Biological Journal of the Linnean Society* 33:95–203.

Ackery, P.R., and R.I. Vane-Wright. 1984. *Milkweed butterflies: Their cladistics and biology.* London: British Museum (Natural History) and Ithaca, NY: Cornell University Press, ix + 425 pp.

Ackery, P.R., and R.I. Vane-Wright. 1985. *Patterns of plant utilization by danaine butterflies.* Proceedings of the 3rd Congress of European Lepidopterology, Cambridge 1982, 3–7.

Bates, H.W. 1862. Contributions to an insect fauna of the Amazon Valley. *Transactions of the Linnean Society of London* 23:495–566 (+2 pls.).

Benson, W.W., K.S. Brown, Jr., and L.E. Gilbert. 1976. Coevolution of plants and herbivores: Passion flower butterflies. *Evolution* 29:659–680.

Boppré, M. 1984. Chemically mediated interactions between butterflies. In *The biology of butterflies,* eds. R.I. Vane-Wright and P.R. Ackery, 259–275. London: Academic Press.

Bowers, M.D. 1980. Unpalatability as a defense strategy of *Euphydryas phaeton. Evolution* 34:586–600.

Bowers, M.D. 1981. Unpalatability as a defense strategy of western checkerspot butterflies (*Euphydryas*). *Evolution* 35:367–375.

Bowers, M.D. 1983. Mimicry in North American checkerspot butterflies: *Euphydryas phaeton* and *Chlosyne harrisii* (Nymphalidae). *Ecological Entomology* 8:1–8.

Brower, L.P. 1984. Chemical defence in butterflies. In *The biology of butterflies,* eds. R.I. Vane-Wright and P.R. Ackery, *Symposia of the Royal Entomological Society of London* 11:109–134. London and New York: Academic Press (1989, Princeton, NJ: Princeton University Press).

Brower, L.P., and J.V.Z. Brower. 1964. Birds, butterflies and plant poisons: A study in ecological chemistry. *Zoologica, New York* 49:137–159.

Brown, K.S., Jr. 1973. *A portfolio of neotropical lepidopterology.* Rio de Janeiro: Private production, 28 pp.

Brown, K.S., Jr. 1981. The biology of *Heliconius* and related genera. *Annual Review of Entomology* 26: 427–456.

Brown, K.S., Jr. 1984. Adult-obtained pyrrolizidine alkaloids defend ithomiine butterflies against a spider predator. *Nature* 309:707–709.

Brown, K.S., Jr. 1985. Chemical ecology of dehydropyrrolizidine alkaloids in adult Ithomiinae (Lepidoptera: Nymphalidae). *Revista Brasiliera de Biologia* 44:435–460.

Davis, R.H., and A. Nahrstedt. 1985. Cyanogenesis in insects. In *Comprehensive insect physiology, biochemistry and pharmacology* 11, eds. G.A. Kerkut and L.I. Gilbert, 635–654. Oxford: Pergamon Press.

Drummond, B.A. 1976. Comparative ecology of mimetic relationships of Ithomiine butterflies in eastern Ecuador. Ph. D. thesis. Ann Arbor, MI: University of Michigan.

Edgar, J.A. 1982. Pyrrolizidine alkaloids sequestered by Solomon Island danaine butterflies. The feeding preferences of the Danainae and Ithomiinae. *Journal of Zoology, London* 196:385–399.

Edgar, J.A. 1984. Parsonsieae: Ancestral larval food plants of the Danainae–Ithomiinae? In *The biology of butterflies*, eds. R.I. Vane-Wright and P.R. Ackery, 91–93. London: Academic Press.

Edgar, J.A., C.C.J. Culvenor, and T.E. Pliske. 1974. Coevolution of Danaid butterflies with their hostplants. *Nature (London)* 250:646–648.

Edgar, J.A., P.A. Cockrum, and J.L. Frahn. 1976. Pyrrolizidine alkaloids in *Danaus plexippus* L. and *Danaus chrysippus* L. *Experientia* 32:1535–1537.

Edgar, J.A., M. Boppré, and D. Schneider. 1979. Pyrrolizidine alkaloid storage in African and Australian danaid butterflies. *Experientia* 35:1447-1448.

Eltringham, H. 1910. *African mimetic butterflies.* Oxford: Clarendon Press, 136 pp. (+10 pls. and 1 map).

Futuyma, D.J. 1983. Evolutionary interactions among herbivorous insects and plants. In *Coevolution*, eds. D.J. Futuyma and M. Slatkin, 207-231. Sunderland, MA: Sinauer Associates.

Grimshaw, P. 1983. Test releases of *Heliconius melpomene*, carried out to discover if it is feasible to house this species in conjunction with a large collection of tropical birds. *Newsletter, Exotic Entomology Group. Amateur Entomologists' Society* 1983:9–12.

Harborne, J.B. 1982. *Introduction to ecological biochemistry*, 2nd. ed. London: Academic Press, xvi + 278 pp.

Müller, F. 1879. *Ituna* and *Thyridia.* Ein merkwürdiges Beispiel von Mimicry bei Schmetterlingen. *Kosmos (Liepzig)* 5:100-108. (English translation in *Proceedings of the Entomological Society of London* 1879: xx–xxix.

Nahrstedt, A., and R.H. Davis. 1981. The occurrence of cyanoglucosides, linamarin and lotaustralin, in *Acraea* and *Heliconius* butterflies. *Comparative Biochemistry and Physiology* 68B:575-577.

Nahrstedt, A., and R.H. Davis 1983. Occurrence, variation and biosynthesis of the cyanogenic glucosides linamarin and lotaustralin in species of the Heliconiini. *Comparative Biochemistry and Physiology* 75B:65–73.

Owen, D.F., and J. Owen. 1972. Systematics and bionomics of butterflies seen and collected in the forest region of Sierra Leone. Part 1. Introduction, Papilionidae, Danaidae and Acraeidae. *Revue de Zoologie et de Botanique Africaines, Bruxelles* 85:287–308.

Pierre, J. 1984. Systématique évolutive, cladistique et mimétisme chez les lépidoptères du genre *Acraea* (Nymphalidae). Ph.D. thesis. Paris: University of Paris.

Raubenheimer, D. 1987. Cyanogenesis and the feeding preference of *Acraea horta* (L.) (Lepidoptera: Acraeinae). M.Sc. thesis. University of Cape Town.

Reichstein, T., J. von Euw, J.A. Parsons, and M. Rothschild. 1968. Heart poisons in the monarch butterfly. *Science, New York* 161:861-866.

Rothschild, M. 1972. Secondary plant substances and warning colouration in insects. In *Insect/plant relationships*, ed. H.F. van Emden, 59–83. Oxford: Blackwell.

Rothschild, M., and J.A. Edgar. 1978. Pyrrolizidine alkaloids from *Senecio vulgaris* sequestered and stored by *Danaus plexippus. Journal of Zoology, London* 186:347–349.

Rothschild, M., T. Reichstein, J. von Euw, R. Aplin, and R.R.M. Harman. 1970. Toxic Lepidoptera. *Toxicon* 8:293–299.

Rothschild, M., J. von Euw, and T. Reichstein. 1972. Aristolochic acids stored by *Zerynthia polyxena* (Lepidoptera). *Insect Biochemistry* 2:234–243.

Rothschild, M., N. Marsh, and B. Gardiner. 1978. Cardioactive substances in the monarch butterfly and in *Euploea core* reared on leaf-free artificial diet. *Nature (London)* 275:646-650.

Rothschild, M., R.J. Nash, and E.A. Bell. 1986. Cycasin in the endangered butterfly *Eumaeus atala florida. Phytochemistry* 25:1853–1854.

Seigler, D.S. 1981. Cyanogenic glycosides and lipids: Structural types and distribution. In *Cyanide in biology*, eds. B. Vennesland, E.E. Conn, C.J. Knowles, J. Westley, and F. Wissing, 133-143. London: Academic Press.

Watson, A., and P.E.S. Whalley. 1975. *The dictionary of butterflies and moths in colour.* New York, St Louis, San Francisco, and Toronto: McGraw-Hill Book Company, xiv + 296pp.

A MODEL FOR CARDENOLIDE AND CARDENOLIDE GLYCOSIDE STORAGE BY THE MONARCH BUTTERFLY

Carolyn J. Nelson

ABSTRACT. Laboratory and field studies are reviewed and used to establish a model of cardenolide and cardenolide glycoside storage by the monarch butterfly. Data from feeding four compounds to monarch larvae indicated the quantity stored was linearly related to the logarithm of the quantity consumed. Field studies indicated that for butterflies reared on five *Asclepias* species, most with relatively low concentrations of cardenolide glycosides, the same relationship was observed. For plants with higher concentrations, the butterfly concentrations were relatively constant and independent of plant concentrations. Monarchs selectively stored intermediate and polar cardenolide glycosides, metabolized nonpolar cardenolide glycosides to glycosides of intermediate polarity, and conjugated aglycones producing amphiphilic compounds.

INTRODUCTION

Cardenolides are plant-derived C_{23} steroids that occur as either an aglycone or a glycoside derivative (Fieser and Fieser, 1959). They are used as therapeutic agents for congestive heart failure (Thomas, 1981). Monarch butterflies utilize cardenolide glycoside-containing plants during their larval stages and store these potentially protective compounds (Reichstein et al., 1968; Brower, 1969, 1984; Seiber et al., 1983).

This paper presents a model for the storage of cardenolides and cardenolide glycosides by the monarch butterfly, based on existing data. A model can best be developed when the variables are controlled as in our laboratory study, in which monarch larvae were fed four compounds (Seiber et al., 1980). Studies examining the interrelationships of cardenolide glycosides in monarch butterflies and nine host plants expand the model (Brower et al., 1982, 1984a,b; Lynch and Martin, 1987; Martin and Lynch, 1988; Malcolm et al., 1989; Brower, Seiber, Nelson, Lynch, and Holland, unpubl.; Brower, Seiber, Nelson, and Lynch, unpubl.).

RESULTS AND DISCUSSION

QUANTITATIVE ASPECTS OF STORAGE

Seiber et al. (1980) fed fourth-instar monarch larvae two cardenolide glycosides, uscharidin (**1**) and calotropin (**2**), and two cardenolide aglycones, uzarigenin (**3**) and digitoxigenin (**4**) (Fig. 1) at four concentrations. They found that at low-dose levels the percent of consumed compound stored was greater than at high levels. More cardenolide-derived materials were stored by larvae at all levels when dosed uscharidin and calotropin than uzarigenin and digitoxigenin (Fig. 2).

As part of this study, regression analyses for the absolute amount stored (Y) as a function of the amount consumed (X) were carried out for the following seven relationships: $Y = aX$, $Y = a + bX$, $Y = e^{bX}$, $Y = a + b/X$, $Y = a + b(\log_{10}X)$, $Y = aX^b$, and $Y = X/a + bX$. The logarithmic relationship provided the best fit and is highly significant for all compounds (Table 1a). The graphic representation of the logarithmic relationships, after being transformed back to standard coordinates, is shown in Figure 2. The percent stored or storage efficiency for these compounds, i.e. (amount stored $[Y]$/amount consumed $[X]$)100, decreases as the amount consumed (X) increases. Thus, the slope of the line ($\Delta Y/\Delta X$) describes the change in rate of storage or efficiency of storage with a change in the amount consumed. If the data are transformed, and Y versus $\log_{10}X$ is plotted, the relationship becomes linear with a slope of b ($\Delta Y/\Delta \log_{10}X$). Then b, the storage rate, is a constant for any single compound.

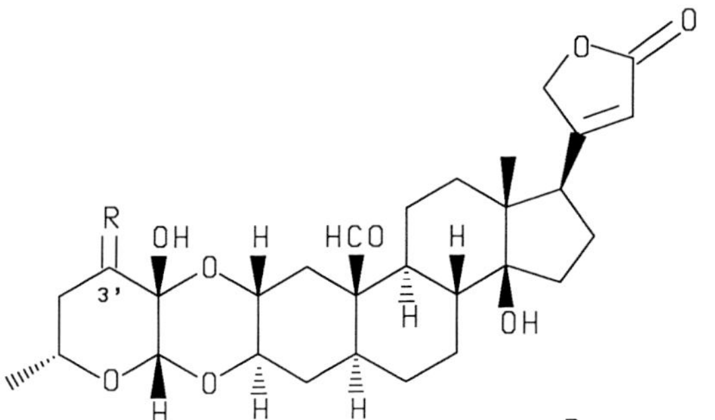

		R
Uscharidin	(1)	O
Calotropin	(2)	α-OH, β-H

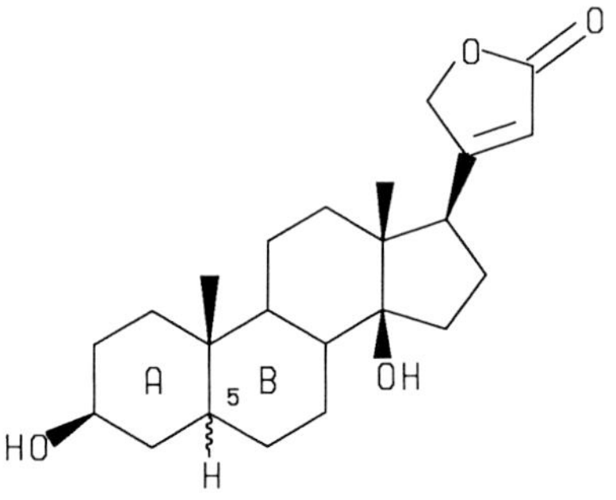

Uzarigenin	(3)	5α-H
Digitoxigenin	(4)	5β-H

Figure 1. Structures of compounds dosed to monarch larvae (Seiber et al., 1980). The cardenolide glycosides **1** and **2** differ only in the functionality at C-3′ and the cardenolide aglyones **3** and **4** differ only in the stereochemistry of the A/B ring junction, as is indicated by the proton at C-5.

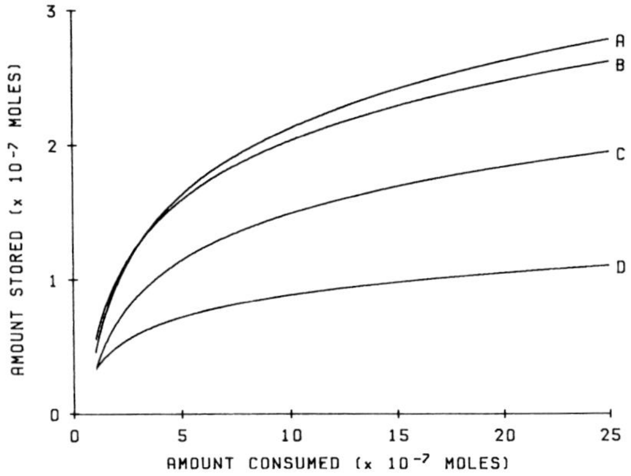

Figure 2. Amount of compound stored by fourth-instar larvae (Y axis) as a function of the amount consumed (X axis). Larvae were fed uscharidin (**A**), calotropin (**B**), uzarigenin (**C**), and digitoxigenin (**D**) (see Fig. 1). The curves are from the regression equations given in Table 1a, $Y = a + b(\log_{10}X)$, transformed back to standard coordinates.

For the cardenolide glycosides uscharidin and calotropin, the storage rates (b values, 1.66 and 1.48; Table 1a) were greater than for the aglycones uzarigenin and digitoxigenin (1.16 and 0.55). This is also indicated by the relative positions of the curves A, B, C, and D in Figure 2. The analysis of variance and covariance (Table 1b) indicates that the slopes among the treatments, and therefore the storage rates, differ significantly. Storage rate and, hence, amount stored varied with the compound dosed.

Since analyses of cardenolide glycoside storage in field-reared monarch butterflies are made on concentration data (Brower et al., 1982, 1984a,b; Lynch and Martin, 1987; Malcolm et al., 1989; Brower, Seiber, Nelson, Lynch, and Holland, unpubl.; Brower, Seiber, Nelson, and Lynch, unpubl.), storage from the dosage of these four compounds was also compared on a concentration basis. Thus, linear regression anlyses of the larval concentration (Y) as a function of diet concentration ($\log_{10}X$) in Table 1c also indicate significant regressions.

Monarch butterfly larvae have also been reared on seven California milkweed species using the technique of pairing a larva with its individual host plant, thus allowing subsequent evaluation of their cardenolide glycoside concentrations (Brower et al., 1982, 1984a,b; Brower, Seiber, Nelson, Lynch, and Holland, unpubl.; Brower, Seiber, Nelson, and Lynch, unpubl.). Using the seven regression relationships listed above, analyses of the cardenolide glycoside concentration in the butterfly (Y) as a function of that in the host plant (X) indicated that a significant correlation ($p < 0.001$) existed with several of the functions for the species *Asclepias cordifolia, A. speciosa,* and *A. californica.* Of these relationships, $Y = a + b(\log_{10}X)$ was chosen as the best fit and these analyses are given in Table 2.

The relative storage efficiencies of butterflies reared on *A. cordifolia, A. speciosa,* and *A. californica,* which show significant logarithmic storage curves, can be compared using the relationships given in Table 2 and an arbitrary plant concentration (X) of 100 μg/0.1 g to calculate the resulting butterfly concentration Y'. Thus, butterflies reared on *A. californica* would be the most efficient accumulators with a resultant concentration of 277 μg/0.1 g at a plant concentration of 100 μg/0.1 g, whereas monarchs on *A. cordifolia* and *A. speciosa* would be less efficient, showing little difference at 218 and 208 μg/0.1 g, respectively.

The higher storage rates for females relative to males (b; Table 2), for butterflies showing a significant logarithmic storage, suggest they were more efficient at storing cardenolide glycosides. This re-

Table 1a. Linear regression analyses of amount stored by the larva (*Y*) versus amount consumed ($\log_{10}X$) for monarch butterfly larvae fed four compounds.[a]

Compound	N	*a*	*b*	*r*	*p*
Uscharidin (**A**)	20	0.463	1.66	0.920	<0.001
Calotropin (**B**)	20	0.554	1.48	0.763	<0.001
Uzarigenin (**C**)	15	0.331	1.16	0.932	<0.001
Digitoxigenin (**D**)	20	0.334	0.55	0.780	<0.001

Table 1b. Analyses of variance and covariance among treatments for *Y* versus $\log_{10}X$.

Test	df	*F*	*p*
ANOVA	3,71	6.502	<0.001
ANCOVA	3,70	18.628	<0.001
F test for common slope	3,67	6.867	<0.001
F test for single regression	3,70	18.628	<0.001
F test for significance of covariate assuming common slope	1,70	134.606	

Table 1c. Linear regression analyses of concentration in the larva (*Y*) versus diet concentration ($\log_{10}X$).[a]

Compound	N	*a*	*b*	*r*	*p*
Uscharidin	20	0.284	0.554	0.927	<0.001
Calotropin	20	0.297	0.442	0.817	<0.001
Uzarigenin	20	0.287	0.384	0.873	<0.001
Digitoxigenin	20	0.176	0.186	0.788	<0.001

[a] Data as moles × 10^{-7} in Table 1a, and milligrams of cardenolide or cardenolide glycoside per gram of sample in Table 1c. Level of significance calculated from critical levels of *r* (Zar, 1974). Analyses based on original data from Seiber et al. (1980).

lationship prescribes that a female butterfly from any *A. californica* and *A. cordifolia* plant of a given concentration would achieve a higher concentration than a male. Because of the values of *a*, this applies only for *A. speciosa* plant concentrations >26 μg/0.1 g. However, the actual mean concentrations reported for male and female butterflies (Table 3) are significantly different for those reared on *A. californica* but not for *A. speciosa* (Brower et al., 1984a,b).

A logarithmic relationship implies that at high plant concentrations of cardenolide glycosides the butterfly concentrations will be relatively constant compared to the plant concentrations; i.e., the butterflies will reach their capacity (Brower et al., 1982), or level of saturation, for these compounds under those specific conditions. Analyses presented in Table 2 indicated that for three species, *A. erosa*, *A. eriocarpa*, and *A. vestita*, the level of significance of the logarithmic relationship is relatively lower than for *A. cordifolia*, *A. speciosa*, and *A. californica*. For these plants, the mean concentrations were >400 μg/0.1 g and the range and mean of the butterfly concentrations were less than those of the respective host plants (Table 3). In contrast, for *A. cordifolia*, *A. speciosa*, and *A. californica*, the mean plant concentrations were <100 μg/0.1 g and the range and mean of butterfly concentrations were greater than those of the respective host plant. The butterflies from larvae reared on *A. erosa*, *A. eriocarpa*, and *A. vestita* approached or reached their level of saturation at mean concentrations of 347, 318, and 250 μg/0.1 g, respectively, with female butterflies achieving a significantly higher mean (20–29%) relative to the males.

Three other studies have evaluated cardenolide glycoside relationships between butterfly–plant pairs. Lynch and Martin (1987) on *A. viridis* and Malcolm et al. (1989) on *A. syriaca* observed a significant logarithmic relationship between the butterfly and plant concentrations. From these relationships, it is calculated that larvae reared on *A. viridis* or *A. syriaca* with 100 μg/0.1 g would produce monarchs with 188 and 376 μg/0.1 g, respectively. However, similar to the milkweeds with higher cardenolide glycoside concentrations, *A. erosa*, *A. eriocarpa*, and *A. vestita* listed in Table 2, Martin and Lynch (1988) found no significant regression relationship between the cardenolide glycoside concentrations of *A. asperula* with a mean concentration of 886 μg/0.1 g and reared monarchs with a mean of 363 μg/0.1 g.

Like *A. cordifolia*, *A. speciosa*, and *A. californica* monarchs from *A. viridis* and *A. syriaca* also had higher means and wider ranges of concentrations relative to their host plants. *Asclepias syriaca*-reared monarchs had a mean of 234 μg/0.1 g (range, 0–792 μg/0.1 g) while the plant had a mean of 50 μg/0.1 g (4–229 μg/0.1 g). *Asclepias viridis*-reared monarchs had a mean of 337 μg/0.1 g (73–591 μg/0.1 g) and their host plants a mean of 245 μg/0.1 g (95–432 μg/0.1 g). For both plant species, female

Table 2. Linear regression analyses ($Y = a + b(\log_{10}X)$) of cardenolide glycoside concentrations (μg/0.1 g dry weight) for monarch butterflies (Y) against their larval host plants ($\log_{10}X$). Values for Y' at an arbitrary value of $X = 100$ are also given where $p \leq 0.001$.

	N	Y′	*a*	*b*	*r*	*p*[a]
A. cordifolia[b]						
Females	60	239	−119	179	0.68	<0.001
Males	52	194	−130	162	0.70	<0.001
Both	112	218	−128	173	0.65	<0.0001
A. speciosa[c]						
Females	45	213	−207	210	0.49	<0.001
Males	66	206	−190	198	0.61	<0.001
Both	111	208	−192	200	0.56	<0.0001
A. californica[d]						
Females	41	280	−3.9	142	0.65	<0.001
Males	44	267	−15.1	141	0.67	<0.001
Both	85	277	−12.9	145	0.65	<0.0001
A. erosa[b]						
Females	34	—	140	88	0.25	<0.15 NS
Males	37	—	141	66	0.23	<0.18 NS
Both	71	—	97	93	0.28	0.018
A. eriocarpa[e]						
Females	80	—	277	44	0.10	<0.05 NS
Males	92	—	154	85	0.23	<0.05
Both	172	—	159	61	0.14	<0.067 NS
A. vestita[f]						
Females	34	—	253	10	0.10	>0.50 NS
Males	29	—	267	−19	0.18	<0.50 NS
Both	63	—	239	4	0.03	>0.80 NS

[a] Level of significance calculated from critical levels of *r* (Zar, 1974). NS = not significant at 5%.

[b] Brower, Seiber, Nelson, and Lynch, unpubl. data.

[c] Brower et al. (1984b).

[d] Brower et al. (1984a).

[e] Data from Brower et al. (1982).

[f] Brower, Seiber, Nelson, Lynch, and Holland (unpubl. data).

butterflies had significantly higher mean concentrations than males by 28 and 34%.

Cardenolide glycoside concentrations for monarch butterflies reared from an additional two New World *Asclepias* species and an African *Gomphocarpus* species are presented in Table 4. Most of these data were not derived from butterflies paired to an individual host plant. Mean concentrations for butterflies reared on *Asclepias curassavica* and *Gomphocarpus physocarpus* were lower than that for the plants. This suggests that butterflies reared on these species reached saturation. For *Asclepias humistrata,* the mean concentrations for plants and butterflies varied widely and overlapped; thus, whether or not the butterflies have reached saturation is questionable. For larvae reared on *A. curassavica* and *A. humistrata,* female butterflies achieved higher mean concentrations than males.

QUALITATIVE ASPECTS OF STORAGE

Seiber et al. (1980) showed by feeding uscharidin (**1**) (Fig. 1) to larvae that it was metabolized mainly to calactin (**6**) (Fig. 3), with minor amount of calotropin (**7**) and trace amounts of a polar metabolite. The conversion of uscharidin to calactin or calotropin requires the reduction of the 3′-oxo of the sugar moiety to a 3′-hydroxyl. Calactin and calotropin are epimers, configurationally isomeric at C-3′. When calotropin (**2**) was fed to larvae, it was stored along with a minor amount of a polar metabolite.

In a preliminary study, Brower et al. (1982) found that both labriformin (**11**) and labriformidin (**12**) were metabolized by monarch larvae to desglucosyrioside (**13**) and a second compound, presumably the C-3′ epimer. This indicates the larva's ability to covert the 3′-thiazoline (**11**) and the 3′-oxo glycosides (**1, 12**) to the corresponding 3′ hydroxyl glycosides (**6, 7, 13**; Fig. 3).

Larvae fed uzarigenin (**3**) stored it in minor amounts, while the major compound stored was a polar metabolite. When the larvae were fed digitoxigenin (**4**), the C-5 epimer of uzarigenin (**3**; Fig. 1), they stored only a polar metabolite and not the dosed aglycone. Adults, from larvae reared on diets containing these aglycones, contained a relatively larger proportion of the aglycones than the polar metabolites. It was suggested that these metabolites were conjugates (Seiber et al., 1980), and recent work has established them as being glucosides of the dosed aglycones (Nelson, 1988; Nelson, Watson, and Seiber, unpubl.).

More cardenolide glycosides with doubly-linked sugars have been isolated from *A. syriaca* and the plants cited in Tables 3 and 4 (Roeske et al., 1976; Brown et al., 1979; Cheung et al., 1980; Seiber et al., 1980, 1982, 1986; Nishio et al., 1982; Nelson, 1988) than those with singly-linked sugars. Monarchs reared on the milkweed species discussed here have all been shown by thin-layer chromatographic analyses to have selective and characteristic storage patterns (Roeske et al., 1976; Brower et al., 1982, 1984a,b; Lynch and Martin, 1987; Martin and Lynch, 1988; Nelson, 1988; Malcolm et al., 1989; Brower, Seiber, Nelson, Lynch, and Holland, unpubl.; Brower, Seiber, Nelson, and Lynch, unpubl.). Monarchs lack the nonpolar plant consitituents, such as the cardenolide glycosides characterized by the following C-3′ functionalities: thiazoline (**11**, uscharin), thiazolidine (voruscharin), oxo (**1, 12**, 5,6-dehydrouscharidin), and acetoxy (**8**, 5,6-dehydroasclepin, 16α-acetoxyasclepin, 16α-acetoxy-5,6-dehydroasclepin). Instead, monarchs store compounds of intermediate polarity, specifically, the 3′-hydroxyl glycosides **6, 7, 10,** and **13** and the more polar compounds **5, 9, 14,** and **15** of Figure 3. The leaves of *A. humistrata* lacked nonpolar cardenolide glycosides, and, as a result, the butterflies reared on this species roughly mirror the cardenolide glycoside content of the plant (Nishio et al., 1983; Cohen, 1985).

Reichstein et al. (1968) isolated the nonpolar uzarigenin (**3**), two glycosides of intermediate polarity, calactin (**6**) and calotropin (**7**), and two more polar compounds, calotropagenin (**5**) and calotoxin (**8**), from butterflies reared on *A. curassavica.* These glycosides (**6, 7,** and **8**) and their aglycone (**5**) are also amoung those isolated from *A. vestita*-reared butterflies (Nelson, 1988). Syriobioside (**14**) and aspecioside (**15**) are two relatively polar compounds that have been identified in butterflies reared from *A. syriaca* and *A. speciosa,* and they were also isolated from overwintering butterflies collected in Mexico (Seiber et al., 1986; Malcolm et al., 1989). Both calotoxin and syriobioside are doubly-linked glycosides with the usual hydroxyls of "stored" glycosides (C-2′, 3′, 14) and with additional hydroxyls on the sugar of calotoxin (C-4′) and on the genin of syriobioside (C-11,12). Aspecioside is a singly-linked glycoside characterized by a 12β-hydroxy and the same glycoside as calotoxin (β-hydroxyls at C-2′, 3′, 4′) (Fig. 3) (Cheung et al., 1986).

Table 3. Cardenolide glycoside concentrations of monarch butterflies and their larval host plants.

	N	Mean concentration (μg/0.1 g) Plant	Butterfly
A. cordifolia[a]			
Females	60	74.4	202
Males	52	71.3	159
Both	112	73.0	182
Range		19–238	38–321
A. speciosa[b]			
Females	45	77.4	175
Males	66	99.1	182
Both	111	90.3	179
Range		19–344	41–547
A. californica[c]			
Females	41	71.5	246
Males	44	61.0	223
Both	85	66.1	234
Range		9–199	59–410
A. erosa[a]			
Females	34	597	381
Males	37	529	316
Both	71	562	347
Range		79–2,102	181–567
A. eriocarpa[d]			
Females	80	408	348
Males	92	433	291
Both	172	421	318
Range		102–919	136–606
A. vestita[e]			
Females	34	753	279
Males	29	746	216
Both	63	750	250
Range		88–1,718	153–352

[a] Brower, Seiber, Nelson, and Lynch, unpubl. data.
[b] Brower et al., 1984b.
[c] Brower et al., 1984a.
[d] Brower et al., 1982.
[e] Brower, Seiber, Nelson, Lynch, and Holland, unpubl. data.

SUMMARY AND A PROPOSED MODEL OF CARDENOLIDE STORAGE

When monarch butterfly larvae were fed two cardenolides and two cardenolide glycosides, they stored these or related compounds in a logarithmic relationship to the amount and concentration consumed. Structure was a variable in determining the storage rate and, hence, the amount stored. Gly-

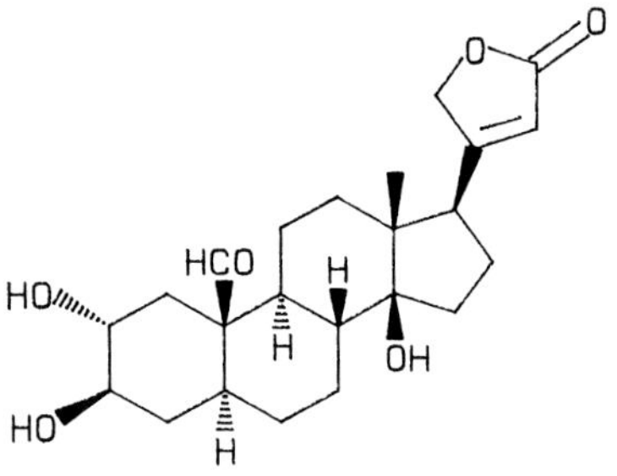

Calotropagenin (5)

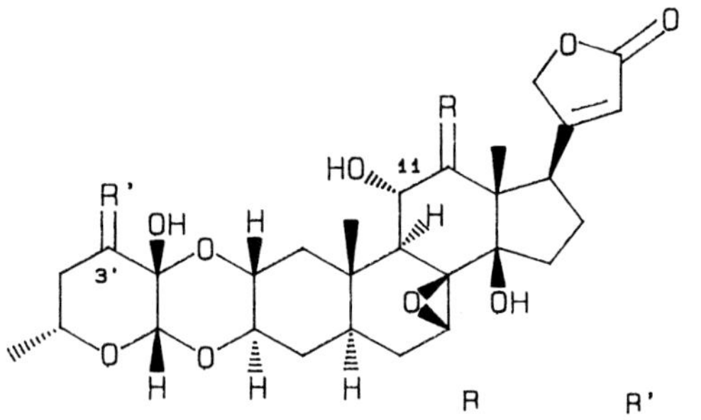

		R	R'
Labriformin	(11)	O	(thiazoline ring)
Labriformidin	(12)	O	O
Desglucosyrioside	(13)	O	β-OH α-H
Syriobioside	(14)	α-OH β-H	β-OH α-H

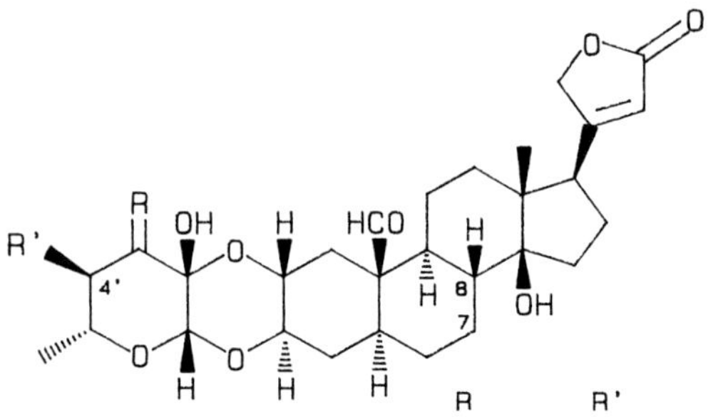

		R	R'
Calactin	(6)	β-OH α-H	H
Calotropin	(7)	α-OH β-H	H
Asclepin	(8)	α-OAc β-H	H
Calotoxin	(9)	β-OH α-H	OH
Humistratin (with $\Delta^{7,8}$)	(10)	β-OH α-H	H

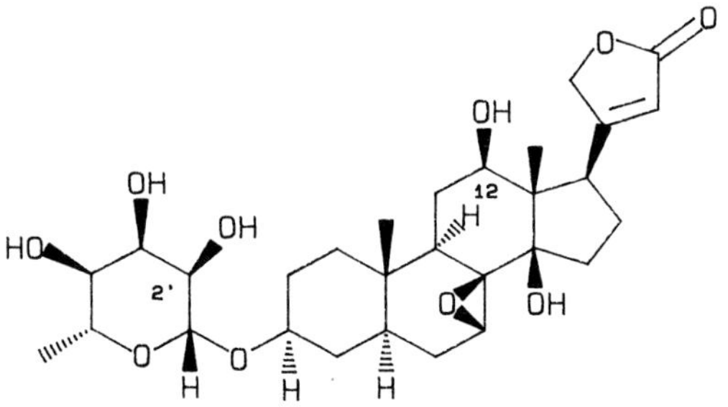

Aspecioside (15)

Figure 3. Some cardenolide glycosides and an aglycone stored or metabolized by the monarch butterfly. Compounds 5, 6, 7, and 9 have been isolated from butterflies of both *A. curassavica-* and *A. vestita*-reared larvae (Reichstein et al., 1968; Nelson, 1988). Compound 8 occurs in both *A. curassavica* (see references in Roeske et al., 1976) and *A. vestita* and is believed to be hydrolyzed to 7 by the larvae (Nelson, 1988). Compound 10 occurs in butterflies from *A. humistrata*-reared larvae (Nishio et al., 1983). Compounds 11 and 12 occur in *A. eriocarpa* and are metabolized by larvae to compound 13 which also occurs in this plant, *A. speciosa,* and *A. syriaca* and in butterflies reared on them. Compounds 14 and 15 occur in butterflies reared on *A. syriaca* and *A. speciosa* (Brower et al., 1982; Seiber et al., 1986; Malcolm et al., 1989).

cosides of low and intermediate polarity were stored at a higher rate than nonpolar aglycones. Glycosides of intermediate polarity were stored intact, while nonpolar glycosides and aglycones were metabolized to more polar compounds and stored.

A highly significant logarithmic relationship for cardenolide glycoside storage, on a concentration basis, was also observed in butterflies reared on five species of *Asclepias,* which had mean concentrations <250 μg/0.1 g. For three additional *Asclepias* species, with mean concentrations >400 μg/0.1 g, the butterfly concentrations were relatively constant and independent of plant concentrations, while a fourth species (*A. erosa*) showed a logarithmic relationship but with a lower level of significance than for species with lower concentrations. The level of saturation for butterflies reared on species with relatively high mean concentrations varied more than threefold, from about 100 μg/0.1 g to 363 μg/0.1 g. Female butterflies reached mean concentrations 6–45% greater than males. Butterflies stored only compounds of intermediate and high polarity.

These results allow formulation of a model based on both quantitative and qualitative storage according to the chemical structure of cardenolides and cardenolide glycosides. Thus, storage by monarch larvae, as reflected in adult butterflies, is linearly related to the $\log_{10}$ of the plant concentration. Therefore, at low plant concentrations this logarithmic relationship is expected to be observable, whereas at high plant concentrations butterfly concentrations are predicted to be relatively constant and essentially independent of plant concentrations. At high plant concentrations, butterflies will reach a saturation level of cardenolide glycosides for that plant. The model also predicts that females

Table 4. Cardenolide glycoside concentrations of three milkweed species and monarch butterflies reared on these species. * pooled sample.

Milkweed species (data source)	Mean concentration (μg/0.1 g)							
	Plant		Butterfly					
	Sample	N	Female	N	Male	N	Mean	N
G. physocarpus								
(Roeske et al., 1976)	331	*					102	*
A. curassavica								
(Roeske et al., 1976)	386	*					319	*
(Brower et al., 1975)			336	15	288	15	312	30
A. humistrata								
(Brower et al., 1975)			374	17	288	12	338	29
(Nishio et al., 1983)	513	2						
	398	3						
	842	3						
(Cohen and Brower, 1982)	492	7						
	441	9						
	476	7						
(Cohen, 1985)			558	6	385	4	489	10
(Malcolm et al., this volume)	389	22	419	5	313	17	337	22

will have a higher storage rate than males, which results in a higher mean concentration when reared on plants of equal concentration.

The model predicts a selective storage pattern for doubly-linked cardenolide glycosides resulting from the metabolism of nonpolar compounds characterized by C-3′ oxo, thiazoline, thiazolidine, and acetoxy functionalities to the analogous 3′-hydroxy compounds of intermediate polarity. Compounds of intermediate and polar character may be stored directly. Relatively nonpolar cardiac aglycones are predicted to be conjugated with glucose to produce amphiphilic compounds. Both the polar metabolite and aglycone are expected to be observed in the adults. The model also anticipates some metabolism of storable cardenolide glycosides of intermediate polarity, which further increases their polarity.

ACKNOWLEDGMENTS

I would like to thank Professors J.N. Seiber and L.P. Brower for access to unpublished data and permission to publish analyses of that data. Their work was supported by National Science Foundation grants to the University of California, Davis, and to Amherst College. I would like to thank Professors T.R. Watson, L.P. Brower, J.N. Seiber, and Dr. H.T.A. Cheung for reviewing the material presented in this manuscript and Neva Nishio for computer assistance.

LITERATURE CITED

Brower, L.P. 1969. Ecological chemistry. *Scientific American* 220(2):22–29.

Brower, L.P. 1984. Chemical defence in butterflies. In *The biology of butterflies*, eds. R.I. Vane-Wright and P.R. Ackery, *Symposia of the Royal Entomological Society of London* 11:109-134. London and New York: Academic Press (1989, Princeton, NJ: Princeton University Press).

Brower, L.P., M. Edmunds, and C.M. Moffitt. 1975. Cardenolide content and palatability of a population of *Danaus chrysippus* butterflies from West Africa. *Journal of Entomology (A)* 42:183–196.

Brower, L.P., J.N. Seiber, C.J. Nelson, S.P. Lynch, and P.M. Tuskes. 1982. Plant-determined variation in the cardenolide content, thin-layer chromatography profiles, and emetic potency of monarch butterflies, *Danaus plexippus* reared on the milkweed, *Asclepias eriocarpa* in California. *Journal of Chemical Ecology* 8:579–633.

Brower, L.P., J.N. Seiber, C.J. Nelson, S.P. Lynch, M.P. Hoggard, and J.A. Cohen. 1984a. Plant-determined variation in cardenolide content and thin-layer chromatography profiles of monarch butterflies, *Danaus plexippus* reared on milkweed plants in California 3. *Asclepias californica. Journal of Chemical Ecology* 10:1823–1857.

Brower, L.P., J.N. Seiber, C.J. Nelson, S.P. Lynch, and M.M. Holland. 1984b. Plant-determined variation in the cardenolide content, thin-layer chromatography profiles, and emetic potency of monarch butterflies, *Danaus plexippus* L. reared on milkweed plants in California: 2. *Asclepias speciosa. Journal of Chemical Ecology* 10:601–639.

Brown, P., J. von Euw, T. Reichstein, K. Stockel, and T.R. Watson. 1979. Cardenolides of *Asclepias syriaca* L., probable structure of syrioside and syriobioside. *Helvetica Chimica Acta* 62:412–441.

Cheung, H.T.A., T.R. Watson, J.N. Seiber, and C.J. Nelson. 1980. 7β,8β-epoxycardenolide glycosides of *Asclepias eriocarpa*. *Journal of the Chemical Society, Perkin Transactions I*, 2169–2173.

Cheung, H.T.A., T.R. Watson, S.M. Lee, M.M. McChesney, and J.N. Seiber. 1986. Structure of aspecioside from the monarch butterfly larvae food plants *Asclepias speciosa* and *A. syriaca*. *Journal of the Chemical Society, Perkin Transactions I*, 61–65.

Cohen, J.A. 1985. Differences and similarities in cardenolide contents of queen and monarch butterflies in Florida and their ecological and evolutionary implications. *Journal of Chemical Ecology* 11:85–103.

Cohen, J.A., and L.P. Brower. 1982. Oviposition and larval success of wild monarch butterflies (Lepidoptera: Danaidae) in relation to host plant size and cardenolide concentration. *Journal of the Kansas Entomological Society* 55:343–348.

Fieser, L.F., and M. Fieser. 1959. *Steroids*. New York: Reinhold, pp. 727–809.

Lynch, S.P., and R.A. Martin. 1987. Cardenolide content and thin-layer chromatography profiles of monarch butterflies, *Danaus plexippus* L., and their larval host-plant milkweed, *Asclepias viridis* Walt., in northwestern Louisiana. *Journal of Chemical Ecology* 13:47–70.

Malcolm, S.B., B.J. Cockrell, and L.P. Brower. 1989. The cardenolide fingerprint of monarch butterflies reared on the common milkweed, *Asclepias syriaca* L. *Journal of Chemical Ecology* 15:819–853.

Martin, R.A., and S.P. Lynch. 1988. Cardenolide content and thin-layer chromatography profiles of monarch butterflies, *Danaus plexippus* L., and their larval host-plant milkweed, *Asclepias asperula* subsp. *capricornu* (Woods.) Woods., in north central Texas. *Journal of Chemical Ecology* 14:295–318.

Nelson, C.J. 1988. *Milkweeds, monarchs and cardenolide glycosides, a study in three parts*. Ph.D. thesis, Sydney: University of Sydney, 320 pp.

Nishio, S., M.S. Blum, J.V. Silverton, and R.J. Highet. 1982. Structure of humistratin: A novel cardenolide from the sand-hill milkweed *Asclepias humistrata*. *Journal of Organic Chemistry* 47:2154–2157.

Nishio, S., M.S. Blum, and S. Takahashi. 1983. Intraplant distribution of cardenolides in *Asclepias humistrata* (Asclepiadaceae), with additional notes on their fates in *Tetraopes melanurus* (Coleoptera: Cerambycidae) and *Rhyssomatus lineaticollis* (Coleoptera: Curculionidae). *Memoirs of the College of Agriculture, Kyoto University* No. 122:43-52.

Reichstein, T., J. von Euw, J.A. Parsons, and M. Rothschild. 1968. Heart poisons in the monarch butterfly. *Science* 161:861–866.

Roeske, C.N., J.N. Seiber, L.P. Brower, and C.M. Moffitt. 1976. Milkweed cardenolides and their comparative processing by monarch butterflies (*Danaus plexippus* L.). *Recent Advances in Phytochemistry* 10: 93–167. New York: Plenum Publishing.

Seiber, J.N., P.M. Tuskes, L.P. Brower, and C.J. Nelson. 1980. Pharmacodynamics of some individual milkweed cardenolides fed to larvae of the monarch butterfly (*Danaus plexippus* L.). *Journal of Chemical Ecology* 6:321–339.

Seiber, J.M., C.J. Nelson, and S.M. Lee. 1982. Cardenolides in the latex and leaves of seven *Asclepias* species and *Calotropis procera*. *Phytochemistry* 21: 2343–2348.

Seiber, J.N., S.M. Lee, and J.M. Benson. 1983. Cardiac glycosides (cardenolides) in species of *Asclepas* (Asclepiadaceae). In *Handbook of natural toxins, Vol. I: Plant and fungal toxins*, eds. R.F. Keeler and A.T. Tu, 43–83. Amsterdam: Marcel Dekker.

Seiber, J.N., L.P. Brower, S.M. Lee, M.M. McChesney, H.T.A. Cheung, C.J. Nelson, and T.R. Watson 1986. Cardenolide connection between overwintering monarch butterflies from Mexico and their larval food plant, *Asclepias syriaca*. *Journal of Chemical Ecology* 12:1157–1170.

Thomas, R.E. 1981. Cardiac drugs. In *Burger's medicinal chemistry*, Part III, ed. M.E. Wolff, 47–192. New York: John Wiley & Sons.

Zar, J.H. 1974. *Biostatistical analysis*. Englewood Cliffs, NJ: Prentice-Hall, 620 pp.

SEQUESTRATION AND STORAGE OF CARDENOLIDES AND CARDENOLIDE GLYCOSIDES BY *DANAUS PLEXIPPUS PLEXIPPUS* AND *D. CHRYSIPPUS PETILIA* WHEN REARED ON *ASCLEPIAS FRUTICOSA*: WITH A REVIEW OF SOME FACTORS THAT INFLUENCE SEQUESTRATION

Carolyn J. Nelson

ABSTRACT. *Asclepias fruticosa* had a mean cardenolide glycoside concentration of 816 μg/0.1 g, *Danaus plexippus* reared on this species contained a mean concentration of 157 μg/0.1 g, and the amounts stored by *Danaus chrysippus* were <14 μg/0.1 g. The major cardenolide glycosides of *A. fruticosa* were nonpolar and were predicted to be metabolized to compounds of intermediate polarity prior to storage by monarch butterflies. Monarch larvae selectively stored cardenolide glycosides from the host plant; however, the selectivity was greater than predicted. *Danaus chrysippus* stored a few very polar cardenolide glycosides. The low level and limited types of cardenolide glycosides stored by the monarch and the possible involvement of amphiphiles in the solubilization of cardenolide glycosides are discussed.

INTRODUCTION

The monarch butterfly, or wanderer as it is known in Australia, *Danaus plexippus plexippus* (L.), utilizes many species of the milkweed family (Asclepiadaceae) as larval host plants from which they sequester and store cardenolides and their glycosides for chemical defense against vertebrate predators (see Roeske et al., 1976; Brower, 1984). A study in which individual compounds were fed to larvae (Seiber et al., 1980) and studies using host plants (Reichstein et al., 1968; Roeske et al., 1976; Brower et al., 1982, 1984a,b) have established both a quantitative and qualitative pattern of cardenolide and cardenolide glycoside storage by the monarch. A model to account for this pattern (Nelson, this volume) predicts that monarch butterflies reared on host plants containing concentrations of cardenolide glycosides >400 μg/0.1 g (dry weight) will reach a plateau of storage. It also predicts a selective storage pattern for doubly-linked cardenolide glycosides that results from metabolism of nonpolar compounds and direct storage of compounds of intermediate polarity.

Asclepias fruticosa L., sometimes referred to as *Gomphocarpus fruticosus* (L.) R. Br., is a common plant in the state of New South Wales (Australia) and serves as a host plant for both *D. plexippus* and *Danaus chrysippus petilia* (Stoll) (the lesser wanderer) (Common and Waterhouse, 1982). Twelve doubly-linked cardenolide glycosides and one aglycone have been isolated from the leaves of this plant (Cheung et al., 1983). It was anticipated from the model (Nelson, this volume) that four compounds of intermediate polarity would be directly stored and that the majority of the compounds, being nonpolar, would require metabolic conversion prior to storage.

This paper reports the quantitative variation of cardenolides and cardenolide glycosides among *A. fruticosa* plants and butterflies (*D. plexippus* and *D. chrysippus*) reared as larvae on them. Both the quantitative and qualitative aspects of storage are examined for the monarch within the framework of the model (Nelson, this volume) and for *D. chrysippus* by comparison to previous studies. Details regarding the isolation and identification of compounds from the monarch butterflies are reported in a separate paper (Cheung et al., 1988).

METHODS AND MATERIALS

COLLECTION OF PLANTS AND BUTTERFLIES

Upper and middle leaves were collected from 24 *A. fruticosa* plants and dried overnight at 60°C. The collection was made in May 1986 near Wisemans Ferry, New South Wales, the same site used by Cheung et al. (1983). Fifth-instar larvae of *D. plexippus* and *D. chrysippus*, and stalks of *A. fruticosa*, were collected at the same site in October 1983. The larvae were reared on the leaves at 26°C, with a 16:8 hr light : dark cycle. After pupation the adults were frozen within 24 hr of emergence and similarly dried.

EXTRACTION AND GROSS CARDENOLIDE GLYCOSIDE CONTENT

Plant (N = 24) and butterfly materials (12 female and 12 male *D. plexippus*, and 5 female and 4 male *D. chrysippus*) were individually ground in a mortar and pestle and individually extracted in 75% aqueous ethanol (0.2 g/10 ml for plants; 0.1 g/5 ml for *D. plexippus*; and the entire butterfly for *D. chrysippus*, ranging form 0.0366 to 0.0854 g/5 ml) for 1 hr at about 75°C.

Cooled extracts were assayed for cardenolides and cardenolide glycosides using a Perkin-Elmer Lambda 5 dual beam UV/VIS spectrophotometer and a procedure similar to that of Brower et al. (1982), but with the following changes: All extracts and reference solutions were prepared in 75% aqueous ethanol (refluxed with sodium hydroxide and distilled); 0.1 *M* sodium hydroxide was an aqueous solution; and the 0.15% 2,2′4,4′-tetranitrodiphenyl (TNDP) solution was prepared by dissolving it in acetonitrile (10%) and then adding distilled ethanol (90%). Plastic semimicro cuvettes (Kartell) were used and correction factors were established for each set by measuring their absorption while empty. The sample cuvette contained 0.5 ml sample extract or reference solution, 0.4 ml TNDP solution, and 0.2 ml sodium hydroxide; the reference cuvette contained 0.5 ml sample extract or reference solution, 0.4 ml acetonitrile-ethanol 1:9, and 0.2 ml sodium hydroxide. The contents were mixed using a glass Pasteur pipette and the absorbance read at 620 nm every 18 sec until it reached a maximum (about 18–20 min).

Regression of absorbance (Y) as a function of concentration (X) ($M \times 10^{-5}$), for a standard curve using 0.3×10^{-5} to 18.0×10^{-5} M solutions of digitoxin, gave $Y = 0.103X + 0.100$, $r^2 = 99.7\%$. Plant samples were diluted to absorb within the standard curve. Absorbance values were converted to equivalent micrograms of digitoxin per 0.1 g dry weight of material assayed.

CLEANUP AND QUALITATIVE CHROMATOGRAPHY

The remaining aqueous ethanol extracts from the plant samples were combined and evaporated to dryness on a rotatory evaporator under vacuum. The residue was dissolved in 30 ml ethanol and 60 ml 10% aqueous lead acetate, left on ice for 20 min, and centrifuged. The precipitate was rinsed, centrifuged, and then discarded. The combined rinse and supernatant was saturated with ammonium sulfate and centrifuged. After discarding the precipitate, the supernatant was extracted three times with 30 ml dichloromethane, which was dried with anhydrous sodium sulfate and evaporated under vacuum.

The extracts of *D. plexippus* and *D. chrysippus* were separately combined and evaporated to dryness under vacuum, defatted with petroleum ether, using for the *D. plexippus* extract 30 ml and for the *D. chrysippus* extract 15 ml, and centrifuged, and then the petroleum ether was removed and discarded. The residues were dissolved in water (20 ml and 10 ml) and dichloromethane (15 ml and 10 ml); the aqueous solutions were extracted three times with the above amounts of dichloromethane. The dichloromethane solutions were dried with anhydrous sodium sulfate and evaporated as above. The aqueous fractions were evaporated, dissolved in methanol, and checked for cardenolide glycosides with TNDP and shown to be negative.

Cleaned-up plant and butterfly extracts, containing approximately a total of 100 and 50 μg, respectively, of cardenolide glycosides, were chromatographed using 0.20 mm Merck silica gel 60 F-254 thin-layer chromatography (TLC) plates and distilled solvents. Two different solvent systems were used—(1) developed twice in ethyl acetate and (2) developed three times in chloroform–methanol–formamide (90:10:1)—with the following reference compounds for comparison: digitoxin, digitoxigenin, gomphoside, 3′-*epi*-gomphoside, 3′-*epi*-gomphoside 3′-acetate, afroside, 3′-*epi*-afroside, 3′-*epi*-afroside 3′-acetate, calactin, calotropin, asclepin, 19-deoxyuscharin, uscharin, uscharidin, 3′-didehydrogomphoside, and uzarigenin. Digitoxin and digitoxigenin were obtained from a commercial source while the remaining standards were obtained from T.R. Watson, H.T.A. Cheung, and J.N. Seiber. Cardenolides and their glycosides were detected by spraying with 0.4% TNDP in toluene followed by 10% potassium hydroxide in methanol–water (1:1). The lower detectable limit for this reagent is approximately 1–5 μg.

ISOLATION AND IDENTIFICATION OF CARDENOLIDE GLYCOSIDES FROM *D. PLEXIPPUS*

Butterflies used for cardenolide glycoside isolation were of the same origin as above. Details of the solvent extraction and partition cleanup; isolations involving column chromatography, high-performance liquid chromatography (HPLC), and TLC; and identifications by chromatographic comparisons, derivatization, methane-chemical ionization/mass spectral (CI/MS) and 400 MHz ^{1}H nuclear magnetic resonance (NMR) analyses are reported elsewhere (Nelson, 1988; Cheung et al., 1988).

RESULTS

QUANTITATIVE CARDENOLIDE GLYCOSIDE CONCENTRATION

The mean cardenolide glycoside concentration for leaf samples was 816 μg/0.1 g dry weight with a range of 323–1,500 (Table 1). In terms of percentage of dried plant mass, the range was from 0.3 to 1.5%. The range of concentrations for *D. plexippus* was 99–208 μg/0.1 g, with a mean of 157. Female butterflies had a higher mean concentration than the males (176 vs. 138). The concentrations in the nine *D. chrysippus* butterflies were below the lowest standard (3×10^{-6} M). Since 0.5 ml of a 3×10^{-6} M solution of digitoxin contains 1.15 μg and the most concentrated *D. chrysippus* extract was 0.0854 g/5 ml, then all samples contained <14 μg/0.1 g.

Table 1. Cardenolide glycoside concentrations of *A. fruticosa* leaves and butterflies reared on the plant.

	Concentrations[a]		
	Mean	SD	Range
A. fruticosa (N = 24)	816	322	323–1,500
D. plexippus			
Females (N = 12)	176	25	109–208
Males (N = 12)	138	26	99–192
All (N = 24)	157	32	99–208
D. chrysippus (N = 9)	<14		

[a] Concentrations are expressed as μg (equivalent to digitoxin)/0.1 g dry weight butterfly or plant material.

TLC ANALYSIS OF PLANTS AND BUTTERFLIES

The redrawn TLC plates (Fig. 1) indicate five discernible spots in the plant, four of them relatively intense. Considering the R_f values of reference compounds, the five plant spots could have contained the following cardenolide glycosides or aglycone: 3′-didehydrogomphoside, 3′-*epi*-gomphoside 3′-acetate, 3′-*epi*-afroside 3′-acetate, asclepin, uscharidin, uzarigenin, 19-deoxyuscharin, uscharin, gomphoside, and *epi*-afroside. Uscharin appeared to be present in the highest concentration, then 3′-*epi*-gomphoside 3′-acetate, followed by 19-deoxyuscharin, 3′-*epi*-afroside 3′-acetate, and 3′-*epi*-afroside. The two major compounds appearing in the *D. plexippus* extract had R_f values that matched calactin and calotropin; minor amounts of more polar compounds were also present. Only trace amounts of polar compounds were detected in *D. chrysippus.*

IDENTIFICATION OF ISOLATED CARDENOLIDE GLYCOSIDES AND AN AGLYCONE FROM *D. PLEXIPPUS*

Structures for the eight cardenolide glycosides and the aglycone isolated from the monarch butterflies reared on *A. fruticosa* are shown in Figure 2. The identification of calactin (**1**), calotropin (**2**), afroside (**3**), and calotoxin (**4**) was made by direct comparison of TLC and HPLC retention values and CI/MS data with those of reference compounds. Frugoside (**7**) was identified by similar comparisons of chromatographic data and confirmed by CI/MS and NMR analyses after conversion to the tetra-acetate. The remaining four compounds, calotropagenin (**5**), gofruside (**6**), gomphogenin 3-β-D-glucopyranoside (**8**), and calotropagenin 3-β-D-glucopyranoside (**9**), were identified by NMR and CI/MS analyses after conversion to the peracetates. Calotropagenin 2,3-diacetate was identified by direct comparison of chromatographic and CI/MS data with the acetylated reference compound. The β-D-glucopyranoside linkage in **8** and **9** was substantiated by hydrolysis using β-D-glucoside glucohydrolase to generate the corresponding aglycones, calotropagenin (**5**) and gomphogenin. Full details are presented elsewhere (Nelson, 1988; Cheung et al., 1988).

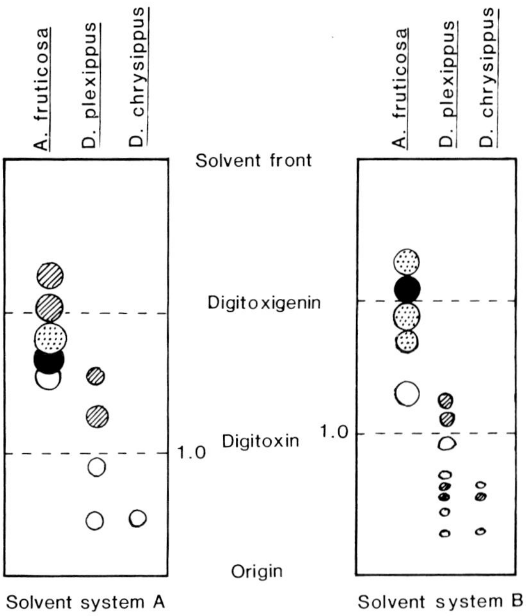

Figure 1. Drawings that represent thin-layer chromatographic plates of extracts from *A. fruticosa* and from *D. plexippus* and *D. chrysippus* reared on *A. fruticosa* and indicate a selective storage by the butterflies of plant cardenolide glycosides. The nonpolar compounds predominant in the plant were absent from the butterflies, which preferentially stored compounds of medium and high polarity, which were either metabolites of the nonpolar compounds or minor components of the plant. Plate **A** (left) was developed twice in ethyl acetate and plate **B** (right) was developed three times in chloroform–methanol–formamide (90:10:1). Shading of spots reflects intensities (black > hatched > dotted > closed) in the original chromatograms. $R_{digitoxin}$ values for solvent system A (and B) for 3′-didehydrogomphoside were 2.38 (and 2.00); 3′-*epi*-gomphoside 3′-acetate, 2.38 (2.18); 3′-*epi*-afroside 3′-acetate, 2.13 (1.82); digitoxigenin, 2.13 (1.89); asclepin, 2.08 (1.75); uscharidin, 2.08 (1.61); uzarigenin, 1.96 (1.67); 19-deoxyuscharin, 1.88 (2.00); gomphoside, 1.79 (1.57); uscharin, 1.67 (1.61); calactin, 1.58 (1.07); 3′-*epi*-gomphoside, 1.58 (1.68); afroside, 1.58 (0.89); calotropin, 1.25 (1.18); 3′-*epi*-afroside, 1.25 (1.29).

DISCUSSION

QUANTITATIVE VARIATION OF CARDENOLIDES AND THEIR GLYCOSIDES IN PLANTS AND BUTTERFLIES

Asclepias fruticosa leaves are among the highest and most variable in cardenolide and cardenolide glycoside concentration for *Asclepias* species (see

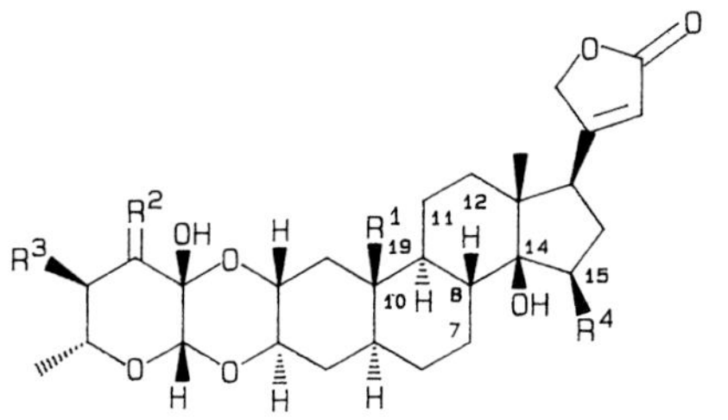

		R^1	R^2	R^3	R^4
Calactin	(1)	CHO	β-OH, α-H	H	H
Calotropin	(2)	CHO	α-OH, β-H	H	H
Afroside	(3)	CH_3	β-OH, α-H	H	OH
Calotoxin	(4)	CHO	β-OH, α-H	OH	H

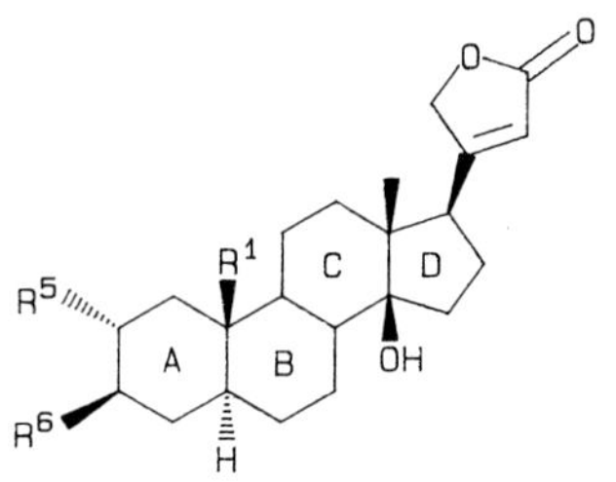

		R^1	R^5	R^6
Calotropagenin	(5)	CHO	OH	OH
Gofruside	(6)	CHO	H	6-deoxy-β-D-allose
Frugoside	(7)	CH_2OH	H	6-deoxy-β-D-allose
Gomphogenin 3-β-D-glucopyranoside	(8)	CH_3	OH	β-D-glucose
Calotropagenin 3-β-D-glucopyranoside	(9)	CHO	OH	β-D-glucose

Figure 2. Structures of cardenolide glycosides (**1**, **2**, **3**, **4**, **6**, **7**, **8**, and **9**) and an aglycone (**5**) isolated from *D. plexippus* reared on *A. fruticosa* (Cheung et al., 1988). Compounds **1**, **2**, **3**, **4**, **5**, **6**, and **7** have all been isolated from *A. fruticosa* or other species of *Asclepias*. Their presence in the butterflies may have resulted from direct sequestration and storage. The glucopyranosides, **8** and **9**, probably represent metabolites derived from gomphogenin and calotropagenin glycosides and/or the aglycones themselves (see text for references and details).

Seiber et al., 1983) and the second highest of the 12 species studied as host plants for the monarch butterfly (see Nelson, this volume). The range observed for this plant, 323–1,500 μg/0.1 g, is similar to *A. asperula*, at 341–1,616 μg (Martin and Lynch, 1988).

In spite of the very high concentrations in *A. fruticosa*, the mean concentration for *A. fruticosa*-reared monarchs was at 157 μg/0.1 g, only 19% of the plant mean, the second lowest mean among the 12 species studied, as cited above. Comparing in the same manner the 12 other relationships, this relationship is closest to *Gomphocarpus physocarpus* (Roeske et al., 1976). Both *A. fruticosa* and *Asclepias physocarpa* (or *G. physocarpus*) are similar in description and native to South Africa and are naturalized in Australia with similar distribution and habitat (Everist, 1981).

Because the monarchs in this study were not matched to individual plants, there could be no evaluation of the relationship between plant and butterfly concentrations. By examining the data in Table 1, it is evident that (1) *A. fruticosa* had high mean concentration (i.e., >400 μg/0.1 g), (2) the mean butterfly concentration was significantly lower than the mean plant concentration, and (3) the concentration range for butterflies was less than for host plants. These three characteristics are observed in relationships where butterflies reach their level of saturation (Nelson, this volume). Assuming saturation was achieved, the level for *A. fruticosa*-reared monarchs is only 43% of that observed in butterflies reared on *A. asperula* (Martin and Lynch, 1988). The mean concentrations for *A. asperula*-reared butterflies were among the highest of the 12 species summarized by Nelson (this volume). In the present study, the mean concentration for the female butterflies was 27% greater than for the males. Similarly, the differences were between 17 and 30% for butterflies reared on *A. curassavica*, *A. eriocarpa*, *A. erosa*, *A. vestita*, and *A. humistrata* (see Nelson, this volume).

Adult *D. chrysippus* contained no measurable quantity of cardenolides or cardenolide glycosides; however, TLC analyses of the combined extracts indicated the presence of some very polar compounds, which were also present in *D. plexippus* (see Fig. 1). Compared to *D. plexippus*, *D. chrysippus* was considerably less efficient and stored only the more polar steroidal compounds.

Brower et al. (1975) found that *D. chrysippus* reared on *Calotropis procera* and *A. curassavica* contained 31 and 44%, respectively, of the cardenolide and cardenolide glycoside concentration in *D. plexippus* adults. When reared on *A. fruticosa*, the difference was greater; *D. chrysippus* contained <9% of that in *D. plexippus* adults. Ferm (1977), comparing storage by *D. plexippus* and *D. chrysippus* on *A. curassavica* and five African species of milkweeds, found *D. plexippus* to have a higher mean concentration than *D. chrysippus* for all six species. TLC comparisons showed *D. plexippus* butterflies selectively stored compounds of medium polarity, whereas *D. chrysippus* butterflies generally stored more polar compounds. Rothschild et al. (1975) also concluded that *D. chrysippus* stored less and chiefly polar compounds compared to *D. plexippus*.

OCCURRENCE OF CARDENOLIDE GLYCOSIDES IN *A. FRUTICOSA* AND THE RELATIONSHIP BETWEEN THEIR PREDICTED AND OBSERVED STORAGE BY *D. PLEXIPPUS*

TLC indicated that cardenolide glycosides of highest concentration in *A. fruticosa* were relatively

Table 2. Uzarigenin and cardenolide glycosides occurring in the leaves of *A. fruticosa* (from Cheung et al., 1983) and prediction of their storage and/or metabolism prior to storage by the monarch butterfly.

Cardenolide glycoside or aglycone of the plant[a]	Stored or metabolized by the butterfly[b]	Metabolized to and stored[a]
Most abundant		
Uscharin (**17**)	M	Calactin (**0**), Calotropin (**2**)
3′-*epi*-Gomphoside 3′-acetate (**10**)	M	3′-*epi*-Gomphoside (**19**)
3′-*epi*-Afroside 3′-acetate (**11**)	M	3′-*epi*-Afroside
Substantial quantities		
3′-Didehydrogomphoside (**13**)	M	Gomphoside (**18**), 3′-*epi*-Gomphoside (**19**)
Uscharidin (**15**)	M	Calactin (**1**), Calotropin (**2**)
Asclepin (**12**)	M	Calotropin (**2**)
Gomphoside (**18**)	S	
19-Deoxyuscharin (**16**)	M	Gomphoside (**18**), 3′-*epi*-Gomphoside (**19**)
Minor constituents		
3′-Didehydroafroside (**14**)	M	Afroside (**3**), 3′-*epi*-Afroside
Calactin (**1**)	S	
3′-*epi*-Gomphoside (**19**)	S	
Afroside (**3**)	S	
Uzarigenin	M/S	Polar metabolite

[a] Bold numbers in parentheses refer to structures in Figures 2 and 3.
[b] S = stored, M = metabolized.

nonpolar. The content of these leaves appeared similar to those collected at the same site in 1982 (Cheung et al., 1983), which contained the compounds listed in the first column of Table 2. These cardenolide glycosides are derived from three different genins: gomphogenin (Column 1, Fig. 3), afrogenin (Column 2), and calotropagenin (Column 3). Uzarigenin was the only cardenolide aglycone isolated. Variations among the five different sugars (Column 4) are at C-3′, i.e. α-acetates (**10**, **11**, and **12**), ketones (**13**, **14**, and **15**), thiazolines (**16** and **17**), β-hydroxyls (**18**, **3**, and **1**), and α-hydroxyl (**19**). The sugars are doubly-linked to the genins through an acetal bond (1′ to 3) and a hemiacetal bond (2′ to 2).

Glycosides of gomphogenin and afrogenin have been isolated only from *A. fruticosa* whereas those of calotropagenin have also been isolated from *A. subulata*, *A. vestita*, *A. cordifolia*, *A. californica*, *A. curassavica*, *C. procera*, and *Pergularia extensa* (see Roeske et al., 1976; Seiber et al., 1980, 1982; Jolad et al., 1986; Nelson, 1988). Uzarigenin has been reported present in many *Asclepias* species (see Roeske et al., 1976; Nelson et al., 1981; Nishio et al., 1983; Nelson, 1988). Many other cardiac glycosides isolated from *A. subulata*, *A. syriaca*, *A. eriocarpa*, *A. vestita*, *A. californica*, *A. cordifolia*, and *A. humistrata* contain common sugars with those from *A. fruticosa* (Brown et al., 1979; Cheung et al., 1980; Nishio et al., 1982; Jolad et al., 1986; Nelson, 1988).

The model for cardenolide glycoside storage (Nelson, this volume) predicts butterflies will store directly intermediate and high-polarity compounds and convert nonpolar compounds to intermediate polarity prior to storage. Thus, it was predicted that of the compounds occurring in relatively high concentrations (Table 2) only gomphoside (**18**) would be directly stored. The others, all nonpolar due to variation at C-3′, would be metabolized to 3′-hydroxyl compounds (Column 3, Table 2). Cardenolide glycosides of intermediate polarity (i.e., **1**, **19**, and **3**) and uzarigenin should be stored directly with minor amounts of polar metabolites.

The selective storage of cardenolides and their glycosides, as shown by TLC (Fig. 1) and confirmed by isolation (see Fig. 2), was greater than predicted by the model (see Table 2). It predicted the primary storage of calactin, calotropin, 3′-*epi*-gomphoside, 3′-*epi*-afroside, gomphoside, and afroside; however, of these only calactin and calotropin and minor amounts of afroside were stored, which suggests the original model was inadequate. The apparent inability of the monarchs to store 3′-*epi*-gomphoside, 3′-*epi*-afroside, and gomphoside, and their apparently low storage efficiency of afroside, were probably factors contributing to the low level of saturation observed for these butterflies.

POSSIBLE FACTORS INFLUENCING THE LEVEL OF SATURATION AND SELECTIVITY OF CARDENOLIDE GLYCOSIDE STORAGE OBSERVED IN BUTTERFLIES REARED ON *A. FRUTICOSA*

Storage of cardenolides and cardenolide glycosides reflects complex interacting kinetic processes. The

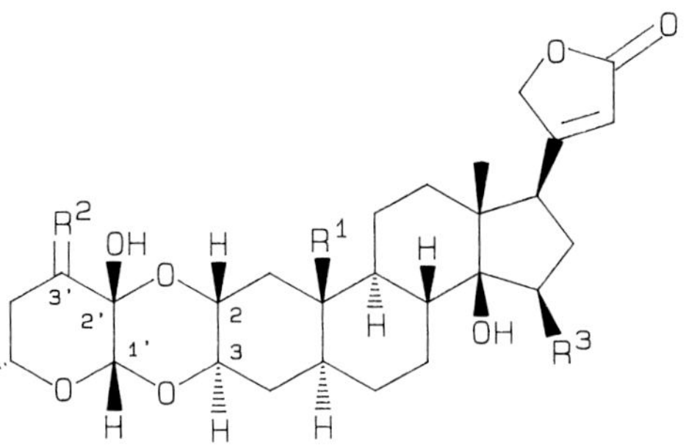

R^1=CH$_3$ R^3=H	R^1=CH$_3$ R^3=OH	R^1=CHO R^3=H	R^2
3'-epi-Gomphoside 3'-acetate (10)	3'-epi-Afroside 3'-acetate (11)	Asclepin (12)	β-H, α-OAc
3'-Didehydro-gomphoside (13)	3'Didehydro-afroside (14)	Uscharidin (15)	O
19-Deoxyuscharin (16)		Uscharin (17)	
Gomphoside (18)	Afroside (3)	Calactin (1)	β-OH, α-H
3'-epi-Gomhposide (19)			β-H, α-OH

Figure 3. Cardenolide glycosides isolated from the leaves of *A. fruticosa* (Cheung et al., 1983); listed in approximate order of increasing polarity, down and left to right. Only those below the line are predicted to be stored without prior metabolism by the monarch butterfly (see Table 2).

fate of these steroidal compounds, from ingestion by the larva to storage in the adult butterfly, is a result of both the inherent properties of the compounds (structure, shape, size, solubility, and polarity) and their interaction with the organism (via gut lumen and tissue solubility, metabolism, and sequestration into various tissues). Some of the inherent properties of cardenolides are shared by other lipids such as triglycerides, fatty acids, and sterols. It is possible that some of the interactions between these lipids and insects are similar to those of the cardenolides.

All insects lacking steroidogenic symbiotes require an exogenous source of dietary sterol for lipid biostructures and as steroid hormone precursors (see Dadd, 1985). Due to this requirement, insects may have developed highly effective absorption and transport mechanisms for sterols.

An insect's alimentary canal is an aqueous system in which small, neutral, hydrophilic molecules dissolve and are absorbed. The presence of an unstirred water layer is a barrier for lipid absorption (Turunen, 1985). Vertebrates solubilize lipids with bile salts by the formation of mixed micelles. Solubilized lipids can either be excreted or penetrate the unstirred water layer and be absorbed from the micellar solution by the intestinal cells (Wilson et al., 1971; Carey and Small, 1972). The biochemical process(es) that insects utilize to solubilize lipophilic compounds may involve phospholipids (Weintraub and Tietz, 1973), lysophospholipids (Turunen, 1983), or fatty acyl-amino acid conjugates (Collatz and Mommsen, 1974).

The levels of cardenolide glycoside saturation observed in monarch butterflies reared on various host plants (see Nelson, this volume) could have resulted from differences in the processes involved in larval sequestration and storage. For the purpose of this discussion two major areas are delineated: (1) the fate of the cardenolide glycosides prior to and during initial absorption and (2) the fate of the cardenolide glycosides after initial absorption until the ultimate storage in the adult butterfly.

Some processes concerning the first area are the mechanisms for (a) dispersion, (b) metabolism, and (c) initial absorption. If initial solubilization was limited, then the amount available for absorption would be limited. If the mechanism and/or rate of absorption was dependent on certain physical–chemical characteristics, then the level of saturation could be affected by the concentration of compounds with the "favorable" characteristics and/or the larva's ability to transform "unfavorable" compounds to "favorable" ones. If the mechanism for initial absorption became saturated, then storage would again be limited.

Some processes concerning the second area are the characteristics of the (a) storage sites, (b) transport mechanism(s), and (c) metabolism. If storage sites are limited, then the system could become saturated. If the type or capacity of sites and/or the mechanism or rate of transport from the gut tissue to the site(s) of deposition varied with the physical–chemical characteristics of the compounds, then storage variation could result. The extent to which absorbed cardenolide glycosides are metabolized, facilitating their transport to storage sites, storage or excretion, or their metabolism to noncardenolides, would influence the level of saturation.

Lipid-soluble compounds (fatty acids, partial glycerides, cholesterol, lysophospholipids, and lipid-soluble vitamins) are believed to be absorbed by passive diffusion (Turunen, 1985). A carrier-mediated, facilitated diffusion mechanism would result in a logarithmic accumulation pattern (Wolfe, 1981) such as that observed for cholesterol absorption in the roach *Eurycotis floridana* (Clayton et al., 1964) and characteristic of cardenolide glycoside storage by the monarch (see Nelson, this volume).

Structural features of the cardenolide aglycones and glycosides are so far the only variable identified in determining their storage rate and therefore the level of saturation (see Nelson, this volume). By comparing the compounds from *A. curassavica* and *A. eriocarpa,* which produced high levels of saturation, with those from *A. fruticosa,* some differences may be observed that may have contributed to the low level of saturation for *A. fruticosa* (Roeske et al., 1976; Brower et al., 1982; Nelson, this volume). One difference is the preponderance of nonstorable nonpolar glycosides in *A. fruticosa*; both *A. curassavica* and *A. eriocarpa* have about equal concentrations of nonstorable glycosides (i.e. labriformin, labriformidin, uscharin, and uscharidin) and storable glycosides (i.e., desglucosyrioside, calactin, and calotropin). If the larvae reared on *A. fruticosa* were incapable of transforming nonstorable glycosides (i.e., **17**, **10**, **11**, **13**, **15**, **12**, and **16**) to glycosides of intermediate polarity (as indicated in Table 2), then the quantity of glycosides of intermediate polarity available for sequestration and storage would be only a small fraction of the total. The capacity for these transformations probably exists, as is indicated by the following: Deacetylation readily occurred during the isolation of asclepin, 16α-acetoxyasclepin, 16α-hydroxyasclepin, and their 5,6-dehydro analogues from *A. vestita.* For example, asclepin deacetylated to calotropin during attempts to recover it from aqueous methanol, reverse-phase HPLC fractions by evaporating to dryness under vacuum at 40°C. This and other situations in which the above glycosides deacetylated suggest that the 3′α-acetoxy cardenolide glycosides are particularly labile to hydrolysis at C-3′ (Nelson, 1988). Basic conditions are known to promote the hydrolysis of esters (Morrison and Boyd, 1966). Therefore, it is likely that any dissolved asclepin in the aqueous alkaline environment of the larval gut (Wigglesworth, 1972) would be hydrolyzed to calotropin. 3′-Keto and 3′-thiazoline compounds are transformed to 3′-hydroxyl compounds prior to storage (Seiber et al., 1980; Brower et al., 1982; Marty and Krieger, 1984).

Danaus chrysippus larvae may lack the metabolic capability to convert nonpolar cardenolide glycosides to compounds of greater polarity. If so, this could contribute to their consistently lower storage ability compared to *D. plexippus* and their very low storage capability on *A. fruticosa.*

Seiber et al. (1980) fed monarch larvae uscharidin and concluded its reduction to calactin and calotropin was accomplished rapidly in the gut. Monarchs through all developmental stages showed the same selective storage of intermediate and polar cardenolide glycosides (Brower, 1984). The supposition that the transformations from nonpolar to intermediately polar compounds take place in the intestinal lumen and/or intestinal tissues prior to storage is consistent with these observations. For chemical hydrolysis and/or enyzmatic transformation, the nonpolar glycosides must be soluble or emulsified to facilitate interaction between the medium or enzyme and substrate. For example, differences in the enzymatic lipolysis of triolein and tripalmitin of *Locusta migratoria* have been attributed to differences in their emulsification in the intestinal lumen (Weintraub and Tietz, 1973).

Structural differences among the cardenolide glycosides of *A. curassavica, A. eriocarpa,* and *A. fruticosa* lie in the genin moiety. The stored 3′-hydroxy glycosides tend to be more polar than the nonstored compounds due to additional oxygen functions. Calotropagenin (**5**) (Fig. 2) contains an aldehyde at C-10 (19-oxo) and the genin of desglucosyrioside is an 11-hydroxy, 12-oxo, and 7,8-epoxy compound. Gomphogenin (**8** where R^6=OH) has no additional oxygen functions, and afrogenin (see **3**) has a hydroxyl group at C-15. These functionalities may subtly affect the glycoside's polarity, amphiphilicity, solubility, partition coefficients, shape, size, and/or other possible physical–chemical characteristics and subsequently the processes of sequestration and storage.

For the housefly (*Musca domestica*) and the hide beetle (*Dermestes maculatus*), the storage of cholesterol was greater than campesterol or β-sitosterol and the storage of campesterol was greater than β-sitosterol (Thompson et al., 1963; Katz et al., 1971). Structural differences among these sterols, which may influence their absorption, are primarily in the alkyl side chain attached at C-17. It would be anticipated that such differences would slightly affect the shape and size of the molecule, rather than significantly changing its polarity or other inherent characteristics.

Katz et al. (1971) observed with *D. maculatus* larvae that phytosterols added to the diet decreased the storage of cholesterol. This could have resulted from sterols competing for solubilization, absorption, and/or retention. Specificity is another characteristic of carrier-mediated, facilitated diffusion; thus, substances with structures closely related to the normally transported molecule can inhibit the rate of transport of that molecule (Wolfe, 1981). Similarly, for monarch larvae reared on *A. fruticosa,* cardenolide glycosides with low absorption and/or retention characteristics such as gomphoside, afroside, and their 3′-*epi* analogues could perhaps compete and/or interfere with glycosides

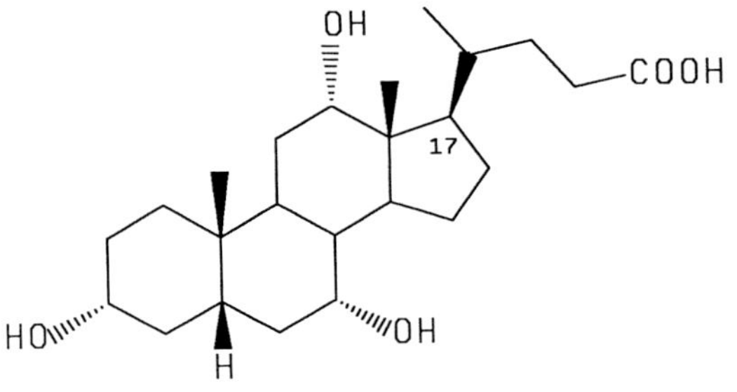

Cholic acid (20)

Figure 4. Sodium and potassium salts of the glycine and taurine conjugates of cholic acid (**20**) are common bile salts in humans. Bile salts are important in solubilizing lipids such as cholesterol in the aqueous medium of the intestine, which is necessary for both their absorption and excretion (Carey and Small, 1972).

having higher absorption and/or retention characteristics such as calactin and calotropin, thereby reducing the amount of calactin and calotropin stored.

Ito and Nakasone (1967) found soybean fatty acids increased the efficiency of β-sitosterol utilization in the silkworm (*Bombyx mori*) and suggested an active participation of fatty acids in sterol utilization. Dadd and Kleinjan (1984) observed that small additions of lecithin allowed the mosquito (*Culex pipiens*) to utilize several phytosterols to the same degree as cholesterol and suggested it was due to improved absorption. The sterol requirements have been reduced by olive oil in *D. maculatus* larvae (Budowski et al., 1967), by lecithin and kephalin in *Aedes aegypti* larvae (Golberg and De Meillon, 1948), and by α-glycerophosphates, oleic acid and oleate-containing compounds, and phosphatides in the ciliate *Tetrahymena corlissi* (Holz et al., 1961). One characteristic of all these compounds is that they are amphiphilic (or amphipathic). Amphiphiles possess distinct nonpolar and polar regions that allow them to form molecular aggregates such as micelles under certain conditions (Carey and Small, 1972). When added to a diet, amphiphilic compounds may increase the concentration of amphiphiles to a critical level that allows the formation of mixed micelles or other molecular aggregates. Or, they may change the characteristics of the existing molecular aggregates, increasing the solubilization of sterols, thereby promoting their absorption.

A comparison (Nelson, 1988) of cardenolide glycoside storage by *D. plexippus* from *A. curassavica* and *A. vestita*, which contain a qualitatively similar array of cardenolide glycosides (Roeske et al., 1976; Brower, Seiber, Nelson, Lynch, and Holland, unpubl.), suggests that the observed different levels of saturation must be influenced by factors not yet investigated. The processes for solubilization, absorption, and/or retention of cardenolide glycosides may be influenced by other phytochemicals.

The ability of the monarch larvae to utilize the cardenolide glycosides in a particular host plant may also be related to their ability to utilize other dietary compounds. This would be the case if the processes for cardenolide glycoside acquisition had a mechanistic commonality with those operative with certain essential lipids such as sterols and fatty acids. Brower and Moffitt (unpubl., as cited in Roeske et al., 1976) found that monarch larvae reared on *A. curassavica* were more efficient in converting ingested food to biomass and storing cardenolide glycosides than *G. physocarpus*-reared larvae. Dixon et al. (1978) found that larvae reared on *Asclepias fascicularis* were more efficient in converting ingested and digested food to biomass and attained a significantly higher mean concentration of cardenolide glycosides than the butterflies from *G. fruticosus*-reared larvae.

OCCURRENCE OF CARDENOLIDE GLYCOSIDES AND AGLYCONES IN BUTTERFLIES NOT REPORTED AS OCCURRING IN *A. FRUTICOSA*

Six of the nine compounds isolated from the butterflies were not noted as occurring in, or were predicted metabolites from, the cardenolide glycosides present in the larval host plant (Fig. 2 and 3 and Table 2). These compounds were calotoxin (**4**), calotropagenin (**5**), gofruside (**6**), frugoside (**7**), and the two glucose conjugates of gomphogenin (**8**) and calotropagenin (**9**).

Calotoxin (**4**), 4′β-hydroxycalactin (Cheung et al., 1983), has occurred coincidently with other calotropagenin glycosides in *A. curassavica, A. vestita, A. cordifolia, C. procera,* and *P. extensa* (see Roeske et al., 1976; Seiber et al., 1982; Nelson, 1988) and has been isolated from butterflies reared on *A. curassavica* and *A. vestita* (Reichstein et al., 1968; Nelson, 1988). Calotoxin probably occurred in low concentration in *A. fruticosa* (i.e., <41 μg/0.1 g dried plants, based on the detectable limit of TLC analysis and total cardenolide glycoside concentration) and was selectively concentrated in the butterflies.

The presence of calotropagenin (**5**) in the butterflies could have been due to (a) direct sequestration from the plant, (b) hydrolysis of doubly-linked calotropagenin glycosides (**1** or **2**), and/or (c) from hydrolysis of the glucoside (**9**). Calotropagenin has been isolated from *A. curassavica* and *A. vestita* (Tschesche et al., 1958, 1959; Nelson, 1988); hence, it may also have occurred in minor quantities in *A. fruticosa*. Analyses (Nelson, 1988) of monarch larvae fed calotropin and uscharidin (Seiber et al., 1980) indicated that they hydrolyzed these glycosides to produce calotropagenin. The results of Seiber et al. (1980) in conjunction with the findings of Nelson (1988) suggest that the glucose conjugates produced by the larvae are, to a fairly large extent, hydrolyzed during the metamorphosis to the adult butterfly. This aglycone was also isolated from butterflies reared on *A. curassavica* and *A. vestita* (Reichstein et al., 1968; Nelson, 1988).

Gofruside (**6**) and frugoside (**7**) have been isolated from the seeds of *A. fruticosa* (Keller and Reichstein, 1949; Hunger and Reichstein, 1952a,b), probably occurring in the leaves in low quantities (i.e. <41 μg/0.1 g dried plant), and were concentrated by the larvae to detectable levels. These two cardenolide glycosides have also been isolated from *A. vestita* and the butterflies reared on this species (Nelson, 1988).

Singly-linked cardenolide glycosides, such as gofruside (**6**) and frugoside (**7**), have been isolated less frequently from *Asclepias* spp. and generally appear to occur in lower concentrations than doubly-linked glycosides. Aspecioside, which contains the same carbohydrate moiety and linkage as **6** and **7**, has been isolated from *A. syriaca* and *A. speciosa*, and from wings of monarch butterflies collected at an overwintering site in Mexico (Cheung et al., 1986). It was the major cardenolide glycoside in butterflies reared on *A. speciosa* and *A. syriaca* (Malcolm et al., 1989; Seiber et al., 1986; Brower et al., 1984b).

The presence of calotoxin, gofruside, and frugoside in *A. fruticosa*-reared butterflies suggests a higher storage efficiency for these compounds than for afroside, and perhaps even higher than calactin and calotropin. Relative storage efficiencies for cardenolide glycosides from *A. speciosa* can be estimated by comparing mean spot intensities (Brower et al., 1984b; Seiber et al., 1986). These butterflies and plants had about equal concentrations of desglucosyrioside (spot 19). The butterflies enriched syriobioside (spot 11) 1.8 times relative to the plants, and aspecioside (spot 10) 2.3 times. Assuming each spot was the single glycoside, this suggests the relative order of storage efficiencies as aspecioside → syriobioside → desglucosyrioside. These follow in decreasing polarity, which may imply that the more polar compounds (aspecioside and syriobioside) had a higher storage efficiency than the compound of medium polarity (desglucosyrioside).

The basis for a greater enrichment of aspecioside relative to desglucosyrioside (from *A. speciosa*), and of gofruside and frugoside relative to afroside (from *A. fruticosa*), may reside in amphiphilic differences that affect their absorption. In vertebrates, amphiphilic bile salts, such as those derived from cholic acid (**20**, Fig. 4), play a critical role in the solubilization of lipids (Carey and Small, 1972). Both types of amphiphiles are similar in size, and for both the steroid moiety provides the nonpolar region. For the bile salts, the polar region is mainly due to the side chain attached to the D-ring (C-17), whereas for the cardenolide conjugates the polar glycoside, attached at the opposite end of the molecule to the A-ring (C-3), serves the same function. The amphiphilic character of cardenolide glycosides, determined in part by the number, type, and location of oxygen and other functionalities, may determine their participation in the formation of mixed micelles or other molecular aggregates. Singly-linked cardenolide glycosides, such as aspecioside, gofruside, and frugoside, have greater flexibility in the orientation of the hydrophilic sugar moiety than the doubly-linked glycosides, which may enhance their lipophilic or hydrophilic properties. Solubility, a general requirement for absorption (Turunen, 1985), can be achieved for lipids by the formation of molecular aggregates. The role of cardenolide glycosides, if any, in the formation of molecular aggregations may vary depending on both their amphiphilic characteristics and those of other amphiphilic phytochemicals in the plant. Duffey et al. (1978) proposed a physical mode of cardenolide glycoside sequestration in *Oncopeltus fasciatus* based on a phase-formation phenomenon.

Glucosides of gomphogenin (**8**) and calotropagenin (**9**) probably resulted from the hydrolysis of the doubly-linked glycosides, followed by conjugation with glucose and by direct conjugation if the aglycones occurred in the plant. The glucose conjugate of calotropagenin was present in larvae fed calotropin and uscharidin, and the very polar metabolites of digitoxigenin and uzarigenin were also identified as glucose conjugates (Seiber et al., 1980; Nelson, 1988). Some cardenolides and cardenolide glycosides, like other secondary plant products (Hosel, 1981), occur in the plant as glucosides (Fieser and Fieser, 1959). Of the cardenolides isolated from the *Asclepias* spp., only glucosides of uzarigenin have been noted (see Roeske et al., 1976; Seiber et al., 1983). The occurrence of glucosides of calotropagenin and gomphogenin in the butterflies are unlikely to be the result of direct larval sequestration.

FURTHER CIRCUMSTANTIAL EVIDENCE IMPLYING AN INFLUENCE OF NONCARDENOLIDE COMPOUNDS IN DETERMINING CARDENOLIDE GLYCOSIDE STORAGE AND EMETICITY

Considerable data have been published on the cardenolide glycoside-related interactions, (a) between the monarch butterfly larvae and their host plants and (b) between the butterflies and the blue jays (*Cyanocitta cristata bromia*) used to bioassay emeticity. I hypothesize that amphiphiles may affect the organisms' responses to ingested cardenolides and cardenolide glycosides and, thus, the interactions between the different trophic levels.

Emulsification of nonpolar cardenolide glycosides in the larval intestinal lumen could facilitate their conversion to storable compounds of intermediate polarity and thus affect the plant–larva interaction. Subsequent storage would reflect a relatively higher proportion of medium- to higher-polarity compounds. Conversely, the inability to emulsify nonpolar glycosides would impede their enzymic and chemical conversion. Under these conditions, the polar glycosides would probably have greater solubility in the larval intestinal lumen than the compounds of less polarity. Polar compounds, with higher concentrations in the solubilized contents and a higher permeability of the unstirred water layer, may thereby exist at a relatively higher concentration in the proximity of the gut wall. Thus, even though the penetration of polar cardenolide glycosides through the gut wall may be slower than less polar compounds (Yoder et al., 1976; Scudder and Meredith, 1982), their ultimate concentration within the larva may be greater.

The effective emulsification of cardenolide glycosides in the jays' intestinal lumen would similarly increase their absorption, reducing the amount required for emeticity (ED_{50}). Amphiphilic phytochemicals other than cardenolide glycosides, varying in type and/or quantity among the monarch's host plants, may result in amphiphilic differences among larval frass and adult butterflies.

Turunen and Kastari (1979) noted the conversion of dietary polar lipids, phosphatidyl cholines (lecithins) to lysophosphatidyl cholines (lysolecithins) by *Pieris brassicae* and suggested this might contribute to lipid emulsification in the intestinal lumen, thus facilitating the hydrolysis and absorption of other dietary lipids. Turunen (1983, 1985) suggested that the dietary lipid composition is important for the biological availability of esterified fatty acids and that the lipid composition of photosynthetic tissues may have considerable influence on lipid absorption in phytophagous insects. Harwood (1980) stated that lipids are mainly present as triacylglycerols (triglycerides) in seeds or glycolipids (glycoglycerolipids, glycosylglycerides) and phospholipids (glycerophospholipids, phosphoglycerides) in leaves, roots, or shoots. These are all amphiphiles or pro-amphiphiles and would be expected to show variation in quantity and type among the monarch's host plants and variation in larval utilization. The term "pro-amphiphile(s)," analogous to pro-drug, is used here to designate chemicals that are capable of being biologically modified to enhance their amphiphilic character. Triacylglycerols, for example, when hydrolyzed by gut lipases, liberate amphiphilic fatty acids, which may serve as precursors for amphiphilic phosphoacygylcerols (Weintraub and Tietz, 1973) or may be both sequestered and excreted (Turunen, 1983).

Dietary amphiphiles/pro-amphiphiles would primarily affect absorption and/or pre-absorption metabolism of cardenolides and cardenolide glycosides, which are only two of many processes determining their storage or emeticity. The following application of the amphiphile/pro-amphiphile hypothesis to the data from two plants (Table 3) provides alternative concepts to those we have previously expressed (Brower and Glazier, 1975; Roeske et al., 1976; Brower et al., 1988). Assume *G. physocarpus* represents a plant that contained a relatively low level of, or ineffectively utilized, amphiphiles/pro-amphiphiles, and *A. curassavica* represents the contrary situation. Assume the monarch larvae ingesting *A. curassavica* converted the pro-amphiphiles to amphiphiles, which aided in the solubilization process but were not completely absorbed; thus, some were excreted in the frass.

The following differences in storage and emeticity (see Table 3) could be in part due to different levels of emulsification of cardenolides and their glycosides. *Gomphocarpus physocarpus*-reared monarchs stored mainly polar compounds (77%) at a lower efficiency (3 vs. 13%), obtaining a lower mean concentration (102 vs. 319 μg/0.1 g) than *A. curassavica*-reared monarchs, which stored mostly medium-polar compounds (64%). The *G. physocarpus*-reared larvae appear less efficient at converting the nonpolar cardenolide glycosides to ones of medium polarity, as is evident from the greater relative proportion of nonpolar compounds in the frass. *Asclepias curassavica* contained 34% nonpolar compounds and larval frass 10%, thus reducing the amount relative to the plant by 70%, whereas for *G. physocarpus* the reduction was 38%.

The relative proportions of nonpolar, medium-polar, and polar cardenolide glycosides in *G. physocarpus* were quite different for plant, butterfly, and frass samples (Table 3); however, the amount of cardenolide glycoside contained in an emetic dose (the reciprocal of emetic potency; Roeske et al., 1976) for each sample differed by <10% (198–181 μg, maximum variation). The emetic doses, and hence the emetic potencies, appear to be relatively consistent and independent of qualitative cardenolide glycoside variation. The amphiphilic content present in plant, butterfly, and frass materials may have limited the solubility of the cardenolide glycosides and, hence, their absorption and emeticity.

Asclepias curassavica and derived samples also varied in cardenolide glycoside polarity (Table 3), and again there was little difference in the emeticity of plant and butterfly samples (68 and 70 μg/ED_{50} dose). The emetic potency for *A. curassavica* plant and butterfly materials was relatively greater than for similar *G. physocarpus* materials (68 vs. 198 μg/ED_{50} dose, plant; 70 vs. 181 μg/ED_{50} dose, butterfly). Host-plant amphiphile variation reflected in the adult butterfly may have contributed to these differences. Larval frass from *A. curassavica*-reared monarchs was more emetic than either butterfly or plant material (44 vs. 68 and 70 μg/ED_{50} dose). This would be expected if some of the cardenolide glycosides were solubilized in molecular aggregations and excreted, and/or if the amphiphiles produced (from the pro-amphiphiles) were excreted.

Differences in emetic potency among samples showing little qualitative cardenolide glycoside variation may be due to qualitative or quantitative variation in amphiphiles. The emetic potency differences among abdomen, wings, and thorax material for butterflies reared on *A. curassavica* (Table 3) (Brower and Glazier, 1975) could not be attrib-

Table 3. Results of cardenolide glycoside interactions at three trophic levels observed for monarch butterflies reared on two milkweed species.

	A. curassavica			*G. physocarpus*		
	Plant[a]	Butterfly[a]	Frass[a]	Plant[a]	Butterfly[a]	Frass[a]
Mean cardenolide glycoside concentration[b]	386	319	325	331	102	277
Storage efficiency of ingested cardenolide glycosides		13%			3%	
Polarity of cardenolide glycosides[c]						
Nonpolar	34%	0%	10%	29%	0%	18%
Medium polar	43%	64%	59%	38%	23%	54%
Polar	23%	36%	31%	33%	77%	28%
Emetic dose (whole butterflies)[d]	68	70	44	198	181	182
Emetic dose (butterfly parts)[e]		Male	Female			
Wings		236	229			
Abdomen		54	43			
Thorax		205	183			
Whole		84	67			

[a] Data and calculated values are based on data from Roeske et al. (1976).

[b] Concentrations are expressed as μg (equivalents to digitoxin)/0.1 g dry weight of plant or butterfly material.

[c] Nonpolar compounds are in regions A and A′, corresponding to the R_f region of uscharidin; medium polarity, B, C, B′, and C′, in the region of calactin and calotropin; polar, all regions below digitoxin; see Roeske et al. (1976), for details.

[d] Micrograms of cardenolide glycoside (equivalents to digitoxin) in an ED_{50} dose for a 84.25-g blue jay. Emetic potency is the reciprocal of emetic dose, i.e. number of ED_{50} units per unit of cardenolide.

[e] Calculated values as in [d] based on data of Brower and Glazier (1975) for *A. curassavica* only.

uted to differences in the cardenolide glycoside content, and so we proposed that these differences were due to cardenolide glycoside adsorption or binding to chitin, protein, or other structural materials (Brower et al., 1988). The emetic potency of abdomen material was four to five times greater than other body parts (43 and 54 vs. 183–236 μg/ED_{50} dose). These differences may have been in part due to a higher lipid concentration in the abdomen, which probably included amphiphiles such as phospholipids or lysophospholipids. Brower and Glazier (1975) observed that the emeticity of the whole butterfly was greater than the emeticity of the sum of the parts by 13 and 17%. If the differences in emetic potency among body parts were due to differences in cardenolide glycoside binding to structural material, as previously proposed, there would be no anticipated change in the amount or degree of binding when mechanically mixing various butterfly parts; therefore, the anticipated emeticity of the whole would be the summation of the contributing parts. However, if the differences in emetic potency of the body parts were due to differences in amphiphiles, as currently proposed, then conceivably the emeticity of the whole could be greater than the summation of contributing parts, because relatively amphiphilic "rich" parts could potentiate the emeticity of cardenolide "rich"-amphiphilic "poor" parts, thereby increasing the overall emetic potency. Similarly, differences in lipid content between migrating and freshly emerged monarchs may have contributed to differences in the emetic potency of monarchs with the same "fingerprint," i.e. *A. syriaca*-reared (Brower, 1969; Brower and Moffitt, 1974; Roeske et al., 1976; Malcolm et al., 1989). Amphiphiles may have also contributed to the difference in the emetic potency between *A. curassavica* and *A. vestita* and monarchs reared on these two species (Nelson, 1988; Roeske et al., 1976; Brower, Seiber, Nelson, Lynch, and Holland, unpubl.).

The emetic potencies of female butterfly parts were consistently greater than for male parts (Table 3). The greatest difference (20%) was between abdomen materials, which is probably the site of greatest lipid storage. Among overwintering populations females had a greater quantity of lipids than did males (Tuskes and Brower, 1978; Brower and Calvert, 1985). Conceivably, female larvae could have slightly different lipid storage capacities or rates of metabolism and/or sequestration than males, resulting in qualitative and/or quantitative differences in adults. The metabolism and sequestration of lipids, which may involve their conversion from pro-amphiphiles to amphiphiles (Turunen, 1983), may affect the efficiency of cardenolide glycoside absorption. This is a possible mechanistic reason underlying the higher storage rate for cardenolide glycosides in females compared to males (see Nelson, this volume).

Results of this and other studies indicated *D. chrysippus* butterflies contained a lower concentration and more polar cardenolide glycosides than did *D. plexippus* (Rothschild et al., 1975; Ferm, 1977). The amphiphile/pro-amphiphile hypothesis would suggest that this could have resulted from nonpolar cardenolide glycosides being less efficiently solubilized by *D. chrysippus* than *D. plexippus* larvae. Differences between the species in their ability to utilize and/or metabolize plant amphiphiles or pro-amphiphiles could conceivably contribute to these differences of cardenolide glycosides storage.

This amphiphile/pro amphiphile hypothesis has been developed and discussed for the cardenolide glycoside-based interactions among the monarch butterflies, their host plants, and the blue jays. It deals with one of the first interactions between an organism and lipophilic compounds encountered by ingestion, i.e., absorption. Because it is dealing with a physical aspect of xenobiotic sequestration, if verified, it may have wider ramifications in the understanding and practical applications of plant–insect and prey–predator interactions than the limited interactions on which it was developed.

SUMMARY

Asclepias fruticosa leaves varied in cardenolide glycoside concentration from 323 to 1,500 μg/0.1 g of dried material with a mean of 816. *Danaus plexippus* reared on this species ranged in concentration from 99 to 208 μg/0.1 g with a mean of 157 and with males having 78% of that found in the females. Samples of *D. chrysippus* butterflies contained <14 μg/0.1 g.

TLC of the plant extracts indicated the major cardenolide glycosides were nonpolar and, most likely, uscharin, 3′-*epi*-gomphoside 3′-acetate, 19-deoxyuscharin, 3′-*epi*-afroside 3′-acetate, and 3′-*epi*-afroside. Chromatographic analyses of *D. plexippus* indicated compounds of intermediate and high polarity whereas *D. chrysippus* showed only polar compounds present. Calactin and calotropin were the major compounds isolated from *D. plexippus* along with substantial amounts of gofruside and frugoside and traces of afroside, calotoxin, calotropagenin, gomphogenin 3-β-D-glucoside, and calotropagenin 3-β-D-glucoside (Cheung et al., 1988).

The results of this study concur with the model (Nelson, this volume) in that high plant concentrations of cardenolide glycosides resulted in lower and reduced variability in butterfly concentrations, that female butterflies had a higher concentration than male butterflies, and that the nonpolar constituents of the plant were absent from the butterfly. The level of saturation achieved by *D. plexippus* butterflies and the quantity stored by *D. chrysippus* were comparatively lower than on other host plants.

The model predicted monarch larvae reared on *A. fruticosa* would store six different 3′-hydroxyl glycosides; however, only three (calactin, calotropin, and afroside) were isolated from the adults. Calotropagenin and the 3-β-D-glucosides of calotropagenin and gomphogenin probably resulted from hydrolysis of doubly-linked glycosides and conjugation with glucose. Calotoxin, gofruside, and frugoside may have resulted from accumulation of minor quantities in the plant.

The bases for the relatively low level of cardenolide glycoside saturation and the greater-than-anticipated selectivity observed in monarchs reared on *A. fruticosa* is not known. It may have resulted from *A. fruticosa* having high concentrations of cardenolide glycosides, which exhibited low storage characteristics, and low concentrations of cardenolide glycosides, which exhibited relatively high storage characteristics. Or, the absorption of cardenolide glycosides with high storage characteristics may have been decreased by competition from glycosides with less favorable storage characteristics. The greater selectivity might suggest that the model was inadequate for this host. However, a review of some aspects of sterol utilization indicates that it would be simplistic to assume the inadequacies were totally due to endogenous or metabolic differences arising from the structurally different cardenolide glycosides present. The larvae's ability to metabolize, absorb, and store cardenolide glycosides may be primarily and directly related to their ability to disperse or solubilize them in the intestinal lumen. This ability may be affected by the presence, and amphiphilic potential, of other phytochemicals. Thus, a further possibility is that the absorption of all cardenolide glycosides may have been decreased due to the phytochemical composition of the host plant (apart from the cardenolide glycosides).

The processes governing quantitative and qualitative storage of cardenolide glycosides in the monarch are conceivably parallel and, possibly, integral to nutritional processes and may not be completely understood if regarded as an independent aspect. Co-ingested amphiphiles and pro-amphiphiles may affect cardenolide glycoside-related interactions among three tropic levels: plants, herbivores, and predators. The effectiveness of a chemical for defense may depend not only on its presence but also on its presentation.

ACKNOWLEDGMENTS

I would like to thank Professor T.R. Watson and Dr. H.T.A. Cheung for their advice and support and for reviewing the work. I would also like to thank Professors L.P. Brower and J.N. Seiber for reviewing the study and Drs. Watson, Cheung, and Seiber for reference compounds. I gratefully acknowledge the financial assistance from the University of Sydney from the award of the A.E. and F.A.Q. Stephens Postgraduate Research Fellowship, the L.M. Pattinson Scholarship, and a pharmacy department tutorship.

LITERATURE CITED

Brower, L.P. 1969. Ecological chemistry. *Scientific American* 220:22–29.

Brower, L.P. 1984 Chemical defence in butterflies. In *The biology of butterflies*, eds. R.I. Vane-Wright and P.R. Ackery, *Symposia of the Royal Entomological Society of London* 11:109–134. New York: Academic Press (1989, Princeton, NJ: Princeton University Press).

Brower, L.P., and W.H. Calvert. 1985. Foraging dynamics of bird predators on overwintering monarch butterflies in Mexico. *Evolution* 39:852–868.

Brower, L.P., and S.C. Glazier. 1975. Localization of heart poisons in the monarch butterfly. *Science* 188: 19–25.

Brower, L.P., and C.M. Moffitt. 1974. Palatability dynamics of cardenolides in the monarch butterfly. *Nature* 249:280–283.

Brower, L.P., M. Edmunds, and C.M. Moffitt. 1975. Cardenolide content and palatability of a population of *Danaus chrysippus* butterflies from West Africa. *Journal of Entomology (A)* 49:183–196.

Brower, L.P, J.N. Seiber, C.J. Nelson, S.P. Lynch, and P.M. Tuskes. 1982. Plant-determined variation in the cardenolide content, thin-layer chromatography profiles, and emetic potency of monarch butterflies, *Danaus plexippus* reared on the milkweed, *Asclepias eriocarpa* in California. *Journal of Chemical Ecology* 8:579–633.

Brower, L.P., J.N. Seiber, C.J. Nelson, S.P. Lynch, M.P. Hoggard, and J.A. Cohen. 1984a. Plant-determined variation in cardenolide content and thin-layer chromatography profiles of monarch butterflies, *Danaus plexippus* reared on milkweed plants in California 3. *Asclepias californica*. *Journal of Chemical Ecology* 10:1823–1857.

Brower, L.P., J.N. Seiber, C.J. Nelson, S.P. Lynch, and M.M. Holland. 1984b. Plant-determined variation in the cardenolide content, thin-layer chromatography profiles, and emetic potency of monarch butterflies, *Danaus plexippus* L. reared on milkweed plants in California 2. *Asclepias speciosa*. *Journal of Chemical Ecology* 10:601–639.

Brower, L.P., C.J. Nelson, J.N. Seiber, L.S. Fink, and C. Bond. 1988. Exaptation as an alternative to coevolution in the cardenolide-based chemical defense of monarch butterflies (*Danaus plexippus* L.) against avian predators. In *Chemical mediation of coevolution*, ed. K.C. Spencer, 447–475. New York: Academic Press.

Brown, P., J. von Euw, T. Reichstein, K. Stockel, and T.R. Watson. 1979. Cardenolides of *Asclepias syriaca* L., probable structure of syrioside and syriobioside. *Helvetica Chimica Acta* 62:412–441.

Budowski, P., I. Ishaaya, and M. Katz. 1967. Growth inhibition of *Dermestes maculatus* by phytosterols. *Journal of Nutrition* 9:201–207.

Carey, M.C., and D.M. Small. 1972. Micelle formation by bile salts. *Archives of Internal Medicine* 130:506–527.

Cheung, H.T., T.R. Watson, J.N. Seiber, and C.J. Nelson. 1980. 7β,8β-epoxycardenolide glycosides of *Asclepias eriocarpa*. *Journal of the Chemical Society, Perkin Transactions I*, 2169–2173.

Cheung, H.T.A., F.C.K. Chiu, T.R. Watson, and R.J. Wells. 1983. Cardenolide glycosides of the Asclepiadaceae. New glycosides from *Asclepias fruticosa* and stereochemistry of uscharin, voruscharin and calotoxin. *Journal of the Chemical Society, Perkin Transactions I*, 2827–2835.

Cheung, H.T.A., T.R. Watson, S.M. Lee, M.M. McChesney, and J.N. Seiber. 1986. Structure of aspecioside from the monarch butterfly larvae food plants *Asclepias speciosa* and *A. syriaca*. *Journal of the Chemical Society, Perkin Transactions I*, 61–65.

Cheung, H.T.A., C.J. Nelson and T.R. Watson. 1988. New glucoside conjugates and other cardenolide glycosides from the monarch butterfly reared on *Asclepias fruticosa* L. *Journal of the Chemical Society, Perkin Transactions I*, 1851–1857.

Clayton, R.B., P.C. Hinkle, D.A. Smith, and A.M. Edwards. 1964. The intestinal absorption of cholesterol, its esters and some related sterols and analogues in the roach, *Eurycotis floridana*. *Comparative Biochemical Physiology* 11:333–350.

Cohen, J.A. 1985. Differences and similarities in the cardenolide contents of queen and monarch butterflies in Florida and their ecological and evolutionary implications. *Journal of Chemical Ecology* 11:85–103.

Collatz, K.G., and T. Mommsen. 1974. Die Struktur der emulgierenden Substanzen verschiedener Invertebraten. *Journal of Comparative Physiology* 94:339–352.

Common, I.F.B., and D.F. Waterhouse. 1982. *Butterflies of Australia*. Sydney: Angus and Robertson Publishers, 434 pp.

Dadd, R.H. 1985. Nutrition: Organisms. In *Comprehensive insect physiology biochemistry and pharmacology, Vol. 4: Regulation: Digestion, nutrition, excretion*, eds. G.A. Kerkut and L.I. Gilbert, 313–390. Sydney: Pergamon Press.

Dadd, R.H., and J.E. Kleinjan. 1984. Lecithin-dependent phytosterol utilization by larvae of *Culex pipiens* (Diptera:Culicidae). *Annals of the Entomological Society of America* 77:518–525.

Dixon, C.A., J.M. Erickson, D.N. Kellett, and M. Rothschild. 1978. Some adaptations between *Danaus plexippus* and its food plant, with notes on *Danaus chrysippus* and *Euploea core* (Insecta: Lepidoptera). *Journal of Zoology, London* 185:437–467.

Duffey, S.S., M.S. Blum, M.B. Isman, and G.G.E. Scudder. 1978. Cardiac glycosides: A physical system for their sequestration by the milkweed bug. *Journal of Insect Physiology* 24:639–645.

Everist, S.L. 1981. *Poisonous plants of Australia*. Sydney: Angus and Robertson, 966 pp.

Ferm, B. 1977. *A comparative study of cardiac glycoside sequestering by* Danaus plexippus *and* Danaus chrysippus. Senior honors thesis. Amherst, MA: Amherst College, 64 pp.

Fieser, L.F., and M. Fieser. 1959. *Steroids*. New York: Reinhold, pp. 727–809.

Goldberg, L., and B. De Meillon. 1948. The nutrition of the larva of *Aedes aegypti* Linnaeus—III. Lipid requirements. *Biochemistry Journal* 43:372–379.

Harwood, J.L. 1980. Plant acyl lipids: Structure, distribution, and analysis. In *The biochemistry of plants, Vol. 4: Lipids: Structure and function,* ed. P.K. Stumpf, 1–55. New York: Academic Press.

Holz, G.G., Jr., B. Wagner, and J. Erwin. 1961. Sterol requirements of a ciliate *Tetrahymena corlissi* Th-X.—I. A nutritional analysis of the sterol requirements of *T. corlissi* Th-X. *Comparative Biochemistry and Physiology* 2:202–217.

Hosel, W. 1981. Glycosylation and glycosidases. In *The biochemistry of plants,* Vol. 7, ed. E.E. Conn, 725–753. New York: Academic Press.

Hunger, A., and T. Reichstein. 1952a. Frugosid, ein zweites kristallisiertes Glykosid aus den Samen von *Gomphocarpus fruticosus* (L.) R. Br. Glykoside und Aglycone, 91. Mitteilung. *Helvetica Chimica Acta* 35:429–433.

Hunger, A., and T. Reichstein. 1952b.Die Konstitution von Gofrusid und Frugosid. *Helvetica Chimica Acta* 35:1073–1103.

Ito, T., and S. Nakasone. 1967. Nutrition of the silkworm, *Bombyx mori.* XV. Utilization of sterol in the presence of dietary fatty acids. *Journal of Insect Physiology* 13:281–287.

Jolad, S.D., R.B. Bates, J.R. Cole, J.J. Hoffmann, T.J. Siahaan, and B.N. Timmermann. 1986. Cardenolides and a lignan from *Asclepias subulata. Phytochemistry* 25:2581–2590.

Katz, M., P. Budowski, and A. Bondi. 1971. The effect of phytosterols on the growth and sterol composition of *Dermestes maculatus. Journal of Insect Physiology* 17:1295–1303.

Keller, M., and T. Reichstein. 1949. Gofrusid, ein krystallisiertes Gylkosid aus dem Samen von *Gomphocarpus fruticosus* (L.) R. Br. *Helvetica Chimica Acta* 32:1607–1612.

Malcolm, S.B., B.J. Cockrell, and L.P. Brower. 1989. The cardenolide fingerprint of monarch butterflies reared on the common milkweed, *Asclepias syriaca* L. *Journal of Chemical Ecology* 15:819–853.

Martin, R.A., and S.P. Lynch. 1988. Cardenolide content and thin-layer chromatography profiles of monarch butterflies, *Danaus plexippus* L., and their larval host-plant milkweed, *Asclepias asperula* subsp. *capricornu* (Woods.) Woods., in north central Texas. *Journal of Chemical Ecology* 14:295–318.

Marty, M.A., and R.I. Krieger. 1984. Metabolism of uscharidin, a milkweed cardenolide, by tissue homogenates of monarch butterfly larvae, *Danaus plexippus* L. *Journal of Chemical Ecology* 10:945–965.

Morrison, R.T., and R.N. Boyd. 1966. *Organic chemistry.* Boston: Allyn and Bacon, 1204 pp.

Nelson, C.J. 1988. *Milkweeds, monarchs and cardenolide glycosides, a study in three parts.* Ph.D. thesis. Sydney: University of Sydney, 320 pp.

Nelson, C.J., J.N. Seiber, and L.P. Brower. 1981. Seasonal and intraplant variation of cardenolide content in the California milkweed, *Asclepias eriocarpa,* and implications for plant defense. *Journal of Chemical Ecology* 7:981–1010.

Nishio, S., M.S. Blum, J.V. Silverton, and R.J. Highet. 1982. Structure of humistratin: A novel cardenolide from the sandhill milkweed *Asclepias humistrata. Journal of Organic Chemistry* 47:2154–2157.

Nishio, S., M.S. Blum, and S. Takahashi. 1983. Intraplant distribution of cardenolides in *Asclepias humistrata* (Asclepiadaceae), with additional notes on their fates in *Tetraopes melanurus* (Coleoptera: Cerambycidae) and *Rhyssomatus lineaticollis* (Coleoptera: Curculionidae). *Memoirs of the College of Agriculture, Kyoto University* No. 122:43–52.

Reichstein, T., J. von Euw, J.A. Parsons, and M. Rothschild. 1968. Heart poisons in the monarch butterfly. *Science* 161:861–866.

Roeske, C.N., J.N. Seiber, L.P. Brower, and C.M. Moffitt. 1976. Milkweed cardenolides and their comparative processing by monarch butterflies (*Danaus plexippus* L.). *Recent Advances in Phytochemistry* 10: 93–167. New York: Plenum Publishing.

Rothschild, M., J. von Euw, T. Reichstein, D.A.S. Smith, and J. Pierre. 1975. Cardenolide storage in *Danaus chrysippus* (L.) with additional notes on *D. plexippus* (L.). *Proceedings of the Royal Society of London B* 190:1–31.

Scudder, G.G.E., and J. Meredith. 1982. The permeability of the midgut of three insects to cardiac glycosides. *Journal of Insect Physiology* 28:689–694.

Seiber, J.N., P.M. Tuskes, L.P. Brower, and C.J. Nelson. 1980. Pharmacodynamics of some individual milkweed cardenolides fed to larvae of the monarch butterfly (*Danaus plexippus* L.). *Journal of Chemical Ecology* 6:321–339.

Seiber, J.N., C.J. Nelson, and S.M. Lee. 1982. Cardenolides in the latex and leaves of seven *Asclepias* species and *Calotropis procera. Phytochemistry* 21: 2343–2348.

Seiber, J.N., S.M. Lee, and J.M. Benson. 1983. Cardiac glycosides (cardenolides) in species of *Asclepias* (Asclepiadaceae). In *Handbook of natural toxins, Vol. I: Plant and fungal toxins,* eds. R.F. Keeler and A.T. Tu, 43–83. Amsterdam: Marcel Dekker.

Seiber, J.N., L.P. Brower, S.M. Lee, M.M. McChesney, H.T.A. Cheung, C.J. Nelson, and T.R. Watson. 1986. Cardenolide connection between overwintering monarch butterflies from Mexico and their larval food plant, *Asclepias syriaca. Journal of Chemical Ecology* 12:1157–1170.

Thompson, M.J., S.J. Louloudes, W.E. Robbins, J.A. Waters, J.A. Steele, and E. Mosetting. 1963. The identity of the major sterol from houseflies reared by the CSMA procedure. *Journal of Insect Physiology* 9: 615–622.

Tschesche, R., D. Forstmann, and V.K.M. Rao. 1958. Zur Kenntnis der Cardenolid-Inhaltsstoffe von *Asclepias curassavica* L. *Chemische Berichte* 91:1204–1211.

Tschesche, R., G. Snatzke, and G. Grimmer. 1959. Calotropagenin aus *Asclepias curassavica* L. *Naturwissenschaften* 46:263–264.

Turunen, S. 1983. Absorption and utilization of essential fatty acids in lepidopterous larvae: Metabolic implications. In *Metabolic aspects of lipid nutrition in insects*, eds. T.E. Mittler and R.H. Dadd, 57–71. Boulder, CO: Westview Press.

Turunen, S. 1985. Absorption. In *Comprehensive insect physiology, biochemistry and pharmacology, Vol. 4: Regulation: Digestion, nutrition, excretion*, eds. G.A. Kerkut and L.I. Gilbert, 241–277. New York: Pergamon Press.

Turunen, S., and T. Kastari. 1979. Digestion and absorption of lecithin in larvae of the cabbage butterfly, *Pieris brassicae. Comparative Biochemistry and Physiology* 62A:933–937.

Tuskes, P.M., and L.P. Brower. 1978. Overwintering ecology of the monarch butterfly, *Danaus plexippus* L., in California. *Ecological Entomology* 3:141–153.

Weintraub, H., and A. Tietz. 1973. Triglyceride digestion and absorption in the locust, *Locusta migratoria. Biochimica et Biophysica Acta* 306:31–41.

Wigglesworth, V.B. 1972. *The principles of insect physiology*. London: Chapman and Hall, 827 pp.

Wilson, F.A., V.L. Sallee, and J.M. Dietschy. 1971. Unstirred water layers in intestine: Rate determinant of fatty acid absorption from micellar solutions. *Science* 174:1031–1033.

Wolfe, S.L. 1981. *Biology of the cell*. Belmont, CA: Wadsworth, 544 pp.

Yoder, C.A., D.E. Leonard, and J. Lerner. 1976. Intestinal uptake of ouabain and digitoxin in the milkweed bug, *Oncopeltus fasciatus. Experientia* 32: 1549–1550.

MILKWEED HOST PLANT UTILIZATION AND CARDENOLIDE SEQUESTRATION BY MONARCH BUTTERFLIES IN LOUISIANA AND TEXAS

Steven P. Lynch and Ronald A. Martin

ABSTRACT. Data synthesized from two previous studies on the cardenolide profiles of monarch butterflies, *Danaus plexippus* (L.), reared on the milkweeds *Asclepias viridis* Walt. and *Asclepias asperula* subsp. *capricornu* (Woods.) Woods. are reviewed, with new information on larval host plant utilization of these two species and *Asclepias amplexicaulis* Sm. and *Asclepias viridiflora* Raf. in Texas (TX) and Louisiana (LA). Initial counts of monarch larvae in early or mid-April were higher than at other times in all populations examined and appear correlated with the arrival of remigrating, egg-laying monarchs from Mexican overwintering sites. Breeding in the southern states is short-lived, and by early to mid-May immatures are generally absent and adults are rarely seen. High larval mortality was observed, particularly in the first three instars. Of 822 immatures found, only 3% had reached the third instar and between 22 and 59% of the fifth-instar larvae collected for pairwise analysis were parasitized by tachinid larvae. Crab spider and ant predation of small larvae was observed but was weakly correlated with numbers of immature monarchs. Other phytophagous insects, including lygaeid bugs, were uncommon during the period in which monarch larvae utilized the host plants. Both *A. viridis* and *A. asperula* were found to contain moderate to high levels of cardenolides with mean cardenolide concentrations of 881 μg/0.1 g for *A. asperula* at Denton, TX (DN), 245 μg/0.1 g for *A. viridis* at Wallace Lake, LA (WL), and 498 μg/0.1 g for *A. viridis* growing sympatrically with *A. asperula* at DN. Butterflies reared on each species were also found to be cardenolide-rich with mean concentrations of 362 μg/0.1 g for the DN *A. asperula*-reared butterflies, 337 μg/0.1 g for WL *A. viridis*-reared butterflies, and 315 μg/0.1 g for DN *A. viridis*-reared butterflies. Female monarchs in each population contained significantly greater cardenolide concentrations than did males even though the males were significantly larger. The range in total cardenolide for all 139 butterflies analyzed was 146–1,105 μg. Cardenolide uptake conformed to an inverse logarithmic relationship in WL *A. viridis*-reared butterflies with saturation occurring in the butterflies with increasing host plant concentrations. No correlation in cardenolide concentrations was found between the DN *A. asperula*-reared butterflies and their host plants. Butterfly dry weights were not correlated with cardenolide concentration when differences between sexes in size and cardenolide concentration were accounted for. Although distinguishable from each other and from previously published profiles for other *Asclepias* species, the overall thin-layer chromatography (TLC) patterns for *A. viridis* and *A. asperula* were similar, with the butterflies predominantly storing cardenolide metabolites more polar than those obtained from their host plants. The unique identity of the TLC profiles for the two species support the use of fingerprints to make ecological predictions concerning larval host plant utilization. *Asclepias viridis* and *A. asperula* are the first abundant species of the milkweed biomass available to monarchs as they reach temperate North America each spring. We propose most early-generation monarchs are reared on these two species and that their high levels of toxicity are instrumental in educating naive predators as the species continues northward to establish summer breeding patterns.

INTRODUCTION

While other danaid butterflies are restricted to subtropical and tropical climates and food resources, the monarch, *Danaus plexippus* (L.), has evolved a migratory dispersal strategy, which may reduce competition for larval host plants by utilizing the abundant temperate milkweeds in the genus *Asclepias* L. (Young, 1982). Each autumn millions of monarchs in the eastern United States leave their North American breeding grounds and fly southward to the Gulf Coast and to the volcanic mountains of eastern Michoacán, in central Mexico (Urquhart and Urquhart, 1976, 1978; Brower et al., 1977). Monarchs that breed in the western United States comprise an entirely separate population long known to overwinter along the California coast (see Lane, this volume; Nagano et al., this volume; Wenner and Harris, this volume).

Since the discovery of the Mexican overwintering sites in 1974 (Urquhart, 1976, 1987), much effort has been made to document the arrival, overwintering behavior, and spring breakup of these spectacular colonies (Brower et al., 1977; Calvert and Brower, 1981, 1986; Calvert et al., 1979, 1983; Fink and Brower, 1981). Until recently, however, little has been known about the ecology of these monarchs as they recolonize their temperate breeding grounds in eastern North America.

Monarch larvae have been reported to feed on a wide variety of milkweeds in the temperate United States. Malcolm and Brower (1986) compiled a list of 28 species serving as hosts of monarch butterflies in nature. To this list we can add *Asclepias cryptoceras* (California and Nevada; Lynch, unpubl.), *Asclepias engelmanniana* (Oklahoma and Kansas; Lynch, unpubl.), *Asclepias latifolia* (Texas; Lynch, unpubl.), *Asclepias perennis* (Louisiana; Lynch, unpubl.), *Asclepias solanoana* (California; Lynch, unpubl.), and *Asclepias viridiflora* (Texas; Lynch and Martin, unpubl.). In addition, we have observed monarch larvae on three species of milkweed vine in the south-central United States including *Sarcostemma crispum* and *Sarcostemma cynanchoides* in Texas and *Cynanchum laeve* in Louisiana (Lynch, unpubl.).

Two abundant spring-flowering species, *Asclepias viridis* and *Asclepias asperula,* appear to be particularly important to early spring monarchs in the south. Brower (1961, 1962), Cohen and Brower (1982), Cohen (1985), and Malcolm and Brower (1986) have reported selective oviposition by monarchs on *Asclepias humistrata* and *Asclepias curassavica* in Florida, but since most monarchs appear to remigrate through Texas and Louisiana, *A. viridis* and *A. asperula* are likely to be most important at the start of their annual breeding cycle.

In this paper we synthesize data from two previous studies on *A. viridis* (Lynch and Martin, 1987) and *A. asperula* (Martin and Lynch, 1988) with new information on host plant utilization and cardenolide chemistry in these two species. Also included are data on monarch host plant utilization of *Asclepias amplexicaulis* Sm. and *A. viridiflora* Raf. in east Texas. Our purposes are to describe the timing and extent of milkweed host plant utilization by monarchs and other arthropods, discuss monarch larval behavior and survival, and further establish the quantitative and qualitative variation in cardenolide contents of butterflies as they relate to their host plants *A. viridis* and *A. asperula.* We then discuss the temporal, geographical, and chemical roles these two species play in the re-establishment of monarch summer breeding populations in temperate North America.

METHODS AND MATERIALS

HOST PLANT MATURATION AND MONARCH UTILIZATION

In April and May of 1985 populations of *A. viridis, A. asperula* subsp. *capricornu, A. amplexicaulis,* and *A. viridiflora* were selected in eastern Texas and western Louisiana, and repeated observations were made of their growth and reproductive phenology. In April and May of 1987 one additional population of *A. asperula* subsp. *capricornu* was observed in central Texas. At each site transects were drawn across the population and the first 50 plants of each species encountered were identified and marked. Data were collected on the number of stems, the number of umbels in bud, at anthesis (open), and past anthesis (past), and the number of umbels with developing fruit >1 cm.

During each visit, we recorded immature monarchs found on the same 50 host plants. Each was identified as an egg or a specific instar, and whether it was located on the stem, a leaf, or an umbel. For all populations in 1985, the numbers of spiders, adult *Lygaeus* and *Oncopeltus* seed bugs (Lygaeidae), and other arthropods present were also counted, while ants and immature lygaeids were recorded as present or absent.

Four sites were established for *A. viridis.* These were Wallace Lake (WL), 0.5 mi north of Wallace Lake on Wallace Lake Road, Caddo Parish, Louisiana; Lake Charles (LC), 1.8 mi west of Jefferson Parish on US 190, Beauregard Parish, Louisiana; Livingston (LN), on Farm Road 942W, 2.2 mi east of US 59, Polk County, Texas; and Denton (DN), 0.8 mi west of I-35 on US 380, Denton County, Texas. Observations were made at WL on 2, 9 17, and 24 April and 1, 9, and 16 May, at LC on 5 and 21 April and 7 May, at LN on 6 April, and at DN on 19 April and 7 and 20 May.

A population of *A. asperula* subsp. *capricornu* growing sympatrically with *A. viridis* was also sampled at DN in 1985. Counts on *A. asperula* were made on 20 April and 7 and 20 May, 1985. A second population for *A. asperula* growing allopatrically from *A. viridis* was sampled in 1987 northwest of Buda (BT) in Hays County, Texas, 7.3 mi south of Texas Farm Road 967. Counts and collections were made on 5 and 18 April and 2 and 28 May and 5 July 1987. *Asclepias amplexicaulis* and *A. viridiflora* populations were sampled in 1985 at Ore City (OC), 1.0 mi east of Ore City on Farm Road 450, Upshur County, Texas. Counts were made on each species on 19 and 28 April and 7 and 20 May.

PAIRWISE COLLECTIONS FOR CARDENOLIDE ANALYSIS

Methods for the pairwise collection and qualitative and quantitative cardenolide analyses of adult monarchs reared as larvae on *A. viridis* and *A. asperula* host plants are given in Lynch and Martin (1987) and Martin and Lynch (1988). In addition to the 60 pairwise samples examined for *A. viridis* and 41 for *A. asperula,* 38 additional butterflies were reared on these two species at DN in 1985, 11 on *A. asperula,* 18 on *A. viridis,* and 9 on putative hybrids morphologically intermediate to the two parental species. These butterflies were analyzed following procedures established for *A. asperula* (Martin and Lynch, 1988). For the hybrid study, each thin-layer chromatography (TLC) plate included both sexes of each species and the various combinations were intentionally placed in different regions on each plate to reduce or eliminate error due to plate position. The two standards, digitoxin and digitoxigenin, were run as a mixture in the outermost positions and in the central position of each TLC plate, and pooled samples of *A. viridis* and *A. asperula* were run immediately flanking the central channel.

STATISTICAL ANALYSES

Statistical analyses were made on an IBM 3330 computer housed at Louisiana State University in Baton Rouge utilizing SAS statistical software, Version 5, and on a Compaq Portable Personal Computer using SAS-PC, Version 6.03. Statistical procedures generally followed those in Lynch and Martin (1987).

RESULTS

MONARCH UTILIZATION OF MILKWEED HOST PLANT RESOURCES

Since 1981 our first sightings of monarch adults in northwest Louisiana have occurred between 28 March and 5 April (Lynch and Martin, unpubl.). In this region we consistently find numerous early instars on *A. viridis* by early April (see also Riley, this volume). In Table 1 we list the mean number of immature monarchs (including eggs) observed on host plants at each site included in this study. The highest immature monarch density was found on *A. viridis* at WL on 9 April, with a total of 3.96 eggs and larvae per plant. Of these, 67.2% were eggs, 20.7% were first instars, 10.6% were second instars, and the remaining 1.5% were third instars. On 17 April at WL, only 1.1 eggs and larvae were found per plant. Of these, 54.6% were eggs, first and second instars each comprised 16.4% of the total, 9.1% were third instars, and the remaining 3.6% were fourth instars. On 24 April, only 0.44 immatures were found per plant with eggs comprising 22.7% and fourth instars 32% of the total. On 1 May, only three eggs, one first instar, and one second instar were found on the 50 plants. No immature monarchs were observed on either 9 or 16 May at WL.

High initial numbers of eggs and immatures were also observed in early April 1985 on *A. viridis* at LC and at LN. Of 1.16 immatures per plant observed at LC on 5 April, 38.3% were eggs, 48.5% were first instars, and the remaining 13.2% were second instars. At LN on 6 April, 81.0% of 1.7 immatures per plants were eggs and the remaining 19.0% were first instars. Counts at DN, BT, and OC were not initiated until mid-April due to the later emergence and development of the milkweeds at those two sites. Both *A. viridis* and *A. asperula* hosted low numbers of monarchs at DN with initial counts of 0.38 and 0.10 eggs and immatures per plant, respectively, on 19 and 20 April. Of these, 50.0% were eggs, 20.8% were first instars, 12.5% were second instars, 12.5% were third instars, and one (4.2%) was a fourth-instar larva. A high of 1.88 immatures was observed on *A. asperula* plants at BT on 18 April, including 74.5% eggs, 20.2% first instars, and 5.3% second instars. Initial counts made at OC on 19 April were also high, with 1.96 monarch immatures on *A. amplexicaulis* and 0.60 on *A. viridiflora*. Here, 74.9% were eggs, 19.8% were first-instar larvae, and 5.3% were second.

At WL the emergence of *A. viridis* stems occurred within 1 wk of our peak count of immature larvae at that site. In Table 2 the percent of stems with flower buds, open flowers (at anthesis), past flowering (no fruit), and developing fruit >1 cm in length is given for the 50 WL *A. viridis* host plants examined for monarch larvae over seven observation periods. On 2 April, no monarchs (adult or immature) were observed at the site and either the plants were dormant or new growth had emerged <5 cm. On 9 April, when peak numbers of immature monarchs were observed, the plants were still emerging and all flowering stems (89.9% of the total stems) were in bud only. By 24 April, a peak of 46.3% of the total stems were at anthesis, 23.8% of the stems were in bud, and 3.9% of the stems were past anthesis. Flowering dropped off to 9.6% of the stems by 9 May, but then increased to 10.9% by 16 May. This extended flowering period may be due largely to new growth initiated after stem damage caused by the hooves of cattle grazing in the fields. In contrast to flowering, numbers of immature monarchs at WL had decreased to 1.1 per plant on 17 April and to only 0.44 per plant at our measured peak flowering on 24 April. No immatures were found after 9 May, even though 22.7% of the observed stems on that date and 16.8% of the stems on 16 May remained in bud or at anthesis. Many plants remained in flower and all the host plants remained green through the maturation and dehiscence of their fruits in late June.

Counts were also made on the number of stems at anthesis for each milkweed population on each observation date included in this study (Table 2). The peak flowering periods for *A. viridis* varied considerably among the sites. At LC, peak flowering was observed during our initial counts on 5 April, over 2 wk earlier than the observed 24 April peak at WL, and >1 mo earlier than the observed peak flowering for *A. viridis* at DN on 7 May. Some flowering extended past our last observation date at all four *A. viridis* sites. *Asclepias viridis* plants at LN had not yet reached anthesis on 6 April.

The observed peak flowering data for *A. asperula* at DN was 7 May, coinciding with that of *A. viridis* at the same location. However, since only 22.2% of the flowers were at anthesis, 32.5% were already past anthesis, and 6.0% already carried developing fruit, peak flowering probably occurred earlier than our observed date. In 1987 at BT, *A. asperula* peak flowering was observed on 2 May. At both sites all

Table 1. Mean number of monarch eggs and larvae per host plant observed on four *Asclepias* species at six sites in Louisiana and Texas. Data are taken from 50 individual host plants for each population at each site.

Site and species		Egg	Larval instar					Total
			First	Second	Third	Fourth	Fifth	
(1) Wallace Lake								
A. viridis	2 Apr 85	0	0	0	0	0	0	0
	9 Apr 85	2.66	0.82	0.42	0.06	0	0	3.96
	17 Apr 85	0.60	0.18	0.18	0.10	0.04	0	1.10
	24 Apr 85	0.10	0.08	0.06	0.04	0.14	0.02	0.44
	1 May 85	0.06	0.02	0.02	0	0	0	0.10
	9 May 85	0	0	0	0	0	0	0
	16 May 85	0	0	0	0	0	0	0
(2) Lake Charles								
A. viridis	5 Apr 85	0.38	0.54	0.24	0	0	0	1.16
	21 Apr 85	0.22	0.30	0.06	0	0	0	0.58
	7 May 85	0.06	0	0	0	0	0.02	0.08
(3) Livingston								
A. viridis	6 Apr 85	1.70	0.62	0	0	0	0	2.32
(4) Denton								
A. viridis	19 Apr 85	0.16	0.08	0.06	0.06	0.02	0	0.38
	7 May 85	0	0	0	0.02	0	0.02	0.04
	20 May 85	0	0	0	0	0	0	0
A. asperula	20 Apr 85	0.08	0.02	0	0	0	0	0.10
	7 May 85	0	0	0	0	0	0	0
	20 May 85	0	0	0	0	0	0	0
(5) Buda								
A. asperula	5 Apr 87	0	0	0	0	0	0	0
	18 Apr 87	1.40	0.38	0.10	0	0	0	1.88
	2 May 87	0	0	0	0.02	0.04	0.10	0.16
	28 May 87	0	0	0	0	0	0	0
	5 July 87	0	0	0	0	0	0	0
(6) Ore City								
A. amplexicaulis	19 Apr 85	1.56	0.36	0.04	0	0	0	1.96
	28 Apr 85	0.08	0.42	0.34	0.10	0.06	0.04	1.04
	7 May 85	0	0	0.02	0.02	0.04	0.10	0.18
	20 May 85	0	0	0	0	0	0	0
A. viridiflora	19 Apr 85	0.48	0.08	0.04	0	0	0	0.60
	28 Apr 85	0	0.06	0.10	0.08	0.04	0	0.28
	7 May 85	0	0	0	0	0.02	0.06	0.08
	20 May 85	0	0	0	0	0	0	0

A. asperula stems were past anthesis by 22 May. Peak flowering was observed for *A. amplexicaulis* at OC on 28 April. Nine days earlier all the flowering stems were in bud (46.0% of the total). Grazing cattle damaged several *A. amplexicaulis* stems lowering the percentages listed in Table 2 for 7 and 20 May. All *A. amplexicaulis* stems were past flowering by 20 May. *Asclepias viridiflora* did not produce any visible buds until 7 May and remained in bud throughout the study.

Our data indicate a high mortality rate for immature monarchs, particularly in the first three instars. Cohen and Brower (1982; pers. comm.) and Malcolm et al. (1987) report similarly low survivorships in monarchs feeding on Florida milkweeds. Of 822 immatures counted in our study, 477 (58.0%) were eggs, 198 (24.1%) were first instars, 84 (10.2%) were second instars, and only 25 (3.0%) were third instars. Fourth- and fifth-instar larvae were observed nearly as often as third instars, at 20 (2.4%) and 18 (2.3%), respectively.

Many small-instar larvae were found in developing inflorescences surrounded by buds and/or pedicels. We suggest this location affords them protection from crab spiders and other arthropod predators. For *A. viridis*, *A. asperula*, and *A. amplexicaulis* the number of eggs laid on leaves versus inflorescences was not significantly different from

Table 2. **Reproductive condition of stems observed on four *Asclepias* species at six sites in Louisiana and Texas. Data are percent of total stems for 50 individual host plants for each population at each site.**

Site and species		Condition of stem (%) Bud	Anthesis	Past	Fruit[a]
(1) Wallace Lake					
A. viridis	2 Apr 85	0	0	0	0
	9 Apr 85	89.9	0	0	0
	17 Apr 85	74.0	9.8	0	0
	24 Apr 85	23.8	46.3	3.9	3.4
	1 May 85	12.6	35.6	15.0	14.2
	9 May 85	13.1	9.6	14.7	41.5
	16 May 85	5.9	10.9	10.0	61.6
(2) Lake Charles					
A. viridis	5 Apr 85	51.4	43.6	0	0
	21 Apr 85	14.2	20.1	40.4	28.6
	7 May 85	7.3	2.4	29.7	52.4
(3) Livingston					
A. viridis	6 Apr 85	98.0	0	0	0
(4) Denton					
A. viridis	19 Apr 85	69.6	0	0	0
	7 May 85	27.3	34.7	3.8	0
	20 May 85	11.1	24.0	11.7	35.8
A. asperula	20 Apr 85	77.4	0	0	0
	7 May 85	5.1	22.2	32.5	6.0
	20 May 85	0	0.4	14.1	47.7
(5) Buda					
A. asperula	5 Apr 87	0	0	0	0
	18 Apr 87	80.4	9.2	0	0
	2 May 87	43.0	28.6	14.5	4.7
	28 May 87	0	0	31.4	64.2
(6) Ore City					
A. amplexicaulis	19 Apr 85	46.0	0	0	0
	28 Apr 85	23.4	22.6	0	0
	7 May 85	9.2	7.3	22.7	0
	20 May 85	0	0	7.6	26.8
A. viridiflora	19 Apr 85	0	0	0	0
	28 Apr 85	0	0	0	0
	7 May 85	22.0	0	0	0
	20 May 85	35.4	0	0	0

[a] Percent total stems with developing fruit >1 cm long.

the assumption of a 50–50 distribution. However, for each species first- and second-instar larvae were found in significantly greater numbers in the inflorescences than in the leaves. Of the 449 eggs found on the three species, 207 were laid in the inflorescences and 242 were laid on leaves ($\chi^2 = 1.36$, 1 df, $p > 0.20$). Of 258 first- and second-instar larvae, 161 were found on the inflorescences and only 97 were found on leaves ($\chi^2 = 7.94$, 1 df, $p < 0.01$). In a 2-×-2 contingency table comparing the distribution of eggs with first- and second-instar larvae on leaves and inflorescences we also found a significantly greater than expected number of the larvae in the inflorescences ($\chi^2 = 17.44$, df $= 2$, $p < 0.01$). *Asclepias viridiflora* at OC did not develop inflorescences while immature monarchs were present and is not included in our calculations.

We also recorded the frequency of occurrence of lygaeid bugs, spiders, and ants present on the host plants that may compete for food and space with immature monarchs or prey on them (Fig. 1). At WL on 9 April, 86% of the 50 study plants possessed at least one immature monarch, but just one spider was observed and only one plant hosted any ants. No lygaeids were found on 9 April. By 24 April only 32% of the plants hosted monarch larvae, while ants were found on 42% of the plants, spiders on 26%, and lygaeids on 6%. The spider

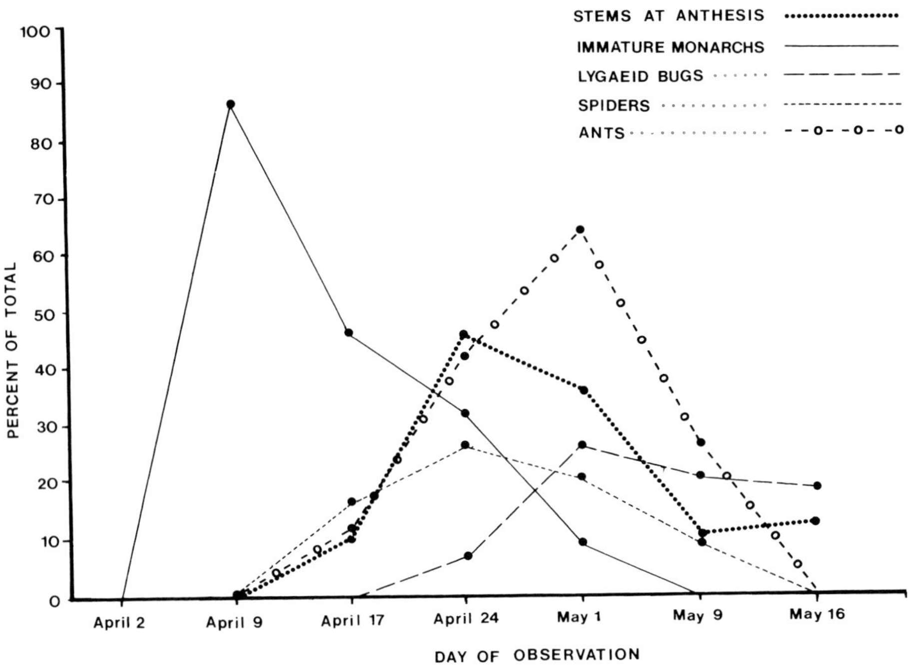

Figure 1. Percent of the total number of stems at anthesis and the frequency of occurrence of immature monarchs, lygaeid bugs, crab spiders, and ants on 50 *A. viridis* host plants at Wallace Lake. Individual host plants and observation dates correspond to those in Table 1.

population peaked on 24 April and no spiders were found on study plants by 16 May. The ant population peaked soon after the peak of anthesis in *A. viridis,* present on 64% of the plants on 1 May. No ants were observed on 16 May.

Crab spiders are known to ambush a variety of insects, and may feed primarily on floral visitors (Morse and Fritz, 1982). We observed predation of small-instar monarch larvae by crab spiders on numerous occasions and suggest that they represent a major factor in the mortality of immature monarchs. However, we could find only a weak negative correlation between the presence of immature monarchs and spiders on all the host plants and dates included in the study ($F = 4.57$, $p = 0.03$, $r^2 = 0.0046$). There were no significant correlations for any specific date at WL. It may be that individual spiders are highly mobile and periodic observations are ineffective indicators of spider activity. Also, examining our data at WL (see Fig. 1), we find that the spider population roughly follows the pattern of flowering of *A. viridis.*

Ant predation of monarch larvae occurred at all instars, even when the larvae were enclosed in nylon net bags. On *A. viridis* at WL, foraging lines of ants would chew small holes through the bags to gain entrance. Even large fifth-instar larvae were devoured. However, the presence of ants on the host plants was not significantly correlated with the presence or absence of immature monarchs on *A. viridis* at WL and was only weakly negatively correlated with the presence of immatures on all the host plants and dates used in this study ($F = 9.30$, $p = 0.002$, $r^2 = 0.0092$). Ant population patterns appear to follow the flowering pattern of the host plants, and their presence on the plants appears primarily associated with the availability of nectar.

Lygaeid bugs, including *Oncopeltus fasciatus,* and *Lygaeus kalmii* were frequent visitors on *A. viridis* at WL and on other milkweeds in our region but appear to develop their populations well after most monarch larvae have left their host plants. At WL, their peak frequency was observed on 1 May. Although the number of plants hosting lygaeids decreased slightly through 16 May, their total numbers were actually increasing as many plants hosted numerous immatures on that date. In all cases where developing fruits were present there was a significant correlation between the presence of lygaeids and the number of developing fruits. Other plant feeders were infrequent and sporadic, and their presence was not significantly correlated with immature monarchs, predators, or lygaeids. Thus, there appears to be very little competition for milkweed host plant resources between other plant feeders and monarchs in the south-central United States.

The actual number of monarchs that survive to pupate may actually be higher than that indicated by our data because later-instar larvae may spend some time away from their host plants. In contrast to western United States monarchs, which nearly always form their chrysalids upon their host plant

Table 3. Cardenolide concentrations in monarch butterflies and their host-plant milkweeds, *A. viridis* and *A. asperula* subsp. *capricornu*. Data are μg (equivalent to digitoxin)/0.1 g dry weight of butterfly or plant material.

		Plant material			Butterfly material		
	N	Mean	SD	Range	Mean	SD	Range
Wallace Lake, LA							
A. viridis[a]							
Males	29	229	64	95–432	286	98	73–482
Females	31	259	72	130–432	384	89	216–591
Both	60	245	70	95–432	337	105	73–591
Denton, TX							
A. viridis							
Males	5	578	129	373–730	290	31	254–325
Females	13	468	175	251–779	324	71	217–466
Both	18	498	167	250–779	315	64	217–466
A. asperula subsp. *capricornu*[b]							
Males	25	863	272	341–1,572	324	53	231–431
Females	27	897	260	341–1,616	397	80	285–529
Both	52	881	264	341–1,616	362	77	231–529
A. asperula × *viridis*							
Males	5	655	357	388–1,272	241	44	198–310
Females	4	772	265	388–972	378	28	346–407
Both	9	707	306	388–1,272	302	81	198–407

[a] Data taken with permission from Lynch and Martin (1987).

[b] Data include 41 pairwise samples taken with permission from Martin and Lynch (1988).

(Brower et al., 1982, 1984a, 1984b), no chrysalids were found on unbagged plants. Even bagged monarchs tended to spin their pads on the nylon net bags rather than on the leaves of the plant. Unbagged third- and fourth-instar larvae were also observed to leave their host plant occasionally, possibly in preparation for molting.

However, after pupating, many additional monarchs fail to emerge as adults. Twenty-one of the 74 unbagged monarch fifth-instar larvae collected from *A. viridis* for the pairwise analyses in Lynch and Martin (1987) were parasitized by tachinid larvae. The tachinids, as many as 15 in a single monarch, characteristically emerged to pupate either as the monarchs were hanging in preparation for pupation or during the chrysalid stage. No parasitized monarchs survived to emerge as adults. Six of 14 unbagged fifths collected off *A. asperula* for the Martin and Lynch (1988) study died from tachinid and hymenopteran parasitoids. Even though bagging small-instar larvae on the plants in nylon net bags afforded considerable protection, 9 of 131 (6.9%) bagged *A. viridis* and *A. asperula* larvae were also lost to parasitoids. On *A. amplexicaulis,* 22 of 81 (27.2%) of the fifth instars we collected were parasitized by tachinids. Comparable collections of unbagged monarchs reared from several milkweeds in California often suffered mortality rates >50% from tachinid parasitoids (Lynch and Brower, unpubl. data).

CARDENOLIDE CONTENT OF PLANTS AND BUTTERFLIES

Table 3 lists dry weight cardenolide concentrations equivalent to digitoxin for *A. viridis* and *A. asperula* and for monarch adults reared as larvae on these two host plants. We found *A. viridis* host plants at WL (Lynch and Martin, 1987) to contain moderate amounts of cardenolide. The mean dry weight concentration for all 60 *A. viridis* plants measured at WL was 245 μg/0.1 g, ranging from 95 to 432 μg/0.1 g. Butterflies reared on *A. viridis* at WL were found to be cardenolide-rich with a mean of 337 μg/0.1 g dry weight.

Asclepias asperula host plants at DN contained high concentrations of cardenolides, with a mean dry weight concentration of 881 μg/0.1 g for 52 host plants, ranging from 341 to 1,616 μg/0.1 g. Butterflies reared on *A. asperula* contained a mean of 362 μg cardenolide/0.1 g dry weight. Eighteen *A. viridis* pairwise samples and nine plants morphologically intermediate to *A. viridis* and *A. asperula* were also analyzed from the DN site. *Asclepias viridis* at DN was significantly higher in cardenolide content than *A. viridis* at WL, with a mean of 498 μg/0.1 g, ranging from 250 to 779 μg/0.1 g. The putative hybrids contained concentrations intermediate to the parental populations with a mean of 707 μg/0.1 g and a range of 388–1,272 μg/0.1 g.

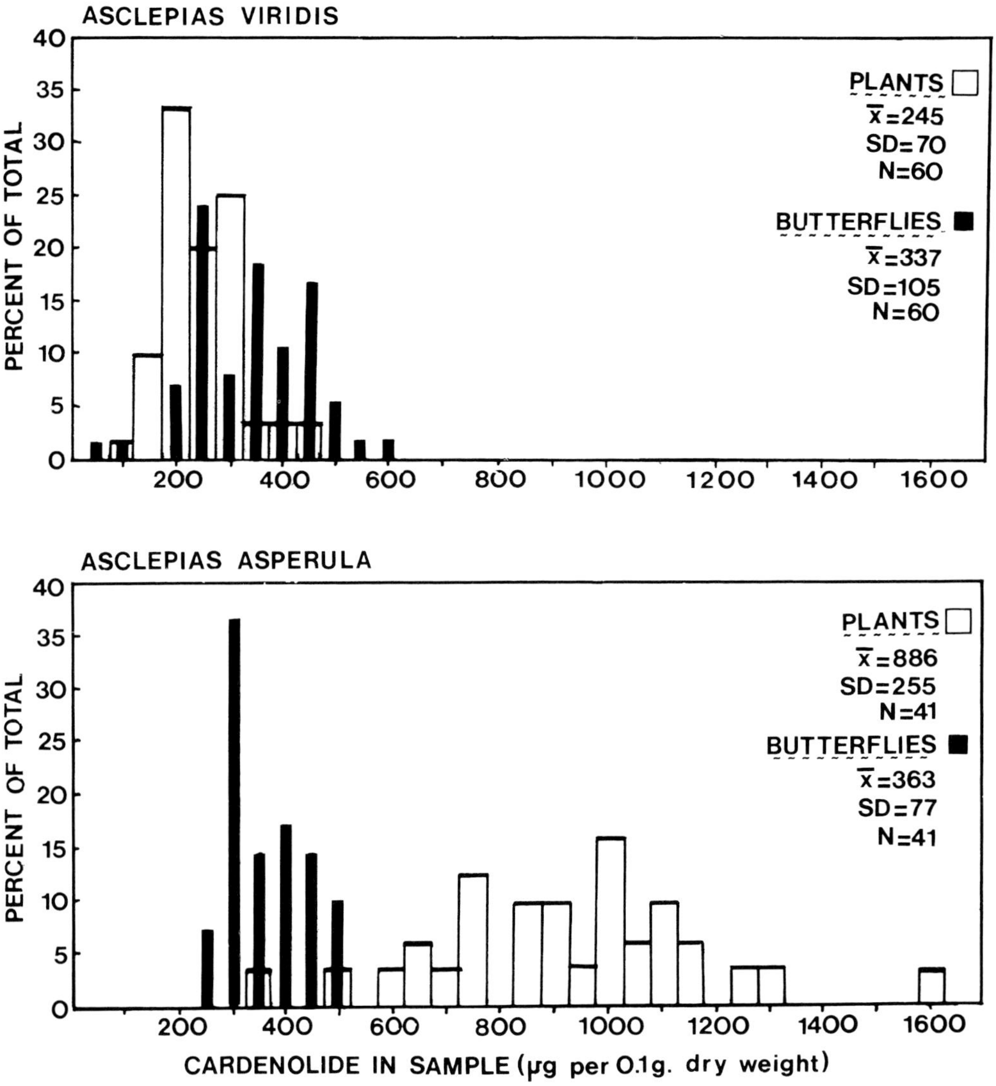

Figure 2. Frequency distributions of the gross cardenolide content of 60 *A. viridis* and 41 *A. asperula* host plants and monarch butterflies reared thereon. Concentrations are expressed as μg/0.1 g dry weight, equivalent to digitoxin. The width of each bar is 50 μg.

The patterns of variation in cardenolide concentrations of 60 WL *A. viridis* pairwise samples (Lynch and Martin, 1987) and 41 DN *A. asperula* pairwise samples (Martin and Lynch, 1988) are shown in Figure 2. The distribution of cardenolide concentrations in the *A. viridis* plants was positively skewed so that the majority of plants contained less than the mean concentration of cardenolide. The concentration distribution for the *A. asperula* plants, although negatively kurtotic, was approximately normal. Conversely, the pattern of variation in cardenolide concentration in the *A. viridis* butterflies was normally distributed, while the concentration distribution in the *A. asperula* butterflies was not. For the *A. viridis* plants and the *A. asperula* butterflies, the $\log_{10}$ transformed data did not depart significantly from normality.

The mean cardenolide concentration in the *A. viridis* butterflies was 1.4 times greater than that of their respective plants and the overall range of variation in the butterflies was also greater than that in the plants so that the lowest and the highest concentrations were found in the butterflies. In *A. asperula* we observed the opposite pattern. The mean cardenolide concentration in the *A. asperula* plants was 2.4 times greater than that of their respective butterflies, and the overall range of variation in the plants was also greater than that found in the butterflies.

Figure 3 illustrates butterfly concentrations (*Y*,

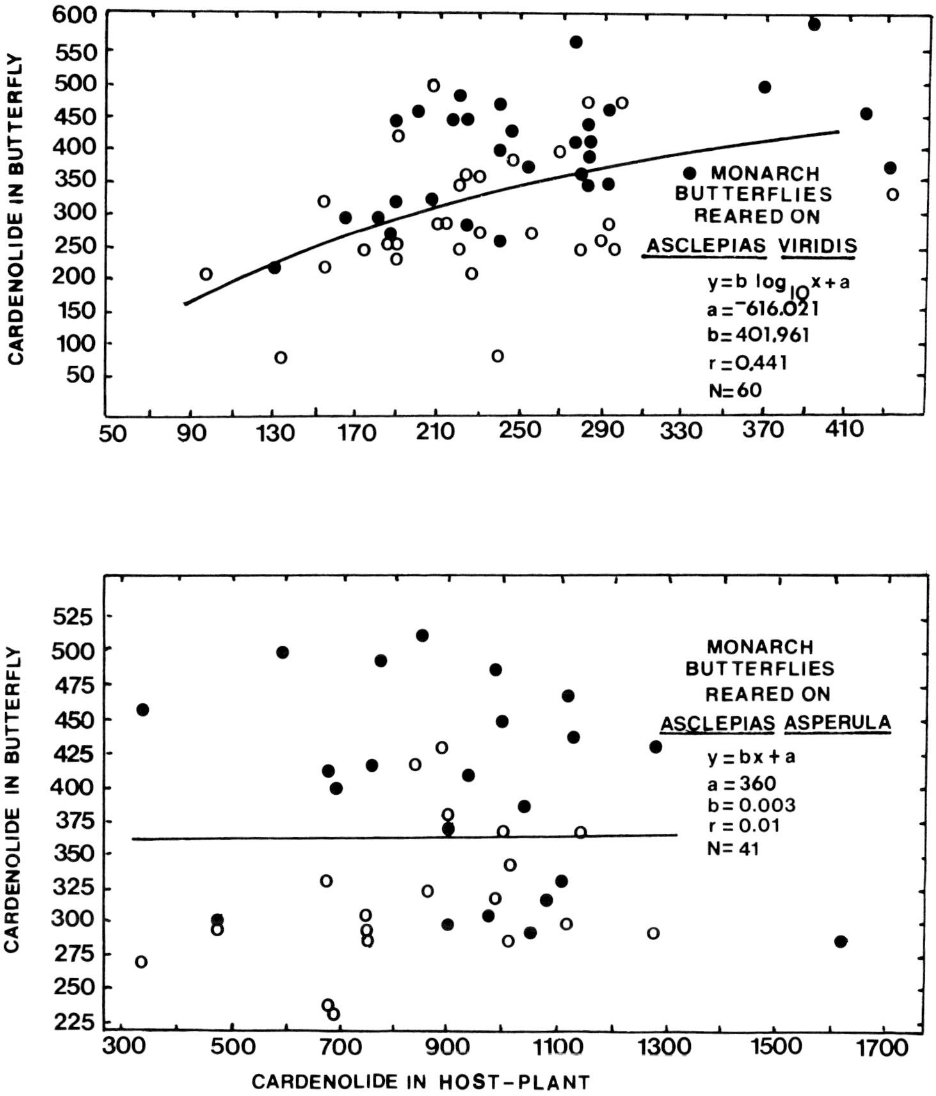

Figure 3. Cardenolide concentrations of monarch butterflies (*Y* axis) as a function of the cardenolide concentrations of their larval food plants, *A. viridis* and *A. asperula* (*X* axis). Concentrations are expressed as μg/0.1 g dry weight, equivalent to digitoxin. Each of the data points represents one plant–butterfly pair. Open circles are males and solid circles are females.

the dependent variable) plotted against respective plant concentrations (X, the independent variable) for the 60 *A. viridis* and the 41 *A. asperula* plant–butterfly pairs. In *A. viridis*, the main regression model confirmed the logarithmic relationship $Y = 401.961 \log_{10}X - 616.021$ ($r^2 = 0.441, p < 0.0001$). Both the slope and the intercept were significantly different from 0. Plotting of the residual values in both the linear and logarithmic regressions substantiated this logarithmic relationship. In *A. asperula* the regression of butterfly concentrations on their respective plant concentrations was not a significant fit to either a linear or log plot. For graphic simplicity we plotted the "best fit" linear model $Y = 0.003X + 360$ in Figure 3. The calculated regression model, $\log_{10}Y = 0.00001X + 2.539$ was not a significantly better fit ($r^2 = 0.036$, $p < 0.82$). In the latter regression the intercept was significantly different from 0 and the slope was not.

Table 3 also shows mean cardenolide concentrations of *A. viridis* and *A. asperula* plants fed on by male versus female butterflies. Differences in host plant cardenolide concentration according to the the sex of the butterfly were not significant within the *A. viridis, A. asperula,* or hybrid plant populations. Differences in mean cardenolide concentrations for male and female butterflies were significant for *A. viridis* at WL ($F = 15.98$, $p <$

Table 4. Summary of dry weights and total cardenolide content in butterflies reared on *A. viridis* Walt. and *A. asperula* subsp. *capricornu* (Woods.) Woods.

	N	Dry weight (g)			Total cardenolide (μg)		
		Mean	SD	Range	Mean	SD	Range
Wallace Lake, LA							
A. viridis[a]							
Males	29	0.185	0.026	0.126–0.232	523	179	146–990
Females	31	0.172	0.028	0.110–0.212	655	167	411–973
Both	60	0.178	0.027	0.110–0.232	591	184	146–990
Denton, TX							
A. viridis							
Males	5	0.218	0.019	0.198–0.247	631	69	543–735
Females	13	0.193	0.033	0.128–0.246	631	208	377–1,105
Both	18	0.200	0.032	0.128–0.247	631	178	377–1,105
A. asperula subsp. *capricornu*[b]							
Males	25	0.211	0.032	0.134–0.262	684	134	324–860
Females	27	0.188	0.026	0.133–0.242	745	165	417–1,084
Both	52	0.199	0.032	0.133–0.262	715	153	324–1,084
A. asperula × *viridis*							
Males	5	0.208	0.009	0.195–0.217	501	94	413–637
Females	4	0.185	0.011	0.168–0.192	702	88	581–779
Both	9	0.198	0.015	0.168–0.217	591	136	413–779

[a] Data are taken with permission from Lynch and Martin (1987).

[b] Data include 41 pairwise samples taken with permission from Martin and Lynch (1988).

0.0002), for *A. asperula* at DN ($F = 14.47$, $p < 0.0004$), and for the hybrids at DN ($F = 29.25$, $p < 0.001$) but were not significant for the 5 male and 18 female *A. viridis* butterflies at DN.

Table 4 lists the dry weights and total cardenolide contents of butterflies reared at WL and DN. The dry weights and total cardenolide contents for butterflies in each population were normally distributed. We also found no significant differences in the variances in dry weight or total cardenolide between sexes for butterflies reared on either species. The average male butterfly was slightly heavier than the female in all cases. Differences in dry weight according to sex were significant in the DN *A. asperula* butterflies ($F = 7.66$, $p < 0.007$) and the DN hybrids ($F = 12.13$, $p < 0.01$), but in the WL *A. viridis* butterflies ($F = 3.62$, $p > 0.062$) and the DN *A. viridis* butterflies ($F = 1.01$, $p > 0.33$) they were not. The mean total cardenolide content was generally greater in female butterflies than in male butterflies, but the difference was significant only in the WL *A. viridis* butterflies ($F = 7.84$, $p < 0.007$) and the DN hybrids ($F = 10.67$, $p < 0.014$).

Butterfly dry weights were not significantly correlated to their host plant cardenolide concentrations for *A. viridis* at WL ($F = 1.06$, $p > 0.30$), for *A. asperula* at DN ($F = 0.62$, $p > 0.43$), or for the DN hybrids ($F = 0.21$, $p > 0.63$). However, dry weights for *A. viridis*-reared butterflies at DN were negatively correlated to their host plant concentrations ($F = 13.73$, $p < 0.002$). Within each population, when we compared butterfly dry weights with butterfly cardenolide concentrations alone, we found significant negative correlations in *A. viridis* at WL ($r^2 = 0.11$, $p < 0.011$), and for *A. asperula* at DN ($r^2 = 0.11$, $p < 0.016$). However, in a second set of regressions, which included sex and the plant concentration as covariates, none of the independent variables or their interactions were significant predictors of butterfly dry weight within populations using type II SS (all p values > 0.09).

Although the DN *A. asperula*-reared butterflies did not contain a significantly higher mean concentration of cardenolides than those reared on the WL *A. viridis*, we found they were significantly heavier than the latter ($F = 14.33$, $p < 0.0003$) and contained significantly greater total cardenolide ($F = 13.60$, $p < 0.0004$). We found butterfly total cardenolides were highly dependent on the plant concentration in *A. viridis* ($r^2 = 0.43$, $p < 0.0006$) and that both sexes individually showed similar dependence. Based on the equation $Y = 637.33(\log_{10}X) - 920.32$, the overall model predicted that butterflies that fed on plants containing 95 μg/0.1 g cardenolide (corresponding to that in the lowest host plant measured) would contain approximately 340 μg, those that fed on 245-μg plants (the mean for all *A. viridis* host plants) would contain 602 μg, and those that fed on 432-μg plants (the highest *A. viridis* host plant measured) would contain 759 μg. In *A. asperula* we found no significant dependence of butterfly total cardenolide on the plant concentration ($p < 0.626$) and that both sexes similarly showed no dependence.

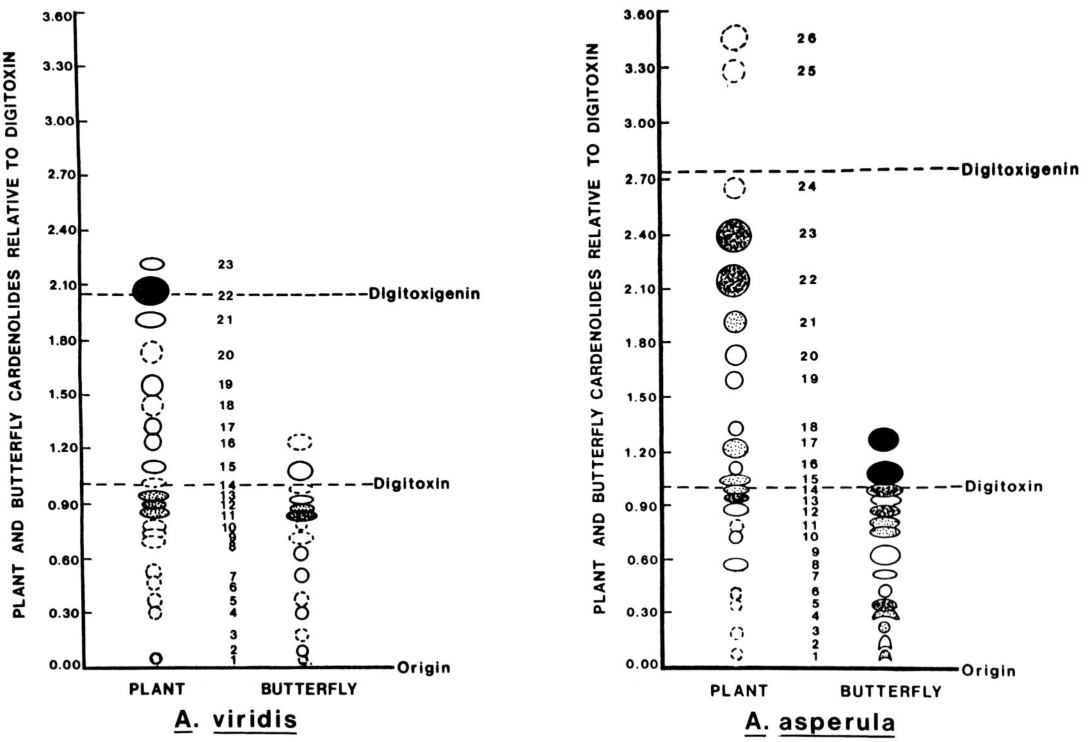

Figure 4. The mean cardenolide profiles for 25 monarch butterflies and their individual *A. viridis* and *A. asperula* host plants based on the CMF TLC solvent system. Shown are the mean R_d and SI values for the plant and butterfly cardenolide spots. SI are represented as follows: 1.00 < 1.25, dotted circle; 1.25 < 2.00, closed circle; 2.00 < 3.00, light stipple; 3.00 < 4.00, dark stipple; 4.00 < 5.00, black circle. The combined figure was produced, with permission, from individual profiles and data presented in Lynch and Martin (1987) and Martin and Lynch (1988).

TLC CARDENOLIDE PROFILES OF PLANTS AND BUTTERFLIES

Summary tables listing the means and standard deviations for cardenolide migration distances relative to digitoxin (R_d), spot intensity (SI) values, and probability of occurrence (PO) for all the spots in the 24 plant and butterfly samples separated by TLC in the chloroform–methanol–formamide (CMF) solvent system are presented for *A. viridis* in Lynch and Martin (1987) and for *A. asperula* in Martin and Lynch (1988). Figure 4 represents the mean cardenolide profiles for the 24 plants and butterflies each utilized in those TLC analyses of *A. viridis* and *A. asperula*. The relative positions and darkness of the spots were calculated from the mean R_d and SI values. The shapes and sizes of the spots were drawn from representative TLC plates and generalized in the figure.

Twenty-three cardenolides were resolved in the CMF system for *A. viridis*. Twenty-one cardenolides were discerned in the plants and 15 in the butterflies. Spots 2 and 3 were not present in any of the *A. viridis* host plants while spots 6 and 17–23 were absent in their respective butterflies. Seven plant spots (nos. 1, 4, 5, and 7–10) had a PO < 0.50, leaving 14 spots (nos. 6 and 11–23) as diagnostic. Nine spots (43%) had an R_d > 1.0 (above digitoxin) but only spot 23 was consistently above digitoxigenin. Spot 22, with an R_d essentially the same as digitoxigenin, had the highest SI. In the *A. viridis*-reared butterflies four spots had a PO < 0.50 (nos. 3, 9, 10, and 14), leaving 11 as diagnostic. The butterflies only stored cardenolides within the lowest 55% of the R_d range of cardenolides present in the plants. The mean R_d value for all plant spots was 1.04 compared with 0.62 for all butterfly spots. Only two butterfly spots (nos. 15 and 16) had an R_d > 1.0. The mean SI value for all plant spots was 1.53 compared with 1.45 for all butterfly spots.

In *A. asperula,* 16 spots were resolved in the CMF system. Twenty-two spots were discerned in the plants and 15 in the butterflies. Spots 2, 4, 7, and 9 were not present in any of the *A. asperula* host plants while spots 8, 16, and 18–26 were absent in the butterflies. In the plants there were six spots (nos. 1, 6, 14, and 24–26) that had a PO < 0.50, leaving 16 spots as diagnostic (nos. 3, 5, 8, 10–13, and 15–23). Twelve spots (55%) had an R_d > 1.0. Spots 25 and 26 were consistently above digitoxigenin, although spot 24 occasionally migrated slightly above digitoxigenin's position. Spots 8, 13, 15, 17, and 21–23 generally had the highest spot intensities. In the *A. asperula*-reared butterflies no spots had a PO < 0.50, leaving all spots detected as diagnostic. Only two spots had an $R_d > 1.0$ (13%, spots 15 and 17) while the remaining spots (87%) were below digitoxin. The butterflies only stored cardenolide within the lowest 36% of the R_d range of cardenolides present in the plants. The mean R_d value for all plant spots was 1.29 compared with 0.64 for all butterfly spots detected. The mean SI value for all plant spots was 1.96 compared with 3.15 for all butterfly spots.

DISCUSSION

PATTERNS OF HOST PLANT UTILIZATION

Milkweed species occurring in tropical or subtropical climates may remain active throughout the year. However, all *Asclepias* species that occur naturally in temperate North America are herbaceous perennials that die back in autumn and produce new stems from a subterranean hypocotyl as weather permits in spring. The date of emergence of these temperate species is critical to their exploitation by monarchs in spring. Our data on host plant utilization followed a consistent pattern at all our study sites. Initial counts of immature monarchs on *A. viridis* and *A. asperula* were highest in early or mid-April (see also Riley, this volume). These immatures are most likely to have been produced by adults remigrating from Mexican overwintering sites. At this time, most of the northern United States milkweeds are still dormant.

For those monarchs remigrating through Texas and Louisiana, *A. viridis* and *A. asperula* subsp. *capricornu* represent the first abundant and widespread milkweeds they encounter. Other milkweed species in our region are either less common or do not begin their development until after most migratory monarchs have passed northward. *Asclepias amplexicaulis* and *A. viridiflora* serve as host plants for first-generation monarchs but are relatively uncommon in our region. We have also observed large-instar monarch larvae feeding on *A. longifolia* Michx. in Louisiana but have not yet observed fourth or fifth instars on other species in Louisiana or eastern Texas.

The success of monarchs in the south-central United States appears more dependent on the number of host plants and less on survival per plant. While only a small percentage of the larvae appear to reach the fourth- and fifth-instar stages, the abundance of *A. viridis* and *A. asperula* subsp. *capricornu* suggests that larvae survive to pupate and emerge as adults over a very wide area. We believe that the majority of first-generation monarchs that continue the summer reproductive cycle of the monarch (see Malcolm et al., this volume; Cockrell et al., this volume) utilize these two species as their larval food plant.

Although we were unable to establish any negative correlation between the presence of spider and ant predators, and monarch larvae on individual host plants, we still suggest that these two groups of organisms could account for much of the decrease in numbers we observed through the first three larval instars. Few other animals were found on the plants, and ants and crab spiders increased in numbers soon after monarch eggs were laid on the plants. The significantly greater than expected numbers of first- and second-instar larvae found in the inflorescences adds support to our hypothesis that these small larvae gain some advantage from predators by feeding in hidden spaces among the young flower buds. Young umbels may also reduce abiotic mortality by providing shelter from rain and low temperatures. Plants with young, tightly packed inflorescences would therefore serve as more effective host plants than those past anthesis or in fruit. Whether there is simply a higher larval survival rate for eggs laid in the inflorescences or the larvae tend to crawl to the inflorescences after hatching on the leaves is unclear.

The breeding cycle in the southern states is short-lived. Monarch larval populations decreased rapidly at all our study sites (Table 1), disappearing entirely at LC by 7 May, at WL by 9 May, at DN and OC by 20 May, and at BT by 22 May. By early May we observed freshly eclosed adults in increasing numbers (see also Malcolm et al., this volume). In Louisiana and Texas we have seldom seen adult monarchs after late May, and immature stages are generally absent, even though, as documented in this study, many plants remain green through most of June (see also Riley, this volume).

Malcolm et al. (1987) report that two spring monarch generations are produced on milkweed host plants in Florida. No second generation was produced on host plants at our study sites. A second, and possibly a third, generation of monarchs may be produced in the Central Plains. We have observed monarchs in central Oklahoma in late May searching and mating in fields of *A. viridis* and *A. asperula* and in central Kansas during early June laying eggs on *A. speciosa* and *A. syriaca*. These adults, fresh in appearance, likely fed as larvae in Texas and Louisiana since suitable host plants farther north would not have emerged in time to produce a second generation that far north by the end of May. Although our rearing methodology did not provide exact information on when eggs were laid, nearly all the first-instar larvae we bagged on plants emerged as adults between 28 and 33 days later. *Asclepias viridis* and *A. asperula* in DN did not emerge for egg laying until mid-April. We believe the low numbers of monarchs observed on both *A. viridis* and *A. asperula* at this site are due to the emergence of the host plants after the majority of remigrants laid their eggs farther south and died.

By mid-summer, monarchs are widely dispersed throughout much of the northeastern United States and southeastern Canada. *Asclepias syriaca* and *A. speciosa* appear to be the major food source for monarch larvae during the summer breeding season (Urquhart, 1960; Fink and Brower, 1981; Seiber et al., 1986; Malcolm et al., 1987, 1989, this volume; Malcolm and Brower, 1989; Cockrell et al., this volume), although several other milkweed species are abundant and may serve as monarch host plants, particularly along the western margins of their distribution in the eastern United States (Woodson, 1954). By late summer, monarchs may expand their breeding to reinclude areas in the south-central United States since we have observed monarchs breeding on *Cynanchum laeve* in northwest Louisiana in early September (Lynch and Martin, unpubl.). By this time, however, most southern milkweeds, including *A. viridis, A. asperula, A. amplexicaulis,* and *A. longifolia,* are generally dormant and unsuitable as host plants.

QUANTITATIVE RELATIONSHIPS OF PLANT AND BUTTERFLY CARDENOLIDES

Seiber et al. (1980), Malcolm and Brower (1989), and Nelson (this volume) suggest that monarchs sequester cardenolides more efficiently when present in low concentrations, with uptake generally conforming to an inverse logarithmic relationship: $Y = a(\log_{10}X) + b$. The positive logarithmic relationship we observed between individual plant and butterfly concentrations in *A. viridis* (Fig. 3) is consistent with this hypothesis. Similar patterns were found by Brower et al. (1984a,b) in *A. californica* and *A. speciosa* butterfly–plant pairs. Both species have a lower mean cardenolide concentration than that of *A. viridis.*

That we were unable to confirm statistically any relationship in *A. asperula* between cardenolide concentrations in monarchs and their respective host plants is predictable through the same model. All the *A. asperula* plants examined contained high enough levels of cardenolide that our samples represented only the saturation portion of the curve. The calculated slopes were so small that concentrations and total cardenolide in the butterflies approximated the intercept. In *A. eriocarpa,* another cardenolide-rich milkweed, Brower et al. (1982) also found no significant correlation between butterfly and plant cardenolide concentrations.

While the mechanisms for cardenolide uptake in monarchs are unknown, these patterns could indicate that monarch cardenolide sequestration is limited by their physical capacity to store them (but see Nelson, this volume). Duffey et al. (1978) and Scudder et al. (1986) describe cardenolide sequestration in *Oncopeltus,* a common milkweed bug, as a nonsaturable, passive diffusion process in proportion to the cardenolide content in the food source and the hemolymph. That saturation does not occur in *Oncopeltus* even with cardenolide-rich host plants apparently reflects the storage capacity of a specialized inner epithelium (the dorsolateral space).

The higher mean concentration of cardenolide we observed in female butterflies compared to that in male butterflies reared on either *A. viridis* or *A. asperula* is consistent with data on butterflies reared on several other *Asclepias* species (Brower and Glazier, 1975; Brower et al., 1975, 1982, 1984b) and in wild summer breeding populations in the eastern United States (Brower et al., 1972; Brower and Moffitt, 1974). These data may largely reflect differences in the reproductive morphology between males and females. Brower (1984) cites work in which eggs produced by butterflies reared on *A. curassavica* were found to contain higher levels of cardenolide (0.6% by dry weight) than adults (0.4%). A female may lay as many as 100 (Erickson, 1973) to 400 eggs (Urquhart, 1960). Therefore, significant amounts of cardenolide could be stored in the female reproductive tract. Dixon et al. (1978) found that monarch females that had laid all their eggs were less emetic to pigeons than freshly eclosed females.

There does not appear to be a significant physiological cost for monarchs in sequestering cardenolides. Brower et al. (1972) generated considerable controversy when, using size reduction as a potential indicator, they first suggested that monarchs incur a physiological cost in sequestering cardenolides. More recent work by Brower et al. (1982, 1984a,b) and the results published in Lynch and Martin (1987) and Martin and Lynch (1988) do not directly support this hypothesis. Higher levels of sequestration in females in comparison to the significantly larger males initially confused the issue. Also, female monarchs tend to feed on host plants with higher cardenolide concentrations (Brower et al., 1982, 1984a,b; Lynch and Martin, 1987; Martin and Lynch, 1988). In these five studies butterfly dry weights were negatively correlated to their cardenolide concentrations. However, when the larger size of the males, the higher cardenolide concentrations present in females, and the generally higher host plant cardenolide concentrations of females were corrected for in the regression model, no significant correlations were obtained. Neither butterfly concentration, butterfly sex, plant concentration, nor their interactions were significant predictors of butterfly weight using type II SS. The lack of correlation may be explained by the presence of cardenolide-resistant Na,K-ATPases in the monarch (Vaughan and Jungreis, 1977), which may reduce or eliminate physiological costs of cardenolide sequestration. Moore and Scudder (1985) report similar ouabain-resistant Na,K-ATPases in *Oncopeltus,* which also incur no detectable physiological costs from the sequestration of cardenolides.

QUALITATIVE ANALYSES OF PLANT AND BUTTERFLY CARDENOLIDES

Although distinguishable, the overall TLC patterns of *A. viridis* and *A. asperula* plant and butterfly cardenolides were generally similar and also conformed to those observed by Brower et al. (1982, 1984a,b) in *A. eriocarpa, A. speciosa,* and *A. californica.* Butterfly cardenolides were sequestered in a more polar range than their host plant cardenolides, providing a pattern of cardenolides in the adult butterflies with lower R_d values.

While it is important to compare and contrast the identities of individual spots between the two species and to individual spots in profiles reported in other species, many problems exist. Relatively few milkweed cardenolide structures have been established and most cardenolides found in established TLC fingerprint profiles remain uncharacterized. In addition, different compounds may have similar polarities and fail to separate in a particular TLC solvent system [e.g., calactin and calotropin in the CMF system (chloroform–methanol–formamide, 90:6:1) and desglucosyrioside and labriformin in the EM system (ethylacetate–methanol, 97:3)]. We have also had difficulty standardizing the R_d values of digitoxigenin on our plates. Brower et al. (1982, 1984a,b) had similar difficulty. While this does not invalidate the profiles, it does cause some problems in comparing the R_f values of spots between plates, particularly those with values above digitoxin. We have reduced the discrepancies in digitoxigenin R_f values in our most recent work through better control of atmospheric humidity in the chromatographic environment.

The presence of hybridization and the subsequent introgression of *A. asperula* genes into *A. viridis* in areas of sympatry also presents a problem. At present we are unable to distinguish TLC profiles of the two species from the DN site. We have observed hybridization between *A. viridis* and *A. asperula* at other sites in Texas and Oklahoma as well (Lynch, unpubl. field records). Hybridization also occurs between other species of *Asclepias* important to the life history of the monarch. In central Kansas, *A. speciosa* and *A. syriaca* clearly hybridize, forming a variety of intermediate forms (Lynch, pers. obs.; Adams et al., 1987). Hybridization between *A. syriaca* and *A. exaltata* has been reported by Kephart et al. (1988). Robert Wyatt (pers. comm.) has also observed hybridization between *A. syriaca* and *A. purpurascens.* Although these hybrid complexes have yet to be examined in detail for cardenolides, they may actually improve our ability to predict patterns of monarch host plant utilization.

In addition to the TLC plates reported in Lynch and Martin (1987) and Martin and Lynch (1988), we have run additional plates in both the CMF and EM solvent systems, which included pooled samples of *A. viridis* and *A. asperula* plants and butterflies with the cardenolide standards desglucosyrioside, labriformin, and labriformidin. CMF plant spots 26 and 25 in *A. asperula* had similar actual and reported R_f values to labriformidin and labriformin, respectively. However, the EM system firmly establishes that desglucosyrioside is absent in *A. asperula*-reared butterflies, suggesting the absence of this entire series of epoxy cardenolides in *A. asperula.* Carolyn Nelson (pers. comm.) has suggested that acetate derivatives of various cardenolides also have high R_f values and could be mistaken for labriformin and/or labriformidin. The situation in *A. viridis* is not quite as clear because we were unable to detect either labriformidin or labriformin in the plants but both the plants and the butterflies possess spots with similar R_f values to those of desglucosyrioside in both the CMF (spot 16) and EM systems.

In Lynch and Martin (1987) we reported that the coincidence of R_f values in pairwise *A. viridis* plants and butterflies strongly indicated that the CMF spots 10 and 14 were calotropagenin and calotropin, respectively, and that spot 14 may represent calactin. We further suggested that *A. viridis* plant spots 19–22 appeared to be uscharin, uscharidin, uzarigenin, and vorusharin, respectively. All these cardenolides are members of a series derived from the genin calotropagenin (Brüschweiler et al., 1969; Roeske et al., 1976; Seiber et al., 1983, 1984). Marty and Krieger (1984) and Brower et al. (1984b) cite evidence that calactin and calotropin are the principal cardenolides stored when monarchs feed on plants containing calotropagenin glycosides and that the 3′-keto glycoside uscharidin and its 3′-spiro derivatives uscharin and voruscharin are not stored by monarchs but are metabolized by larvae to the isomeric 3′-OH derivatives calactin and calotropin.

The situation became less clear after we analyzed pairwise samples of *A. asperula* (Martin and Lynch, 1988) and then subsequently compared pooled TLC plates of both species. *Asclepias asperula* plant and butterfly profiles also contain cardenolide spots that have similar values to those in the calotropagenin series. However, in the CMF system the two most diagnostic high R_f plant spots in *A. viridis* (spots 22 and 21) each occupied different positions than the nearest plant spots (23, 22, or 21) in *A. asperula.* In the EM system the plant spots appeared more congruent between the two species but the butterfly profiles were quite different. Butterflies reared on both species produce two large and intense spots above digitoxigenin. However, these two spots in *A. viridis* consistently have different R_f values than the two in *A. asperula.*

In the absence of additional standards on our plates, we choose not to speculate further on the identity of the cardenolides in either *A. viridis* or *A. asperula.* No specific cardenolides have been isolated and identified from the *Asclepias* subgenus *Asclepiodora* and both *A. viridis* and *A. asperula* may well contain previously unidentified cardenolides. Work is currently underway in our laboratory to isolate and purify cardenolide fractions for structural identification.

CARDENOLIDE PROFILES AND MONARCH MIGRATION

Asclepias viridis and *A. asperula* subsp. *capricornu* are heavily utilized by early spring monarchs in western Louisiana and eastern Texas. We believe many second generation monarchs are also reared on *A. viridis* and *A. asperula* in Oklahoma and Kansas. Brower (1961, 1962), Cohen and Brower (1982), and Malcolm et al. (1987) report that the common sandhill milkweed, *A. humistrata* Walt., is heavily utilized by monarch remigrants in central and northern Florida. If remigrating monarchs do in fact oviposit predominantly on these three southern milkweeds as they re-enter the temperate United States, the majority of late spring and early summer breeding monarchs seen in the central and northeastern United States and Canada would possess their characteristic TLC profiles (see Malcolm et al., this volume).

All three milkweeds are relatively rich in cardenolides and produce adult monarchs that contain high levels of these compounds. These toxic first-generation monarchs could be instrumental in educating naive predators as they continue northward, establishing visual rejection by would-be predators of later-generation monarchs regardless of their taste or toxicity due to cardenolide content (Brower, 1985; Lynch and Martin, 1987; Martin and Lynch, 1988; Malcolm and Brower, 1989; Malcolm et al., this volume).

The butterflies we analyzed from *A. viridis* host plants at WL each contained between 146 and 990 μg of cardenolide with a mean of 591 μg. Those *A. viridis* butterflies analyzed from DN contained between 377 and 1,105 μg with a mean of 631 μg. *Asclepias asperula*-reared butterflies at DN contained even greater amounts of cardenolide with a mean of 715 μg and a range of 324–1,084 μg. Compared with the range of ED_{50}'s between 76 μg established for *A. curassavica* butterflies (Brower et al., 1975) and 36–39 μg established for *A. eriocarpa* butterflies (Brower et al., 1982), the average WL *A. viridis* butterfly would contain between 7.8 and 16.4 emetic units, the average DN *A. viridis* butterfly between 8.3 and 17.5 emetic units, and the average DN *A. asperula* butterfly between 9.4 and 19.9 emetic units. For all 139 butterflies analyzed for either species, the least toxic butterfly (146 μg) would contain at least 1.9 emetic units and the most toxic (1,084 μg) could contain up to 30.1 emetic units. Brower (1969) calculated that monarchs reared on *A. humistrata* contained an average of 8 emetic units.

The cardenolide contents of butterflies reared on *A. viridis, A. asperula,* and *A. humistrata* (Martin et al., in prep.) contrast sharply with contents of butterflies examined at the Michoacán, Mexico, overwintering site. Fink and Brower (1981) showed that only 10% of the butterflies sampled at the Michoacán overwintering site contained sufficient cardenolide to produce emesis in blue jays. Seiber et al. (1986) have shown that 85% of all monarchs sampled from there contain similar patterns of epoxy cardenolide glycosides, which suggests that *A. syriaca* and *A. speciosa* are the late summer food plants of monarchs in eastern North America. Both species produce monarchs containing low to moderate amounts of cardenolide (Brower et al., 1984b; Seiber et al., 1986; Malcolm et al., 1989), and the predominant glycosides in each are the polar syrioside and aspecioside, which have a significantly lower emetic potency than cardenolides of intermediate polarity often sequestered by monarchs in California (Brower et al., 1982, 1984b).

Differences in the ecological chemistry between the eastern monarchs overwintering in Michoacán and western monarchs overwintering along the California coast further emphasize the importance of cardenolides in the protection of adult monarchs. Brower and Moffitt (1974) showed that 49% of the overwintering adults in California contain more than 1 emetic dose of cardenolide. More extensive avian and rodent predation occurs at the Mexican overwintering site than occurs in overwintering sites along the California coast (Calvert et al., 1979; Fink and Brower, 1981; Fink et al., 1983; Brower et al., 1984b; Brower and Calvert, 1985; Glendinning et al., 1988; Glendinning, this volume; Arellano G. et al., this volume).

Calvert et al. (1979) argue that individuals in large overwintering colonies are better protected against birds than those in small ones. They further suggest that bird predation on the periphery of the colony would produce intense competition for nonperipheral positions, contributing to the evolution of the dense packing in the colony. By packing together into small areas in their remote, high-elevation sites, the huge overwintering colonies seem able to insulate themselves from undue predation, thus partially circumventing the need for chemical protection. Western monarchs produce smaller, more numerous colonies but maintain a higher level of chemical protection.

In contrast to the cardenolide contents of the Michoacán overwintering monarchs, Brower and Moffitt (1974) showed that nearly 55% of their September sample population of Massachusetts monarch adults were emetic. Brower et al. (1972) earlier found that approximately 40% of autumn migrating monarchs in Ontario and Massachusetts contained emetic concentrations of cardenolide but that migrating monarchs in Maryland and Florida contained significantly less cardenolide (Malcolm and Brower, 1989). Malcolm and Brower (1989) suggest that geographical differences in emetic potencies of late summer monarchs in eastern North America may reflect cardenolide loss in metabolism and excretion as they migrate southward. It is also possible that large numbers of adults reared as larvae on *A. viridis* and *A. asperula* are still present in the summer northern populations but have a lower autumn survival rate than late summer generation butterflies as they emigrate southward.

These hypotheses can be verified by matching cardenolide fingerprint profiles of wild-caught northern butterflies against established cardenolide fingerprints of butterflies reared on the predominant southern milkweeds (see Malcolm et al., this volume). As additional monarch–milkweed pairwise analyses are completed, individual fingerprints from the spring and summer breeding populations can be compared with those from autumn emigrating and Mexican overwintering populations with increasing accuracy. This should provide key data to our understanding of the monarch butterfly's remarkable life history. Determinations of the structural identity of individual cardenolides will further clarify ecological predictions based on these TLC profiles and may provide valuable insight into the infrageneric relationships of the more than 108 North American species of *Asclepias.*

LITERATURE CITED

Adams, R.P., A.S. Tomb, and S.C. Price. 1987. Investigation of hybridization between *Asclepias speciosa* and *A. syriaca* using alkanes, fatty acids, and triterpenoids. *Biochemical Systematics and Ecology* 15(4): 395–399.

Brower, L.P. 1961. Studies on the migration of the monarch butterfly I. Breeding populations of *Danaus plexippus* and *D. gilippus berenice* in south central Florida. *Ecology* 42(1):76–83.

Brower, L.P. 1962. Evidence for interspecific competition in natural populations of the monarch and queen butterflies, *Danaus plexippus,* and *D. gilippus berenice* in south central Florida. *Ecology* 43(3):549–552.

Brower, L.P. 1969. Ecological chemistry. *Scientific American* 220(2):22–29.

Brower, L.P. 1984. Chemical defence in butterflies. In *The biology of butterflies,* eds. R.I. Vane-Wright and P.R. Ackery, *Symposia of the Royal Entomological Society of London* 11:109–134. London and New York: Academic Press (1989, Princeton, NJ: Princeton University Press).

Brower, L.P. 1985. New perspectives on the migration biology of the monarch butterfly, *Danaus plexippus* (L.) In *Migration: Mechanisms and adaptive significance,* ed. M.A. Rankin, 748–785. Austin: University of Texas, *Contributions in Marine Science,* Supplement, 27.

Brower, L.P., and W.H. Calvert. 1985. Foraging dynamics of bird predators on overwintering monarch butterflies in Mexico. *Evolution* 39:852–868.

Brower, L.P., and S.C. Glazier. 1975. Localization of heart poisons in the monarch butterfly. *Science* 188: 19–25.

Brower, L.P., and C.M. Moffitt. 1974. Palatability dynamics of cardenolides in the monarch butterfly. *Nature* 249:280–283.

Brower, L.P., P.B. McEvoy, K.L. Williams, and M.A. Flannery. 1972. Variation in cardiac glycoside content of monarch butterflies from natural populations in eastern North America. *Science* 177:426–429.

Brower, L.P., M. Edmunds, and C.M. Moffitt. 1975. Cardenolide content and palatability of a population of *Danaus chrysippus* butterflies from West Africa. *Journal of Entomology (A)* 49(2):183–196.

Brower, L.P., W.H. Calvert, L.E. Hedrick, and J. Christian. 1977. Biological observations on an overwintering colony of monarch butterflies (*Danaus plexippus,* Danaidae) in Mexico. *Journal of the Lepidopterists' Society* 31:232–242.

Brower, L.P., J.N. Seiber, C.N. Roeske, and S.P. Lynch. 1982. Plant-determined variation in the cardenolide content, thin-layer chromatography profiles, and emetic potency of monarch butterflies, *Danaus plexippus* L., reared on the milkweed, *Asclepias eriocarpa,* in California. *Journal of Chemical Ecology* 8:579–633.

Brower, L.P., J.N. Seiber, C.J. Nelson, S.P. Lynch, M.P. Hoggard, and J.A. Cohen. 1984a. Plant-determined variation in the cardenolide content, thin-layer chromatography profiles, and emetic potency of monarch butterflies, *Danaus plexippus,* reared on the milkweed, *Asclepias californica,* in California. *Journal of Chemical Ecology* 10:1823–1857.

Brower, L.P., J.N. Seiber, C.J. Nelson, S.P. Lynch, and M.M. Holland. 1984b. Plant-determined variation in the cardenolide content, thin-layer chromatography profiles, and emetic potency of monarch butterflies, *Danaus plexippus,* reared on the milkweed, *Asclepias speciosa,* in California. *Journal of Chemical Ecology* 10:601–639.

Brüschweiler, F., K. Stöckel, and T. Reichstein. 1969. *Calotropis*-Glykoside, vermutliche Teilstruktur. *Helvetica Chimica Acta* 52:2276–2303.

Calvert, W.H., and L.P. Brower. 1981. The importance of forest cover for the survival of overwintering monarch butterflies (*Danaus plexippus,* Danaidae). *Journal of the Lepidopterists' Society* 35:216–225.

Calvert, W.H., and L.P. Brower. 1986. The location of monarch butterfly (*Danaus plexippus* L.), overwintering colonies in Mexico in relation to topography and climate. *Journal of the Lepidopterists' Society* 40(3):164–187.

Calvert, W.H., L.E. Hedrick, and L.P. Brower. 1979. Mortality of the monarch butterfly (*Danaus plexippus* L.): Avian predation at five overwintering sites in Mexico. *Science* 204:847–850.

Calvert, W.H., W. Zuchowski, and L.P. Brower. 1983. The effect of rain, snow, and freezing temperatures on overwintering monarch butterflies in Mexico. *Biotropica* 15:42–47.

Cohen, J.A. 1985. The differences and similarities in cardenolide contents of queen and monarch butterflies in Florida and their ecological and evolutionary implications. *Journal of Chemical Ecology* 11:85–103.

Cohen, J.A., and L.P. Brower. 1982. Oviposition and larval success of wild monarch butterflies (Lepidoptera: Danaidae) in relation to host plant size and cardenolide concentration. *Journal of the Kansas Entomological Society* 55(2):343–348.

Dixon, C.A., J.M. Erickson, D.N. Kellett, and M. Rothschild. 1978. Some adaptations between *Danaus plexippus* and its food plant, with notes on *D. chrysippus* and *Euploea core* (Insecta: Lepidoptera). *Journal of Zoology, London* 185:437–467.

Duffey, S.S., M.S. Blum, M.B. Isman, and G.G.E. Scudder. 1978. Cardiac glycosides: A physical system for their sequestration by the milkweed bug. *Journal of Insect Physiology* 24:639–645.

Erickson, J.M. 1973. The utilization of various *Asclepias* species by larvae of the monarch butterfly *Danaus plexippus*. *Psyche* 80:230–244.

Fink, L.S., and L.P. Brower. 1981. Birds can overcome the cardenolide defence of monarch butterflies in Mexico. *Nature* 291:67–70.

Fink, L.S., L.P. Brower, R.B. Waide, and P.R. Spitzer. 1983. Overwintering monarch butterflies as food for insectivorous birds in Mexico. *Biotropica* 15: 151–153.

Glendinning, J., A. Alonso Mejía, and L.P. Brower. 1988. Behavioral and ecological interactions of foraging mice (*Peromyscus melanotis*) with overwintering monarch butterflies (*Danaus plexippus*) in Mexico. *Oecologia* 75:222–227.

Kephart, S.R., R. Wyatt, and D. Parella. 1988. Hybridization in North American *Asclepias* I. Morphological evidence. *Systematic Botany* 13:456–473.

Lynch, S.P., and R.A. Martin. 1987. Cardenolide content and thin-layer chromatography profiles of monarch butterflies, *Danaus plexippus* L. and their larval host-plant milkweed, *Asclepias viridis* Walt., in northwestern Louisiana. *Journal of Chemical Ecology* 13:47–70.

Malcolm, S.B., and L.P. Brower. 1986. Selective oviposition by monarch butterflies (*Danaus plexippus* L.) in a mixed stand of *Asclepias curassavica* L. and *A. incarnata* L. in south Florida. *Journal of the Lepidopterists' Society* 40(4):255–263.

Malcolm, S.B., and L.P. Brower. 1989. Evolutionary and ecological implications of cardenolide sequestration in the monarch butterfly. *Experientia* 45:284–295.

Malcolm, S.B., B.J. Cockrell, and L.P. Brower. 1987. Monarch butterfly voltinism: Effects of temperature constraints at different latitudes. *Oikos* 49:77–82.

Malcolm, S.B., B.J. Cockrell, and L.P. Brower. 1989. The cardenolide fingerprint of monarch butterflies reared on the common milkweed, *Asclepias syriaca* L. *Journal of Chemical Ecology* 15(3):819–853.

Martin, R.A., and S.P. Lynch. 1988. Cardenolide content and thin-layer chromatography profiles of monarch butterflies, *Danaus plexippus* L. and their larval host-plant milkweed, *Asclepias asperula* subsp. *capricornu* (Woods.) Woods., in north central Texas. *Journal of Chemical Ecology* 14:295–318.

Marty, M.A., and R.I Krieger. 1984. Metabolism of uscharidin, a milkweed cardenolide, by tissue homogenates of monarch butterfly larvae, *Danaus plexippus* L. *Journal of Chemical Ecology* 10:945–956.

Moore, L.V., and G.G.E. Scudder. 1985. Selective sequestration of milkweed (*Asclepias* spp.) cardenolides in *Oncopeltus fasciatus*. *Journal of Insect Physiology* 1:27–33.

Morse, D.H., and R.S. Fritz. 1982. Experimental and observational studies of patch choice at different scales by the crab spider *Misumena vatia*. *Ecology* 63:172–182.

Roeske, C.N., J.N. Seiber, L.P. Brower, and C.M. Moffitt. 1976. Milkweed cardenolides and their comparative processing by monarch butterflies (*Danaus plexippus* L.). *Recent Advances in Phytochemistry* 10: 93–167.

Scudder, G.G.E., L.V. Moore, and M.B. Isman. 1986. Sequestration of cardenolides in *Oncopeltus fasciatus*: Morphological and physiological adaptations. *Journal of Chemical Ecology* 12:1171–1187.

Seiber, J.N., P.M. Tuskes, L.P. Brower, and C.J. Nelson. 1980. Pharmacodynamics of some individual milkweed cardenolides fed to larvae of the monarch butterfly (*Danaus plexippus* L.). *Journal of Chemical Ecology* 6:321–339.

Sieber, J.N., S.M. Lee, and J.M. Benson. 1983. Cardiac glycosides (cardenolides) in species of *Asclepias* (Asclepiadaceae). In *Handbook of natural toxins, Vol. 1. Plant and Fungal Toxins*, eds. R.F. Keeler and A.T. Tu, 43–83. Amsterdam and New York: Marcel Dekker.

Seiber, J.N., S.M. Lee, and J.M. Benson. 1984. Chemical characteristics and ecological significance of cardenolides in *Asclepias* (milkweed) species. In *Isopentenoids in plants: Biochemistry and function*, eds. W.D. Nes, G. Fuller, and L. Tsai, 563–588. Amsterdam and New York: Marcel Dekker.

Seiber, J.N., L.P. Brower, S.M. Lee, M.M. McChesney, H.T.A. Cheung, C.J. Nelson, and T.R. Watson. 1986. Cardenolide connection between overwintering monarch butterflies from Mexico and their larval food plant, *Asclepias syriaca*. *Journal of Chemical Ecology* 12:1157–1170.

Urquhart, F.A. 1960. *The monarch butterfly*. Toronto: University of Toronto Press, 361 pp.

Urquhart, F.A. 1976. Found at last: The monarch's winter home. *National Geographic* 150:160–173.

Urquhart, F.A. 1987. *The monarch butterfly: International traveler*. Chicago: Nelson-Hall, 232 pp.

Urquhart, F.A., and N.R. Urquhart. 1976. The overwintering site of the eastern population of the monarch butterfly (*Danaus p. plexippus*: Danaidae) in southern Mexico. *Journal of the Lepidopterists' Society* 30:153–158.

Urquhart, F.A., and N.R. Urquhart. 1978. Autumnal migration routes of the eastern population of the monarch butterfly (*Danaus p. plexippus* L.; Danaidae; Lepidoptera) in North America to the overwintering site in the neovolcanic plateau of Mexico. *Canadian Journal of Zoology* 56:1759–1764.

Vaughan, G.L., and A.M. Jungreis. 1977. Insensitivity of lepidopteran tissues to ouabain: Physiological mechanisms for protection from cardiac glycosides. *Journal of Insect Physiology* 23:585–589.

Woodson, R.E., Jr. 1954. The North American species of *Asclepias* L. *Annals of the Missouri Botanical Garden* 41:1–211.

Young, A.M. 1982. An evolutionary-ecological model of the evolution of migratory behavior in the monarch butterfly and its absence in the queen butterfly. *Acta Biotheoretica* 31:214–237.

REJECTION OF *APOCYNUM ANDROSAEMIFOLIUM* AND *A. SIBIRICUM* (APOCYNACEAE) AS FOOD PLANTS BY LARVAE OF *DANAUS PLEXIPPUS*: REFUTATION OF EARLY ACCOUNTS

Susan Sullivan Borkin

ABSTRACT. A preliminary investigation was conducted to examine the acceptability of *Apocynum androsaemifolium* and *Apocynum sibiricum* (Apocynaceae) as larval food plants for *Danaus plexippus* (L.). For the purpose of this study, an acceptable larval food plant was defined as a species on which larvae feed and develop normally to produce viable adults. In the laboratory *D. plexippus* larvae, with and without prior plant feeding experience, were offered one of the two *Apocynum* spp. in no-choice feeding experiments. The results of the feeding tests showed that all larvae tested totally rejected the apocynaceous plants they were offered in the experiments. This finding refutes various published accounts that *Apocynum* spp. serve as larval food plants for *D. plexippus.*

INTRODUCTION

Life-history accounts for the monarch butterfly, *Danaus plexippus* (L.), date back to the 1820s or earlier (Tietz, 1972). In one account, Saunders (1873) states that *D. plexippus* lay their eggs "on the bitter root (*Apocynum androsaemifolium*)" in addition to milkweeds (*Asclepias* spp.), and in a later paper, he also reports "the food plant of the caterpillar is Milkweed, usually the common species (*Asclepias syriaca*) but often other kinds of Milkweed and even Dogbane" (Saunders, 1932).

Bitter root (*A. androsaemifolium* L.), also called spreading dogbane, is a member of the Apocynaceae, a family closely related to the Asclepiadaceae, the milkweed family (Cronquist, 1981). Both plant groups are characterized as having milky latex and distasteful, poisonous properties due to the presence of highly toxic secondary compounds, particularly cardiac glycosides (Gleason, 1968; Gibbs, 1974). The geographic range of *A. androsaemifolium* extends from Mexico throughout most of North America, much the same as for *D. plexippus.*

Dogbane (*A. androsaemifolium*) or dogbanes (*Apocynum* spp.) have been included in numerous references as larval food plants of *D. plexippus* (Scudder, 1889; Klots, 1951; Tietz, 1952; Brown et al., 1954–1957; Ehrlich and Ehrlich, 1961; Mitchell and Zim, 1962; Kimball, 1965; Pyle, 1981; Ackery and Vane-Wright, 1984; Opler and Krizek, 1984), apparently based on the early reports by Saunders. Additional observations substantiating that *D. plexippus* either oviposit or feed as larvae on these plants have not been recorded in the literature.

The purpose of this preliminary study was to examine the acceptability of *A. androsaemifolium* and *Apocynum sibiricum* Jacq., sometimes called Indian hemp, as larval food plants for *D. plexippus.* The term "acceptability" is used in this study to mean both that larvae feed on a plant species and that the plants support normal larval development to produce viable adults.

MATERIALS AND METHODS

Preliminary no-choice feeding experiments were conducted with a sample of 10 *D. plexippus* larvae in various instars collected from *A. syriaca* L. in the vicinity of Milwaukee, Wisconsin, U.S.A., during June and July of 1985. Each larva was placed individually into a clear plastic bag and supplied with a fresh slip of *A. androsaemifolium* with the cut stem in water. All bags were kept under similar laboratory conditions at room temperature with natural, indirect lighting. Each bag was examined daily for feeding, evidenced by leaf damage and fresh feces, for as long as a larva survived.

More extensive no-choice feeding experiments were conducted in June through August of 1986 with 241 *D. plexippus* larvae. In these tests, each larva was offered one of two plants, *A. androsaemifolium* or *A. sibiricum,* for a time period of between 24 (minimum) and 48 hr (maximum) using the same plastic bag technique and laboratory conditions as in 1985. As a control, after the no-

Table 1. No-choice tests of *D. plexippus* larvae for feeding on *A. androsaemifolium* and *A. sibiricum*.

Plant offered	No. larvae per instar tested						Total	No. larvae that fed
	1*	1	2	3	4	5		
A. androsaemifolium (June–July 1985)	0	4	1	2	1	2	10	0
A. androsaemifolium (June–August 1986)	54	32	16	6	14	50	172	0
A. sibiricum (June–August 1986)	20	14	5	2	8	20	69	0
Total	74	50	22	10	23	72	251	0

* Collected as eggs; no prior feeding on *Asclepias*.

choice test was completed, each larva was maintained in the laboratory and provided with fresh cuttings of both *A. syriaca* and *A. androsaemifolium* until development was completed. Larvae of various instars were tested, including 74 newly hatched larvae that had no prior plant feeding experience. Each larva was used in a no-choice test only once, and data from parasitized larvae were excluded. Eggs and larvae used in these experiments were collected from *A. syriaca, A. incarnata* L., or *Asclepias verticillata* L. in southeastern Wisconsin. In addition, numerous *A. androsaemifolium* and *A. sibiricum* plants were searched to determine if *D. plexippus* eggs or larvae were also present on these plants under natural conditions.

RESULTS

In all of the no-choice feeding tests, no evidence of feeding could be detected upon close examination of the *A. androsaemifolium* and *A. sibiricum* cuttings offered to the *D. plexippus* larvae, and no fresh fecal material was found in any of the bags containing these plants (Table 1). Results were identical for larvae with prior feeding experience on *Asclepias* and for those with no prior plant feeding. Typically, larvae would actively wander over the surfaces of the plants offered and the inner walls of the plastic bags, eventually coming to rest on one or the other substrate with no obvious preference.

In the 1985 experiments, all 10 *D. plexippus* larvae died within 5 days after the start of each test. In 1986, under the control conditions, larvae began to feed immediately upon contact with *A. syriaca* leaves. In one instance, a larva fed first on a drop of milky sap that it encountered before reaching the *A. syriaca* plant cutting. In this two-choice situation, *D. plexippus* larvae completed development successfully feeding on *A. syriaca* and, as in the no-choice experiments, they totally rejected the *A. androsaemifolium* provided. No *D. plexippus* eggs or larvae were found on *Apocynum* spp. at sites where plant material was collected for use in the experiments, although adults were observed feeding at the blossoms.

DISCUSSION

In nature, the larval food plants utilized by *D. plexippus* are selected by the females' choice of plants for oviposition, by larval acceptance or rejection of plants for feeding, and by the suitability of plants for sustaining complete, normal development of *D. plexippus*. Many factors determine and affect each of these components.

Although no eggs or larvae of *D. plexippus* were found on the *Apocynum* plants searched, the question of whether or not female *D. plexippus* will oviposit on *Apocynum* spp., and, if so, with what frequency, would be worth further investigation. Dixon et al. (1978) found that under crowded greenhouse conditions *D. plexippus* females initially chose a suitable plant for oviposition, but then laid eggs on plants and soil in the vicinity without further testing. In contrast, Dethier (1959) found *D. plexippus* females to be highly precise in selecting *Asclepias* for oviposition under natural conditions.

Studies of the mechanisms used by lepidopterous larvae to recognize their food plants have shown that neither nutritive plant compounds nor individual secondary plant chemicals are responsible for determination of larval food-plant recognition (Dethier, 1982). Olfactory and gustatory sensory receptors have been identified and studied, but precisely how the myriad of structural and chemical properties of a plant are perceived and evaluated by larvae remains largely unknown (Schoonhoven, 1973; Hanson, 1983; Miller and Strickler, 1984). No chemical analysis was made of the plant material used in this study; however, Gibbs (1974) reported cardiac glycosides to be present in *Apocynum* spp. as well as in species of *Asclepias*. Assuming that cardiac glycosides were present in the two *Apocynum* spp. used in the feeding experiments, the results confirm that presence of these secondary chemicals is not sufficient to elicit a feeding response in *D. plexippus*.

Prior larval feeding experience on one food-plant species may, in some herbivores, induce preference for that plant over others that are equally or more suitable (Hanson, 1983). In the present study, no difference was found between first-instar *D. plexippus* larvae that had fed on *Asclepias* spp. prior to testing and those that had no prior feeding experience. Both groups rejected the two *Apocynum* spp. offered. Similar results were obtained by L.P. Brower (pers. comm.) in the early 1970s at Amherst, Massachusetts, U.S.A. He placed 100 *D. plexippus* eggs into containers with *A. androsaemifolium*, and all of the resulting larvae died in the first instar without feeding.

Some danaid butterflies, especially the Euploeini, are known to utilize apocynaceous larval food plants (Ackery and Vane-Wright, 1984). In feeding experiments conducted by Dixon et al. (1978), first- and fourth-instar *D. plexippus* larvae refused to feed on *Nerium oleander* L., an apocynaceous food plant of *Euploea core* (Cr.). However, Young (1981) demonstrated with just four larvae in a no-choice feeding test using *Prestonia* sp. (Apocynaceae) that *D. plexippus* larvae possess the "chemo-behavioral sensory mechanisms" to feed on an apocynaceous plant, but since the larvae did not complete development, he suggests they may have evolutionarily lost the ability to process this plant material. According to Edgar (1982), *Prestonia* and other members of the tribe Parsonsieae may be similar to the ancestral danaid food plant.

Care must be taken to distinguish between plants that are used only for oviposition, or those that larvae will feed but not survive on, and plants that are acceptable larval food plants, i.e., a food resource capable of sustaining normal larval development to produce reproductively viable adults in nature (Shields et al., 1969; Kitching and Zalucki, 1983). It is unclear what Saunders (1873, 1932) observed that led him to conclude that *A. androsaemifolium* was a food plant of *D. plexippus.* For example, his record provides no information on whether he found larvae feeding on *A. androsaemifolium* or witnessed oviposition on that plant. If Saunders observed *D. plexippus* females ovipositing on *A. androsaemifolium,* results from the present study indicate that it is unlikely the resulting larvae would feed on the plants. Since *Apocynum* spp. sometimes grow in close proximity to *Asclepias* spp., discounting the possibility of an oviposition error, Saunders could have encountered a *D. plexippus* larva that wandered from an *Asclepias* plant onto a neighboring *A. androsaemifolium. Danaus plexippus* larvae are mobile and may leave plants on which they have been feeding, particularly in the last instar (Rawlins and Lederhouse, 1981; Borkin, 1982). It is also possible that the populations of *A. androsaemifolium* at Saunders' study sites differed in some critical aspect from the Wisconsin plants used in this study and the Massachusetts plants used by Brower in the 1970s, making them acceptable to *D. plexippus* larvae.

Comparative chemical analysis or feeding experiments with *A. androsaemifolium* plants from the Allegany State Park area, observed by Saunders (1932), would be informative. However, given the broad geographic range of *A. androsaemifolium* and numerous studies of *D. plexippus* by amateurs and professionals, at least one additional published account documenting the use of *A. androsaemifolium* as an oviposition or larval food plant of *D. plexippus* might be expected. None was found in a search of the literature conducted for this study. Presumably, later authors included *Apocynum* spp. as larval food plants of *D. plexippus* in their works based on the early reports by Saunders. Results from the simple feeding experiments conducted in this study question the validity of *A. androsaemifolium* and *A. sibiricum* as acceptable food plants for *D. plexippus* larvae.

SUMMARY

This study has shown that *D. plexippus* larvae reject *A. androsaemifolium* and *A. sibiricum* as food plants, even in a no-choice situation where the alternatives are feed or starve to death. The results refute earlier published accounts, which have been carried through the secondary literature, that these apocynaceous plants serve as larval food plants for *D. plexippus.*

ACKNOWLEDGMENTS

This study was funded by Friends of the Museum, Inc., of the Milwaukee Public Museum with support from Dr. M.K. Starr. A portion of the plants and animals used were collected at the University of Wisconsin–Milwaukee Field Station. Dr. M. Ficken and Dr. J. Reinartz kindly provided access to the Field Station, and A. Brower and R. Sullivan helped with collecting. I gratefully acknowledge Dr. L. Brower for discussions stimulating the development of this project and for sharing his unpublished observations. Special thanks also go to Dr. A. Young for generous assistance, including review of the manuscript.

LITERATURE CITED

Ackery, P.R., and R.I. Vane-Wright. 1984. *Milkweed butterflies: Their cladistics and biology.* London: British Museum (Natural History); and Ithaca, NY: Cornell University Press, 425 pp.

Borkin, S.S. 1982. Notes on shifting distribution patterns and survival of immature *Danaus plexippus* (Lepidoptera: Danaidae) on the food plant *Asclepias syriaca. The Great Lakes Entomologist* 15:199–206.

Brown, F.M., D. Eff, and B. Rotger. 1954–1957. *Colorado butterflies.* Denver, CO: Denver Museum of Natural History, *Proceedings* 3–7, 368 pp.

Cronquist, A. 1981. *An integrated system of classification of flowering plants.* New York: Columbia University Press, 1262 pp.

Dethier, V.G. 1959. Egg-laying habits of lepidoptera in relation to available food. *The Canadian Entomologist* 91:554–561.

Dethier, V.G. 1982. Mechanism of host-plant recognition. *Entomologica Experimentalis et Applicata* 31:49–56.

Dixon, C.A., J.M. Erickson, D.N. Kellett, and M. Rothschild. 1978. Some adaptations between *Danaus plexippus* and its food plant, with notes on *Danaus chrysippus* and *Euploea core* (Insecta: Lepidoptera). *Journal of Zoology, London* 185:437–467.

Edgar, J.A. 1982. Pyrrolizidine alkaloids sequestered by Solomon Island Danaine butterflies. The feeding preferences of the Danainae and Ithomiinae. *Journal of Zoology, London* 196:385–399.

Ehrlich, P.R., and A.H. Ehrlich. 1961. *How to know the butterflies.* Dubuque, IA: Wm.C. Brown Company, 262 pp.

Gibbs, R.D. 1974. *Chemotaxonomy of flowering plants,* 4 vols. Montreal and London: McGill-Queen's University Press, 2372 pp.

Gleason, H.A. 1968. *The new Britton and Brown illustrated flora of the northeastern United States and Canada,* Vol. 3. New York: Hafner Publishing, 595 pp.

Hanson, F.E. 1983. The behavioral and neurophysiological basis of food plant selection by lepidopterous larvae. In *Herbivorous insects,* ed. S. Ahmad, 3–23. New York: Academic Press.

Kimball, C.P. 1965. *The Lepidoptera of Florida.* Gainesville: Florida Department of Agriculture, 363 pp.

Kitching, R.L., and M.P. Zalucki. 1983. A cautionary note on the use of oviposition records as larval food plant records. *Australian Entomological Magazine* 10:64–66.

Klots, A.B. 1951. *A field guide to the butterflies of North America, east of the Great Plains.* Boston: Houghton Mifflin Company, 349 pp.

Miller, J.R., and K.L. Strickler. 1984. Finding and accepting host plants. In *Chemical ecology of insects,* eds. W.J. Bell and R. T. Cardé, 127–157. Sunderland, MA: Sinauer Associates.

Mitchell, R.T., and H.S. Zim. 1962. *Butterflies and moths.* New York: Golden Press, 160 pp.

Opler, P.A., and G.O. Krizek. 1984. *Butterflies east of the Great Plains, an illustrated natural history.* Baltimore, MD: The Johns Hopkins University Press, 273 pp.

Pyle, R.M. 1981. *The Audubon Society field guide to North American butterflies.* New York: Alfred A. Knopf, 916 pp.

Rawlins, J.E., and R.C. Lederhouse. 1981. Developmental influences of thermal behavior on monarch caterpillars (*Danaus plexippus*): An adaptation for migration (Lepidoptera: Nymphalidae: Danainae). *Journal of the Kansas Entomological Society* 54: 387–408.

Saunders, A.A. 1873. On some of our common insects. Paper No. 1. *The Canadian Entomologist* 5:4–8.

Saunders, A.A. 1932. *The butterflies of Allegany State Park.* New York State Museum Handbook 13. Albany: The University of the State of New York, 270 pp.

Schoonhoven, L.M. 1973. Plant recognition by lepidopterous larvae. In *Insect/plant relationships,* ed. H.F. van Emden, 87–99. New York: John Wiley & Sons.

Scudder, S.H. 1889. *The butterflies of the eastern United States and Canada with special reference to New England,* 3 vols. Cambridge, MA: publ. by author, 1958 pp.

Shields, O., J.F. Emmel, and D.E. Breedlove. 1969. Butterfly larval records and a procedure for reporting foodplants. *The Journal of Research on the Lepidoptera* 8:21–36.

Tietz, H.M. 1952. *The Lepidoptera of Pennsylvania: A manual.* University Park: Pennsylvania Agricultural Experiment Station, Pennsylvania State College School of Agriculture, 194 pp.

Tietz, H.M. 1972. *An index to the described life histories, early stages and hosts of the Macrolepidoptera of the continental United States and Canada,* 2 parts. Sarasota, FL: A.C. Allyn, 1041 pp.

Young, A.M. 1981. On the evolutionary distance between asclepiadaceous-feeding danaida and apocynaceous-feeding ithomiids. *The Journal of Research on the Lepidoptera* 18:251–254.

A REASSESSMENT OF THE MIMICRY RELATIONSHIP AMONG VICEROYS, QUEENS, AND MONARCHS IN FLORIDA

David B. Ritland and Lincoln P. Brower

ABSTRACT. This paper outlines our continuing reassessment of a classic example of Batesian mimicry—the case of the viceroy butterfly, *Limenitis archippus.* Viceroys are traditionally viewed as edible, Batesian mimics of two purportedly distasteful danaids, the monarch (*Danaus plexippus*) and the Florida queen (*Danaus gilippus berenice*). However, existing evidence suggests that the viceroy is not completely palatable; furthermore, queens may be less protected than previously believed. To clarify the nature of the mimicry relationship between these two butterflies, we directly investigated their comparative palatability in a laboratory experiment using captive red-winged blackbirds (*Agelaius phoeniceus*). Our preliminary data indicate that both viceroys and Florida queens are somewhat unpalatable to birds, thereby challenging the traditional view that viceroys are Batesian mimics of queens.

Instead, the mimicry relationships among viceroys, queens, and monarchs in Florida may be dynamic, shifting along a continuum from Batesian to Müllerian mimicry. Three factors could contribute to this flexible mimicry situation: (1) queens and monarchs feed on a variety of food plants, and therefore vary in palatability both seasonally and geographically; (2) the relative frequencies of both danaines and viceroys change spatiotemporally; and (3) viceroys in Florida populations vary in wing color from light orange to dark mahogany-brown; some resemble monarchs, some resemble queens, and some are intermediate. We hypothesize that these three factors significantly influence the way in which predators perceive and react to commingling populations of viceroys, monarchs, and queens, thereby creating a complex and labile mimicry situation.

INTRODUCTION

The viceroy butterfly, *Limenitis archippus* (Cramer), is the premier example of Batesian mimicry among North American butterflies. Viceroys vary geographically in wing color and resemble distinct danaid butterflies in different regions. The familiar, bright orange northern viceroy, *Limenitis archippus archippus,* ranges over most of the eastern United States and is thought to gain protection against avian predators by its striking resemblance to the widespread monarch, *Danaus plexippus* (L.). In the southeastern United States, the viceroy's wing color darkens along a latitudinal cline, culminating in the deep mahogany-brown hue of the Florida viceroy, *Limenitis archippus floridensis* Strecker, which is presumed to mimic the chocolate-brown Florida queen, *Danaus gilippus berenice* (Cramer). The change in viceroy phenotype in the southeastern United States is conventionally attributed to a switch in mimicry, due to the greater abundance and consistent availability of the queen in the southern states (e.g., Poulton, 1909). Brower (1958a,b) demonstrated the effectiveness of viceroy mimicry; she concluded that captive birds rejected light northern viceroys after experience with monarchs and, likewise, avoided dark Florida viceroys after previous exposure to queens.

Historically, viceroys have been referred to as palatable Batesian mimics of the two danaids. Both queens and monarchs have been presumed to be distasteful to birds (e.g., Poulton, 1909; Brower, 1958b), and in fact both can sequester toxic cardiac glycosides (cardenolides) from their larval milkweed food plants (Asclepiadaceae) (Brower et al., 1975; Roeske et al., 1976). However, Florida queens may be less well defended than previously thought, partly because they are less efficient than monarchs at sequestering cardenolides (Brower et al., 1975), and partly because many Florida queen populations appear to feed primarily on low-cardenolide larval food plants (e.g., Kimball, 1965). Moreover, experimental feeding studies have hinted that the vice-

Table 1. **Comparative acceptability of northern viceroys (*L. a. archippus*) and swallowtails (*Papilio* spp.) to four wild-captured scrub jays (data from J. Brower, 1958a).**

Scrub jay behavior	Northern viceroy	Swallow-tail
% attacked	90.0	100.0
% eaten (of butterflies attacked)	66.7	100.0
Mean time to attack (sec)	8.5	3.8
Aversion behavior frequency	25%	0%

roy is not entirely palatable to captive birds (Brower, 1958a,b; Platt et al., 1971). These findings cast doubt on the classic interpretation of viceroy–danaid mimicry, particularly the relationship between viceroys and queens in Florida.

Therefore, we have initiated laboratory and field studies directed at reanalyzing the Florida viceroy's mimicry relationships; this paper presents preliminary findings from our continuing investigation. After reviewing published work on viceroy and queen palatability, we present data from a new laboratory experiment designed to assess the comparative acceptability of the two butterflies to captive birds. We then briefly discuss the chemical basis of distastefulness in these butterflies. Finally, we offer a new interpretation of a *dynamic* mimicry relationship among viceroys, queens, and monarchs in Florida, in which viceroys may play a changing role in the mimicry relationship, switching from Müllerian co-models to Batesian mimics in different locales and seasons.

PREVIOUS EXPERIMENTAL EVIDENCE OF VICEROY DISTASTEFULNESS

Brower (1958a) reported that northern viceroys (*L. a. archippus*) were somewhat distasteful to captive Florida scrub jays, *Aphelocoma coerulescens* (Bosc). Her data (Table 1) indicate that four birds (which had no prior laboratory experience with either monarchs or viceroys) readily attacked bright orange northern viceroys but rejected 33% after tasting them. In addition, the scrub jays did not attack viceroys as quickly as they did palatable swallowtail butterflies (*Papilio* spp.), and they exhibited significantly more feather-ruffling behavior and agitation when presented with viceroys.

Brower (1958b) also reported the reactions of three of these same scrub jays to Florida viceroys (*L. a. floridensis*). These data (Table 2) must be interpreted with caution—because the birds had previous experience with monarchs and northern viceroys, their rejection of Florida viceroys could have been based on visual cues (wing coloration patterns) rather than palatability per se. However, the scrub jays ate only 7% of those viceroys that they tasted, they attacked them more hesitantly than the palatable *Papilio* species, and they frequently exhibited feather-ruffling behavior in response to viceroys. (Brower reported that one additional scrub jay, which did not have prior laboratory exposure to monarchs or viceroys, treated *L. a. floridensis* as a palatable butterfly, eating 97% of those attacked). Thus, although hampered by the problem of previous experience, Brower's data do suggest that neither *L. a. archippus* nor *L. a. floridensis* are completely acceptable to scrub jays.

Table 2. **Comparative acceptability of Florida viceroys (*L. a. floridensis*) and swallowtails (*Papilio* spp.) to three captive scrub jays with previous monarch and northern viceroy experience (data from J. Brower, 1958b). These data must be interpreted cautiously, because previous experience may have caused the birds to reject Florida viceroys based on visual cues and not unpalatability.**

Scrub jay behavior	Florida viceroy	Swallow-tail
% attacked	12.3	100.0
% eaten (of butterflies attacked)	7.1	100.0
Mean time to attack (sec)	14.0	2.0
Frequency of feather-ruffling behavior	39%	0%

In another experiment, Platt et al. (1971) offered northern viceroys (*L. a. archippus*) that they had reared on weeping willow (*Salix babylonica* L.) to captive blue jays (*Cyanocitta cristata* Oberholser). They found that 15 blue jays (which had no previous experience with monarchs or viceroys in the laboratory) rejected 47% of northern viceroys on sight. In addition, they found (A. Platt, pers. comm.) that the birds rejected nearly half of the viceroys that they did attack after tasting them (Table 3). Like Brower's scrub jays, these blue jays commonly responded to viceroys with head-shaking and feather-ruffling.

PALATABILITY AND THE CARDENOLIDE DEFENSE OF THE FLORIDA QUEEN

Queens, like monarchs and other danaids, have long been considered to be protected against avian predators (e.g., Poulton, 1909). However, laboratory investigations have suggested that Florida queens are less defended than monarchs. Data extracted from J. Brower's mimicry studies (1958a,b) indirectly compare the reactions of two groups of captive scrub jays presented with monarchs and queens (Table 4). The scrub jays attacked queens more frequently and more quickly than they did monarchs; in addition, they ate more of the queens that they did attack and exhibited fewer behavioral expressions of unacceptability (e.g., feather-ruffling) in response to queens. Differences in the previous experience of the two bird groups preclude strong inference, but the data do suggest that queens were less protected than were monarchs.

Queens are apparently less distasteful than mon-

Table 3. Comparative acceptability of northern viceroys and three other butterflies to captive blue jays (unpubl. data, from study of Platt et al., 1971).

Butterfly	No. of birds	No. of butterflies presented	% eaten (of those attacked)
Northern viceroy (*L. a. archippus*)	15	30	56.2
Red-spotted purple (*L. a. astyanax*)	8	16	100.0
Tiger swallowtail (*P. glaucus*)	31	341	95.8
Monarch (*D. plexippus*)	8	80	15.2

archs because they are less efficient at storing toxic cardenolides sequestered from their larval milkweed food plants. Brower et al. (1975) and Cohen (1985) showed that queens reared on the toxic sandhill milkweed (*Asclepias humistrata*) had mean cardenolide *concentrations* only 67–77% those of monarchs reared on the same milkweed (hence, they are presumably less bitter tasting than monarchs). In addition, because queens are smaller than monarchs, they have a lower cardenolide *content*—only 30–42% as much as monarchs (Table 5); evidently, therefore, queens are also less emetic than monarchs. The possibility that queens compensate for these reduced quantities by storing a more potent complement of cardenolides than monarchs was eliminated by Cohen (1985), who reared queens and monarchs on the same food-plant species and discovered via chromatography that they stored virtually identical complements of cardenolides.

Cohen (1985) further suggested that natural queen populations in three Florida locales were poorly defended—less than one-third of the butterflies were estimated to contain sufficient cardenolide to induce emesis in a blue jay predator. In fact, 17–57% of the butterflies (depending on locale) contained no measurable cardenolide (Cohen, 1985). Mathematical models developed by Brower et al. (1970) and Pough et al. (1973) predict that such scarcity of emetic individuals will lower the protection level of the entire local queen population (however, queens may rely on other defensive chemicals in addition to cardenolides; see "Perspectives on Chemical Defense in Queens and Viceroys," below). The queen's relatively low cardenolide-storing capacity is compounded by the fact that throughout much of Florida, queen larvae appear to feed principally on low-cardenolide asclepiad vines (*Sarcostemma* and *Cynanchum* spp.; Kimball, 1965; Ackery and Vane-Wright, 1984). In addition, it has been suggested that even when cardenolide-rich food plants are available, competition with the monarch forces queens onto milkweeds containing low titers of cardenolide (e.g., *Asclepias tuberosa*) (Brower, 1961, 1962; Brower and Glazier, 1975). The utilization of these cardenolide-poor food plants would automatically reduce the queen's cardenolide defense.

Table 4. Comparative acceptability of monarchs (*D. plexippus*) and queens (*D. g. berenice*) to captive scrub jays (data from J. Brower, 1958a,b).

Scrub jay behavior	Queen	Monarch
% attacked	23.7	39.3
% eaten (of butterflies attacked)	7.8	0.0
Mean time to attack (sec)	13.0	21.0
Frequency of feather-ruffling behavior (%)	22.7	49.5

A NEW INVESTIGATION OF COMPARATIVE VICEROY AND QUEEN PALATABILITY

Together, the summarized feeding studies and chemical assays suggest that the traditional view—that the Florida viceroy is a classic Batesian mimic of the queen—is inaccurate. However, the nature of the relationship remains unclear, because the rejection of viceroys and queens in Brower's (1958a,b) experiments may have resulted from the birds' previous experience with the wing coloration patterns of monarchs and viceroys. We therefore designed a bird feeding experiment that would directly assess the palatability of Florida viceroys and queens and thereby determine the mode of mimicry between them (Batesian, Müllerian, or a combination). The problem of learned aversions to specific wing patterns was eliminated by presenting only the abdomens of the butterflies to captive birds; therefore, the birds had to accept or reject butterflies based primarily on gustatory cues, not on visual (wing coloration) cues.

METHODS

Male red-winged blackbirds (*Agelaius phoeniceus* (L.); Icteridae) collected in decoy traps in north-central Florida

Table 5. Comparative cardenolide concentration and total cardenolide content of monarchs (*D. plexippus*) and queens (*D. g. berenice*) reared on *A. humistrata*. Data are extracted from Brower et al. (1975) and Cohen (1985) and represent means of male and female butterfly values.

	Monarch	Queen	Queen : monarch ratio
Mean cardenolide concentration (equiv. μg digitoxin/0.1 g dry wt.)			
Brower et al. (1975)	329	221	0.67
Cohen (1985)	472	364	0.77
Total cardenolide content (equiv. μg digitoxin/butterfly)			
Brower et al. (1975)	527	160	0.30
Cohen (1985)	650	270	0.42

were obtained from the Denver Wildlife Research Center in Gainesville, Florida. These birds are ecologically relevant predators, because they forage in marshy areas where queens and particularly viceroys are found (Orians, 1985), they include a substantial proportion of animal matter in their diet (Bent, 1965), and they have been observed attacking butterflies in the wild (Brower and Brower, 1962; A.M. Shapiro, pers. comm.). Birds were housed in specialized experimental cages containing a motorized, rotating lazy-susan device equipped with 12 feeding cups (see Coppinger, 1970, for a detailed description). One-way glass on the front of the cage allowed observation of the birds with minimal disturbance. The feeding experiments were conducted from 8 July to 4 August 1986.

After learning to eat poultry feed and mealworms from the rotating cups, the birds were trained over a 3-day period to eat severed abdomens of three species of swallowtail butterfly (*Papilio troilus* L., *Papilio glaucus* L., and *Papilio palamedes* Drury; Papilionidae) that had been previously shown to be palatable to scrub jays (Brower, 1958a). We included these three visually distinctive butterflies in the training period to familiarize the birds with a variety of abdomen coloration patterns, thereby reducing any tendency of the birds to reject novel abdomens presented during the subsequent experiment. To qualify for the experiment, a bird had to eat all six swallowtail abdomens presented during the final training period; 10 of 12 birds met this criterion.

The ensuing experiment compared the reactions of these 10 birds to the abdomens of three experimental butterflies in relation to the abdomens of control butterflies (*Papilio* spp.). The three experimental butterflies were Florida viceroys (*L. a. floridensis*), Florida queens (*Danaus gilippus berenice*), and red spotted purples (*Limenitis arthemis astyanax* (Fabr.)). The red-spotted purple, a congener of the viceroys, is reportedly palatable (Platt et al., 1971); we included this butterfly because birds may reject novel prey items simply due to unfamiliar taste or appearance (neophobia; Coppinger, 1970). As long as the birds accepted the abdomens of the unfamiliar red-spotted purple, we could safely assume that rejection of viceroys or queens was based on distastefulness and not neophobia. Each bird received every type of butterfly abdomen and therefore served as its own control, as opposed to comparing the reactions of one set of "control" birds and one set of "experimental" birds that received different butterfly species.

The viceroys used in the experiment were reared under ambient conditions in the laboratory on their local larval food plant, coastal plain willow (*Salix caroliniana* Michx.), which is the only food plant available to viceroys through much of peninsular Florida (Wunderlin, 1982). The queens were collected as adults in Collier County, Florida, in March–April 1986; the predominant larval host plants in the area were *Sarcostemma clausum* (Jacq.) and *Asclepias incarnata* L.

Birds were deprived of food for 14 hr prior to testing, and experimental trials were conducted from 1000 to 1200 hours daily. Severed abdomens from frozen butterflies were thawed and presented in the rotating feeding dishes. In the case of large butterflies (e.g., female *Papilio* spp.), less than the entire abdomen was presented, so that each food item the birds received was approximately the same size (pilot trials had shown that splitting an abdomen had no effect on acceptability). Abdomens were presented individually for 25 sec at 15-min intervals.

Every bird participated in a series of 24 feeding trials over a 6-day period (four trials were conducted per day). Each "trial" consisted of presenting the bird with two butterfly abdomens in succession: one control (*Papilio*) and one experimental species (viceroy, queen, or red-spotted purple). Within each trial, the presentation order of control and experimental abdomens was randomized. Each bird received abdomens of a single experimental species each day, in addition to controls. Over the 6-day experimental period, different birds received the three experimental species in different orders. This randomization of presentation order prevented any bias that might be introduced by using a single order of viceroy and queen presentations. Subsequent statistical analysis indicated that presentation order per se did not affect the birds' reactions to the butterflies; there was no difference in number of viceroys eaten before versus after queens (Wilcoxon signed-rank test, $p = 0.30$); likewise, queens presented before and after viceroys were treated similarly (Wilcoxon, $p = 0.17$). By the end of the experiment, every bird had received eight viceroys, eight queens, and eight red-spotted purples, in addition to control (swallowtail) abdomens.

STATISTICAL ANALYSIS

For each of the four butterfly types, we recorded the following data to quantify palatability of the abdomens:

1. Overall proportion of abdomens eaten by each bird.
2. Manipulation time (time elapsed between attack and swallowing) of those abdomens that the birds ate.
3. "Response category": an ordinal measurement indicating the bird's treatment of each individual abdomen presented (0 = not touched; 1 = pecked only; 2 = partially eaten; 3 = totally eaten). This index is similar to Brower's (1958a) "reaction category," but we make no inference as to whether or not a given attack would kill a butterfly in the wild.
4. "Behavior score": a numerical value reflecting the incidence of excessive drinking (score = 1), head-shaking (minimum of 3 head shakes) or feather-ruffling (score = 2), and agitation (alarm calls, jumping around the cage; score = 3) in response to an abdomen presentation. This ranking scheme assumes that agitation represents the greatest distress, and drinking the least. Birds received a *total* behavior score for each butterfly species, based on the frequency with which each type of behavior was observed. For example, if a bird exhibited four episodes of feather-ruffling and two agitations during its eight viceroy presentations, its total "viceroy behavior score" was 14 (4 ruffle incidents at 2 pts. each + 2 agitations at 3 pts. each = 14).

Data for proportion eaten, response score, and behavioral score were not normally distributed, so nonparametric Friedman two-way analyses were employed to test whether or not a significant difference existed among the four butterfly types. Wilcoxon signed-rank tests were then performed to determine pairwise differences in the pro-

Table 6. **Percentage of abdomens eaten (in eight presentations) of four butterfly types by 10 captive red-winged blackbirds. Values indicated by different letters are statistically different at the 5% significance level (determined by pairwise Wilcoxon signed-rank tests).**

Bird #	Swallow-tails (*Papilio* spp.)	Red-spotted purple (*L. a. astyanax*)	Viceroy (*L. a. floridensis*)	Queen (*D. g. berenice*)
1	100	100	88	38
2	100	100	25	75
3	100	88	100	88
4	100	100	88	62
5	100	100	100	0
6	100	100	88	50
7	100	100	100	100
8	100	100	100	12
9	100	100	75	88
10	100	88	62	75
Mean	100	97.6	82.6	59.0
Median	100	100	88.0	68.5
Wilcoxon grouping	a	a	b	b

portion eaten, response score, and behavioral score of each butterfly type.

Manipulation time was only recorded when an abdomen was eaten; because birds ate differing proportions of the four butterfly types, this created an unbalanced data set that was intractable to statistical analysis. Therefore, for each individual bird, we calculated two *mean time values* per butterfly type. One value was the mean manipulation time for the first four presentations of the particular butterfly species, and the other was the mean time for the second four presentations of that butterfly. This operation yielded eight values per bird (two for each butterfly type) and created a balanced data set for analysis, eliminating the problem of differentially weighted individual bird responses. The mean time data were then subjected to a square-root transformation to achieve homogeneity of variances (Steel and Torrie, 1960). A parametric two-way ANOVA was performed on the transformed data, and the Student-Newman-Keuls multiple comparison test determined differences among mean manipulation times for the four butterfly types. One bird (#8) was excluded from the manipulation time analysis because it did not eat sufficient queens to yield two mean time values.

RESULTS

While the birds ate virtually every swallowtail and red-spotted purple abdomen presented to them, they ate only 83% of the viceroys and 68% of the queens offered (Table 6). The difference between viceroys and queens is not significant ($p = 0.13$), but both butterflies were eaten significantly less frequently than the controls ($p < 0.01$ in both cases). The birds evidently rejected viceroys and queens because they were distasteful and not simply because they were unfamiliar, since they readily ate the unfamiliar red-spotted purple.

An analysis of mean manipulation time for those butterflies that the birds did eat indicates that birds were significantly more hesitant to swallow viceroys than the control swallowtails (Table 7). The longer manipulation times for viceroys resulted because the birds often threw the abdomens about their cage or submerged them in their water bottle before eating them. Queen handling time was significantly longer than all of the others, because birds frequently removed and ate the abdominal contents completely before eating the exoskeleton (where much of the cardenolide is concentrated; Brower and Glazier, 1975). Manipulation times for red-spotted purples were intermediate between swallowtails and viceroys.

Table 8 indicates that viceroys and queens received significantly lower "response category scores" than red-spotted purples and swallowtails ($p < 0.02$ in both cases), but there was no difference between viceroys and queens ($p = 0.16$). This again indicates the relative unpalatability of viceroys and queens. Similarly, the total "behavior scores" (Table 9) indicate that viceroys and queens elicited significantly more aversion behavior (feather-ruffling, etc.) than did swallowtails and red-spotted purples ($p < 0.05$), but again there was no difference between viceroys and queens. All of the measures of acceptability were strongly correlated with one another (Table 10), as Brower (1958a,b) noted with scrub jays.

The sex of butterflies had no effect on bird reactions; there were no significant differences in proportion eaten, manipulation time, response score,

Table 7. **Mean manipulation time of abdomens of four butterfly types by 10 captive red-winged blackbirds. Values indicated by different letters are statistically different at the 5% significance level (determined by Student-Newman-Keuls (SNK) multiple comparison of transformed means).**

Butterfly	Manipulation time (sec) Mean ± SE	Range	n	SNK grouping
Swallowtails (*Papilio* spp.)	5.4 ± 0.9	1–35	80	a
Red-spotted purple (*L. a. astyanax*)	9.3 ± 2.3	1–43	78	ab
Viceroy (*L. a. floridensis*)	10.3 ± 2.2	1–42	66	b
Queen (*D. g. berenice*)	25.6 ± 4.0	1–85	38	c

Table 8. Mean response scores (as described in text) of 10 captive red-winged blackbirds in response to abdomens of four butterfly types. Values indicated by different letters are statistically different at the 5% significance level (determined by Friedman two-way ANOVA and pairwise Wilcoxon signed-rank tests).

Bird #	Swallow-tails (*Papilio* spp.)	Red-spotted purple (*L. a. astyanax*)	Viceroy (*L. a. floridensis*)	Queen (*D. g. berenice*)
1	3.0	3.0	2.8	1.9
2	3.0	3.0	1.9	2.8
3	3.0	2.9	3.0	2.8
4	3.0	3.0	2.9	2.1
5	3.0	3.0	3.0	1.0
6	3.0	3.0	2.9	2.2
7	3.0	3.0	3.0	3.0
8	3.0	3.0	3.0	0.9
9	3.0	3.0	2.5	2.9
10	3.0	2.9	2.2	2.5
Mean	3.00	2.98	2.70	2.20
Median	3.00	3.00	2.90	2.35
Wilcoxon grouping	a	a	b	b

Table 9. Total behavioral scores (as described in text) of 10 captive red-winged blackbirds in response to abdomens of four butterfly types. Values indicated by different letters are statistically different at the 5% significance level (determined by Friedman two-way ANOVA and pairwise Wilcoxon signed-rank tests).

Bird #	Swallow-tails (*Papilio* spp.)	Red-spotted purple (*L. a. astyanax*)	Viceroy (*L. a. floridensis*)	Queen (*D. g. berenice*)
1	0	1	1	9
2	0	3	6	7
3	0	2	3	4
4	0	3	8	8
5	0	1	1	6
6	2	0	5	10
7	1	2	3	5
8	0	0	12	6
9	0	3	8	8
10	0	4	5	0
Mean	0.3	1.9	5.2	6.3
Median	0.0	2.0	5.0	6.5
Wilcoxon grouping	a	a	b	b

or behavior score for males versus females of any butterfly species.

While viceroys were evidently somewhat *distasteful* to the birds, they did not appear to be *toxic*; birds that ingested them neither vomited nor exhibited depressed activity levels. Likewise, birds that ate several queens exhibited no obvious toxicosis or emesis. However, subtle toxic effects cannot be ruled out for either butterfly (Brower, 1984).

The experimental birds varied in their treatment of viceroys and queens; different individuals ate from 25 to 100% of the viceroys offered them, and from 12 to 100% of the queens. Birds that rejected viceroys did not necessarily reject queens (e.g., bird #2), and vice versa (e.g., bird #8) (Table 6). In fact, a separate experiment demonstrated that certain birds would readily devour abdomens that had been previously rejected by other individuals (unpubl. data). Whether this variation in bird behavior is due to individual differences in ability to *taste* the butterflies' defensive compounds or differential *tolerance* of bitter compounds by individual birds is unclear (cf. Brower, 1984).

DISCUSSION

The data on proportion eaten do not by themselves provide compelling evidence that viceroys are unpalatable (since 83% of them were eventually eaten). However, taken together, the concordant trends in proportion eaten, handling time, response category score, and behavior score do indicate that our captive red-winged blackbirds found viceroys significantly less acceptable than swallowtails and red-spotted purples. Even when very hungry, as evidenced by continual attempts to locate food during the experimental trials and very fast attack times (0.5–3.0 sec), most birds were hesitant to swallow viceroys. (Subsequent experiments using additional birds have provided much stronger evidence that viceroys are indeed unacceptable to these predators; Ritland, 1991a,b; in press a; Ritland and Brower, 1991).

Our Collier County queen sample was also relatively unacceptable to the experimental birds, although surprisingly, the queens were not significantly better protected than the viceroys. These results contradict the traditional hypothesis that viceroys are edible Batesian mimics of queens. Rather, our data indicate that these samples of viceroys and queens are relatively weak Müllerian comodels.

Individual queens appeared to vary considerably in palatability; some were eaten quickly after they were tasted, but others were eaten only after long delays, or not at all. This verifies for the first time that a palatability spectrum (Brower et al., 1968) exists in wild queen populations, presumably due to variation in cardenolide content (Cohen, 1985) or other chemical defenses (see following section). Further work using reared queens is planned to clarify the extent to which acceptability of individual queens is related to defensive chemicals obtained from larval or adult food plants.

Our experiment demonstrated that substantial variation exists in the ability of individual predators to perceive or tolerate distasteful compounds. Similar findings reported by J. Brower (1958a,b) and Platt et al. (1971) illustrate that no chemical defense is absolute. In L. Brower's (1969) terminology, some individual predators are relatively indiscriminate ("gourmands"), whereas others are very selective ("gourmets"), with regard to their reactions to chemically defended prey. An indiscriminate predator represents a very different selective force than a choosy one; hence, predator populations with different proportions of gourmets and gourmands must place different selective pressures on the butterfly populations they exploit. The variability in predator sensitivity must play a central role in the continuing coevolution of predators and their chemically defended prey (Ritland, 1991a).

The Müllerian mimicry relationship we suggest for viceroys and Collier County queens may not apply to every queen and viceroy population. As will be discussed below ("The Dynamic Mimicry Complex in Florida"), the butterflies probably rely on different food plants in different areas; their palatability, and thus the nature of their mimicry relationship, must vary in time and space. Future experimentation will attempt to clarify the palatability relationship between Florida viceroys and queens in different locales and assess the acceptability of various Florida monarch populations in relation to viceroys and queens.

PERSPECTIVES ON CHEMICAL DEFENSE IN QUEENS AND VICEROYS

Investigations into the queen's chemical defense have concentrated on the emetic properties of cardenolides, and alternative possibilities have received little experimental attention. Cardenolides in subemetic dosages or bitter, noncardenolide milkweed phytochemicals (Hegnauer, 1964) may also contribute to queen distastefulness. Many organisms innately avoid bitter substances (Garcia and Hankins, 1975), and bitterness per se can be sufficient to promote learned aversions in predators (Alcock, 1970; review in Brower, 1984). Even cardenolide-poor milkweeds (e.g., *A. tuberosa*) contain bitter compounds (Punyarajun, 1965), which might be stored by the adult queen, conferring substantial protection in the absence of emetic or toxic dosages of cardenolides (Ritland, 1991a).

In addition, pyrrolizidine alkaloids (PAs) may contribute to unpalatability in queens. PAs are bitter vertebrate toxins (Bull et al., 1968; Deinzer et al., 1977) and have been implicated as defensive chemicals in several insects (Aplin and Rothschild, 1972; Rothschild et al., 1970, 1979; Boppré, 1984). Adult danaids sequester and store PAs from flowers and withered plants of several genera (e.g., *Crotalaria, Senecio, Eupatorium;* Edgar et al., 1979; Edgar, 1982; Boppré, 1984; Schneider, this volume).

Table 10. Correlation matrix indicating Spearman rank correlations among values for the four acceptability indices used in red-winged blackbird feeding study.

Index	Index: % eaten	Manipulation time	Response score	Behavior score
% eaten	—	1.0	1.0	1.0
Manipulation time	1.0	—	1.0	1.0
Response score	1.0	1.0	—	1.0
Behavior score	1.0	1.0	1.0	—

It has been suggested that in addition to the established role of PA metabolites as male pheromone components (e.g., Brower et al., 1965; Pliske and Eisner, 1969; Boppré, 1981) these alkaloids contribute to the distastefulness of the insect (Edgar et al., 1976; Conner et al., 1981; Boppré, 1978, 1984; Eisner, 1980; Edgar, 1982; Brower, 1984; but cf. Glendinning, this volume). The hypothesis that PAs are defensive compounds in danaids is perhaps supported by the fact that even monarchs, which apparently do not require PAs for courtship (Schneider, this volume), store these compounds (Edgar et al., 1971, 1976; Dixon et al., 1978; Rothschild and Edgar, 1978; Kelley et al., 1987).

Experimental studies have demonstrated that PAs deter invertebrate predators, apparently via taste (Brown, 1984; Vasconcellos-Neto and Lewinsohn, 1984). However, there are no published studies demonstrating whether the naturally occurring PAs or PA metabolites that are sequestered by danaids are effective in deterring vertebrate predators. To investigate the possibility of a dual defense system (PA + cardenolide) in the Florida queen, queens of four types (PA + cardenolide, PA only, cardenolide only, and neither) are being reared in the laboratory for presentation to captive birds. This should make it possible to assess the significance of these two toxic compounds in queen defense.

The viceroy's chemical ecology has not been adequately investigated, and whether its distastefulness is due to sequestration and storage of phytochemicals or to *de novo* synthesis of defensive chemicals by the butterfly is unclear. The viceroy's larval food plants (willows and poplars; Salicaceae) contain a diversity of bitter-tasting phenylglycosides (e.g., Pearl and Darling, 1968; Bryant et al., 1985) that could potentially be sequestered by viceroy larvae and stored by the adult as defensive chemicals. A precedent for the defensive role of phenylglycosides exists, since Pasteels et al. (1982, 1983) have shown that willow-feeding leaf beetles (Coleoptera: Chrysomelidae) sequester salicin from their host plant and employ a metabolite, salicylaldehyde, in their defensive secretions. Also, viceroys, like queens, might sequester alkaloids or other compounds from their *adult* food sources. Since lab-reared viceroys that have not fed as adults are in fact somewhat distasteful (e.g., Platt et al., 1971, this paper), adult-obtained deterrents are apparently not critical; however, they may enhance the protection level of the butterflies through additive or synergistic effects. Alternatively, viceroys may synthesize defensive compounds *de novo*. Evidence is accumulating that butterflies are capable of producing an array of compounds with putative defensive roles (Ackery, 1987, this volume). The hypothesis that viceroys synthesize defensive compounds *de novo* gains some support from the fact that although viceroys rely on many of the same food plants as other *Limenitis* species (Howe, 1975; Platt, 1983), they are the only distasteful species among those that have been investigated.

THE DYNAMIC MIMICRY COMPLEX IN FLORIDA

The nature of a butterfly mimicry relationship depends on how predators perceive and react to the different butterfly species. For mimicry to be effective, predators must encounter enough unpalatable models to develop an aversion to the model's color pattern, and they must then avoid mimics (or co-models) because they are unable or unwilling to discriminate them visually from models. These aspects of predator learning and behavior, and thus the *dynamics* of mimicry, are affected by many factors, ranging from the relative frequencies of models and mimics to the similarity in model and mimic color patterns (Sheppard, 1959). But perhaps the most important determinant of mimicry dynamics is variable unpalatability.

Queens and monarchs breeding in different parts of Florida probably rely on distinct arrays of larval and adult food plants (e.g., Brower, 1961; Cohen, 1985). These regional floras provide differing complements of cardenolides and PAs, and the degree to which danaid populations are chemically defended must therefore vary geographically. Thus, Florida monarch and queen populations presumably exhibit a geographic palatability spectrum similar to that demonstrated for monarchs from other parts of the United States (Brower et al., 1972; Malcolm et al., this volume). In fact, we have demonstrated that queens reared on different larval food plants differ significantly in acceptability to laboratory birds (Ritland, 1991a), and Cohen (1985) documented geographic variation in queen cardenolide content within Florida.

In addition, there may be *temporal* variation in the chemical protection level of queens and monarchs, due to the seasonal succession of milkweed species and PA-containing plants (e.g., Brower, 1961; Wunderlin, 1982; Clewell, 1985; Malcolm et al., this volume). Temporal (seasonal) fluctuation in palatability is probably greatest in monarchs, which feed on a variety of milkweeds in different seasons due to their migratory life-style. Thus, monarchs breeding in north Florida in the spring appear to rely primarily on the cardenolide-rich sandhill milkweed, *A. humistrata* (Malcolm et al., 1987); larvae feeding on this plant produce highly emetic adults (Brower, 1969). By contrast, the autumn influx of migrant monarchs into north Florida consists largely of butterflies that fed as larvae upon a cardenolide-poor northern milkweed, *Asclepias syriaca* (Seiber et al., 1986; Malcolm et al., 1989). Although these migrant monarchs may also contain PAs (Kelley et al., 1987), they are only moderately protected against bird predators (Ritland, 1991a; Ritland and Brower, 1991). Thus, even within a single region, monarch palatability varies greatly from season to season.

Queen palatability must also vary seasonally due to the changing milkweed flora; in addition, queens may be more strongly influenced by seasonal changes in the availability of PA-containing plants, since queens rely more on PA sources than monarchs do (Pliske, 1975). Finally, viceroys themselves may vary in palatability, as a result of either variation in phenylglycoside content in willows (including seasonal fluctuation; Pearl and Darling, 1968; Rowell-Rahier, 1984; Smiley et al., 1985) or genetic variation among viceroys in the ability either to sequester or to synthesize defensive chemicals.

We hypothesize that such seasonal and geographic variation in chemical defense creates a dynamic "palatability hierarchy" among viceroys, queens, and monarchs. The nature of the mimicry association must therefore differ from population to population, shifting along a continuum from Batesian to Müllerian mimicry. In addition, variation in queen and monarch palatability will yield populations containing varying proportions of edible "automimics" (Brower, 1969) that essentially mimic better-protected individual *conspecifics*. Changing the proportion of palatable automimics must affect the protection level of the local danaid population (Brower et al., 1970; Pough et al., 1973) and thereby compromise the umbrella of protection extending to the entire mimicry complex.

Thus, variable unpalatability (probably augmented by variation in abundance and phenotypic similarity of the butterflies) creates an unexpectedly complex mimicry relationship, in which viceroys may be either mimics, co-models, or models of monarchs and queens in different locales and seasons.

CONCLUSION

Our experimental feeding studies, together with the work of Brower (1958a,b) and Platt et al. (1971), indicate that viceroys are relatively unpalatable to red-winged blackbirds, scrub jays, and blue jays. Moreover, we have confirmed that in spite of a less efficient cardenolide defense system than monarchs, queens also enjoy some protection from bird predators. Our findings refute the conventional hypothesis that viceroys are Batesian mimics of queens; rather, we conclude that queens and viceroys in

Florida are involved in a complex and labile Müllerian association.

The evident unacceptability of viceroys compels a closer look at the ecology and evolution of the classic viceroy–danaid mimicry relationship. We have proposed that in the ecological sense, viceroys may be either "mimics" or "models," depending upon their palatability and abundance relative to the danaids. In the phylogenetic sense, though, the viceroy is undoubtedly the mimic (R.I. Vane-Wright, pers. comm.), having diverged dramatically from the ancestral *Limenitis* coloration pattern of white-on-black to adopt the orange and brown hues of the danaids. A.P. Platt (1983, pers. comm.), noting that other *Limenitis* species are palatable, has suggested that viceroys first evolved as Batesian mimics of danaids, and then developed distastefulness subsequently. Apparently, viceroys are currently evolving from Batesian into Müllerian mimics of their danaid models. Because the ecological and evolutionary consequences for models, mimics, and predators differ in Batesian and Müllerian modes (e.g., Huheey, 1984), the evolution of viceroy distastefulness must place the entire mimicry assemblage, as well as their predators, in a very different selective milieu. Detailed study of the viceroy's chemical ecology and mimicry dynamics may provide insights into the evolution of distastefulness, the development of Müllerian mimics from Batesian ancestors, and the putative existence of a Batesian–Müllerian mimicry continuum. These poorly studied phenomena remain a focal point of debate in the mimicry literature (e.g., Huheey, 1984; Turner, 1984).

The mimicry relationship among monarchs, queens, and viceroys in Florida is presumably mediated by their chemical ecology, their seasonally changing abundance, and their phenotypic resemblance. We have postulated that spatiotemporal variation in these factors produces an unexpectedly dynamic mimicry system; because local populations of the three butterflies exist in different ecological contexts, they must differ in the nature of their mimicry relationships. In some instances, the traditional status of the queen or monarch as a distasteful model and the viceroy as a Batesian mimic may in fact be reversed.

A complex combination of environmental and organismal characteristics dictates the butterflies' protection level, their apparency to predators, and the ability of predators to discriminate visually among them, thus guiding the continuing coevolution of models, mimics, and predators. Studying mimicry dynamics at the population level therefore bridges the gap between ecological circumstance and evolutionary result and is crucial to deciphering the evolutionary factors that create and shape mimetic butterfly association.

Note added in proof: Many of the predictions generated herein have been tested and upheld in subsequent experiments (Ritland, 1991a,b; Ritland, in press a,b; Ritland and Brower, 1991).

ACKNOWLEDGMENTS

We are grateful to the U.S. Department of Agriculture Denver Wildlife Research Laboratory in Gainesville, Florida, for providing red-winged blackbirds. The comments of Drs. S.B. Malcolm, A.P. Platt, and R.I. Vane-Wright, as well as those of R.L. Ritland, have been very helpful in the revision of the manuscript. This work has been financially supported by the National Geographic Society (grant #3417-86) and a Sigma Xi Grant-in-Aid of Research.

LITERATURE CITED

Ackery, P.R. 1987. Hostplants and classification: A review of nymphalid butterflies. *Biological Journal of the Linnean Society* 33(2):95–203.

Ackery, P.R., and R.I. Vane-Wright. 1984. *Milkweed butterflies: Their cladistics and biology.* London: British Museum (Natural History); and Ithaca, NY: Cornell University Press, 425 pp.

Alcock, J. 1970. Punishment levels and the response of black-capped chickadees (*Parus atricapillus*) to three kinds of artificial seeds. *Animal Behavior* 18:592–599.

Aplin, R.T., and M. Rothschild. 1972. Poisonous alkaloids in the body tissues of the garden tiger moth (*Arctia caja* L.) and the cinnabar moth (*Tyria* (=*Callimorpha*) *jacobeae* L.) (Lep.). In *Toxins of animal and plant origin,* eds. A. deVries and E. Kochuapp, 579–595. London: Gordon and Breach.

Bent, A.C. 1965. *Life histories of North American blackbirds, orioles, tanagers, and allies.* New York: Dover Publications, 549 pp.

Boppré, M. 1978. Chemical communication, plant relationships and mimicry in the evolution of Danaid butterflies. *Entomologia Experimentalis et Applicata* 24: 64–77.

Boppré, M. 1981. Adult lepidoptera 'feeding' at withered *Heliotropium* plants (Boraginaceae) in East Africa. *Ecological Entomology* 6:449–452.

Boppré, M. 1984. Chemically mediated interactions between butterflies. In *The biology of butterflies,* eds. R.I. Vane-Wright and P.R. Ackery, *Symposia of the Royal Entomological Society of London* 11:259–275. London and New York: Academic Press (1989, Princeton, NJ: Princeton University Press).

Brower, J.V.Z. 1958a. Experimental studies of mimicry in some North American butterflies. I. The monarch, *Danaus plexippus* and the viceroy, *Limenitis archippus. Evolution* 12:32–47.

Brower, J.V.Z. 1958b. Experimental studies of mimicry in some North American butterflies. III. *Danaus gilippus berenice* and *Limenitis archippus floridensis. Evolution* 12:273–285.

Brower, L.P. 1961. Studies on the migration of the monarch butterfly. I. Breeding populations of *Danaus plexippus* and *D. gilippus berenice* in south-central Florida. *Ecology* 42:76–83.

Brower, L.P. 1962. Further evidence for interspecific competition in natural populations of the monarch and queen butterflies, *Danaus plexippus* and *D. gi-*

lippus berenice in south central Florida. *Ecology* 43: 549–552.

Brower, L.P. 1969. Ecological chemistry. *Scientific American* 220:22–29.

Brower, L.P. 1984. Chemical defence in butterflies. In *The biology of butterflies,* eds. R.I. Vane-Wright and P.R. Ackery, *Symposia of the Royal Entomological Society of London* 11:109–134. London and New York: Academic Press (1989, Princeton, NY: Princeton University Press).

Brower, L.P., and J.V.Z. Brower. 1962. Investigations into mimicry. *Natural History* 71:8–19.

Brower, L.P., and S.C. Glazier. 1975. Localization of heart poisons in the monarch butterfly. *Science* 188: 19–25.

Brower, L.P., J.V.Z. Brower, and F.P. Cranston. 1965. Courtship behaviour of the queen butterfly, *Danaus gilippus berenice* (Cramer). *Zoologica* 50:1–39.

Brower, L.P., W.N. Ryerson, L.L. Coppinger, and S.C. Glazier. 1968. Ecological chemistry and the palatability spectrum. *Science* 161:1349–1351.

Brower, L.P., F.H. Pough, and H.R. Meck. 1970. Theoretical investigations of automimicry. I. Single trial learning. *Proceedings of the National Academy of Sciences* 66:1059–1066.

Brower, L.P., P.B. McEvoy, K.L. Williamson, and M.A. Flannery. 1972. Variation in cardiac glycoside content of monarch butterflies from natural populations in eastern North America. *Science* 177:426–429.

Brower, L.P., M. Edmunds, and C.M. Moffitt. 1975. Cardenolide content and palatability of a population of *Danaus chrysippus* butterflies from West Africa. *Journal of Entomology (A)* 49:183–196.

Brown, K.S. 1984. Adult-obtained pyrrolizidine alkaloids defend ithmoniine butterflies against a spider predator. *Nature* 309:707–709.

Bryant, J.P., G.D. Wieland, T. Clausen, and P. Kuropat. 1985. Interactions of snowshoe hare and feltleaf willow in Alaska. *Ecology* 66:1564–1573.

Bull, L.B., C.C.J. Culvenor, and A.T. Dick. 1968. *The pyrrolizidine alkaloids.* Amsterdam: North-Holland Publishing, 293 pp.

Clewell, A.F. 1985. *Guide to the vascular plants of the Florida panhandle.* Tallahassee: University Presses of Florida and Florida State University Press, 605 pp.

Cohen, J.A. 1985. Differences and similarities in cardenolide content of queen and monarch butterflies in Florida and their ecological and evolutionary implications. *Journal of Chemical Ecology* 11:85–103.

Conner, W.E., T. Eisner, R.K. Vander Meer, A. Guerrero, and J. Meinwald. 1981. Precopulatory sexual interaction in an arctiid moth (*Utetheisa ornatrix*): Role of a pheromone derived from dietary alkaloids. *Behavioral Ecology and Sociobiology* 9:227–235.

Coppinger, R.P. 1970. The effect of experience and novelty on avian feeding behavior with reference to the evolution of warning coloration in butterflies. II. Reactions of naive birds to novel insects. *American Naturalist* 104:323–335.

Deinzer, M.L., P.A. Thomson, D.M. Burgett, and D.L. Isaacson. 1977. Pyrrolizidine alkaloids: Their occurrence in honey from Tansy Ragwort (*Senecio jacobeae* L.). *Science* 195:497–499.

Dixon, C.A., D.N. Kellett, and M. Rothschild. 1978. Some adaptations between *Danaus plexippus* and its food plant, with some notes on *Danaus chrysippus* and *Euploea core* (Insecta: Lepidoptera). *Journal of Zoology* 185:437–467.

Edgar, J.A. 1982. Pyrrolizidine alkaloids sequestered by Solomon Island danaine butterflies. The feeding preferences of the Danainae and Ithomiinae. *Journal of Zoology* 196:385–399.

Edgar, J.A., C.C.J. Culvenor, and L.W. Smith. 1971. Dihydropyrrolizine derivatives in the hair pencil secretions of danaid butterflies. *Experientia* 27:761–762.

Edgar, J.A., P.A. Cockrum, and J.L. Frahn. 1976. Pyrrolizidine alkaloids in *Danaus plexippus* and *Danaus chrysippus. Experientia* 32:1535–1537.

Edgar, J.A., M. Boppré, and D. Schneider. 1979. Pyrrolizidine alkaloid storage in African and Australian danaid butterflies. *Experientia* 35:1447–1448.

Eisner, T. 1980. Chemistry, defence and survival: Case studies and selected topics. In *Insect biology in the future,* eds. M. Locke and D.S. Smith, 847–878. New York: Academic Press, 977 pp.

Garcia, J., and W.G. Hankins. 1975. The evolution of bitter and the acquisition of toxiphobia. In *Fifth International Symposium on Olfaction and Taste,* eds. D.A. Denton and J.P. Coghlan, 39–45. New York: Academic Press, 460 pp.

Hegnauer, R. 1964. *Chemotaxonomie der Pflanzen,* Band 3. Basel and Stuttgart: Birkhauser e Verlag.

Howe, W.H. 1975. *The butterflies of North America north of Mexico.* Garden City, New York: Doubleday and Co., 633 pp.

Huheey, J.E. 1984. Warning coloration and mimicry. In *Chemical ecology of insects,* eds. W.J. Bell and R.T. Cardé, 257–297. London and New York: Chapman and Hall, 524 pp.

Kelley, R.B., J.N. Seiber, A.D. Jones, H.J. Segall, and L.P. Brower. 1987. Pyrrolizidine alkaloids in overwintering monarch butterflies (*Danaus plexippus*) from Mexico. *Experientia* 43:943–946.

Kimball, C.P. 1965. *Lepidoptera of Florida* (*Arthropods of Florida and neighboring land areas,* Vol. 1). Gainesville, FL: Division of Plant Industry, 363 pp.

Malcolm, S.B., B.J. Cockrell, and L.P. Brower. 1987. Monarch butterfly voltinism: Effects of temperature constraints at different latitudes. *Oikos* 49:77–82.

Malcolm, S.B., B.J. Cockrell, and L.P. Brower. 1989. Cardenolide fingerprint of monarch butterflies reared on common milkweed, *Asclepias syriaca* L. *Journal of Chemical Ecology* 15:819–853.

Orians, G.H. 1985. *Blackbirds of the Americas.* Seattle: University of Washington Press, 163 pp.

Pasteels, J.M., J.C. Braekman, and D. Daloze. 1982. Chemical defence in chrysomelid larvae and adults. *Tetrahedron* 38:1891–1897.

Pasteels, J.M., M. Rowell-Rahier, J.C. Braekman, and A. Dupont. 1983. Salicin from host plant as precursor of salicylaldehyde in defensive secretion of chrysomeline larvae. *Physiological Entomology* 8:307–314.

Pearl, I.A., and S.F. Darling. 1968. Studies of the leaves of the family Salicaceae. XI. The hot water extractives of the leaves of *Populus balsamifera*. *Phytochemistry* 7:1845–1849.
Platt, A.P. 1983. Evolution of North American admiral butterflies. *Bulletin of the Entomological Society of America* 29:10–22.
Platt, A.P., R.P. Coppinger, and L.P. Brower. 1971. Demonstration of the selective advantage of mimetic *Limenitis* butterflies presented to caged avian predators. *Evolution* 25:692–701.
Pliske, T.E. 1975. Attraction of Lepidoptera to plants containing pyrrolizidine alkaloids. *Environmental Entomology* 4:455–473.
Pliske, T.E., and T. Eisner. 1969. Sex pheromone of the queen butterfly: Biology. *Science* 164:1170–1172.
Pough, F.H., L.P. Brower, H.R. Meck, and S.R. Kessell. 1973. Theoretical investigations of automimicry: Multiple trial learning and the palatability spectrum. *Proceedings of the National Academy of Sciences* 70:2261–2265.
Poulton, E.B. 1909. Mimicry in the butterflies of North America. *Annals of the Entomological Society of America* 2:203–242.
Punyarajun, S. 1965. Chemical investigations of *Asclepias tuberosa*. Ph.D. dissertation. West Lafayette, IN: Purdue University.
Ritland, D.B. 1991a. Ecological dynamics of viceroy butterfly (*Limenitis archippus*) mimicry in Florida. Ph.D. dissertation, University of Florida, Gainesville, Florida.
Ritland, D.B. 1991b. Revising a classic butterfly mimicry scenario: Demonstration of Müllerian mimicry between Florida viceroys (*Limenitis archippus floridensis*) and queens (*Danaus gilippus berenice*). *Evolution* 45:918–934.
Ritland, D.B. In press a. Unpalatability of viceroy butterflies (*Limenitis archippus*) and their purported mimicry models, Florida queens (*Danaus gilippus*).
Ritland, D.B. In press b. Palatability of aposematic queen butterflies (*Danaus gilippus*) feeding on *Sarcostemma clausum* (Asclepiadaceae) in Florida. *Journal of Chemical Ecology* 17(8).
Ritland, D.B., and L.P. Brower. 1991. The viceroy butterfly is not a batesian mimic. *Nature* 350:497–498.
Roeske, C.N., J.N. Seiber, L.P. Brower, and C.M. Moffitt. 1976. Milkweed cardenolides and their comparative processing by monarch butterflies (*Danaus plexippus* L.). *Recent Advances in Phytochemistry* 10: 93–167.
Rothschild, M., and J.A. Edgar. 1978. Pyrrolizidine alkaloids from *Senecio vulgaris* sequestered and stored by *Danaus plexippus*. *Journal of Zoology* 186:347–349.
Rothschild, M., T. Reichstein, J. von Euw, R. Alpin, and R.R.M. Harman. 1970. Toxic lepidoptera. *Toxicon* 8:293–299.
Rothschild, M., R.T. Alpin, P.A. Cockrum, J.A. Edgar, P. Fairweather, and R. Lees. 1979. Pyrrolizidine alkaloids in arctiid moths with a discussion on host plant relationships and the role of these secondary plant substances in the Arctiidae. *Biological Journal of the Linnean Society* 12:305–326.
Rowell-Rahier, M. 1984. The presence or absence of phenolglycosides in *Salix* (Salicaceae) leaves and the level of dietary specialisation of some of their herbivorous insects. *Oecologia* 62:26–30.
Seiber, J.N., L.P. Brower, S.M. Lee, M.M. McChesney, H.T.A. Cheung, C.J. Nelson, and T.R. Watson. 1986. Cardenolide connection between overwintering monarch butterflies from Mexico and their larval foodplant, *Asclepias syriaca*. *Journal of Chemical Ecology* 12:1157–1170.
Sheppard, P.M. 1959. The evolution of mimicry; a problem in ecology and genetics. *Cold Spring Harbor Symposia Quantitative Biology* 24:131–140.
Smiley, J.T., J.M. Horn, and N.E. Rank. 1985. Ecological effects of salicin at three trophic levels: New problems from old adaptations. *Science* 229:649–651.
Steel, R.G.D., and J.H. Torrie. 1960. *Principles and procedures of statistics*, 2nd ed. New York: McGraw-Hill, 481 pp.
Turner, J.R.G. 1984. Mimicry: The palatability spectrum and its consequences. In *The biology of butterflies*, eds. R.I. Vane-Wright and P.R. Ackery, *Symposia of the Royal Entomological Society of London* 11:141–161. London and New York: Academic Press (1989, Princeton, NJ: Princeton University Press).
Vasconcellos-Neto, J., and T.M. Lewinsohn. 1984. Discrimination and release of unpalatable butterflies by *Nephila clavipes*, a neotropical orb-weaving spider. *Ecological Entomology* 9:337–344.
Wunderlin, R.P. 1982. *Guide to the vascular plants of central Florida*. Tampa: University Presses of Florida, 472 pp.

6

PHYSIOLOGICAL ECOLOGY AND THE ANNUAL CYCLE

A male monarch with its wings opened to the sun. Such basking behavior enables monarchs to heat up their bodies sufficiently to fly, especially in the cool ambient temperatures characteristic of the overwintering sites. Photograph by Julian P. Donahue.

ENDOCRINOLOGY OF THE MONARCH BUTTERFLY

William S. Herman

ABSTRACT. The juvenile hormone (JH), a lipid produced by the corpora allata, is the most important hormone of the adult monarch. The use of allatectomies, neck ligatures, hormone injections, and JH titer determinations have shown that JH plays a very important role in the stimulation of reproductive tract development in both sexes. More specifically, JH accelerates the growth of the ovaries and reproductive glands of the female and the growth of the accessory glands, tubular glands, and ejaculatory ducts of the male. JH is required for both the posteclosion growth and the postdiapause growth of the above reproductive organs in both sexes. Comparable experiments suggest that high titers of JH accelerate senescence in adult monarchs and that low levels of JH production provoke both reproductive diapause and southward migration.

Two additional hormones, which appear to be neuropeptides produced by the brain–corpora cardiaca complex, are also known to play important roles in adult monarchs. The monarch adipokinetic hormone acts to elevate circulating diglycerides, and thus provide fuel for flight metabolism; it may be of major importance during migration and overwintering. This monarch hormone appears to be a member of a large family of arthropod neuropeptides, all of which resemble the prawn red pigment-concentrating hormone. The monarch diuretic hormone acts for a short time after adult eclosion to accelerate water loss and assist in the establishment of the appropriate body weight for monarch flight. The diuretic hormone, which has not been fully characterized, may also regulate other monarch activities.

Our current understanding of the biology of these three hormones in adult monarchs will be discussed, and the endocrinology of the monarch will be compared with that of other adult Lepidoptera.

It is now well documented that juvenile hormone (JH) is a regulator of major importance in *Danaus plexippus* (L.), the monarch butterfly. Initial studies on the role of this hormone in monarchs demonstrated that posteclosion ovarian maturation was dependent upon the presence of JH (Pan and Wyatt, 1971, 1976; Barker, 1976; Barker and Herman, 1973). Subsequent studies showed that JH also stimulated posteclosion growth of the colleterial glands of females and the accessory glands, tubular glands, and ejaculatory ducts of males (Herman, 1975a,b; Johnson, 1979). More recent work indicates JH regulation of both the receptacle gland and the bursa copulatrix of females (Herman, 1982). All of the above organs exhibit little or no growth in the absence of JH, but all undergo significant wet and dry weight increases when the hormone is present. The important role of JH in posteclosion growth of the reproductive tracts in both sexes is supported by measurements of hemolymph JH titers in posteclosion animals (Herman et al., 1981), which demonstrate a correlation between reproductive tract growth and JH levels in both males and females. JH production in posteclosion females is sensitive to environmental conditions (Barker and Herman, 1976), and mating enhances reproductive tract growth, and by inference JH production, in females (Herman and Barker, 1977). Little is known about the specific molecules produced by monarch reproductive tracts in response to JH, but Pan and Wyatt (1971, 1976) showed that the hormone induces vitellogenin production by the female, and Herman and Peng (1976) reported the production of a sperm activator by the male ejaculatory duct in response to JH. JH I and JH II appear to be equally effective in adult monarchs, while JH III has significantly less biological activity (Lessman et al., 1982). Recent studies indicate that JH II is the principal JH produced by adult monarchs (Lessman, Herman, and Schooley, unpubl.), but JH I and JH III are also present in adult hemolymph. A JH esterase and JH binding protein play important roles in the regulation and transport of JH in both sexes (Lessman, 1979).

JH also appears to play an important role in the regulation of the annual reproductive cycle of male and female monarchs. JH levels in the hemolymph are low in southward migrating and overwintering animals (Lessman and Herman, 1983) when reproductive tract growth is reduced (Herman, 1985), but tract growth rapidly follows JH administration (Herman, 1973, 1985). Apparently, the absence of sufficient JH is at least one reason for the lack of reproductive tract development in such adults.

Other research has outlined JH involvement in three periods of the adult monarch annual cycle, i.e., prediapause, diapause, and postdiapause (Herman, 1981). Prediapause animals (emerging from June to August in Minnesota) respond to summer-like conditions with pronounced reproductive tract development, whereas diapause animals (captured in September in Minnesota, or from September to January in California) exhibit significantly reduced responses to such conditions. By contrast, postdiapause adults obtained from the Californian overwintering colonies after the completion of diapause possess underdeveloped reproductive tracts comparable to those of diapause adults, but they respond to summer conditions with tract growth comparable to that of prediapause animals. Diapause therefore appears to be a period of reduced ability to produce effective JH titers in response to summer conditions. The major importance of diapause may be that it prevents futile reproductive effort when larval food plant is generally in short supply and thereby conserves valuable energy reserves for use during overwintering. Recent work has shown that the reproductive dormancy of Australian monarchs differs from that of those in North America in several respects (James, 1982, 1983; James and Hales, 1983).

Other data suggest that JH may also accelerate senescence in adult monarchs (Herman, 1985), perhaps by elevating metabolic rate and oxygen consumption (Herman, unpubl.). High JH titers may therefore be at least partially responsible for the relatively short life-spans of reproductively active monarchs, whereas the low titers present in overwintering monarchs could enhance longevity in such insects.

The role of JH, if any, in the southward migration of monarchs is as yet undefined. It is known that JH titers during such flights are lower than those of summer reproductives (Lessman and Herman, 1983), and such titer reductions are believed to be of importance in stimulating migratory behavior in other insects (Rankin, 1985). It has been reported that flight may actively reduce JH levels in monarchs by enhancing the activity of JH esterase (Lessman and Herman, 1981), and that JH may increase flight behavior in postdiapause northward migrants (Rankin, 1985).

Monarch adults of both sexes also use peptide hormones to regulate important activities. Dallman and Herman (1978) have presented evidence for an adipokinetic hormone (AKH) released from the monarch cephalic neuroendocrine system that acts to elevate hemolymph diglycerides. Presumably, this hormone plays a vital role in the mobilization of nutrients that will be used to provide fuel for monarch flight. The action of the monarch hormone is duplicated by locust AKH and crustacean red pigment-concentrating hormone (RPCH) (Dallman et al., 1981), indicating that the monarch hormone is a member of the growing family of RPCH/AKH peptides found in arthropods. The possibility exists that the monarch hormone may be identical to the AKH peptides recently isolated from both *Heliothis* (Jaffe et al., 1986) and *Manduca* (Ziegler et al., 1985). Additional roles of monarch AKH, such as the inhibition of protein synthesis reported for AKH in locusts (Carlisle and Loughton, 1986) have yet to be fully documented, but Rankin (1985) has obtained preliminary evidence that locust AKH can stimulate monarch flight behavior.

Monarch adults also produce a peptide hormone, the diuretic hormone (DH), that acts to accelerate water loss in posteclosion adults (Dores et al., 1979; Dores and Herman, 1981). The DH is released immediately after eclosion, and its action results in the excretion of excess body fluid subsequent to wing expansion. The DH does not appear to be important in postprandial diuresis in monarchs (Dores et al., 1979), but since it is apparently present in the neuroendocrine system throughout adult life it may play additional roles in this species. The chemical nature of the monarch DH has not yet been elucidated, but the available evidence strongly indicates that the monarch DH is not the monarch AKH.

The above studies have demonstrated the presence and importance of at least three hormones, JH, AKH, and DH, in adult *D. plexippus*, and the currently available data indicate that such hormones are also of importance in at least some other Lepidoptera. JH clearly plays as important role in *Nymphalis* (Herman and Bennett, 1975), *Vanessa* (Herman and Dallmann, 1981), and *Manduca* (Sasaki and Riddiford, 1984), and it seems to be important in at least some other species of the order (Engelmann, 1983; Sims, 1984). AKH has been demonstrated in two additional Lepidoptera, i.e., *Manduca* (Zeigler et al., 1985) and *Heliothis* (Jaffe et al., 1986), and DH is known to be present not only in *Danaus* but also in *Pieris* and *Papilio* (Nicolson and Millar, 1983). Three additional hormones, i.e., the eclosion hormone, bursicon, and a peptide(s) regulating heart activity, have also been reported in other adults of the order (Reynolds, 1983; Reynolds and Truman, 1983; Tublitz and Truman, 1985), but their possible presence in monarchs has not been examined.

ACKNOWLEDGMENTS

The contributions of both William and William L. Sherwood to the above discussed research are very much appreciated, as are the constructive editorial remarks of B. Weidel.

LITERATURE CITED

Barker, J.F. 1976. On the neurendocrinology of reproduction and migration in the monarch butterfly, *Danaus plexippus.* Ph.D. thesis. St. Paul: University of Minnesota.

Barker, J.F., and W.S. Herman 1973. On the neuroendocrine control of ovarian development in the monarch butterfly. *Journal of Experimental Zoology* 183: 1–10.

Barker, J.F., and W.S. Herman. 1976. Effect of photoperiod and temperature on reproduction of the monarch butterfly, *Danaus plexippus. Journal of Insect Physiology* 22:1565–1568.

Carlisle, J., and B.G. Loughton. 1986. The inhibition of protein synthesis in *Locusta migratoria* by adipokinetic hormone. *Journal of Insect Physiology* 32: 573–578.

Dallmann, S.H., and W.S. Herman. 1978. Hormonal regulation of hemolymph lipid concentration in the monarch butterfly, *Danaus plexippus. General and Comparative Endocrinology* 36: 142–150.

Dallmann, S.H., W.S. Herman, J. Carlsen, and L. Josefsson. 1981. Adipokinetic activity of shrimp and locust peptide hormones in butterflies. *General and Comparative Endocrinology* 43:256–258.

Dores, R.M., and W.S. Herman. 1981. Effect of age, ambient temperature, and exposure to hormone on the posteclosion diuretic response of the monarch butterfly. *Experientia* 37:529–530.

Dores, R.M., S.H. Dallmann, and W.S. Herman. 1979. The regulation of posteclosion and postfeeding diuresis in the monarch butterfly, *Danaus plexippus. Journal of Insect Physiology* 25:895–901.

Engelmann, F. 1983, Vitellogenesis controlled by juvenile hormone. In *Endocrinology of insects,* eds. G.H. Downer and H. Laufer, 259–270. New York: Alan R. Liss.

Herman, W.S. 1973. The endocrine basis of reproductive inactivity in monarch butterflies overwintering in central California. *Journal of Insect Physiology* 19:1883–1887.

Herman, W.S. 1975a. Endocrine regulation of posteclosion enlargement of the male and female reproductive glands in monarch butterflies. *General and Comparative Endocrinology* 26:534–540.

Herman, W.S. 1975b. Juvenile hormone stimulation of tubular and accessory glands in male monarch butterflies. *Comparative Biochemistry and Physiology* 51(A):507–510.

Herman, W.S. 1981. Studies on the adult reproductive diapause of the monarch butterfly, *Danaus plexippus. Biological Bulletin* 160:89–106.

Herman, W.S. 1982. Endocrine regulation of the bursa copulatrix and receptacle gland of *Danaus plexippus* L. (Lepidoptera:Danaidae) *Experientia* 38:631–632.

Herman, W.S. 1985. Hormonally mediated events in adult monarch butterflies. In *Migration: Mechanisms and adaptive significance,* ed. M.A. Rankin, 779–815. Austin: University of Texas, *Contributions in Marine Science,* Supplement, 27.

Herman, W.S., and J.F. Barker. 1977. Effect of mating on monarch butterfly oogenesis. *Experientia* 33:688–689.

Herman, W.S., and D.C. Bennett. 1975. Regulation of oogenesis, female specific protein production, and male and female reproductive gland development by juvenile hormone in the butterfly, *Nymphalis antiopa. Journal of Comparative Physiology* 99:331–338.

Herman, W.S., and S.H. Dallmann. 1981. Endocrine biology of the painted lady butterfly, *Vanessa cardui. Journal of Insect Physiology* 27:163–168.

Herman, W.S., and P. Peng. 1976. Juvenile hormone stimulation of sperm activator production in male monarch butterflies. *Journal of Insect Physiology* 22:579–581.

Herman, W.S., C.A. Lessman, and G.D. Johnson. 1981. Correlation of juvenile hormone titer changes with reproductive tract development in the posteclosion monarch butterfly. *Journal of Experimental Zoology* 218:387–395.

Jaffe, H., A.K. Raina, C.T. Riley, B.A. Fraser, G.M. Holman, R.M. Wagner, R.L. Ridgway, and D.K. Hayes. 1986. Isolation and primary structure of a peptide from the corpora cardiaca of *Heliothis zea* with adipokinetic activity. *Biochemical and Biophysical Research Communications* 135:622–628.

James, D.G. 1982. Ovarian dormancy in *Danaus plexippus* (L.) (Lepidoptera:Nymphalidae)—Oligopause not diapause. *Journal of the Australian Entomological Society* 21:31–35.

James, D.G. 1983. Induction of reproductive dormancy in Australian monarch butterflies, *Danaus plexippus* (L.). *Australian Journal of Zoology* 31:497–504.

James, D.G., and D.F. Hales. 1983. Sensitivity to juvenile hormone is not reduced in clustering monarch butterflies, *Danaus plexippus,* in Australia. *Physiological Entomology* 8:273–276.

Johnson, G.C.D. 1979. The accessory reproductive glands of the male monarch butterfly, *Danaus plexippus:* Their development, fine structure, and biochemistry. Ph.D. thesis. St. Paul: University of Minnesota.

Lessman, C.A. 1979. The biology of juvenile hormone in the adult monarch (*Danaus p. plexippus:* Lepidoptera). Ph.D. thesis. St. Paul: University of Minnesota.

Lessman, C.A., and W.S. Herman. 1981. Flight enhances juvenile hormone inactivation in *Danaus plexippus plexippus* (L.) (Lepidoptera: Danaidae). *Experientia* 37:599–601.

Lessman, C.A., and W.S. Herman. 1983. Seasonal variation in hemolymph juvenile hormone of adult monarchs (*Danaus p. plexippus:* Lepidoptera). *Canadian Journal of Zoology* 61:88–94.

Lessman, C.A., L. Rollins, and W.S. Herman. 1982. Effects of juvenile hormones I, II, and III on reproductive tract growth in male and female monarch butterflies. *Comparative Biochemistry and Physiology* 71A:141–144.

Nicolson, S.W., and R.P. Millar. 1983. Effects of biogenic amines and hormones on butterfly malphighian tubules: Dopamine stimulates fluid secretion. *Journal of Insect Physiology* 29:611–615.

Pan, M.L., and G.R. Wyatt. 1971. Juvenile hormone induces vitellogenin synthesis in the monarch butterfly. *Science* 174:503–505.

Pan, M.L., and G.R. Wyatt. 1976. Control of vitellogenin synthesis in the monarch butterfly by juvenile hormone. *Developmental Biology* 54:127–134.

Rankin, M.A. 1985. Endocrine influence of flight behavior. In *Migration: Mechanisms and adaptive significance,* ed. M.A. Rankin, 817–841. Austin: University of Texas, *Contributions in Marine Science,* Supplement, 27.

Reynolds, S.E. 1983. Bursicon. In *Endocrinology of insects,* eds. G.H. Downer and H. Laufer, 235–250. New York: Alan R. Liss.

Reynolds, S.E., and J.W. Truman. 1983. Eclosion hormone. In *Endocrinology of insects,* eds. G.H. Downer and H. Laufer, 217–234. New York: Alan R. Liss.

Sasaki, M., and L.M. Riddiford. 1984. Regulation of reproductive behaviour and egg maturation in the tobacco hawk moth, *Manduca sexta. Physiological Entomology* 9:315–327.

Sims, S.R. 1984. Reproductive diapause in *Speyeria* (Lepidoptera: Nymphalidae). *Journal of Research on the Lepidoptera* 23:211–216.

Tublitz, N.J., and J.W. Truman. 1985. Intracellular stimulation of an identified neuron evokes cardioacceleratory peptide release. *Science* 228:1013–1014.

Ziegler, R., K. Eckart, H. Schwarz, and R. Keller. 1985. Amino acid sequence of *Manduca sexta* adipokinetic hormone eluciated by combined fast atom bombardment (FAB)/tandem mass spectrometry. *Biochemical and Biophysical Research Communications* 133:337–342.

TEMPERATURE AND THERMOREGULATION IN THE MONARCH BUTTERFLY

Alan R. Masters

ABSTRACT. The monarch butterfly, *Danaus plexippus* (L.), has a complex life history in North America including several generations that breed in the temperate zone, the last of which migrates to overwintering areas in Mexico or California and then back again in the spring. The significance of temperature and thermoregulation to the monarch reflects its complex migratory life cycle. Breeding individuals use thermoregulation both to increase the duration of their activity period and to avoid overheating. Overwintering monarchs thermoregulate to escape lower lethal temperatures and economize necessary activity to avoid lipid depletion. The resulting thermal conflicts in the monarch appear to be resolved through morphological compromise, physiological acclimation, and behavioral plasticity.

INTRODUCTION

Ambient temperatures in natural habitats influence the evolution of morphology, physiology, and behavior of organisms and ecologically limit their range, phenology, and activity. Insects are particularly subject to temperature effects because of their small size and weak endothermic capabilities (May, 1979). Patterns of heat exchange, mechanisms of thermoregulation, and the ecological and evolutionary significance of thermoregulation in insects have been investigated (see reviews by Cloudsley-Thompson, 1970; Heinrich, 1974, 1981; May, 1979, 1984; Kingsolver, 1985), and it is evident that many insects can alter their body temperature from that of their environment in adaptive ways.

Until recently (Masters et al., 1988), studies of temperature regulation in butterflies have been restricted to temperate species active only a portion of the year (Kingsolver, 1985). During this time, such butterflies thermoregulate to increase activity periods (Clench, 1966; Rawlins, 1980; Kingsolver, 1985). While natural selection may favor increasing activity periods for most butterflies, this is not always the case for the monarch (*Danaus plexippus*). Recent range expansion from the tropics allows the monarch to exploit abundant sources of milkweed (Asclepiadaceae) in the temperate zone (Ackery and Vane-Wright, 1984; Brower, 1985), but migration and overwintering have imposed new and varied thermal constraints. Adaptations for increasing activity during the breeding season may be offset by selection for conserving fixed energy reserves while overwintering (Masters et al., 1988).

The monarch provides an ideal opportunity to explore the ecological and evolutionary effects of temperature regulation in a butterfly exposed to a wide range of thermal environments. Here, I integrate studies of butterfly thermal biology with monarch natural history. In doing so, I examine the ecological influence of temperature on all life stages of the monarch and the often opposing temperature constraints acting on summer breeding and overwintering/recolonizing adults. I also explore how thermoregulation aids the monarch in reducing the influence of temperature, and how the evolution of thermoregulation in the monarch reflects its complex life history.

TEMPERATURE REGULATION IN BUTTERFLIES

MODES OF HEAT TRANSFER

Insect body temperature is the result of differential rates of heat gain and loss. An inactive insect may gain heat ectothermally from solar radiation or from direct contact with a warm surface via conduction (Digby, 1955; Church, 1960). Active insects may generate enough heat endothermically from muscle metabolism to raise their body temperature above ambient (Krogh and Zeuthen, 1941; Heinrich, 1974).

Heat loss occurs when heat is radiated from the insect to the surrounding environment and via conduction when in direct contact with a cooler substrate, including air. Convective cooling is also an important avenue of heat loss for butterflies, especially at high wind speeds (Digby, 1955; Kingsolver, 1983), and evaporative heat loss may occasionally occur in insects with concurrent water loss (Church, 1960).

ECTOTHERMY VERSUS ENDOTHERMY

Butterflies are almost entirely ectothermic, with direct solar radiation providing the major source of external heat. Although little myogenic heat is generated in crawling, both shivering (preflight warm-up) and powered flight generate enough heat to raise the thoracic temperature above ambient (Heath and Heath, 1982; Alonso M. et al., this volume). Shivering occurs in many moths (Oosthuizen, 1939; Heath and Adams, 1965; Kammer, 1968; Casey and Joos, 1983) and may raise thoracic temperatures >10°C (Kammer, 1968). Shivering is uncommon in butterflies and is considered unimportant for the temperate species examined (Kingsolver, 1985), with the exception of overwintering monarchs (Masters et al., 1988). Flight also generates heat and under low insolation raises thoracic temperatures 10–20°C in moths (Heath and Adams, 1965; Heinrich, 1971; Heinrich and Casey, 1973; Bartholomew and Epting, 1975) but only 3–6°C in the black swallowtail, *Papilio polyxenes asterius* (Rawlins, 1980). Experimental evidence from other animals (Bartholomew, 1981) suggests that slender body shape, small body size, and poor insulation make endothermy impossible for most butterflies.

MORPHOLOGY

Rates of ectothermic heat gain and loss are influenced by butterfly morphology. A large thorax decreases the rate of heat loss by increasing thermal inertia, a result of lower surface to volume ratio (Bartholomew, 1981). Hence, large butterflies heat and cool more slowly and maintain higher body temperatures (Douglas, 1978). Coloration also influences the rate of ectothermic heating but not cooling (Watt, 1968; Casey, 1981). The darker but similarly sized *Vanessa atalanta rubria* heats more rapidly and to a higher thoracic temperature than *Vanessa virginiensis* when placed under a heat lamp but cools at the same rate (Douglas, 1978). In addition, insulation in the form of hairlike scales on the thorax and basal wing surface is important in restricting heat loss (Church, 1960). Butterflies with insulating scales can be active at lower air temperatures and at greater wind speeds (Kingsolver, 1983).

WINGS AS SOLAR PANELS

Wings do not act as solar collecting panels as earlier maintained by Clench (1966). This is because hemolymph warmed in the wings circulates to the body too slowly to explain the much faster rise in thoracic temperature from radiation; moreover, dead butterflies show the same rates of heating as do live ones (Arnold, 1964; Kammer and Bracchi, 1973; Wasserthal, 1975, 1984; Douglas, 1978). However, the basal third of a basking butterfly's wings provides a greater mass around the thorax for heat retention and produces a boundary layer of still air that reduces convective heat loss (Kingsolver and Koehl, 1985). The wings of some pierids may also provide a reflective surface from which light is deflected to the thorax (Kingsolver, 1985).

BEHAVIORAL THERMOREGULATION

Basking postures that alter the amount of ectothermic heat gain are classified by wing position (Clench, 1966; Kingsolver, 1985): Dorsal baskers position their dorsal wing surface perpendicular to the sun, lateral baskers keep wings closed and position their lateral surface to the sun, and body baskers open their forewings at acute angles to gain both direct insolation and that reflecting from the wing surface. Experimental studies with dead butterflies of several species indicate that dorsal basking is the quickest way to gain solar heat and maintain higher thoracic temperatures for large butterflies, but dorsal and lateral basking result in similar heating rates and thoracic temperatures for smaller butterflies (Douglas, 1978). This may explain the relationship between basking posture and taxonomy suggested by Clench (1966) and Kingsolver (1985): Most large nymphalids (Longstaff, 1906; Vielmetter, 1958) and swallowtails (Rawlins, 1980) are dorsal baskers. Smaller butterflies, including pierids, lycaenids, and satyrs, tend to be lateral or body baskers.

Butterfly larvae also use a variety of thermoregulatory behaviors to control body temperature (Scriber and Lederhouse, 1983). At low and high temperatures outside their normal activity range, larvae of *Colias* (Sherman and Watt, 1973), *Euphydryas aurinia* (Porter, 1982), *Hemileuca oliviae* (Capinera et al., 1980), and the sphingid *Hyles lineata* (Casey, 1976) descend from the upper leaves of their host plants and seek refuge at the base of the plant or in nearby leaf litter. At high temperatures, *Colias* larvae also move to the underside of leaves (Sherman and Watt, 1973). At low ambient temperatures with strong insolation, basking in the sun may raise larval body temperature several degrees above ambient (Porter, 1982). Gregarious behavior in larvae may increase body temperature by reducing exposure to wind and, hence, the cooling effect of convection (Seymour, 1974).

THE ADAPTIVE SIGNIFICANCE OF THERMOREGULATION

Thermoregulation allows butterflies to increase the range of ambient temperatures in which activity and survival are possible. Larvae active for longer periods or at warmer body temperatures grow and

Table 1. Temperature constraints for survival and activity in the monarch butterfly (*D. plexippus*).

Life stage	Category	Thoracic temperature (°C)	Source
Adult	Lower lethal	−14 to −1.5	Anderson and Brower, this volume
	Crawling	4–12	Alonso M. et al., this volume
	Shivering	6–24	Masters et al., 1988; Alonso M. et al., this volume
	Flying	12–41	Douglas, 1978; Masters et al., 1988; Alonso M. et al., this volume
	Upper lethal	>41	Douglas 1978
Immatures			Rawlins and Lederhouse, 1981; Zalucki, 1982
	Lower lethal	<10	
	Development	12–35	
	Upper lethal	>35	

develop more quickly (Porter, 1982; Scriber and Lederhouse, 1983). Adult butterflies able to fly at a broader range of ambient temperatures may have an increased opportunity to feed, mate, and oviposit (Stern and Smith, 1960; Gossard and Jones, 1977; Kingsolver, 1983; Shreeve, 1984). For example, high-altitude *Colias philodice* may be able to lay only 20–50% of their eggs because of temperature-limited flight time (Kingsolver, 1983). An increased range of ambient temperatures for activity may also allow exploitation of new habitat. A temperature-induced polyphenism for basal wing color in *Nathalis iole* allows melanic individuals to fly farther north (Douglas and Grula, 1978), and similar differences in basal wing color allow higher-altitude *Colias meadii* to fly at temperatures 5°C lower than the lower-altitude, lighter-winged *C. philodice* (Kingsolver and Watt, 1983). In addition, thermoregulation may decrease the possibility of death or stress from lethal temperature extremes. Swallowtails often flutter their wings while foraging under high ambient temperatures and insolation levels, possibly to avoid overheating (Douglas, 1978; Rawlins, 1980).

THE INFLUENCE OF TEMPERATURE IN MONARCH BIOLOGY

TEMPERATURE LIMITS FOR SURVIVAL AND ACTIVITY

Survivorship in monarchs is restricted to a range of body temperatures (see Table 1). Temperatures below −14°C, sometimes reached near overwintering areas, are sufficient to kill all monarchs and some death occurs at −1.5°C (Calvert et al., 1982, 1983; Anderson and Brower, this volume). Body temperatures >41°C, possible throughout the summer, create aberrant behavior in monarchs (Douglas, 1978). Immatures begin to show temperature-related mortality below 10°C and above 31°C (Rawlins and Lederhouse, 1981; Zalucki, 1982).

Activity and development are possible within a narrower temperature range: Crawling is possible at ambient temperatures >4°C, shivering occurs >6°C, and flight becomes possible between 11.7 and 16°C (Masters et al., 1988; Alonso M. et al., this volume) until flight ceases between 35 and 41°C (Douglas, 1978; James, this volume). Development of immatures cannot proceed below 12°C and above 31–35°C (Rawlins and Lederhouse, 1981; Zalucki, 1982). The temperature constraints an individual monarch faces differ not only with the stage of development but also with the portion of the monarch's annual cycle in which it finds itself (see Fig. 1).

TEMPERATURE AND THE ANNUAL CYCLE

Each spring, adult monarchs return from overwintering sites in Mexico and California and recolonize the southern United States (see Brower, 1985; Brower and Calvert 1985; Calvert and Brower 1986; Seiber et al., 1987; Cockrell et al., this volume; Malcolm et al., this volume; Nagano et al., this volume). Subsequent generations fly from the south to recolonize most of temperate North America (Malcolm et al., 1987, this volume). Two to five generations are produced throughout the summer before monarchs migrate south in the autumn (Malcolm et al., 1987; Cockrell et al., this volume). Monarchs overwinter in Mexico approximately 4 months before returning north in the spring (Calvert and Brower, 1986). The monarch annual cycle can be divided into two major stages with regard to temperature and its constraints on survival and activity: summer breeding and overwintering. In each case, individuals are faced with very different thermal problems.

SUMMER BREEDING GENERATIONS

Monarchs resulting from the eggs of recolonizing females begin the summer breeding generations, and their thermal problems are similar to other temperate butterflies studied. These individuals spend their lives in temperate North America throughout the summer. They are warm enough to fly most daylight hours and are not threatened by low temperature extremes (Malcolm et al., 1987). The mean thoracic temperature for 45 field-captured flying monarchs in Michigan from April to November was 35.5°C (SD = 3.69) and these monarchs were active at ambient temperatures from 18.0 to 34.5°C (Douglas, 1978). Overheating may be a problem for both adults and immatures. At lower, warmer latitudes, ambient temperatures commonly exceed upper lethal extremes for immatures. Adult monarchs and immatures disappear from north Florida after only a few days with ambient temperatures consistently >33°C (Malcolm et al., 1987; Cockrell et al., this volume) and similarly disappear from all the Gulf states (Lynch and Martin, this volume).

OVERWINTERING

Temperature is a major determinant of overwintering success (Masters, 1985; Masters et al., 1988; Anderson and Brower, this volume; Alonso M. et al., this volume; Brower et al., this volume). Overwintering monarchs are commonly faced with lower temperature extremes while daytime ambient (shade) temperatures fall below flight threshold until all but the very end of the overwintering period (Fig. 1). Activity, including crawling, shivering, and flight, is necessary to avoid low lethal temperatures and mouse predation if individuals are dislodged from tree clusters and land on the ground (Calvert and Cohen, 1983; Calvert et al., 1983; Brower et al., 1985; Glendinning et al., 1988; Glendinning, this volume; Alonso M. et al., this volume). In addition, flight is necessary to reach sources of water, to avoid bird predation, and to follow large-scale repositioning of the colony in response to changing climatic conditions (Brower et al., 1977; Calvert et al., 1979, 1983; Brower and Calvert, 1985; Calvert and Lawton, this volume). Ambient temperatures are high enough to break diapause but mating does not occur until the end of the overwintering period (Herman, 1985; Herman et al., 1989), and upper lethal extremes for survival are never attained (Fig. 1). However, even though lethal extremes are not obtained, heat avoidance behaviors are important to keep body temperatures low. Because sources of nectar are severely limited in overwintering areas, increasing body temperature for activity is offset by the need to conserve fixed lipid stores (Masters et al., 1988). A large percentage of overwintering monarchs die from depleting lipid reserves, and elevated body temperatures, particularly through endothermy, greatly increase this risk (Chaplin and Wells, 1982; Masters et al., 1988; Brower et al., this volume). Overwintering monarchs are therefore faced with the challenge of activity at the lowest possible energetic cost. Heat-avoidance behaviors such as shade-seeking and gliding at cold ambient temperatures aid in keeping thoracic temperatures low.

Migrants returning from overwintering sites in Mexico to temperate North America in the spring are often exposed to temperatures that are too low for flight, as well as extremely cold temperatures that may cause permanent damage or death to the adult (Urquhart, 1960; Malcolm et al., 1987; Cockrell et al., this volume; Anderson and Brower, this volume; Fig. 1). Furthermore, recolonizing females may be faced with temperatures that are too cold for ovarian development and egg maturation (Zalucki, 1981). Both males and females benefit from increasing body temperatures above ambient temperatures because they may avoid low lethal extremes and increase time spent foraging, mating, and ovipositing. Immature stages would also benefit from increasing body temperatures above ambient because they are exposed to lethally low temperatures or temperatures prolonging development throughout the first generation at a given latitude (Malcolm et al., 1987; Cockrell et al., this volume). Heat avoidance is less necessary for recolonizers because ambient temperatures during recolonization do not reach upper lethal extremes (Malcolm et al., 1987).

TEMPERATURE REGULATION IN THE MONARCH BUTTERFLY

MORPHOLOGY AND TEMPERATURE REGULATION

In general, the monarch's morphology is better suited to heat gain and heat retention than most temperate butterflies. Its relatively large body size, large dark thorax, heavy coat of hairlike scales on and around the thorax, and large wing mass near the thorax promote rapid heating and greater differences between ambient and body temperatures (Table 2). However, ectothermic heat gain is slower in larger butterflies like the monarch and overheating is more likely (Douglas, 1978). So, while monarch morphology favors activity at cooler temperatures, the intensity and duration of insolation must be greater to rely on ectothermic heat gain. Endothermic heat, including that generated while shivering and flying, is more difficult to dissipate, making endothermy effective for obtaining or maintaining activity at low ambient temperatures (see Bartholomew, 1981; Heath and Heath, 1982) but problematic at higher ones.

PHYSIOLOGY AND TEMPERATURE REGULATION

Monarchs are capable of endothermy through myogenic heat generation but are largely ectothermic, receiving the majority of their body heat directly from insolation or, less commonly, from warm substrates in their environment (Masters et al., 1988). When crawling, shivering, or flying, thoracic muscles are generating heat (Masters et al., 1988; Alonso M. et al., this volume). Such myogenic heat generated while crawling is probably insufficient to raise body temperature (Alonso M. et al., this volume). Heat generated through shivering, wing-flapping, or flying liberates larger amounts of heat capable of raising monarch body temperatures

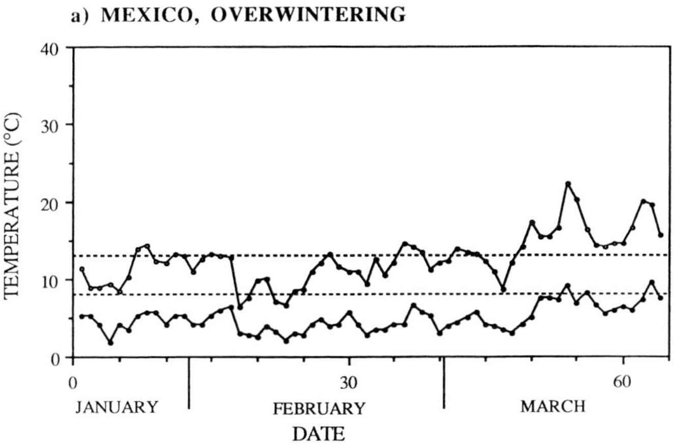

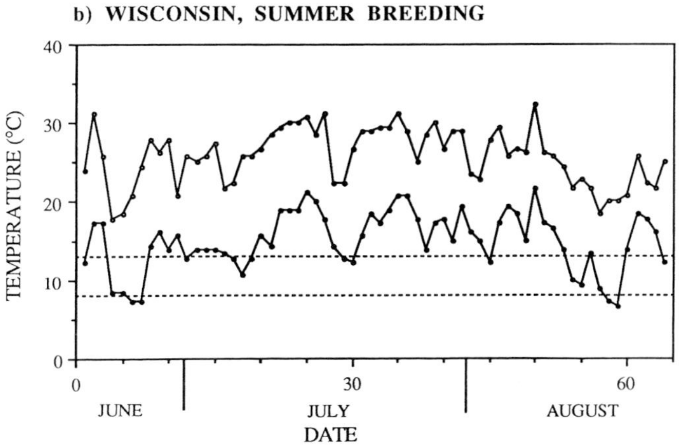

Figure 1. Daily maximum and minimum temperatures faced by monarchs overwintering in Mexico and monarchs breeding in Wisconsin. Temperatures from Mexico originally reported in Brower and Calvert (1985) are measurements for 1979 taken within a colony. Temperatures in Wisconsin originally reported in Malcolm et al. (1987) are taken from N.O.A.A., U.S. Environmental Data Service, National Climatic Center, Asheville, North Carolina for 1979 at St. Croix Falls in northwest Wisconsin. The upper dashed line represents the ambient (shade) temperature at which flight becomes possible, and the lower dashed line represents the ambient temperature at which shivering becomes possible (from Alonso M. et al., this volume). Temperature-related adult mortality occurs below −1.5°C (Anderson and Brower, this volume) and above 40°C (Douglas, 1978).

several degrees above ambient (Kammer, 1981; Masters et al., 1988).

The physiology of shivering in the monarch has been studied by Kammer (1968, 1970, 1981). During flight, antagonistic flight muscles are excited by a short burst of nerve impulses and contract alternately. While shivering, these elevator and depressor muscles contract simultaneously, producing heat but not mobility. For monarchs held at an ambient temperature of 15.5°C, shivering increases the thoracic temperature at the same rate as flying and with the same metabolic costs (Kammer, 1970). At low levels of insolation and ambient temperature, conditions commonly found in Mexican overwintering areas, shivering is a mechanism that increases thoracic temperature for more rapid crawling (Alonso M. et al., this volume) and for attaining thoracic temperatures necessary for flight (Masters et al., 1988).

At higher thoracic temperatures and insolation, flight is possible through ectothermic heat gain alone (Kammer, 1970; Masters et al., 1988). As thoracic temperature increases, so does power output, resulting in faster flight with more control (Heath and Heath, 1982). At high ambient temperatures, there is a possibility of overheating during flight because heat generated by flight muscles exceeds heat loss to the atmosphere. Possible mechanisms for avoiding such overheating include reducing endothermy through cessation of powered flight by landing or

Table 2. Rates of heat gain and magnitude of thoracic excess in butterflies of different body size as measured in thoracic diameter. Maximum heating rates were calculated and the maximum thoracic temperature excess (thoracic − ambient temperature) is reported from heating curves generated by Douglas (1978) using freshly killed specimens in dorsal basking posture with insolation 0.84 cal/cm²/min at an ambient temperature of 20°C.

Species	Diameter of thorax (mm)	Heating rate (°C/min)	Thoracic excess (°C)
Celestrina pseudargiolus (Lycaenidae)	2.96	15	12
Eurema lisa (Pieridae)	3.68	10	15
Colias eurytheme (Pieridae)	5.66	12	23
Limenitis archippus (Nymphalidae)	6.70	5	26
Danaus plexippus (Nymphalidae)	8.86	4	28

increasing time spent gliding or soaring (Masters et al., 1988).

BEHAVIOR AND TEMPERATURE REGULATION

Ovipositing females preferentially place their eggs on more exposed milkweeds in the sun (Urquhart, 1960; Zalucki, 1982). Monarch larvae select portions of milkweed plants receiving direct sunlight and orient their bodies to maximize the effects of insolation and to minimize convective heat loss (Rawlins and Lederhouse, 1981). Larval posturing in fourth- and fifth-instar monarchs can increase body temperature by as much as 8°C, potentially reducing the duration of late instars by 10–50% (Rawlins and Lederhouse, 1981). The ability of monarch larvae to bask in direct sunlight on the upper surface of leaves may be a result of their cardenolide-based defense mechanism (Rawlins and Lederhouse, 1981; Brower, 1984). Heat-avoidance behaviors include descending the milkweed and resting in nearby leaf litter, or crawling to the undersides of leaves (Urquhart, 1960; Rawlins and Lederhouse, 1981).

Like other large butterflies, the adult monarch is ordinarily a dorsal basker (Clench, 1966; Douglas, 1978; Kingsolver, 1985; Masters et al., 1988). By opening its wings with the dorsal surface perpendicular to the sun, monarchs can ectothermically raise their thoracic temperatures rapidly under conditions of low ambient temperature and high insolation (Fig. 2). Monarchs that bask with wings closed over the thorax or those with wings ablated have lower rates of ectothermic heat gain and smaller differences between thoracic and ambient temperatures (Douglas, 1978; Table 3). Dorsal basking gives way to resting with wings closed in monarchs at high ambient temperatures and levels of insolation (Douglas, 1978). In addition, monarchs may cease activity altogether and rest in shady locations when air temperatures become too great (Urquhart, 1960; Douglas, 1978).

EVOLUTIONARY CONFLICTS FROM ECOLOGICAL DIFFERENCES

The importance of temperature as a selective force in the evolution of monarch morphology, physiology, and behavior is difficult to ascertain, but it is clear that summer breeding and overwintering/recolonizing individuals are under different temperature-related selective pressures. Natural selection should favor traits in summer breeding monarchs that allow activity at higher temperatures, unlike overwintering individuals that benefit from activity at lower temperatures while avoiding lower lethal extremes. Such evolutionary conflicts may be resolved in a number of ways: one best overall phenotype that is not optimal in any one generation, environmental control of certain traits to produce different morphs depending on environmental conditions, and behavioral thermoregulation. All are probably occurring in the monarch (see also James, this volume).

MORPHOLOGICAL ADAPTATIONS AND COMPROMISE

As a whole, the monarch's adult morphology is better suited to heat gain and heat retention than most temperate butterflies. Summer breeding, tropical, and subtropical populations of monarchs that are rarely exposed to cool temperatures and often faced with upper lethal temperatures might be expected to have smaller bodies, lighter coloration, and less insulation than overwintering monarchs. Color and insulation have not been examined for these monarch populations, nor have measures of body size.

To determine if body size differences occur, I measured 40 monarchs (20 of each sex) from each of three groups: June 1985 south Florida nonmigratory monarchs, July 1985 summer breeding individuals from Wisconsin, and January 1986 overwintering monarchs from Mexico. I measured maximum thoracic width, height, and length of the thorax and calculated thoracic volume by multiplying these parameters. Thoracic volume is an important thermal measure because it reflects surface area exposed to insolation and rate of heat loss through thermal inertia. I found that the thoracic volumes of overwintering monarchs ($\bar{X} = 665.35$ mm³, SD = 67) were 10% larger than breeding individuals ($\bar{X} = 603.92$ mm³, SD = 42). The means were significantly different by Student's t test at $p < 0.05$. Likewise, Wisconsin breeding monarchs

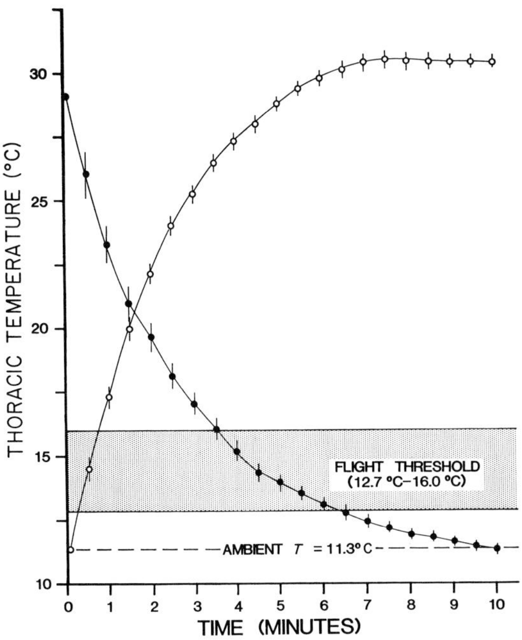

Figure 2. Changes in thoracic temperatures of resting overwintering monarchs ($\bar{X} \pm SE$) when moved from the sun to the shade and vice versa. ○ = heating curve, ● = cooling curve (from Masters et al., 1988).

had thoracic areas 8% larger than nonmigratory monarchs from south Florida ($\bar{X}$ = 556.03 mm^3, SD = 29), and the means were significantly different (Student's *t* test, $p < 0.05$). Smaller body size in south Florida monarchs may be a result of selection for lowering thoracic temperature excess. Wisconsin butterflies may not be as small because selection for small size is offset during the overwintering period, presenting a compromise between optimum size for breeding and overwintering. However, while these measures support the idea of a thermally based evolutionary compromise, they may be explained equally well in terms of other hypotheses (e.g., migrating ability). It is also possible that such size differences reflect a polyphenism where temperature determines adult size. Smaller wing size was found in first-generation monarchs that grew up in the southeastern United States, compared to later, more northern populations (Herman, 1985).

PHYSIOLOGICAL ADAPTATIONS AND ACCLIMATION

There is considerable evidence for environmental control over monarch reproduction and migration (Herman, 1985; James, this volume). Short photoperiod and cool temperatures induce reproductive diapause and migratory behavior through changes in juvenile hormone titers (Herman, 1985, this volume). Similarly, these cues bring on physiological changes that elicit clustering behavior (Herman, 1985). Clustering is important to overwintering because individuals outside of clusters are subject to greater temperature extremes (Calvert and Brower, 1986) and lethal freezing temperatures (Calvert and Cohen, 1983; Calvert et al., 1983).

Table 3. The effects of body and wing orientation on the rate of heating and thoracic temperature maintained for individual freshly killed monarch butterflies (*D. plexippus*) with thoracic diameter of 8.86 mm. Heating rates and thoracic temperatures are reported here from figures in Douglas (1978), where individuals were pinned into orientations and placed under a light source producing 0.84 cal/cm²/min at an ambient temperature of 20°C.

Orientation of thorax to light source	Maximum heating rate (°C/min)	Thoracic temperature (°C)
Dorsal, wings open	4.0	44
Lateral, wings closed	3.7	42
Dorsal, wings ablated	2.7	35
Lateral, wings ablated	2.2	33
Dorsal, wings closed	1.2	28

Temperature controls physiological characters that are important to monarch thermoregulation through acclimation. Acclimation may explain why overwintering monarchs can withstand colder temperatures, before freezing damages or kills, than monarchs from summer populations (Anderson and Brower, this volume). Likewise, endothermy becomes possible at lower temperatures once monarchs have been kept for several days at cooler ambient temperatures (Kammer, 1971). More monarchs kept in the laboratory at 4–5°C shiver at ambient temperatures of 15–16°C than those kept at 23–24°C (Kammer, 1971), and overwintering monarchs in Mexico display shivering behavior at ambient temperatures as low as 6°C (Alonso M. et al., this volume) compared to a minimum of 13°C for warmer, California overwintering monarchs (Kammer, 1970). Like other temperate butterflies, summer breeding monarchs may have no need for endothermic shivering as dorsal basking is an effective means of rapid ectothermic heat gain and no shivering in breeding monarchs has been reported. Conversely, overwintering monarchs on the ground, near flight threshold, shiver because climatic factors, such as low ambient temperatures and/or low levels of insolation, may make ectothermic heat gain ineffective for hours or days (Masters et al., 1988). Only if these monarchs shiver can they attain body temperatures necessary for activity to escape lethal temperatures on the ground, such as when dislodged from clusters. The monarch is the only species to date providing evidence for the adaptiveness of shivering in butterflies.

Temperatures at which other activities occur also differ between breeding and overwintering generations, further suggesting acclimation to temperature. Breeding individuals in Michigan are active at a mean thoracic temperature of 35.5°C and are flying, foraging, mating, and ovipositing, but show

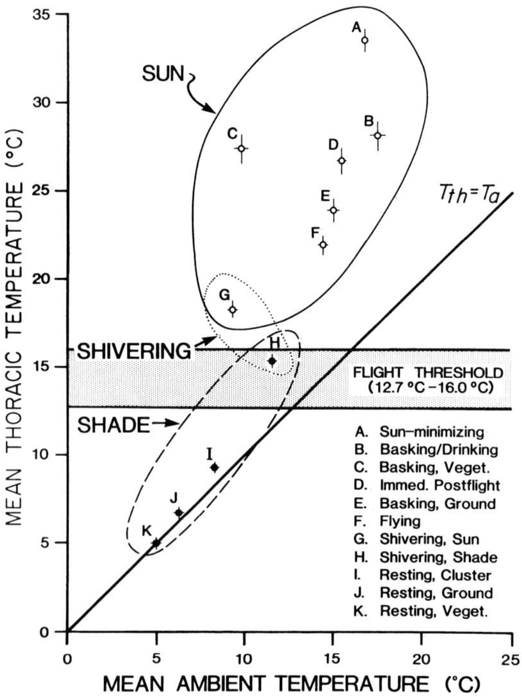

Figure 3. Relationships of mean and ambient shade temperatures (T_a) and mean thoracic temperatures (T_{th}) of 440 overwintering monarch butterflies engaged in various behaviors in either sun or shade during January 1983. Each point and standard error represents 40 butterflies and 40 corresponding ambient temperatures (from Masters et al., 1988).

heat stress at body temperatures near 41°C (Douglas, 1978). Overwintering monarchs, on the other hand, are active at mean thoracic temperatures between 22.0 and 28.2°C before activity ceases and heat avoidance behavior is shown at 33.6°C (Masters et al., 1988; Fig. 3). Such acclimation is adaptive for overwintering monarchs because activity occurs at lower ambient temperatures and with lower metabolic costs, despite a possible loss of power output and control in flight.

BEHAVIORAL PLASTICITY

While morphology and physiology control the range of possible thoracic temperatures, behavior allows temperature regulation. It has already been noted that monarchs differ from most temperate butterflies in using endothermic shivering to warm to flight temperatures. Ectothermic heat gain and avoidance through changes in behavior, particularly posturing with respect to the sun, provide a wide choice of thoracic temperatures at a given ambient temperature. In the laboratory monarch thoracic temperature ranges 16°C through orientation alone (Douglas, 1978) and overwintering monarchs in the sun showed body temperatures of 9.4–37.2°C, a range of >27°C, over a much smaller 12.3°C range of ambient temperatures (Fig. 3). The ability to thermoregulate is not universal in butterflies. Some tropical species, for instance, show no behavioral thermoregulation (Heinrich, 1972). The monarch's recent range expansion into the temperate zone from the tropics almost certainly required behavioral thermoregulation, as it does today. More observations of temperature regulation in tropical monarchs are needed to determine whether existing behaviors aided in range expansion and migration, in which case they are exaptations, or many of the thermoregulatory behaviors reported here are relatively recent, specific adaptations to overwintering and temperate breeding.

SUMMARY

Unlike most temperate butterflies that restrict activity to warm seasons, the North American monarch may be active throughout the year using a system of southern migration and overwintering to escape freezing temperatures in the north. This complex life-history strategy exposes a single population of monarchs to very different temperature constraints. Temperature regulation is important in all phases of the monarch life cycle. Larvae can optimize development time by basking in cool weather and guard against overheating by seeking shade. Breeding monarchs are rarely faced with cool temperatures throughout the summer, but may avoid overheating through heat-avoidance behaviors. Overwintering monarchs reduce the likelihood of freezing by remaining in clusters and escaping from the ground by either basking in sunlight or shivering at temperatures just below flight threshold. The monarch has responded to differing temperature-related selection pressures at both phases in the life cycle with a possible compromise in body size, a series of physiological acclimations, and behavioral thermoregulation.

ACKNOWLEDGMENTS

I would like to thank Steve Malcolm, Myron Zalucki, and Lincoln Brower for help and comments on this manuscript. I give special thanks to Lincoln Brower for introducing me to the monarch and for providing support, guidance, and patience. I would also like to thank Lincoln and Andy Brower for providing monarch collections.

LITERATURE CITED

Ackery, P.R., and R. I. Vane-Wright. 1984. *Milkweed butterflies: Their cladistics and biology.* London: British Museum (Natural History); and Ithaca, NY: Cornell University Press, 425 pp.

Arnold, J.W. 1964. Blood circulation in insect wings. *Memoirs of the Entomological Society of Canada* 38:1–48.

Bartholomew, G.A. 1981. A matter of size: An examination of endothermy in insects and terrestrial vertebrates. In *Insect thermoregulation,* ed. B. Heinrich, 45–78. New York: Wiley.

Bartholomew, G.A., and R.J. Epting. 1975. Allometry of postflight cooling rates in moths: A comparison with vertebrate homeotherms. *Journal of Experimental Biology* 63:603–613.

Brower, L.P. 1984. Chemical defence in butterflies. In *The biology of butterflies*, eds. R.I. Vane-Wright and P.R. Ackery, *Symposia of the Royal Entomological Society of London* 11:109–134, London and New York: Academic Press (1989, Princeton, NJ: Princeton University Press).

Brower, L.P. 1985. New perspectives on the migration biology of the monarch butterfly, *Danaus plexippus* L. In *Migration: Mechanisms and adaptive significance*, ed. M.A. Rankin, 748–785. Austin: University of Texas, *Contributions in Marine Science*, Supplement, 27.

Brower, L.P., and W.H. Calvert. 1985. Foraging dynamics of bird predators on overwintering monarch butterflies in Mexico. *Evolution* 39:852–868.

Brower, L.P., W.H. Calvert, L.E. Hedrick, and J. Christian. 1977. Biological observations on an overwintering colony of monarch butterflies (*Danaus plexippus*, Danaidae) in Mexico. *Journal of the Lepidopterists' Society* 31:232–242.

Brower, L.P., B.E. Horner, M.M. Marty, C.M. Moffitt, and B. Villa R. 1985. Mice (*Peromyscus maniculatus labecula, P. spicilegus*, and *Microtus mexicanus*) as predators of overwintering monarch butterflies (*Danaus plexippus*) in Mexico. *Biotropica* 17:89–99.

Calvert, W.H., and L.P. Brower. 1986. The location of monarch butterfly (*Danaus plexippus* L.) overwintering colonies in Mexico in relation to topography and climate. *Journal of the Lepidopterists' Society* 35:216–225.

Calvert, W.H., and J.A. Cohen. 1983. The adaptive significance of crawling up onto foliage for the survival of grounded overwintering monarch butterflies (*Danaus plexippus*) in Mexico. *Ecological Entomology* 8:471–474.

Calvert, W.H., L.E. Hedrick, and L.P. Brower. 1979. Mortality of the monarch butterfly (*Danaus plexippus* L.): Avian predation at five overwintering sites in Mexico. *Science* 204:847–851.

Calvert, W.H., W. Zuchowski, and L.P. Brower. 1982. The impact of forest thinning on microclimate in monarch butterfly (*Danaus plexippus* L.) overwintering areas of Mexico. *Boletín de la Sociedad Botánica de Mexico* 42:11–18.

Calvert, W.H., W. Zuchowski, and L.P. Brower. 1983. The effect of rain, snow and freezing temperatures on overwintering monarch butterflies in Mexico. *Biotropica* 15:42–47.

Capinera, J.L., L.F. Wiener, and P.R. Anamosa. 1980. Behavioral thermoregulation by late instar range caterpillar larvae *Hemileuca oliviae* Cockerell (Lepidoptera: Saturniidae). *Journal of the Kansas Entomological Society* 53:631–638.

Casey, T.M. 1976. Activity patterns, body temperature, and thermal ecology in two desert caterpillars (Lepidoptera: Sphingidae). *Ecology* 52:485–497.

Casey, T.M. 1981. Behavioral mechanisms of thermoregulation. In *Insect thermoregulation*, ed. B. Heinrich, 79–114. New York: Wiley.

Casey, T.M., and B.A. Joos. 1983. Morphometrics, conductance, thoracic temperature, and flight energetics of noctuid and geometrid moths. *Physiological Zoology* 56:160–173.

Chaplin, S.B., and P.H. Wells. 1982. Energy reserves and metabolic expenditures of monarch butterflies overwintering in southern California. *Ecological Entomology* 7:249–256.

Church, H.K. 1960. Heat loss and the body temperature of flying insects. II. Heat conduction within the body and its loss by radiation and convection. *Journal of Experimental Biology* 37:187–212.

Clench, H.K. 1966. Behavioral thermoregulation in butterflies. *Ecology* 47:1021–1034.

Cloudsley-Thompson, J.L. 1970. Terrestrial invertebrates. In *Comparative physiology of thermoregulation*, Vol. 1, Invertebrates and nonmammalian vertebrates, ed. G.C. Whittow, 15–77. New York: Academic Press.

Digby, P.S.B. 1955. Factors influencing the temperature excess of insects in sunshine. *Journal of Experimental Biology* 32:279–298.

Douglas, M.M. 1978. The behavioral and biophysical strategies of thermoregulation in temperate butterflies. Ph.D. thesis. Lawrence: University of Kansas.

Douglas, M.M., and J. Grula. 1978. Thermoregulatory adaptations allowing ecological range expansion by the pierid butterfly, *Nathalis iole*. *Evolution* 32:776–783.

Glendinning, J.I., A. Alonso Mejía, and L.P. Brower. 1988. Behavioral and ecological interactions of foraging mice (*Peromyscus melanotis*) with overwintering monarch butterflies (*Danaus plexippus*) in Mexico. *Oecologia* 75:222–227.

Gossard, T., and R. Jones. 1977. The effects of age and weather on egg-laying in *Pieris rapae* L. *Journal of Animal Ecology* 14:65–71.

Heath, J.E., and P.A. Adams. 1965. Temperature regulation in the sphinx moth during flight. *Nature* 205: 309–310.

Heath, J.E., and M.S. Heath. 1982. Energetics of locomotion in endothermic insects. *Annual Review of Physiology* 44:133–143.

Heinrich, B. 1971. Temperature regulation of the sphinx moth, *Manduca sexta* I. Flight energetics and body temperature regulation during free and tethered flight. *Journal of Experimental Biology* 54:141–152.

Heinrich, B. 1972. Thoracic temperatures of butterflies in the field near the equator. *Comparative Biochemistry Physiology* 43A:459–467.

Heinrich, B. 1974. Thermoregulation in endothermic insects. *Science* 185:747–756.

Heinrich, B. 1981. *Insect thermoregulation*. New York: Wiley.

Heinrich, B., and T.M. Casey. 1973. Metabolic rate and endothermy in sphinx moths. *Journal of Comparative Physiology* 82:195–206.

Herman, W.S. 1985. Hormonally mediated events in adult monarch butterflies. In *Migration: Mechanisms and adaptive significance*, ed. M.A. Rankin,

799–815. Austin: University of Texas, *Contributions in Marine Science,* Supplement, 27.

Herman, W.S., L. P. Brower, and W.H. Calvert. 1989. Reproductive tract development in monarch butterflies overwintering in California and Mexico. *Journal of the Lepidopterist's Society* 43:50–58.

Kammer, A.E. 1968. Motor patterns during flight and warm-up in Lepidoptera. *Journal of Experimental Biology* 48:89–109.

Kammer, A.E. 1970. Thoracic temperature, shivering, and flight in the monarch, *Danaus plexippus* (L.). *Zeitschrift für Vergleichende Physiologie* 68:334–344.

Kammer, A.E. 1971. Influence of acclimation temperature on the shivering behavior of the butterfly *Danaus plexippus* (L.). *Zeitschrift für Vergleichende Physiologie* 72:364–369.

Kammer, A.E. 1981. Physiological mechanisms of thermoregulation. In *Insect thermoregulation,* ed. B. Heinrich, 115–158. New York: Wiley.

Kammer, A.E., and J. Bracchi. 1973. Role of wings in the absorption of radiant energy by a butterfly. *Comparative Biochemistry and Physiology* 45A:1057–1063.

Kingsolver, J.G. 1983. Ecological significance of flight activity in *Colias* butterflies: Implication for reproductive strategy and population structure. *Ecology* 64:546–551.

Kingsolver, J.G. 1985. Butterfly thermoregulation: Organismic mechanisms and population consequences. *Journal of the Lepidopterists' Society* 24:1–20.

Kingsolver, J.G., and M.A.R. Koehl. 1985. Aerodynamics, thermoregulation, and evolution of insect wings: Differential scaling and evolutionary change. *Evolution* 39:448–504.

Kingsolver, J.G., and W.B. Watt. 1983. Thermoregulatory strategies in *Colias* butterflies: Thermal stress and the limits to adaptation in temporally varying environments. *American Naturalist* 121:32–55.

Krogh, A., and E. Zeuthen. 1941. The mechanism of flight preparation in some insects. *Journal of Experimental Biology* 18:1–10.

Longstaff, G.B. 1906. Some rest-attitudes of butterflies. *Transactions of the Royal Entomological Society of London* 1:97–118.

Malcolm, S.B., B.J. Cockrell, and L.P. Brower. 1987. Monarch butterfly voltinism: Effects of temperature constraints at different latitudes. *Oikos* 49:77–82.

Masters, A.R. 1985. The influence of temperature on the overwintering biology of the monarch butterfly (*Danaus plexippus*). M.S. thesis. Gainesville: University of Florida.

Masters, A.R., S.B. Malcolm, and L.P. Brower. 1988. Monarch butterfly (*Danaus plexippus*) thermoregulatory behavior and adaptations for overwintering in Mexico. *Ecology* 69:458–467.

May, M.L. 1979. Insect thermoregulation. *Annual Review of Entomology* 24:313–349.

May, M.L. 1984. Thermoregulation. In *Comprehensive insect physiology, biochemistry, and pharmacology,* Vol. 4, eds. G.A. Kerbut and L.I. Gilbert, 507–552. New York: Pergamon.

Oosthuizen, M.J. 1939. The body temperature of *Samia cecropia* Linn. (Lepidoptera, Saturniidae) as influenced by muscular activity. *Journal of the Entomological Society of South Africa* 2:63–73.

Porter, K. 1982. Basking behaviour in larvae of the butterfly *Euphydryas aurinia. Oikos* 38:308–312.

Rawlins, J.E. 1980. Thermoregulation by the black swallowtail butterfly *Papilio polyxenes* (Lepidoptera: Papilionidae). *Ecology* 61:345–357.

Rawlins, J.E., and R.C. Lederhouse. 1981. Developmental influences of thermal behavior on monarch caterpillars (*Danaus plexippus*): An adaptation for migration (Lepidoptera: Nymphalidae: Danainae). *Journal of the Kansas Entomological Society* 54: 387–408.

Scriber, J.M., and R.C. Lederhouse. 1983. Temperature as a factor in the development and feeding ecology of tiger swallowtail caterpillars, *Papilio glaucus* (Lepidoptera). *Oikos* 40:95–102.

Seiber, J.N., L.P. Brower, S.M. Lee, M.M. McChesney, H.T.A. Cheung, C.J. Nelson, and T.R. Watson. 1987. The cardenolide connection between overwintering monarch butterflies from Mexico and their larval foodplant, *Asclepias syriaca. Journal of Chemical Ecology* 12:1157–1170.

Seymour, R.S. 1974. Convective and evaporative cooling in sawfly larvae. *Journal of Insect Physiology* 20: 2447–2457.

Sherman, P.W., and W.B. Watt. 1973. The thermal ecology of some *Colias* butterfly larvae. *Journal of Comparative Physiology* 83:25–40.

Shreeve, T.G. 1984. Habitat selection, mate location, and microclimatic constraints on the activity of the speckled wood butterfly, *Pararge aegeria. Oikos* 43: 371–377.

Stern, V., and R.F. Smith. 1960. Factors affecting egg production and oviposition in populations of *Colias philodice eurytheme* Boisduva (Lepidoptera: Pieridae). *Hilgardia* 29:411–454.

Urquhart, F.A. 1960. *The monarch butterfly.* Toronto: University of Toronto Press, 361 pp.

Vielmetter, W. 1958. Physiologie der Verhaltens zur Sonnenstrahlung bei dem Tagfalter *Argynnis paphia* L. I. Untersuchungen im Freiland. *Journal of Insect Physiology* 2:13–37.

Wasserthal, L.T. 1975. The role of butterfly wings in regulation of body temperature. *Journal of Insect Physiology* 21:1921–1930.

Wasserthal, L.T. 1984. Hemolymph flows in the wings of butterflies visualized by vital staining. *Journal of Comparative Physiology* 83:45–49.

Watt, W.B. 1968. Adaptive significance of pigment polymorphism in *Colias* butterflies. I. Variation in melanin pigment in relation to thermoregulation. *Evolution* 22:437–458.

Zalucki, M.P. 1981. The effects of age and weather on egg laying in *Danaus plexippus* L. (Lepidoptera: Danaidae). *Researches in Population Ecology* 23: 318–327.

Zalucki, M.P. 1982. Temperature and rate development in *Danaus plexippus* L. and *D. chrysippus* L. (Lepidoptera: Nymphalidae). *Journal of the Australian Entolomological Society* 21:241–246.

COLD-HARDINESS IN THE ANNUAL CYCLE OF THE MONARCH BUTTERFLY

James B. Anderson and Lincoln P. Brower

ABSTRACT. Almost the entire eastern North American population of the monarch butterfly, *Danaus plexippus* (L.), migrates each autumn to overwinter at high-altitude sites in central Mexico. Although these sites are usually cool and stable through the winter, occasional episodes of freezing weather occur. Since most overwintering butterflies survive, we suspected some mechanism of freeze-protection. Experiments determined that migrating and overwintering monarchs supercool to lower temperatures than freshly eclosed or summer reproductive monarchs or butterflies from two other closely related Danaine species. The seasonal increase in monarch cold-hardiness, probably a physiological adaptation to the overwintering environment, is moderate compared with other, more boreal insect species but appears adequate for all but extreme low temperatures at the overwintering sites. The importance of an intact, closed-canopy forest in protecting the overwintering monarchs from cold extremes is discussed.

INTRODUCTION

MONARCH OVERWINTERING

Among insects, the monarch butterfly, *Danaus plexippus* (L.), is one of the most striking and well-studied examples of seasonal migration to an overwintering site (Tauber et al., 1986). Each autumn, American monarchs east of the Rocky Mountains abandon their northern breeding ranges and migrate southwestward to form dense overwintering aggregations high in the transvolcanic mountains of central Mexico (Urquhart and Urquhart, 1976, 1977; Brower, 1977, 1985; Calvert and Brower, 1986). Within limited boreal forests of oyamel fir (*Abies religiosa* H.B.K.) only found on the higher mountains, monarch colonies form each winter at sites on southwestern slopes between altitudes of 2,300 and 3,100 m (Calvert and Brower, 1986). At these elevations, ambient temperatures are sufficiently cool and stable throughout the winter to allow most of the butterflies to conserve lipid reserves (Masters et al., 1988) and to remain in a state of reproductive diapause until the spring emigration (Barker and Herman, 1976; Herman, 1981, 1985). In general, the balance of moisture, insolation, and temperature, which allows movement to water as well as repositioning within the clusters, seems ideally suited for monarch survival and for maintenance of the dense colonies (Calvert and Brower, 1986; Masters et al., 1988; Alonso M. et al., this volume).

However, extreme temperature fluctuations sometimes occur at these altitudes, and despite the ameliorating influence of the forest canopy (Calvert and Brower, 1981), the overwintering monarchs are occasionally exposed to severe cold. During January, the coldest month, freezing temperatures are frequent at night, especially close to the ground in open areas (Calvert et al., 1982, 1983; Calvert and Cohen, 1983; Alonso M. et al., this volume), and episodes of freezing weather occur during or immediately after winter storms, which have been observed in most years since the discovery of the Mexican overwintering sites in 1974 (Urquhart and Urquhart, 1976; Brower, 1985; Calvert et al., 1989). Although many monarchs succumb during these periods of cold, 80% survived the most intense storm yet documented (Calvert et al., 1983, 1989; Calvert and Brower, 1986). Moreover, Urquhart (1960) reported that some adult monarchs that had been acclimated to 12-hr low temperature regimes in the laboratory could survive temperatures as low as −6.8°C. Hence, we suspected that overwintering monarchs must have some mechanism of freeze-protection.

INSECT COLD-HARDINESS

Many insects possess physiological mechanisms that protect them from subzero temperatures (Dankø, 1978; DeVries, 1983; Tauber et al., 1986). Only a few species can survive being frozen, a lethal event for most organisms, by tolerating ice within their cells (Salt, 1961) or by directing ice formation away from intracellular tissues with extracellular nucleating agents (Miller, 1982; Zachariassen and Hammel, 1976). Most insect cold-hardiness, however, is due to increased concentrations of antifreeze biochemicals, which help prevent the formation of ice anywhere within the body. These chemicals can function in three ways: (1) by lowering the freezing point of body fluids, (2) by lowering the supercooling point of body fluids, or (3) by inhibiting the growth of ice crystals within the body, producing an effect called "thermal hysteresis." For clarity, we will briefly define these concepts.

The *freezing point* of a solution is the highest temperature at which an ice crystal will grow during gradual cooling, and the *melting point* is the temperature at which an ice crystal begins to shrink during gradual warming. The two are usually the same. Freezing point depression is directly related to the number of solute molecules dissolved, irrespective of solute, each mole of which lowers the freezing point of an aqueous solution by 1.86°C. Because high concentrations of solute are required to provide relatively small freezing-point depressions, this mechanism is biologically impractical (although it is the principle of automobile antifreeze).

The *supercooling point* (SCP) of a solution is the highest temperature at which spontaneous freezing occurs during gradual cooling when no seed crystal is present (Salt, 1961; Zachariassen, 1980; Somme, 1982). In the absence of ice crystals or nucleating agents, pure water and aqueous solutions will commonly supercool, or persist in the liquid phase at a lower temperature than the freezing point. This state is unstable, with decreasing temperature increasing the probability of some water molecules coming into organization as a microscopic seed crystal, after which freezing rapidly progresses by crystal growth. This sudden transition is the SCP, and can be conveniently recognized by a sharp increase in the temperature of the solution as the latent heat of fusion of water is released. Small molecular weight polar compounds such as simple sugars or small polyhydric alcohols are very effective at producing a low SCP, probably by disrupting the formation of initial seed crystals.

A third type of antifreeze mechanism involves a "*thermal hysteresis*," or difference between a solution's freezing point and melting point. This effect is produced by special proteins or glycoproteins that adsorb to incipient ice crystals, inhibiting further spread of ice within the body (Duman, 1979; DeVries, 1983; Duman and Horwath, 1983). Since these proteins inhibit ice crystal growth, they cause the measured freezing point to be several degrees lower than the melting point and, thus, are called thermal hysteresis proteins (THPs). When ice crystals finally grow at the lowered freezing point, they form unusual and characteristic needlelike shapes as a result of the protein inhibitor. Unlike supercooling antifreezes, THPs function in the presence of ice crystals and seem particularly effective at preventing ice crystals encountered in the environment from radiating by inoculative freezing across the skin or exoskeleton into body fluids (DeVries, 1971, 1980; Duman et al., 1982).

MONARCH COLD-HARDINESS

Variations on these physiological themes of cold-hardiness have been found in many insects, other poikilotherms, and plants and may have evolved many times. The monarch butterfly is the only one of 157 species of the tropical subfamily Danainae to migrate extensively into the temperate zone (Ackery and Vane-Wright, 1984), and it appears to have evolved some protection against freezing. In the present study, we explored two questions: (1) Does the hemolymph of adult monarch butterflies show the characteristic thermal hysteresis indicative of THPs? These proteins would probably be the optimal antifreeze for climatic conditions at the overwintering sites because they would prevent external ice (snow, sleet, freezing rain, frost) from radiating into the bodies of exposed monarchs. (2) Do overwintering monarchs have lower SCPs compared with those of monarchs tested in the summer and those of other danaid species that do not extend their ranges into northern areas?

METHODS

MEASUREMENT OF THERMAL HYSTERESIS

Hemolymph from overwintering and migratory monarchs was tested for thermal hysteresis. Each sample of hemolymph (2–3 μl) was extracted by abdominal puncture into a 10-μl capillary tube, which was then sealed (one end by flame and the other end by injection of a layer of mineral oil) and arranged so that the fused end projected from the bottom of a rubber stopper that could be plugged into a small insulated chamber. Ethylene glycol from a Haake cold temperature bath (model A80) was circulated through the chamber, and a dissecting microscope was positioned at plexiglass viewports so that crystal growth and shrinkage could be closely observed. Polarizing filters at the viewports allowed a clearer image of the ice crystal. Hemolymph temperature was estimated with a YSI telethermometer connected to a small surface thermistor (YSI, 400 series) affixed in the ethylene glycol within 5 mm of the hemolymph sample. After an ice crystal was induced in the hemolymph by spraying the capillary tube with freon spray (Cryoquik, American Scientific), the tube and stopper were quickly inserted into the chamber and the bath temperature raised or lowered at 0.2°C/min. The freezing point was taken as the temperature at which the seed crystal began to grow during temperature decrease, and the melting point was the temperature at which the seed crystal began to melt during temperature increase.

MEASUREMENT OF SUPERCOOLING

For determination of the SCPs of live monarchs, the Haake cooling bath was set to lower the temperature of the ethylene glycol solution from just above zero to −25°C, at a rate of 0.44°C/min. During this slow drop in temperature, several 25-ml vials (usually seven at a time), each plugged with a rubber stopper and containing a live monarch, were submerged (except for the top centimeter) in the bath liquid. A small YSI surface thermistor was secured with masking tape to the anterior ventral surface of the abdomen, and body temperatures were recorded at 1-min intervals using the YSI telethermometer. We previously determined that this measurement of body temperature was identical to that taken within the thorax by a needle thermistor (YSI series 500, 34-gauge hypodermic probe). The monarchs, then, were not in contact with the bath liquid, but with the air inside the immersed vials (a capillary tube through the stopper allowed pressure equilibrium). When freezing occurred at the SCP, we recorded this point as the lowest body temperature reached before the sharp elevation in body temperature caused by release of the latent heat of fusion as ice radiated within the butterfly (Salt, 1961; Somme, 1964).

OVERWINTERING SAMPLES

In 1984, SCPs were initially tested in the laboratory at the University of Florida on a small sample of butterflies sent back alive on ice in glassine envelopes from one of the Mexican overwintering sites (Table 1, sample 8). Subsequently, on-site studies were deemed necessary because any cold-hardiness mechanism involved might be altered during the time of shipment. We established a field laboratory, powered by a Honda 3,000-watt generator, near the large overwintering site on the Sierra Chincua in Michoacán, Mexico, from January through March 1985 and 1986 (Table 1, samples 9 and 10). During the winter in 1985, the THP and SCP determinations were done on monarchs collected during four activities, i.e. hanging in clusters (Table 1, sample 8), nectaring at flowers, flying, and drinking at streams (Table 1, sample 9), whereas in 1986 all were collected from clusters (Table 1, sample 10).

SPRING, SUMMER, AND AUTUMN SAMPLES

To compare overwintering cold-hardiness with that of summer monarchs and two other Danaine species, SCP tests were done on several further groups of butterflies. One comparison was of freshly eclosed monarchs tested in July 1986 in Gainesville, Florida (Table 1, sample 1). These butterflies (from eggs laid by monarchs captured in north Florida) were raised in the laboratory on *Asclepias curassavica* at about 20°C, exposed to natural photoperiods, and tested 24 hr after emergence. The next group consisted of migratory monarchs captured on the Gulf Coast of northern Florida during both spring (Table 1, sample 6) and autumn (Table 1, sample 7), at Cape San Blas in April 1984 and at St. Mark's National Wildlife Refuge in October 1985. Both samples were brought back to the laboratory and tested for SCP within 7 hr of capture. In November 1985, a sample of reproductively active, migrant monarchs was collected from an experimental milkweed patch (*A. curassavica*) at the University of Florida (Table 1, sample 5). We also tested a group of reproductively active, nonmigrant butterflies in south Florida. Monarchs and two closely related Danaine species, *Danaus gilippus berenice* and *Danaus eresimus* (Table 1, samples 2–4) were collected in August 1985 from several areas immediately west of Miami, Florida. SCPs were determined within 7 hr of capture.

LETHALITY OF THE FREEZING EVENT

Until recently, it has been assumed in SCP studies that the subject remained alive during cooling until the actual freezing event occurred. Knight et al. (1986), however, showed that prefreezing mortality occurs in some insects before the SCP is recorded. In testing adult grain aphids *Sitobion avenae* at a cooling rate of 1°C/min, they found a mean SCP of −24.2°C. However, mortality was observed at −5°C and none of the aphids survived temperatures of −15°C. To check for prefreeze mortality in the present study, 15 control monarchs were removed from the cooling bath at temperatures below −10°C before their SCP occurred and then placed in nylon mesh cages for 24 hr at ambient temperatures. Their survival was compared with that of 70 monarchs that had been cooled until SCPs were recorded and also subsequently held in observation cages for 24 hr.

RESULTS

THERMAL HYSTERESIS

Freezing and melting points of hemolymph from 20 overwintering monarchs in Mexico and of 10 autumn migrants in Florida are given in Table 2. Although there is a significant difference between the means of freezing and melting points in both the overwintering and migrant samples, this difference (Table 2) is very small, <0.1°C in both comparisons. Formation of needlelike crystals characteristic of solutions containing THPs did not occur during the growth of these crystals.

SUPERCOOLING

A summary of mean SCPs of the different samples is presented in Table 1. Because there were no significant differences in SCP with regard to sex in any of the samples (one-factor ANOVA), both sexes were lumped in further analyses. The distributions of SCPs in all samples tested were normal (Kolmogornov-Smirnov tests for normality, two-tailed, all $p > 0.05$) and shifted from higher SCPs in reproductive to lower SCPs in migratory and overwintering monarchs (Fig. 1). Sample means were compared using a one-factor ANOVA and Scheffé F test. The mean SCPs of overwintering samples

Table 1. Mean SCPs (°C), standard deviations, and ranges of various samples and species of *Danaus* butterflies tested. Samples designated A are significantly different from B ($p < 0.05$; one-factor ANOVA, Scheffé F test).

Sample number	Collection site	Date	Sample size	Condition	SCP (°C) Mean	SCP (°C) SD	Range below zero	Significance grouping
Summer								
1. (*D. plexippus*)	North Florida	July 1986	11	Freshly eclosed	−4.1	1.5	2–7	A
2. (*D. plexippus*)	South Florida	August 1985	23	Summer fields	−5.1	1.6	1–8	A
3. (*D. gilippus*)	South Florida	August 1985	49	Summer fields	−4.9	2.4	1–10	A
4. (*D. eresimus*)	South Florida	August 1985	22	Summer fields	−5.3	2.3	2–10	A
Autumn/spring								
5. (*D. plexippus*)	North Florida	November 1985	29	Autumn breeders	−4.4	1.5	1–7	A
6. (*D. plexippus*)	North Florida	April 1984	7	Spring migrants	−6.6	0.9	5–8	AB
7. (*D. plexippus*)	North Florida	October 1985	41	Autumn migrants	−7.6	2.1	2–13	B
Winter								
8. (*D. plexippus*)	Mexico	January 1984	8	Winter, clusters	−8.2	0.8	4–10	B
9. (*D. plexippus*)	Mexico	January–March 1985	407	Winter, various activities	−7.7	2.5	1–14	B
10. (*D. plexippus*)	Mexico	January–March 1986	412	Winter, clusters	−8.2	2.4	2–13	B

were all significantly lower than mean SCPs of reproductively active monarchs and closely related Danaine butterflies (Table 1). Autumn migrants also had significantly lower SCPs than breeding monarchs.

SCP AND ACTIVITY

Mean SCPs of overwintering butterflies caught flying, nectaring, or drinking were not different from each other but were significantly different from the mean SCP of clustering butterflies (Table 3).

SCP AND LETHALITY

In Mexico, all 15 clustering butterflies removed from their cool-down period in the SCP test before freezing showed complete recovery within 1 hr in the nylon cages at ambient temperature with no observable ill effects. These monarchs had experienced body temperatures of −10 to −12°C during the test. Another 70 monarchs taken through the same cool-down period and showing evidence of freezing at their SCP were subsequently observed for 24 hr under the same ambient conditions. None recovered, and most showed no signs of movement.

DISCUSSION

Our experiments disclose an annual increase in monarch supercooling ability, with autumn migrants and overwintering samples freezing lower than those in the summertime reproductive state. SCPs of the summer monarchs were not statistically different from SCPs of two close congeners, *D. gilippus* and *D. eresimus,* collected and tested at the same time in south Florida. At the Mexican overwintering sites, clustering monarchs were more resistant to freezing than were monarchs engaged in other activities. We found freezing, not cooling, to be the lethal event. Although a THP mechanism would probably be very effective under the high-altitude winter conditions, we found only a slight hysteresis, <0.1°C, in monarch hemolymph (Table 2). In other studies where THPs have been found, the thermal hysteresis was much greater, ranging from 2.2 to 5.6°C (Duman et al., 1982), and char-

Table 2. Melting points (A), freezing points (B), and thermal hystereses (A − B) of hemolymph (°C) from adult monarchs collected in north Florida during the 1985 autumn migration and during the 1984–1985 overwintering season in Mexico (means ± SD). The differences between mean melting and mean freezing points were analyzed by paired t tests.

	N	A (melting point)	B (freezing point)	A − B (thermal hysteresis)	A versus B (paired t value)	p
Autumn (Florida)	20	−0.92 ± 0.04	−0.98 ± 0.06	0.06 ± 0.05	3.67	0.005
Winter (Mexico)	10	−0.94 ± 0.05	−1.02 ± 0.05	0.07 ± 0.07	4.68	0.0002

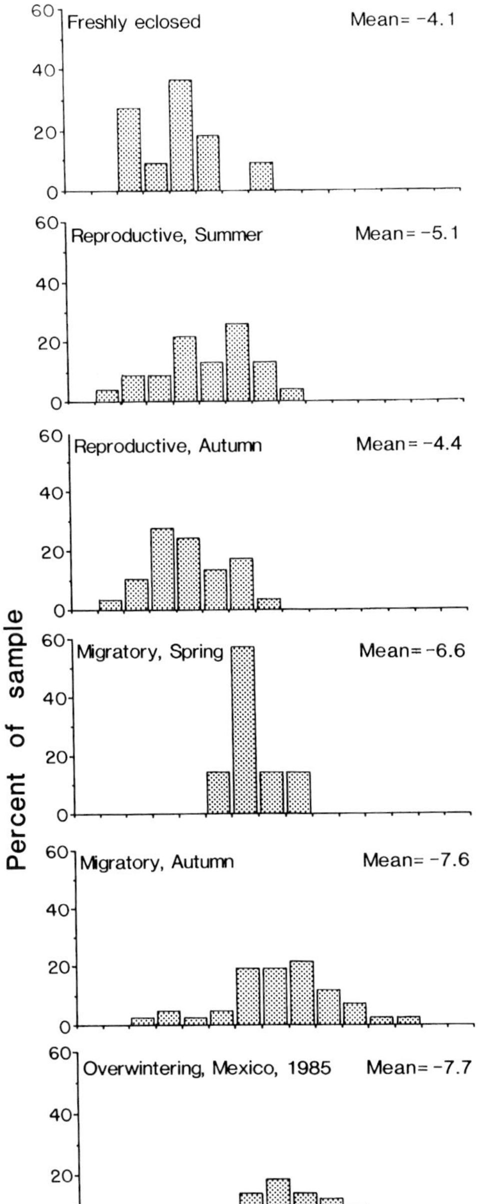

Figure 1. Frequency distributions of SCPs (°C) for samples of monarch butterflies during various phases of their annual cycle (see Table 1 for more details). Samples of freshly eclosed, summer reproductive, autumn reproductive, spring migrant, autumn migrant, and overwintering monarchs from Mexico in 1985 and 1986 are samples 1, 2, 5, 6, 7, 9, and 10, respectively, of Table 1.

acteristic crystal spires (not seen in our study) were observed. We conclude that, while there was a measurable hysteresis, evidence for functional THPs in overwintering monarch butterflies is lacking.

While this freeze-protection may be adequate for overwintering conditions in Mexico, it is certainly moderate compared with the supercooling ability of nonmigrant insect species that inhabit boreal climates. Many of these spend the winter as eggs or larvae, which can withstand ultralow temperatures. Mean SCPs as low as −51°C were found in eggs of the tortricid *Zeiraphera diniana* from inland Norway (Bakke, 1969), and eggs of the aphid *Pterocomma smithia* from Alberta could survive to −42°C (Somme, 1964). In larvae of the bark beetle *Scolytus multistriatus,* a mean SCP of −53°C was recorded (Lozina-Lozinskii, 1974), and gall-forming midge larvae (*Rhabdophaga* sp.) from the Northwest Territories had a mean SCP as low as −61°C (Ring and Tesar, 1980). In contrast, insects that overwinter as adults are much less cold-resistant, with only a few having mean SCPs below −25°C. There are three reports of adult butterfly cold-hardiness, all in the Nymphalinae. In interior Alaska, mourning cloaks (*Nymphalis antiopa*) and a species of *Polygonia* had mean SCPs of −20 and −25°C, respectively (Miller, 1982). *Vanessa urticae* from Leningrad showed a mean SCP of −21°C (Lozina-Lozinskii, 1974). These boreal butterflies are understandably more cold-protected than monarchs, in relation to past selection pressures of colder temperatures. Being able to endure the harsh winter *in situ* is part of an evolutionary strategy that allows extended access to favorable resources and relates to the type of host plant utilized. Butterflies that overwinter as adults in northern areas are frequently associated with woody host plants (Hayes, 1982). The alternate strategy of the monarch, migration over great distances, probably evolved primarily in relation to the availability pattern of its (herbaceous) milkweed hosts (Brower, 1985; Cockrell et al., this volume; Malcolm et al., this volume). Avoidance of severe northern cold may be of secondary importance. Our results suggest that a potential for the evolution of more extreme cold-hardiness is contained in the natural variation of SCP of overwintering monarchs.

The monarch migration/diapause syndrome is a physiological state cued by environmental stimuli, mediated by endocrine mechanisms and involving several behavioral and physiological changes in the adult butterflies (Herman, 1981, 1985; Tauber et al., 1986; James, this volume). Some of these changes are cessation of reproduction, migratory behavior, lipid storage, and low levels of juvenile hormone

Table 3. Mean SCPs (°C), standard deviations, and ranges of overwintering monarchs in Mexico, 1985, collected during four different activities (subset of sample 9, Table 1). Clustering monarchs had significantly lower SCPs than the other three samples ($p < 0.05$; one-factor ANOVA, Scheffé F test).

		SCP			
Activity	N	Mean	SD	Range	Test group
Clustering	157	−8.9	2.3	−3.5 to −14.0	A
Nectaring	40	−7.6	2.0	−1.5 to −12.0	B
Flying	28	−7.2	2.1	−4.2 to −12.9	B
Drinking	123	−6.5	2.2	−2.5 to −12.8	B

(Herman, 1985, this volume). In our study, autumn migrants show increased cold-hardiness, but reproductively active monarchs in autumn have the same SCP as reproductively active summer samples. These autumn reproductives were probably migrants that had broken diapause in response to warm temperatures and encounters with cultivated milkweed host plants in Florida. If low juvenile hormone titers mediate the migration/diapause state and high JH levels lead to reproduction (Herman, 1985), then the seasonal changes in SCP shown in this study probably reflect a cold-hardiness physiology also controlled by JH levels. This leads to the testable prediction that injection of JH will not only cause overwintering monarchs to break diapause, as Herman (1985) found, but also result in higher SCPs. The fact that more active overwintering monarchs in 1985 had higher SCPs could mean that their physiologies were shifted toward the reproductive mode by the warmer daytime temperatures.

If this moderate supercooling ability of the monarch is in fact a physiological adaptation to overwintering in Mexico, it must relate to current and past minimum temperatures at the altitude and latitude of the current overwintering sites, which now appear limited to an area of 60 × 30 km (Calvert and Brower, 1986; Calvert et al., 1989). While no temperature records from actual overwintering sites at their average altitude of 3,000 m existed before their discovery in 1974, data from nearby weather stations are available. Table 4 shows average minimum and extreme minimum temperatures over 10–26 yr from nine weather stations >2,500 m during January, the coldest month. The average lows of January for these stations ranged from 2.4 to −1.7°C, and the extreme lows ranged from −5 to −14°C. A temperature as low as −14°C would kill virtually all overwintering monarchs (Fig. 1). However, weather stations are usually situated in clearings, which are colder and less protected than within the forest where monarchs cluster (Calvert and Brower, 1981; Calvert et al., 1982, 1989).

We have additional evidence that an intact forest canopy functions to protect the butterfly clusters from potentially lethal cold temperatures, especially at night. We found that body temperatures of monarchs exposed to the night sky were an average of 4°C cooler than either the surrounding air or other monarchs under overhead cover. We also found that the presence of water on the body greatly increases a monarch's risk of freezing, thwarting any supercooling ability. Overhead cover in the form of forest canopy or other butterfly clusters reduces wetting from rain or dew (Anderson and Brower, in prep.). Other studies have demonstrated that a greater proportion of monarchs freeze at night on

Table 4. Average January minimum and extreme low temperatures (°C) recorded at nine, high-altitude weather stations near the monarch overwintering sites in the states of México and Michoacán (10–26 yr of data from *Normales Climatologicas, Periodo 1941–1970, Mexico, 1982*).

Weather station	Latitude (°N)	Longitude (°W)	Altitude (m)	Average January low	Extreme minimum
México					
Almoloya del Río	19°09′	99°29′	2,670	−1.7	−9.0
Atlacomulco	19°48′	99°53′	2,526	1.5	−9.0
Río Frío	19°21′	98°40′	3,000	−0.3	−14.0
San Felipe del Progreso	19°42′	99°58′	2,550	2.0	−9.0
Tenango del Valle	19°06′	99°35′	2,660	1.1	−9.0
Texcoco	19°33′	98°53′	2,800	−0.5	−11.5
Michoacán					
Villa Victoria	19°27′	99°59′	2,608	−0.7	−8.9
Presa Pucuato	19°38′	100°42′	2,505	2.0	−6.0
Presa Sabaneta	19°38′	100°41′	2,515	2.4	−5.0

the ground in open areas than those beneath understory vegetation (Calvert et al., 1986) or within the monarch colony (Calvert and Cohen, 1983). Temperatures as low as the extremes of Table 4 are probably experienced within the sheltered forests of the overwintering sites very infrequently, which may well account for the highly restricted area in which the butterflies are known to cluster. An intact canopy is clearly protective, and since the oyamel forests are subject to extensive harvesting (Snook, this volume), our results are important evidence in consideration of future conservation and management strategies.

ACKNOWLEDGMENTS

This study was partially funded by grants from the World Wildlife Fund, Sigma Xi, and National Science Foundation Grant BSR-850041 to L.P. Brower. The Zoology Department of the University of Florida provided invaluable assistance and ambience. For aid in the struggle against entropy while camping in Michoacán and for stimulating conversation, we thank John Glendinning, Tonya Van Hook, Bill Calvert, Alfonso Alonso M. and Alfredo Arellano G.

LITERATURE CITED

Ackery, P.R., and R.I. Vane-Wright. 1984. *Milkweed butterflies: Their cladistics and biology*. London: British Museum (Natural History); and Ithaca, NY: Cornell University Press, 425 pp.

Bakke, A. 1969. Extremely low supercooling points in eggs of *Zeiraphera diniana* (Guenée) (Lepidoptera: Tortricidae). *Norwegian Journal of Entomology* 16: 81–83.

Barker, J.F., and W.S. Herman. 1976. Effect of photoperiod and temperature on reproduction of the monarch butterfly *Danaus plexippus*. *Journal of Insect Physiology* 22:1565–1568.

Brower, L.P. 1977. Monarch migration. *Natural History* 86:40–53.

Brower, L.P. 1985. New perspectives on the migration biology of the monarch butterfly, *Danaus plexippus* L. In *Migration: Mechanisms and adaptive significance*, ed. M.A. Rankin, 748–785. Austin: University of Texas, *Contributions in Marine Science*, Supplement, 27.

Calvert, W.H., and L.P. Brower. 1981. The importance of forest cover for the survival of overwintering monarch butterflies (*Danaus plexippus*, Danaidae). *Journal of the Lepidopterists' Society* 35:216–255.

Calvert, W.H., and L.P. Brower. 1986. The location of monarch butterfly (*Danaus plexippus* L.) overwintering colonies in Mexico in relation to topography and climate. *Journal of the Lepidopterists' Society* 40:164–187.

Calvert, W.H., and J.A. Cohen. 1983. The adaptive significance of crawling up onto foliage for the survival of grounded overwintering monarch butterflies (*Danaus plexippus*) in Mexico. *Ecological Entomology* 8:471–474.

Calvert, W.H., M. Hyatt, and N.P. Mendoza. 1986. The effects of understory vegetation on the survival of overwintering monarch butterflies (*Danaus plexippus* L.) in Mexico. *Acta Zoologica Mexicana* 18:1–17.

Calvert, W.H., W. Zuchowski, and L.P. Brower. 1982. The impact of forest thinning on microclimate in monarch butterfly (*Danaus plexippus* L.) overwintering areas of Mexico. *Boletín de la Sociedad Botánica de México* 42:11–18.

Calvert W.H., W. Zuchowski, and L.P. Brower. 1983. The effect of rain, snow and freezing temperatures on overwintering monarch butterflies in Mexico. *Biotropica* 15(1):42–47.

Calvert, W.H., S.B. Malcolm, J.I. Glendinning, L.P. Brower, M.P. Zalucki, T. Van Hook, J.B. Anderson, and L.C. Snook. 1989. Conservation biology of monarch butterfly overwintering sites in Mexico. *Vida Silvestre Neotropical* 2(1):38–48.

Dankø, H.V. 1978. Modes of seasonal adaptation in the insects: Winter survival. *Canadian Entomologist* 110:1167–1205.

DeVries, A.L. 1971. Glycoproteins as biological antifreeze agents in antarctic fishes. *Science* 172:1152–1155.

DeVries, A.L. 1980. Biological antifreezes and survival in freezing environments. In *Animals and environmental fitness*, ed. R. Giles, 583–607. Oxford: Pergamon Press.

DeVries, A.L. 1983. Biological antifreeze agents. *Annual Review of Physiology* 25:805–810.

Duman, J.G. 1979. Thermal-hysteresis factors in overwintering insects. *Journal of Insect Physiology* 25: 805–810.

Duman, J.G., and K. Horwath. 1983. The role of hemolymph proteins in the cold tolerance of insects. *Annual Review of Physiology* 45:261–270.

Duman, J.G., K.L. Horwath, A. Tomchaney, and J.L. Patterson. 1982. Antifreeze agents of terrestrial arthropods. *Comparative Biochemistry and Physiology* 73A:545–555.

Hayes, J.L. 1982. A study of the relationships of diapause phenomena and other life history characters in temperate butterflies. *The American Naturalist* 120:160–170.

Herman, W.S. 1981. Studies on the adult reproductive diapause of the monarch butterfly, *Danaus plexippus*. *Biological Bulletin* 160:89–106.

Herman, W.S. 1985. Hormonally mediated events in adult monarch butterflies. In *Migration: Mechanisms and adaptive significance*, ed. M.A. Rankin, 799–815. Austin: University of Texas, *Contributions in Marine Science*, Supplement, 27.

Knight, J.D., J.S. Bale, F. Franks, S.F. Mathias, and J.G. Baust. 1986. Insect cold hardiness: Supercooling points and pre-freeze mortality. *Cryo-Letters* 7:194–203.

Lozina-Lozinskii, L.K. 1974. *Studies in cryobiology*. New York: John Wiley and Sons, pp. 69–84.

Masters, A.R., S.B. Malcolm, and L.P. Brower. 1988. Monarch butterfly (*Danaus plexippus*) thermoregulatory behavior and adaptations for overwintering in Mexico. *Ecology* 69:458–467.

Miller, K. 1982. Cold-hardiness strategies of some adult and immature insects overwintering in interior Alaska. *Comparative Biochemistry and Physiology* 73A: 595–604.

Ring, R.A., and D. Tesar. 1980. Cold-hardiness of the arctic beetle *Phyto americanus* Kirby Coleoptera, Pythidae (Salpingidae). *Journal of Insect Physiology* 26:763–774.

Salt, R.W. 1961. Principles of insect cold hardiness. *Annual Review of Entomology* 6:55–74.

Somme, L. 1964. Effects of glycerol on cold hardiness in insects. *Canadian Journal of Zoology* 42:87–101.

Somme, L. 1982. Supercooling and winter survival in terrestrial arthropods. *Comparative Biochemistry and Physiology* 73A:519–543.

Tauber, M.J., C.A. Tauber, and S. Masaki. 1986. *Seasonal adaptations of insects.* New York and Oxford: Oxford University Press, 411 pp.

Urquhart, F.A. 1960. *The monarch butterfly.* Toronto: University of Toronto Press, 361 pp.

Urquhart, F.A., and N.R. Urquhart. 1976. The overwintering site of the eastern population of the monarch butterfly (*Danaus plexippus,* Danaidae) in southern Mexico. *Journal of the Lepidopterists' Society* 30:153–158.

Urquhart F.A., and N.R. Urquhart. 1977. Overwintering areas and migratory routes of the monarch butterfly (*Danaus plexippus* L.; Danaidae; Lepidoptera) in North America with special reference to the western population. *Canadian Entomologist* 109:1583–1589.

Zachariassen, K.E. 1980. The role of polyols and nucleating agents in cold-hardy beetles. *Journal of Comparative Physiology* 140:227–234.

Zachariassen, K.E., and H.T. Hammel. 1976. Nucleating agents in the hemolymph of insects tolerant to freezing. *Nature* 262:285–287.

THE COMPARATIVE BIOLOGY AND ECOLOGY OF THE AUSTRALIAN DANAINES

R.L. Kitching and E. Scheermeyer

ABSTRACT. *Danaus plexippus* is a newcomer to Australia. When it arrived in the late 1800s, it joined a suite of other danaines, long established naturally on the continent. This paper examines the biogeography, food-plant associations, history, population ecology, and conservation of this group of 12 species, focusing on the similarities and differences between the native species and the monarch and the impact of 200 years of European occupation on the natural history of the group.

With the exception of *D. plexippus,* which is a transpacific invader, Australia's danaines are Asian in origin, some essentially Papuan, others Timorese. Of the 11 genera of danaines, Australia possesses but three, *Danaus, Euploea,* and *Tirumala.* Within these genera, the distributions of most reflect those of their larval food plants, although the ranges and/or abundance of *Danaus chrysippus, Euploea core,* and *D. plexippus* have been greatly increased by accidental and deliberate introductions of additional host plants—principally weeds in the genus *Asclepias* and ornamental *Nerium oleander.* In the case of *D. plexippus,* such weed introductions probably set the scene for its successful invasion during the last century. The range of food plants for each species is reviewed, specialist and generalist species are identified, and the interaction between food-plant range and distribution is examined.

Danaus plexippus presents differing life-history strategies in different climatic zones—ranging from overwintering aggregation followed by enormous summer expansions of range to year-round breeding and foraging activity. Other species are compared with this pattern. Some of the native danaines aggregate during unfavorable situations of coldness or aridity and these include several species of *Euploea* and *Tirumala hamata* but others do not. Tied in with this behavior may be long-distance movements (although nothing on the scale of *D. plexippus* in North America has been recorded), reinvading the full summer range of the species annually or only occasionally.

Conservation of some Australian danaines is of concern as their moist forest habitats dwindle. Others, in contrast, show all the vigor of pest species in their artificially expanded ranges.

INTRODUCTION

In this paper, we review the biogeography, food-plant relationships, life-history strategies, and conservation status of the Australian Danainae. Lest this should sound pretentious, we add that we are just beginning to scratch the surface in our understanding of the biology of the danaines in Australia. The work of researchers such as Zalucki, James, and Scheermeyer over the last few years, however, make it an apposite moment to attempt such a review and, thereby place the Australian biology of *Danaus plexippus* in a wider context. Before proceeding, it must be acknowledged that only the existence of such excellent general works as those of Common and Waterhouse (1981), Ackery and Vane-Wright (1984), and Tsukada (1985) allow such regional reviews: Like other workers in the field, we are deeply indebted to them for their sterling efforts.

BIOGEOGRAPHY

GENERAL PATTERNS

Most groups of Australian butterflies show considerable endemism with centers of local radiation associated principally with the southeast of the continent. Within the Satyrinae and Hesperiidae some groups of genera that occur around the Victorian and New South Wales highlands are found nowhere else in the world, and within the Lycaenidae a comparable degree of specialization is found at the tribal level (Evans, 1949; Eliot, 1973; Common

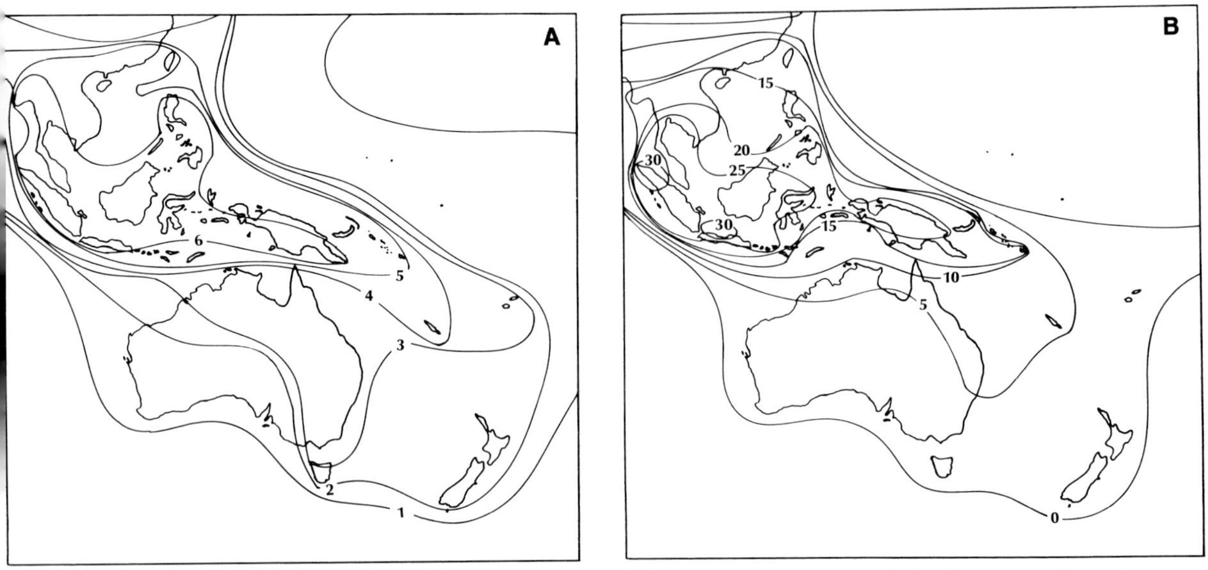

Figure 1. Isoclines for the numbers of Danainae occurring in Australasia and Southeast Asia: (**A**) genera, (**B**) species. Note that the generic isoclines are based on a unit increment, and those for species on an increment of five.

and Waterhouse, 1981; Kitching, 1981). These centers of endemism are usually explained as resulting from radiations early in the history of the groups concerned and consequent upon their long isolation. Some claim Gondwanaland connections for selected groups (e.g., key genera within the Pieridae; A.G. Orr, pers. comm.).

Within the Danainae, however, no such endemism can be claimed in Australia (apart from one or two debatable subspecies) and the subfamily remains a group of tropical expatriates (or temporary residents) with even a human-assisted settler. Australia lies just to the south of one of the great centers for the group. Within 1,000 km of Australia's shores are found 8 of the 11 genera recognized on a worldwide basis with up to 35 species occurring in particularly rich sites. Figure 1A and B, based on information collated from the works of Tsukada (1985) and Ackery and Vane-Wright (1984), shows the distribution isoclines for genera and species. At the generic level, maximum diversity is found throughout the Malayan/Philippine/Indonesian/New Guinea region with a rapid decline away from this central area. The six Indo-Malayan genera *Parantica, Ideopsis,* and *Idea,* together with *Danaus, Euploea,* and *Tirumala,* produce the peak in generic diversity. The extremely local endemic genera *Protoploea* from New Guinea and *Tiradelphe* from the Solomons complete the generic list. At the species level, maximum diversity is found in parts of Java and Sumatra, where more than 30 species fly. Again there is a gradual fall off from these regions, to the northern limits of the group in China and the southern limits in Australasia. These are of course New World connections and westward connections to the rest of Asia. Within the Indo-Australian region, the great peaks of species diversity can be accounted for largely by the distribution of the genus *Euploea* (Fig. 2A) with peaks of up to 16 species in parts of Java, Sumatra, and northern New Guinea. Even the most diverse of the other genera (*Parantica;* Fig. 2D) does little more than smooth out the irregularities in the *Euploea* map!

Turning then to the Australian fauna, it is perhaps surprising given the large size, longevity, and vagility of the group that, of these eight genera, only three have become established in Australia. There is no great barrier between the northern parts of Australia and the faunistically much richer islands of New Guinea and Indonesia. Indeed, across the Torres Straits there is no sea-gap greater than a few tens of kilometers with more or less appropriate vegetation on both sides of the Straits. Nevertheless, the genera *Danaus, Tirumala,* and *Euploea* are all we have, making up a total of 12 species. At the generic level, our *Danaus* species are simply just part of the cosmopolitan distribution of the genus (Fig. 2B) with two widespread species (*chrysippus* (L.) and *plexippus* (L.)) and two species that are widespread in southeast Asia (*genutia* (Cramer) and *affinis* (F.)). The Australian species of *Euploea* reflect, as it were, the southern "tail" of the enormous northern diversity (Fig. 2A). *Tirumala hamata* (Macleay) is the most widespread species of the Indo-Malayan genus (Fig. 2C).

At the level of the species within Australia, it is generally acknowledged that the various danaines have probably entered Australia by different routes and that these may be inferred with some confidence from their present intra- and extra-Australian distributions (Common and Waterhouse, 1981).

"TIMORESE" SPECIES

There is a clearly identifiable group of species, the affinities of which lie to the northwest of the con-

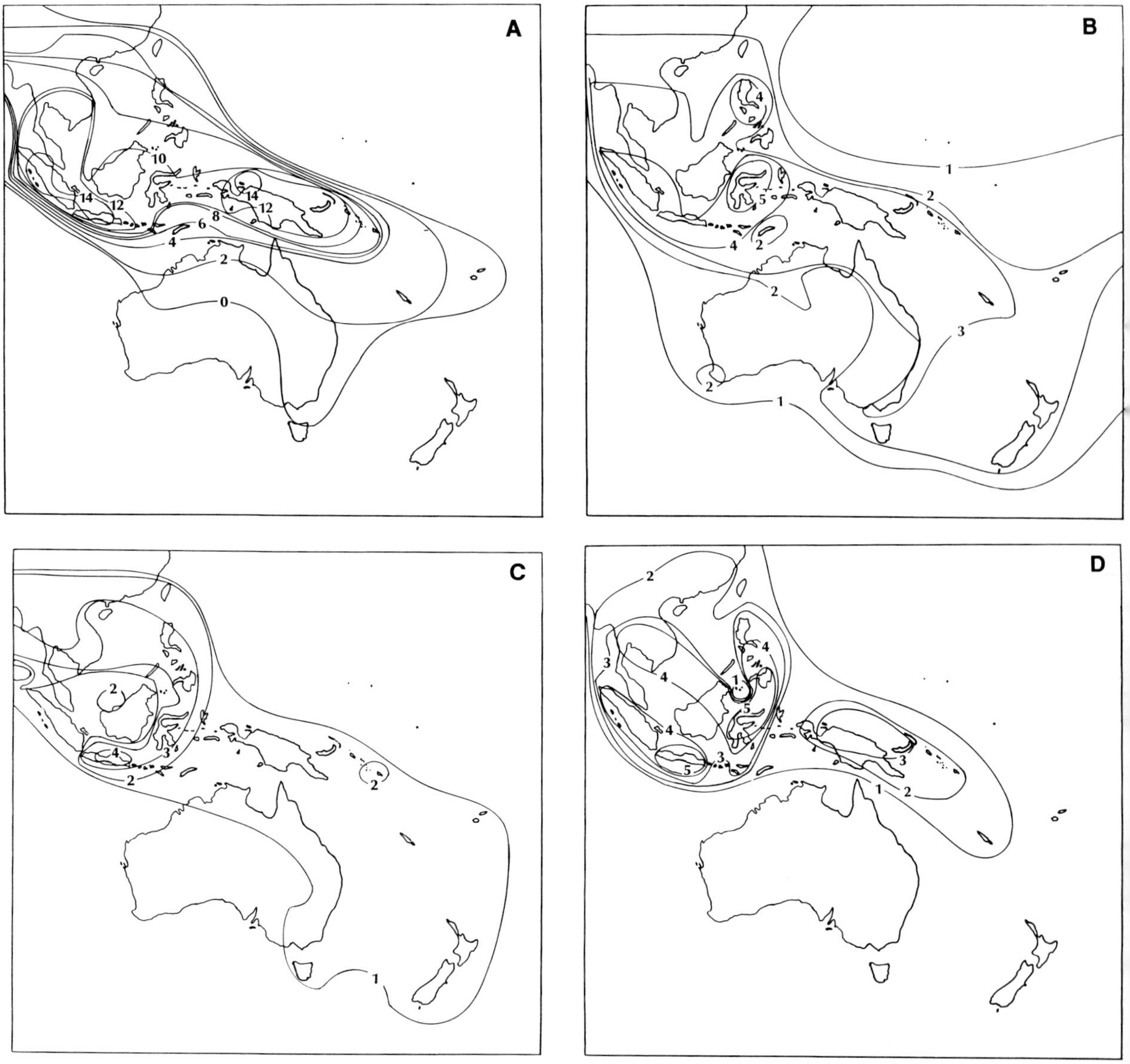

Figure 2. Isoclines for the numbers of species within selected genera of Danainae within Australasia and Southeast Asia: (**A**) *Euploea*, (**B**) *Danaus*, (**C**) *Tirumala*, (**D**) *Parantica*. All isocline increments are one, except for A in which the increment is two.

tinent, connecting to their extra-Australian distributions through the island of Timor. These species are *Danaus genutia* (Fig. 3A), *Euploea climena* (Stoll), and *Euploea darchia* (Macleay) (Fig. 3B). *Euploea core* (Cramer) (Fig. 3C) has a distribution that also suggests a northwestern origin (although bypassing Timor), but its intra-Australian distribution is now greatly enhanced by the presence of people (see below). Ackery and Vane-Wright (1984) express some reservations about the homogeneity of this species, which is polytypic. Its somewhat discontinuous distribution, which bypasses New Guinea but includes the Solomons (Fig. 3C), may reflect this.

CAPE YORK SPECIES

A number of species are included in the Australian fauna on the basis of their occurrence close to the tip of Cape York or on the, politically Australian, Torres Strait Islands. Two of these are basically New Guinea species, *Euploea usipetes* Hewitson (Fig. 3D) and *Euploea batesii* (Felder and Felder), and the third, *Euploea algea* (Godart), is a widespread species, the eastward extension of the range of which stretches as far as New Caledonia and Fiji and just coincides with the Australian continent in so doing. Some of these species have only marginal credentials as breeding members of the Australian fauna.

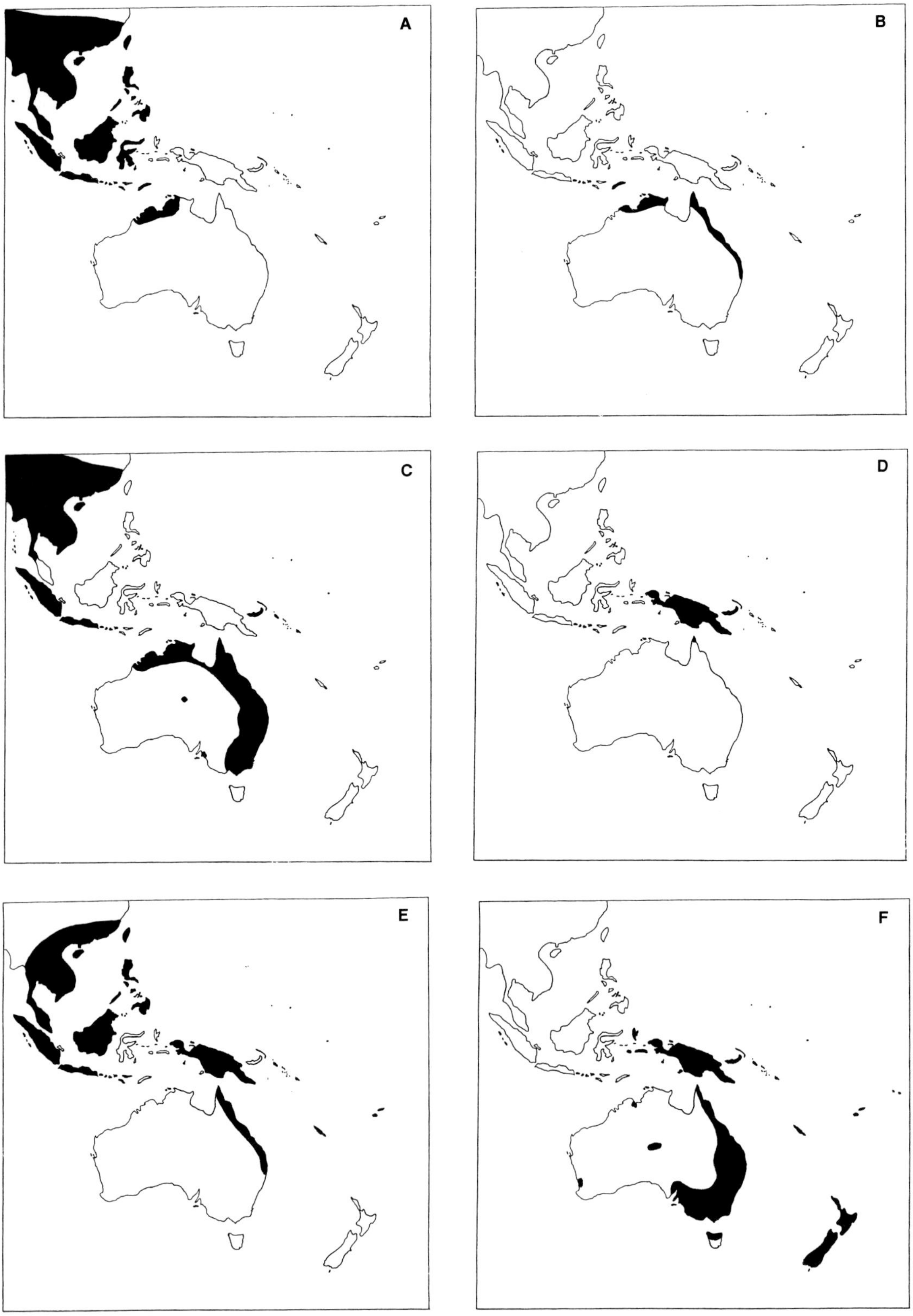

Figure 3. Maps of the distribution of selected species in Australasia and Southeast Asia: (**A**) *Danaus genutia,* (**B**) *Euploea darchia,* (**C**) *Euploea core,* (**D**) *Euploea usipetes,* (**E**) *Euploea tulliolus,* (**F**) *Danaus plexippus.*

GENERAL TROPICAL ENTRY SPECIES

Six species have distributions that indicate an indeterminate tropical route of entry, which may have been originally via Cape York or the northwest or both. Indeed, their present distributions and vagility probably imply an ongoing exchange between their Australian and extra-Australian ranges. The species involved are *T. hamata, Danaus affinis, Danaus chrysippus, Euploea tulliolus* (F.) (Fig. 3E), *Euploea alcathoe* (Godart), and *Euploea sylvester* (F.).

HUMAN-ASSISTED SPECIES

Danaus plexippus, of course, is the only species in this category. It arrived in eastern Australia originally from the New World sometime before 1871, when it was first reported by a south Queensland newspaper (Common and Waterhouse, 1981). Not only is its present distribution (Fig. 3F) human-assisted, but it is doubtful if its initial establishment would have been possible at all without the earlier accidental introduction by Europeans of appropriate species of *Asclepias* (but see Vane-Wright, this volume).

Of course, the distribution of any butterfly species reflects, *inter alia,* the distribution of its food plants and, in considering the present occurrence of danaines in Australia, we must also consider the range and location of these plants.

FOOD-PLANT RANGE AND HISTORY

RECORDS AND GENERAL PATTERNS

We have some information about the larval food plants of 11 of the 12 Australian danaines although, in a couple of cases, this is by extension from observations made by workers in New Guinea. All of the available information is summarized in Tables 1 and 2. These tables are based on data presented by Ackery and Vane-Wright (1984) and Common and Waterhouse (1981) and additional observations on *E. core* by Scheermeyer and Zalucki (1985). As we have pointed out elsewhere (Kitching and Zalucki, 1983), some care must be exercised in accepting records of food plants based, often of necessity, on casual observations and, in particular, oviposition records. Some of the information included in the tables raises such doubts (e.g., *Asclepias* as a food plant for *E. core,* or *Araujia hortorum* as a food plant for *D. plexippus*) but rather than exclude these few records we have left them in so that the tables reflect virtually all of the published information on the topic.

The first observation that can be drawn from the tables is an obvious one—all the food plants are from the expected families, Asclepiadaceae, Apocynaceae, and Moraceae, with the more specialized "milkweed" genera, *Danaus* and *Tirumala,* being restricted to the Asclepiadaceae. Of more interest is the information summarized in Table 2. From this we note some species are far more catholic in their range of food plants than are others. *Euploea core* has 20 recorded food plants, whereas the highly specialist *D. affinis* breeds only on the asclepiadaceous vine *Cynanchum carnosum,* and *E. tulliolus* breeds only on *Malaisia scandens.* Second, we note that some species have adopted introduced food plants with much greater facility than others. Percentages in this regard may be a little misleading, numbers varying greatly as they do. *Euploea core* has 6 of 20 food plants in the introduced category, *D. chrysippus* 5 of 10, and, of course, *D. plexippus* 5 of 5. The strictly tropical species *E. alcathoe* (=*E. eichhorni* Staudinger in much of the Australian literature) is recorded as breeding on two introduced plants out of a total of seven known food plants. The single record of an introduced host for *E. batesii* is of no great consequence in the present context, because this butterfly is a vagrant or, at best, a very rare resident at the tip of Cape York only. The patterns evident in Table 2 may provide the key to understanding the changes of distribution that have occurred in some species in recent times.

HUMAN-INDUCED DISTRIBUTION MODIFICATIONS

Of the Australian danaines, most have distributions that have changed little since records began. Some may have retreated a little as their coastal rainforest habitats were cleared, but their overall geographic ranges remain little affected. This includes some relatively widespread and common species such as *D. affinis, E. tulliolus,* and *T. hamata.* These species, for the most part, have not adopted non-native food plants and presumably remain constrained by the restricted distributions of their hosts. The present distributions or abundance of three other species, however, are substantially greater than previously recorded and it is instructive to examine each of these in turn.

Euploea core

The so-called "common crow" occupies a range of habitats in the north of its range from moist rainforests to open savannahs. In contrast, at the southern and inland limits of its current distribution, it is more often than not a garden insect associated with ornamental plantings of oleander and, to a lesser extent, species of *Hoya, Mandevilla, Trachelospermum,* and *Stephanotis.* Inland, and in much of the northeastern tropics, it thrives, commonly also, on *Cryptostegia grandiflora,* a naturalized alien (Scheermeyer, 1985). Although *C. grandiflora* is only a mediocre host plant in terms of developmental rate and adult size of the butterfly, its common presence and frequent fresh foliage throughout the year probably account for its success as a host of *E. core* (Scheermeyer, 1987; Scheermeyer et al., 1990). The ability to thrive on garden plantings of oleander, a very widely used ornamental in Australia, and on the wide-ranging *C. grandiflora* may well have assisted in producing the very widespread distribution reported by Common and Waterhouse

Table 1. List of larval food plants recorded for Australian danaines. Those indicated by an asterisk are introduced to the continent. See text for sources.

Plant species	Butterfly species
Asclepiadaceae	
*Asclepias fruticosa**	*D. plexippus, D. chrysippus*
*A. curassavica**	*D. plexippus, D. chrysippus*
*A. rotundifolia**	*D. plexippus, D. chrysippus*
*A. semilunata**	*D. plexippus*
Asclepias "spp."*	*E. alcathoe, E. core*
*Calotropis gigantea**	*D. plexippus, D. chrysippus*
*C. procera**	*D. chrysippus*
*Araujia hortorum**	*D. plexippus*
Leichardtia australis	*D. chrysippus, E. core*
Cynanchum carnosum	*D. chrysippus, D. affinis, E. core, T. hamata*
C. floribundum	*D. chrysippus*
Pentatropis atropurpurea	*D. chrysippus*
P. quinquepartita	*D. chrysippus*
Secamone elliptica	*E. core, T. hamata*
Hoya australis	*E. core, E. alcathoe*
Gymnanthera nitida	*E. core, E. alcathoe*
Marsdenia leichardtiana	*E. alcathoe*
Sarcostemma australe	*E. core*
Stephanotis sp.*	*E. core*
*Cryptostegia grandiflora**	*E. core*
Apocynaceae	
*Nerium oleander**	*E. core, E. batesii, E. alcathoe*
Mandevilla sp.*	*E. core*
Trachelospermum sp.*	*E. core*
Parsonsia straminea	*E. core*
P. eucalyptophylla	*E. core*
Ichnocarpus frutescens	*E. core, E. algea*
Ichnocarpus sp.	*E. sylvester*
Carissa ovata	*E. core*
Moraceae	
Ficus eugenioides	*E. alcathoe*
F. muelleri	*E. alcathoe*
F. platypoda	*E. core*
F. coronata	*E. core*
F. benjamina	*E. core*
F. obliqua	*E. core*
F. racemosa	*E. sylvester*
Ficus sp.	*E. algea*
Malaisia scandens	*E. darchia, E. tulliolus*
M. tortuosa	*E. darchia*

(1981). This includes at least periodic breeding occurrences in central Australia and on the Spenser Gulf in south Australia (Fig. 3C) (Scheermeyer, 1985). The tendency of the species to participate in mass movements (see e.g., Smithers, 1983) also will have contributed to this opportunistic range spread. However, the demonstrated ability of the species to exploit a wide range of food plants seems of paramount importance as other widespread Australian danaines show the overwintering and migratory tendencies without having achieved such widespread distributions (see below).

Danaus chrysippus

The "lesser wanderer," as it is known in Australia, is one of the most widespread of Australian butterflies, with a recorded distribution that encompasses the whole of the continent, plus occasionally northern Tasmania. It has a range of native and exotic food plants (Table 1) throughout its distribution. This range may not actually have been increased by the activities of Europeans. However, we have certainly provided the species with the opportunity to become much more abundant. This

Table 2. An analysis of food-plant range and origin for selected species of Australian Danainae.

Butterfly species	Asclepiadaceae			Apocynaceae			Moraceae			% exotic
	Native	Exotic	Total	Native	Exotic	Total	Native	Exotic	Total	
D. plexippus	0	6	6	0	0	0	0	0	0	100
D. chrysippus	5	5	10	0	0	0	0	0	0	50
D. affinis	1	0	1	0	0	0	0	0	0	0
T. hamata	2	0	2	0	0	0	0	0	0	0
E. batesii	0	0	0	0	1	1	0	0	0	(100)
E. alcathoe	3	1	4	0	1	1	2	0	2	28.5
E. core	6	2	8	4	4	8	4	0	4	30
E. algea	0	0	0	1	0	1	1	0	1	0
E. sylvester	0	0	0	1	0	1	1	0	1	0
E. darchia	0	0	0	0	0	0	2	0	2	0
E. tulliolus	0	0	0	0	0	0	1	0	1	0

has been, principally, by the accidental introduction of a range of species of *Asclepias* that, as weeds of old fields, have become its most frequently used food plant in many regions (see e.g., Zalucki and Kitching, 1982). This broadening of the species' resource base may have been repeated in the more tropical regions by the introduction of *Calotropis gigantea* and *Calotropis procera*. This butterfly shares both *Asclepias* and *Calotropis* with *D. plexippus* and their larvae may co-occur on the same individual plants. Active competition through resource depletion no doubt occurs from time to time. The two species do, however, have different temperature relations (Zalucki, 1982) and different patterns of mortality due to predators and parasitoids (Zalucki and Kitching, 1982) and these will tend to minimize competitive interaction.

Danaus plexippus

The "wanderer," or monarch as it is also called, has a different history from any other Australian danaine. It is an exotic that has only become established following the equally artificial introduction of appropriate larval food plants, species of *Asclepias*. The mechanism by which the monarch arrived in Australia is a matter for unresolvable debate. It has been variously suggested that it simply flew across the Pacific reflecting the extremely high vagility of the species, that it island-hopped across the ocean establishing interim populations en route, and that it came as a passenger on a sailing ship (see Vane-Wright, this volume). Some combination of all of these mechanisms was probably involved at one time or another. We strongly expect that a series of arrivals was involved rather than the traditional single gravid female! Furthermore, it is quite possible that one way or another monarchs had been arriving in Australia for an extended period before establishment, the actual colonization taking place only when suitable food plants had themselves become established. *Asclepias* seed probably came in as seed-grain contaminants from South Africa and the Americas.

In the hundred or so years since its first introduction, the species has spread throughout much of the continent. We have reconstructed this spread as well as we could from a series of accounts from that time, although freely admitting to some necessary interpretation of some of the sketchier accounts in the sources. Figure 4 shows approximate minimal distributions for the species for 1871 (from the original newspaper report), for 1907 (from Rainbow, 1907), for 1951 (from Barrett and Burns, 1951), and for 1981 (from Common and Waterhouse, 1981). The species was first noted in southern Queensland in 1871 (although it may well have established some considerable time before this date) and was recorded from Melbourne, over 1,000 km to the south, by the following year (Barrett and Burns, 1951). The high rate of initial spread is evidenced by the extensive east coast distribution suggested by Rainbow as early as 1907. Since that time it has continued to spread, albeit much more slowly. It has established outlier colonies in Western Australia, central Australia, and the northwest within the last 15 yr or so and is, presumably, still spreading. It remains but an occasional visitor to northern Tasmania (Fig. 4D).

At this stage the monarch is approaching the supposed limits of range of the introduced species of *Asclepias*, its principal food plant, and *Calotropis*, a newly adopted host in northern Queensland. Other native asclepiads occur throughout the continent (as evidenced by the distribution of *D. chrysippus*) and, occasionally, food plants in other genera are recorded for non-Australian individuals (Ackery and Vane-Wright, 1984), so further spread may be possible if the species adopts other hosts in Australia.

COMPARATIVE POPULATION ECOLOGY

Other authors in this volume (James, Scheermeyer, Zalucki) deal with aspects of the population biology of selected danaines in Australia in much more detail than we attempt here. We use this opportunity to collate a few comparative observations on the group.

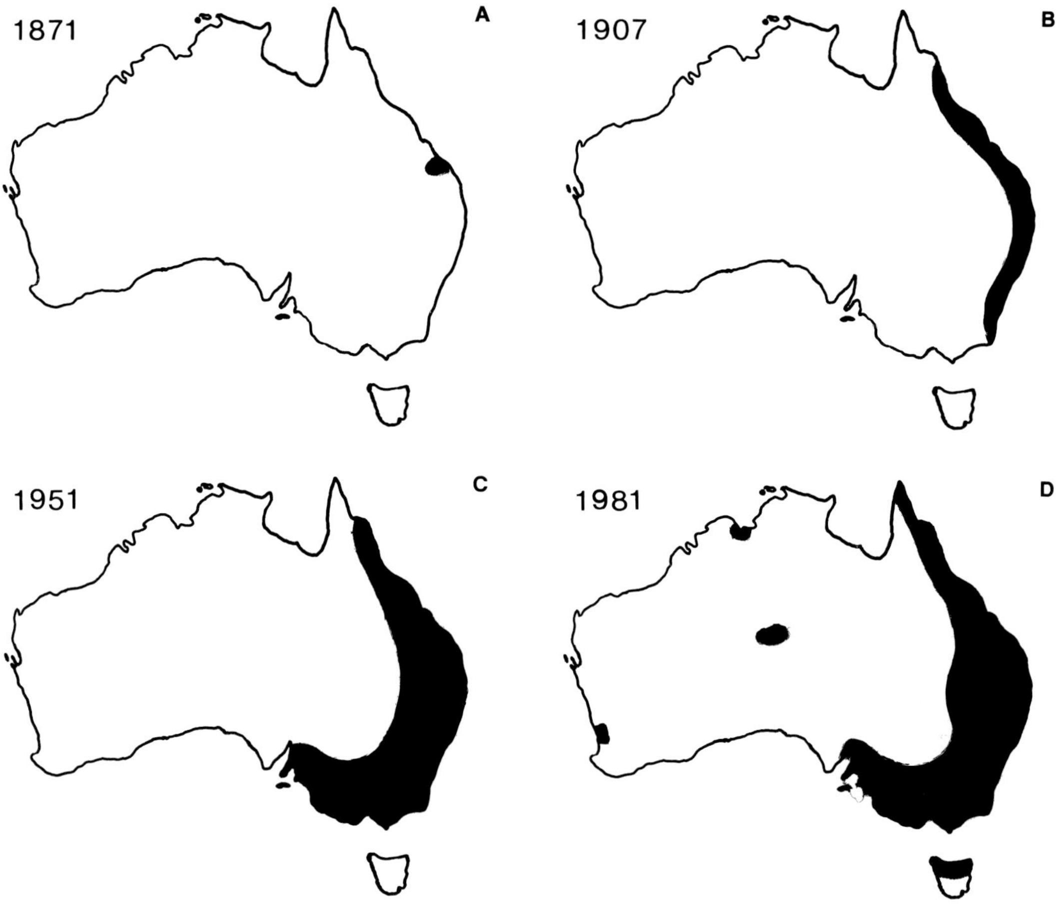

Figure 4. Maps of the supposed distributions of *Danaus plexippus* in Australia since the arrival of the species on the continent (**A**) 1871 (**B**) 1907 (**C**) 1951 (**D**) 1981. See text for sources (also see Vane-Wright, this volume).

As we stated at the outset, few species of danaines in Australia have been studied ecologically. The exceptions are *D. plexippus,* which has been examined by James, Smithers, and their collaborators in the Sydney region and Zalucki, Kitching, and occasional collaborators in the Brisbane region. *Euploea core* has been studied by Scheermeyer, Zalucki, Orr, and Kitching in the Brisbane area, and Scheermeyer has made extensive comparative observations on *E. core, E. tulliolus,* and *T. hamata* principally in the Townsville and Burdekin River regions. In addition there have been a few useful observations on migratory and overwintering behavior in other species (see below). Some species exhibit substantially different life histories and behavior in different parts of their ranges. *Danaus plexippus* provides an excellent example of this. In southern Queensland, part of the subtropical and tropical Torresian biogeographic subregion, the species breeds year-round and there are no records of either aggregations or mass movements (Zalucki, 1986). In contrast, in the Sydney region and the associated parts of the species' range, part of the seasonal Bassian subregion, there are overwintering aggregations (although, sadly, not as spectacular as the Californian and Mexican versions) with only occasional breeding in winter in favored locations (Smithers, 1977; James, 1979, 1981, 1982, this volume). From these centers the full southern range of the species is reinvaded each spring. There is some evidence of a return but the data are few. Sites in the Adelaide region in South Australia may provide a similar winter refuge although no records of winter breeding have been made.

Two features, across species, indicate to us the general potential of danaines for exploiting wide ranges and regions that are marginal climatically: reproductive dormancy, often in aggregations, to circumvent periods of adverse conditions for breeding, be they winter or dry seasons; and mass movements. Table 3 summarizes such records for the Australian danaines with respect to these important ecological characteristics.

Seven of our 12 danaines are recorded as aggregating, which is frequently associated with reproductive dormancy. In one case, however, this is

Table 3. **The occurrence of aggregating behavior and mass movements in Australian danaines.**

Butterfly species	Aggregations?	Mass movements?
D. plexippus	Yes, regularly in south[1]	Yes, probably annual[2]
D. chrysippus	Occasionally in north[6]	Yes, many records[2,3]
D. affinis	Occasionally in north[4,6]	Doubtful[2]
T. hamata	Yes[5,6]	Yes, many records[6]
E. core	Yes, many records[5,7]	Yes, but few records[2,6]
E. tulliolus	Yes[5,6]	Yes, but few records[2]
E. sylvester	Yes[2,5,6]	No records
E. alcathoe (=*E. eichhorni*)	Yes[8]	Yes, single record[8]
E. algea	Yes, in Papua New Guinea[9]	Yes, in Papua New Guinea[9]
E. darchia	No records	No records
E. batesii	No records	No records
E. usipetes	No records	No records

References: [1] James (1979), [2] Smithers (1983), [3] Williams (1930), [4] R. Kitching (pers. obs.), [5] Monteith (1982), [6] Scheermeyer (1987), [7] Kitching and Zalucki (1981), [8] Straatman (1963), [9] Ackery and Vane-Wright (1984).

based on a single observation (*E. alcathoe*), and in another on a New Guinea report (*E. algea*). Two other species, *D. affinis* and *D. chrysippus,* may also form aggregations in the tropical part of their range albeit for short periods only (E. Scheermeyer, unpubl. results). In the case of the tropical and subtropical species, these are dry-season aggregations (Scheermeyer, this volume), whereas in the south, as already mentioned, overwintering is involved. In one detailed study of overwintering in *E. core,* Kitching and Zalucki (1981) found that the gullies in which aggregations occurred were slightly more equable climatically than the surrounding region. The number of individuals in such aggregations may vary between tens and several thousand individuals and fluctuate considerably over time (Kitching and Zalucki, 1981; Scheermeyer, 1987).

Studies on overwintering aggregations of *D. plexippus* in the Sydney region have been made by James (1981, 1982). He found that there were both long-term overwintering sites and transient camps, possibly of individuals en route to the major aggregating sites, a situation closely reminiscent of the other very well-known Australian migratory and aggregating lepidopteran, the bogong moth (*Agrotis infusa* (Boisd.)) (Common, 1954). James recorded a gradual decline in condition during the aggregating period and a switch from female- to male-dominated sex ratio during the period. James' (1981) estimate of the maximum number of adults aggregating was approximately 350 individuals, although numbers up to 100,000 have been noted for some sites (C.N. Smithers, pers. comm.). Nevertheless, this is a far cry from the millions estimated for some American situations.

Turning to records of mass movements, seven species have been noted as forming "migratory" swarms. In four cases there are many corroborating records suggesting that such movements form an integral and regular part of the species' life history, at least in some locations. Again note that one of these four records (for *E. algea*) is from New Guinea rather than continental Australia. For the remaining three species, records are few and corroboration is needed. An eighth species, *D. affinis,* has been suggested as a migrant (Smithers, 1983) but this awaits confirmation. One might hope that any patterns that appear within Table 3 would aid in interpreting the observations, presented earlier, concerning those danaines that have extended their range and become at least semidomesticated. Sadly this is not the case. Certainly *T. hamata, E. core, D. plexippus,* and, to some extent, *D. chrysippus* form aggregations and go in for mass movements. However, both *E. tulliolus* and *E. alcathoe* form aggregations and migrate, but both remain restricted to tropical and subtropical, forested regions. The only thing these data show is that both aggregation and migration are evolutionary options for which danaines are preadapted (which we knew anyway) and that the habitat mosaic represented by a land mass such as Australia, or any part of it, presents an evolutionary challenge to its inhabitants for which there are sets of solutions, not unique answers—the "habitat templet" of Southwood (1977) revisited!

CONSERVATION ISSUES

The conservation of butterflies in Australia is in its infancy but, even so, it presents different problems from those associated with the monarch and its habitats in North America.

Certainly we have situations in which individual species can be classed as endangered through having extremely limited distributions that come under threat of development by man. Cases in point are lycaenids such as *Acridipsas illidgei* (Waterhouse and Lyell), restricted to a few mangrove patches close to the metropolis of Brisbane; *Paralucia spinifera* (Edwards and Common), a copper found only in a tiny area of southcentral New South Wales; the nymphaline *Argyreus hyperbius* (Butler), the habitats of which are scattered over a small area of southern Queensland and northern New South Wales in densely farmed land; and *Euschemon raf-*

flesia (W.S. Macleay), a magnificent skipper that is now highly restricted in distribution as are the patches of lowland rainforest with which it is specifically associated. No danaines come into this category.

More common, as in other parts of the world, is the problem of habitat destruction. Some of the danaines are primarily wet forest species (e.g., *E. alcathoe, E. tulliolus, E. darchia*), whereas *D. affinis* is a wetland species associated with swamp/woodland interfaces. There has been and remains a constant pressure upon such habitats in Australia (as elsewhere) and this presents a threat to the whole suite of butterfly species that occurs in them, including the danaines. Fortunately some areas have been conserved either as national parks or state forests, but the need remains for constant vigilance as conflicting interests continue to interact with respect to Australian forests.

Lastly, there is of course the point that some danaines, especially *D. plexippus*, can appropriately be considered as unwanted exotics in Australia and that under no circumstances should any conserving steps be taken in their direction! We would be overstating the case if we said that any real body of opinion was pressing this viewpoint in Australia and, in any case, the species could be seen as at least a partial controller of the much more undesirable aliens, the milkweeds. Two species, the wanderer (monarch) and the common crow, remain among our most familiar garden butterflies in much of the eastern part of the continent. They would be the species most preschoolers would attempt to draw if asked to portray a "butterfly." They are not endangered in any sense of the word and, through their adoption of essentially weedy habits, probably have a more secure future than the vast majority of the rest of our insect fauna.

ACKNOWLEDGMENTS

We have already indicated the intellectual debt owed to those past and present graduate students who have worked on danaine biology: Myron Zalucki and Bert Orr. We reiterate our gratitude to them now. In addition, we thank Stuart Jackson, without whose very able help this paper could not have been prepared.

LITERATURE CITED

Ackery, P.R., and R.I. Vane-Wright. 1984. *Milkweed butterflies: Their cladistics and biology.* London: British Museum (Natural History); and Ithaca, NY: Cornell University Press, 425 pp.

Barrett, C., and A.N. Burns. 1951. *Butterflies of Australia and New Guinea.* Melbourne: N.H. Seward, 187 pp.

Common, I.F.B. 1954. A study of the adult bogong moth, *Agrotis infusa* (Boisd.) (Lepidoptera: Noctuidae) with special reference to its behaviour during migration and aestivation. *Australian Journal of Zoology* 2:223–263.

Common, I.F.B., and D.F. Waterhouse. 1981. *Butterflies of Australia,* 2nd ed. Sydney: Angus and Robertson, 682 pp.

Eliot, J.N. 1973. The higher classification of the Lycaenidae (Lepidoptera): A tentative arrangement. *Bulletin of the British Museum (Natural History) Entomology* 28:373–505.

Evans, W.H. 1949. *A catalogue of the Hesperiidae from Europe, Asia and Australia in the British Museum (Natural History).* London: British Museum, 502 pp.

James, D.G. 1979. Observations on two overwintering clusters of *Danaus plexippus* (L.) (Lepidoptera: Nymphalidae) in the Sydney area during 1978. *Australian Entomological Magazine* 5:81–85.

James, D.G. 1981. Studies on a winter breeding population of *Danaus plexippus* (L.) (Lepidoptera: Nymphalidae) at Spenser, New South Wales. *General and Applied Entomology* 13:47–53.

James, D.G. 1982. A transient non-breeding population of *Danaus plexippus* (L.) (Lepidoptera: Nymphalidae) near Picton, New South Wales. *Australian Entomological Magazine* 8:96–99.

Kitching, R.L. 1981. The geography of the Australian Papilionoidea. In *Ecological biogeography in Australia,* ed. A. Keast, 979–1005, The Hague: Junk.

Kitching, R.L., and M.P. Zalucki. 1981. Observations on the ecology of *Euploea core corinna* (Nymphalidae) with special reference to an overwintering population. *Journal of the Lepidopterists' Society* 35: 106–119.

Kitching, R.I., and M.P. Zalucki. 1983. A cautionary note on the use of oviposition records as larval food plant records. *Australian Entomological Magazine* 10:64–66.

Monteith, G.B. 1982. Dry season aggregations of insects in Australian monsoon forests. *Memoirs of the Queensland Museum* 20:533–543.

Rainbow, W.J. 1907. *A Guide to the study of Australian butterflies.* Melbourne: T.C. Lothian, 272 pp.

Scheermeyer, E. 1985. Some factors affecting the distribution of *Euploea core corinna* (W.S. Macleay) (Lepidoptera: Danainae). *Australian Journal of Zoology* 33:339–348.

Scheermeyer, E. 1987. *Seasonality or opportunism in reproduction of Australian danaine butterflies: Euploea core, E. tulliolus and Tirumala hamata (Lepidoptera).* Ph.D. thesis. Brisbane: Griffith University.

Scheermeyer, E., and M.P. Zalucki. 1985. Food plant records of *Euploea core corinna* (W.S. Macleay) with some notes on larval colouration. *Australian Entomological Magazine* 11:87–90.

Scheermeyer, E., R.L. Kitching, and R.E. Jones. 1990. Host plants, temperature and seasonal effects on immature development and adult size of *Euploea core corinna* (Lepidoptera: Danainae). *Australian Journal of Zoology* 37:299–308.

Smithers, C.N. 1977. Seasonal distribution and breeding status of *Danaus plexippus* (L.) (Lepidoptera: Nymphalidae) in Australia. *Journal of the Australian Entomological Society* 16:175–184.

Smithers, C.N. 1983. Migration records in Australia. 3

Danainae and Acraeinae (Lepidoptera: Nymphalidae). *Australian Entomological Magazine* 10:21–27.

Southwood, T.R.E. 1977. Habitat, the templet for ecological strategies? *Journal of Animal Ecology* 46: 337–365.

Straatman, R. 1963. Observations on migration in certain Australian Lepidoptera. *Tijdschrift voor Entomologie* 106:197–199.

Tsukada, E. (ed.). 1985. *Butterflies of the South-east Asian Islands.* Vol. 2. *Pieridae, Danaidae.* Tokyo: Plapac, 623 pp.

Williams, C.B. 1930. *The migration of butterflies.* London: Oliver & Boyd, 473 pp.

Zalucki, M.P. 1982. Temperature and rate of development in *Danaus plexippus* and *D. chrysippus* (Nymphalidae). *Journal of the Australian Entomological Society* 21:241–246.

Zalucki, M.P. 1986. The monarch butterfly—A nonpest exotic animal. In *The ecology of exotic animals and plants: Some Australian case-histories,* ed. R.L. Kitching, 128–143. Brisbane: Jacaranda-Wiley.

Zalucki, M.P., and R.L. Kitching. 1982. Temporal and spatial variation of mortality in field populations of *Danaus plexippus* L. and *D. chrysippus* L. (Lepidoptera: Nymphalidae). *Oecologia (Berlin)* 53:201–207.

7

MIGRATION

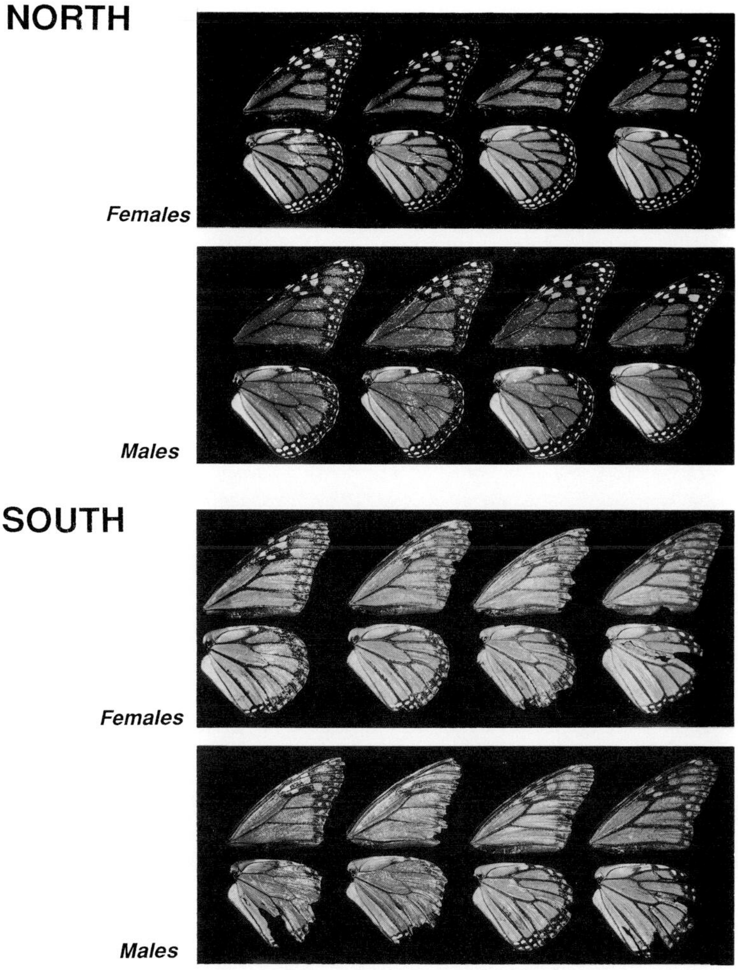

Wings of the earliest spring monarch arrivals caught in the continental United States. The southern butterflies were taken near Austin, Texas on 5–9 April 1985; the northern butterflies were caught in east-central Wisconsin and Door County, Wisconsin on 5 and 9 June 1985. The southern butterflies are battered and worn, whereas the northern butterflies are in good condition. This suggests that northern butterflies are the offspring of early spring arrivals in the southern United States. See Malcolm, Cockrell, and Brower (this volume). Photographs by Tonya Van Hook and Lincoln P. Brower.

THE COLUMBUS HYPOTHESIS: AN EXPLANATION FOR THE DRAMATIC 19TH CENTURY RANGE EXPANSION OF THE MONARCH BUTTERFLY

Richard I. Vane-Wright

ABSTRACT. The dramatic 19th century colonization of Pacific and Atlantic islands by the monarch, *Danaus plexippus plexippus* (Linnaeus), can be well documented. Although the monarch has very recently become established in the Mascarene Islands and gained a toe-hold in Spain, this major range expansion was essentially complete well before 1900. The butterfly has never been recorded reliably from Africa or the Indian subcontinent, and it failed to establish itself west of Wallace's Line in Southeast Asia.

Available records of the North American overwintering colonies of the monarch in California, Florida, and Mexico all postdate the known start of the Pacific and Atlantic expansion. The two phenomena may be linked to the late 18th and 19th century deforestation of North America caused by European colonists, which probably favored a population explosion of this open-ground insect. Before reinforcing selection could lead to the evolution of the highly coordinated migration cycle seen in North America today, the population explosion could have triggered the range expansion. The spectacular monarch roosts of Mexico may be of very recent origin, brought into being, together with the 19th century expansion, as direct consequences of forest destruction in North America by European settlers—the Columbus Hypothesis.

INTRODUCTION

This paper presents an idea, the Columbus Hypothesis, to explain the present global distribution of the monarch butterfly. It suggests that the annual coordinated migration and massed overwintering cycle is a very recent phenomenon and that this, together with the major 19th century range expansion of the monarch in the Atlantic and Pacific, was brought about by the destruction of North America's forests since the arrival of European colonists.

HOW LONG HAVE MIGRATORY MONARCHS BEEN MIGRATING?

The monarch—or the North American variant at least—is a very special butterfly. Its annual migration cycle is the most spectacular (Malcolm, 1987) in the insect world, and it has been described as "a unique biological phenomenon" (Brower, 1985). According to Ward (1987), "the monarch's two-way migration may be unique among the estimated 30 million species of insects on earth." The monarch is unusual in other ways too—including its mating system (e.g., Boppré, this volume; Schneider, this volume; Van Hook, this volume) and the use of sequestered cardenolides as a potent means of defense (e.g., Rothschild and Marsh, 1978; Brower, 1984; see also Ackery, this volume). The subspecies occurring in North America, *Danaus plexippus plexippus* (Linnaeus), has been studied extensively with respect to its migratory habits (e.g., Brower, 1985; Cockrell et al., this volume; Malcolm et al., this volume) as well as its mimicry by the viceroy *Limenitis archippus* (Cramer) (e.g., Brower, 1958a,b; Ritland and Brower, 1991, this volume).

However, in many parts of Central America and the Caribbean, and the whole of South America, the endemic races of the monarch are not involved in mimicry, nor do they undergo migrations on a scale comparable to those found in North America (Haber, this volume). Although the southern monarch, *Danaus erippus,* is recorded as undergoing seasonal migrations in Argentina (Hayward, 1969), Lamas (pers. comm.) indicates that the autumnal migration of *erippus* is southerly (to mirror the North American monarch it would need to be northerly), and that there is no evidence of a return migration. This lack of spectacular migrations in the tropical Americas need not surprise us. The whole point of the complex migratory behavior is to permit larval exploitation of the rich, weedy milkweed flora of the United States and Canada—which the monarch can only accomplish during the frost-free months of the northern summer (Young, 1982).

Kitching et al. (this volume; see also Young, 1982) have suggested that the most likely origin of the North American monarch, in the medium term of about 2 myr, is South America, and that it reached the northern continent via the Pliocene Darien Gap connection. If so, then in the shorter term its origin must have been Central America or Mexico (Young, 1982). Eighteen thousand years ago North America was in the maximum grip of the Wisconsinan Glaciation, which brought permafrost south of Chicago and cool boreal forests to Arkansas. Virtually all of the southern United States was forested, right down to the Gulf Coast, including the area now covered by the prairies (Duplessy and Ruddiman, 1984). Even as far back as 40,000 yr B.P., the climatic conditions were almost as harsh, and the vegetation (Delcourt and Delcourt, 1981) would almost certainly have been largely unsuitable for the monarch. As a result, I think we can take an age of about 10,000 yr, at the beginning of the Holocene, postglacial, or mesolithic periods (Sutcliffe, 1985), as an upper bound for the age of the present North American monarch migration phenomenon. At this time the prairies were beginning to form in the rainshadow to the east of the Rockies (Delcourt and Delcourt, 1981).

But before pursuing this line of argument further, I first wish to illustrate details of the equally spectacular, but far less well-known, monarch phenomenon, the 19th century colonization of the Atlantic and Pacific oceans.

THE MONARCH'S DOUBLE INVASION OF THE OLD WORLD

Prior to about 1840, there is no record of the monarch outside the Americas and Caribbean. A decade later we have reliable records for Bermuda (Jones, 1859; Walker, 1914) and Hawaii (Scudder and Gulick, 1875; Scudder, 1888–1889) and a doubtful account for New Zealand's North Island (Walker, 1914; Gibbs, 1980). By 1860 we have additional records for the Caroline Islands (Scudder and Gulick, 1875; Scudder, 1888–1889) and Marquesas (Walker, 1914) and a doubtful record for New South Wales (Olliff, 1889). During the following 10 yr Australia is confirmed (first at Brisbane, in 1870; Marks, 1963), together with several records for central Pacific islands (Walker, 1914) and Flores and Fayal in the Atlantic Azores group (in 1864; Godman, 1870). As early as 1873, Semper gives an account of its occurrence in the Manado area of northern Sulawesi (where it can evidently still be found, but not elsewhere on that island; Morishita, 1981; R.I.V.-W., pers. obs.), and by 1880 there are records for New Guinea (BMNH collection), Cape York (BMNH collection), Tasmania (Couchman and Couchman, 1977), South Island of New Zealand (Walker, 1914), France (Baret, 1878), and England (Llewellyn, 1876; Williams, 1942). By 1890 the picture for both the Atlantic and Pacific is consolidated (Walker, 1914), with records for Spain and the Canary Islands, the Solomons and New Caledonia, the Moluccas (MNH collection, Leiden), and even the Straits of Malacca in Malaysia (de Nicéville and Martin, 1896). The next 15 yr sees records for Hong Kong, Taiwan, Java, Borneo, the Philippines, Guam, and a host of other small islands (Walker, 1914; Ackery and Vane-Wright, 1984; BMNH and MNH collections).

By the turn of the century most entomologists were convinced of two things about the monarch: its extraordinary spread (Fig. 1) had been brought about by trade, either directly (Scudder, 1888–1889) or in the wake of the introduction of *Asclepias* (Walker, 1914), and it would continue to advance and eventually encircle the globe (Distant, 1877; Holland, 1893; Walker, 1914). However, during the ensuing 80 yr, practically nothing has happened. The butterfly has become common in the Mascarene Islands in the last 10 yr (J.R. Williams, pers. comm.; Vane-Wright, 1986; Paré, 1987), it has bred for a few years since about 1980 in Spain (Bretherton, 1984; Edwards, 1984, 1988), it has been recorded once from Denmark (Toft, 1980), and it eventually reached the west coast of Australia (Common and Waterhouse, 1972). However, it has never been reported again from the Greater Sunda Islands (Corbet and Pendlebury, 1978), it has failed to colonize the rest of Sulawesi (Morishita, 1981; pers. obs.), it is extinct in the Philippines (pers. obs.; C.G. Treadaway, pers. comm.), and it is probably extinct in mainland China (M. Bascombe, pers. comm.). There has never been a single reliable record for the whole of Africa or the subcontinent of India (Fig. 2).

When you look at the orderly spread reflected by the maps (Figs. 1 and 2), based as they necessarily are on uncoordinated and haphazardly reported sightings, one could speculate that this could have been brought about directly through the introduction of fast trading ships (the Scudder Hypothesis). The paddle steamer *Savannah* crossed the Atlantic from west to east in 1819 and, following the voyage of the *Sirius* from Cork to New York in 18 days in 1837, the Cunard Company introduced a regular Atlantic steamer service in 1840. But in light of this, and the distribution of entomologists, if the monarch was directly transported by shipping, then we would expect records for coasts of Europe or Australia before we received reports from the Azores, Marquesas, or Carolines. The latter information was received not days, or months, but *years* in advance. And why did the monarch run out of steam by 1905, or even 1880 (see Fig. 1)?—the world's ships certainly hadn't!

If not directly introduced to the Pacific and Atlantic by fast ships, what of the alternative—that the monarch merely used its migratory ability to track the spread of *Asclepias* brought by man's trade (the J.J. Walker Hypothesis; see also Malcolm and Brower, 1987)? If so, we must then ask why it stopped short of Africa and India. One possibility is that the monarch has been inhibited by Old World *Danaus* species capable of out-competing it—or as a result of disease (as suggested by T.J. Walker, in litt.). But there is no species of *Danaus* in Africa or India with which the monarch does not come into contact in the eastern Indo-Australian region (see Ackery and Vane-Wright, 1984, table 31). In Africa, *Danaus chrysippus* is particularly abundant, but so it is in the Canary Islands, where the two species have now coexisted for well over 100 yr. In Mauritius, where the monarch has recently established itself, it has to co-exist with *chrysippus*. In the recent colonization of Spain by both species, the monarch is still present, but *chrysippus* may have already disappeared (Edwards, 1988).

Whereas the monarch's colonization of the Pacific and Atlantic is known to have taken place so recently, and has been such a short-lived phenomenon in terms of rapid advance, the annual migration cycle in North America is usually assumed to be a relatively ancient phenomenon—about 10,000 yr in terms of the picture given above, and it could have occurred even earlier, during previous interglacials. Young (1982) suggests that the presence of both the queen and the monarch in North America represents different responses of these butterflies to a post-Pleistocene invasion of Mexico and Central America by the rich *Asclepias* flora evolved in North America. Kitching et al. (this volume) discuss evidence for the monarch reaching North America during the last 2 myr, since closure of the Darien Gap. Genetic markers, such as the enzyme variations discussed by Kitching et al. (this volume), if studied sufficiently, should eventually enable us to make better estimates of the timing and sequence of various events.

WHITE MONARCHS—NEW MUTANTS OR OLD COLONISTS?

Danaus plexippus plexippus shows little phenotypic variation. One exception is the white or 'nivosus' form, in which the normal bright orange ground-color of the wings is replaced by a faintly pinkish or grayish white. The white monarchs of Oahu, Hawaii, first noted in 1965 (Mitchell, 1966), comprise about 4% of the local monarch population (Clarke and Rothschild, 1980; Stimson and Meyers, 1984), currently rising to even 10% (J. Stimson, pers. comm.).

Form 'nivosus' is a simple autosomal recessive (Stimson and Meyers, 1984), possibly involving the loss of ability to produce the normal orange pigment. According to Stimson and Meyers, 'nivosus' was unknown on Hawaii until 1965 and must therefore be the product of a new mutation. However, the Rothschild Collection (BMNH) includes a single white monarch labeled 'Hawaii, Palmer'. This specimen, together with "regular" Hawaiian monarchs, was evidently received by Walter Rothschild in the mid-1890s. Thus 'nivosus' has probably persisted on Hawaii since the end of the last century, and the allele responsible may well have been present in the founder population of ca. 1845 (Scudder, 1888–1889).

Elsewhere, white monarchs have been reported from Washington, D.C., in 1896 (Clark, 1932); Missouri, 1908 (Gunder, 1927); Pennsylvania, 1921 (Gunder, 1927); the coast of California, ca. 1980 (John Lane, pers. comm., in Vane-Wright, 1987); Brisbane, 1980 (De Baar, 1982); North Island, New Zealand, 1985 (G. Bulow, pers. comm.); and Ambon and Seram (in the Moluccas), 1906–1916 (BMNH collection).

Although most of the 11 recognized species of *Danaus* (see Kitching et al., this volume) are invariably orange-pigmented, white races do occur in the Austro-Oriental *Danaus melanippus*, and all races of the Wallacean species *Danaus ismare* lack orange wing coloration (see Ackery and Vane-Wright, 1984). Thus, it would not be surprising if 'nivosus' mutants had arisen independently in *plexippus* more than once. However, it is also possible that the disjunct Hawaiian, New Zealand, Australian, and Moluccan white forms have been produced by recombination of a rare but homologous allele, inherited directly from their North American ancestors. Hybridization experiments involving 'nivosus' of different origin could shed some light on this problem. In particular, if double heterozygous Hawaii × non-Hawaii crosses produced only typical F1 monarchs, we could reject the single-origin hypothesis. But if typical : 'nivosus' segregated 3:1 in the F1, the American origin hypothesis would receive corroboration—although this would not, of course, reject the multiple-origin possibility.

Genetically controlled phenotypic traits such as the 'nivosus' form, the less well-known 'fumosus' trait of the Pacific and North America, and the 'leucogyne' variants of the Caribbean, together with other genetic variations, offer some hope of getting a more detailed picture of the spread of the monarch during the last century. But such phenomena as the white monarchs of Hawaii also remind us of fundamental issues of interpretation in historical biogeography (e.g., Humphries and Parenti, 1986). The need to understand cause and effect and to separate those events that are historically connected from those that have arisen independently are paramount.

EARLY RECORDS OF THE OVERWINTERING COLONIES

The extraordinary Mexican overwintering sites, although well known to local people at the time of their "discovery" in the mid-1970s, were not reported in the literature until that period (Urquhart,

1850

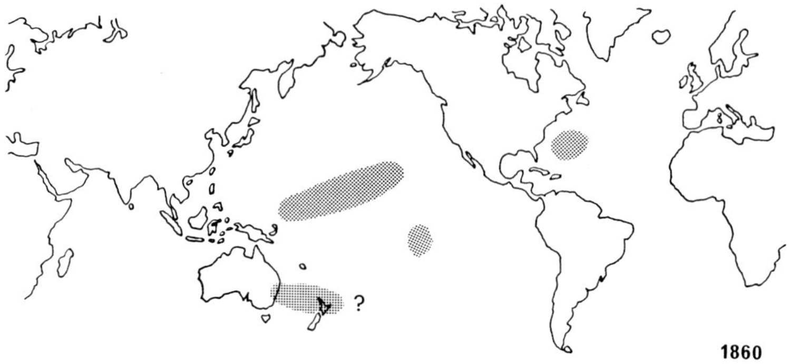

1860

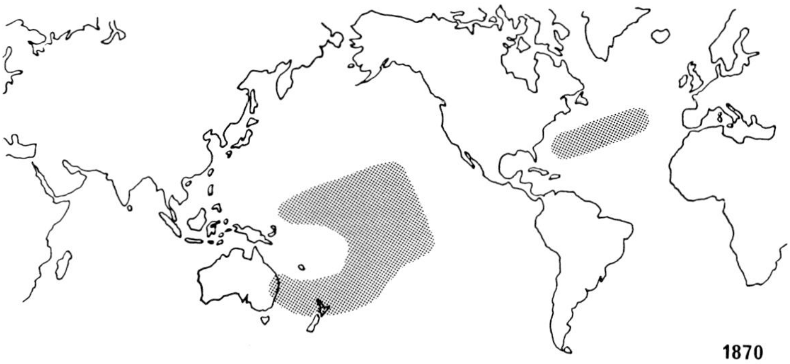

1870

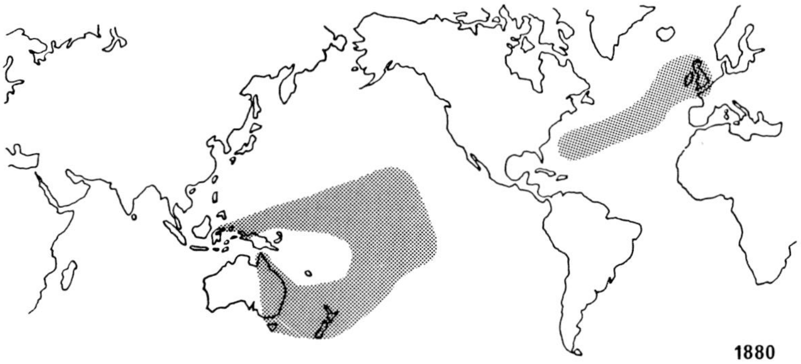

1880

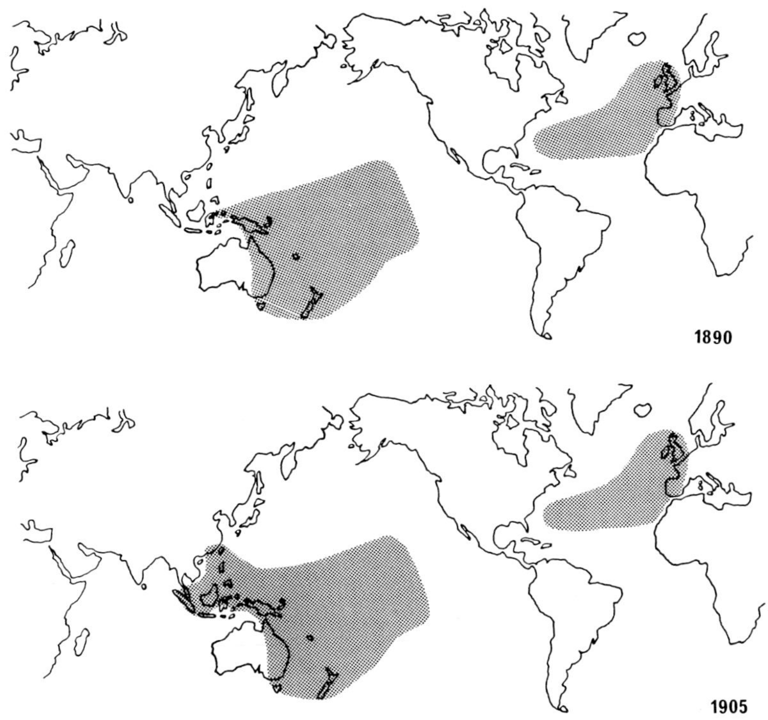

Figure 1. Cumulative maps of the recorded distribution of *D. p. plexippus* outside the Americas, up to 1850, and in 1860, 1870, 1880, 1890, and 1905. Based mainly on information in Walker (1914); see also text.

1976; Brower, 1977). The first published observations of massed overwintering colonies in Florida (at Apalachicola) and at Pacific Grove in California that I have been able to trace are those by Thaxter (1880) and Bush (1881), respectively (also see Lane, this volume). Bush states emphatically that "A lady resident informed me that for the twelve years she had lived there the appearance [of the massed monarchs on the pine trees] had been the same," thus documenting the phenomenon in California back to about 1869.

Lucia Shepardson, in her charming yet curious booklet published in 1914, states that "The earliest authentic information as to this annual migration [at Pacific Grove] dates back fifty years [1864], when the monarchs were seen upon the pines just as they are to-day." (This reference probably relates to Bush, but I am not certain.) She goes on, "Previous to that, no mention has been found of this interesting phenomenon . . . . The early Spanish chronicles and traditions make no mention of it, although Monterey, a scant three miles distant, was gay with life when the last century came in . . . even David Douglas, the world-famed botanist, and the keenest-eyed of all the strangers who came here, is silent regarding it." Douglas, who died a sudden and tragic death in 1834, visited Monterey from December 1830 to August 1832 (Harvey, 1947). During this time he discovered the Monterey pine (*Pinus radiata* Don; *Pinus insignis* Douglas), noted by Shepardson as *the* tree to which the Pacific Grove monarchs flocked.

There are few, if any, accounts of the autumnal clustering and migratory behavior of the monarch before the 1860s, when the then burgeoning North American literature suddenly starts to report observations by the dozen—together with argument and speculation as to their meaning. If we accept Shepardson's account literally, the earliest historical record of massed overwintering appears to have been made at about the same time, in 1864 (but see Lane, this volume). Douglas, the indefatigable fir tree collector, appears to have made no mention of the phenomenon in 1830–1832, despite spending two winters at Monterey. As outlined above, range expansion by the monarch into the Pacific and Atlantic is known to have started exactly midway between these dates. Are the two phenomena linked historically? If so, which is the cause and which the effect?

THE COLUMBUS HYPOTHESIS

The monarch is an open-country, frost-intolerant insect (Anderson and Brower, this volume) that exploits the rich North American asclepiad flora by means of the annual remigration cycle (Young, 1982). In the tropical Americas, from Nicaragua southward, the monarch is also largely restricted to open areas but only makes more limited local

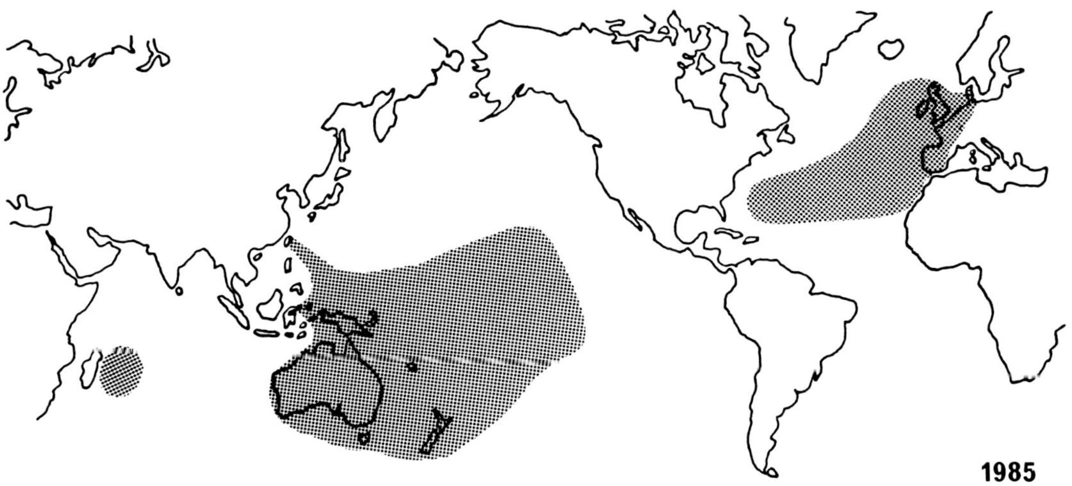

Figure 2. Current (1985) known range of *D. p. plexippus* outside the Americas. [Note: Apart from the east coast region of Spain (Edwards, 1988) monarchs are not known to breed in the British Isles or continental Europe, but they still occur as more or less occasional vagrants on western coasts (e.g. Bretherton and Chalmers-Hunt, 1982).]

migrations (such as the east–west, dry–wet season movements observed in Costa Rica: Haber, this volume; note also Young, 1982). Before Europeans destroyed extensive areas of woodlands and forest, the weedy, disturbed-ground asclepiads of North America must have been much less common than they are now. Before this, in the natural prairies, monarchs may have behaved much as they now do in Australia or New Zealand, or as *D. erippus* appears to do in the Argentinian pampas (Williams, 1942; Hayward, 1969).

Could the deforestation of North America, which started in earnest about 200 yr ago (Delcourt and Delcourt, 1981), have led to a 19th century population explosion of milkweed plants and, in turn, of the native monarch, from which the overspill colonized the Atlantic and Pacific? Is it possible that in succeeding autumns, millions of monarchs flew east and west as well as south, until the locally adaptive, southerly, massed and orderly, "out-and-back" annual migrations and overwintering roosts rapidly evolved through reinforcing selection? Did the anticipated global colonization run out of steam simply because, by about 1880, nearly all North American monarchs had been programmed in this way to fly south in autumn (those that weren't simply never came back)? Do the monarchs of Australia and New Zealand (Common and Waterhouse, 1972; Wise, 1980; Gibbs, 1980) fail to show this precise north–south behavior pattern because they are derived from an ancestral population in which it had not fully evolved?

I am not suggesting that migration and clustering is limited to North American monarchs. Such behavior is of course widespread not only in the monarch as a whole, but also in many other danaines as well (see for example review in Ackery and Vane-Wright, 1984; and recent detailed work of Scheermeyer, 1987; Wang and Emmel, 1991; Kitching and Scheermeyer, this volume; Scheermeyer, this volume). What I do want to suggest is that the *spectacular annual migration cycles,* found only in the populations of North American monarchs, have evolved *in their present form* as a result of the cataclysmic ecological changes wrought by European colonists—the Columbus Hypothesis. According to Dasmann (1965), for example, extensive coastal lumber extraction in northern California started in 1851, shortly after the Gold Rush, and by "1870 most of the redwoods in the immediate vicinity of Humboldt Bay were cut, and the hills behind Eureka and its sister town of Arcata were left deforested." More generally for North America, Williams (1989) states that "300 million acres of forest land had been cleared by the beginning of the twentieth century." It is perhaps ironic to contemplate the idea that the Mexican winter roosts, which, together with the Californian sites, were among the first "Threatened Phenomena" to be designated in the history of international conservation (Wells et al., 1983), are less than 200 yr old—and a product of man's destruction!

The only monarch populations to have had their migration and dormancy cycles extensively studied are those of the United States and Australia. They exhibit a number of differences (Herman, 1981, this volume; James and Hales, 1983; James, 1984, this volume). Recently it has been suggested that in the Australian monarchs, over the last 115 yr, "both diapause-mediated dormancy and migration behavior have been altered" (Tauber et al., 1986: 243). Altered in relation to what? An equal possibility is that the "standard" North American populations have changed. As in the case of the white monarchs, the Australian populations may have merely retained characteristics of their early Victorian ancestors—able to make local migrations between breeding and overwintering sites, but not necessarily capable of the extreme coordinated behavior seen in North America. The universal need to distinguish primitive from specialized traits (Hennig, 1966), just as with the white monarchs of the Pacific, means that comparative physiologists and ecologists dealing with historical events can no more safely ignore Hennig's principles than can other systematists.

FEW ANSWERS, MORE QUESTIONS

I am well aware that these suggestions raise far more questions than they resolve. In general, it is certain that some form of migration occurred in North American monarch populations, before the arrival of Europeans, to take advantage of the naturally occurring prairie milkweeds. Did deforestation merely cause quantitative enhancement of this phenomenon, or has there been some qualitative change in behavior as well? The fact that monarchs, particularly in the southern parts of North America, operate a sort of "mixed strategy" (Malcolm, 1987), with some individuals joining the mass migrations south while others remain to form more or less continuously breeding colonies, may throw some light on this problem.

More particular questions include, amongst others: What *is* the earliest record of massed overwintering in California? If indeed it is only 1864 or 1869, is this significant, or does it merely reflect the late European settlement of western North America? (The Spanish established Monterey as the capital of California in 1770, the Mexicans took over in 1821, and it was seized by Americans in 1846 and ceded to the United States in 1848.) Can anyone make an estimate of the progress of North American deforestation up to about 1880 (see Williams, 1989), and its likely impact on native milkweed plants? Does the continuing arrival of monarchs along the western coasts of Europe, notably in peak years such as 1933 (Williams, 1958) and 1981 (Bretherton and Chalmers-Hunt, 1982), tell against the Columbus Hypothesis—or have many of these latter-day wanderers only strayed from closer established colonies, such as those on the Canary Islands?

In addition to answers to such questions, we require a thorough investigation of the monarch's genetics, particularly its molecular variation. Before speculation outruns the supply of essential facts, we really need a worldwide survey of this beautiful—and still mysterious—insect. We also need to devise critical tests or experiments, in the hope of eliminating the Columbus Hypothesis—or its rivals.

ACKNOWLEDGMENTS

In addition to those friends and correspondents noted in the text, I would like to thank P.R. Ackery, L.P. Brower, J. Dayton, M. De Baar, J. Diment, P. Forey, L. Gilbert, J. Hodson, C.J. Humphries, J. Lane, S. Malcolm, L.A. Mound, M. Rothschild, A.J. Sutcliffe, and T.J. Walker for their help. I am also deeply indebted to Julian P. Donahue and the organizers of Moncon-2, who made it possible for me to attend the Conference.

LITERATURE CITED

Ackery, P.R., and R.I. Vane-Wright. 1984. *Milkweed butterflies: Their cladistics and biology*. London: British Museum (Natural History), and Ithaca, NY: Cornell University Press, 425 pp.

Baret, [ ]. 1878. Note sur la *Danais archippus. Petites Nouvelles Entomologiques* 2:253–254.

Bretherton, R.F. 1984. Monarchs on the move—*Danaus plexippus* (L.) and *D. chrysippus* (L.). *Proceedings and Transactions of the British Entomological and Natural History Society* 17:65–66.

Bretherton, R.F., and J.M. Chalmers-Hunt. 1982. The monarch or milkweed butterfly (*Danaus plexippus* L.) in 1981. *Entomologist's Record and Journal of Variation* 94:81–87, 141–146.

Brower, J.V.Z. 1958a. Experimental studies of mimicry in some North American butterflies. Part I. The monarch, *Danaus plexippus,* and the viceroy, *Limenitis archippus archippus. Evolution* 12:32–47.

Brower, J.V.Z. 1958b. Experimental studies of mimicry in some North American butterflies. Part III. *Danaus gilippus berenice* and *Limenitis archippus floridensis. Evolution* 12:273–285.

Brower, L.P. 1977. Monarch migration. *Natural History* 84:40–53.

Brower, L.P. 1984. Chemical defence in butterflies. In *The biology of butterflies,* eds. R.I. Vane-Wright and P.R. Ackery, *Symposia of the Royal Entomological Society of London* 11:109–134, London and New York: Academic Press (1989, Princeton, NJ: Princeton University Press).

Brower, L.P. 1985. New perspectives on the migration biology of the monarch butterfly, *Danaus plexippus* L. In *Migration: Mechanisms and adaptive significance,* ed. M.A. Rankin, 748–785. Austin: University of Texas, *Contributions in Marine Science,* Supplement, 27.

Bush, A.E. 1881. Trees attractive to butterflies. *American Naturalist* 15:572.

Clark, A.H. 1932. The butterflies of the district of Columbia and vicinity. *Smithsonian Institution United States National Museum Bulletin,* 157, 337 pp. (+64 pls.).

Clarke, C.A., and M. Rothschild. 1980. A new mutant of *Danaus plexippus* ssp. *erippus* (Cramer). *Journal of the Lepidopterists' Society* 34:224–229.

Common, I.F.B., and D.F. Waterhouse. 1972. *Butterflies of Australia.* Sydney: Angus & Robertson, 498 pp. (+41 pls.).

Corbet, A.S., and H.M. Pendlebury. 1978. *The butterflies of the Malay Peninsula,* 3rd ed., revised by J.N. Eliot. Kuala Lumpur: Malay Nature Society, 578 pp. (+36 pls.).

Couchman, L.E., and R. Couchman. 1977. The butterflies of Tasmania. *Tasmanian Year Book* 11:66–96 (+8 pls.).

Dasmann, R.F. 1965. *The destruction of California.* New York: Macmillan, 247 pp.

De Baar, M. 1982. An aberrant wanderer butterfly. *Entomological Society of Queensland News Bulletin* 10:25–26.

Delcourt, P.A., and H.R. Delcourt. 1981. Vegetation maps for eastern North America: 40,000 yr B.P. to the present. In *Geobotany II,* ed. R.C. Romans, 123–165. New York: Plenum Press.

de Nicéville, L., and L. Martin. 1896. A list of the butterflies of Sumatra with especial reference to the species occurring in the north-east of the island. *Journal of the Asiatic Society of Bengal* 64(2):357–555.

Distant, W.L. 1877. The geographical distribution of *Danais archippus. Transactions of the Royal Entomological Society of London* 10:93–104.

Duplessy, J.-C., and W.F. Ruddiman. 1984. La fonte des calottes glaciares. *La Recherche* (156):806–818.

Edwards, P.J. 1984. Monarchs resident in Spain (*Danaus plexippus* (L.)). *Proceedings and Transactions of the British Entomological and Natural History Society* 17:84.

Edwards, P.J. 1988. Monarchs resident in Spain. *British Journal of Entomology and Natural History* 1:59–60.

Gibbs, G.W. 1980. *New Zealand butterflies: Identification and natural history.* Auckland: W. Collins, 207 pp.

Godman, F.D. 1870. *Natural history of the Azores, or Western Islands.* London: J. van Voorst, 358 pp. (+2 maps).

Gunder, J.D. 1927. New transition forms or "abs" and their classification (Lepid. Rhopalocera). *Entomological News* 38:129–138 (+1 pl.).

Harvey, A.G. 1947. *Douglas of the fir.* Cambridge, MA: Harvard UP, 290 pp.

Hayward, K.J. 1969. Notes on butterfly migration in Argentina between April, 1966, and May, 1968. *The Entomologist* 102:10–11.

Hennig, W. 1966. *Phylogenetic systematics.* Urbana: Illinois UP, 263 pp.

Herman, W.S. 1981. Studies on the adult reproductive diapause of the monarch butterfly, *Danaus plexippus. Biological Bulletin, Marine Biological Laboratory, Woods Hole* 160:89–106.

Holland, W.J. 1893. Notes and queries. *Annual Report of the Entomological Society of Ontario* 24:53–55.

Humphries, C.J., and L.R. Parenti. 1986. *Cladistic biogeography.* Oxford: Clarendon Press, 98 pp.

James, D.G. 1984. Population and general biology of non-reproductive colonies of the monarch butterfly, *Danaus plexippus* (L.) (Lepidoptera: Danaidae) in New South Wales. *Australian Journal of Zoology* 32:663–670.

James, D.G., and D.F. Hales. 1983. Sensitivity to juvenile hormone is not reduced in clustering monarch butterflies, *Danaus plexippus,* in Australia. *Physiological Entomology* 8:273–276.

Jones, J.M. 1859. *The naturalist in Bermuda.* London: Reeves & Turner, 200 pp. (+1 map).

Llewellyn, J.T.D. 1876. A foreign visitor (*Danais archippus*). *Entomologist's Monthly Magazine* 13:107–108.

Malcolm, S.B. 1987. Monarch butterfly migration in North America: Controversy and conservation. *Trends in Ecology and Evolution* 2:135–138.

Malcolm, S.B., and L.P. Brower. 1987. White monarchs. *Antenna* 11:2–3.

Marks, E.N. 1963. The first record of *Danaus plexippus* L. from Australia. *Journal of the Entomological Society of Queensland* 2:8.

Mitchell, W.C. 1966. [Note on] *Danaus plexippus* (L.). *Proceedings of the Hawaiian Entomological Society* 19:129.

Morishita, K. 1981. Danaidae. In *Butterflies of the Southeast Asian islands,* ed. E. Tsukada, 2:439–628 (+ pls. 85–162). Tokyo: Plapac.

Olliff, A.S. 1889. *Australian butterflies.* Sydney: Natural History Association of N.S.W., 49 pp.

Paré, R. 1987. A note from Zimbabwe. *News of the Lepidopterists' Society* Nov./Dec. 1987:80.

Ritland, D.B. and L.P. Brower. 1991. The viceroy butterfly is not a batesian mimic. *Nature* 350:497–498.

Rothschild, M., and N. Marsh. 1978. Some peculiar aspects of danaid/plant relationships. *Entomologia Experimentalis et Applicata* 24:637–650.

Scheermeyer, E. 1987. Seasonality or opportunism in reproduction of Australian danaine butterflies: *Euploea core, E. tulliolus* and *Tirumala hamata* (Lepidoptera). Ph.D. thesis. Brisbane: Griffith University.

Scudder, S.H. 1888–1889. *The butterflies of the eastern United States and Canada with special reference to New England,* 3 vols. Cambridge, MA: publ. by author, 1958 pp. (+89 pls.) (in 3 vols.).

Scudder, S.H., and L.H. Gulick. 1875. The introduction of *Danaida plexippus* into the Pacific islands. *Psyche* 1:81–84.

Semper, G. 1873. Die Wanderung von *Danais erippus,* Cramer, nach dem Südseeinseln, Australien und Celebes. *Journal des Museum Godeffroy* 1:293–295.

Shepardson, L. 1914. *The butterfly trees.* San Francisco: J.H. Barry, 32 pp. (+4 pls.).

Stimson, J., and L. Meyers. 1984 [1985]. Inheritance and frequency of a color polymorphism in *Danaus plexippus* (Lepidoptera: Danaidae) on Ohahu [*sic!*], Hawaii. *Journal of Research on the Lepidoptera* 23: 153–160.

Sutcliffe, A.J. 1985. *On the track of ice age mammals.* London: British Museum (Natural History), 224 pp.

Tauber, M.J., C.A. Tauber, and S. Masaki. 1986. *Seasonal adaptations of insects.* Oxford: Oxford University Press, 411 pp.

Thaxter, R. 1880. Swarming of *archippus. Canadian Entomologist* 12:38–39.

Toft, S. 1980. First Danish record of the monarch butterfly (*Danaus plexippus* Linn) (Lepidoptera: Danaidae). *Entomologiske Meddelelser* 47:115–116.

Urquhart, F.A. 1976. Found at last: The monarch's winter home. *National Geographic* 150:160–173.

Vane-Wright, R.I. 1986. White monarchs. *Antenna* 10: 117–118.

Vane-Wright, R.I. 1987. [Reply to Malcolm and Brower.] *Antenna* 11:3.

Walker, J.J. 1914. The geographical distribution of *Danaida plexippus*, L. (*Danais archippus*, F.) with especial reference to its recent migrations. *Entomologist's Monthly Magazine* 50:181–193, 233–237 (+1 map).

Wang, H.Y., and T.C. Emmel. 1991. Migration and overwintering aggregations of nine danaine butterfly species in Taiwan (Nymphalidae). *Journal of the Lepidopterists' Society* 44:216–288.

Ward, B. 1987. The mystery of monarch migration. *Sky* (Sept. 1987):12–23.

Wells, S.M., R.M. Pyle, and N.M. Collins. 1983. *The IUCN invertebrate red data book*. Gland, Switzerland: IUCN, 632 pp.

Williams, C.B. 1942. In C.B. Williams, G.F. Cockbill, M.E. Gibbs, and J.A. Downes. Studies in the migration of Lepidoptera. *Transactions of the Royal Entomological Society of London* 92:101–283 (+2 pls.).

Williams, C.B. 1958. *Insect migration*. London: Collins, xiii + 235 pp.

Williams, M. 1989. *Americans and their forests*. Cambridge: Cambridge University Press, xii + 599 pp.

Wise, K.A.J. 1980. Monarch butterfly dispersal in New Zealand. *Record of the Auckland Institute and Museum* 17:157–173.

Young, A.M. 1982. An evolutionary-ecological model of the evolution of migratory behavior in the monarch butterfly, and its absence in the queen butterfly. *Acta Biotheoretica* 31:219–237.

MIGRATION BIOLOGY OF THE MONARCH BUTTERFLY IN AUSTRALIA

David G. James

ABSTRACT. Summer breeding populations of the monarch butterfly occur throughout eastern Australia. Late summer and autumn generations south of latitude 34° generally become nonreproductive and migrate to coastal overwintering sites. Reproductive dormancy and migratory behavior are induced by cool (<18°C), cloudy conditions during early adult life. Migration in a north to northeasterly direction for up to 380 km has been recorded and is stimulated by warm (>20°C) days and cold (<10°C) nights. Female migrants show reproductive development when brought into the laboratory and exposed to warmth or injected with juvenile hormone. They fail to respond to oogenesis-inducing stimuli in the field. The energy reserves of migrants are extensive and are maintained at high levels until mid–late winter. Lipogenesis and lipid conservation are enhanced by hot (>30°C) or cool (<18°C) temperatures but inhibited by warm (20–30°C) conditions. In the absence of rigid physiological control (diapause), it is suggested that reproductive dormancy and maintenance of energy reserves in migrants result from behavioral modification of body temperature (T_b). Behavior such as sustained flight, basking, and nocturnal/cool weather inactivity create a T_b regime inhibitory to reproduction and stimulatory to lipogenesis/lipid conservation (i.e., >30°C or <20°C). During autumn, migrants form male-dominated transient roosts of 10–1,000 individuals. With the advent of cooler (<20°C) day temperatures in late autumn–early winter, stable overwintering colonies are formed in the Sydney basin. Colonies disperse in mid–late winter and limited data indicate the possibility of a return, southwesterly migration.

INTRODUCTION

The monarch butterfly, *Danaus plexippus* (L.), was first recorded in Australia in 1870 (Miskin, 1871; Marks, 1963) and now occurs throughout most of eastern and in some parts of Western Australia. During summer its distribution is extensive, ranging from northern Queensland to Tasmania, South Australia and Western Australia (Fig. 1). Severe limitations are placed on its abundance in many areas by the relative scarcity of introduced milkweeds (Asclepiadaceae). All milkweeds used by the monarch in Australia [(*Asclepias curassavica* L.; *Asclepias fruticosa* L., (also known as *Gomphocarpus fruticosus* L. (R. Br.)); *Gomphocarpus physocarpus* (L.); and *Calotropis procera* (R. Br.)] are restricted in their distribution and abundance by intolerance to dry conditions and frost. Thus, the largest monarch and milkweed populations in Australia occur in coastal areas, particularly in Queensland and New South Wales (Smithers, 1977). The occurrence of milkweeds in inland areas of Australia is sporadic and often cyclical with respect to drought periods. When they do occur in these areas, milkweeds are invariably confined to moist situations such as ditches and along watercourses. During late summer and autumn the monarch's range contracts, so that by mid-winter three major populations occur: in the southern Queensland–northern New South Wales coastal strip, the Sydney basin–Hunter Valley region, and the Adelaide area (Smithers, 1977). Small populations also exist throughout the year near Perth in Western Australia, near Alice Springs in the Northern Territory, and in north Queensland, although little information is available (Common and Waterhouse, 1981) (Fig. 1). Prior to the 1980s, little was known about the seasonal biology and ecology of the monarch in Australia except for the studies of Smithers (1965, 1972, 1973a,b, 1977). In 1978 a program of research was initiated to investigate the seasonal biology of the monarch in New South Wales. To a large extent, this work has established the ecological and physiological fundamentals of monarch overwintering in Australia

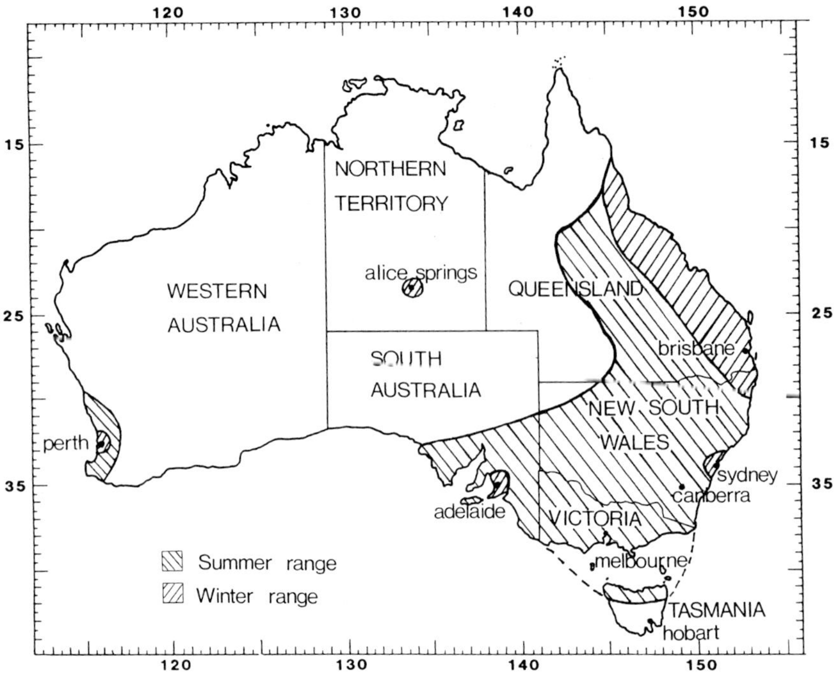

Figure 1. Summer and winter distribution of the monarch butterfly in Australia.

(James, 1982b, 1983, 1984b, 1986a). In addition, our knowledge of the ecology of continuously breeding monarch populations in Australia has been much improved (Zalucki, 1981a,b, 1982; Zalucki and Kitching, 1982a,b,c; Bull et al., 1985; Suzuki and Zalucki, 1986). A major handicap to studies on the seasonal movement of the monarch in Australia is the relatively small population of people and butterflies, particularly in inland areas, compared with that found in North America. Limited monarch populations effectively dilute the visual impact of seasonal behavior such as migration. This problem is exacerbated by the fact that the populations with the greatest propensity to migrate are the small inland populations (James, 1984b). The inland areas of Australia are also characterized by their low human population densities! Thus, it was clear from the outset that an intensive monarch tagging program in Australia, such as that undertaken by Urquhart and cooperators in North America (Urquhart and Urquhart, 1978, 1979), would be fraught with problems. Such an approach was initiated in the early 1960s by the Australian Museum in Sydney under the guidance of Dr. Courtenay Smithers. To date there is no evidence for regular long-distance movement of the monarch in Australia (Smithers, pers. comm.). The aim of this paper is to review recent biological studies on monarch migration and overwintering, which have been conducted principally in the Sydney region of New South Wales. Salient ecological and physiological features of monarch overwintering in this area are presented in Figures 2 and 3, respectively.

AUTUMN POPULATIONS OF THE MONARCH IN SOUTHEASTERN AUSTRALIA

Early studies investigated the incidence of overwintering monarch butterflies in the Sydney area of New South Wales and provided evidence of coexistence of substantial reproductive and nonreproductive populations (James, 1979, 1981, 1984b). This is an interesting situation that does not appear to occur on the same scale in North America (Urquhart, 1960). Summer breeding populations of the monarch in Sydney and surrounding areas show considerable variation in size from year to year. Field observations indicate that three to four summer adult generations occur in most years (James, 1984b). Divergence of monarch populations into reproductive and nonreproductive components occurs (Figs. 2 and 3) first in the tableland and highland areas to the west and southwest of Sydney during March (early autumn). Field observations over a number of years showed that reproductive dimorphism of tableland populations invariably followed short periods of cool, cloudy weather (James, 1983). Field experiments confirmed the hypothesis that low temperatures (<18°C) following adult eclosion were the primary cue for induction of reproductive dormancy and migratory behavior (James, 1983, 1984b). Photophase after emergence and during immature development did not affect reproductive status or migratory behavior. Field observations also suggest that temperature and photophase experienced during larval and pupal development are un-

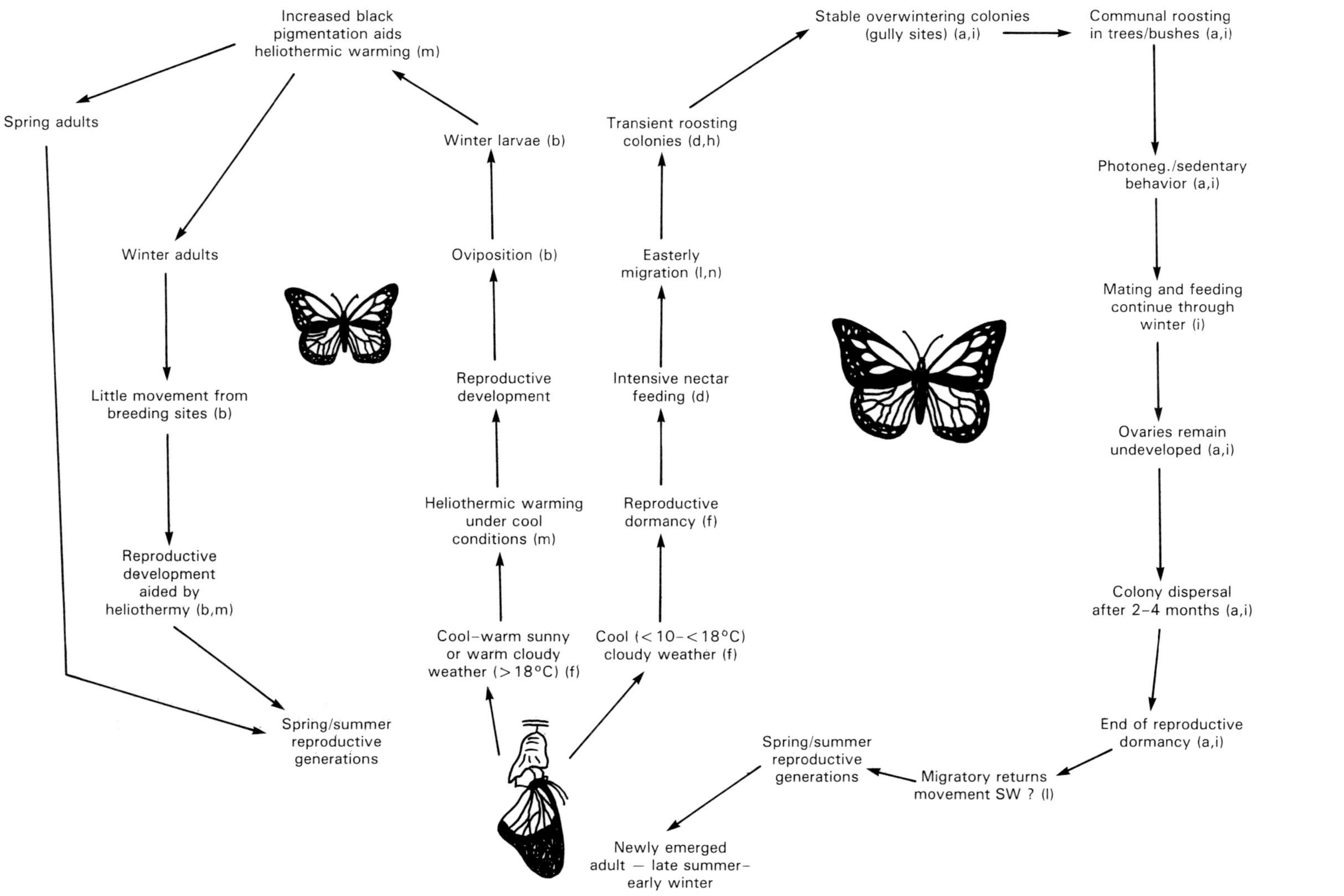

Figure 2. Salient ecological features of monarch overwintering in the Sydney area of New South Wales. References from James: **a,** 1979; **b,** 1981; **d,** 1982b; **f,** 1983; **h,** 1984a; **i,** 1984d; **l,** 1986b; **m,** 1986c; **n,** 1988.

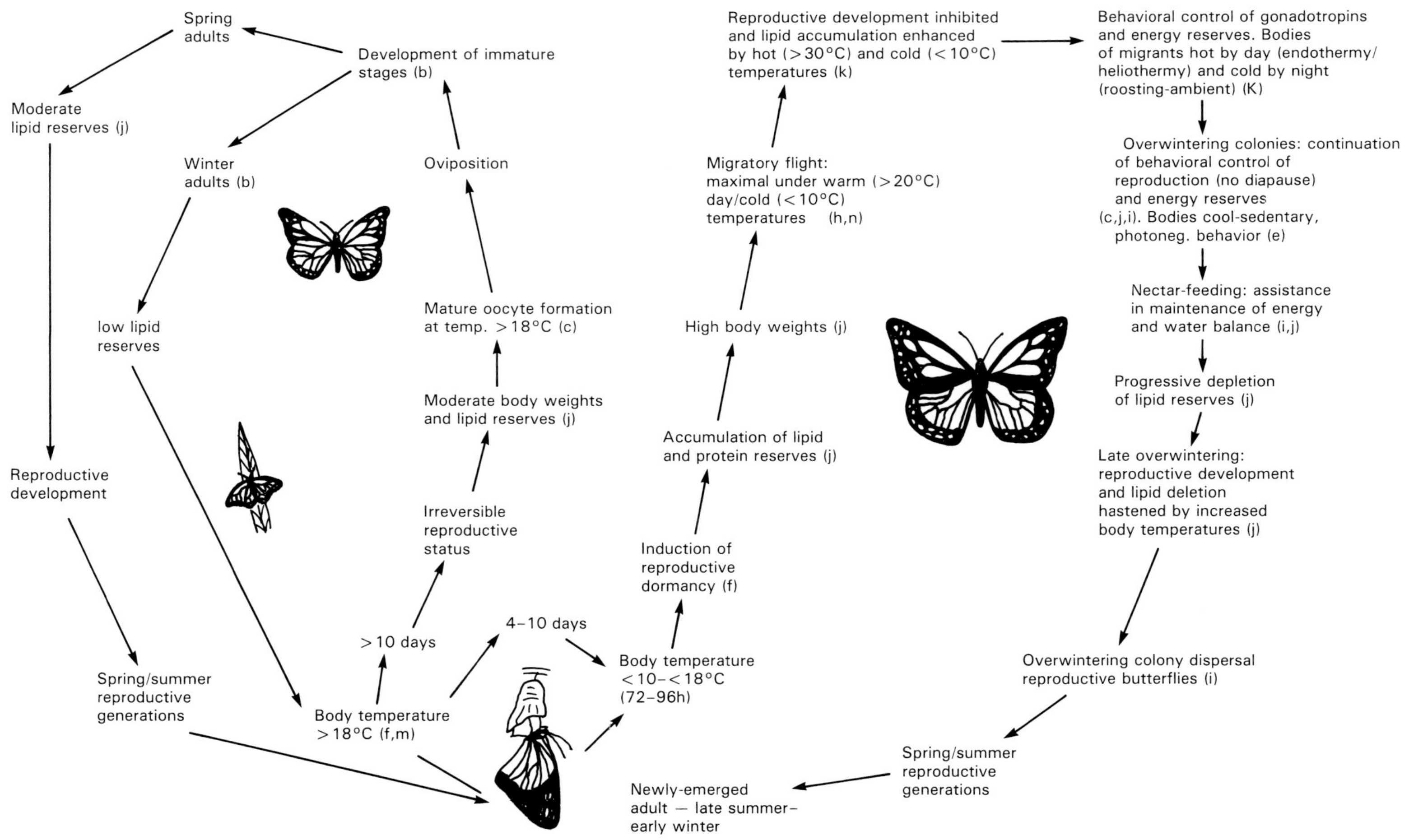

Figure 3. Salient physiological features of monarch overwintering in the Sydney area of New South Wales. References from James: **b,** 1981; **c,** 1982a; **e,** James and Hales, 1983; **f,** 1983; **h,** 1984a; **i,** 1984d; **j,** 1984a; **k,** 1986a; **m,** 1986c; **n,** 1988.

important in determining adult reproductive status, since breeding can occur following immature development during autumn and/or winter (James, 1984b). During autumn, cold fronts, following days or weeks of fine weather with increasing temperatures, periodically affect coastal and tableland areas of New South Wales, generating clouds and lowering temperatures. On the tablelands, temperatures usually remain <15–18°C, and the overcast skies prevent butterflies from elevating body temperatures (T_b's) by basking (James, 1984b, 1986c). Persistence of these conditions for 2–4 days is sufficient to induce reproductive dormancy in newly emerged monarchs, as well as individuals that are up to a week or so old (James, 1986a). Monarchs that emerge during periods of sunny, settled weather become reproductive and form an autumn breeding tableland generation. The proportion of reproductive to nonreproductive individuals is probably dependent upon the frequency of cold fronts and the degree of persistence of subsequent south to southeasterly winds and associated cloudy weather. It is also undoubtedly a function of latitude and generation phenology, with populations further south more likely to be entirely nonreproductive. Generally, autumn generations south of latitude 34° become nonreproductive with the Sydney area representing a transition zone between predominantly reproductive and nonreproductive populations (James, 1984b).

AUTUMN MONARCH MIGRATION IN SOUTHEASTERN AUSTRALIA

Data collected over 5 yr indicate that nonreproductive autumn monarchs in New South Wales migrate coastward in a general easterly or northeasterly direction (James, 1984b, 1986b, 1988). Information on monarch migration has been derived from two main sources: field observations and recapture of tagged individuals. The latter yielded data that for the first time demonstrated the occurrence of long-distance autumn migration of the monarch in Australia with five individuals released near Canberra in the Australian Capital Territory (Fig. 1) recaptured in the Sydney area, up to 380 km to the northeast. Other releases of tagged monarchs in the Blue Mountains highland area immediately west of Sydney also resulted in substantial movement into the Sydney basin (James, 1988). Some of these individuals were recovered from overwintering colonies. These observations indicate that overwintering nonreproductive monarchs in the Sydney area are derived from inland areas to the west and southwest. It is likely the majority of individuals originate in the tableland and highland areas of the Great Dividing Range, 200–300 km from Sydney. Some may fly greater distances, but in most years these will be in the minority due to the generally low monarch populations in western New South Wales. The possibility of migration from the southerly area of Victoria cannot be discounted, but again this area does not usually harbor substantial monarch populations (Smithers, 1977; Common and Waterhouse, 1981).

PHYSIOLOGY OF AUTUMN MIGRANTS

Most data on the physical condition and physiology of autumn migrants have been obtained from studies on transient roosting populations (James, 1982b, 1984c) (Fig. 3). As in North America, autumn migrants in Australia are characteristically young adults in good to excellent physical condition. The sex ratio of butterflies in transient colonies is always substantially biased in favor of males. Females rarely account for >10–15% of these populations (James, 1982b). Normal 1:1 sex ratios usually occur in overwintering colonies, so it is likely that a behavioral difference between the sexes during migration accounts for this sex ratio imbalance. Migrants do not differ in size from nonmigrants as judged by forewing length, but they are significantly heavier (James, 1984c). Wet, dry, and lean dry weights are substantially greater than those of reproductive monarchs (James, 1984c). The increased weights of migrants are primarily due to the accumulation of extensive reserves of fat (see also Calvert and Lawton, this volume). A lipid content of 100–200 mg is commonly found in autumn migrants, compared with an average value of around 20 mg for reproductive individuals (James, 1984c). The high lean dry weights found in Australian migrant and overwintering monarchs appear to reflect increased soluble protein content (James, 1984c). This may occur as the result of synthesis and storage of protein, which in reproductive individuals is presumably utilized in spermatogenesis and vitellogenesis. Lean dry weights of winter reproductive monarchs in Sydney, which show reduced reproductive development in a suboptimal temperature regime, are also greater than those of summer reproductives (James, 1984c). Lower lean dry weights are found in early colonizers of overwintering populations than in transient roosting colonies, suggesting that some protein may be utilized as an energy source during migration (James, 1984c).

Laboratory studies on the effect of temperature on lipid accumulation in nonreproductive monarchs show that high (>35°C) and low (<18°C) temperatures enhance storage of fat reserves, whereas medium (20–30°C) temperatures do not (James, 1986a). These results are compatible with the known physiology and endocrinology of nonreproductive monarchs (see Herman, this volume). Reproductive development and lipid utilization occurs in the presence of high juvenile hormone (JH) titers induced by warm temperatures (Barker and Herman, 1976; Lessman and Herman, 1983). Under cold conditions JH titers are low or undetectable, reproductive development is prevented, and fat body hypertrophy occurs (Barker and Herman, 1976; Dallman and Herman, 1978; Lessman and Herman,

1983). Temperatures > 28–30°C inhibit reproductive development because of an increase in a JH-specific esterase (JHE) (Lessman, 1980), which is expected to result in lowered JH titers and increased lipid content. In the laboratory, substantial storage of lipid occurs in monarchs exposed to posteclosion temperature treatments known to cause low–moderate levels of JH (e.g., cool–cold temperatures), and then transferred to temperatures known to inhibit synthesis (low) or promote degradation (high) of JH (James, 1986a). Reproductively active monarchs, with high JH titers after exposure to warm posteclosion conditions, were unable to accumulate lipid even if subsequently transferred to low temperatures (James, 1986a).

Reproductive dormancy in Australian migrating and overwintering monarchs is not an example of diapause (James, 1982a; James and Hales, 1983) (Fig. 3). Diapause is usually characterized by suppression of development in the period following induction of dormancy, regardless of environmental conditions (Tauber and Tauber, 1976). Migrating and overwintering monarchs in North America show depressed reproductive growth when exposed to summer conditions of temperature and photophase (Herman, 1981). This period, when females show a reproductive response that averages 35% of that obtained with summer individuals, was considered by Herman to represent the refractory phase of reproductive diapause. Results obtained from similar experiments on migrating and overwintering individuals show no evidence of a refractory phase of reproductive dormancy in Australian monarchs (James, 1982a, 1984b). Laboratory and field data clearly indicate the flexible nature of reproductive dormancy. Full reproductive development may be stimulated at any time during adult life if appropriate environmental conditions are present (James, 1982b). Similarly, there is no period of reduced JH sensitivity in migrating or overwintering Australian monarchs, as there is in diapause individuals in North America (Lessman, 1980; Herman, 1981; James and Hales, 1983; James, 1984b). Comparisons with Herman's (1981) data show that Australian nonreproductive monarchs respond to JH by producing approximately three times as many eggs as American diapause monarchs (James and Hales, 1983). Reproductive dormancy in male Australian monarchs has not been investigated in detail, although it is apparently less intense than in females, as is the case with North American diapause monarchs (Herman, 1981). Mating activity is common even in transient colonies, and virgin females never account for >30% of such populations (James, 1982b, 1984d).

The flexible nature of reproductive dormancy in monarchs in the Sydney area raises an intriguing question: How is reproductive dormancy and associated energy reserve storage maintained in the face of often quite warm conditions, which, as noted previously, permit significant winter breeding by some populations? This question is addressed in the following section.

BEHAVIOR OF MIGRANTS

Migrating monarchs in New South Wales are generally not conspicuous, especially in the low-population western areas. In years of high abundance, funneling of migrants through valleys in the Great Dividing Range west and southwest of Sydney produces noticeable migratory activity in the Sydney basin (James, 1984b, 1986b). Most migratory flights recorded over a 4-yr period occurred during sunshine and temperatures of 20–25°C (James, 1984b, 1986b).

Feeding is a major activity of migrants and is especially noticeable at transient roosting sites which often have an abundance of nectar sources such as *Lantana camara* (L.) (Verbenaceae) (James, 1982b). Courtship flights and mating also occur frequently at transient roosting sites. In some years these sites are only occupied for a few weeks, whereas in others they are used for 2–3 mo with a continuous stream of migrants passing through (James, 1984b). Colonies of short duration usually occur in years of low monarch abundance. Density of butterflies within a colony and intracolony roosting patterns are strongly influenced by temperature and wind strength (James, 1984b).

The physiology of Australian migrant and overwintering monarchs seemingly presents a paradox (Fig. 3). These individuals are able to maintain reproductive dormancy and associated energy reserves without diapause, in conditions known to be favorable for reproductive development. Circumstantial and some experimental evidence indicate that control of nonreproductive physiology in Australian monarchs may be achieved through adoption of behavior that ensures T_b's remain favorable for retention of reproductive dormancy and energy reserve storage. Ambient temperatures (T_a's) during monarch migration in New South Wales range from <10 to >35°C, although for most of the time diurnal temperatures vary between 20 and 30°C (James, 1984b). According to the data obtained on the effect of temperature on energy reserve storage, unless T_b's were <18 or >30°C, monarchs would have high JH titers and be incapable of effective lipid storage (James, 1986c). However, normal behavior of autumn migrants (sustained flight and basking/feeding in sunshine) ensure high T_b and thus lipogenesis. The T_b of a monarch exposed to winter sunshine in New South Wales at air temperatures >10°C is usually >30°C (James, 1986c). Similarly, thorax temperature (T_{th}) in sustained flight is 6–10°C above ambient (Kammer, 1970). These behaviors at T_a's of 20–25°C will promote increased JHE titers and JH inactivation (Lessman and Herman, 1981). During late autumn and early winter, T_a's in the Sydney area usually remain <20°C and migration ceases (James, 1984a,b). According to the data of Kammer (1970), migrants at this temperature would be unable to achieve T_{th} >30°C and lipid storage would be inhibited. The clustering behavior of monarchs in late autumn and early

Table 1. Summary of the apparent relationship between temperature and migration in the monarch butterfly.

Monarch colony behavior	Temperature: Mean daily maximum	Mean daily minimum	Observations[a]
Nil–minor instability	<20°C	<10°C	Camden, May/June 1981 Wallacia and Camden, June 1982
Moderate instability	>20°C	>10°C	Wallacia and Camden, April 1982
Substantial instability	>20°C	<10°C	Picton, April/May 1981–1983 Wallacia and Camden, May 1982

[a] Complete data contained in James, 1984b.

winter (Smithers, 1965; James, 1979, 1984d) prevents endothermic or heliothermic warming, so lipid storage is favored (James, 1986a,c). Migration of Australian monarchs occurs preferentially at warm (20–25°C) T_a's (James, 1984a, 1986b, 1988), which result in hot (>30°C) T_b, conducive to lipid accumulation. Ambient temperatures of 20–25°C are normally associated with reproductive development and energy reserve usage in the monarch (Barker and Herman, 1976; James, 1982a). However, the characteristic behavior of migrating monarchs ensures that T_b favors retention of reproductive dormancy and energy reserve storage. A similar hypothesis of behavioral control of JH gonadotrophic levels in migrants has also been proposed by Lessman and Herman (1981) for North American monarchs. Much of the above is speculative. Preliminary studies on captive migrants have provided some evidence for behavioral control of energy reserves. Unfed migrants maintained in a glasshouse and displaying at least some natural behavior are more successful in maintaining energy reserves than restrained individuals under the same conditions (James, unpubl. data). Similarly, fed migrants maintained in small cages at 25–30°C, unable to exhibit much natural behavior, lost 6–9 mg of fat daily, compared with free-flying individuals that lost 3 mg a day in a glasshouse. At lower temperatures (15–20°C), when migrants characteristically show little flight activity, all individuals lose 1–2 mg fat daily, whether caged or not (James, unpubl. data). The importance of feeding in maintaining energy reserves during migration and overwintering is indicated by data on unfed migrants stored at 5°C, which showed a lipid loss of 2 mg a day reducing fat reserves to low levels after only 10 wk. This is a more rapid depletion of energy reserves than is found in wild nonreproductive monarchs in Sydney, which are exposed to considerably higher temperatures and yet lose only around 0.5 mg per day (James, unpubl. data). Similar relationships with time and temperature were also found for lean dry weights and by inference, protein storage (James, unpubl. data). Distinct behavioral differences have been observed in captivity and in the field, between flight responses of nonmigrant (reproductive) and migrant (nonreproductive) monarchs to temperatures >25°C. Reproductive butterflies generally continue normal flight behavior at T_a's up to 33–35°C, whereupon they adopt shade-seeking behavior. Migrants show flight inhibition and photonegativity at much lower temperatures (25–26°C) (James, unpubl. data). This behavior is progressive with increasing temperature, so that flight by migrants is often completely inhibited at temperatures >30°C. Sustained flight at high (>30°C) temperatures would clearly result in excessive and possibly deleterious T_b. Migrants forced to fly at 38–40°C adopt a strange undulating movement by interposing active flight with free-fall dives. If left alone, ground roosting positions are sought (James, unpubl. data). Reproductive monarchs are less constrained by high temperatures because their trivial flight pattern usually incorporates frequent pauses, which could be used for cooling. At temperatures <20°C, flight activity of migrants and overwintering nonreproductives is inhibited in comparison with winter reproductives in the Sydney area (James, unpubl. data). Inactivity, feeding, and basking (usually in a delta-shape posture; see James, 1986c) are prevalent behaviors among migrants in cool conditions (James, 1979, 1984b; Alonso M. et al., this volume). Even though basking enables a T_b in excess of 30°C to be attained, the inability of flight activity in air temperatures <20°C to maintain this high T_b restricts migration. Instead, migrants opt for the only low JH-producing behaviors available to them: basking or shade inactivity.

ECOLOGY OF MIGRANTS

Sites used by autumn migrants for transient roosts and by winter monarchs for extended roosting are similar in configuration, and the same sites may be used by both groups of butterflies (James, 1984a,b). The sites are generally lightly wooded gullies or valleys, which are often blind-ended. In most instances, sites are situated in major breeding grounds, resulting in the co-existence of reproductive and nonreproductive winter populations (James, 1979, 1984b) (Fig. 2). Sites used by migrant and overwintering populations are distributed throughout the Sydney basin, although they are most common in southwestern districts. Most are inland (20–80 km), although a few are within meters of the sea (James, 1984b). Some sites have been recorded along the coast up to 200 km south of Sydney, while a large number have been recorded in the Newcastle area 200 km to the north (Smithers, 1965; James, unpubl. data). Migratory populations invariably appear first in March near Picton, a site on the southwestern perimeter of the Sydney basin (James, 1982b). Major characteristics of migratory populations at this and other sites are low recapture rates of tagged individuals, a high percentage of butterflies in excellent condition, and low numbers of females (James, 1982b, 1984b). At the Picton site, migrants congregate at the southern end of a north–south oriented valley at an altitude of about 250 m. Protection from predominantly southerly and westerly winds is given by the valley sides, which rise to 440 m, at the southern end. A group of small trees and bushes are used by the butterflies for roosting and consist mainly of *Melaleuca styphelioides* S.m. (Myrtaceae), prickly leaved tea tree, and *L. camara*. The area of occupation is about 30 m long and 10 m wide (James, 1982b). Numbers of migrants at transient roosting sites during 1980–1986 ranged from <10 to close to 1,000. In addition to the "typical" sites, a number of other locations in the Sydney metropolitan area regularly contain migrant and overwintering monarchs. These are usually parks or wasteland that lack an obvious gully-type configuration. Populations at these sites are invariably small (<50) but otherwise display all characteristics of nonreproductive migrant or overwintering monarchs (James, 1984b; unpubl. data). These populations may represent individuals that become "stranded" in the metropolitan area after reaching the coast.

The timing of migration and clustering of nonreproductive monarchs with respect to environmental stimuli has been little studied, although observations in North America, without actually defining the factors involved in inducing reproductive diapause, have long indicated the importance of decreasing temperature and photoperiod in stimulating migration (Urquhart, 1960; Barker and Herman, 1976). Climatic conditions following induction of reproductive dormancy are likely to be important influences on the phenology of migration and clustering in the monarch. Studies in New South Wales over a number of years have shown substantial migration to be associated with warm (20–25°C), sunny days and cold (<10°C) nights (James, 1984a,b). When nights are mild (>10°C), migration decreases. Migration ceases under a climatic regime of cool (<20°C) days and cold or mild nights. In these conditions the formation of stable overwintering colonies occurs (James, 1984a). The apparent relationship between temperature and migration in the monarch is summarized in Table 1. Prolonged overcast weather during autumn results in reduced migratory activity (James, 1984b; unpubl. data). Observations in North America indicate similar relationships between climate and monarch migration (Urquhart, 1960). A winter transient roosting colony in southern California was reported by Tuskes and Brower (1978), who suggested that it was the result of frequently clear and warm weather. It seems likely that weather conditions affecting butterflies following induction of reproductive dormancy will determine the extent, duration, and end-point of migration. If there is a rapid deterioration of weather during autumn, then migratory activity is relatively short-lived, resulting in the formation of sizeable overwintering colonies in the Sydney area. However, if the seasonal transition from autumn to winter is protracted, as it was in 1982 and 1986, optimal conditions for migration (warm days/cold nights) persist for 2–3 mo (James, 1984a; unpubl. data). This results in migrants continuing on an extended quest for overwintering sites and "overshooting" the Sydney area. Sites that harbor stable overwintering populations during cool–cold autumns contain large, unstable migratory populations when the arrival of cool winter conditions is late (James, 1984a,b). Ultimately, most individuals reach the coast and probably continue flying northward, increasing the likelihood of early reproductive maturation. It is significant that many instances of monarchs washed up on the shoreline are reported in years of extended, warm autumn weather (James, unpubl. data). Stable overwintering colonies are invariably small in these years (James, 1984b). Such principles of monarch migration may also be applicable in North America. In their southern movement, migratory populations may "track" the retreat of optimal weather for migration and terminate movement when such conditions disappear. Early migrants may "overshoot" overwintering areas if migration-terminating stimuli are not yet present, and late migrants may become "trapped" if such stimuli envelop them before reaching wintering grounds. In Australia, the small latitudinal shift necessary for migrants to reach overwintering sites, the thermally restricted northerly range of stable winter colonies, and frequent extended autumns must result in many butterflies being lost from nonreproductive populations. It is clear that the dynamics of monarch migration with respect to spatial, temporal, and climatic factors are complex. Detailed studies of such relationships are necessary and would undoubtedly contribute to a greater understanding of the phenology of monarch migration.

The reasons for choice of sites by migrating monarchs for temporary roosts or overwinter colonies remain obscure. Some sites in the Sydney area have been used annually for more than 20 yr (Smithers, 1965; James, 1984b), while adjacent similar situations are ignored. The presence of milkweed and reproductive monarch populations are invariable, common denominators at roosting sites (James, 1984b). Such sites may simply represent suitable refuges, located along well-defined migration routes, which might result from funneling of migrants through tableland valleys and passes. The chemical or visual perception of milkweed and/or breeding monarchs may provide the initial cue for colonization.

WINTER POPULATIONS OF THE MONARCH IN SOUTHEASTERN AUSTRALIA

By mid-winter in southeastern Australia (south of latitude 34°) adult monarch populations may be found only in the Adelaide and Sydney areas (Smithers, 1977). These locations are alone in this part of the continent in having an average winter temperature that does not fall <10°C. Although no published data are available, overwintering populations in the Adelaide area are reportedly nonreproductive, forming colonies in hills close to the city (Common and Waterhouse, 1981). Nothing is known regarding the origin of the Adelaide overwintering populations. In the Sydney area, as already described, winter populations of reproductive and nonreproductive monarchs occur (Smithers, 1965; James, 1979, 1981) (Fig. 2). Winter reproductive populations are more frequently found in the northern areas of the Sydney basin and are often in sheltered, frost-free river valleys (James, 1984b).

Since 1978 the largest nonreproductive overwintering colony in the Sydney area consisted of approximately 3,500 individuals (James, 1984d). Some transient roosting sites have been estimated to have had >10,000 migrants passing through them during autumn (James, 1984a,b; unpubl. data). Stable overwintering colonies in some years consist of <500 individuals. Overwintering colonies in contemporary times have been much smaller than those reported in the early 1960s by Smithers (1965), when populations of up to 40,000 individuals were common. However, annual inconsistency in population size, as seen over the past few years, was also a feature of the populations studied by Smithers (1965, 1972). Annual variation in cluster population size, as discussed previously, may be attributable to climatic influences upon the extent and duration of migration. Other major limiting factors, particularly those affecting growth and viability of summer breeding populations, must also be considered. The incidence of drought and parasitism, major constraints upon successful development of monarch populations (Smithers, 1973a; Zalucki, 1981b; James, 1984b), will have an important effect on the size of overwintering populations. It is clear, however, that monarch populations in New South Wales have declined over the past two decades, and we can only speculate on the causes. It is unquestioned that the availability of suitable breeding habitats has declined substantially. The possibility of stock poisoning by milkweed has led to its virtual eradication from farmland, where formerly it often flourished.

Early overwintering populations at all sites are dominated by males. While this dominance continues in migratory colonies, winter populations develop a reasonably balanced sex ratio once the formation phase has ended. This is usually short-lived as males again dominate the population during colony dispersal (James, 1984b). Male dominance of overwintering colonies in formation and dispersal phases suggests that females arrive later and leave earlier. Early dispersal of females has also been reported from the Californian colonies (Urquhart, 1960; Tuskes and Brower, 1978). The wing condition of butterflies in clusters does not deteriorate markedly during the winter. As noted previously, migratory colonies contain a significantly higher percentage of individuals in an excellent or recently emerged condition than do winter colonies. However, butterflies in a good condition show little deterioration during overwintering (James, 1984b). Wing damage in Sydney overwintering monarchs is rare, presumably because of the relatively short time spent in overwintering colonies. Monarchs dispersing from Californian colonies may be 6–7 mo old, while butterflies leaving clusters in Sydney will rarely be >4 mo old. Mortality in overwintering populations is rare, with direct evidence found only on three occasions during 1978–1986. In 1978, prolonged wet and windy weather appeared to cause high mortalities among a number of populations. Predation by birds was suspected to be the cause of significant mortality at one site in 1982 and 1986 (James, 1984b; unpubl. data). Smithers (1973a) identified a number of birds seen attacking Australian monarchs and also reported the possibility of a small nocturnal mammal predator operating among clustering butterflies.

SPRING POPULATIONS IN SOUTHEASTERN AUSTRALIA

The break-up of winter colonies does not appear to be signaled by any specific environmental stimulus. Dispersal usually begins in the absence of any discernible change in temperature (James, 1984a), or before daylength starts increasing (James, 1984b). There is also no evidence of mass mating prior to dispersal as described for Californian monarchs (Hill et al., 1976). Most females in Sydney colonies are mated soon after colony formation (James, 1984b). Abandonment of overwintering sites by Australian monarchs is probably a function of physiological condition. All dispersing females examined show some reproductive development, and most contain immature and/or mature oocytes (James, 1979, 1984b). Lipid content of dispersing monarchs is variable but usually averages 40 mg, about twice that found in reproductive butterflies (James, 1984c).

Overwintering populations in the Sydney area begin dispersing soon after the arrival of the winter solstice in June. In some years, complete dispersal occurs before days begin lengthening in mid-July, whereas in other years some individuals may still be present in colonies in early September (James, 1984b; unpubl. data). Thus, stable overwintering colonies occur for only 2–4 mo, which contrasts with the 4–5 mo spent overwintering by North American monarchs. The dispersal of Sydney colonies during winter, rather than after it, does not pose many problems to survival and reproduction, providing the butterflies remain in the Sydney area or fly northward. Some individuals have been recorded moving northward after leaving clusters (Common and Waterhouse, 1981), but generally our knowledge of postoverwintering dispersal is poor. Limited data on the recovery of tagged butterflies during the past few years suggests that there is at least some return south–southwestward migration (James, 1984b, 1986b). Such individuals may face greater problems in survival because their flight may take them into tableland areas where winter conditions are generally still inhospitable. If this is the case, then selection may reduce the tendency for postcluster remigration, in favor of continued residence in the milder Sydney basin. Spring expansion of the monarch's range in southeastern Australia is slow and a significant southward component has not been noted (Smithers, 1977). More work is needed to clarify the importance of postoverwintering migration in Australian monarchs.

CONCLUSIONS

The monarch butterfly in New South Wales appears to be remarkably well adapted to its environment. Whether the contemporary monarch populations in New South Wales had tropical or temperate origins, some modification of biology has clearly occurred, remembering that the species has been present in Australia for only a little over 100 yr. Comparative biological studies on Australian, North American, and Pacific island populations are needed to clarify the extent to which adaptive evolution has occurred.

The most significant feature of monarch overwintering in New South Wales is the absence of reproductive diapause. The elimination of winter reproductive development in overwintering monarchs in the Sydney area would not have the positive survival value that such a strategy has in North America. Winter environmental conditions can support monarch breeding populations and Australian monarchs have responded by either (1) continuous reproductive development in resident Sydney basin monarchs or (2) adoption by inland populations of a flexible winter dormancy state, which is not under strict physiological control (James, 1982a, 1984b; James and Hales, 1983). The latter strategy appears to be an example of "bet-hedging," where reproductive dormancy can be maintained should conditions be suboptimal, but the capability for early maturation is retained should environmental conditions become favorable. In the marginal winter climate of the Sydney area, this is clearly an appropriate reproductive strategy. In North America, reproductive diapause allows effective removal of populations from the greater part of the continent in which winter survival and reproduction would be impossible. However, some individuals may possess the ability for early reproductive maturation (Herman, 1973; Barker and Herman, 1976) and join the small continuously breeding populations in some southern areas of the United States (Brower, 1961; Funk, 1968; Urquhart et al., 1968).

The evidence to date strongly suggests that reproductive dormancy in Australian monarchs is maintained by the repertoire of behaviors associated with this physiological status (James, 1984b, 1986a; unpubl. data). Although behavioral control of gonadotropins has been suggested for North American monarchs by Lessman and Herman (1981), it is expected that butterflies in reproductive diapause would show less dependence upon behavior as a means to ensure reproductive inactivity. More studies are needed on the role of behavior in regulating the nonreproductive status of Australian and North American monarchs.

Although there are no published data to confirm it, photoperiod has frequently been implicated in the induction of reproductive diapause in adult North American monarchs (Urquhart, 1960). The suggestion has also been made that diapause may be programmed in an earlier developmental stage (Herman, 1981). Australian monarchs enter reproductive dormancy as a consequence of exposure, as newly emerged adults, to cool temperatures. Photoperiod is unimportant, as are conditions affecting developmental stages (James, 1983, 1984b). The possibility, therefore, exists that the dormancy induction mechanism differs from that of North American individuals. The less critical requirement for dormancy in Australian monarchs may have reduced the importance of photoperiodic cues, which are often associated with induction of insect reproductive diapause in temperate regions (Tauber and Tauber, 1976). A number of other differences in the migration and winter biology of the monarch in Australia are also apparent, although many are related to the fundamental physiological difference in reproductive dormancy (e.g. the relative scarcity of virgin females and absence of mass mating in overwintering colonies) (James, 1984d).

The story of the monarch has been one of considerable adaptation and this is amply demonstrated by its successful exploitation of the Australian environment. Future studies should further enhance our knowledge of this remarkable insect in Australia. It is also to be hoped that similar work in other regions, apart from North America, will be attempted to provide a more global view of this most successful and adaptable organism.

LITERATURE CITED

Barker, J.F., and W.S. Herman. 1976. Effect of photoperiod and temperature on reproduction of the monarch butterly, *Danaus plexippus*. *Journal of Insect Physiology* 22:1565–1568.

Brower, L.P. 1961. Studies on the migration of the monarch butterfly. 1. Breeding populations of *Danaus plexippus* and *D. gillipus berenice* in south central Florida. *Ecology* 29:76–83.

Bull, C.M., M.P. Zalucki, Y. Suzuki, D.A. Mackay, and R.L. Kitching. 1985. An experimental investigation of patch use by female monarch butterflies, *Danaus plexippus* (L.) *Australian Journal of Ecology* 10:391–398.

Common, I.F.B., and D.F. Waterhouse. 1981. *Butterflies of Australia.* Sydney: Angus and Robertson.

Dallman, S.H., and W.S. Herman. 1978. Hormonal regulation of hemolymph lipid concentration in the monarch butterfly, *Danaus plexippus. General and Comparative Endocrinology* 36:142–150.

Funk, R.S. 1968. Overwintering of monarch butterflies as a breeding colony in southwestern Arizona. *Journal of the Lepidopterists' Society* 22:63–64.

Herman, W.S. 1973. The endocrine basis of reproductive inactivity in monarch butterflies overwintering in central California. *Journal of Insect Physiology* 19:1883–1887.

Herman, W.S. 1981. Studies on the adult reproductive diapause of the monarch butterfly, *Danaus plexippus. Biological Bulletin* 160:89–106.

Hill, H.F., A.M. Wenner, and P.H. Wells. 1976. Reproductive behavior in an overwintering aggregation of monarch butterflies. *American Midland Naturalist* 95:10–19.

James, D.G. 1979. Observations on two overwintering clusters of *Danaus plexippus* (L.) (Lepidoptera: Nymphalidae) in the Sydney area during 1978. *Australian Entomological Magazine* 5:81–85.

James, D.G. 1981. Studies on a winter breeding population of *Danaus plexippus* (L.) (Lepidoptera: Nymphalidae) at Spencer, New South Wales. *General and Applied Entomology* 13:47–53.

James, D.G. 1982a. Ovarian dormancy in *Danaus plexippus* (L.) (Lepidoptera: Nymphalidae)—Oligopause not diapause. *Journal of the Australian Entomological Society* 21:31–35.

James, D.G. 1982b. A transient non-breeding population of *Danaus plexippus* (L.) (Lepidoptera: Nymphalidae) near Picton, New South Wales. *Australian Entomological Magazine* 8:96–99.

James, D.G. 1983. Induction of reproductive dormancy in Australian monarch butterflies, *Danaus plexippus. Australian Journal of Zoology* 31:497–504.

James, D.G. 1984a. Migration and clustering phenology of *Danaus plexippus* (L.) (Lepidoptera: Nymphalidae) in Australia. *Journal of the Australian Entomological Society* 23:199–204.

James, D.G. 1984b. Overwintering biology of the monarch butterfly, *Danaus plexippus* in New South Wales. Ph.D. thesis. Sydney: Macquarie University.

James, D.G. 1984c. Phenology of weight, moisture and energy reserves of Australian monarch butterflies, *Danaus plexippus. Ecological Entomology* 9:421–428.

James, D.G. 1984d. Population and general biology of non-reproductive colonies of the monarch butterfly, *Danaus plexippus* (L.) (Lepidoptera: Nymphalidae) in New South Wales. *Australian Journal of Zoology* 32:663–670.

James, D.G. 1986a. Effect of temperature upon energy reserves of the monarch butterfly, *Danaus plexippus* (L.) (Lepidoptera: Danaidae). *Australian Journal of Zoology* 34:373–379.

James, D.G. 1986b. Studies on the migration of *Danaus plexippus* (L.) (Lepidoptera: Nymphalidae) in the Sydney area. *Australian Entomological Magazine* 13:27–31.

James, D.G. 1986c. Thermoregulation in *Danaus plexippus* (L.) (Lepidoptera: Nymphalidae)—Two cool climate adaptations. *General and Applied Entomology* 18:43–47.

James, D.G. 1988. Migration and behavior of non-reproductive *Danaus plexippus* (L.) (Lepidoptera: Nymphalidae) in the Blue Mountains, New South Wales. *Australian Entomological Magazine* 15:50–55.

James, D.G., and D.F. Hales. 1983. Sensitivity to juvenile hormone is not reduced in clustering monarch butterflies, *Danaus plexippus* in Australia. *Physiological Entomology* 8:372–376.

Kammer, A.E. 1970. Thoracic temperature, shivering and flight in the monarch butterfly. *Zeitschrift für Vergleichende Phsyiologie* 68:334–344.

Lessman, C.A. 1980. Biology of hemolymph juvenile hormone in *Danaus plexippus.* Ph.D. thesis. Minneapolis: University of Minnesota.

Lessman, C.A., and W.S. Herman. 1981. Flight enhances juvenile hormone inactivation in *Danaus plexippus plexippus* (L.) (Lepidoptera: Danaidae). *Experientia* 37:599–601.

Lessman, C.A., and W.S. Herman. 1983. Seasonal variation in hemolymph juvenile hormone of adult monarchs (*Danaus p. plexippus:* Lepidoptera). *Canadian Journal of Zoology* 61:88–94.

Marks, E.N. 1963. The first record of *Danaus plexippus* (L.) from Australia. *Journal of the Entomological Society of Queensland* 2:8.

Miskin, W.H. 1871. Occurrence of *Danaus archippus* in Queensland. *Entomologists' Monthly Magazine* 8:17.

Smithers, C.N. 1965. A note on overwintering in *Danaus plexippus* (Linnaeus) (Lepidoptera: Nymphalidae) in Australia. *Australian Zoologist* 13:135–136.

Smithers, C.N. 1972. Observations on a breeding population of *Danaus plexippus* (L.) (Lepidoptera: Nymphalidae) at Camden, New South Wales. *Australian Zoologist* 17:142–148.

Smithers, C.N. 1973a. A note on natural enemies of *Danaus plexippus* (L.) in Australia. *Australian Entomological Magazine* 1:37–40.

Smithers, C.N. 1973b. A note on length of life of some Australian butterflies. *Australian Entomological Magazine* 1:62–66.

Smithers, C.N. 1977. Seasonal distribution and breeding status of *Danaus plexippus* (L.) (Lepidoptera: Nymphalidae) in Australia. *Journal of the Australian Entomological Society* 16:175–184.

Suzuki, Y., and M.P. Zalucki. 1986. The influence of sex ratio on female dispersal in *Danaus plexippus*

(L.) (Lepidoptera: Danaidae). *Journal of the Australian Entomological Society* 25:31–35.

Tauber, M.J., and C.A. Tauber. 1976. Insect seasonality: Diapause maintenance, termination and postdiapause development. *Annual Review of Entomology* 21:81–107.

Tuskes, P.M., and L.P. Brower. 1978. Overwintering ecology of the monarch butterfly, *Danaus plexippus* L., in California. *Ecological Entomology* 3:141–153.

Urquhart, F.A. 1960. *The monarch butterfly*. Toronto: University of Toronto Press, 361 pp.

Urquhart, F.A., and N.R. Urquhart. 1978. Autumnal migration routes of the eastern population of the monarch butterfly (*Danaus p. plexippus*) (L.) (Danaidae: Lepidoptera) in North America to the overwintering site in the neovolcanic plateau of Mexico. *Canadian Journal of Zoology* 56:1754–1764.

Urquhart, F.A., and N.R. Urquhart. 1979. Vernal migration of the monarch butterfly (*Danaus p. plexippus*, Lepidoptera: Danaidae) in North America from the overwintering site in the neovolcanic plateau of Mexico. *Canadian Entomologist* 111:15–18.

Urquhart, F.A., N.R. Urquhart, and F. Munger. 1968. A study of a continuously breeding population of *Danaus plexippus* in southern California compared to a migratory population and its significance in the study of insect movement. *Journal of Research on the Lepidoptera* 7:169–181.

Zalucki, M.P. 1981a. The effects of age and weather on egg-laying in *Danaus plexippus* L. (Lepidoptera: Danaidae). *Researches on Population Ecology* 23: 318–327.

Zalucki, M.P. 1981b. Temporal and spatial variation of parasitism in *Danaus plexippus* (L.) (Lepidoptera: Nymphalidae). *Australian Entomological Magazine* 8:3–8.

Zalucki, M.P. 1982. Temperature and rate of development in *Danaus plexippus* L. and *D. chrysippus* L. (Lepidoptera: Nymphalidae). *Journal of the Australian Entomological Society* 21:241–246.

Zalucki, M.P., and R.L. Kitching. 1982a. The analysis and description of movement in adult *Danaus plexippus* L. (Lepidoptera: Danainae). *Behaviour* 80:174–198.

Zalucki, M.P., and R.L. Kitching. 1982b. Dynamics of oviposition in *Danaus plexippus* (Insecta: Lepidoptera) on milkweed, *Asclepias* spp. *Journal of Zoology* 198:103–116.

Zalucki, M.P., and R.L. Kitching. 1982c. Temporal and spatial variation of mortality in field populations of *Danaus plexippus* L. and *D. chrysippus* L. larvae (Lepidoptera: Nymphalidae). *Oecologia* 53:201–207.

SEASONAL MIGRATION OF MONARCHS AND OTHER BUTTERFLIES IN COSTA RICA

William A. Haber

ABSTRACT. Butterflies flying eastward through passes at Monteverde, Costa Rica, were sampled by hand-netting for 1.5 yr to determine the frequency, flight time, sex ratio, and reproductive condition of migrants. Of 508 species recorded, 268 (53%) are considered migrants. Both sexes migrated, but sex ratios were usually female-biased. Dissected females of all but two species were reproductively active during migration, carrying mature eggs and spermatophores, and those with widespread host plants oviposited during migration. The exceptions were two species of *Anteos* (Pieridae) that migrated as virgins or in reproductive diapause.

Unlike the north–south migration of North American monarchs, migration of Costa Rican butterflies, including monarchs, was oriented east–west across the country between seasonally dry, semideciduous Pacific habitats and evergreen Atlantic habitats. Emigration from dry forest peaked in mid-dry season (February), while monarch migration peaked in late dry season (March–April). A few species left the dry forest in July, after only one wet season generation, and others emigrated in both wet and dry seasons. Return migration to the Pacific dry forest peaked during April and May, at the transition from dry to wet season.

INTRODUCTION

Seasonal migration is now recognized as an important component in the natural history of many neotropical animals, especially birds and insects (Janzen, 1983). For example, many Costa Rican birds migrate altitudinally, usually breeding in montane cloud forest while passing the nonbreeding season in lowland wet forest (Stiles, 1983).

Similar interhabitat movements of neotropical Lepidoptera have often been reported anecdotally (Williams, 1930; Beebe, 1949, 1950; Haber, 1978; Johnson, 1978; DeVries, 1983; Janzen, 1983, 1984) but have received little serious study until recently (Smith, 1982, 1983; Janzen, 1987a,b; Haber and Frankie, 1989). DeVries (1983) listed only six presumed migrant butterfly species in Costa Rica, and Janzen (1983) specifically excluded the monarch, *Danaus plexippus* (L.), as a migrant in Costa Rica. However, DeVries (1987) later suggested that migration may occur in most species of Pieridae and Ithomiinae and in a few species in other taxa.

This study was motivated by observations of ithomiine butterflies that were apparently migrating through mountain passes near Monteverde, Costa Rica, during 1973–1974 (Haber, 1978) and similar observations of other taxa (including the monarch) repeated in succeeding years. Here I summarize the temporal patterns of butterfly migration, the taxa and sex ratios of the migrants, and the reproductive condition of migrant and nonmigrant females.

STUDY SITES AND METHODS

STUDY SITES

The principal study site was the Monteverde Cloud Forest Reserve and the adjoining Monteverde community straddling the Cordillera de Tilarán, a north–south mountain chain in north-central Costa Rica (10°18′N, 84°48′W). The upper slopes and continental divide (1,300–1,600 m elevation) support a species-rich cloud forest described by Lawton and Dryer (1980). A highly seasonal, premontane moist forest grows on the lower Pacific slope at 800–1,300 m. Premontane rainforest occurs on the Atlantic slope at the same elevations. Other sites included a 60-ha fragment of tropical moist forest 1 km north of Cuatro Cruces (10°8′N, 84°48′W) at 200 m elevation in the Pacific lowlands and a 2-km^2 patch of tropical wet forest (10°8′N, 84°32′W) at 50 m elevation in the Atlantic lowlands 15 km east of Fortuna. These sites lie approximately 30 km south and 40 km northeast of Monteverde, respectively.

The Pacific slope and lowlands experience a 6-mo dry

Table 1. Numbers of resident and migrant butterfly species recorded at Monteverde, Costa Rica. Taxonomic arrangement follows DeVries (1987).

Family Subfamily	Total species	Migrant species	% migrants
Papilionidae	16	4	25
Pieridae	50	32	64
Dismorphiinae	11	4	36
Pierinae	19	9	47
Coliadinae	20	20	100
Nymphalidae	222	116	52
Charaxinae	14	0	0
Apaturinae	5	3	60
Libytheinae	1	1	100
Nymphalinae	69	42	61
Acraeinae	2	2	100
Heliconiinae	16	13	81
Melitaeinae	19	10	53
Danainae	6	5	83
Ithomiinae	52	35	67
Morphinae	3	0	0
Brassolinae	5	0	0
Satyrinae	30	5	17
Riodinidae	35	6	17
Lycaenidae	80	50	62
Hesperiidae	105	60	57
Totals	508	268	53

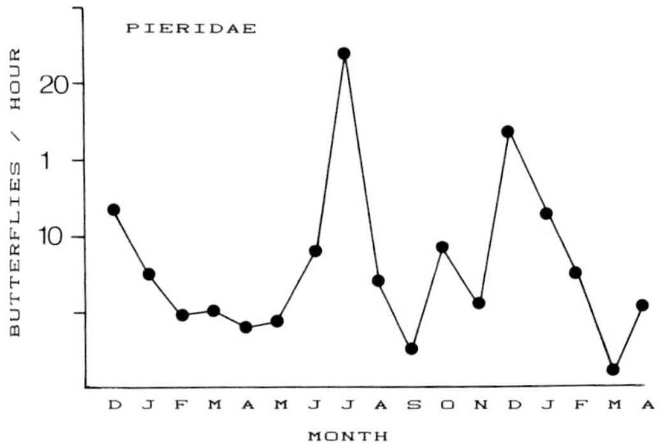

Figure 1. Eastward migration times of Pieridae from Pacific to Atlantic habitats sampled from December 1984 to April 1986, as numbers of individuals of all migrant species caught per hour during each month in flyways at Monteverde, Costa Rica.

season extending from mid-November to mid-May. The upper Pacific slope and continental divide near Monteverde receive moisture during early dry season (November–January) from mist blown across the continental divide by the trade winds. The Atlantic slope and lowlands have a much less pronounced dry season from January to April.

METHODS

Sampling Migrants

Eastbound migrants were sampled in two windy road cuts that served as passes or flyways at Monteverde: one at 1,500 m on the Pacific slope at the entrance to the reserve, the other in the reserve at 1,600 m on the continental divide. The strong northeast trade winds (10–20 km/hr) forced most butterflies migrating eastward toward the Atlantic to fly near the ground in the flyways where they were easily netted. Butterflies were collected for identification, sex determination, and dissection during clear weather (usually between 0900 and 1300 hours C.S.T.) for 2–6 hr on up to 12 days each month from December 1984 to April 1986, the number of sampling days per month varying with weather conditions.

Identifying Migrants

Butterflies were identified as migrants if they met either of two criteria: first, those that moved directly east through the flyways against the trade winds and second, those caught in the flyways that belonged to species known to breed only in lowland habitats and whose host plants did not occur anywhere near the sampling sites (DeVries, 1987; W.A.H., unpubl. data).

Sampling Nonmigrants

Butterflies were sampled approximately every 2 mo on the Atlantic and Pacific slopes below Monteverde (800–1,200 m) and two to three times in both wet and dry seasons in the lowland sites described above. Sampling consisted of netting as many individuals and species as possible during each site visit.

Reproductive Condition

Reproductive status of females was assessed by inspection of abdomens dissected under a binocular microscope, using scoring methods similar to those developed independently by Jones and Rienks (1987). Immature and mature eggs were recorded as present or absent and spermatophores (including remnant tails and shells) were counted. Age was estimated by scoring wing wear and size of the fat body on a scale of 0–3.

RESULTS

MIGRANT TAXA AND SEX RATIOS

Of 508 butterfly species recorded at Monteverde (DeVries, 1983; plus author's records), 268 (53%) species were classed as migrants during this study (Table 1). Migrants were prominently represented among the Pieridae, Nymphalidae, Lycaenidae, and Hesperiidae, but a few species of Papilionidae and Riodinidae were collected in the flyways. All but four of the nymphalid subfamilies included >50% migrant species. I found no evidence for migration among species of Charaxinae, Morphinae, and Brassolinae. *Cissia hermes* (Fabr.), *Euptychia jesia* Butler, and *Manataria maculata* (Hopffer) were the only common migrants among Satyrinae.

Although both sexes were well represented among migrants, sex ratios were usually female-biased. For example, the sex ratio for all migrant species sam-

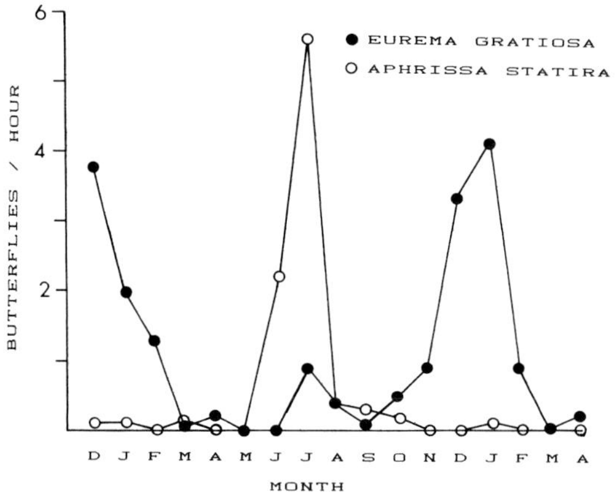

Figure 2. Numbers of migrating *Eurema gratiosa* and *Aphrissa statira* (Pieridae) at Monteverde (data as in Fig. 1).

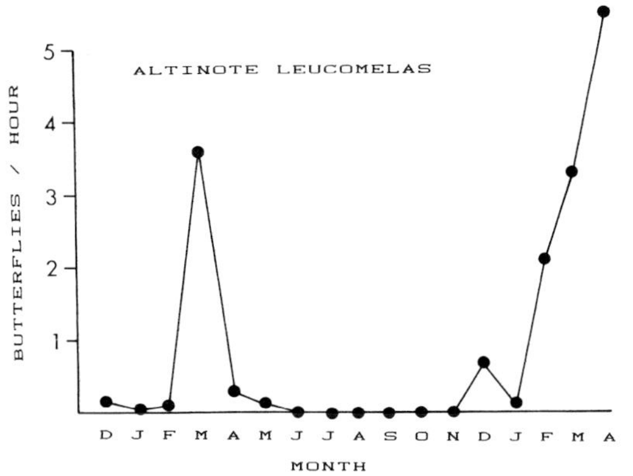

Figure 4. Numbers of migrating *Altinote leucomelas* (Acraeinae) at Monteverde (data as in Fig. 1).

pled during December 1984 to January 1985 was 60.5% female (n = 1,057). On 16 of 19 (84%) days during this period the samples contained more females than males.

TIME OF EASTWARD MIGRATION

Eastward migration was heaviest from October to April, beginning in the last month of the wet season and continuing throughout the dry season. However, some migration occurred in every month. The numbers of most migrant species showed distinct seasonal peaks, although timing of the peaks differed widely among species. The following examples cover the main temporal patterns observed.

The numbers of all Pieridae showed two distinct peaks, one in December–January (early dry season) and another in July (mid-wet season) (Fig. 1). The numbers of some species, such as *Eurema gratiosa* (Doubleday & Hew.), reached a peak in the dry season, whereas the numbers of others, such as *Aphrissa statira* (Cramer), reached a peak in the wet season (Fig. 2). It should be noted that small numbers of *E. gratiosa* were also involved in movements through Monteverde during July, suggesting that a second emigration from dry forest areas had occurred, this time by wet season adults.

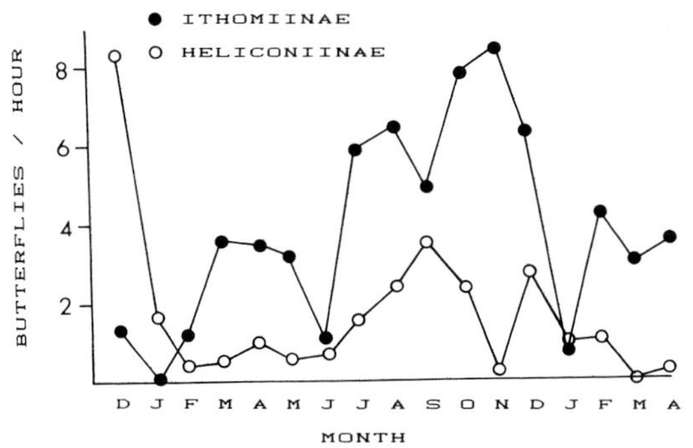

Figure 3. Numbers of migrating Ithomiinae and Heliconiinae (Nymphalidae) sampled at Monteverde (data as in Fig. 1).

Migration of Heliconiinae and Ithomiinae was spread out more than in Pieridae (Fig. 3). This resulted from some species migrating during several months as well as from differences in flight times among species. Numbers of migrating ithomiines were highest at the end of the wet season (October–November). The stepped peaks of the ithomiines in April, August, and November also suggest that some adults from successive wet season generations emigrated. The Hesperiidae also showed this pattern, although a few species migrated in each month.

Altinote leucomelas (Bates) (Acraeinae) [traditionally placed in the genus *Actinote,* where DeVries (1987) retained it, from which Potts (1943) segregated *Altinote* from congeners of *leucomelas*], illustrates a seasonal pattern in which the number of migrants peaked in March–April near the end of the dry season (Fig. 4). In 1985 almost every individual passed through Monteverde during 2 wk in March, while in 1986 migration extended over 3 mo, indicating the high variation between years that was typical for many species at Monteverde. *Danaus plexippus* also showed this pattern, although some individuals also migrated during May–June and September–October (Fig. 5). *Danaus gilippus* Bates migrated from the lowlands to the moist Pacific slope (1,000–1,400 m) where they remained reproductively active throughout the dry season, but, in contrast to monarchs, they rarely crossed the continental divide.

The Lycaenidae illustrate an eastbound flight pattern seen in few other species (Fig. 6). Beginning in the late dry season, migration peaked sharply in June (early wet season).

TIME OF WESTWARD MIGRATION

The westward migration (back to the Pacific side) peaked from April to June. Large butterflies, such

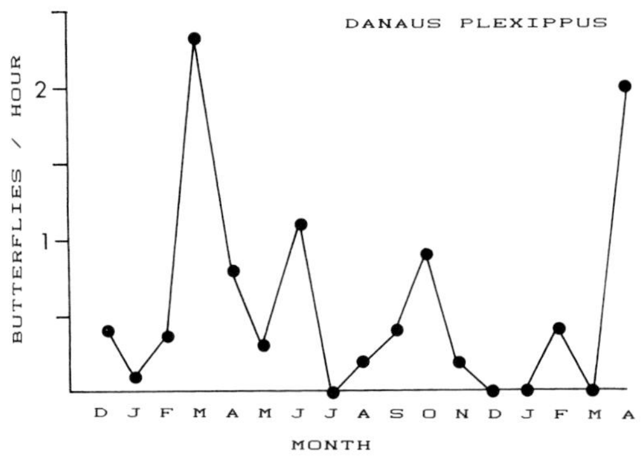

Figure 5. Numbers of migrating *Danaus plexippus* (Danainae) at Monteverde (data as in Fig. 1).

as species of Heliconiinae and Pieridae, were particularly conspicuous, while small, dark hesperiids and lycaenids were hardly noted. Most of the migrants flew fast, usually 1–2 m above ground, and passed straight over obstacles in their path such as forest stands or buildings. However, ithomiines sometimes flew through forest understory instead of over the canopy and monarchs flew in a slow, gliding manner near the ground, frequently stopping to investigate host plants.

Migration was most pronounced during warm, calm, clear weather. This behavior was observed on the Atlantic slope, where the butterflies flew uphill toward the continental divide, and on the Pacific slope near Monteverde, where the butterflies flew downhill over forest and pasture or along rivers, as well as along the Inter-American Highway in the Pacific lowlands. While easy to observe, the westward migration was difficult to quantify because the trade winds had calmed and the butterflies were not concentrated in flyways where systematic sampling was possible. However, the relative abundance of these returning species appeared to be generally lower than during eastward migration. No data were recorded on sex ratios of the westbound migrants.

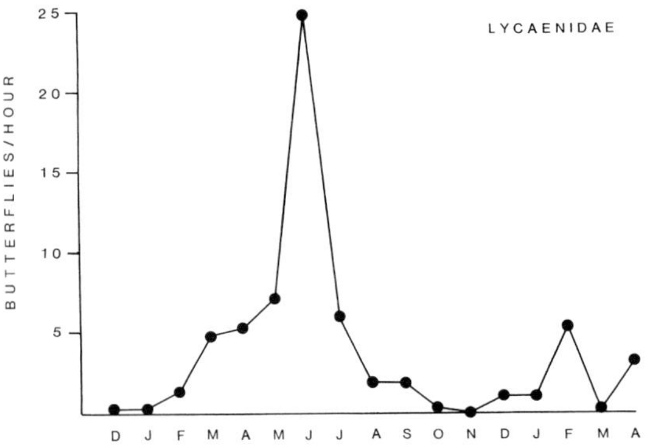

Figure 6. Numbers of migrating Lycaenidae (all species) at Monteverde (data as in Fig. 1).

REPRODUCTIVE CONDITION OF EASTBOUND MIGRANTS

With only two exceptions, the females of all migrant species checked were reproductively active; i.e. they carried eggs and spermatophores (e.g. see Tables 2 and 3). In contrast, all dissected females of the pierids *Anteos clorinde* Godart and *Anteos maerula* Fabr. were virgin, had large fat bodies, and carried no immature or mature eggs.

Females of several species were observed to oviposit during migration. Most of these had weedy host plants with broad elevational ranges [e.g. *A. leucomelas* on *Liabum* spp. (Asteraceae), *D. plexippus* on *Asclepias curassavica* L. (Asclepiadaceae), and *Leptophobia aripa* (Boisd.) on *Lepidium virginicum* L. (Brassicaceae)]. Minor wing wear, a well-developed fat body, and good condition of the spermatophores indicated that most migrants were recently emerged, though at least a few days old and not teneral.

BEHAVIOR AND REPRODUCTIVE STATUS OF NONMIGRANTS

Over 80% of the lowland dry forest species migrated (though the proportion was lower when montane species were included). Those remaining passed the dry season as adults that continued to feed but were otherwise largely inactive. Dissection of resident dry forest females revealed that nearly all species were reproductively inactive and presumably diapausing. None carried mature eggs, few had spermatophores, and most had large fat bodies. While oocytes were visible in the upper portion of the ovarioles, these were apparently absorbed on reaching a tissue plug located above the oviduct (see also Jones and Rienks, 1987). No evidence of reproductive diapause was found in the monarch (Table 3) or the other five danaid species (see Herman, this volume; James, this volume).

Several species included both migrant and diapausing individuals, such as some *Eurema* species and *Siproeta stelenes biplagiata* (Fruhstorfer) (Nymphalinae). The contrasting reproductive status of migrant and nonmigrant *Eurema daira* is demonstrated in Table 2. Nonmigrants collected in November (beginning dry season) were reproductive. Samples from the same area in the dry season (January–April) were nonreproductive. In contrast, 78% of migrants collected at Monteverde in January were reproductively active.

DISCUSSION

In this study it was determined that in northwestern Costa Rica, >80% of the lowland dry forest butterfly species emigrated at some time of year to wet forest in the northeastern lowlands. Within the limits of topography, migration covered distances of from as little as 10 km to a maximum of just over 100 km in straight-line flights between lowland habitats on opposite sides of the central mountain range.

Table 2. Reproductive condition of *Eurema daira* collected (A) on the Pacific slope below Monteverde at 1,000 m elevation at the end of the wet season on 27 November 1984, (B) migrating through Monteverde during January 1985, also from the Pacific dry forest during the dry season on (C) 13 January, (D) 18 February, (E) 12 March, (F) 24 April, and (G) 13 May (after the first wet season rains), and (H) from the Atlantic lowlands on 28 March 1985.

	Wings[a]	Fat[b]	Immature eggs[c]	Mature eggs	Spermatophores[d]	% reproduction[e]	n[f]
A	1.33	1.25	0.75	0.76	1.78 (1.07)	100	63
B	0.67	2.17	0.72	0.61	1.22 (0.88)	78	18
C	0.19	2.97	0	0	0.31 (0.47)	0	86
D	1.18	2.92	0.04	0	0.69 (0.47)	4	48
E	1.00	3.00	0.12	0	0.79 (0.42)	12	24
F	0.91	3.00	0	0	0.91 (0.42)	0	23
G	2.83	2.42	0.87	0.58	1.03 (0.60)	84	31
H	2.15	0.92	0.85	1.0	1.62 (0.52)	100	13

[a] Mean wing wear: 0 = teneral; 1 = new, little, or no scale loss; 2 = obvious scale loss; 3 = heavy scale loss and wings tattered.

[b] Mean fat body: 0 = no fat present to 3 = maximum fat present.

[c] Mean immature and mature eggs recorded as absent (0) or present (1).

[d] Number of spermatophores per female: mean (SD).

[e] Females with either immature or mature eggs were counted as reproductive.

[f] Total number of females sampled.

Although migrants were observed all along their flight route, they were most conspicuous (and easiest to sample) when flying eastward across the continental divide in the Monteverde Cloud Forest Reserve. Among the 508 butterfly species collected at Monteverde, over half (53%) were collected during this study as they migrated eastward through the community and reserve (Table 1). These migrants included many lowland species as well as butterflies resident on the upper Pacific slopes near Monteverde.

The eastward migration was strongly concentrated during early to mid-dry season, with the result that lowland Pacific forests were left nearly empty of butterflies in the second half of the dry season (February–April). Species that passed the dry season in dry forest sites were, with few exceptions, nonreproductive. Absorption of oocytes within the ovarioles, and sedentary adult behavior, suggested that these females entered a physiological state of reproductive diapause (see Jones and Rienks, 1987). Nevertheless, females of some species (e.g., *E. daira*) mated during this period but did not develop eggs (Table 2). Females of the few species that remained reproductively active during the dry season, such as the monarch (Table 3), used host plants that continued to grow in wet microhabitats until later in the dry season. Janzen (1987a) reported an identical pattern among Sphingidae in Costa Rican dry forest.

Females migrating eastward were nearly all reproductively active and some species, such as *D. plexippus* and *A. leucomelas*, oviposited during the trip, which contrasts sharply with the general pattern of other insects that migrate while in diapause (Johnson, 1978). Most individuals of species sampled from the Atlantic slope and lowlands east of Monteverde (the presumed destination of the eastbound migrants) were reproductive throughout the dry season. Together, these observations suggest that emigration and reproductive diapause serve as alternative strategies for coping with the dry season when reproduction is inhibited by leafless host plants or other inimical conditions (Janzen, 1983, 1987a; Jones and Rienks, 1987).

A subset of butterfly species also emigrated from the dry forest during the wet season, especially some species of Pieridae (e.g., *Aphrissa* and *Eurema*) and many species of Lycaenidae. Since host plants were not leafless at this time, factors such as changes in nutritional quality, levels of secondary compounds, or predator and parasitoid interactions may be important (Janzen, 1987a,b). The diversity of migrant species and patterns of emigration from dry forest

Table 3. Reproductive condition of female *Danaus plexippus* sampled during the dry season (A) while migrating through Monteverde and (B) from the Altantic slope, nonmigrant population. Data as in Table 2.

	Wings	Fat	Immature eggs	Mature eggs	Spermatophores	% reproduction	n
A	1.64	1.73	0.91	1.0	2.36 (1.36)	100	11
B	1.33	1.67	1.00	1.0	1.22 (0.44)	100	9

suggest that a variety of factors invoke migratory behavior in different species.

The emigration of lycaenids from dry forest in June was exceptional compared with that of other taxa. However, larvae of many dry forest lycaenids feed on buds and flowers of mass-flowering trees and lianas that bloom from February to April (Frankie et al., 1974, 1983). Lycaenid adults presumably emigrated to the Atlantic side soon after emerging in May and June.

Why do more females than males migrate? Extremely ragged males of some species (e.g., *S. stelenes*) were conspicuous in Pacific sites after peak migration through Monteverde. These males, perhaps lacking adequate resources for migration, may have stayed behind to mate with late-emerging females before they abandoned the dry forest. In this way, males could participate in continued reproduction during the dry season through their offspring.

It is especially peculiar that peak eastward migration of monarchs (along with *A. leucomelas* and a few other species) occurred at the end of dry season. Two possible explanations fit these observations.

1. *Danaus plexippus* (and *D. gilippus*) originating in the Pacific lowlands moved upslope to moist forest valleys at 800–1,400 m elevation, where healthy individuals of *A. curassavica,* the only host known in this area, survived in wet microhabitats. Here, the monarchs and queens continued to lay eggs as the dry season progressed. When these plants finally succumbed to drought, the monarchs continued their migration over the mountains to the Atlantic slope, where milkweeds persisted throughout the dry season, producing the peak in Figure 6. Offspring that emerged as late as May and June probably returned to the Pacific lowlands. In contrast, *D. gilippus* remained abundant and reproductive on the upper pacific slopes (1,000–1,400 m) during the entire dry season. Only a few individuals were caught in flyways at Monteverde (>1,400 m).

2. Small numbers of monarchs migrated through Monteverde during early to mid-dry season, having oviposited on host plants encountered on the way. Offspring emerging into dry conditions during March–April followed their parents over the mountains to create the late dry season peak seen in Figure 6.

Westward migration was much more difficult to observe than eastward flights and could not be quantified by systematic sampling. However, on certain warm, clear, calm days, usually following a day of rain, hundreds of butterflies could be seen flying downhill through Monteverde clearly heading for the Pacific lowlands. Once it was realized that these butterflies were migrating the phenomenon was obvious, but to most people the butterflies appeared to be simply abundant and active. Westward migration was occasionally observed at other times, such as a flight of thousands of *Marpesia chiron* (Fabr.) observed during 2 days in February 1985. Westward migration could perhaps be documented best by periodic sampling of the dry forest sites, as observed by Janzen (1987a) for the sudden arrival of moths at lights in Santa Rosa National Park.

This study provides three points toward understanding the migration of Costa Rican butterflies.

1. Migration is a predictable seasonal occurrence, not a curious and haphazard anomaly. Costa Rican butterflies can be observed all along their migratory route and the sites of origin and destination are accessible for study.

2. A diversity of migratory patterns was observed, suggesting that different adaptations exist and that several hypotheses may be needed to explain them. Although certain patterns were consistent even at the family level, some species migrated in both wet and dry seasons, and others included both migrant and resident individuals.

3. We have long known the importance of conserving habitats at both ends of the migration route for vertebrates. It now seems clear that multihabitat protection is also necessary for preserving the species diversity of tropical butterflies and many other migrating insects. Extensive clearcutting in the lowlands of Costa Rica has undoubtedly affected the relative abundance of migrating butterflies already, perhaps making weedy, second-growth species more common, while deep forest butterflies have moved toward rarity and extinction.

ACKNOWLEDGMENTS

I wish to thank my field assistants K. Hoffmann and J. Wolfe and the staff of the Monteverde Cloud Forest Reserve. Also, I appreciate the encouragement of P. Feinsinger, G. Frankie, D. Futuyma, L. Gilbert, and D. Gill. D. Janzen and P. Opler shared their ideas and unpublished data. Funding was provided in part by The National Geographic Society and the Missouri Botanical Garden.

LITERATURE CITED

Beebe, W. 1949. Insect migrations at Rancho Grande, northcentral Venezuela. *Zoologica* 34:107–110.

Beebe, W. 1950. Migration of Danaidae, Ithomiidae, Acraeidae and Heliconidae (butterflies) at Rancho Grande, north-central Venezuela. *Zoologica* 35:57–68.

DeVries, P.J. 1983. Checklist of butterflies. In *Costa Rican natural history,* ed. D.H. Janzen, 654–678. Chicago: University of Chicago Press.

DeVries, P.J. 1987. *The butterflies of Costa Rica and their natural history.* Princeton: Princeton University Press, 327 pp.

Frankie, G.W., H.G. Baker, and P.A. Opler. 1974. Comparative phenological studies of trees in tropical wet and dry forests in the lowlands of Costa Rica. *Journal of Ecology* 62:881–919.

Frankie, G.W., W.A. Haber, P.A. Opler, and K. Bawa. 1983. Characteristics and organization of the large bee pollination system in the Costa Rican dry forest. In *Handbook of experimental pollination biology,* eds. C.E. Jones and R.J. Little, 411–447. New York: Van Nostrand Reinhold.

Haber, W.A. 1978. Evolutionary ecology of tropical mimetic butterflies (Lepidoptera: Ithomiinae). Ph.D. thesis. Minneapolis: University of Minnesota, 227 pp.

Haber, W.A., and G.W. Frankie. 1989. A tropical hawkmoth community: Costa Rican dry forest Sphingidae. *Biotropica* 21:155–172.

Janzen, D.H. 1983. Insects. In *Costa Rican natural history,* ed. D.H. Janzen, 619–645. Chicago: University of Chicago Press.

Janzen, D.H. 1984. Two ways to be a tropical moth: Santa Rosa saturniids and sphingids. *Oxford Surveys in Evolutionary Biology* 1:85–140.

Janzen, D.H. 1987a. How moths pass the dry season in a Costa Rican dry forest. *Insect Science and Its Application* 8:489–500.

Janzen, D.H. 1987b. When, and when not to leave. *Oikos* 49:241–243.

Johnson, C.G. 1978. *Migration and dispersal of insects by flight.* London: Methuen, 763 pp.

Jones, R.E., and J. Rienks. 1987. Reproductive seasonality in the tropical genus *Eurema* (Lepidoptera: Pieridae). *Biotropica* 19:7–16.

Lawton, R., and V. Dryer. 1980. The vegetation of the Monteverde Cloud Forest Reserve. *Brenesia* 18:101–116.

Potts, R.W.L. 1943. Systematic notes concerning American Acraeinae (Lepidoptera: Nymphalidae). *The Pan-Pacific Entomologist* 19:31–32.

Smith, N. 1982. Population eruptions and periodic migrations in the day-flying moth, *Urania fulgens.* In *The ecology of a tropical forest,* eds. E. Leigh, A. Rand, and D. Windsor, 331–344. Washington, D.C.: Smithsonian Institution Press.

Smith, N. 1983. *Urania fulgens.* In *Costa Rican natural history,* ed. D.H. Janzen, 775–777. Chicago: University of Chicago Press.

Stiles, F.G. 1983. Birds. In *Costa Rican natural history,* ed. D.H. Janzen, 502–530. Chicago: University of Chicago Press.

Williams, C.B. 1930. *The migration of butterflies.* Edinburgh: Oliver and Boyd, 473 pp.

DO CALIFORNIA MONARCHS UNDERGO LONG-DISTANCE DIRECTED MIGRATION?

Adrian M. Wenner and Ann M. Harris

ABSTRACT. Californian summers have previously been considered unsuitable for survival of monarch larvae; consequently, coastline, winter aggregations of monarch butterflies in the state have been believed to be composed of individuals produced to the north and east, beyond the Sierra Nevada and Cascade ranges (the Great Basin region and more northern states). Studies on monarch oviposition and larval biology in Santa Barbara County documented year-round incidence of adults and larvae. Oviposition in spring on local milkweed species, coupled with the existence of successive generations on those plants, assures the continued existence of local populations. Autumn movements of adults appear to consist almost entirely of individuals flying against ocean breezes, followed by a concentration of monarchs along the coastline and later aggregation. An alternative to the directed migration theory is thus plausible for California: Monarchs, largely a warm region species, may be experiencing a range extension into the north and east in spring and contraction of that range in autumn.

INTRODUCTION

The monarch butterfly is present year-round in California; as a consequence, many residents have advocated naming it the California state butterfly. Monarchs are especially obvious in Santa Barbara. In the autumn hundreds can be seen each day, flying against the gentle ocean breezes prevalent at that time of year. By mid-November more than 25 winter aggregations can be found along the coastline, along with perhaps 10 others, which are only temporary (Harris, 1986; Lane, this volume; Nagano et al., this volume). Throughout the winter, individuals fly out from their aggregations and drink water or nectar on warm days. By mid-February, aggregations begin dispersing (Hill et al., 1976); oviposition on local milkweed plants occurs shortly thereafter (Harris, 1986).

California has long been considered principally a winter destination of monarch butterflies that have migrated south from Washington, Oregon, Idaho, Montana, and the western provinces of Canada. In that connection Williams wrote (1958: 108):

> . . . the butterflies which breed in the north move south in the autumn, spend the winter in a state of semi-hibernation—most certainly not breeding—in Southern California and similar latitudes, and then return to the north in the spring . . . . What is so extraordinary is that they should fly a thousand miles to find a suitable place for hibernation.

Urquhart et al. (1970) felt that the monarch flight behavior represented "... a movement similar in many respects to that of a species of migratory bird." That sentiment was repeated later by Rankin (1978), when she wrote:

> The monarch flies up to 2000 miles from Canada and the northern United States to overwintering sites in southern California, Florida, and Mexico and migrates back again to the north the following spring.

However, several features of the monarch presence in the Santa Barbara region do not correspond with what one might expect from the above-described, long-distance "migration" theory as applied to monarch butterflies in the western states (see Harris, 1986, for a more complete review). The spring dispersal in late February normally occurs long before any stage of these butterflies (including adults) can survive the low temperatures prevailing at that time of year in the extensive San Rafael and Sierra Madre Mountain ranges north of the immediate vicinity (within 25 km) of Santa Barbara. Snowstorms often strike those mountains weeks after aggregation dispersal, and the average date of last killing frost in that mountain range is 20 March

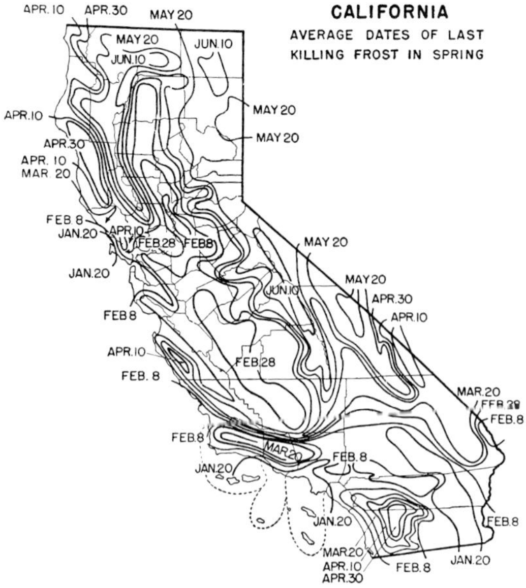

Figure 1. Isotherm map showing average dates of last killing frost in California (from Sprague, 1941). An atlas should be consulted to reveal the exact position of Santa Barbara County and Santa Cruz Island in the lower left-hand corner.

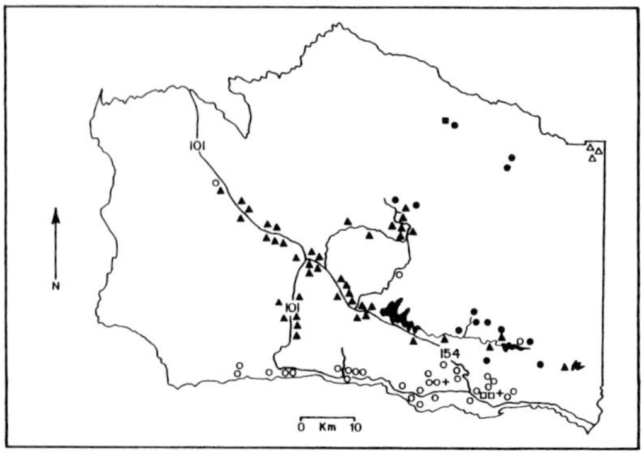

Figure 2. Location of known milkweed patches in Santa Barbara County in accessible areas; highways 101 and 154 are labeled for ease of reference. Symbols represent the species of *Asclepias* as follows: closed circles, *A. californica;* open circles, *A. fascicularis;* closed triangles, *A. eriocarpa;* open triangles, *A. vestita;* closed squares, *A. erosa;* open squares, cultivated *A. curassavica* plus *A. fruticosa.* The black patches (left to right) represent lakes Cachuma, Gibralter, and Juncal, respectively. The Pacific Ocean is both to the south and to the west of the county.

(Fig. 1; from Sprague, 1941). Even if adults could cross the San Rafael and Sierra Madre barrier, they would face the more formidable barrier of the Sierra Nevada and contiguous Cascade Mountain ranges, where passage would likely not be possible, on average, until the end of May (Fig. 1).

Neither is flight in the springtime unidirectional; adults can be seen flying in all directions. Adults caught and released tend to fly toward the highest point on the horizon, rather than against the wind as they do in the autumn (unpubl. data). Also, females lay a great number of eggs on local plants during and after their dispersal from aggregations (Harris, 1986), often with the result that newly emerging plants are stripped of most of their vegetation. Furthermore, old adults (the faded and tattered ones) disappear from the Santa Barbara region, apparently perishing after ovipositing, before more northern regions become suitable for survival.

Anomalies also exist in the autumn. The monarch's leisurely flight is primarily against the wind and occurs only during midday, hardly a sufficient rate for traveling the great distances required of the long-distance navigation hypothesis. Numerous trips to California's Central Valley during September have failed to reveal the presence of migrating adults. Neither has it been possible to find larvae on the milkweed plants at that time of year in that vast area (unpubl. data).

Specimens aggregating at the overwintering sites exhibit no evidence of long-distance migration. Most individuals are "mint-condition" specimens, as if they have recently emerged from the chrysalis, and arrive in the winter aggregations well supplied with lipid energy reserves (Chaplin and Wells, 1982). That is not what one might expect after a lengthy migratory flight from Idaho or Montana at that time of year, due to the vast amount of desert to be traversed in the Great Basin (making fueling en route difficult). By contrast, monarchs disperse from the Santa Barbara aggregations in an energy-depleted state, despite available nectar supplies in winter (Chaplin and Wells, 1982).

Finally, no evidence is available that might indicate what percentage of individuals found in Santa Barbara aggregations might have flown in from more northerly regions. It is for all of the above reasons that the question arose: Is it possible that overwintering monarch butterflies along the Santa Barbara coastline spend their larval life in that county or in regions adjacent to Santa Barbara County rather than immigrate from distant regions?

Investigators of the western monarch population have directed their attention to the biology of adults rather than to that of the larvae (e.g., Williams et al., 1942; Urquhart et al., 1970; Hill et al., 1976; Urquhart and Urquhart, 1977; Tuskes and Brower, 1978; Fadem and Shapiro, 1979; Brower et al., 1982, 1984; Chaplin and Wells, 1982; but see Cockrell et al., this volume). We felt it appropriate to investigate the possibility that the Santa Barbara County region is a suitable habitat for summer populations of monarch larvae (see above quotation from Williams, 1958: 108). If so, it follows that these local regions could supply most of the adults that aggregate along that coastline during the winter.

To gain a better understanding of the biology of

monarch butterflies living in the western states, our laboratory has conducted a series of studies, beginning in 1961. General observations have continued in each year, but in 1963, 1966–1968, 1970–1972, 1975, and 1984–1986 more intensive studies were done on monarch oviposition, larval abundance, sex ratios, parasitism, orientation, courtship, and/or overwintering behavior. We report here the results of those studies that are relevant to the long-distance migration hypothesis.

METHODS AND NOTES

During and after aggregation dispersal in late February, and before aggregation formation in early November, various accessible plant communities in Santa Barbara County have been visited intermittently during the past 25 yr in a search for milkweed patches and to inspect plants for the presence of monarch eggs and larvae. In the spring of 1970, members of a College of Creative Studies class at the University of California in Santa Barbara participated in that project at a more formal level. Fred Hill and one of us (A.M.W.) continued that formal study in 1971 and 1972. Dan Claffey elucidated the food web associated with milkweek plants in 1975. Flight orientation studies were done in 1970 by a graduate level animal behavior class at the University of California, Santa Barbara.

A more comprehensive study of monarch life history was begun in 1984 by one of us (Harris, 1986). This latter study continued through the 1985 and 1986 seasons and included an intensive search for aggregation sites and for additional localities of milkweed occurrence. In 1985–1986 (as in 1970–1972), all plant clumps within a patch were checked carefully on each visit for condition of plants (e.g., time of emergence, blooming season, number of stems per plant) and for presence of monarch eggs and larvae (see Harris, 1986, for details).

During the broader time period (1966–1986), we also inspected milkweed patches (*Asclepias fascicularis* Dcne.) and observed adult behavior on Santa Cruz Island (a part of Santa Barbara County) on numerous occasions. That island has many of the topographical features that exist in the remainder of the county.

Several factors weigh against any attempt to estimate the total number of monarch butterflies raised in Santa Barbara County. The annual fire closure of the Los Padres National Forest in Santa Barbara County's back country excludes most of the northeast portion of the county (nearly one-half of the county) for studies. A large portion of the southwest corner (nearly one-fourth of the county) is also closed, either because it is within the Vandenberg Air Force Base or because it is within one of the private Spanish Land Grant ranches.

Nevertheless, habitats in the remaining accessible Santa Barbara County area had characteristics similar to those plant communities found in the inaccessible area (Harris, 1986) and were studied when time permitted. Much of the area that might be considered accessible, however, may have no highways or may be too rugged for inspection in the time available for research. In addition, caterpillars may be abundant in some patches in some years but be absent in those same patches in other years, whereas, conversely, other patches may have caterpillars every year.

Neither was it possible to visit the patches on a regular schedule in all of the 4 yr, nor do milkweed plants (settling as they do primarily in disturbed areas) occur in all of the apparently suitable habitats. Also, the above-ground portion of individual plants in any one patch emerge throughout their growing season, leading to a constantly changing number of suitable oviposition plants within each patch.

The above three factors preclude an estimation of total milkweed plant abundance in the county. The patch locations shown in Figure 2 are thus only partially representative of the total distribution and abundance of the various milkweed species and represent those patches accessible by road. It is for all of the above reasons that we gathered data on relative (not absolute) distribution of monarch immatures to illustrate that they could be found in abundance throughout the summer months. Only data for eggs are presented here.

A more complete exposition of methods and biological information can be found in Hill et al. (1976) and in Harris (1986).

RESULTS

MILKWEED PLANTS AND MONARCH OVIPOSITION PATTERNS

Five species of milkweed plants are native to Santa Barbara County: *Asclepias californica* Greene, *Asclepias fascicularis*, *Asclepias eriocarpa* Benth., *Asclepias vestita* Hook. & Arn., and *Asclepias erosa* Torr. The first three were found in abundance in the accessible portion of the southern half of the county (Fig. 2); the last two are known from herbarium records (Harris, 1986). In addition to the native species, several exotics can be found in gardens in Santa Barbara, notably *Asclepias curassavica* L. and *Asclepias fruticosa* L. (Fig. 2).

Asclepias californica is prostrate and abundant on steep south-facing ridgetops among yucca plants on those chapparal-covered slopes (Fig. 2; see also Woodson, 1954, for distributions and habitats of this and other milkweed species). The time of emergence from the ground for this species is variable (e.g., Smith, 1976; Fig. 3) and is apparently correlated with the onset of the rainy season. The above-ground portions of these plants usually died 1 or 2 mo after cessation of the rainy season, except during one very wet year (1970).

Monarch eggs were found on *A. californica* soon after its emergence from the ground in the Santa Ynez Valley during 1970 and 1985 (e.g., Figs. 2 and 3), subsequent to adult aggregation dispersal along the Santa Barbara coastline (Harris, 1986). The duration of larval presence corresponded with the time necessary for one generation, in agreement with the low night temperatures in the area (approximately 0–10°C) during the early portion of that period (March–April). The egg counts shown are only relative, not absolute, because the number

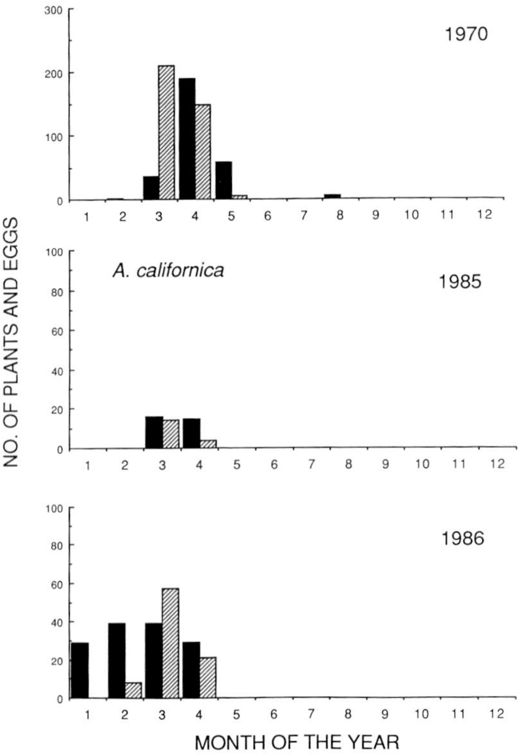

Figure 3. Relative (not absolute) abundance of *A. californica* and monarch oviposition incidence (see text) as a function of time for the years 1970, 1985, and 1986, with data grouped by month. The solid bars represent the mean number of plants inspected for the one or more different trips within each month. The hatched bars represent the total number of eggs found during all visits in each month. (The same eggs could not be counted on each visit, but the plants could.)

of plants, field trips, and investigators necessarily varied from year to year (see Methods and Notes).

Asclepias fascicularis plants usually emerged from the ground in mid-March or early April along the coastal slope (Fig. 2), but the exact time was difficult to determine due to the manner in which these plants blended into the surrounding vegetation. However, the plants persisted until late autumn or even winter (Fig. 4). The 1985 study contained few records of these plants for the first part of the year due to the destruction of our traditional study patches by landowners prior to that year (because of conversion to farming or because of weed control). The members of this species that were found were largely restricted to the coastal slope, but its cryptic habit obscured the full extent of its distribution. (Herbarium records indicate a presence away from the coastal slope.)

Oviposition on narrow-leaf milkweed, *A. fascicularis,* began as soon as new vegetation appeared above ground (Fig. 4). The lengthy oviposition period during April and May in 1970 (actually bimodal in a probit analysis of raw data) corresponded with the time necessary for production of two spring generations, before the onset of school terminated the project that year. In addition, in each of the autumn seasons in which plants have been inspected, there was sufficient time for the production of one (1971) or two (1985) generations of eggs and larvae (e.g., Fig. 4).

At approximately the same time that aboveground portions of *A. californica* die in the spring, the erect *A. eriocarpa* plants emerge from the ground in Santa Barbara County (Figs. 2 and 5). This latter species occurs near stream and river courses at all elevations and is particularly noticeable along highway rights-of-way (e.g., Fig. 2).

By visiting patches of this milkweed species at various elevations, we found plants that were suitable for oviposition at any time during the summer reproductive season (Fig. 5), due to the staggered time of their emergence from the ground (see Methods and Notes). Likewise, eggs (Fig. 5) and larvae were found on this species during that same period. Sufficient time elapsed for the production of four or five generations of larvae on this plant species.

SANTA CRUZ ISLAND AS A MICROCOSM OF THE SANTA BARBARA REGION

Asclepias fascicularis is the only milkweed species reported from Santa Cruz Island (Wallace, 1985; Harris, 1986), but it is widely distributed there (Fig. 6). As in the remainder of Santa Barbara County, monarch butterflies occur on the island all year, but the number is very small by comparison. We found eggs and/or larvae on plants in spring, summer, and autumn on every inspection trip. Adults were also seen flying or in aggregations on every visit.

The island has a diverse topography and is similar to Santa Barbara County as a whole, with its inland valleys and corresponding climate. The island differs from the main part of the county in that its principal streams drain in all directions from the center of the island. When autumn arrives, adult monarchs can be seen flying leisurely against the ocean breezes, just as in Santa Barbara County proper. This against-the-wind flight pattern results in their arrival on the coastline of the island. However, that behavior takes them in no particular direction; aggregations form at the heads of each of the major canyons (Fig. 6) on whatever vegetation exists there, but only one canyon mouth opens on the north shore (at Prisoner's Harbor). In contrast to the large aggregations on the mainland, aggregations on the island numbered at most 100 or 200 individuals at each of the canyon mouths studied.

Santa Rosa Island lies 10 km west of Santa Cruz Island, is about the same size, and also has a diverse topography. However, Santa Rosa Island lacks milkweed plants. During several trips to that island, we have seen only a single monarch adult. Also, we have not found any aggregations at any of the canyon mouths.

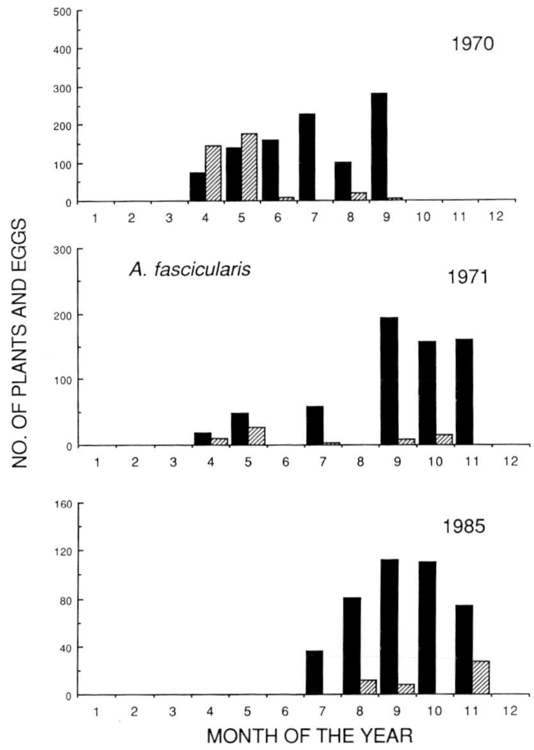

Figure 4. Relative abundance of *A. fascicularis* and monarch oviposition incidence during the years 1970, 1971, and 1985, with data grouped by month. Bar representations are the same as those in Figure 3.

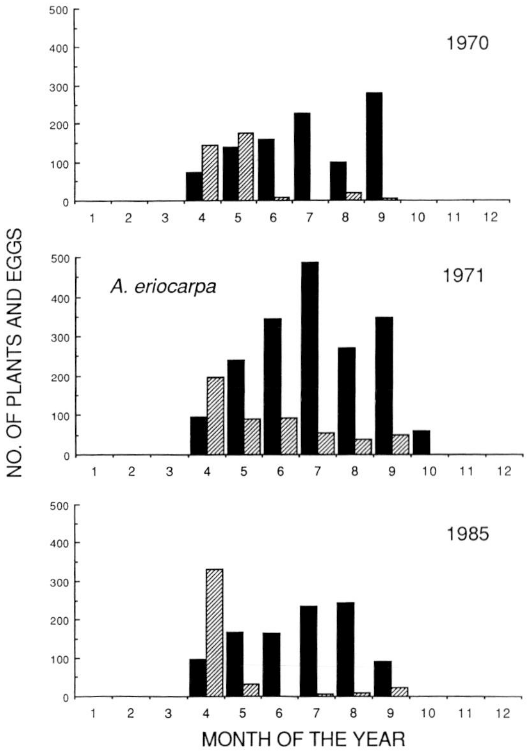

Figure 5. Relative abundance of *A. eriocarpa* and monarch oviposition incidence during the years 1970, 1971, and 1985, with bars indicating the same variables as those in Figure 3.

DISCUSSION

FATE AND ORIGIN OF AGGREGATING MONARCHS

Two questions can be asked regarding aggregating monarchs in Santa Barbara County: (1) Where do the adults go after aggregation dispersal in the spring? and (2) Where do adults come from prior to aggregation formation along the coastline in the autumn? The present study focused on the possibility that various species of milkweed plants in the county could provide a food supply for sequential generations of monarch larvae between the time of aggregation dispersal and aggregation formation. The results obtained apply to interpretations of monarch butterfly movements in the west.

With regard to the first question, it is evident from the data gathered and from climatic considerations that aggregation dispersal occurs too soon in the spring (late February) for an effective birdlike migration to distant northern regions. Severe frosts often occur in regions immediately inland from the California coast after the aggregations have already dispersed (Fig. 1).

Furthermore, it is evident that adults need not travel far to the north or east from the aggregation sites (see Harris, 1986) before ovipositing on milkweed plants (Fig. 2). At the time of aggregation dispersal or shortly thereafter (depending on the exact time of plant emergence from the ground), one species of milkweed is available along the coastline (*A. fascicularis*) and another (*A. californica*) is present on south-facing ridges behind the Santa Ynez Mountains, which lie immediately to the north of Santa Barbara's south coast.

Oviposition can be intense on newly emerging plants during and immediately after cluster dispersal in the spring, and plants are often stripped of their leaves. *Asclepias californica* normally has above-ground vegetation long enough to support only a single generation of larvae, but suitable *A. fascicularis* plants are available for at least the first two generations of monarch larvae after aggregation dispersal. In approximately mid-May the coast becomes fog-shrouded, and adults more frequently can be found behind the Santa Ynez Mountains. Some adults remain along the coastal slope after the first or second generations, as is evident by the continued presence of larvae on cultivated, exotic milkweed plants.

By April, the third species of native milkweed, *A. eriocarpa*, has emerged from the ground at low elevations throughout the interior of the county (Fig. 5). Other patches emerge at higher elevations

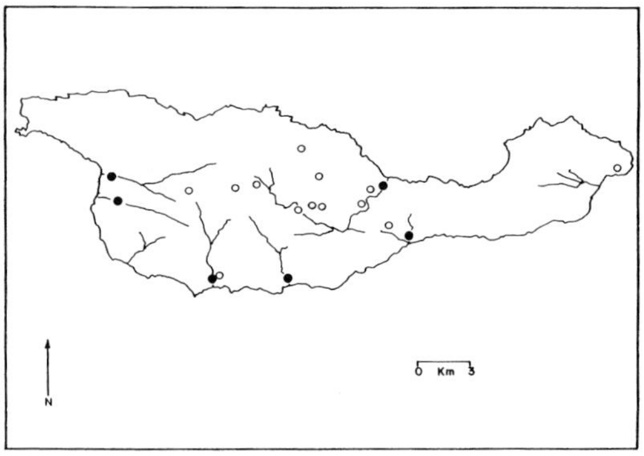

Figure 6. Known localities of *A. fascicularis* (open circles) and of monarch aggregations (closed circles) on Santa Cruz Island, in relation to major stream drainages. (Prisoner's Stream is the only stream that reaches the northern coastline.)

later. All of this area is exceedingly hot and dry during the summer months; however, subsequent generations of monarchs obviously mature in these regions; a sequential incidence of egg laying and larval feeding occurs on plants throughout the county, representing four to five generations (Harris, 1986).

Consider now the second of the above questions: Where do the monarchs come from during aggregation formation in Santa Barbara? Earlier, J.A. Downes (in Williams et al., 1942: 160) raised the same question we raise here with regard to the situation in California. He wrote:

> . . . the precise point of origin of the autumn migration, and the course of both the autumn and spring flights, seems to be largely unknown.

That assessment by Downes seems to hold true today (but see Nagano et al., this volume).

In late summer and autumn in the Santa Barbara area, adults can be seen flying against the onshore ocean breezes so prevalent at that time of year. This flight pattern is perhaps triggered by a change in photoperiod, a cold snap, or a drying up of the larval food plant in the interior valleys. This behavior takes them to the coastline, where the narrow-leaf milkweed (*A. fascicularis*) is common. Many of the patches of these plants have been free of larvae during the previous 3 mo and have had an opportunity to recover from the intensive larval feeding prevalent during spring. One then finds oviposition and large numbers of larvae after mid-August and into October (Fig. 4).

Santa Barbara County, of course, is not a geographically defined area. Much of the Coast Range of California has features similar to those of Santa Barbara County with respect to vegetation, climate, and topography. Thus, results similar to those found in this study of milkweed plant occurrence and oviposition in Santa Barbara County would likely be obtained in other counties along the central California coastline. The production of monarch adults along the stream drainages and in the mountains of Santa Barbara County is likely repeated to the east in Ventura County and to the north in San Luis Obispo County. The Santa Clara, Ventura, and Sespe-Piru rivers drain vast areas of the back country of Ventura County. All of that area undoubtedly produces large numbers of monarchs during the summer months.

In late summer, the adults produced in these back-country areas, by merely flying against ocean breezes, will reach the coastline. After reaching the ocean, flight against the afternoon prevailing westerlies would bring them to the Santa Barbara aggregation areas. The mint condition of the vast majority of specimens in newly formed aggregations partly supports that interpretation. Downes (in Williams et al., 1942: 161) also recognized that one need not assume a purposeful southward flight behavior. He wrote:

> In reality, however, the situation is much simpler . . . It would seem . . . that the butterflies reach any point on this long coast line and then collect at, or discover, practically all the groups of trees, remaining in numbers proportional to the suitability of the sites found. The migration is therefore merely to the coast, and not to particular small places.

ON THE QUESTION OF "LONG-DISTANCE MIGRATION"

The interpretation proposed above is not substantially contradicted by results from mark–recapture studies. The design of those studies has been influenced by the assumption that long-distance migration does indeed occur, as was perhaps best stated by Williams (1958: 201):

> . . . we must continue to try (to mark and recapture individual butterflies), whatever the difficulties; for only by recovery at a distance can we *prove* that a single individual really makes the whole journey. Apart from such evidence, long distance migration of insects is an inferential theory, perhaps the only one that will explain the known fact, but not "proved" to the unbeliever!

Only a few such mark–recapture studies have been done with monarchs in the west. Urquhart (1960, fig. 79) reported on one such study. Adults marked in aggregation sites along the California coastline were searched for in the remainder of the west (see also Nagano et al., this volume). However, most of the results indicated that a relatively short distance had been traversed between marking and recapture. (The time elapsed between marking and later recapture was not published, but see Nagano et al., this volume.)

Urquhart et al. (1970) repeated that study with release of monarchs from a point source at Whittier, California. Again, the results did not support a single hypothesis; marked individuals were caught in all directions. Although there was some seasonal variation in recapture direction, prevailing wind directions were not reported. They concluded:

> It would appear . . . that the monarch butterfly exhibits two types of movement; one extending over long distances as in the case for the Eastern North American Population . . . and others in which the movement is restricted to shorter distances.

Urquhart and Urquhart (1977) later conducted the only published West Coast experiment in which tagged monarchs were released in the north and searched for in more southerly localities. They sent an unspecified number of "migrants from areas of great abundance" by airmail to volunteers in British Columbia, Oregon, Idaho, and Nevada. Only 10 of the marked individuals that had been released in the north (out of an unspecified number) were caught in California (presumably the only area searched for monarchs).

Unfortunately, the Urquharts' report did not include the site of capture, the number of monarchs released, the time of release (other than "autumn"), or the time between release and capture. The experimental design thus did not distinguish between a "directed" long-distance migration and a dispersal from the point of release.

To understand why the supportive evidence is insufficient confirmation of a "birdlike" migration when applied to the western population of monarchs, one must refer to Williams' (1958: 3) definition. He wrote:

> "Migration" for the purpose of this book is considered to be a continued movement in a more or less definite direction, in which both movement and direction are under the control of the animal concerned.

Urquhart (1960: 319) obviously worked within those same assumptions when he described what he felt must occur during the spring "migration." He wrote:

> . . . the flight is direct, north or northeast, rather than meandering . . . Once the spring migration is under way the female must reach the breeding grounds as quickly as possible, because the eggs begin to develop rapidly and must be deposited.

The evidence gathered to date in the west does not support the impressions of either Williams or Urquhart.

Return now to the opening questions in this section. If monarchs migrate north immediately after aggregation dispersal in the spring, one would expect a marked depletion of them from Santa Barbara County. However, both adults and larvae are present in the coastal area through at least two generations in the spring, can be found in the remainder of the county through successive generations in the summer, and produce at least one more generation in the coastal area in the autumn. Once again, Downes (in Williams et al., 1942: 164) reported that same phenomenon earlier:

> . . . the whole of California is repopulated by the middle of (April). Many butterflies, however, are still at Pacific Grove in mid-April, and . . . some are still there as late as 8th May. It is indeed possible that the spring migration is never quite complete.

On the other hand, monarchs are known as "wanderer butterflies" in many parts of their range (e.g., Smithers, 1977). Although none of their life-history stages can survive deep frost (Anderson and Brower, this volume; Masters, this volume), some small percentage of the overwintering adults may well reach the more northern regions of the west after the majority has laid eggs in the area near aggregation sites and once the few survivors can clear the formidable mountain ranges ringing California in late May. Future research may reveal what that percentage of long-distance "wanderers" may be (see Nagano et al., this volume).

Similarly, if most of the monarchs in winter aggregations had migrated into California from more northern regions, one would expect a high percentage of the adults in aggregations to have traveled great distances. As of now, however, little documentation supports long-distance directed migration by a majority of aggregating monarchs. Furthermore, nearly all of the individuals are in excellent condition at the time of aggregation (as mentioned above), despite the paucity of nectar-bearing plants at that time of year in the Great Basin deserts. Neither would one expect the existing spurt of autumn oviposition and larval rearing on Santa Barbara milkweed to occur if the "migrating forms" had previously entered an arrested state of reproduction (assuming all individuals are at the same stage of oviposition).

THE RANGE EXPANSION AND CONTRACTION MODEL

All of the results provided by these Santa Barbara County studies have led us to generate an alternative hypothesis for the life history of monarch butterflies in California (Harris, 1986). One can consider monarchs to be a warm region (tropical and subtropical) species that thrives in the more equable regions of California. At the time of aggregation dispersal, mating is intense (Hill et al., 1976), with oviposition soon thereafter on those milkweed plants that are not in regions subject to severe frost. The first generation thus develops on local milkweed and provides the next generation of adults, some of which can expand the monarch range into the north, east, and south when conditions permit.

Under this model, one would then expect a sequence of generations of larvae on Santa Barbara County plants, an event that does occur. With the advent of autumn, the monarch range would contract to the coastal regions once again, with most of the more northern forms perhaps perishing. Individuals from the more southern and western populations of the range would be sufficiently abundant and could once again form aggregations, supplemented to a great extent by the one or two autumn generations of adults produced on milkweed growing relatively near the aggregation sites.

We are not the first to propose such a model; the Australian situation near Sydney (e.g., James, 1979, 1981, this volume; Smithers, 1965, 1972, 1977) is remarkably similar to the pattern we have observed in the Santa Barbara region (and likely elsewhere along the California coast). For example, Smithers (1965) wrote:

> . . . some aspects of the pattern of behaviour of this species do not correspond with that of the same species in North America, as indicated by Urquhart (1960), at least in the areas of Australia in which most of the investigations have been made . . . . Strong tendencies to migrate have not yet been detected; only one confirmed report of obvious unidirectional flight is available; all known clustering sites, whether overwintering or not, are adjacent to areas with good summer growth of the host plant (*Asclepias* sp.; breeding can continue well into winter in protected situations . . . even though it may have ceased several weeks earlier in colder areas only a few miles distant.

Smithers (1977: 183) later summarized his views, which correspond well with our conclusions for Santa Barbara monarchs. He wrote:

> A perusal of the series of maps indicates a withdrawal of *D. plexippus* from an extensive summer distribution to a restricted temperature-controlled winter one . . . . Release and extension of range comes only with the rising temperatures of spring and the availability of larval host plants.

In a series of papers, James (e.g., 1979, 1981) provided additional input bearing on the findings obtained by Smithers. James (1979) concurred with the conclusion of Smithers by writing, "Although population movements do occur in Australia, they appear to be mainly related to an annual extension and contraction of range (Smithers, 1977)." James also obtained results similar to those we obtained (e.g., Harris, 1986) and wrote, "Fortnightly sampling of the two areas of *G. fruticosus* revealed an abundance of ova and larvae through the winter months." James also found aggregations that contained some long-distance migrants from farther south in Australia (see James, this volume).

The data gathered on spring behavior of monarchs in Santa Barbara indicate that they do not behave as described by Urquhart (1960: 319), and the weather at that time of year would likely cause their demise if they did. The range expansion model published earlier for some Australia monarch populations fits the results we have obtained and is now open to test. Although the flight of monarchs is clearly not wind-controlled, neither is it ". . . a continued movement in a more or less definite direction."

The spring behavior of monarchs in Santa Barbara appears to fall somewhat within Johnson's Class IE category ("migration" in the broad sense of mass movement). In spring the monarchs in Santa Barbara disperse ". . . by controlling their track more or less independently of the wind, also make one-way journeys, and oviposit and die soon after reaching their destination" (as in Johnson, 1969: 25). That destination, however, may be no more than a few kilometers, for most monarchs, from the aggregation sites.

Monarch flight behavior in late summer and autumn differs from that observed in the spring. On the basis of more than 20 yr of observation in the Santa Barbara region, we conclude that Johnson's (1969: 33) statement—that there is ". . . not much evidence to suggest that the displacement is other than largely wind-controlled . . ." applies to monarch adults during the autumn in Santa Barbara County. This behavior thus places the autumn mass movement of monarchs only partially within Johnson's Class IIIB category (Johnson, 1969: 31). However, because Johnson listed round trips as a requisite for the Class III categories, it perhaps appears more reasonable to consider the autumn mass movement of monarchs in Santa Barbara to be only a special case of range contraction by flight against the wind. That behavior would parallel the phenomenon of "withdrawal" from a larger range, as described by Smithers (1977).

SANTA CRUZ ISLAND AND TESTS OF THE TWO COMPETING HYPOTHESES

Two tests of the "migration" hypothesis against the range expansion and contraction hypothesis are feasible. One of them, an indirect test, has already been done in part by nature. An examination of population data gathered on Santa Cruz Island lends no support to the migration theory. The other test, a comparison of cardenolide "fingerprinting" of adults raised on various species of milkweed (e.g., Brower et al., 1982, 1984), to the chemical composition of adults found in winter aggregations in California, remains to be done (see Malcolm et al., this volume).

The Santa Cruz Island microcosm has provided a convenient opportunity to compare application of the "migration" and "range contraction and expansion" hypotheses. If monarchs were migrating to the north after aggregation dispersal in the spring, they should be absent from Santa Cruz Island during summer and should again appear only after southward migration in the autumn. Instead, a small population of adults can be found year-round on that island.

Neither do the adults aggregate only on the south shore of Santa Cruz Island; they aggregate at the heads of *each* of the major existing canyons (only one canyon drains to the north shore). The largest aggregation ever found on the island (perhaps 250 adults) was recorded by Patrick Wells in December 1980, near the Christi Ranch house at the west end of the island. The second largest aggregation regularly forms at the head of the Prisoner's Stream, which flows to the north shore of the island.

Similarly, if monarchs were migrating south from more northerly regions in the autumn, large numbers of them would be expected to overshoot the coastline (see Schmidt-Koenig, this volume) and aggregate on both Santa Cruz and Santa Rosa islands immediately to the south of the Santa Barbara coastline. Instead, we have found monarch aggregations on Santa Cruz Island only at the mouths of canyons and only in the numbers that could have been produced on the amount of local milkweed found on that island. Occasionally an adult can be seen flying on Santa Rosa Island, but no aggregations are formed there in the autumn, despite suitable protection in Lobos Canyon. On the other hand, under the range expansion and contraction interpretation, the adult population levels necessary for aggregations do not exist and are not to be expected on Santa Rosa Island, because no milkweed species exist on that island.

A more direct comparative test of the two hypotheses is now within our grasp. California has relatively few species of milkweed plants (Woodson, 1954; figs. 4, 16, 55, 70, 71, 74, 75, 80, 92, 98, 99, and 110), only some of which are prevalent in wide areas of California. A cardenolide analysis can be done on adults that have been raised on the more prominent species, as Brower and co-workers have shown is feasible (e.g., Brower et al., 1982, 1984; Malcolm et al., this volume). A similar analysis can then be executed on adults found in the aggregation areas in Santa Barbara. If a high percentage of monarchs found in the Santa Barbara aggregations exhibit "cardenolide-prints" that differ from those expected on the basis of what is found in the milkweed species found in Santa Barbara, then they must have come from outside the county. If they differ from those raised on any of the milkweed plants found in the state, then they must have come from outside that boundary.

On the other hand, if almost all of the adults found in the winter aggregations along the Santa Barbara coastline had obviously fed only upon those milkweed species that may be found in Santa Barbara or neighboring counties, there would be no need to invoke the long-distance migration hypothesis to explain the origin of monarchs along that coastline. A similar set of comparisons could also be run for aggregation areas and milkweed species throughout the state.

We conclude that sufficient data on milkweed distribution and expertise now exist to test the relative predictive abilities of the migration hypothesis and the range expansion and contraction hypothesis. We also look forward to the possibility that future rigorous experimental tests of these two hypotheses against one another (see Wenner, 1989) will supplement the supportive types of study done heretofore on monarch butterfly movements.

ACKNOWLEDGMENTS

We thank Al and Russ Vail of the Vail and Vickers Company (Santa Rosa Island), the late Carey Stanton of the Santa Cruz Island Company, Rancho Dos Pueblos, and the U. S. National Forest Service for providing access to remote areas for this research. H. Fred Hill, Larry Friesen, Patrick Wells, and many graduate and undergraduate students helped gather and analyze data through the years. Hugh Dingle, Mark Page, Maynard Moseley, and Patrick Wells read and provided comments on the manuscript. We also thank the anonymous referees who provided valuable input. The Chevron Oil Corporation provided partial financial assistance to A.M.H. for the project.

LITERATURE CITED

Brower, L.P., J.N. Seiber, C.J. Nelson, S.P. Lynch, and P.M. Tuskes. 1982. Plant-determined variation in the cardenolide content, thin-layer chromatography profiles, and emetic potency of monarch butterflies, *Danaus plexippus* reared on the milkweed, *Asclepias eriocarpa* in California. *Journal of Chemical Ecology* 8:579–633.

Brower, L.P., J.N. Seiber, C.J. Nelson, S.P. Lynch, M.P. Hoggard, and J.A. Cohen. 1984. Plant-determined variation in cardenolide content and thin-layer chromatography profiles of monarch butterflies, *Danaus plexippus* reared on milkweed plants in California 3. *Asclepias californica*. *Journal of Chemical Ecology* 10:1823–1857.

Chaplin, S.B., and P.H. Wells. 1982. Energy reserves and metabolic expenditures of monarch butterflies overwintering in southern California. *Ecological Entomology* 7:249–256.

Fadem, C.M., and A.M. Shapiro. 1979. Notes on winter roosting by monarchs (Lepidoptera: Danaidae) at an inland site in California. *Pan-Pacific Entomologist* 55:309–310.

Harris, A. 1986. Natural history of the monarch butterfly (*Danaus plexippus*) in the Santa Barbara region. M.A. thesis. Santa Barbara: University of California, 33 pp. (+figs. and appendices).

Hill, H.F., A.M. Wenner, and P.H. Wells. 1976. Reproductive behavior in an overwintering aggregation of monarch butterflies. *American Midland Naturalist* 95:10–19.

James, D.G. 1979. Observations on two overwintering clusters of *Danaus plexippus* (L.) (Lepidoptera: Nymphalidae) in the Sydney area during 1978. *Australian Entomological Magazine* 5:81–85.

James, D.G. 1981. Studies on a winter breeding population of *Danaus plexippus* (L.) (Lepidoptera: Nymphalidae) at Spencer, New South Wales. *General and Applied Entomology* 13:47–53.

Johnson, C.G. 1969. *Migration and dispersal of insects by flight.* London: Methuen, 763 pp.

Rankin, M.A. 1978. Hormonal control of insect migratory behavior. In *Evolution of insect migration and diapause,* ed. H. Dingle, 5–32. New York: Springer-Verlag.

Smith, C.F. 1976. *A flora of the Santa Barbara region, California.* Santa Barbara, California: Santa Barbara Museum of Natural History, 331 pp.

Smithers, C.N. 1965. A note on overwintering in *Danaus plexippus* (Linnaeus) (Lepidoptera: Nymphalidae) in Australia. *Australian Zoologist* 13:135–136.

Smithers, C.N. 1972. Observations on a breeding population of *Danaus plexippus* (L.) (Lepidoptera: Nymphalidae) at Camden, New South Wales. *Australian Zoologist* 17:142–148.

Smithers, C.N. 1977. Seasonal distribution and breeding status of *Danaus plexippus* (L.) (Lepidoptera: Nymphalidae) in Australia. *Journal of the Australian Entomological Society* 16:175–184.

Sprague, M. 1941. Climate of California. In *Climate and man: Yearbook of Agriculture,* 783–797. Washington, D.C.: United States Department of Agriculture, United States Government Printing Office.

Tuskes, P.M., and L.P. Brower. 1978. Overwintering ecology of the monarch butterfly, *Danaus plexippus* L., in California. *Ecological Entomology* 3:141–153.

Urquhart, F.A. 1960. *The monarch butterfly.* Toronto: University of Toronto Press, 361 pp.

Urquhart, F.A., and N.R. Urquhart. 1977. Overwintering areas and migratory routes of the monarch butterfly (*Danaus p. plexippus,* Lepidoptera: Danaidae) in North America with special reference to the western population. *Canadian Entomologist* 109:1583–1589.

Urquhart, F.A., N.R. Urquhart, and F. Munger. 1970. A study of a continuously breeding population of *Danaus plexippus* in southern California compared to a migratory population and its significance in the study of insect movement. *Journal of Research on the Lepidoptera* 7:169–181.

Wallace, G.D. 1985. *Vascular plants of the Channel Islands of Southern California and Guadalupe Island, Baja California, Mexico.* Contributions in Science, no. 365, 136 pp. Natural History Museum of Los Angeles County.

Wenner, A.M. 1989. Concept centered versus organism centered biology. *American Zoologist* 29:1177–1197.

Williams, C.B. 1958. *Insect migration.* London: Collins, 235 pp.

Williams, C.B., G.F. Cockbill, M.E. Gibbs, and J.A. Downes. 1942. Studies in the migration of Lepidoptera. *Transactions of the Royal Entomological Society of London* 92:101–283.

Woodson, R.E., Jr. 1954. The North American species of *Asclepias* L. *Annals of the Missouri Botanical Garden* 41:1–261.

SPRING MIGRATION OF MONARCH BUTTERFLIES IN CALIFORNIA

Christopher D. Nagano, Walter H. Sakai, Stephen B. Malcolm, Barbara J. Cockrell, Julian P. Donahue, and Lincoln P. Brower

ABSTRACT. The evidence for migration by monarch butterflies in California is incomplete and has been difficult to interpret. This has led to the belief that these large, mobile butterflies move randomly between overwintering sites and breeding areas and do not engage in directed, seasonal migrations in California. Here we publish mark–release–recapture data that refute the random movement hypothesis to provide the first unequivocal evidence for directed, seasonal monarch migration in California.

In the spring of 1986 and 1987, monarchs left overwintering sites along the coast of central and southern California to head in one of two directions. The two headings were approximately northwesterly and easterly with mean angles of 326.7 and 99.2°, respectively. We interpret these nonrandom headings as evidence that monarch departures from their coastal overwintering sites were migrations into, and along, the Coast and Transverse mountain ranges in search of early-spring milkweed host plants. Later departures from the overwintering sites moved farther inland on longer migratory flights. These flights would allow monarchs to exploit milkweeds that appear later in spring in the San Joaquin Valley and the Sierra Nevada. The bimodal migratory headings may also be related to coastal orientation and the location and extent of the Mojave Desert.

INTRODUCTION

Since 1937, F.A. Urquhart has worked from Toronto in Canada with associates across most of North America to tag monarch butterflies, *Danaus plexippus* (L.), in an effort to describe their annual patterns of migration (Urquhart, 1960, 1966, 1976, 1987; Urquhart et al., 1965, 1970; Urquhart and Urquhart, 1976a,b,c, 1977, 1978, 1979a,b). This patient research effort culminated with the publication, almost 40 years later, of the discovery that the majority of monarchs east of the Rocky Mountains overwinter each year, tightly aggregated in high-altitude fir forests west of Mexico City (Urquhart, 1976). Since 1976, much has been learned about the biology of monarch overwintering in Mexico (reviews in Brower, 1985; Calvert and Brower, 1986; Urquhart, 1987). Likewise, much has been learned about both autumn migration to the overwintering sites (Urquhart and Urquhart, 1976b,c, 1977, 1978, 1979a; Brower, 1985; Seiber et al., 1986) and spring migration back to breeding areas in the United States and Canada (Urquhart and Urquhart, 1979c; Malcolm et al., 1987; Malcolm and Brower, 1989; and Lynch and Martin, Cockrell et al., Malcolm et al., Riley, this volume).

In contrast, although monarchs of the population found west of the Rocky Mountains have been recorded in California since at least 1816 (Lane, this volume), and in overwintering aggregations since approximately 1864 (Lane, this volume; Vane-Wright, this volume; Nagano and Sakai, 1989), little is known about migration of these western monarchs. Nevertheless, the accessibility of monarchs overwintering in aggregations in groves along the coast of California, has stimulated research on many aspects of their biology, including estimates of population size, mating behavior, hormonal control of diapause, thermoregulation, fat utilization, and cardenolide storage (Williams, 1930; Williams et al., 1942; Beall, 1948; Johnson, 1969; Kammer, 1970; Herman, 1973, 1975, 1981, 1985; Brower and Moffitt, 1974; Barker and Herman, 1976; Hill et al., 1976; Roeske et al., 1976; Herman and Barker, 1977; Tuskes and Brower, 1978; Chaplin and Wells, 1982; Herman et al., 1989). However, much remains to be learned about monarch migration in California (see Wenner and Harris, this volume).

Until now, the primary evidence that monarchs do migrate in California has been the published

Table 1. Numbers of monarch butterflies marked and released at the 14 overwintering locations shown in Figure 1 and listed in Table 2 with the recaptures of these tagged butterflies.

Location and date of mark–release event	Number of visits	Number marked			Recaptures	
		Males	Females	Total	N	%
1. Morro Bay State Park, San Luis Obispo County						
December 1985	2	898	615	1,513	5	0.33
January 1986	1	93	91	184	0	0
December 1986	2	405	328	733	0	0
2. North Campground, Pismo State Beach, San Luis Obispo County						
January 1986	1	47	50	97	0	0
December 1986	1	114	101	215	0	0
January 1987	2	1,178	815	1,993	5	0.25
February 1987	2	262	154	416	0	0
3. Ellwood, Goleta, Santa Barbara County						
November 1985	3	561	409	970	0	0
December 1985	7	2,234	1,418	3,652	4	0.11
January 1986	19	2,765	2,027	4,792	7	0.15
February 1986	7	937	649	1,586	0	0
March 1986	2	27	11	38	0	0
November 1986	5	4,691	4,071	8,762	15	0.17
December 1986	2	96	66	162	0	0
January 1987	4	118	126	244	0	0
February 1987	1	128	85	213	0	0
4. Summerland, Santa Barbara County						
December 1985	2	430	297	727	2	0.28
5. Carpinteria, Santa Barbara County						
October 1986	3	91	51	142	0	0
November 1986	1	10	9	19	0	0
January 1987	3	60	7	67	1	1.49
6. Harmon Barranca, Arrundell Barranca, and Camino Real Park (closely adjacent sites in Ventura, Ventura County)						
Harmon Barranca						
November 1986	4	16	8	24	1	4.17
December 1986	11	39	10	49	0	0
January 1986	4	24	5	29	0	0
Arrundell Barranca						
November 1986	1	175	62	237	0	0
December 1986	1	612	388	1,000	2	0.20
January 1987	2	1,425	571	1,996	7	0.35
Camino Real Park						
December 1986	1	41	8	49	0	0
January 1987	2	1,848	810	2,658	6	0.23
7. Big Sycamore Canyon, Point Mugu State Park, Ventura County						
January 1986	1	10	0	10	0	0
February 1986	1	2	0	2	0	0
September 1986	7	267	36	303	1	0.33
October 1986	23	2,352	531	2,883	5	0.17
November 1986	12	617	250	867	3	0.35
December 1986	8	282	95	377	0	0
January 1987	4	59	26	85	0	0
February 1987	1	2	0	2	0	0
March 1987	2	3	0	3	0	0
8. Little Sycamore Canyon, Ventura County						
December 1985	2	94	41	135	1	0.74
January 1986	4	214	40	254	0	0
February 1986	1	2	0	2	0	0
November 1986	5	1,232	821	2,053	8	0.39

Table 1. Continued.

Location and date of mark–release event	Number of visits	Number marked			Recaptures	
		Males	Females	Total	N	%
December 1986	7	1,177	649	1,826	6	0.33
January 1987	3	188	105	293	3	1.02
February 1987	6	142	34	176	0	0
March 1987	1	2	0	2	0	0
9. Leo Carrillo State Beach, Los Angeles County						
December 1985	1	119	55	174	3	1.72
January 1986	1	4	0	4	0	0
September 1986	3	13	6	19	0	0
October 1986	3	50	9	59	0	0
November 1986	9	733	386	1,119	1	0.09
December 1986	5	129	82	211	0	0
January 1987	3	80	46	126	0	0
10. Zuma Canyon, Los Angeles County						
January 1986	1	5	1	6	0	0
November 1986	1	64	41	105	1	0.95
December 1986	1	99	30	129	0	0
January 1987	1	35	16	51	0	0
11. Malibu Canyon, Malibu, Los Angeles County						
December 1985	1	5	1	6	0	0
December 1986	3	3,444	1,223	4,667	8	0.17
12. Rustic Canyon, Los Angeles County						
October 1986	1	75	18	93	0	0
November 1986	7	1,071	321	1,392	2	0.14
December 1986	1	144	36	180	1	0.56
13. Santa Monica, Los Angeles County						
December 1985	1	5	1	6	0	0
February 1986	2	4	2	6	0	0
March 1987	1	6	0	6	1	16.67
14. Ballona Wetlands, Los Angeles County						
December 1985	3	81	25	106	1	0.94
October 1986	6	220	14	234	0	0
November 1986	4	123	21	144	0	0
December 1986	8	76	12	88	0	0
Totals	251	32,555	18,216	50,771	100	0.20

recaptures of monarchs tagged by Urquhart and his colleagues (Urquhart, 1960; Urquhart et al., 1965, 1970; Urquhart and Urquhart, 1977). Unfortunately, the data are difficult to interpret for two reasons: (1) it is unclear which of the tagged butterflies were reared artificially and then released from locations chosen by Urquhart and his colleagues, and which were wild, field-tagged butterflies, and (2) mark, release, and recapture data necessary for detailed interpretation were omitted, including the numbers marked, the dates marked, proportions of the two sexes, effort expended in tagging monarchs, and the number of recaptures. Since Urquhart's published data are not complete enough to allow critical interpretation, it is not surprising that Wenner and Harris (this volume) have speculated that monarchs do not migrate in California. Instead, Wenner and Harris consider that because they observed monarchs breeding throughout the year in Santa Barbara County in southern California, these monarchs were not migrants. Rather, they argue that seasonal movements can be described by random range expansion and contraction of monarch populations and not by migration (also see Brower et al., 1984; Malcolm, 1987, for a discussion of monarch migration in California).

To address the question of whether or not monarchs migrate in California we used Urquhart's tagging method to mark and release monarchs overwintering naturally at a series of locations along the coast of central and southern California. Nonrandom movement of monarchs away from the

Table 2. The locations and dates of recapture, with the distances and overall compass headings flown by each of the 100 monarch butterflies after mark and release on the dates indicated at the 14 overwintering sites in California. Sites are numbered as in Table 1 and Figure 1.

Mark–release event		Recapture event				
Date	Butter-fly sex	Location	County	Date	Distance flown (km)	Compass heading (°)
1. Morro Bay State Park, San Luis Obispo County (N = 5)						
8 Dec 85	f	Three Rivers, Sequoia National Park	Tulare	10 Mar 86	216	55
8 Dec 85	m	Vandenberg Air Force Base	Santa Barbara	15 Mar 86	82	156
8 Dec 85	f	Woodland	Sacramento	23 Mar 86	434	350
8 Dec 85	f	Carmel	Monterey	Mar 86[a]	173	324
8 Dec 85	m	Lemore	Kings	Mar 86[a]	143	44
2. North Campground, Pismo State Beach, San Luis Obispo County (N = 5)						
6 Jan 87	m	San Luis Obispo	San Luis Obispo	14 Jan 87	19	353
6 Jan 87	f	San Luis Obispo	San Luis Obispo	26 Mar 87	19	353
6 Jan 87	m	Yosemite National Park	Mariposa	6 Apr 87	299	17
6 Jan 87	m	Aptos	Santa Cruz	20 Apr 87	238	332
7 Jan 87	f	Santa Barbara	Santa Barbara	16 Mar 87	118	134
3. Ellwood, Goleta, Santa Barbara County (N = 26)						
14 Dec 85	m	Chula Vista	San Diego	Apr 86[a]	343	127
15 Dec 85	m	Carpinteria	Santa Barbara	"spring" 86[a]	35	96
28 Dec 85	f	Santa Maria	Santa Barbara	14 Apr 86	81	318
28 Dec 85	m	Atascadero	San Luis Obispo	14 Apr 86	144	327
3 Jan 86	m	Pismo State Beach	San Luis Obispo	2 Feb 86	109	318
4 Jan 86	f	Oceano	San Luis Obispo	28 Feb 86	105	317
4 Jan 86	f	Canyon Country	Los Angeles	20 Mar 86	119	88
14 Jan 86	f	Ventura	Ventura	"spring" 86[a]	47	108
14 Jan 86	f	Visalia	Fresno	20 Mar 86	221	14
25 Jan 86	f	Lost Hills	Kern	21 Apr 86	139	8
29 Jan 86	f	San Jose	Santa Clara	5 Apr 86	376	332
8 Nov 86	m	Santa Barbara	Santa Barbara	16 Mar 87	19	100
8 Nov 86	m	Carpinteria	Santa Barbara	20 Mar 87	41	96
8 Nov 86	m	Bishop	Inyo	23 Apr 87	350	23
15 Nov 86	f	Elk Grove	Sacramento	8 May 87	465	345
22 Nov 86	f	Montecito	Santa Barbara	27 Mar 87	24	85
22 Nov 86	f	Pacific Palisades	Los Angeles	20 Feb 87	143	108
28 Nov 86	f	Hope Ranch	Santa Barbara	8 Mar 87	14	94
28 Nov 86	f	Valencia	Los Angeles	15 Mar 87	124	91
28 Nov 86	m	Ojai	Ventura	16 Mar 87	61	89
28 Nov 86	m	Santa Barbara	Santa Barbara	16 Mar 87	16	100
28 Nov 86	m	Santa Barbara	Santa Barbara	18 Mar 87	20	90
28 Nov 86	f	Carpinteria	Santa Barbara	20 Mar 87	41	96
28 Nov 86	f	San Luis Obispo	San Luis Obispo	29 Mar 87	124	331
28 Nov 86	f	San Jose	Santa Clara	31 Mar 87	376	332
28 Nov 86	m	San Luis Obispo	San Luis Obispo	29 Apr 87	127	330
4. Summerland, Santa Barbara County (N = 2)						
3 Dec 85	f	Pismo State Beach	San Luis Obispo	20 Feb 86	167	309
3 Dec 85	f	La Puente	Los Angeles	26 Feb 86	167	106
5. Carpinteria, Santa Barbara County (N = 1)						
14 Jan 87	m	Chino	Riverside	20 Apr 87	180	103
6. Harmon Barranca, Arrundell Barranca, and Camino Real Park (closely adjacent sites in Ventura, Ventura County [N = 16])						
Harmon Barranca						
12 Nov 86	f	Carpinteria	Santa Barbara	19 Mar 87	32	296

Table 2. Continued.

Mark–release event		Recapture event				
Date	Butter-fly sex	Location	County	Date	Distance flown (km)	Compass heading (°)
Arrundell Barranca						
4 Dec 86	m	Canyon Country	Los Angeles	7 Mar 86	64	73
4 Dec 86	f	Fillmore	Ventura	1 Apr 87	37	69
2 Jan 87	f	Santa Paula	Ventura	10 Mar 87	21	67
2 Jan 87	f	El Segundo	Los Angeles	17 Mar 87	91	118
19 Jan 87	f	Placentia	Orange	1 Apr 87	144	173
19 Jan 87	f	Laguna Hills	Orange	15 Feb 87	162	121
19 Jan 87	m	Santa Barbara	Santa Barbara	16 Apr 87	47	290
19 Jan 87	m	Ojai	Ventura	8 Jun 87	22	3
20 Jan 87	m	Glendora	Los Angeles	4 Apr 87	139	98
Camino Real Park						
2 Jan 87	m	Little Sycamore Canyon	Ventura	20 Jan 87	40	138
2 Jan 87	m	Fresno	Fresno	7 Mar 87	286	349
2 Jan 87	m	Santa Barbara	Santa Barbara	26 Mar 87	47	290
2 Jan 87	m	Santa Paula	Ventura	1 Apr 87	21	67
2 Jan 87	m	Johnson Valley	San Bernardino	9 Apr 87	261	85
2 Jan 87	m	Ojai	Ventura	2 May 87	22	3
7. Big Sycamore Canyon, Point Mugu State Park, Ventura County (N = 9)						
26 Sep 86	m	Pacific Palisades	Los Angeles	7 Feb 87	50	97
7 Oct 86	m	Pismo State Beach	San Luis Obispo	11 Feb 87	183	308
12 Oct 86	m	Torrance	Los Angeles	7 Feb 87	74	115
12 Oct 86	m	Ventura	Ventura	26 Feb 87	35	318
13 Oct 86	m	Arroyo Grande	San Luis Obispo	14 Apr 87	180	310
24 Oct 86	f	Ventura	Ventura	28 Jan 87	35	318
5 Nov 86	m	Ventura	Ventura	25 Jan 87	35	318
15 Nov 86	m	Camarillo	Ventura	8 Jan 87	22	348
25 Nov 86	f	Ventura	Ventura	26 Jan 87	35	318
8. Little Sycamore Canyon, Ventura County (N = 18)						
26 Dec 85	m	Paradise Cove	Los Angeles	20 Apr 86	19	103
15 Nov 86	f	West Covina	Los Angeles	15 Feb 87	102	89
15 Nov 86	m	Newbury Park	Ventura	20 Apr 87	16	37
15 Nov 86	f	Malibu	Los Angeles	4 Feb 87	32	93
19 Nov 86	f	Oxnard	Ventura	24 Feb 87	28	307
19 Nov 86	m	Burbank	Los Angeles	1 Feb 87	68	79
21 Nov 86	f	Carpinteria	Santa Barbara	14 Feb 87	60	308
21 Nov 86	f	Carpinteria	Santa Barbara	14 Feb 87	60	308
27 Nov 86	m	Beaumont	Riverside	4 Apr 87	192	94
14 Dec 86	f	Santa Barbara	Santa Barbara	10 Apr 87	74	300
20 Dec 86	f	Oxnard	Ventura	17 Feb 87	28	307
20 Dec 86	f	Oxnard	Ventura	17 Feb 87	28	307
20 Dec 86	m	Camarillo	Ventura	19 Feb 87	26	348
20 Dec 86	m	Camarillo	Ventura	19 Feb 87	26	348
20 Dec 86	m	Santa Paula	Ventura	19 Feb 87	32	346
17 Jan 87	f	Santa Paula	Ventura	8 Feb 87	32	346
20 Jan 87	f	Santa Paula	Ventura	8 Feb 87	32	346
20 Jan 87	m	Malibu	Los Angeles	2 Mar 87	32	93
9. Leo Carillo State Beach, Los Angeles County (N = 4)						
26 Dec 85	m	Santa Cruz Island	Santa Barbara	5 Feb 86	91	268
26 Dec 85	m	Santa Cruz Island	Santa Barbara	5 Feb 86	73	267
26 Dec 85	m	Ventura	Ventura	20 Mar 86	44	303
2 Nov 86	f	Ventura	Ventura	1 Feb 87	44	303
10. Zuma Canyon, Los Angeles County (N = 1)						
15 Nov 86	f	Long Beach	Los Angeles	11 Jan 87	71	115

Table 2. Continued.

Mark–release event		Recapture event				
Date	Butter-fly sex	Location	County	Date	Distance flown (km)	Compass heading (°)
11. Malibu Canyon, Malibu, Los Angeles County (N = 8)						
26 Dec 86	m	Getty Museum	Los Angeles	22 Feb 87	11	92
26 Dec 86	m	Gaviota State Beach	Santa Barbara	10 Mar 87	135	288
26 Dec 86	m	Moreno Valley	Riverside	"spring" 87[a]	147	95
27 Dec 86	f	Woodland Hills	Los Angeles	31 Mar 87	15	29
27 Dec 86	m	Thousand Oaks	Ventura	3 May 87	19	305
28 Dec 86	f	Fillmore	Ventura	22 Feb 87	54	333
28 Dec 86	m	Tustin	Orange	3 Mar 87	90	112
28 Dec 86	m	Zuma Canyon	Los Angeles	13 Mar 87	12	266
12. Rustic Canyon, Los Angeles County (N = 3)						
16 Nov 86	m	Long Beach	Los Angeles	5 Feb 87	46	121
30 Nov 86	f	San Bernardino	San Bernardino	14 Feb 87	123	86
16 Dec 86	f	2 km NW of Little Sycamore Canyon	Ventura	1 Mar 87	46	278
13. Santa Monica, Los Angeles County (N = 1)						
16 Mar 87	m	Temple City	Los Angeles	28 Mar 87	42	76
14. Ballona Wetlands, Los Angeles County (N = 1)						
22 Dec 85	m	Harbor City	Los Angeles	4 Mar 86	25	148

[a] Data incomplete or unrecorded.

wooded overwintering habitat of mark sites was considered to satisfy generally accepted definitions of migration and migratory behavior (Kennedy, 1961, 1985; Johnson, 1969; Southwood, 1977; Baker, 1978; Dingle, 1980; Rogers, 1983; Taylor and Taylor, 1983).

METHODS

During the winters of 1985–1986 and 1986–1987, >50,000 monarch butterflies were captured and tagged at 62 overwintering sites in southern and central California out of the 129 colonies verified state-wide (Sakai et al., 1989). Recaptured monarchs that moved >10 km were recorded from only 14 overwintering areas and are listed in Tables 1 and 2 and are shown in Figure 1. The 14 groups of sites ranged from Morro Bay in San Luis Obispo County in the north, on the coast at latitude 35°21′N, to Ballona Creek, on Santa Monica Bay (Los Angles County) in the south, at latitude 33°52′N. Monarchs were collected randomly in long-handled nets from clusters at varying heights (7.5 cm to 25 m) within overwintering aggregations, or they were netted while basking or flying. Butterflies were tagged on the right forewing with self-adhesive paper labels, 6 × 10 mm in size, printed by Walter Goodman, Jr., of Dallas, Texas. Each label was printed with instructions to return the specimen to the Natural History Museum of Los Angeles County and with a code number, according to the method of Urquhart (1960, 1987). The locality, date, sex, and label number were recorded for each butterfly, and immediately after tagging, butterflies were either returned to clusters on tree branches or allowed to fly free within the overwintering habitat. All recaptured monarchs and original data have been deposited in the Migratory Monarch Special Collection at the Natural History Museum of Los Angeles County.

Both linear and circular statistical analyses of the data followed Batschelet (1981) and Zar (1984) with levels of significance set at 5%. Tests for normality of linear data used the Kolmogorov-Smirnov one-sample *D* statistic. Although tagged monarchs that moved <10 km were also recaptured, this paper only considers recaptures of butterflies that moved >10 km from their mark site.

RESULTS

A total of 50,771 monarchs were tagged and released at the 14 overwintering locations during 251 visits between November 1985 and March 1986 and between September 1986 and March 1987 (Table 1 and Fig. 1). One hundred of these, which moved >10 km, were recovered in the late winter or spring of 1986 and 1987. Thus, 0.20% of the tagged monarchs were recaptured after having moved >10 km from their mark–release sites (Tables 1 and 2 and Fig. 1). Most recaptures (26) had been marked at Ellwood (site 3, Tables 1 and 2), followed by 18 at Little Sycamore (site 8), 16 at Ventura (three locations in the town were grouped as site 6 of Tables 1 and 2 and Fig. 1), 9 at Big Sycamore (site 7), and 8 at Malibu (site 11). Only five or fewer butterflies marked at the remaining

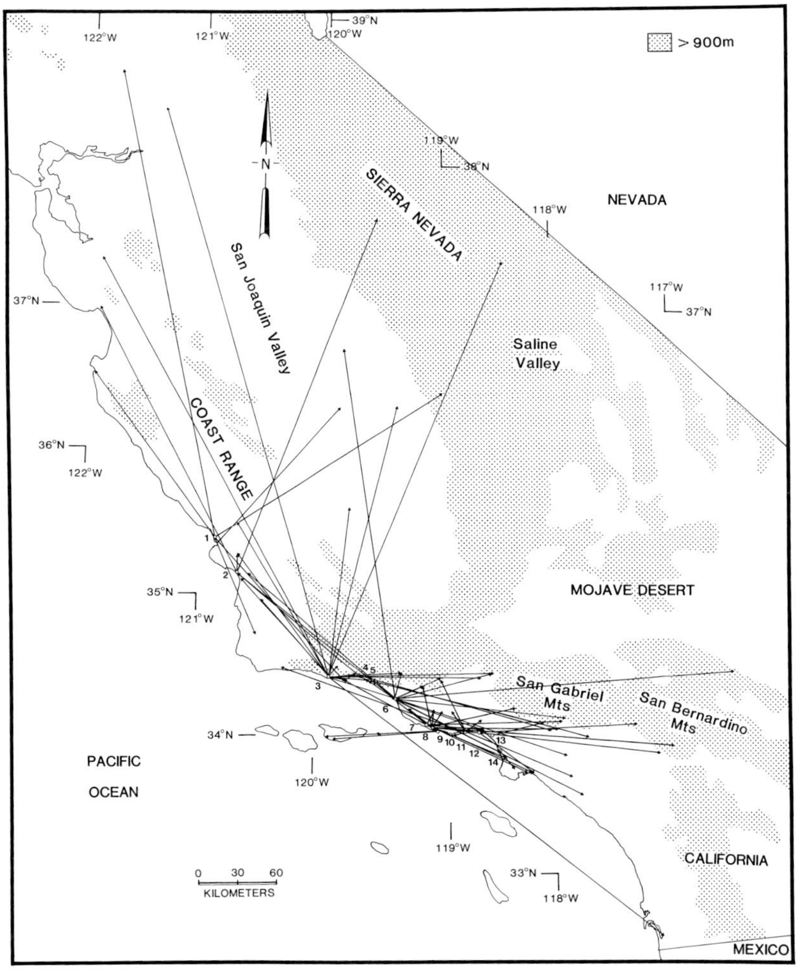

Figure 1. Distribution of the 14 numbered mark–release sites in California (see Tables 1 and 2) with lines drawn between release and recapture points for 100 monarchs. Arrowheads represent the recapture locations. Stippled areas represent principal mountain ranges >900 m.

sites were recaptured. The data include information on overall angular headings, distances traveled by recaptured monarchs, and dates of recapture.

ANGULAR HEADINGS

The headings of butterflies from mark–release sites were distinctly bimodal, with most recaptures either to the northwest or east (Fig. 2). Since the two heading distributions separate naturally at ≷40° on a von Mises plot (Fig. 3), we followed Batschelet (1981) and separated the data into two normal distributions (<40°, D = 0.11, $p = 0.21$; >40°, D = 0.12, $p = 0.20$). Headings of 266–37° were predominantly northwesterly, and those of 44–173° were predominantly easterly (Table 3A).

No recaptures were made of butterflies that may have moved between headings of 173 and 266°. Despite two seaward recaptures on Santa Cruz Island (Table 2 and Fig. 1) in the Channel Islands, more than one-quarter of possible compass headings were not represented by monarch recaptures. This quadrant of 93° is oriented southwest toward the sea, and the lack of seaward recaptures may reflect either the absence of observers or that monarchs do not fly out to sea from the coast.

For the two normal distributions, which spanned a clockwise arc from 266 to 173°, all 54 northwesterly headings were oriented at a mean angle of 326.7° and the remaining 46 easterly headings showed a mean angle of 99.2° (Table 3A). The difference between these mean angles is thus 132.5°. The two heading distributions were both significantly nonrandom (Rayleigh test; Table 3A) and they were significantly different from each other (Watson-Williams two-sample test; Table 3B). Similarly, all male and female headings, within each of the two heading directions, were significantly nonrandom (Table 3A). However, there were no sig-

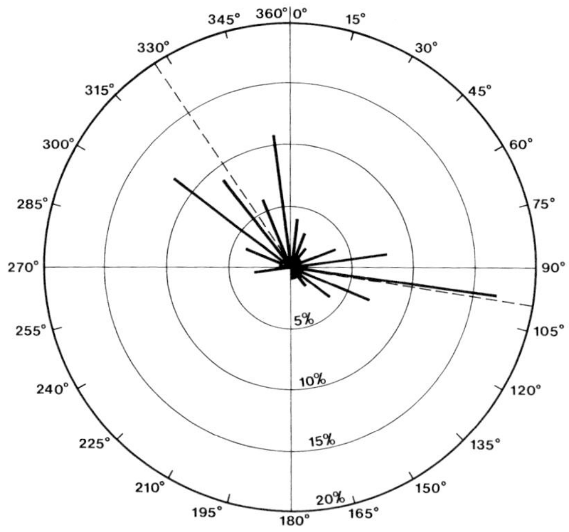

Figure 2. Angular frequency distribution (%) of release–recapture headings for the 100 monarchs listed in Table 2 and illustrated in Figure 1. The dashed lines represent the mean angular headings of 326.7° for northwesterly headings and 99.2° for easterly headings (Table 3). Angular intervals are set at 15°.

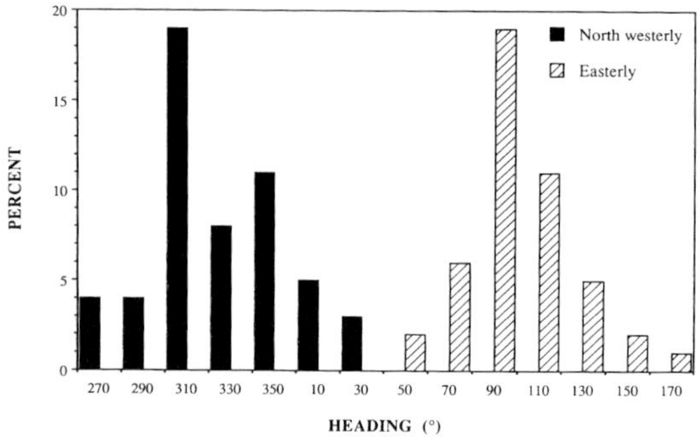

Figure 3. von Mises frequency distribution of the 100 recaptured monarch headings. The data separate naturally at 40° into two distributions: northwesterly and easterly recaptures. No headings were recorded between 173 and 266°. Both distributions were normal with a mean angle of 326.7° (s = 30.2°, N = 54), over a range of 266–37° for northwesterly headings, and a mean angle of 99.2°. (s = 23.8°, N = 46), over a range of 44–173° for easterly headings.

nificant differences between male and female northwesterly headings, or between male and female easterly headings (Table 3B).

Since 77 of the 100 recaptures were marked at just 5 of the 14 sites, the heading data for the recaptured monarchs that departed from these 5 sites are shown in Table 4A. There were no differences between the numbers of northwesterly and easterly recaptures derived from the five sites (Table 4A; contingency table $\chi^2 = 5.84$, df = 4, $p = 0.21$ NS). There were also no significant differences among the headings of either northwesterly or easterly recaptures derived from these five sites (Table 4B). Northwesterly headings of monarchs from the five sites had mean angles that ranged between 313 and 339°, and easterly headings ranged between 92 and 106° (Table 4A), like those of the entire recapture sample, at 327 and 99°, respectively (Table 3A).

DISTANCES TRAVELED AND DATES OF RECAPTURE

For all recaptured monarchs (N = 100) the mean distance traveled was 99.8 km (SD = 99.7, range = 11–465 km), but the data were heavily skewed to the right with a high proportion of short recapture distances recorded (Fig. 4), resulting in a significantly non-normal distribution (D = 0.187, $p = 0.03$). Males (N = 54) and females (N = 46) showed no significant difference in the distances traveled before recapture (Mann–Whitney U test, $Z = 1.15$, NS). There were no significant differences in the distances traveled by recaptured monarchs among the five sites that generated most recaptures (Table 5; Kruskal-Wallis ranked ANOVA, $H_4 = 8.9$, $p = 0.10 > 0.05$ NS).

If monarchs do migrate away from overwintering sites, their recapture distances should be related to the dates of recapture, with longest distances recorded for the latest recaptures. Such a trend is illustrated in Figure 5. Because the distance data were not normally distributed, they were logarithmically transformed to a distribution that did not differ significantly from normality (D = 0.099, $p = 0.16$ NS). The regression of transformed distances against recapture date (days since 1 January)—*ln distance = 0.007(recapture date) + 3.70*—was significant ($r = 0.20$, $F_{95} = 4.08$, $p = 0.046$).

In California, monarchs start to leave their overwintering sites in January, and by the beginning of March most monarchs have dispersed from their dense overwintering aggregations, although some sites last until the end of March (Tuskes and Brower, 1978). At the colonies that persist so late, nearly all the remaining butterflies are males. Before 1 March monarchs moved a mean of 69.7 km (SE = 8.1, N = 39) over a range of 11–183 km (Fig. 5). After this date they almost doubled their mean recapture distances to 120.8 km (SE = 16.0, N = 57) over a greater range of distances from 12 to 465 km (Fig. 5). Monarchs grouped into distance categories of <50 km, 50–200 km, and >200 km (Fig. 6) showed significant differences in the distances moved before and after 1 March (contingency table $\chi^2 = 9.35$, df = 2, $p = 0.009$). Before 1 March, 50.0% of monarchs moved <50 km, and after this date, 44.8% of monarchs moved <50 km (Fig. 6). Thus, more monarchs move short distances of <50 km before 1 March than those recaptured after this date. However, after 1 March, fewer monarchs (34.5%) moved intermediate distances of 50–

Table 3. A. Mean angular headings of 100 recaptured monarchs separated into northwesterly (266–37°) and easterly (44–173°) distributions and by sex. The angle ranges, angular deviations, mean vectors, and the Rayleigh Z values with probabilities of randomness are also given. B. Watson-Williams two-sample tests of differences between pairs of heading groups.

A

	Northwesterly			Easterly		
Estimate	All	Male	Female	All	Male	Female
Mean angle ($\phi°$)	326.7	326.9	326.5	99.2	99.3	99.0
Angular deviation (s°)	30.2	34.3	25.4	23.8	23.0	24.8
Range (°)	266–37	266–37	278–29	44–173	44–156	55–173
Mean vector (r)	0.86	0.82	0.90	0.91	0.92	0.91
N	54	27	27	46	27	19
Rayleigh Z	40.1	18.2	22.0	38.4	22.8	15.6
p	<0.001	<0.001	<0.001	<0.001	<0.001	<0.001

B

Comparison	F	v_1	v_2	p
Northwesterly versus easterly	487.3	1	98	<0.001
Northwesterly: males versus females	0.005	1	52	>0.75 NS
Easterly: males versus females	0.010	1	44	>0.75 NS

200 km than those before 1 March (50.0%). No monarchs moved >200 km before 1 March, but 20.7% moved >200 km after 1 March. Thus, it is only after 1 March that long-distance movements were detected among recaptures (Figs. 5 and 6).

DISCUSSION

The movement of monarch butterflies away from their coastal overwintering sites in central and southern California can be described by three measurements: direction, distance, and time. For clarity we consider each of these measures in turn.

DIRECTION

The mean angular headings of 326.7 and 99.2° flown by recaptured monarchs (Table 3A) and the lack of a difference between the numbers of monarchs that headed in these directions show that equal numbers of monarchs left their 14 overwintering sites in either northwesterly or easterly directions.

Table 4. A. Mean angular headings ($\phi°$) of the 77 recaptured monarchs tagged at the five mark–release sites that produced most recaptures. The data for each site are separated into northwesterly (266–37°) and easterly (44–173°) distributions. The angular deviations (s°), mean vectors (r), and Rayleigh Z values with probabilities (p) of randomness are also given. Sites are numbered as in Tables 1 and 2 and Figure 1. * = the sample size is too small to determine significance from tables of r or Z. B. Watson-Williams multisample tests among headings of recaptured monarchs that were tagged at the five mark–release sites that produced most recaptures. Comparisons are separated into northwesterly and easterly headings.

A

	Northwesterly						Easterly					
Site	N	$\phi°$	s°	r	Z	p	N	$\phi°$	s°	r	Z	p
3. Ellwood	12	339	22	0.93	10.3	<0.001	14	98	10	0.98	13.5	<0.001
6. Ventura	6	325	33	0.83	4.2	0.009	10	99	44	0.70	4.9	0.005
7. Big Sycamore	7	320	12	0.98	6.7	<0.001	2	106	9	0.99	2.0	*
8. Little Sycamore	12	330	27	0.89	9.5	<0.001	6	92	7	0.99	5.9	<0.001
11. Malibu	5	313	40	0.75	2.8	0.05	3	100	9	0.99	2.9	*

B

Five-site comparison	F	v_1	v_2	p
Among northwesterly recaptures	0.90	4	37	>0.25 NS
Among easterly recaptures	0.22	4	30	>0.75 NS

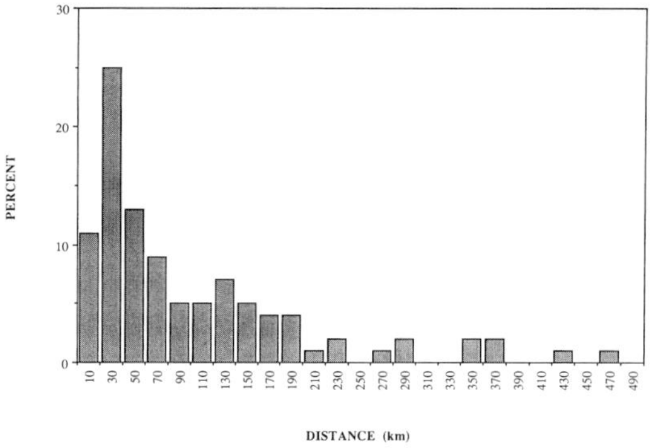

Figure 4. Percent frequency distribution of the distances moved by the 100 recaptured monarchs of Figure 1 and Table 2. Distances are the mid-points of 20-km intervals from 0 to 500 km.

Table 5. Mean distances traveled by the 77 recaptured monarchs tagged at the five mark-release sites that produced most recaptures. Mark sites are ranked from north to south, and the site numbers correspond to those in Tables 1 and 2 and Figure 1. Distances traveled do not vary significantly among the five sites (Kruskal-Wallis ANOVA, $H_4 = 8.9$, $p = 0.10 > 0.05$ NS).

	Distance (km)			
Site	N	Mean	SD	Range
3. Ellwood	26	140.9	131.9	14–465
6. Ventura	16	89.8	85.6	21–286
7. Big Sycamore	9	72.1	63.7	22–183
8. Little Sycamore	18	49.3	42.2	16–192
11. Malibu	8	60.4	56.7	11–147

The significant consistency of these two headings for males, females, and butterflies from the five sites that generated most recaptures (Tables 3B and 4B) refutes any possibility that these monarchs left their mark–release sites in random directions. Since recaptures were not aligned with coastal orientation, it is most likely that the northwesterly orientation of the Coast Range of mountains and the easterly orientation of the Transverse Range of mountains (including the San Gabriel and San Bernardino mountains) influence the directionality of recaptured monarchs in spring (Fig. 1). Nevertheless, monarchs may also use the coastline orientation as a cue for short distance movements during the overwintering period to locate additional, wooded overwintering sites.

Although it is apparent from the distribution of recaptures in Figure 1 that monarchs can fly through the Coast and Transverse ranges, most early spring migrants will move initially into the foothills of these mountains to locate milkweed host plants (see Wenner and Harris, this volume). Subsequent movement by overwintered butterflies and their offspring could continue the migration to exploit milkweeds throughout California, including the Sierra Nevada, and further east into the Rocky Mountains and north into Oregon. This hypothesis is supported by the first observations of spring monarchs in these areas.

The northwesterly and easterly headings do not include any seaward recaptures over a southwesterly quadrant of 93° from 173 to 266°. Despite the presence of a number of islands at accessible distances from overwintering sites, only two recaptures were made on these islands (Fig. 1 and Table 2). This may indicate a lack of observers to return recaptures or that monarchs do in fact avoid seaward headings during spring migration. Monarchs have been observed flying over the Santa Barbara channel and the San Pedro channel to and from the Channel Islands each autumn from 1985 to 1988. However, selection against seaward movement is likely to be significant, but it would be interesting to mark and release the monarchs that overwinter on the islands (Nagano and Lane, 1985; Sakai et al., 1989; Wenner and Harris, this volume) to determine whether or not they will fly to the mainland to exploit milkweeds in spring.

The bimodal headings also exclude long-distance recaptures through a northeasterly arc of approximately 60°, which coincides with the Mojave Desert (Fig. 1). The absence of tagged monarchs may reflect a lack of observers, rather than avoidance of the region by spring migrants. With more data it could be useful to compare the distribution and timing of recaptures with a random movement model that incorporated predictions based on the distribution of potential observers in all directions and distances from overwintering sites. Nevertheless, spring migrants that may enter the Mojave Desert probably do not linger. Instead, they are likely to move rapidly across the region because temperatures would become lethally hot for any offspring, before completion of larval development (Rawlins and Lederhouse, 1981; Zalucki, 1981, 1982; Malcolm et al., 1987). Although monarchs have been observed breeding in the low desert during the winter (Funk, 1968), it is unlikely that they will exploit the xeric milkweeds characteristic of the Mojave Desert (Woodson, 1954) in the hotter months of the year.

DISTANCE

The frequency distribution of distances moved by the recaptured butterflies is strongly skewed (Fig. 4), with many recaptures made after comparatively short movements—despite our consideration of data for recaptures of >10 km only. The fact that more short distances were traveled at the start of the breakup of the overwintering colonies than after the main breakup of the colonies, when all long-distance recaptures were made (Figs. 5 and 6), indicates that monarchs do migrate away from the coastal sites. This shows that monarchs do move

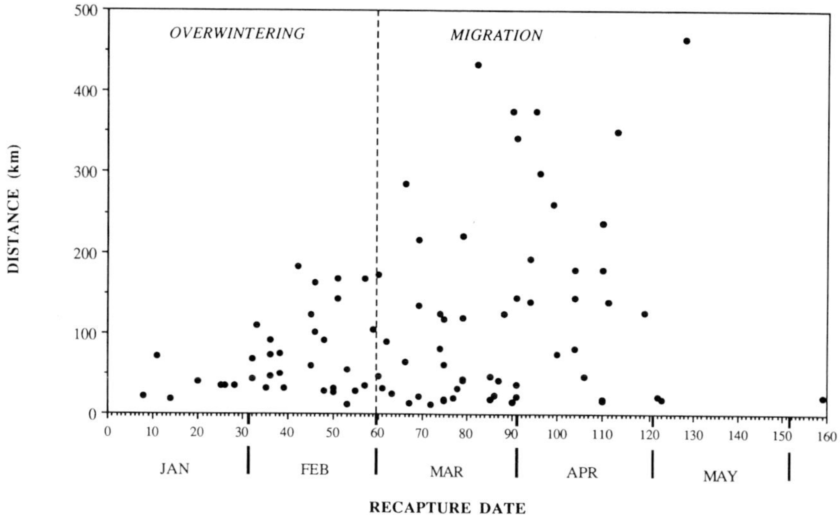

Figure 5. Scatterplot of straight-line, release–recapture distances against recapture date (since 1 January) for the 100 monarchs listed in Table 2 and illustrated in Figure 1. The dashed line at 1 March represents the transition from overwintering aggregations to migration away from the overwintering sites. Movements of >200 km all occurred after 1 March.

during the overwintering period, as well as among overwintering sites in response to changing environmental conditions or poor quality overwintering habitat. Such movements suggest that the colonies should not be viewed as discrete populations. Instead, numerous adjacent sites could form a *metacolony* much like the metapopulation concept of Levins (1970) discussed by Hanski (1989). Thus, the dynamics of monarch movement among the sites encompassed within a metacolony could be an important determinant of overwintering success. Future analyses of our short-distance (<10 km) recapture data should illustrate the extent of intercolony movements during the overwintering period and help to define the dynamics, extent, and duration of the metacolonies.

TIME

The distribution of recaptures with time suggests that January is the end of the overwintering period when the first short-distance movements were made away from overwintering sites (Fig. 5). During February an increasing number of longer-distance movements were recorded, followed by increasing numbers of long-distance recaptures as monarchs dispersed from colony sites after 1 March. Although local, short-distance movement from overwintering roosts to nearby breeding habitat is a likely explanation for the high frequency of shorter distance recaptures, these movements are not random with respect to direction. If the movements are between different habitats and not within a generalized life-history range, then such movements are consistent with migration, just as are long-distance, directed movements (Kennedy, 1961, 1985; Johnson, 1969; Southwood, 1977; Rogers, 1983; Taylor and Taylor, 1983).

SUMMARY

Our spring recapture data show that monarchs leave overwintering sites to move into the Coast Range north of central California and into the San Gabriel and San Bernardino mountains, of the Transverse Range, northeast and east of Los Angeles (Fig. 1). Longer-distance recaptures show movements into the San Joaquin Valley of central California and farther east into the Sierra Nevada. It is also possible that both the Coast Range and the Transverse Range

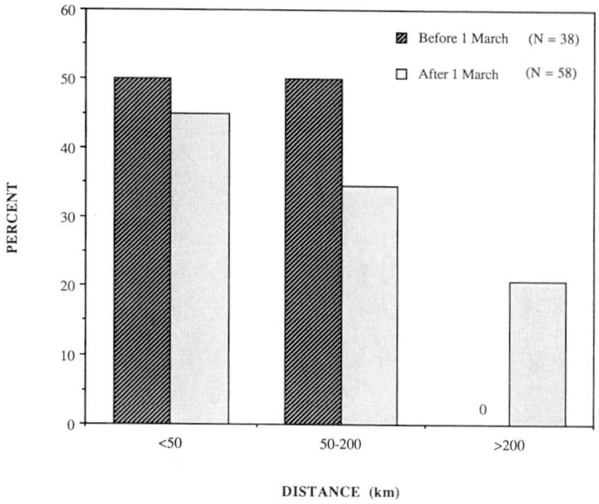

Figure 6. Percentages of release–recapture distances for short- (<50 km), medium- (50–200 km), and long- (>200 km) distance monarch recaptures before and after 1 March.

(San Gabriel and San Bernardino mountains) channel monarchs around the Mojave Desert to account for the absence of recaptures in the desert. The desert may in fact be a comparatively cool and benign environment in early spring. However, it is unlikely that spring monarchs will linger in such habitats and produce offspring.

Earlier recaptures show that, like the distance data, the directional data were dominated by local, short-distance orientations along the coast at the end of the overwintering period. Only after 1 March, when the majority of overwintering colonies have broken up, did monarchs show orientations away from the coastline to migrate inland. Thus, we interpret both the distances and directions flown by monarchs, recaptured after 1 March, as evidence that monarchs migrate inland from coastal overwintering aggregations in search of breeding habitat. As Harris (1986) and Wenner and Harris (this volume) describe, this breeding habitat is characterized in California by the presence of early-spring milkweeds. Although little is known about the life histories of monarchs in California and the timing of breeding in relation to geographical location west of the Rocky Mountains, our data are fully consistent with the interpretation that monarchs do migrate inland each spring away from overwintering habitat along the coast of California.

ACKNOWLEDGMENTS

We gratefully acknowledge the use of facilities at the Natural History Museum of Los Angeles County and the assistance and cooperation there of Charles L. Hogue, Mary Ann Dunn, Brian P. Harris, Roy R. Snelling, William Powder, Sharon Belkin, and Craig C. Black. Richard and Louise Fall of BioQuip Products, Inc., provided field equipment, and the following people assisted with tagging monarchs and associated fieldwork: Diane Ackerman, Sharon Belkin, Rose Bernaldo, Robert Brownell, David Browning, Slader Buck, Mike and Cherie Chudler, Kathy Donahue, Mary Ann Dunn, Jill Eikenberry, Arthur and Linda Evans, Christyann and Gary Evans, Chris Fall, Louise Fall, Richard Fall, Josh Golden, Phil Goldfarb, Nancy Gorman, Jason Ho, Fiona and Cambria Leigh, Jane Marshall, Momo Nagano, Naomi Nagano, Otome Nagano, Eddie Olmedo, Juventino Ortiz III, Kathy Ralls, Cindy Reeves, Vicki Sakai, Liza Salvatore, Santa Monica College Ecology Club (1985–1987), Jan and Christine Schrumpf, Nancy Schuler-Jones, Richard Simpson, Laurie Singer, and Gary Wolfe. Permission to work on lands administered by the California Department of Parks and Recreation was provided by Gary Fregien. Tim Thomas gave permits for the Santa Monica Mountains National Recreation Area, and Richard Idler allowed access to a colony in southern California. Transportation and housing on Santa Cruz Island was provided by the University of California Natural Reserves System. Support is also gratefully acknowledged from the Natural History Museum of Los Angeles County, the California Department of Parks and Recreation, the World Wildlife Fund–US, and especially Otome and Naomi Nagano. We also thank the many people who have taken the time and effort to return tagged monarchs. The helpful review comments of Thomas Walker, William Calvert, and James Anderson are also gratefully acknowledged. We dedicate this paper to two individuals who have been instrumental in conserving the western North American population of the monarch butterfly: Walter Goodman, Jr., of Dallas, Texas, who supplied 300,000 tags for marking monarchs; and California Assemblyman Byron Sher (D–Palo Alto), who is working to ensure that the State provides the protection necessary for the long-term survival of the monarchs' wintering colonies.

LITERATURE CITED

Baker, R.R. 1978. *The evolutionary ecology of animal migration.* London: Hodder and Stoughton, 1012 pp.

Barker, J.F., and W.S. Herman. 1976. Effect of photoperiod and temperature on reproduction of the monarch butterfly *Danaus plexippus. Journal of Insect Physiology* 22:1565–1568.

Batschelet, E. 1981. *Circular statistics in biology.* London: Academic Press, 371 pp.

Beall, G. 1948. The fat content of a butterfly *Danaus plexippus* Linn. as affected by migration. *Ecology* 29:80–94.

Brower, L.P. 1985. New perspectives on the migration biology of the monarch butterfly, *Danaus plexippus* L. In *Migration: Mechanisms and adaptive significance,* ed. M.A. Rankin, 748–785. Austin: University of Texas, *Contributions in Marine Science,* Supplement, 27.

Brower, L.P., and C.M. Moffitt. 1974. Palatability dynamics of cardenolides in the monarch butterfly. *Nature* 249:280–283.

Brower, L.P., J.N. Seiber, C.J. Nelson, S.P. Lynch, M.P. Hoggard, and J.A. Cohen. 1984. Plant-determined variation in cardenolide content and thin-layer chromatography profiles of monarch butterflies, *Danaus plexippus* reared on milkweed plants in California 3. *Asclepias californica. Journal of Chemical Ecology* 10:1823–1857.

Calvert, W.H., and L.P. Brower. 1986. The location of monarch butterfly (*Danaus plexippus* L.) overwintering colonies in Mexico in relation to topography and climate. *Journal of the Lepidopterists' Society* 40:164–187.

Chaplin, S.B., and P.H. Wells. 1982. Energy reserves and metabolic expenditure of monarch butterflies overwintering in southern California. *Ecological Entomology* 7:249–256.

Dingle, H. 1980. Ecology and evolution of migration. In *Animal migration, orientation, and navigation,* ed. S.A. Gauthreaux, Jr., 1–101. New York: Academic Press.

Funk, R.S. 1968. Overwintering of monarch butterflies as a breeding colony in southwestern Arizona. *Journal of the Lepidopterists' Society* 22:63–64.

Hanski, I. 1989. Metapopulation dynamics: Does it help to have more of the same? *Trends in Ecology and Evolution* 4:113–114.

Harris, A. 1986. Natural history of the monarch butterfly (*Danaus plexippus*) in the Santa Barbara area. M.A. thesis. Santa Barbara: University of California, 33 pp.
Herman, W.S. 1973. The endocrine basis of reproductive inactivity in monarch butterflies overwintering in central California. *Journal of Insect Physiology* 19:1883–1887.
Herman, W.S. 1975. Endocrine regulation of post-eclosion enlargement of male and female reproductive glands in monarch butterflies. *General and Comparative Endocrinology* 26:534–540.
Herman, W.S. 1981. Studies on the adult reproductive diapause of the monarch butterfly, *Danaus plexippus. Biological Bulletin* 160:89–106.
Herman, W.S. 1985. Hormonally mediated events in adult monarch butterflies. In *Migration: Mechanisms and adaptive significance,* ed. M.A. Rankin, 799–815. Austin: University of Texas, *Contributions in Marine Science,* Supplement, 27.
Herman, W.S., and J.F. Barker. 1977. Effect of mating on monarch butterfly oogenesis. *Experientia* 33:688.
Herman, W.S., L.P. Brower, and W.H. Calvert. 1989. Reproductive tract development in monarch butterflies overwintering in California and Mexico. *Journal of the Lepidopterists' Society* 43:50–58.
Hill, H.F., Jr., A.M. Wenner, and P.H. Wells. 1976. Reproductive behavior in an overwintering aggregation of monarch butterflies. *American Midland Naturalist* 95:10–19.
Johnson, C.G. 1969. *Migration and dispersal of insects by flight.* London: Methuen, 763 pp.
Kammer, A.E. 1970. Thoracic temperature, shivering, and flight in the monarch, *Danaus plexippus* (L.) *Zeitschrift für Vergleichende Physiologie* 68:334–344.
Kennedy, J. 1961. A turning point in the study of insect migration. *Nature* 189:785–791.
Kennedy, J. 1985. Migration, behavioral and ecological. In *Migration: Mechanisms and adaptive significance,* ed. M.A. Rankin, 5–26. Austin: University of Texas, *Contributions in Marine Science,* Supplement, 27.
Levins, R. 1970. Extinction. *Lectures on Mathematics in the Life Sciences* 2:75–107.
Malcolm, S.B. 1987. Monarch butterfly migration in North America: Controversy and conservation. *Trends in Ecology and Evolution* 2:135–138.
Malcolm, S.B., and L.P. Brower. 1989. Evolutionary and ecological implications of cardenolide sequestration in the monarch butterfly. *Experientia* 45:284–295.
Malcolm, S.B., B.J. Cockrell, and L.P. Brower. 1987. Monarch butterfly voltinism: Effects of temperature constraints at different latitudes. *Oikos* 49:77–82.
Nagano, C.D., and J. Lane. 1985. A survey of the location of monarch butterfly (*Danaus plexippus* [L.]) overwintering roosts in the state of California, U.S.A.: First year 1984/85. Report to the World Wildlife Fund–US, 106 pp.
Nagano, C.D., and W.H. Sakai. 1989. The monarch butterfly. In *Audubon Wildlife Report 1989/1990,* eds. W.J. Chandler and L. Labate, 367–385. San Diego: Academic Press.
Rawlins, J.E., and R.C. Lederhouse. 1981. Developmental influences of thermal behavior on monarch caterpillars (*Danaus plexippus*): An adaptation for migration (Lepidoptera: Nymphalidae: Danainae). *Journal of the Kansas Entomological Society* 54: 387–408.
Roeske, C.N., J.N. Seiber, L.P. Brower, and C.M. Moffitt. 1976. Milkweed cardenolides and their comparative processing by monarch butterflies (*Danaus plexippus* L.). *Recent Advances in Phytochemistry* 10: 93–167.
Rogers, D. 1983. Pattern and process in large-scale animal movement. In *The ecology of animal movement,* eds. I.A. Swingland and P.J. Greenwood, 160–180. Oxford: Clarendon Press.
Sakai, W.H., J. Lane, A.V. Evans, J. Schrumpf, and M. Monroe. 1989. The wintering colonies of the monarch butterfly (*Danaus plexippus* (L.): Nymphalidae: Lepidoptera) in the state of California, U.S.A. *California Department of Fish and Game Contract Report FG7551.* Sacramento, California, 38 pp.
Seiber, J.N., L.P. Brower, S.M. Lee, M.M. McChesney, H.T.A. Cheung, C.J. Nelson, and T.R. Watson. 1986. Cardenolide connection between overwintering monarch butterflies from Mexico and their larval food plant, *Asclepias syriaca. Journal of Chemical Ecology* 12:1157–1170.
Southwood, T.R.E. 1977. Habitat, the templet for ecological strategies? *Journal of Animal Ecology* 46: 337–365.
Taylor, L.R., and R.A.J. Taylor. 1983. Insect migration as a paradigm for survival by movement. In *The ecology of animal movement,* eds. I.A. Swingland and P.J. Greenwood, 181–214. Oxford: Clarendon Press.
Tuskes, P.M., and L.P. Brower. 1978. Overwintering ecology of the monarch butterfly, *Danaus plexippus* L., in California, *Ecological Entomology* 3:141–153.
Urquhart, F.A. 1960. *The monarch butterfly.* Toronto: University of Toronto Press, 361 pp.
Urquhart, F.A. 1966. A study of the migrations of the Gulf Coast population of the monarch butterfly (*Danaus plexippus* L.) in North America. *Annales Zoologici Fennici* 3:82–87.
Urquhart, F.A. 1976. Found at last: The monarch's winter home. *National Geographic* 150:161–173.
Urquhart, F.A. 1987. *The monarch butterfly: International traveler.* Chicago: Nelson-Hall, 232 pp.
Urquhart, F.A., and N.R. Urquhart. 1976a. Migration of butterflies along the gulf coast of northern Florida. *Journal of the Lepidopterists' Society* 30:59–61.
Urquhart, F.A., and N.R. Urquhart. 1976b. The overwintering site of the eastern population of the monarch butterfly (*Danaus p. plexippus:* Danaidae) in southern Mexico. *Journal of the Lepidopterists' Society* 30:153–158.
Urquhart, F.A., and N.R. Urquhart. 1976c. A study of the peninsular Florida populations of the monarch butterfly (*Danaus p. plexippus:* Danaidae). *Journal of the Lepidopterists' Society* 30:73–87.
Urquhart, F.A., and N.R. Urquhart. 1977. Overwintering areas and migratory routes of the eastern pop-

ulation of the monarch butterfly (*Danaus p. plexippus*, Lepidoptera: Danaidae) in North America, with special reference to the western population. *Canadian Entomologist* 109:1583–1589.

Urquhart, F.A., and N.R. Urquhart. 1978. Autumnal migration routes of the eastern population of the monarch butterfly (*Danaus p. plexippus* L.; Danaidae; Lepidoptera) in North America to the overwintering site in the neovolcanic plateau of Mexico. *Canadian Journal of Zoology* 56:1759–1764.

Urquhart, F.A., and N.R. Urquhart. 1979a. Breeding areas and overnight roosting locations in the northern range of the monarch butterfly (*Danaus plexippus plexippus*) with a summary of associated migratory routes. *Canadian Field Naturalist* 93:41–47.

Urquhart, F.A., and N.R. Urquhart. 1979b. Vernal migration of the monarch butterfly (*Danaus p. plexippus*, Lepidoptera: Danaidae) in North America from the overwintering site in the neo-volcanic plateau of Mexico. *Canadian Entomologist* 111:15–18.

Urquhart, F.A., P. Beard, and R. Brownlee. 1965. A population study of a hibernal roosting colony of the monarch butterfly (*D. plexippus*) in northern California. *Journal of Research on the Lepidoptera* 4:221–226.

Urquhart, F.A., N.R. Urquhart, and F. Munger. 1970. A study of a continuously breeding population of *Danaus plexippus* in southern California compared to a migratory population and its significance in the study of insect movement. *Journal of Research on the Lepidoptera* 7:169–181.

Williams, C.B. 1930. *The migration of butterflies*. Edinburgh: Oliver and Boyd, 473 pp.

Williams, C.B., G.F. Cockbill, M.E. Gibbs, and J.A. Downes. 1942. The migrations and reproductive cycle of the monarch butterfly in California. *Transactions of the Royal Entomological Society of London* 92:160–165.

Woodson, R.E., Jr. 1954. The North American species of *Asclepias* L. *Annals of the Missouri Botanical Garden* 41:1–211.

Zalucki, M.P. 1981. The effects of age and weather on egg laying in *Danaus plexippus* L. (Lepidoptera: Danaidae). *Researches in Population Ecology* 23: 318–327.

Zalucki, M.P. 1982. Temperature and rate of development in *Danaus plexippus* L. and *D. chrysippus* L. (Lepidoptera: Nymphalidae). *Journal of the Australian Entomological Society* 21:241–246.

Zar, J.H. 1984. *Biostatistical analysis*, 2nd ed. Englewood Cliffs, NJ: Prentice-Hall, 718 pp.

TIME, TEMPERATURE, AND LATITUDINAL CONSTRAINTS ON THE ANNUAL RECOLONIZATION OF EASTERN NORTH AMERICA BY THE MONARCH BUTTERFLY

Barbara J. Cockrell, Stephen B. Malcolm, and Lincoln P. Brower

ABSTRACT. Monarch butterflies, *Danaus plexippus* (L.), remigrating into the eastern United States from Mexican overwintering sites may deposit all their eggs on the early-spring milkweeds they encounter in the southern United States. Their offspring then migrate farther north to recolonize later and more northern milkweed resources. Alternatively, overwintered monarchs may deposit only some eggs in the south and continue migrating northward in a "single sweep," depositing eggs on route. These two recolonization strategies have different implications for adult migratory behavior and the number of monarch generations produced annually.

Based on temperature data and generation times expressed in day-degrees, we predict that five monarch generations may be produced per season, originating from the spring migrant monarchs returning from Mexico. If the first two generations mature in the region of latitude 30°N, there is time for three more generations to occur farther north. In contrast, populations of apparently nonmigratory monarchs breed for most of the year in south Florida and can complete at least seven generations.

We censused immature monarchs in the central and eastern United States, from the Gulf Coast to northern Minnesota in 1985, 1986, and 1987. Using day-degrees, determined from the daily maximum and minimum temperatures recorded near to each of the 62 sites at which immatures were found, we back-calculated to when the first (most mature) individual was laid, and also to when the main cohort of immature monarchs was oviposited at each site. We used these dates to describe the temporal distribution of ovipositing female monarchs with latitude. We also determined how long it would take both categories of immatures at each site to complete their development and reach adult maturity. This allowed us to assess the effects of both latitude and oviposition date on maturation time.

Monarch maturation time was extended by cooler temperatures at more northerly latitudes but was reduced by later oviposition because of seasonal warming. Thus, the first individuals oviposited at sites had longer maturation times than the subsequently oviposited main cohort. Over the entire south to north distance, which monarchs recolonized at an average rate of 24 km/day, the effect of latitude outweighed oviposition date, resulting in longer maturation times in the north. At sites south of 41°40′N, a relatively constant generation time of 44 days for the first individuals occurred at a recolonization rate of 17 km/day, and a generation time of 40 days for the main cohort was maintained by a recolonization rate of 20 km/day. Recolonization north of 41°40′N was unpredictable, with no significant relationship between latitude and oviposition date. Maturation times of the immatures in this region were longer and more variable than farther south.

The spatial deposition of eggs before, compared with after, 1 May (the earliest calculated maturation date of the first spring generation) and our observations of adult wing wear indicate that the worn, overwintered Mexican spring remigrants deposit most of their eggs south of latitude 35°30′N. Most oviposition farther north is effected from mid-May by the mature adults of this first generation. The main colonization of the region north of 41°40′N occurs after mid-June. Hence, we propose that Mexican remigrants are the main colonizers south of 35°30′N, that first-generation monarchs are the initial colonizers of the central and northern regions, and that they are subsequently joined in the north by later generations. Thus, monarchs recolonize North America by successive broods of migrants and not by a single sweep of overwintered migrants.

INTRODUCTION

Monarch butterflies migrate south to Mexico each autumn and escape both a predictable gap in their larval milkweed food resources and freezing temperatures (Urquhart, 1960; Brower, 1985). As milkweeds become available again in the spring, the same monarch individuals return to exploit them. Compared with our knowledge of the autumn migration (Urquhart, 1960, 1966, 1976, 1987; Urquhart and Urquhart, 1976a,b,c, 1977, 1978, 1979a,b; Brower, 1985; Gibo, 1986; Malcolm, 1987; Schmidt-Koenig, this volume), we know little of the spring remigration (Brower, 1985; Lynch and Martin, this volume; Malcolm et al., this volume; Riley, this volume). In part, this is because monarchs generally occur singly or in only small groups in the spring (Williams, 1930; Fales, 1984) instead of in the large aggregations that characterize their autumn migration. This difference reflects the different destinations and selection pressures on the migrations in these two seasons. While the autumn migrants must reach a relatively small, well defined area as quickly as possible to escape cold temperatures and the absence of larval food resources, the spring remigrants approach a resurgent milkweed resource distributed over much of the continent (Woodson, 1954). The availability of both individuals and species of these milkweeds changes in time and space throughout the spring and summer.

Two alternative spring recolonization strategies have been suggested (Brower, 1985; Malcolm et al., this volume). Remigrant female monarch butterflies from Mexico may deposit all their eggs on the early-spring milkweeds they encounter in the southern United States. Their successive broods of offspring would then migrate farther north to recolonize later, more northern milkweeds. Alternatively, overwintered remigrant monarchs may deposit only some eggs in the south and continue migrating north, ovipositing on route, eventually to cover the entire recolonizable range. In this paper we describe the spatial and temporal patterns of monarch larval distribution in the eastern United States and relate our results and their ecological implications to the different migratory strategies that have been proposed.

Urquhart used a mark–release–recapture technique to describe monarch migration (Urquhart, 1987), but after 50 years this has yielded little published data on the spring remigration since so few marked butterflies have been recaptured (Malcolm, 1987). Here we use a new approach, recording the presence of the offspring of migrant adults as an assay of monarch movement, rather than attempting to intercept the migrating adults themselves. This method simplifies the problems of encountering migrants in space and time, because immature monarchs are confined to milkweed food plants and so far have been recorded feeding on only 27 North American *Asclepias* species (Malcolm and Brower, 1986). The rate of monarch larval development seems not to be significantly influenced by the host-plant species (Erickson, 1973; see also Malcolm et al., 1987) but it is temperature-dependent (Rawlins and Lederhouse, 1981; Zalucki, 1982). We can back-calculate to when an immature was oviposited and, hence, a female monarch was at a site, from published information of the number of day-degrees required to complete each developmental stage (Zalucki, 1982) and local daily maximum and minimum temperatures prior to a census date (Allen, 1976). Thus, throughout the egg and larval development period, which although temperature-dependent generally takes at least 2 wk, the immatures provide an opportunity to gather information about the former presence of ovipositing females. A series of observations of larvae can build a picture of the monarch's overall spring recolonization pattern through time and space. The technique can only census surviving immatures and so it is a conservative description of the movements of ovipositing females. Nevertheless, in terms of migratory success we are most interested in surviving offspring, and we think this method provides useful insights into the spring remigration, especially when used in concert with other methods such as the reproductive status and chemical fingerprints of captured adult spring migrants (see Malcolm et al., this volume).

Migration allows effective exploitation of the total milkweed resource, but it incurs significant energetic, fecundity, and mortality costs (Baker, 1978; Inglesfield and Begon, 1983), which must be balanced by reproductive benefits (Kennedy, 1961, 1985; Dingle, 1980; Rogers, 1983; Taylor and Taylor, 1983). Selection will favor those monarchs that most efficiently use the dynamic milkweed resource to maximize their production of offspring (Southwood, 1977). Spring monarchs must arrive at a latitude sufficiently early for their offspring to exploit the newly available food resource, but not so early that these immatures are killed, or have overly extended generation times resulting from cool spring temperatures. We used maturation time and the number of successfully completed generations of offspring to evaluate different migratory strategies. Monarch growth and reproduction are mediated by both temperature and daylength (Barker and Herman, 1976; Rawlins and Lederhouse, 1981; Zalucki, 1981, 1982; Herman, 1985; James, this volume). We determined the number and duration of monarch generations at five latitudes by dividing the number of day-degrees available for reproduction by the day-degree duration of a monarch generation. We also determined the expected time for field-censused immatures to reach maturity from minimum and maximum field temperatures recorded after the census date. We compared maturation time of the first monarch eggs laid at a particular latitude with that of the main cohort at that latitude, allowing comparison of generation times among sites and also between early and later arrivals at sites.

Because of the absence of adequate information concerning the changing biotic environment within the recolonizable range, we only made descriptive reference to factors such as the dynamics of food-plant availability and quality, and predator and competitor numbers. Thus, in this paper we evaluate migration strategies primarily in relation to abiotic constraints.

METHODS

VOLTINISM MODEL PREDICTIONS AT FIVE LATITUDES

We calculated the expected duration and number of monarch generations in 1981, 1982, and 1983 at five sites, each separated by approximately 5°N latitude, together spanning latitudes 25–45°N, from southern Florida to northern Wisconsin (Fig. 1a). The sites were (A) Flamingo, Florida (25°10′N), (B) Gainesville, Florida (29°40′N), (C) Asheville, North Carolina (35°30′N), (D) Dayton, Ohio (39°55′N), and (E) St. Croix Falls, Wisconsin (45°25′N). We defined the "window" of time when monarch larvae might be present at each site for each of the 3 yr by relating the number of daylight hours (Beck, 1980) and the daily maximum and minimum temperatures recorded at each site in 1981, 1982, and 1983 [published by the National Oceanic and Atmospheric Administration (NOAA), U.S. Environmental Data Service, National Climatic Center, Asheville, North Carolina] to experimentally determined temperature and daylength constraints on monarch growth and reproduction. Monarch larval development is arrested at ambient temperatures <12°C and >30°C (Rawlins and Lederhouse, 1981; Zalucki, 1982). Temperature-dependent growth is resumed when temperatures return within this range, but temperatures between 31 and 35.5°C are lethal to immature monarchs (Rawlins and Lederhouse, 1981; Zalucki, 1982). Oocyte production is completely inhibited at <15°C (Barker and Herman, 1976), while photophases of <12 hr light lower juvenile hormone titers and inhibit mature oocyte production (Barker and Herman, 1976; Herman, 1985). We included this daylength constraint because, although day lengths of <12 hr do not completely suppress oocyte production, reproduction is significantly reduced. Furthermore, we are principally concerned with the options available to monarchs repopulating the United States from Mexico, and the butterflies usually remain at the Mexican overwintering colonies until mid-March (Calvert and Brower, 1986; Calvert and Lawton, this volume; Van Hook, this volume), with monarch remigrants rarely recorded in the United States before 21 March (Malcolm et al., 1987; Lynch and Martin, this volume; Riley, this volume) when daylight is 12 hr or more in the Northern Hemisphere (Beck, 1980).

At each site the abiotic window ended either when maximum temperatures of <15°C inhibited egg production, marking the end of the breeding season, or when persistent daily maximum temperatures <12°C, or >30°C, permanently arrested larval development. We calculated the number of day-degrees available for larval growth at each site for each year, using Allen's (1976) modified sine-wave method on the daily minimum and maximum temperatures recorded at each site. This day-degree total was then divided by the 397 day-degrees monarchs require to develop from an egg to a reproductively mature adult [352 day-degrees from egg to adult stage (Zalucki, 1982) and 45 for adult maturation (Zalucki, 1981)]. We did not attempt to estimate possible reduction in generation times due to insolation through larval basking (Rawlins and Lederhouse, 1981) because this behavior is so variable among individuals and weather conditions. Thus, we determined the number of generations that could be completed within the favorable window of time available at each latitude in each year. In the absence of basking effects, our calculations are liable to overestimate generation times, but this is counteracted by our determination of generation time to the onset, rather than the peak of reproductive maturity.

EVIDENCE FROM FIELD CENSUSED LARVAE

During the spring and summer of 1985, 1986, and 1987, we counted monarch eggs and larvae on milkweeds at 93 sites at different latitudes, across the eastern United States between longitudes 70° and 105°W (Fig. 1). In 1985 we made a southern transect between these longitudes in April and a northern transect in May and June. In 1986 we made an east–west southern transect from Florida to central Texas during April, a south–north transect around longitude 98°W to northern Minnesota during May, and a west–east transect to eastern Michigan in June. In 1987 we made an eastern transect from north-central Florida to Cape Cod, Massachusetts and then westward to Lake Erie. In addition, spot checks were made in Florida, Louisiana, Texas, and Wisconsin throughout these seasons. At each site we determined the milkweed species that were present and their growth and flowering status and searched either all the available milkweeds or a subsample of between 100 and 1,000 plants. We found immature monarchs in 62 of the 93 sites (Fig. 1 and Tables 1 and 2) and obtained the daily minimum and maximum temperatures recorded during the appropriate year at a NOAA weather station near each site (see Table 2). Applying Allen's (1976) day-degree calculation method to the temperatures recorded prior to the census at each site, together with Zalucki's (1981, 1982) values for the day-degree duration of each instar, we determined for each site (a) the date when the most mature larva originated as an egg, i.e. when the first immature was oviposited at the site, and (b) the date at which the main cohort (predominant immature stage) was oviposited (with the qualification that at least two immature stages must be present for one to predominate). By implication (a) indicates when the first and (b) when the main influx of monarchs recolonized the site, although in the absence of available data these estimates necessarily exclude effects produced by differential mortality between instars. We also calculated when the first individual and the main cohort of immature monarchs would be expected to become reproductively mature, using the daily maximum and minimum temperatures for the period following the census date. Where possible, we collected adults from the census sites and determined their wing wear on a 0.5-interval scale from 1 (perfect) to 5 (extremely worn). Wing wear and cardenolide analyses of some of the 1985 butterfly samples are given in Malcolm et al. (this volume).

RESULTS

VOLTINISM MODEL PREDICTIONS AT FIVE LATITUDES

Spring remigrant monarchs could lay eggs at Flamingo, Gainesville, and Asheville from 21 March

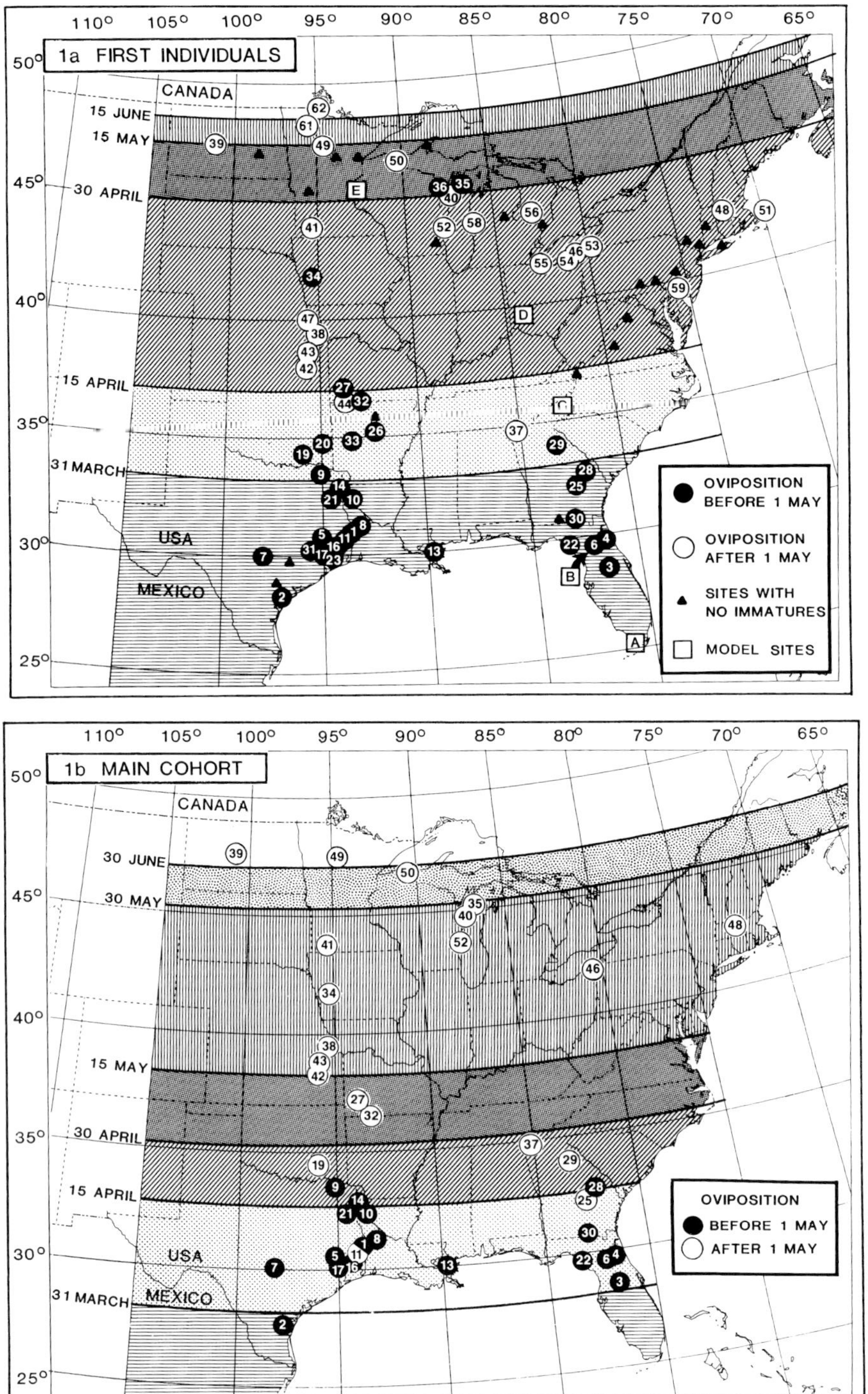

Figure 1. Distribution of larval census sites across the eastern United States in 1985, 1986, and 1987. The northern limits of recolonization with time by (**a**) the first individuals and (**b**) the main cohorts are indicated by dated latitudinal lines. The site numbers represent the chronological sequence of first oviposition dates and correspond with the sample numbers in Tables 1 and 2. (Where a site was sampled in successive years, the lowest sample number is shown.) Certain sites in (**a**) are not represented in (**b**) because there was no main cohort at the time of sampling. Solid circles show where larvae were oviposited before 1 May (probably by Mexican remigrants), and open circles show those oviposited from 1 May (predominantly by first-generation migrants). Sites at which no immatures were found at the time of sampling are shown by solid triangles in (**a**). The locations of the five sites we selected to predict the pattern of monarch voltinism within abiotic constraints are marked as A–E in (**a**).

(Fig. 2). Farther north, maximum daily temperatures <15°C preclude egg production before the third week of April at Dayton and the beginning of May at St. Croix Falls. Figure 2 shows the average outcome of the 3 yr at each site. There was considerable consistency between years, despite minor year-to-year variation. Maturation of the first generation took longer at more northerly latitudes. Over all the sites, generation times decreased with rising summer temperatures and lengthened again at the end of the season (Fig. 2). From the onset of oviposition, the model predicted seven generations at Flamingo, two at Gainesville, four at Asheville, three at Dayton, and three at St. Croix Falls. The predictions do not show a gradual trend from many generations in the south to few generations in the north primarily because of the onset of lethally high summer temperatures in the south at Gainesville in Florida (Malcolm et al., 1987). At Flamingo, on the southern tip of Florida, the cooling influence of the sea maintains summer temperatures lower than those of Gainesville, and because temperatures remain warm and within the upper and lower growth-inhibiting extremes, many generations of short duration are completed. There may even be year-round breeding of a nonmigratory population at this site (Brower, 1961, 1985; Urquhart and Urquhart, 1976c; Malcolm, 1987; see also James, this volume; Wenner and Harris, this volume).

At the two most northern sites, the model indicated that oviposition of a fourth generation would be sometimes inhibited by daylengths <12 hr. When oviposition could occur, completion of this fourth generation would be generally curtailed by persistent maximum daily temperatures of <12°C arresting larval development. Many of these individuals would eventually be killed by the first autumn frosts (Young, 1980).

EVIDENCE FROM FIELD-CENSUSED LARVAE

The locations and dates of our field samples are shown in Table 1, together with the milkweed species at each site and its flowering status, the number of monarch immatures at each stage of development, and the number of worn and fresh adults captured at each location. Plant size and the flowering data indicate that the development of the northern milkweed, *Asclepias syriaca,* is considerably later than that of the southern species, *Asclepias humistrata, Asclepias viridis,* and *Asclepias asperula. Asclepias syriaca* was still prebud in May when seed pods of *A. humistrata* were recorded.

Worn adult monarchs were seen throughout the southern states from the end of March to mid-May (Table 1). Our earliest siting of a new first spring generation monarch was on 21 April 1986 near New Waverly in east Texas, and three more first-generation monarchs were captured on 28 April at Bonham in northeast Texas (Table 1). From this time new monarchs (wing wear <3.5) increased in the adult population until they consistently predominated by the end of the second week in May at sites in Missouri and Kansas (Table 1). Cardenolide analyses (Malcolm et al., this volume) indicate that worn individuals captured at the end of June in Wisconsin (Table 1) are aging first-generation monarchs and not overwintered remigrants. The last worn monarch that had probably overwintered in Mexico was recorded on 22 May in central Iowa (41°30′N; see footnote c, Table 1).

Recolonization Rate

The calculated dates of oviposition and the number of days required for the first and for the main cohort of immatures to reach adult reproductive maturity at each of the censused sites are ranked in chronological order in Table 2. The earliest estimated oviposition date was 15 March at Lakeview, Texas, in 1986. The data at this and three other sites (one in Texas, two in Florida) show that some monarchs may re-enter the United States and start egg laying almost a week earlier than March 21—the date we set for the onset of oviposition in our model based on day-length requirements. The calculated oviposition dates of both the first eggs and the main cohort changed steadily with latitude as shown in Figure 1 and Table 2. Oviposition had occurred as far north as latitude 33°30′ by the end of March, 37°00′ by mid-April, 45°00′ by the end of April, 47°00′ by mid-May, and 48°30′ by mid-June. The main cohort were oviposited at these latitudes approximately 15 days later. The latitudinal lines in Figure 1a show the extreme northernmost limits of recolonization with time, indicating the northward progress of the first adult monarch arrivals. Those in Figure 1b show the northernmost position of the main cohort with time, indicating the northward movement of the majority of recolonizing adults.

There was a significant positive relation between latitude and oviposition date both for the first and for the main cohort of immatures (Fig. 3a and 3b). These predictable relationships between time and the distance covered by ovipositing females is evidence for directed northward migration. It was not surprising to find this result for the first immatures at sites, because they are the offspring of the earliest migrants to reach each site and we would expect these adults to take progressively longer to reach more northerly latitudes (Fig. 3a). By contrast, there is no intrinsic reason to expect oviposition of the main cohort to follow such a predictable pattern. The butterflies that deposited these immatures are not constrained to continue moving north consistently, because milkweeds are simultaneously available for oviposition over a broad area. Furthermore, there was no significant difference in the migration rates of the two groups, as measured by the slopes of the relationships (t_{101} = 0.32, NS). In both cases recolonization rate was approximately 0.21°N/day, or 24 km/day (where 1° latitude = 111 km). The difference in the elevations (t_{101} =

Table 1. **Date, location, and latitude of the 62 sample sites with immatures and the 31 sites at which no immatures were found. Sample numbers correspond with Table 2 and Figure 1. The number of immatures at each stage, the number of milkweeds sampled, the milkweed species, and flowering status are given. The number of old (wing wear >3.5) and new (wing wear <3.5) adults collected at sites on the same day as the immatures were censused is given. Host plant species: amp = *A. amplexicaulis,* asp = *A. asperula,* hir = *A. hirtella,* hum = *A. humistrata,* long = *A. longifolia,* quad = *A. quadrifolia,* spe = *A. speciosa,* syr = *A. syriaca,* var = *A. variegata,* vir = *A. viridis,* o = prebud, b = bud, f = flower, p = pods; + signifies presence but no actual count was made; ? = plant number was not recorded; NF = National Forest.**

				Host plant		
Sample date	Sample no.	Location	Latitude	N	Species	Status
Immature monarchs present						
04 April 1986	3	Tavares, Florida	28°49′	?	hum	b, f
04 April 1986	6	Gainesville, Florida	29°40′	141	hum	o
05 April 1985	13	New Orleans, Louisiana	30°00′	50	vir	b, f
06 April 1985	15	Livingston, Texas	30°42′	50	vir	b, f
06 April 1986	18	Kisatchie NF, Louisiana	31°32′	28	amp	o
07 April 1985	2	Refugio Co., Texas	28°40′	?	asp	b
08 April 1986	1	Lakeview, Texas	31°10′	44	vir	o
09 April 1985	12	Gainesville, Florida	29°40′	241	hum	f
09 April 1986	4	Melrose, Florida	29°42′	126	hum	b, f
09 April 1986	10	Shreveport, Louisiana	32°30′	50	vir	b, f
10 April 1986	5	Livingston, Texas	30°42′	35	vir	f
12 April 1986	23	Dayton, Texas	30°03′	30	vir	b
16 April 1986	7	Pedernales, Texas	30°17′	30	asp	f
17 April 1986	31	Anderson, Texas	30°32′	40	vir	f
18 April 1986	17	New Waverly, Texas	30°33′	25	vir	b, f
19 April 1986	16	S. Houston NF, Texas	30°30′	28	var	b, f
20 April 1985	8	Kisatchie NF, Louisiana	31°32′	67	vir	f
21 April 1986[b]	17	New Waverly, Texas	30°33′	—	—	—
23 April 1986	11	Angelina NF, Texas	31°10′	32	amp	b
26 April 1986	14	McLeod, Texas	33°00′	8	amp	f
27 April 1985	22	Steinhatchee, Florida	29°40′	225	hum	b, f
27 April 1986	21	Lindale, Texas	32°31′	20	vir	b
28 April 1986	9	Bonham, Texas	33°36′	21	vir	b
30 April 1985	24	Shreveport, Louisiana	32°30′	?	vir	b, f
03 May 1986	19	Boggy Depot, Oklahoma	34°23′	50	amp	b, f
05 May 1986	20	Arrowhead, Oklahoma	35°07′	65	vir	f
07 May 1986	25	McCrae, Georgia	32°06′	142	hum	f
07 May 1986	30	Statenville, Georgia	30°51′	148	hum	p
07 May 1986	33	Y City, Arkansas	35°08′	14	vir	b, f
08 May 1986	28	Kite, Georgia	32°41′	177	hum	f
09 May 1986	26	Centerville, Arkansas	35°09′	74	vir	b, f
09 May 1986	29	Athens, Georgia	33°57′	36	amp	b, f
10 May 1986	37	Chicamauga, Georgia	34°52′	31	vir	f
12 May 1986	32	Table Rock, Missouri	36°34′	77	vir	b, f
13 May 1986	44	Neosho, Missouri	36°53′	83	hir	o, b
13 May 1986	27	Carthage, Missouri	37°12′	253	?	?
15 May 1986	42	Flint Hills, Kansas	38°16′	170	syr	o
16 May 1986	43	Reading, Kansas	38°49′	100	vir	b, f
19 May 1986	38	Atchison, Kansas	39°33′	217	syr	o
21 May 1986	47	Indian Cave, Nebraska	40°14′	30	syr	o
23 May 1986[c]	34	Prairie Rose, Iowa	41°39′	104	syr	b
26 May 1986	41	Kilen Woods, Minnesota	43°40′	130	syr	o
28 May 1986	51	Cape Cod, Massachusetts	41°56′	372	syr	o, b
30 May 1985	53	Presque Isle, Pennsylvania	42°07′	?	syr	o, b
31 May 1985	46	Ashtabula, Ohio	41°53′	15	syr	o
31 May 1985	54	Geneva, Ohio	41°54′	?	syr	o
31 May 1986	48	Amherst, Massachusetts	42°23′	335	syr	b
01 June 1985	55	Port Clinton, Ohio	41°30′	?	syr	o, b
01 June 1986	36	Beaver, Wisconsin	45°07′	78	syr	o
02 June 1985	56	Port Austin, Michigan	44°04′	?	syr	o, b

Table 1. Extended.

Immature stage						Adults	
E	1	2	3	4	5	Old	New
12	2	1	4	0	0	1	0
22	10	8	4	1	0	0	0
22	28	0	0	0	0	0	0
94	22	0	0	0	0	4	0
36	0	0	0	0	0	1	0
0	0	0	3	0	1	7	0
40	29	7	17	11	1	62	0
55	8	1	0	0	0	9	0
9	2	0	0	0	2	1	0
133	41	21	3	0	0	12	0
5	0	8	7	15	13	34	0
3	0	0	0	0	0	3	0
1	8	0	2	0	4	0[a]	0
4	0	0	0	0	0	0	0
6	0	0	1	0	0	14	0
0	4	1	0	1	0	4	0
8	8	9	12	4	5	5	0
—	—	—	—	—	—	6	1
7	0	0	0	0	2	0	0
20	11	0	5	0	1	0	0
6	3	3	14	8	2	0	0
9	10	5	2	0	0	0[a]	0
2	1	0	0	0	1	29	3
3	6	11	2	4	2	6	0
16	9	2	0	0	2	26	0
0	0	0	0	0	9	0	0
29	1	6	5	11	14	1	0
0	0	1	9	8	11	0	0
0	0	0	5	0	0	3	10
0	0	1	2	4	5	0	0
3	2	5	6	0	1	9	2
24	3	4	1	0	1	0	0
11	0	1	0	0	0	0	0
2	0	0	0	1	0	0	0
27	0	0	0	0	0	0	0
20	0	0	1	0	1	3	11
3	3	0	0	0	0	9	22
4	2	0	0	0	0	7	21
68	1	2	2	2	0	3	73
6	0	0	0	0	0	2	8
19	0	0	1	1	0	0	20
29	3	0	0	0	0	0	0
2	0	0	0	0	0	0	0
5	0	0	0	0	0	0	7
19	2	0	0	0	0	0	2
6	0	0	0	0	0	0	0
5	11	0	0	0	0	0	0
+	0	0	0	0	0	0	11
0	0	0	0	1	0	0	0
+	0	0	0	0	0	0	29

69.4, $p < 0.001$) indicates a later start by the main group.

Too rapid movement north may result in considerably extended maturation times (Table 2). For example, the first eggs in Newport, Wisconsin, on 28 April 1985, were estimated to require 91 days to mature, compared with 48 days for eggs deposited there 2 mo later. Over the whole data set, the modal development time of the first individuals at sites was 39 days, less than half the time taken by the first Newport individuals. In the time taken to complete just one generation at this time, this far north, two generations could have been completed farther south.

Predicted Recolonization Rate for a Constant Maturation Time of 40 Days. The modal development time of the main cohort (Table 2) was 40 days (compared with 39 days for first individuals). We calculated the earliest predicted oviposition date from which a monarch could mature in this time period at each site in Table 2 using the maximum and minimum temperature data recorded at the weather stations nearby. In 10 cases, all north of 44°N, and several on the south shores of the Great Lakes, temperatures were too low for maturation to occur this rapidly. Latitude is plotted against predicted oviposition date for the remaining 52 sites in Figure 4. The slope of this relationship indicates that by maintaining a constant recolonization rate of 14 km/day, ovipositing females could theoretically maintain a maturation time of 40 days between latitudes 28°40′ and 44°N.

Maturation Time against Latitude and Oviposition Date

First Individuals. The number of days the first individuals at sites took to mature is shown against oviposition date in Figure 5a, against latitude in Figure 5b, and against both these variables together in Figure 5c. Multiple regression analysis indicates that maturation time was significantly extended at higher latitudes and reduced by later oviposition [days to mature = 3.12 latitude − 0.42 oviposition date − 46.7, $p = 0.0001$, with both independent variables significant (Table 3a)]. The standardized beta values shown in Table 3 express the impact of the independent variables (latitude and temperature) on the dependent variable (maturation time) in terms of standard deviations. In this way they eliminate the differences in measurement scale of latitude and oviposition date, allowing a direct comparison of their effects on maturation time. Over all the sites, an increase in one standard deviation of latitude increased maturation time by 1.59 times its standard deviation, while an increase in one standard deviation of oviposition date decreased maturation time by 0.91 times its standard deviation (Table 3a). Thus, over this range of sites and oviposition dates, the maturation time of the first individuals at sites was increased more by colder, more northerly latitudes than it was reduced by later oviposition.

Table 1. Continued.

Sample date	Sample no.	Location	Latitude	Host plant N	Species	Status
02 June 1986	40	Potawatomi, Wisconsin	44°51′	104	syr	o, b
02 June 1986	45	Newport, Wisconsin	45°18′	104	syr	b
03 June 1987	57	Presque Isle, Pennsylvania	42°07′	311	syr	b
04 June 1986	58	Ludington, Michigan	43°58′	104	syr	b
05 June 1986	56	Port Austin, Michigan	44°04′	216	syr	b
06 June 1986	59	Newark, Delaware	39°42′	12	syr	b
16 June 1985	39	Hensler, North Dakota	47°10′	100	syr	o, b
19 June 1985	61	Beltrami Isle, Minnesota	48°42′	?	syr	b
19 June 1985	62	Warroad, Minnesota	48°54′	?	spe	b
20 June 1985	49	Bemidji, Minnesota	47°29′	202	syr	b
21 June 1985	50	Ashland, Wisconsin	46°34′	90	syr	b
24 June 1985	35	Newport, Wisconsin	45°18′	100	syr	b
25 June 1985	52	Harrington Bch., Wisconsin	43°23′	100	syr	o
Immature monarchs absent						
23 March 1985		Gainesville, Florida	29°40′	285	hum	b
05 April 1985		Steinhatchee, Florida	29°40′	141	hum	b
14 April 1986		Weimer, Texas	29°42′	25	vir	f
15 April 1985		Cuero, Texas	29°10′	25	asp	b
16 April 1986		Valdosta, Georgia	30°51′	8	hum	b
06 May 1986		Ouachita NF, Oklahoma	34°48′	33	quad	f
11 May 1986		Rotary Ann, Arkansas	35°41′	50	quad	b, f
18 May 1986		Norfolk, Connecticut	42°00′	?	syr	o
20 May 1986		Saukville, Wisconsin	43°07′	50	syr	o
23 May 1985		Kisatchie NF, Louisiana	31°32′	94	long	f
26 May 1987		Draper, Virginia	37°02′	68	syr	b
26 May 1987		Lexington, Virginia	37°47′	40	syr	b
27 May 1986		Sauk Center, Minnesota	45°43′	104	syr	o
28 May 1986		Bemidji, Minnesota	47°23′	55	syr	o
28 May 1987		Front Royal, Virginia	38°56′	554	syr	b
29 May 1986		Bena, Minnesota	47°18′	26	syr	o
29 May 1986		Proctor, Minnesota	46°45′	50	syr	o
29 May 1987		Harrisburg, Pennsylvania	40°18′	152	syr	o
29 May 1987		Bethel, Pennsylvania	40°28′	55	syr	b
29 May 1987		Allentown, Pennsylvania	40°37′	55	syr	b
29 May 1987		Hainesberg, New Jersey	40°57′	35	syr	b
29 May 1986		Houghton, Michigan	47°06′	104	syr	o
30 May 1987		Orient, New York	41°09′	167	syr	b
31 May 1987		Fishkill, New York	41°32′	220	syr	b
01 June 1987		New Paltz, New York	41°41′	800	syr	b
03 June 1985		Port Austin, Michigan	44°04′	50	syr	b
04 June 1986		N. Bradley, Michigan	43°42′	80	syr	b
05 June 1987		Amherst, Massachusetts	42°23′	367	syr	b
07 June 1987		Bethel, Pennsylvania	40°28′	498	syr	b
07 June 1987		Front Royal, Virginia	38°56′	214	syr	b
13 June 1985		Valley City, North Dakota	46°50′	60	spe	b

[a] At least one adult of the indicated category was seen but not captured.

[b] The first observation of a new-generation monarch. Immatures were censused 3 days previously but not recensused.

[c] The sample taken at this site on 22 May 1986 included one worn, old monarch with 43 new monarchs.

Main Cohort. The data for maturation time of the main cohort is shown against oviposition date in Figure 6a, against latitude in Figure 6b and against both variables in Figure 6c. Multiple regression showed that both latitude and oviposition date were significant (Table 3a). Interestingly, the magnitudes of the standardized beta values closely correspond with those for the first individuals at sites. As before, increasing latitude had a greater influence on increasing maturation time than increasing oviposition date had in decreasing it. Thus, even though oviposition was generally later at more northerly

Table 1. Extended, continued.

Immature stage						Adults	
E	1	2	3	4	5	Old	New
3	0	0	1	0	0	0	1
2	2	5	0	0	0	0	0
6	0	0	0	0	0	0	0[a]
4	0	0	0	0	0	0	0
7	0	0	0	0	0	0	11
10	0	0	0	0	0	0	0
3	0	0	0	0	2	0	3
10	2	0	0	0	0	0	4
+	0	0	0	0	0	0	2
70	69	11	2	0	0	9	57
43	12	29	1	0	0	13	53
20	18	9	8	5	0	50	46
0	3	0	1	1	0	0	0
0	0	0	0	0	0	0	0
0	0	0	0	0	0	0	0
0	0	0	0	0	0	0	0
0	0	0	0	0	0	4	0
0	0	0	0	0	0	0	0
0	0	0	0	0	0	0	0
0	0	0	0	0	0	0	0[a]
0	0	0	0	0	0	0	0
0	0	0	0	0	0	0	0
0	0	0	0	0	0	0	0
0	0	0	0	0	0	0	0
0	0	0	0	0	0	0	0
0	0	0	0	0	0	0	0[a]
0	0	0	0	0	0	0	0
0	0	0	0	0	0	0	0[a]
0	0	0	0	0	0	0	0
0	0	0	0	0	0	0	0
0	0	0	0	0	0	0	0
0	0	0	0	0	0	0	0
0	0	0	0	0	0	0	0
0	0	0	0	0	0	0	0
0	0	0	0	0	0	0	0
0	0	0	0	0	0	0	0
0	0	0	0	0	0	0	0
0	0	0	0	0	0	0	0
0	0	0	0	0	0	0	50
0	0	0	0	0	0	0	1
0	0	0	0	0	0	0	0
0	0	0	0	0	0	0	0
0	0	0	0	0	0	0	0
0	0	0	0	0	0	0	0

latitudes, this did not outweigh the longer generation time farther north.

Successive Generations

Figure 1 describes the recolonization pattern of the eastern United States in 1985, 1986, and 1987, but not necessarily the movement of any individual butterfly. The entire route north to southern Canada appears not to be completed by Mexican remigrants but, instead, by subsequent spring generations. The earliest calculated date of adult maturation from our field samples was 1 May (Refugio County, Texas, in 1985; Table 1). Thus, eggs oviposited on or after this date may be the offspring of the first spring generation of monarchs rather than of Mexican remigrants. This date compares well with our earliest field observations of freshly emerged, first-generation adult monarchs made during the last week of April at two sites in east Texas (Table 1). The three females in this sample all had poorly developed ovaries. Hence, we separated the data shown in Table 2 and Figures 3, 5, and 6 into two groups: eggs laid before 1 May and eggs laid from this date. We do not suggest that all oviposition from 1 May is solely that of first generation monarchs, but our data indicate that from this date we can no longer assume their origin was Mexico.

Oviposition before 1 May. *First Individuals.* At the 36 sites the first eggs were oviposited before 1 May and were thus attributed to Mexican remigrants. Thirty-three of these sites were south of latitude 38°N (Fig. 1a and Table 2), which runs through Kansas, Missouri, Illinois, Indiana, Kentucky, and West Virginia. At the remaining three sites, oviposition occurred in the last few days of April: in central Iowa (latitude 41°39′, one egg on 27 April and one on 30 April 1986) and in northeast Wisconsin (latitude 45°07′, one egg on 30 April 1986, and latitude 45°18′, five eggs on 28 April 1985; see Table 2 and Fig. 1). The standardized beta values (Table 3b) show that over all 36 sites latitude had a greater effect on increasing maturation time than later oviposition date had on reducing it. This was reflected in the considerable range of maturation times from 36 to 91 days with a mean of 47.1 days (Table 4a).

Main Cohort. The main cohort was oviposited before 1 May at 23 sites. All were south of 35°30′N, in the southern half of the United States up to the southern borders of Kansas, Missouri, Kentucky, and Virginia (Fig. 1b and Table 2). Eggs were laid between 25 March and 27 April with a mean oviposition date of 13 April (Table 4a). The mean maturation time of 39.6 days varied little (SD = 2.6), ranging from 36 to 45 days (Table 4a). Neither the simple regressions of days to mature against latitude or oviposition date, nor the multiple regression combining these effects, were significant (Table 3b). This suggests that temperature, and therefore maturation time, was independent of date and latitude from 25 March to 27 April between latitudes 28°40′N and 35°12′N.

Oviposition from 1 May. *First Individuals.* The 26 records where oviposition of first individuals occurred on or after 1 May were all north of 34°50′N (Fig. 1a and Table 2). The number of days to mature ranged from 38 to 77 days, with a mean of 56.7 (Table 4b). Increasing latitude had a considerably greater influence on increasing maturation time than

Table 2. Estimated dates of egg laying (egg date) and the duration to adult maturity (days to mature), calculated using day-degrees (from daily minimum and maximum temperatures recorded at the NOAA weather stations shown) for the first individual and the main cohort of monarchs to occupy each field site at different latitudes across the eastern United States in 1985, 1986, and 1987. Observations are listed in the chronological order of the estimated dates that first eggs were laid at each sample site. Sample numbers correspond with those in Table 1 and Figure 1. – signifies absence of a main cohort at the time of sampling; NF = National Forest, SP = State Park.

First individual		Main cohort							
Egg date	Days to mature	Egg date	Days to mature	Year	Sample No.	Location	State	Latitude	Weather station
15 March	55	08 April	38	1986	1	Lakeview	TX	31°10′	Toledo Bend Dam
18 March	43	25 March	40	1985	2	Refugio Co.	TX	28°40′	Goliad
20 March	49	04 April	42	1986	3	Tavares	FL	28°49′	Lisbon
20 March	50	09 April	44	1986	4	Melrose	FL	29°42′	Gainesville
22 March	43	01 April	38	1986	5	Livingston	TX	30°42′	Livingston
28 March	46	05 April	43	1986	6	Gainesville	FL	29°40′	Gainesville
29 March	39	11 April	38	1986	7	Pedernales	TX	30°17′	Austin
29 March	43	08 April	38	1985	8	Kisatchie NF	LA	31°32′	Natchitoches
29 March	45	27 April	38	1986	9	Bonham	TX	33°36′	Bonham
29 March	47	09 April	45	1986	10	Shreveport	LA	32°30′	Shreveport
30 March	39	07 May	36	1986	11	Angelina NF	TX	31°10′	Sam Rayburn Dam
31 March	43	09 April	40	1985	12	Gainesville	FL	29°40′	Gainesville
01 April	40	01 April	40	1985	13	New Orleans	LA	30°00′	New Orleans
02 April	48	26 April	39	1986	14	McLeod	TX	33°00′	Jefferson
02 April	39	06 April	39	1985	15	Livingston	TX	30°42′	Livingston
04 April	39	14 April	39	1986	16	S. Houston NF	TX	30°30′	Huntsville
06 April	38	18 April	36	1986	17	New Waverly	TX	30°33′	Huntsville
06 April	39	–	–	1986	18	Kisatchie NF	LA	31°32′	Natchitoches
06 April	53	03 May	40	1986	19	Boggy Depot	OK	34°23′	Atoka Dam
06 April	54	–	–	1986	20	Arrowhead	OK	35°07′	Eufaula
07 April	46	22 April	38	1986	21	Lindale	TX	32°31′	Tyler
08 April	41	17 April	36	1986	22	Steinhatchee	FL	29°40′	Steinhatchee
12 April	37	–	–	1986	23	Dayton	TX	30°03′	Houston
13 April	41	22 April	39	1985	24	Shreveport	LA	32°30′	Shreveport
13 April	45	07 May	35	1986	25	McCrae	GA	32°06′	Eastman
13 April	48	27 April	40	1986	26	Centerville	AR	35°09′	Morrilton
13 April	58	13 May	42	1986	27	Carthage	MO	37°12′	Neosho
15 April	44	15 April	44	1986	28	Kite	GA	32°41′	Swainsboro
15 April	49	09 May	39	1986	29	Athens	GA	33°57′	Athens
17 April	41	17 April	41	1986	30	Statenville	GA	30°51′	Valdosta
17 April	36	–	–	1986	31	Anderson	TX	30°32′	Huntsville
24 April	50	12 May	43	1986	32	Table Rock SP	MO	36°34′	Neosho
27 April	40	–	–	1986	33	Y City	AR	35°08′	Morrilton
27 April	56	23 May	40	1986	34	Prairie Rose	IA	41°39′	Harlan
28 April	91	24 June	48	1985	35	Newport	WI	45°18′	Washington Island
30 April	81	–	–	1986	36	Beaver	WI	45°07′	Oconto
01 May	38	10 May	37	1986	37	Chicamauga	GA	34°52′	Chattanooga TN
01 May	46	19 May	38	1986	38	Atchison	KS	39°33′	Atchison
03 May	77	16 June	65	1985	39	Hensler	ND	47°10′	Wilton
06 May	68	02 June	51	1986	40	Potawatomi	WI	44°51′	Sturgeon Bay
09 May	64	26 May	54	1986	41	Kilen Woods	MN	43°40′	Worthington
10 May	42	15 May	41	1986	42	Flint Hills	KS	38°16′	Emporia
11 May	43	16 May	40	1986	43	Reading	KS	38°49′	Emporia
13 May	42	–	–	1986	44	Neosho	MO	36°53′	Neosho
15 May	62	15 May	62	1986	45	Newport	WI	45°18′	Washington Island
20 May	61	31 May	55	1985	46	Ashtabula	OH	41°53′	Ashtabula
21 May	40	–	–	1986	47	Indian Cave	NE	40°14′	Auburn
23 May	56	23 May	56	1986	48	Amherst	MA	42°23′	Amherst
23 May	69	20 June	54	1985	49	Bemidji	MN	47°29′	Bemidji
27 May	68	21 June	56	1985	50	Ashland	WI	46°34′	Ashland Exp. Farm

Table 2. Continued.

First individual		Main cohort							
Egg date	Days to mature	Egg date	Days to mature	Year	Sample No.	Location	State	Latitude	Weather station
28 May	55	–	–	1986	51	Cape Cod	MA	41°56′	Chatham
29 May	56	17 June	48	1985	52	Harrington Beach	WI	43°23′	Port Washington
30 May	55	–	–	1985	53	Presque Isle	PA	42°07′	Erie
31 May	55	–	–	1985	54	Geneva	OH	41°54′	Ashtabula
01 June	50	–	–	1985	55	Port Clinton	OH	41°30′	Sandusky
02 June	69	–	–	1985	56	Port Austin	MI	44°04′	Harbor Beach
02 June	55	–	–	1987	57	Presque Isle	PA	42°07′	Erie
04 June	57	–	–	1986	58	Ludington	MI	43°58′	Ludington
06 June	39	–	–	1986	59	Newark	DE	39°42′	Newark Farm
05 June	67	–	–	1986	60	Port Austin	MI	44°04′	Harbor Beach
07 June	67	–	–	1985	61	Beltrami Isle	MN	48°42′	Baudette
19 June	66	–	–	1985	62	Warroad	MN	48°54′	Warroad

later oviposition date had on reducing it (Table 3c). The multiple regression, standard deviations, and standardized beta values indicate that while the effect of latitude was little changed from that of the pre-1 May group of first individuals, the negative effect of oviposition date had considerably less influence on the maturation time of this later group.

Main Cohort. The main cohort was oviposited from 1 May at 20 sites. These lay between latitudes 31°10′ in east-central Texas and 47°29′ in north Minnesota, spanning almost the entire recolonizable range (Fig. 1b and Table 4b). The mean oviposition date was 25 May, with eggs laid from 3 May to 24 June, and the mean maturation time was 47.2 days, ranging from 35 to 65 days (Table 4b). Multiple regression indicated that oviposition date had no significant effect on maturation time of this sample (Table 3c), and the simple regression of latitude on maturation time gave practically as good a fit, with days to mature = 1.56 latitude − 16.3, $r^2 = 0.69$, $p = 0.0001$. However, we have reservations about concluding that from 1 May latitude alone and not oviposition date influences maturation time, because the data set includes the last 10 observations of an essentially uniform group of sites south of 41°40′N, with 10 observations north of this latitude (Fig. 5b). The egg stage predominated at seven of these sites, while at the remaining three, first instars predominated and only 16 or fewer immatures were present at each site. This suggests that monarch numbers were still in the process of increasing at these sites.

We subdivided both the first individual and the main cohort data sets at latitude 41°40′N, based on the separation of the data into two clusters in Figure 6b and 6c and, to a lesser extent, in Figure 5b and 5c, and because this also divided our adult samples into those that included worn monarchs compared with those that were all new (Table 1). We also subdivided the data describing the butterflies' rate of progress (Fig. 3) at latitude 41°40′N to compare recolonization rates south and north of this latitude.

Oviposition at Latitudes South of 41°40′N. *First Individuals.* This sample covered a northward distance of 1,400 km across 42 sites, ranging from south Texas (28°40′) to central Iowa (41°39′). The dates when the first eggs were oviposited at these sites ranged over 83 days from 15 March through 6 June, and the mean maturation time was 44.4 days, with a range of 22 days (Table 4c). Multiple regression analysis showed that both latitude and oviposition date had a highly significant effect on days to mature (Table 3d). The standardized beta values were of similar magnitude but opposite sign, indicating that this range of latitudes and oviposition dates exerted a comparable but opposite influence on the number of days to mature. The migration rate of 17 km/day (Table 5a) was such that the prolonging effect of northward movement on maturation was almost completely counterbalanced by the reduction due to later oviposition, thus maintaining reasonable consistency in maturation time over this broad range of space and time.

Main Cohort. There was a similar effect in the 33 incidences where the main cohort was recorded at latitudes south of 41°40′N, although here variation in maturation time was even lower. The latitudinal range and mean were similar to those of the first individuals, while the range in oviposition date was 59 days, with the mean on 22 April (Table 4c). Again, multiple regression analysis showed that maturation time was significantly increased at more northerly latitudes, and significantly reduced by later oviposition date (Table 3d). Although there was comparability in the standard deviations of latitude and oviposition date between the main cohort and the first individuals (Table 4c), the standardized beta

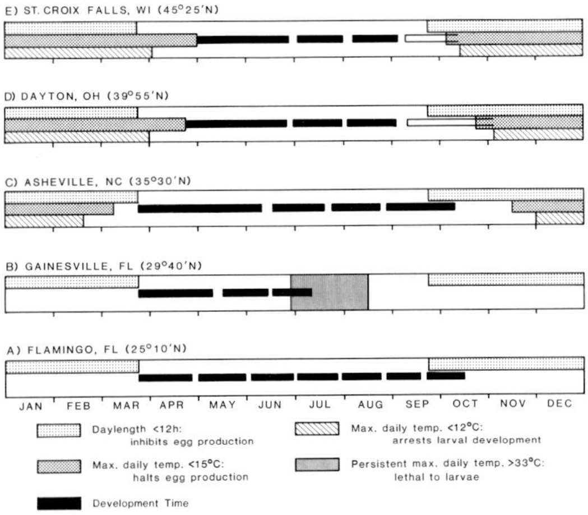

Figure 2. Predicted timing and number of monarch generations possible within the abiotic constraints operating at each of five sites (A–E) from Florida north to Wisconsin (shown in Fig. 1a), each separated by approximately 5°N latitude. Maturation times, combining development from egg to adult emergence (indicated by the dark development time bars) and their subsequent reproductive maturation (spaces between bars) were calculated in day-degrees using Zalucki's (1981, 1982) data and Allen's (1976) modified sine wave method on NOAA daily maximum and minimum temperatures at each site. The durations of the temperature constraints and maturation times were calculated separately for 1981, 1982, and 1983 at each site and then averaged to produce the figure.

values for the main cohort were lower (Table 3d). This suggests that latitude and oviposition date exerted less influence on the maturation time of the main cohort than on the maturation time of the first individuals at sites. The recolonization rate of 20 km/day (Table 5a) maintained a balance between the prolonging influence of increased latitude on maturation time and the reducing influence of later oviposition date and resulted in a mean maturation time of 39.6 days (SD = 2.5), with a range of only 10 days (Table 4c), significantly less than for the first individuals oviposited in this area ($t_{70} = 4.33$, $p < 0.0001$).

Oviposition at Latitudes North of 41°40′N. *First Individuals.* This sample comprised 20 sites from latitude 41°53′N to 48°54′N and covered an oviposition period from 28 April to 19 June. The first individuals at these sites experienced a much longer mean maturation time (65 days) with much greater variability (range of 36 days) than their counterparts farther south (Table 4d). Multiple regression analysis again showed significant and comparable, but opposing, effects of latitude and oviposition date on maturation time (Table 3e). However, in this case there was no significant direct relationship between latitude and oviposition date (Table 5b), suggesting that this far north the recolonization process

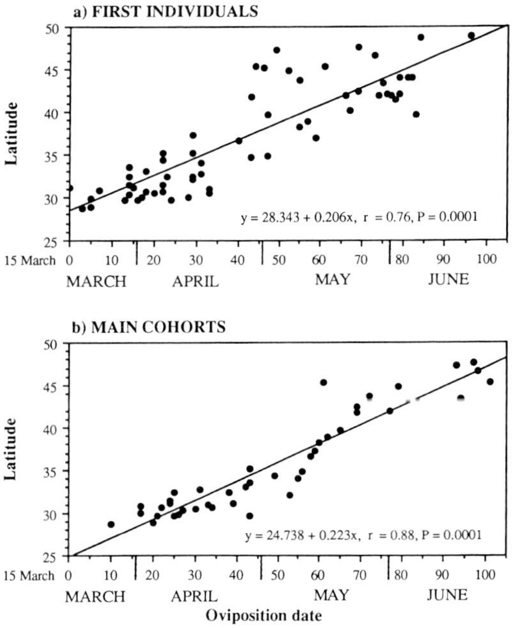

Figure 3. Relationship between site latitudes (°N) and estimated oviposition dates (days from 15 March) of (**a**) the first individuals and (**b**) the main cohorts. Oviposition dates were estimated by back-calculating in day-degrees from the developmental stage of field-censused larvae.

had become much more variable and diffuse. This may signal the cessation of directed northward movement, perhaps in response to environmental cues, or simply reflect variability caused by the different arrival times of overlapping generations moving north from various southerly points.

Main Cohort. The 10 observations in this sample had a latudinal range of almost 6°N, a range in oviposition dates of 40 days and a mean maturation time of 54.9 days (range 40–65 days; Table 4d).

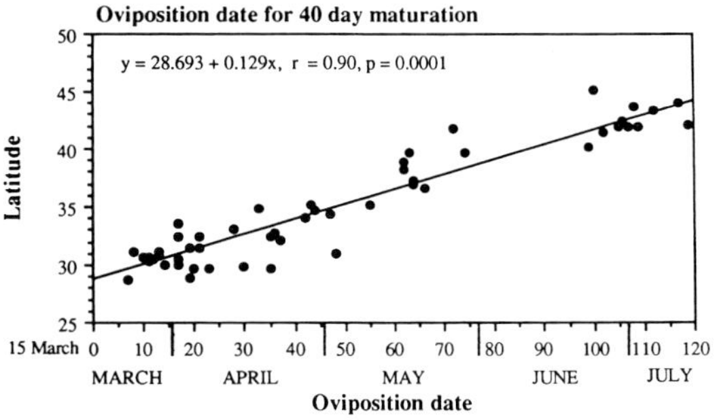

Figure 4. Relationship between site latitude (°N) and the earliest oviposition date (days from 15 March) from which a maturation time of 40 days could theoretically be completed at each of 52 field sites in the year censused. At 10 sites north of 44°N maturation times did not fall this low. The slope indicates a northward movement rate of 14 km/day.

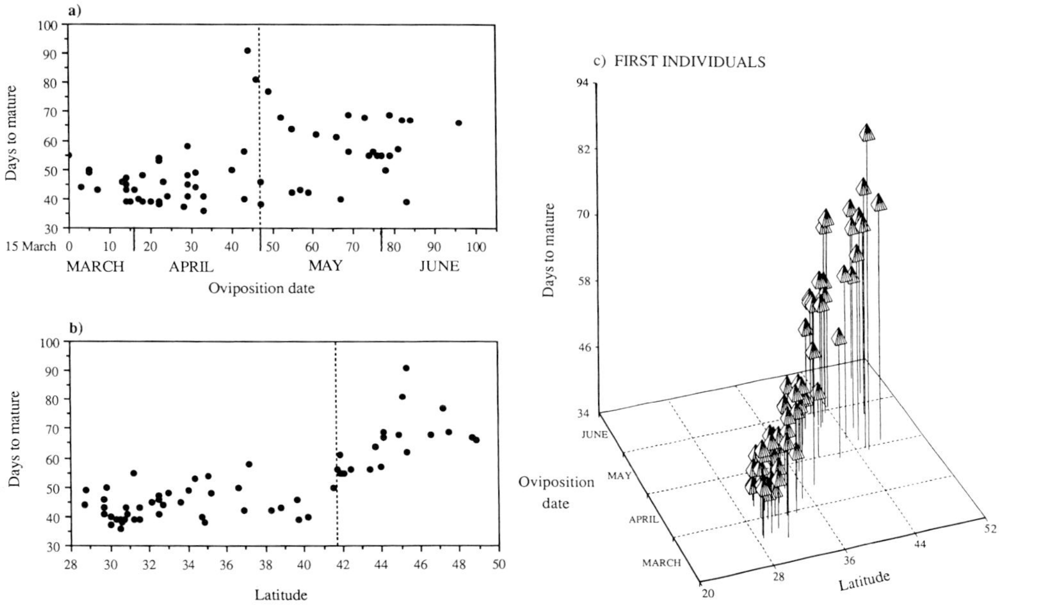

Figure 5. Calculated number of days for the first individuals oviposited at each site to reach maturity against (**a**) oviposition date (days from 15 March), (**b**) latitude (°N), and (**c**) both oviposition date and latitude. Latitude 41°40′N and 1 May are indicated.

Multiple regression analysis showed no significant effects of latitude and oviposition date (Table 3e). This indicates no statistically predictable pattern of maturation time with latitude or oviposition date for this sample, although the standardized beta values (Table 3e) again suggest a balance between the effects of latitude and oviposition date within this data subset. There was no significant relationship between latitude and oviposition date (Table 5b), again suggesting less directed recolonization, and/or less predictable movement, at these latitudes.

DISCUSSION

The abiotic model of the predicted number of monarch generations at 5° intervals between 25° and 45°N, shows that both latitude and oviposition date influence monarch maturation time and hence the number of generations completed. Thus, we would expect monarchs to migrate north each spring at a rate that balances these abiotic constraints (together with as yet largely unquantified biotic constraints) to maximize their offspring production and survival through successive generations. Simulations indicate that monarchs can complete up to seven generations in Flamingo, south Florida. The observed northeasterly flight direction of spring migrating monarchs (Urquhart, 1960, 1966; Malcolm, 1987) and the low numbers of migrants in the southeast (Malcolm et al., this volume) suggest that most remigrants are unlikely to produce the multiple generations we simulated for monarchs in Flamingo, south Florida. These butterflies are probably an isolated, nonmigratory, year-round breeding subpopulation, much like those in southern California (Malcolm, 1987; Wenner and Harris, this volume) or eastern Australia, north of Sydney (James, this volume; Zalucki, this volume).

Spring remigrating monarchs from Mexico reach the Gulf states of the southern United States at the end of March (Riley, 1871; Lynch and Martin, this volume; Riley, this volume), when the early onset of warm temperatures and the abundance of early-spring milkweeds such as *A. humistrata, A. viridis,* and *A. asperula* (Woodson, 1954; Lynch and Martin, 1987; Malcolm et al., 1987; Martin and Lynch, 1988) invite colonization of this region. By mid-June, however, temperatures become too hot for further larval development. Field observations (Malcolm et al., 1987) support our model in showing that monarchs can complete only two generations at Gainesville, Florida (latitude 29°40′N). Lethal summer temperatures occur at similar latitudes across the entire coastal plain from Texas, east to Florida and Georgia, resulting in one (Riley, this volume) or at most two spring generations from this southern region (Malcolm et al., 1987). Our model suggests that by completing two fast-developing southern generations by early June and then ovipositing farther north, it is possible for five generations to be produced during the season, especially if development time is reduced by larval basking (Rawlins and Lederhouse, 1981). The likelihood of five generations per year is corroborated by Malcolm et al. (1987) using data from Florida, together with Borkin's (1982) data from Wisconsin. It is also supported by the field data reported in

Table 3. **Multiple regression analyses (showing slopes, standardized beta values, and significances) of days to mature on latitude, for first individuals and the main cohorts oviposited before or from 1 May, at sites south or north of latitude 41°40′N.**

	First individuals			Main cohort		
	Value	Beta	p	Value	Beta	p
a) All observations						
Intercept	−46.67			−14.51		
Latitude	3.12	1.59	0.0001	1.96	1.59	0.0012
Oviposition date	−0.42	−0.91	0.0001	−0.25	−0.85	0.0001
r^2	0.83		0.0001	0.71		0.0001
N	62			43		
b) Oviposition before 1 May						
Intercept	−51.92			25.75		
Latitude	3.32	1.20	0.0001	0.60	0.38	0.17
Oviposition date	−0.45	−0.48	0.0001	−0.16	−0.60	0.04
r^2	0.83		0.0001	0.13		0.10
N	36			23		
c) Oviposition from 1 May						
Intercept	−67.52			−26.00		
Latitude	3.24	1.00	0.0001	2.13	1.15	0.0001
Oviposition date	−0.21	−0.24	0.0001	−0.19	−0.36	0.14
r^2	0.84		0.0001	0.71		0.0001
N	26			20		
d) Oviposition at latitudes south of 41°40′N						
Intercept	−15.83			22.15		
Latitude	2.13	1.41	0.0001	0.71	0.96	0.009
Oviposition date	−0.37	−1.31	0.0001	−0.15	−0.99	0.007
r^2	0.62		0.0001	0.22		0.02
N	42			33		
e) Oviposition at latitudes north of 41°40′N						
Intercept	−14.45			−12.15		
Latitude	2.44	0.57	0.0005	2.03	0.73	0.06
Oviposition date	−0.43	−0.62	0.0003	−0.28	−0.74	0.05
r^2	0.69		0.0001	0.35		0.09
N	20			10		

this paper, which shows a modal development time of 40 days, allowing five generations in the 200 days between mid-March and the end of September.

Our voltinism model indicates that the least energetically demanding way of producing these offspring is to complete two fast-developing generations in the region of 30°N by early June and then three more at 35°N. It shows no additional advantage from continued movement north of 35°N. The difference between the simple abiotic model predictions and the real situation, where migration continues beyond 35°N, suggests there are biotic selection pressures to proceed north. Although, from the abiotic perspective, no additional generations are gained, migration escapes potential resource competition and depletion, as well as possibly increased predator and parasite pressure, and allows exploitation of North America's most abundant milkweed, *A. syriaca,* whose distribution extends north from latitude 35°N (Woodson, 1954; Malcolm et al., 1989).

Our field data indicate that the recolonization process occurs in at least two stages, with overwintered monarchs from Mexico principally colonizing the south, and their offspring colonizing the north. We base this conclusion both on the spatial deposition of eggs before, compared with after, 1 May (the earliest calculated maturation date of the first spring generation) and on our observations of adult wing wear with latitude. More than 99% (1,021 of 1,029) of all the eggs oviposited before 1 May were laid south of latitude 38°N. The remaining eight were all oviposited after 27 April at three sites north of this limit (Iowa, 41°39′N, one egg on 27 April and one on 30 April 1986; Wisconsin, 45°07′N, one egg on 30 April 1986; Wisconsin, 45°07′N, five eggs on 28 April 1985), and they possibly represent the offspring of only three early first-generation females. The only other evidence that Mexican remigrants may reach these latitudes is an unconfirmed sighting of a wing-tagged Mexican remigrant in New York State at 42°10′N (Urquhart and Ur-

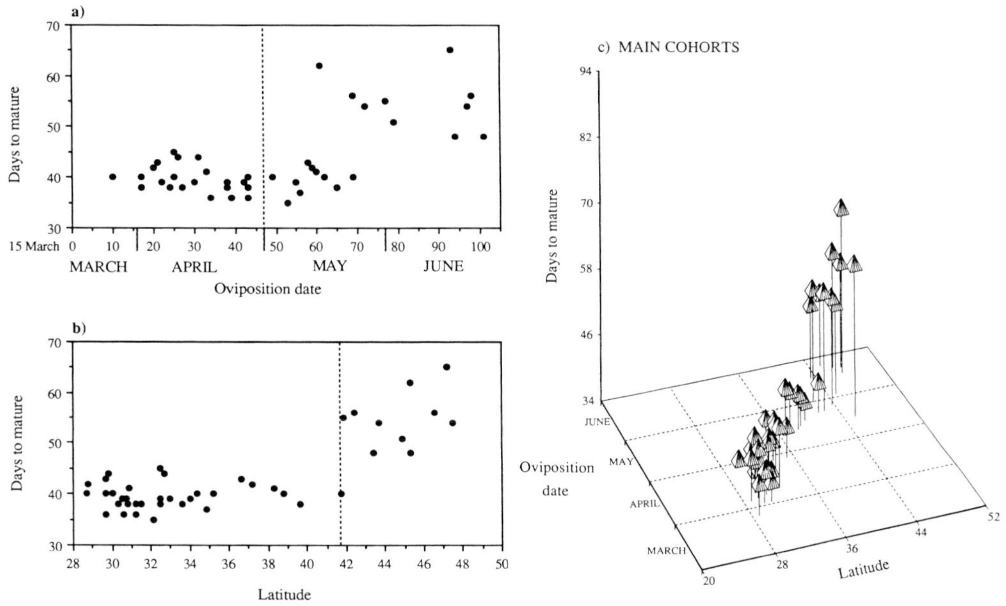

Figure 6. Calculated number of days for the main cohort at each site to reach maturity against (**a**) oviposition date (days from 15 March), (**b**) latitude (°N), and (**c**) both oviposition date and latitude. Latitude 41°40′N and 1 May are indicated.

quhart, 1979b). Any such long-range, single-sweep migration by Mexican remigrants contributes little to the population, as indicated by our data, which showed that before 1 May the north was largely unoccupied. The main cohort was not oviposited at any site north of 35°30′N until after 1 May, while only 5 of the 36 sites (14%) that were colonized before 1 May lay north of 35°30′N. By contrast the southern portion of the continent was effectively colonized before 1 May, oviposition having occurred at all 28 sites south of 34°52′N prior to this date.

After 1 May, oviposition of the main cohort occurred throughout the area east of the Great Plains from Texas to Minnesota. We suggest that this cohort includes offspring of both Mexican remigrant and first-generation monarchs, with new monarchs being increasingly represented in the population from 1 May. Our adult captures changed from being predominantly worn Mexican monarchs on 9 May at 35°09′N in Arkansas, to predominantly pristine, first spring generation individuals on 13 May at 37°12′N in Missouri. The last worn Mexican adult was captured on 22 May at 40°39′N in Iowa (Table 1). Therefore, we suggest that the principal colonizers north of 41°40′N were the first-generation offspring of Mexican remigrants, and this separation of generations is reflected in the differences in our data south and north of 41°40′N. Splitting the data at 1 May produces a less clear division because, since this is the earliest date of maturation, the proportion of new generation adult monarchs will be low initially.

There is strong selection for Mexican remigrants to recolonize the abundant southern milkweed resource and not to advance north before more northern larval food resources are available for oviposition, or when their offspring may be exposed to frosts, or to such cool temperatures that maturation time is extended, even to the point where a potential additional generation is lost (as we report in northeastern Wisconsin at the end of April). Our model does not incorporate any reduction in development time through larval basking (Rawlins and Lederhouse, 1981), which may dampen such extreme differences; nevertheless, the first individuals at sites consistently had significantly longer predicted generation times than later ones, and immatures at more northern sites took statistically longer to mature than more southern ones, despite later oviposition in the north. These factors will select against over-rapid migration rates. Over the entire continent the cooling effect of increasing latitude had a greater effect in prolonging generation times than the warming effect of later oviposition date had in reducing them. In part this is due to the impossibility of producing as short generations in the Great Lakes region and the extreme north, compared with farther south, but it is also because the calculated recolonization rate of 24 km/day was considerably faster than the calculated optimal speed of 14 km/day for producing short 40-day generations. Hence, monarchs were arriving at latitudes too early for them to have received sufficient seasonal warming. We found that south of 41°40′N, recolonization rates of 17 km/day for the first immatures and 20 km/day for the main cohort maintained a balance between the effects of increased

Table 4. **Means, standard deviations, and ranges (minimum and maximum) of latitude, oviposition date, and maturation times of the first individuals and main cohorts of monarchs oviposited before or from 1 May, and south or north of latitude 41°40′N.**

	First individuals			Main cohort		
	Latitude	Oviposition date	Days to mature	Latitude	Oviposition date	Days to mature
a) Oviposition before May 1						
Mean	32°48′	7 April	47.1	31°04′	13 April	39.6
SD	4°02′	11.9 days	11.2	1°36′	9.5 days	2.6
Minimum	28°42′	15 March	36	28°40′	25 March	36
Maximum	45°18′	30 April	91	35°09′	27 April	45
N	36	36	36	23	23	23
b) Oviposition from May 1						
Mean	42°42′	23 May	56.7	40°40′	25 May	47.2
SD	3°03′	13.0 days	11.4	4°05′	16.8 days	8.9
Minimum	34°52′	1 May	38	31°10′	3 May	35
Maximum	48°54′	19 June	77	47°29′	24 June	65
N	26	26	26	20	20	20
c) Oviposition at latitudes south of 41°40′N						
Mean	33°25′	13 April	44.4	32°50′	22 April	39.6
SD	3°52′	19.6 days	5.6	3°20′	16.1 days	2.5
Minimum	28°40′	15 March	36	28°40′	25 March	35
Maximum	41°39′	6 June	58	41°39′	23 May	45
N	42	42	42	33	33	33
d) Oviposition at latitudes north of 41°40′N						
Mean	44°33′	23 May	65.0	44°49′	7 June	54.9
SD	2°27′	14.1 days	10.0	2°00′	6.8 days	5.4
Minimum	41°53′	28 April	55	41°53′	15 May	48
Maximum	48°54′	19 June	91	47°29′	24 June	65
N	20	20	20	10	10	10

latitude and later oviposition date to produce reasonably constant maturation times of 44 and 40 days, respectively.

No matter how the data were divided, latitude and oviposition date had less effect on the maturation time of the main cohort than on the first individuals at sites (Table 3). This, in conjunction with lower and less variable main cohort maturation times (Table 4), suggests that there is a wider band of equivalent options available to the later recolonizing adults. Maturation time of the main group was independent of latitude and oviposition date between 25 March and 27 April, and latitudes 28°40′N and 35°09′N. Within these limits temperatures were sufficiently similar to produce equivalent maturation times, making monarch spatial and temporal choices uncritical.

North of 41°40′N maturation times were considerably extended and variable, and latitude and oviposition date had only marginal significance on the maturation time of the main cohort. At these latitudes there was no significant relationship between latitude and oviposition date, either for the first immatures or the main cohort, suggesting that this far north directed migration either had ceased or was undetectable beneath the variability of overlapping generations of monarchs moving north from various points to the south. Unfortunately we have only 10 observations of a main cohort north of 41°40′N, and, based on the pre-eminence of eggs and first instars at these sites, it seems likely that our northern samples pre-dated most of the main cohort. Later northern main cohorts would have shorter maturation times, like our model simulations for St. Croix Falls, Wisconsin, and Dayton, Ohio, in July (Fig. 2), similar to published information (Borkin, 1982; Malcolm et al., 1987). A lower maturation time may be particularly beneficial if northern larvae are not as well defended against predators and parasites as are southern individuals (Malcolm and Brower, 1989; Malcolm et al., 1989, this volume), requiring them to develop as fast as possible to minimize exposure to natural enemies.

Our paper raises the question of what determines the recolonization rate? The spring migration may be viewed as behavior to track the milkweed resource in space and time, and the recolonization rate should ultimately relate to the availability of larval host plants, mating opportunities, and conditions of offspring development. Butterflies may respond to their immediate temperature and weath-

Table 5. Relationships between latitude and oviposition date in the form of latitude = b(oviposition date) + c, subdivided for the first individuals and main cohorts at sites south and north of latitude 41°40′N. The value of the slope, b, is shown in parentheses as km/day.

	First individuals		Main chort	
	Value	*p*	Value	*p*
a) Oviposition at latitudes south of 41°40′N				
Intercept, c	28.86		25.89	
Slope, b (km/day)	0.157 (17.4)		0.182 (20.2)	
r^2	0.70	0.0001	0.77	0.0001
N	42		33	
b) Oviposition at latitudes north of 41°40′N				
Intercept, c	44.34		38.59	
Slope, b (km/day)	0.003 (0.3)		0.074 (8.2)	
r^2	0.0004	0.93 NS	0.29	0.11 NS
N	20		10	

er conditions by making continuous, gradual physiological and behavioral adjustments to the prevailing climatic conditions, as has been suggested for the autumn migrants (Barker and Herman, 1976; Herman, 1985, this volume; James, this volume). Alternatively, they may respond to correlates of these climatic factors. For example, migration rate could be proportional to the rate of daylength change, and the minimal rates of daylength change around the summer solstice may be the cue to terminate northward movement. Monarchs may also respond to the effects of climatic factors on foodplant and mate availability, both of which increase with rising summer temperatures (Woodson, 1954; Urquhart, 1960; Eanes and Koehn, 1978; Borkin, 1982; Brower, 1985; Malcolm et al., 1987). If monarchs increasingly spend more time mating and ovipositing and less time searching and migrating, their northward progress would slow down. Because milkweed growth rate is temperature-dependent, host-plant availability may provide a sufficient proximate cue to the temperature conditions for immature development.

Urquhart (1960) suggests that monarchs can sustain flight speeds of 18 km/hr, and Gibo (1986) has shown how this can be augmented by major air movements. But Baker (1978) considers that the actual rate of migration is likely to be considerably less than this, and he calculates, from Urquhart's data, an average of approximately 4 km/day for spring remigrants. Our data suggest that the migration rate is not constant over the distance from Mexico to southern Canada, as indicated by the decline in recolonization rate north of 41°40′N. Even this is an oversimplification, for if monarchs were to fly from the overwintering sites in Mexico at the same rate that we record their oviposition progress in the south, then this requires the main cohort to leave the overwintering areas, at approximately latitude 20°N (Calvert and Brower, 1986), by 11 February, and the earliest arrivals to leave by 17 January. This is clearly contrary to the evidence of Calvert and Brower (1986), Calvert and Lawton (this volume), and Van Hook (this volume), which shows that large numbers of monarchs do not leave the overwintering sites until mid-March. Furthermore, if the monarch recolonizes by successive broods, as our data and those of Malcolm et al. (this volume) indicate, then a constant daily recolonization rate by individual butterflies is clearly impossible, because the immatures are immobile during development. Therefore, to maintain the appearance of a constant daily recolonization rate as we have described, the monarchs must show a rapid spurt of migration after adult emergence but before sexual maturity. Thus, we suggest that remigrant monarchs initially fly rapidly from the overwintering sites, over central and northern Mexico, when the dry season is at its peak intensity and there is an absence of spring milkweeds (unpubl. obs.; Stephen P. Lynch, Louisiana State University, pers. comm.). During this migration the butterflies become reproductively mature. When they reach the southern milkweed flora, their progress declines as they encounter the milkweed food resource and begin oviposition. When this generation emerges, they also start their northward flight rapidly (necessary to make up the lost ground implied by the constant daily flight rate), again during a period of reproductive maturation, slowing when their eggs are ready to lay and young, flushing, unoccupied milkweeds become available. We illustrate these possible spurts and plateaus in migration rate for the main cohort data of Figure 3b, replotted in Figure 7. This indicates a migration rate of 96 km/day by monarchs leaving the Mexican overwintering sites at 20°N on 15 March to arrive at Refugio County, Texas, at 28°40′N on 25 March. These Mexican remigrants then slow to a recolonization rate of 12.4 km/day across the sites at which there was oviposition before 1 May. Between this date and 41°40′N, at sites where both old and new generations of monarchs were recorded, the recolonization rate increases again to 52 km/day. North of

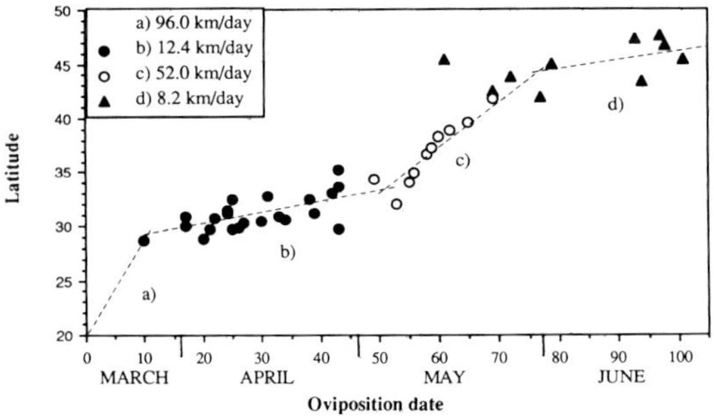

Figure 7. Oviposition date (days from 15 March) of the main cohort against latitudinal position of sites (°N), with fitted regressions showing the variation in recolonization rate for the following subsets of the data: (**a**) the overwintering sites on 15 March at 20°N and Refugio County, Texas, on 25 March at 28°40′N, (**b**) all incidences of oviposition before 1 May [latitude = 0.11(oviposition date) + 27.88, $r^2 = 0.44$, N = 23, $p = 0.0006$], (**c**) oviposition from 1 May and south of 41°40′N [latitude = 0.47(oviposition date) + 9.52, $r^2 = 0.85$, N = 10, $p = 0.0001$], and (**d**) oviposition north of 41°40′N [latitude = 0.07(oviposition date) + 38.59, $r^2 = 0.29$, N = 10, $p = 0.11$].

41°40′N, where only new-generation monarchs were found, the rate falls again to 8.2 km/day, although here the recolonization rate is less predictable and the regression is not statistically significant.

In conclusion, our simulations and field-based analysis strongly support the hypothesis that spring migrating monarchs recolonize their breeding range by successive broods rather than a single sweep strategy. This corroborates the results generated by cardenolide fingerprints of migrating monarchs (Malcolm et al., this volume). We suggest that the early availability of southern milkweeds such as *A. humistrata, A. viridis,* and *A. asperula,* together with the onset of lethal summer temperatures in the south, and the later abundance of *A. syriaca* farther north, provide strong selection for successive brood migration. This strategy has the advantage of utilizing a greater portion of the total milkweed resource in space and time across the continent and produces more monarch generations through shorter maturation times than could the single-sweep strategy.

ACKNOWLEDGMENTS

We gratefully acknowledge funding for this research from National Science Foundation grant BSR-8500416 and from the Division of Sponsored Research, University of Florida. We also thank Susan Borkin, Andrew Brower, Paul Davis, Steve Lynch, Tom Riley, Theodore Sargent, and Tonya Van Hook for their help with field observations.

LITERATURE CITED

Allen, J.C. 1976. A modified sine wave method for calculating degree days. *Environmental Entomology* 5: 388–396.

Baker, R.R. 1978. *The evolutionary ecology of animal migration.* London: Hodder and Stoughton, 1012 pp.

Barker, J.F., and W.S. Herman. 1976. Effect of photoperiod and temperature on reproduction of the monarch butterfly *Danaus plexippus. Journal of Insect Physiology* 22:1565–1568.

Beck, S.D. 1980. *Insect photoperiodism.* New York: Academic Press, 387 pp.

Borkin, S.S. 1982. Notes on shifting distribution patterns and survival of immature *Danaus plexippus* (Lepidoptera: Danaidae) on the food plant *Asclepias syriaca. The Great Lakes Entomologist* 15:199–206.

Brower, L.P. 1961. Studies on the migration of the monarch butterfly I. Breeding populations of *Danaus plexippus* and *D. gilippus berenice* in south central Florida. *Ecology* 42:76–83.

Brower, L.P. 1985. New perspectives on the migration biology of the monarch butterfly, *Danaus plexippus* L. In *Migration: Mechanisms and adaptive significance,* ed. M.A. Rankin, 748–785. Austin: University of Texas, *Contributions in Marine Science,* Supplement, 27.

Calvert, W.H., and L.P. Brower. 1986. The location of monarch butterfly (*Danaus plexippus* L.) overwintering colonies in Mexico in relation to topography and climate. *Journal of the Lepidopterists' Society* 40:164–187.

Dingle, H. 1980. Ecology and evolution of migration. In *Animal migration, orientation, and navigation,* ed. S.A. Gauthreaux, Jr., 1–101. New York: Academic Press.

Eanes, W.F., and R.K. Koehn. 1978. An analysis of genetic structure in the monarch butterfly, *Danaus plexippus* L. *Evolution* 32:784–797.

Erickson, J.M. 1973. The utilization of various *Asclepias* species by larvae of the monarch butterfly, *Danaus plexippus. Psyche* 80:230–244.

Fales, J.H. 1984. Spring occurrence of the monarch butterfly in Maryland. *Maryland Entomologist* 2:76–79.

Gibo, D.L. 1986. Flight strategies of migrating monarch butterflies (*Danaus plexippus* L.) in southern Ontario. In *Insect flight: Dispersal and migration,* ed. W. Danthanarayana, 172–184. Berlin and Heidelberg: Springer-Verlag.

Herman, W.S. 1985. Hormonally mediated events in adult monarch butterflies. In *Migration: Mechanisms and adaptive significance,* ed. M.A. Rankin, 799–815. Austin: University of Texas, *Contributions in Marine Science,* Supplement, 27.

Inglesfield, C., and M. Begon. 1983. The ontogeny and cost of migration in *Drosophila subobscura* Collin. *Biological Journal of the Linnean Society* 19:9–15.

Kennedy, J. 1961. A turning point in the study of insect migration. *Nature* 189:785–791.

Kennedy, J. 1985. Migration, behavioral and ecological. In *Migration: Mechanisms and adaptive significance,* ed. M.A. Rankin, 5–26. Austin: University of Texas, *Contributions in Marine Science,* Supplement, 27.

Lynch, S.P., and R.A. Martin. 1987. Cardenolide content and thin-layer chromatography profiles of monarch butterflies, *Danaus plexippus* L. and their larval host-plant milkweed, *Asclepias viridis* Walt., in northwestern Louisiana. *Journal of Chemical Ecology* 13:47–70.

Malcolm, S.B. 1987. Monarch butterfly migration in North America: Controversy and conservation. *Trends in Ecology and Evolution* 2:135–138.

Malcolm, S.B., and L.P. Brower. 1986. Selective oviposition by monarch butterflies (*Danaus plexippus* L.) in a mixed stand of *Asclepias curassavica* L. and *A. incarnata* L. in south Florida. *Journal of the Lepidopterists' Society* 40:255–263.

Malcolm, S.B., and L.P. Brower. 1989. Evolutionary and ecological implications of cardenolide sequestration in the monarch butterfly. *Experientia* 45:284–295.

Malcolm, S.B., B.J. Cockrell, and L.P. Brower. 1987. Monarch butterfly voltinism: Effects of temperature constraints at different latitudes. *Oikos* 49:77–82.

Malcolm, S.B., B.J. Cockrell, and L.P. Brower. 1989. The cardenolide fingerprint of monarch butterflies reared on the common milkweed, *Asclepias syriaca*. *Journal of Chemical Ecology* 15:819–853.

Martin, R.A., and S.P. Lynch. 1988. Cardenolide content and thin-layer chromatography profiles of monarch butterflies, *Danaus plexippus* L., and their larval host-plant milkweed, *Asclepias asperula* subsp. *capricornu* (Woods.) Woods., in north central Texas. *Journal of Chemical Ecology* 14:295–318.

Rawlins, J.E., and R.C. Lederhouse. 1981. Developmental influences of thermal behavior on monarch caterpillars (*Danaus plexippus*): An adaptation for migration (Lepidoptera: Nymphalidae: Danainae). *Journal of the Kansas Entomological Society* 54: 387–408.

Riley, C.V. 1871. *Third annual report of the noxious, beneficial and other insects of the state of Missouri.* Jefferson City, MO: Horace Wilcox, 179 pp.

Rogers, D. 1983. Pattern and process in large-scale animal movement. In *The ecology of animal movement,* eds. I.A. Swingland and P.J. Greenwood, 160–180. Oxford: Clarendon Press.

Southwood, T.R.E. 1977. Habitat, the templet for ecological strategies? *Journal of Animal Ecology* 46: 337–365.

Taylor, L.R., and R.A.J. Taylor. 1983. Insect migration as a paradigm for survival by movement. In *The ecology of animal movement,* eds. I.A. Swingland and P.J. Greenwood, 181–214. Oxford: Clarendon Press.

Urquhart, F.A. 1960. *The monarch butterfly*. Toronto: University of Toronto Press, 361 pp.

Urquhart, F.A. 1966. A study of the migrations of the Gulf Coast population of the monarch butterfly (*Danaus plexippus* L.) in North America. *Annales Zoologici Fennici* 3:82–87.

Urquhart, F.A. 1976. Found at last: The monarch's winter home. *National Geographic* 150:161–173.

Urquhart, F.A. 1987. *The monarch butterfly: International traveler.* Chicago: Nelson-Hall, 232 pp.

Urquhart, F.A., and N.R. Urquhart. 1976a. Migration of butterflies along the Gulf Coast of northern Florida. *Journal of the Lepidopterists' Society* 30:59–61.

Urquhart, F.A., and N.R. Urquhart. 1976b. The overwintering site of the eastern population of the monarch butterfly (*Danaus p. plexippus:* Danaidae) in southern Mexico. *Journal of the Lepidopterists' Society* 30:153–158.

Urquhart, F.A., and N.R. Urquhart. 1976c. A study of the peninsular Florida populations of the monarch butterfly (*Danaus p. plexippus:* Danaidae). *Journal of the Lepidopterists' Society* 30:73–87.

Urquhart, F.A., and N.R. Urquhart. 1977. Overwintering areas and migratory routes of the eastern population of the monarch butterfly (*Danaus p. plexippus,* Lepidoptera: Danaidae) in North America, with special reference to the western population. *Canadian Entomologist* 109:1583–1589.

Urquhart, F.A., and N.R. Urquhart. 1978. Autumnal migration routes of the eastern population of the monarch butterfly (*Danaus p. plexippus* L.; Danaidae; Lepidoptera) in North America to the overwintering site in the neovolcanic plateau of Mexico. *Canadian Journal of Zoology* 56:1759–1764.

Urquhart, F.A., and N.R. Urquhart. 1979a. Breeding areas and overnight roosting locations in the northern range of the monarch butterfly (*Danaus plexippus plexippus*) with a summary of associated migratory routes. *Canadian Field Naturalist* 93:41–47.

Urquhart, F.A., and N.R. Urquhart. 1979b. Vernal migration of the monarch butterfly (*Danaus p. plexippus,* Lepidoptera: Danaidae) in North America from the overwintering site in the neo-volcanic plateau of Mexico. *Canadian Entomologist* 111:15–18.

Williams, C.B. 1930. *The migration of butterflies.* Edinburgh: Oliver and Boyd, 473 pp.

Woodson, R.E., Jr. 1954. The North American species of *Asclepias* L. *Annals of the Missouri Botanical Garden* 41:1–211.

Young, A.M. 1980. Some observations on the natural history and behaviour of the camberwell beauty (mourning cloak) butterfly, *Nymphalis antiopa* (Linnaeus) (Lepidoptera: Nymphalidae) in the United States. *Entomologists Gazette* 31:7–18.

Zalucki, M.P. 1981. The effects of age and weather on egg laying in *Danaus plexippus* L. (Lepidoptera: Danaidae). *Researches in Population Ecology* 23: 318–327.

Zalucki, M.P. 1982. Temperature and rate of development in *Danaus plexippus* L. and *D. chrysippus* L. (Lepidoptera: Nymphalidae). *Journal of the Australian Entomological Society* 21:241–246.

SPRING RECOLONIZATION OF EASTERN NORTH AMERICA BY THE MONARCH BUTTERFLY: SUCCESSIVE BROOD OR SINGLE SWEEP MIGRATION?

Stephen B. Malcolm, Barbara J. Cockrell, and Lincoln P. Brower

ABSTRACT. Each March, monarch butterflies, *Danaus plexippus* (L.), remigrate north from Mexican overwintering sites. By June they are distributed across most of eastern North America, north to southern Canada. Here we distinguish between two alternative spring recolonization strategies: (1) "single sweep recolonization," in which overwintered individual remigrants from Mexico were hypothesized to colonize the entire eastern North American breeding range, and (2) "successive brood recolonization" in which remigrants from Mexico were hypothesized to lay all of their eggs in the southern United States and then die, leaving their offspring to continue the migration northward and recolonize the northern breeding range.

Cardenolide concentrations, cardenolide "fingerprints," and wing wear of migrant butterflies captured along two latitudinal transects in the spring of 1985 provided three lines of independent support consistent with hypothesis 2. Monarchs captured in the southern United States in late March and early April, which lay eggs on the emergent milkweeds, had (1) low cardenolide concentrations, (2) cardenolide fingerprints identical to those of overwintering monarchs in Mexico, and (3) very worn wings. In contrast, monarchs from the northern United States in May and June had (1) high cardenolide concentrations, (2) fingerprints characteristic of the southern spring milkweed flora, and (3) wings in significantly better condition. Thus, monarchs recolonize their breeding range each spring by migration of successive broods and not by a single sweep of overwintered migrants.

The existence of concurrent migration and reproductive activity of the new spring brood is an exception to the "oogenesis-flight syndrome" of migratory insects, which holds that reproductive activity is suppressed during migration. The new evidence also indicates that chemical defense during the monarch's annual life cycle is dynamic and cycles from poorly protected autumn migrants, which overwinter and then recolonize the southern United States, to highly protected migrants of the first spring generation, which recolonize the northern United States.

INTRODUCTION

The monarch butterfly, *Danaus plexippus* (L.), is perhaps the best-known example of a regular long-range insect migrant. From the long-term research on monarch movement by Urquhart (1960, 1966, 1976, 1987) and Urquhart and Urquhart (1976a,b,c, 1977, 1978; summaries in Brower, 1985; Malcolm, 1987), we know that monarchs that breed east of the Rocky Mountains fly southwest each autumn, from the northern United States and southern Canada to the transvolcanic mountains of central Mexico, where they spend the winter in dense aggregations on high-altitude fir trees (Calvert and Brower, 1986). In spring the same butterflies make the return journey to at least the southern United States, and by mid-summer the species is once again distributed as far north as southern Canada (Urquhart and Urquhart, 1979). The geographical area occupied by monarchs east of the Rocky Mountains (Urquhart, 1960) almost exactly coincides with the distribution of the most common larval host plants in the milkweed genus *Asclepias* (Fig. 1).

Although the monarch's migration and their larvae feeding on milkweeds are familiar phenomena, little is known of the critical spring migration and the annual distribution of monarch generations in time and space across eastern North America. As can be seen in Williams (1958), Urquhart (1960), and Brower (1961, 1985) two hypotheses have emerged to explain how monarchs migrate to recolonize North American milkweeds each spring. (1) The first, which we now formalize as a "single sweep recolonization strategy," holds that overwintered monarchs remigrate from Mexico to de-

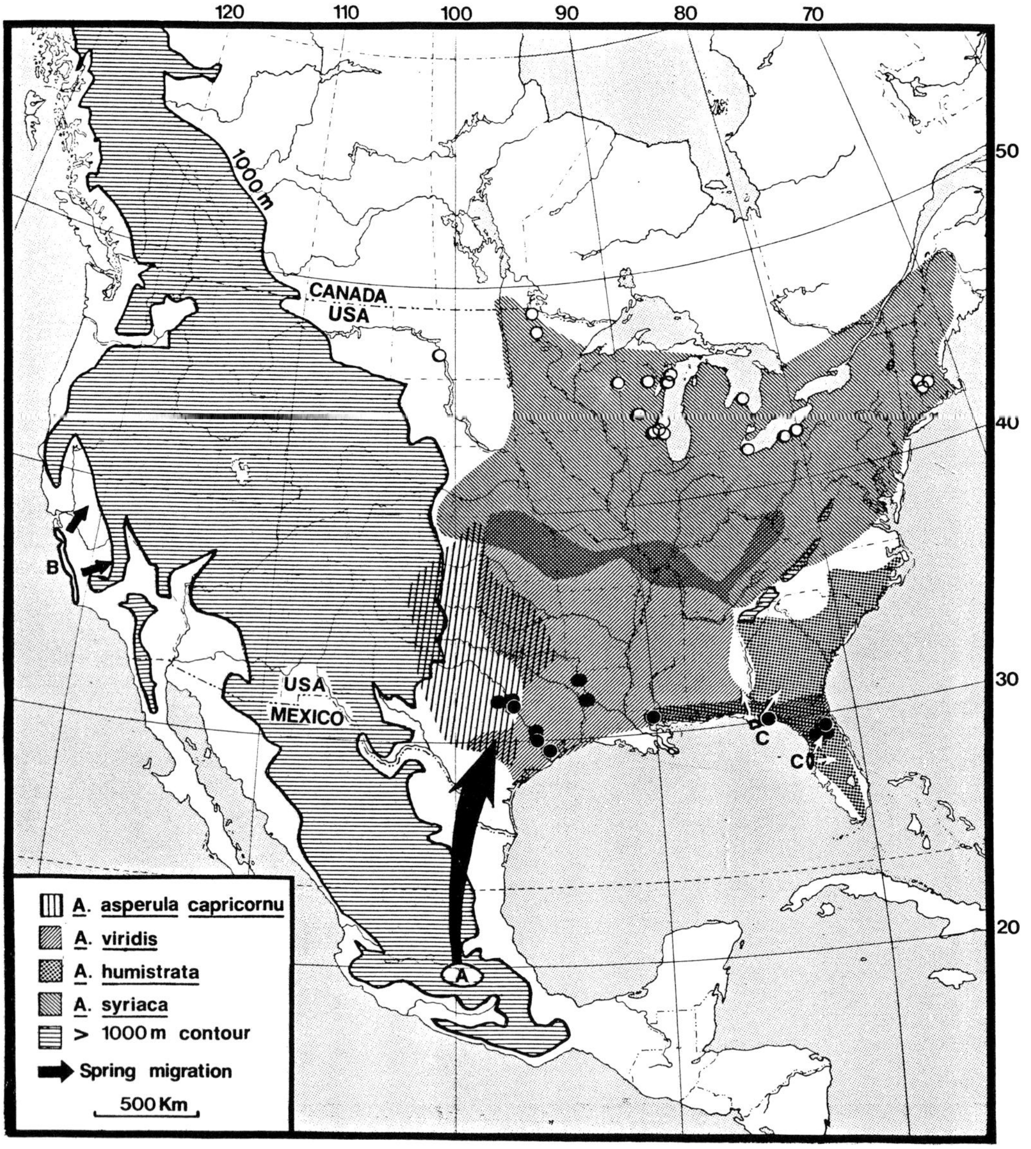

Figure 1. The distribution of the three abundant southern milkweeds, *A. asperula capricornu, A. viridis,* and *A. humistrata,* and the most abundant northern milkweed, *A. syriaca* (after Woodson, 1954). The breeding distribution of migratory eastern monarchs, which overwinter at **A** in Mexico, coincides largely with the distributions of these milkweed species east of the Rocky Mountain barrier. This barrier is indicated by most of the horizontal hatched area within the 1,000-m contour in the western United States and Canada. Monarchs that overwinter at C in Florida may contribute small numbers to the eastern spring migration, whereas monarchs that overwinter at **B** in California appear largely restricted to migration within California. Our 1985 adult monarch sample collection sites in the eastern United States are also shown (black circles = southern samples, white circles = northern samples).

posit eggs on milkweeds over their entire eastern range (Fig. 2A). It is this purported strategy that has prevailed in the literature (Johnson, 1969; Ackery and Vane-Wright, 1984; Brower, 1985; Urquhart, 1987). (2) Alternatively, monarchs could locate milkweed resources over their entire breeding range by a "successive brood recolonization" strategy (Fig. 2B). Overwintered monarchs would deposit all of their eggs on early-spring milkweeds in the southern states of the United States and then die, leaving their offspring to migrate north and recolonize the remainder of the northern range. If recolonization is achieved by the second, successive brood strategy, then the new spring generation may migrate "regionally" (Fig. 2C) or "diffusely" (Fig. 2D) across North America.

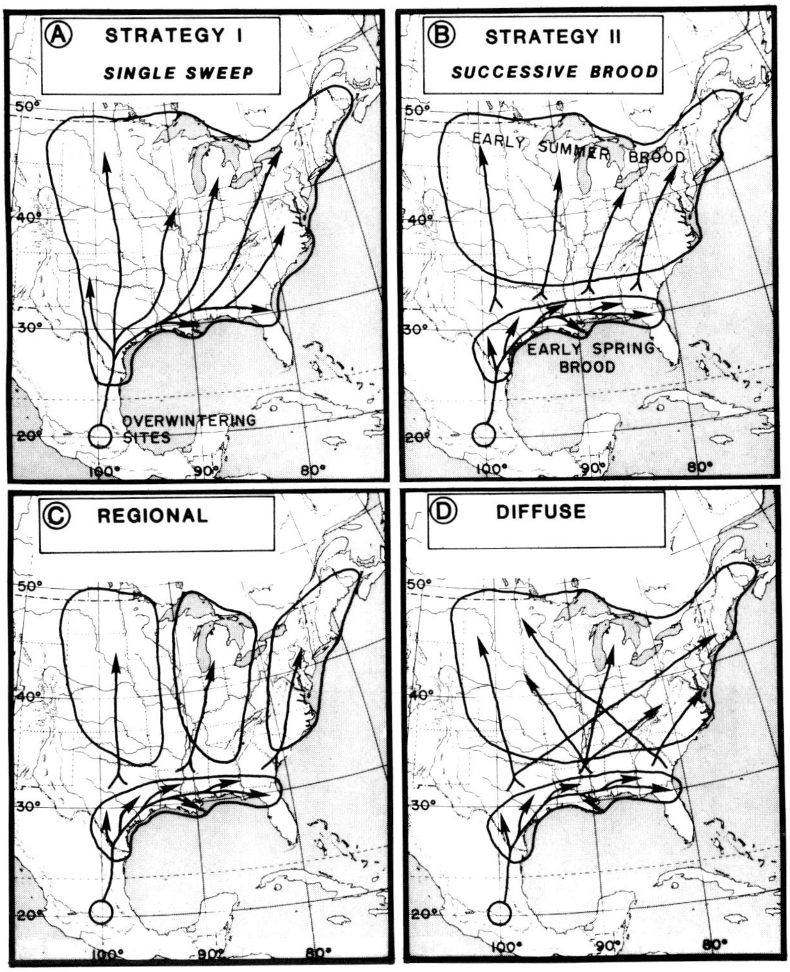

Figure 2. Hypothesized monarch spring migration strategies: (**A**) "single sweep recolonization"—overwintered migrants migrate north and lay eggs on milkweeds over their entire spring and summer range—(**B**) "successive brood recolonization"—overwintered migrants lay all of their eggs on southern United States milkweeds, leaving their offspring to recolonize northern milkweeds. (**C** and **D**) Two alternative ways in which successive brood recolonization could occur by either "regional" migration, directed north along specific corridors, or "diffuse" migration, directed approximately and randomly north.

Urquhart's (1960) pioneering mark–recapture technique with monarch butterflies led to the discovery in 1975 of their overwintering location in Mexico (Urquhart, 1976). However, the small numbers of spring migrant individuals recaptured (Malcolm, 1987) are insufficient to differentiate the possible modes of their remigration. Here we solve this problem by utilizing chemical information stored in the butterflies during their larval stage to assess the spatial and temporal origins of each field-collected adult monarch during their spring migration.

Monarch larvae have been recorded feeding on 27 of the 108 North American milkweed species in the genus *Asclepias* (Malcolm and Brower, 1986), and adult butterflies reflect the characteristic chemical pattern or "fingerprint" of different cardenolides (cardiac-active steroidal glycosides and genins) present in the host-plant species fed upon by their larvae (Roeske et al., 1976; Brower et al., 1982, 1984a,b; Cohen, 1985; Seiber et al., 1986; Lynch and Martin, 1987, this volume; Martin and Lynch, 1988; Malcolm et al., 1989; Nelson, this volume). Since Woodson (1954) gives extensive information on the distribution of *Asclepias* species, the origin of a monarch can be assigned to each milkweed's distribution, samples of which are shown in Figure 1. Furthermore, the common southern milkweeds *Asclepias humistrata*, *Asclepias viridis*, and *Asclepias asperula capricornu* have considerably higher cardenolide concentrations than the extremely abundant northern *Asclepias syriaca* (Table 1). Monarchs reared from these milkweeds reflect these cardenolide concentrations (Table 1), so that individuals originating in the southern United States from southern milkweeds can be expected to have higher cardenolide concentrations than those that fed on milkweeds in the northern United States and Canada. The trend for lower cardenolide contents of northern, *A. syriaca*-derived butterflies is even more pronounced in the March and April remigrants from Mexico because these monarchs lose cardenolide during migration and the overwintering period (Malcolm and Brower, 1989; Malcolm et al., 1989).

In the spring of 1985 we collected monarchs along an early spring, southern latidudinal transect in April and a later, northern latitudinal transect in May and June. We collected the first arrivals of migrating monarchs at these latitudes across the eastern United States, from Texas and Louisiana to Florida in the south, and North Dakota, Minnesota, Wisconsin, Michigan, Ohio, Pennsylvania, and Massachusetts in the north. As measures of monarch host-plant origin, age, activity, and distance moved, we determined the cardenolide concentrations, cardenolide fingerprints, and wing wear of these butterflies.

METHODS

The cardenolide concentrations of ether-defatted monarchs were measured by spectrophotometry at an absorbance wavelength of 622 nm on a Perkin-Elmer 559A spectrophotometer (Brower et al., 1982; Malcolm et al., 1989). Cardenolide concentrations were determined as μg cardenolide/0.1 g dry, defatted weight of butterfly, to control for variation in the fat weights carried by monarchs at different stages in their life history (Brower, 1985). Cardenolide fingerprints were produced by thin-layer chromatographic (TLC) separation of cardenolides in these extracts, after pigments were precipitated with lead acetate (Brower et al., 1975, 1982, 1984a; Malcolm et al., 1989). Silica gel 60 F_{254} TLC plates were run four times in chloroform : methanol : formamide (90:6:1, v:v) at 20°C and cardenolides were visualized as blue spots by spraying with a saturated solution of 2,2′-4,4′-tetranitrodiphenyl in benzene, followed by 10% potassium hydroxide in 50% aqueous methanol (Brower et al., 1982).

To determine the identity of cardenolide fingerprints in wild-caught monarchs, we compared their TLC cardenolide patterns with those of monarchs reared from a selection of known milkweeds. At present our comparative library of fingerprints includes monarchs reared from the *Asclepias* species that occur commonly in various parts of North America, east of the Rocky Mountains—*humistrata, viridis, curassavica, exaltata, speciosa,* and *syriaca* (see Fig. 3 for TLC examples). The fingerprints of monarchs reared on *A. viridis* and *A. asperula capricornu* in Texas and Louisiana have also been determined by Lynch and Martin (1987) and Martin and Lynch (1988). Since the cardenolide fingerprints of spring migrant monarchs were remarkably consistent, with almost all butterflies showing one of two very distinct patterns, we used published accounts of the predominant cardenolides to establish the identity of fingerprints as objectively as possible. The criteria we used for fingerprint identification included comparison of spot intensities and spot mobilities (relative to digitoxin) with published values. For example, the fingerprint of butterflies that fed as larvae on *A. syriaca* is dominated by the cardenolide glycoside aspecioside—the largest and most intense spot that appears approximately halfway between the origin and digitoxin in Figure 3 (see also Malcolm et al., 1989). The only other known fingerprint like this is for monarchs reared from *Asclepias speciosa* (Brower et al., 1984a; Seiber et al., 1986), but because this is a western equivalent of *A. syriaca* we think it is of only marginal significance to monarch breeding east of the Rockies. The other dominant fingerprint in our samples is derived from *A. viridis* and is characterized by more lower polarity cardenolides that move farther up the TLC plate than the cardenolides in *A. syriaca* monarchs. The two obvious sets of paired spots on either side of digitoxin (Fig. 3) are characteristic of *A. viridis*-derived monarchs. The small number of additional fingerprints were always very obviously different from the two predominant patterns. Although our fingerprint identifications do include a degree of subjective pattern comparison we consider that the fingerprints were so consistent within samples that our method is robust (see also Seiber et al., 1986). However, we are developing densitometry measurements of TLC patterns followed by discriminant analyses among sample patterns (Cockrell, Malcolm, and Brower, in prep.).

The location, date, and number of monarchs collected across the two transects, in the spring 1985 samples, are given in Table 2. The collections were designed to sample monarchs east of longitude 105°W in the United States, with western (<35°N, >90°W) and eastern (<35°N, <90°W) divisions of the southern range, and western (>40°N, >80°W) and eastern (>40°N, <80°W) divisions of the northern range (see Fig. 1). Thus, our southern samples are from Texas, Louisiana, and Florida, and the northern samples are from North Dakota east to western Pennsylvania, and Massachusetts (Table 2). The small samples of 19 and 13 from the southeast and northeast in comparison with more western samples of 114 and 633 reflect the much lower numbers of monarchs during the spring in the east. The dates of collection (Table 2) coincided with the known times of arrival of migrant monarchs; late March and early April in the Gulf states and late May and early June in Wisconsin and Minnesota (Brower, 1985; Malcolm et al., 1987; Cockrell et al., this volume; Lynch and Martin, this volume; Riley, this volume).

Cardenolide concentrations were determined from 562 Mexican overwintering monarchs collected in January, February, and March 1978 and 1979. We chromatographed 386 of these butterflies (Table 2). These samples were considered representative of typical overwintering cardenolide contents because Malcolm and Brower (1989) have shown that the cardenolide concentrations and amounts in eight samples, representing 839 monarchs from four overwintering seasons in Mexico, were all extremely similar.

Field-collected monarchs were stored frozen in glassine envelopes. In the laboratory, spermatophores were dis-

Table 1. Cardenolide concentrations in the leaves (μg/0.1 g dry weight) of three of the most abundant southern and the single most abundant northern *Asclepias* species, with the cardenolide concentrations of monarch butterflies reared on these four milkweed species.

Asclepias species	Origin	N[a]	Plant Mean	Plant SD	Plant Range	Butterfly Mean	Butterfly SD	Butterfly Range
Southern								
asperula[b]	TX	41	886	255	341–1,616	363	77	231–515
viridis[c]	LA	60	245	70	95–432	337	105	73–591
viridis[d]	FL	18	376	203	148–972	438	63	337–548
humistrata[d]	FL	22	389	141	71–639	337	65	242–478
Northern								
syriaca[e]	—[e]	158	50	39	4–229	234	150	0–792

[a] Number of plant–butterfly paired samples.

[b] Martin and Lynch (1988).

[c] Lynch and Martin (1987).

[d] S.B. Malcolm, material collected near Gainesville, Florida, 1983 and 1984.

[e] Malcolm et al. (1989); North Dakota east to Vermont and south to Virginia (see Fig. 1).

sected from the bursa copulatrix of females to eliminate possible male contribution to female cardenolides. Overwintering butterflies were not dissected since few females were mated (Brower, 1985; Van Hook, this volume). Wing wear of all butterflies was estimated by the same person on a subjective 0.5 interval scale of 1 (perfect condition) to 5 (extremely worn) based on degree of scale loss, wing fading, and extent of tearing and fraying. All butterflies were photographed to maintain a consistent wing wear estimate and reference record. Butterflies were then dried at 60°C for 16 hr, before fat extraction and cardenolide analysis.

RESULTS

CARDENOLIDE CONCENTRATIONS

Although the five samples of overwintering monarchs from Mexico (Table 2) showed variation among their mean cardenolide concentrations (Kruskal-Wallis ANOVA, $H_4 = 17.4$; $p < 0.01$), only sample 2 (Sierra Chincua, 23 January 1979) had significantly less cardenolide (mean = 63 μg/0.1 g) than samples 1 and 3 (means = 97 and 91 μg/0.1 g, respectively; Student-Newman-Keuls range test at 5%). Samples 4 and 5 (means = 81 and 72 μg/0.1 g, respectively) were not significantly different from any of the other samples. Since sample 2 was the only 1979 sample and the means of all samples were similarly low (between 50 and 100 μg), the differences were considered sufficiently small to permit pooling of the overwintering samples for comparison with wild-caught migrant monarchs.

The mean cardenolide concentrations in monarchs from overwintering sites in Mexico (77 μg/0.1 g) and from the early-spring samples in Florida, Louisiana, and Texas in the southern United States (62 μg/0.1 g) were not significantly different (Table 3; the 1979 Mexican sample described above makes no difference to the significance of this result), and the Mexican and southern United States groups show extremely similar distributions strongly skewed to the right (Fig. 4a). In contrast, the northern United States monarchs show a broad, normal distribution of cardenolide concentrations (Fig. 4b) and, on average, contained more than twice the amount of cardenolide (157 μg/0.1 g; Table 3). The highly significant difference between the Mexican and southern versus the northern data sets is strong evidence that the spring migrants arriving at the southern United States in early April are from the same population of monarchs that leave Mexico in mid- to late March, whereas the northern population fed as larvae on southern milkweeds, which are rich in cardenolides (Lynch and Martin, 1987, this volume; Martin and Lynch, 1988). We conclude that monarchs arriving in the northern United States are the first-generation offspring of the overwintered spring remigrants, i.e. that the successive brood recolonization strategy prevails.

Although the sample sizes for the eastern monarchs were small compared with those for the more western samples, the skewed southern and normal northern patterns of low to high cardenolides remained similar (Fig. 5a and 5b). However, the southern monarchs in the east had significantly higher cardenolide concentrations than the more western ones (Table 4). Most of this difference was caused by three monarchs with >200 μg/0.1 g of cardenolide. These individuals may have fed as larvae on the high cardenolide milkweed *Asclepias curassavica* in south Florida (Cohen, 1985; Malcolm and Brower, 1986), or even on the high cardenolide *Asclepias nivea* from Caribbean islands (Roeske et al., 1976). Alternatively, they may have fed on *A. humistrata* in Florida, but their early presence in north Florida in mid-April required either unusually early larval development or remarkable longevity, because *A. humistrata* is only avail-

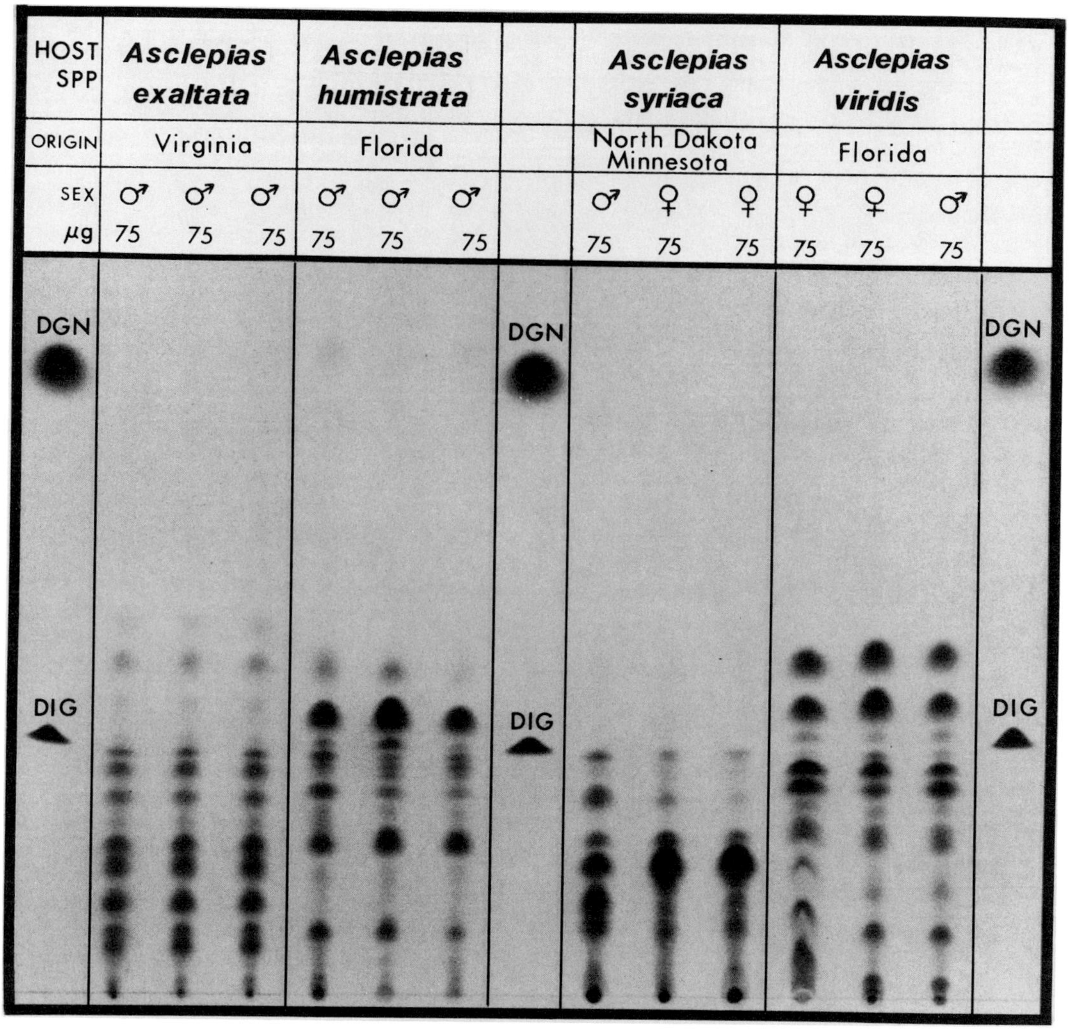

Figure 3. TLC fingerprints of larval host-derived cardenolide spots in adult monarchs reared from the milkweeds *A. exaltata, A. humistrata, A. syriaca,* and *A. viridis.* Each of the 12 monarch extracts were applied to the plate so that equal quantities of cardenolide per butterfly (75 µg) were separated during plate development. The cardenolide standards digitoxin (DIG) and digitoxigenin (DGN) are also shown.

able from early March until mid-July (Malcolm et al., 1987). In contrast, the high cardenolide concentrations of northeastern monarchs were not significantly different from those of the more western samples (Table 4).

CARDENOLIDE FINGERPRINTS

The most striking initial result of examining the TLC patterns of wild-caught monarchs is the small number of different cardenolide fingerprints that we found. In 94% of 1,148 wild overwintering and spring migrant monarchs we identified three common patterns (Table 5). The remaining 6% had zero or indeterminate cardenolide that could not be assigned a fingerprint (Table 5). Of these fingerprints, 92% Mexican and 84% southern United States monarchs had fingerprints that were assignable to *A. syriaca* [Table 5 and Fig. 6, compare the *A. syriaca* pattern of Malcolm et al. (1989), which is also illustrated in Fig. 3, with the Mexican and southern United States monarch patterns]. This fingerprint is easily recognizable by the high polarity of the most concentrated cardenolides such as aspecioside, syrioside, and syriobioside (Seiber et al., 1986; Malcolm et al., 1989). Thus almost all of the overwintered, Mexican butterflies that remigrate to the southern United States in spring must have fed as larvae on *A. syriaca* the previous year in the northern United States (contingency table comparison of fingerprint frequencies, $\chi^2 = 0.3$, df $= 2$, NS). Of the remaining southern monarchs, 9% had

Table 2. The location, date, and numbers of overwintering and migrating monarch adults [males (M), females (F), and totals] collected for analyses of cardenolide concentration, cardenolide fingerprint, and wing wear (see Fig. 1 for the distribution of collection sites).

Location	Date	Sample size M	Sample size F	Sample size Total
Overwintering monarchs in Mexico[a]				
1. Sierra Chincua, Michoacán	15 January 1978	58	47	105 (83)[b]
2. Sierra Chincua, Michoacán	23 January 1979	79	77	156 (97)
3. Cerro Pelón, México	5 February 1978	50	50	100 (80)
4. Cerro Altamirano, Michoacán	15 February 1978	50	50	100 (70)
5. Sierra Chincua, Michoacán	20 March 1978	51	50	101 (56)
Totals		288	274	562 (386)
Migrating monarchs in the United States (all 1985)				
Southern samples				
1. Central Texas (Counties: Travis, Gonzales, Blanco, De Witt, Refugio)	5–25 April	28	17	45
2. West Louisiana (Counties: Natchitoches, Caddo/DeSoto, St. Tammany)	6–28 April	37	24	61
3. Central Texas (County: Blanco)	9–12 May	5	3	8
4. North-central Florida (Counties: Franklin, Alachua, Levy, Putnam)	30 March–5 May	6	13	19
Totals		76	57	133
Northern samples				
1. East Wisconsin (Counties: Ozaukee, Milwaukee, Waukesha)	22–31 May	31	15	46
2. West Pennsylvania (County: Erie)	30 May	4	3	7
3. North Ohio (Counties: Ashtabula, Ottawa)	31 May–1 June	10	3	13
4. East Wisconsin (Counties: Ozaukee, Milwaukee, Door, Langlade, Taylor, Marquette, Waukesha)	1–9 June	221	80	301
5. East Michigan (County: Huron)	2–3 June	56	23	79
6. Central North Dakota (County: Oliver)	14–18 June	15	4	19
7. North Minnesota (County: Beltrami)	19–20 June	42	30	72
8. East Wisconsin (County: Door)	24 June	62	34	96
9. Central Massachusetts (Counties: Hampshire, Franklin)	26 May–22 June	8	5	13
Totals		449	197	646

[a] See Calvert and Brower (1986) for details of overwintering locations in Mexico.

[b] Mexican samples in parentheses were used for TLC analyses; otherwise TLC analyses were performed on all monarchs in a sample.

insufficient cardenolide to establish a fingerprint, and the other 7% may have been collected sufficiently late to be first-generation monarchs that fed as larvae on *A. viridis* and *A. humistrata* (Table 5). In marked contrast, 84% of northern monarchs had a fingerprint characteristic of the southern milkweed *A. viridis,* and 6% had the southeastern *A. humistrata* fingerprint (Table 5 and Fig. 6, compare the *A. viridis* pattern of Figure 3 with the northern United States monarch pattern). Thus, northern monarchs had very different fingerprints from those of southern monarchs (contingency table $\chi^2 = 471.3$,

Table 3. Comparisons of the cardenolide concentrations (μg cardenolide/0.1 g dry lean weight) in Mexican overwintering monarchs (Mexico), southern spring migrants (South), and northern spring migrants (North). Samples are compared by either Mann-Whitney *U* or Wilcoxon two-sample tests with significance set at 5%.

Monarch sample	N	Cardenolide concentration: Mean	SD	Range	Test statistic: *z*	*p*
Mexico	562	79	80	0–366	1.00 (Mexico vs South)	>0.16 NS
South	133	62	65	0–354	12.21 (South vs North)	<0.0001
North	646	157	81	0–395		

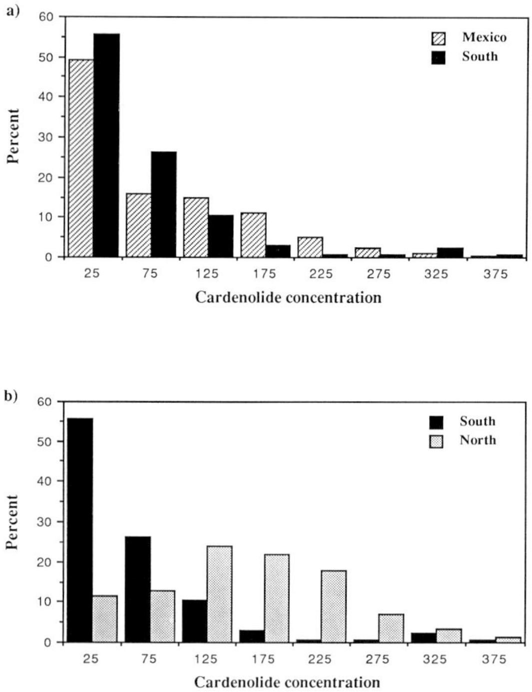

Figure 4. Comparisons of the distributions of cardenolide concentrations in (**a**) Mexican overwintering monarchs (Mexico) and southern United States monarchs (South) and (**b**) southern (South) and northern (North) United States monarchs. See Table 2 for sizes, dates, and locations of samples, and Table 3 for cardenolide concentration data.

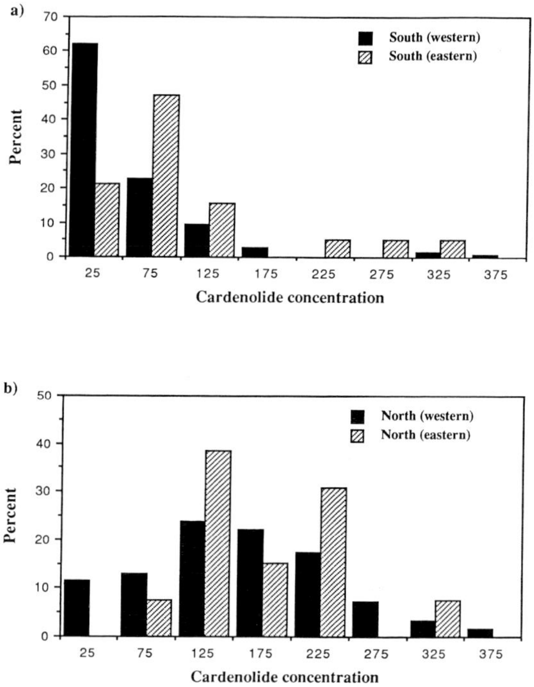

Figure 5. Regional comparisons of the cardenolide concentration distributions in (**a**) western and eastern samples of southern United States monarchs and (**b**) western and eastern samples of northern United States monarchs. Locations and times of collection of these monarchs are given in Table 2.

df = 3, $p = 0.0001$) and most of them must have reached their northern range by a successive brood recolonization strategy, dependent on the abundance of *A. viridis* throughout the southern states—especially Texas. The 6% of *A. syriaca* monarchs in the northern samples may represent either new, first spring generation butterflies reared from *A. syriaca* or overwintered migrants. If these were overwintered migrants, then a small percentage of monarchs may have completed the spring recolonization of their breeding range by a single sweep strategy.

It is interesting to note that although the northeastern sample was small (Table 6) significantly more *A. humistrata* fingerprints (67%) and significantly fewer *A. viridis* fingerprints (17%) were detected than expected in these monarchs, than in northwestern monarchs (contingency table $\chi^2 = 45.8$, df = 3, $p = 0.0001$). Consequently, the converse was also significant in that more *viridis*-derived, and fewer *humistrata*-derived monarchs occurred than expected in more western samples from the north. This regionality suggests that the first-generation spring migrants that fed as larvae on *A. humistrata* in the southeastern coastal plain were largely constrained to the eastern side of the Appalachian mountains in their northward migration, whereas west of the Appalachians monarchs are likely to be derived from *A. viridis*.

Each cardenolide fingerprint of the three southeastern, high cardenolide monarchs described above is most like that derived from *A. humistrata*. Thus, despite their early arrival in Florida (9, 15, and 17 April), we think these monarchs may have been very early first-generation butterflies that fed as larvae on *A. humistrata* in south Florida during March (see Cockrell et al., this volume, for details of breeding constraints in Florida). This may indicate the significance of monarchs overwintering in Florida (Brower, 1985) for early recolonization of southeastern milkweeds.

WING WEAR

The condition of monarch wings is consistent with the cardenolide evidence for successive brood recolonization of North American milkweeds. Monarchs collected in the southern United States in April had wings in significantly worse condition than monarchs from the northern United States in May and June (Fig. 7 and Table 7), suggesting that later, northern arrivals were younger migrants than earlier, southern arrivals.

In regional comparisons, the wings of south-

Table 4. Comparisons of the cardenolide concentrations (μg cardenolide/0.1 g dry lean weight) in spring migrant monarchs from western and eastern subsamples of butterflies collected in the south and north of the United States east of the Rocky Mountains. Samples were compared by Wilcoxon two-sample tests with significance at 5%.

Monarch sample	N	Cardenolide concentration Mean	SD	Range	Test statistic z	p
South						
Western	114	56	60	0–354	2.84	0.005
Eastern	19	99	82	12–314		
North						
Western	633	156	81	0–395	1.04	0.30 NS
Eastern	13	177	65	57–305		

western and southeastern monarchs (Fig. 8a) showed no significant difference in wing wear (Table 8). Despite the overall, similarly worn wings of eastern and more western monarchs from the south, the three southeastern monarchs with the high cardenolide concentrations described above also had wings in good condition with scores of 1.5, 2, and 2.5. This suggests that these butterflies were not overwintered migrants and that they had not flown very far, or for very long, before being caught in north-central Florida.

In contrast, the wings of northwestern monarchs were significantly more worn than those of northeastern monarchs (Fig. 8b and Table 8). The small sample of northeastern monarchs were all in excellent condition, whereas the large northwestern sample of 646 butterflies included 46 battered monarchs, with wing wear scores of 4, 4.5, and 5, caught in June (Fig. 8b). Since 43 of these butterflies showed the *A. viridis* fingerprint and 2 had the *A. humistrata* fingerprint (with one unknown), and 43 of the 46 were late June arrivals, caught on either 20 or 24 June in Minnesota and Wisconsin (3 were caught on 6, 8, and 9 June in Wisconsin), these must have been first-generation monarchs from farther south that had sufficient time to become well worn during migratory and reproductive activities. Interestingly, 80% of these 46, mostly late monarchs were females, which was a significantly higher percentage than the 30% of females in the complete sample of 646 northern butterflies ($\chi^2 = 45.6$, df $= 1$, $p = 0.0001$; Table 2). There were also significantly more females in this worn subsample than in the total sample of 162 late monarchs caught on 20 and 24 June ($\chi^2 = 21.1$, df $= 1$, $p = 0.0001$). Thus, the worn females provide good evidence for the "costs" of activities such as mating and egg laying. In contrast, the male-biased sex ratio of the 162 late, less-worn monarchs did not differ from the total northern sample of 646 monarchs ($\chi^2 = 3.2$, df $= 1$, p = NS).

On first consideration, these worn butterflies suggest the presence of overwintered migrants that used a single sweep recolonization strategy to reach their northern breeding area. However, their remarkably consistent *A. viridis* fingerprint dispels this suggestion. A further group of butterflies, the 6% of northern arrivals with the *A. syriaca* fingerprint (Table 5), could also support the single sweep hypothesis if they showed well-worn wings characteristic of old and overwintered long-distance migrants. However, unlike the group of well-worn butterflies, these 35 *A. syriaca*-derived butterflies had the same sex ratio as the total northern sample ($\chi^2 = 1.3$, df = 1, p = NS) and had wings in relatively good condition with a mean score of 2.4 (SD = 0.5), which was almost identical to that of the total northern sample (Table 7). Most of these *A. syriaca*-derived monarchs were collected in Minnesota, Wisconsin, and Michigan (plus one from Ohio and one from Massachusetts) between 29 May and 9 June (two were collected on 20 June and two on 24 June). Because of their good condition, we conclude that they too are first spring generation individuals which must have exploited early flushing *A. syriaca* in May as the plants first appeared at the southern edge of this milkweed's distribution, from Kansas east to Kentucky (Fig. 1).

Table 5. Percentages of the different *Asclepias* species-specific cardenolide fingerprints assigned by TLC in 1,148 Mexican overwintering monarchs and spring migrant monarchs from the southern and northern United States. Three fingerprint patterns were recognized in 94% of the total samples, and 6% of the total was indeterminate.

Asclepias species	Over-wintering Mexico (%)	Migrants: Southern United States (early spring) (%)	Northern United States (late spring) (%)
viridis	0	3	84
humistrata	0	4	6
syriaca	92	84	6
Indeterminate	8	9	5
Sample size	386	133	629

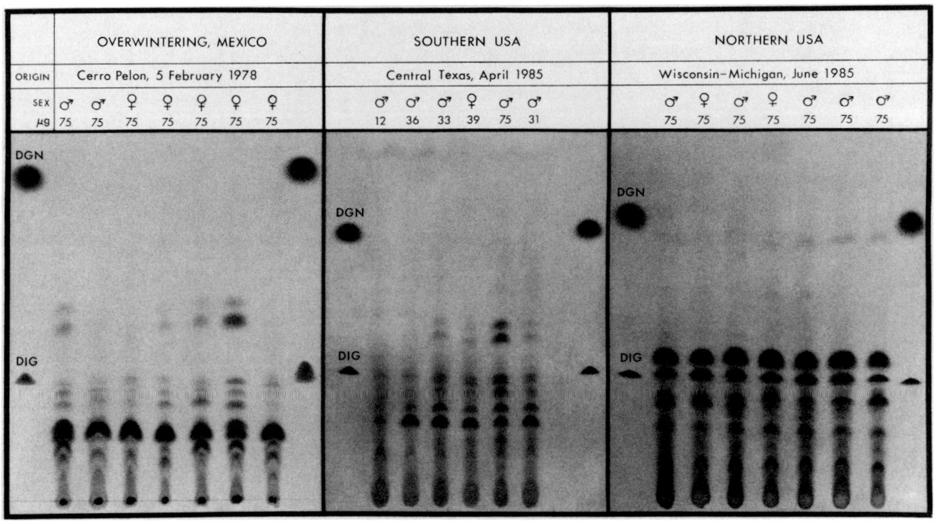

Figure 6. Composite TLC plate of cardenolide fingerprints in monarchs from a Mexican overwintering site (Cerro Pelón), the southern United States (central Texas), and the northern United States (Wisconsin and Michigan). Mexican and southern monarchs have the *A. syriaca* fingerprint shown in Figure 3, and the northern monarchs show the *A. viridis* fingerprint. Where possible each sample was applied to the plate so that equal quantities of cardenolide per butterfly (75 μg) were separated in the solvent system. Only one of the southern monarchs contained sufficient cardenolide, since the other five butterflies had the very low cardenolide concentrations and variable spot intensities, characteristic of southern spring remigrants. The three different TLC plates are aligned by the spot origin and by the cardenolide glycoside standard digitoxin (DIG). The cardenolide genin digitoxigenin (DGN) is also applied to each plate, but variation among plates usually makes it impossible to align plates at both standards. However, via chromatography with additional known plant cardenolides we know that most overwintering monarchs and southern spring arrivals have the same fingerprint pattern, derived from the northern milkweed *A. syriaca* (see Malcolm et al., 1989, for details).

DISCUSSION

Our results show distinct differences in the cardenolide concentrations, cardenolide fingerprints, and wing conditions between the first migrant monarch arrivals in the southern and northern United States. All three sets of evidence show that nearly all of the spring migrant arrivals in the north are the offspring of overwintered remigrants that arrived in the southern United States from Mexico. In Texas and Louisiana these overwintered remigrants laid most of their eggs on the extremely abundant early-spring, southern milkweed *A. viridis*. This milkweed has high concentrations of cardenolide, which produces monarchs with correspondingly high cardenolide concentrations. Although these monarchs appear to lose some of this host-derived cardenolide during their northward, spring migration (Malcolm and Brower, 1989) it is these monarchs that reach the northern United States, with very few, if any, of their overwintered parents reaching as far north as Wisconsin or Michigan. Thus, the spring remigration of monarchs, to recolonize their milkweed larval food resources, is achieved almost exclusively by a successive brood strategy and not by a single sweep of overwintered Mexican migrants. Using completely different methods, which focus on the abiotic constraints faced by monarchs as they migrate north in the spring, Cockrell et al. (this volume) reach the same conclusion.

The monarchs produced in the first spring generation also show some regionality in their continued northward migration and colonization. Thus, those that fed as larvae on the common southeastern milkweed, *A. humistrata*, are later found as far north as Massachusetts. Similarly *A. viridis* is extremely abundant in Texas and Oklahoma and

Table 6. Regional percentages of the different *Asclepias* species-specific cardenolide fingerprints identified by TLC in spring migrant monarchs from the western and eastern subsamples of monarchs from the south and north of the United States, east of the Rocky Mountains.

	South		North	
***Asclepias* species**	**Western (%)**	**Eastern (%)**	**Western (%)**	**Eastern (%)**
viridis	4	0	85	17
humistrata	0	26	5	67
syriaca	87	69	5	17
Indeterminate	10	5	5	0
Sample size	114	19	623	6

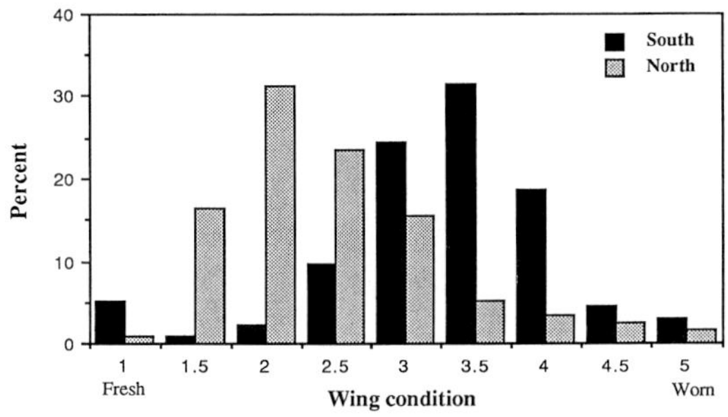

Figure 7. Wing wear distributions of monarch spring arrivals in the southern (South) and northern (North) United States (1 = fresh wings in pristine condition, in 0.5 intervals to 5 = extremely worn and tattered wings). The samples are compared statistically in Table 7.

monarchs with the *A. viridis* fingerprint dominate the monarch samples collected in North Dakota, Minnesota, Wisconsin, Michigan, Ohio, and Pennsylvania, west of the Appalachian Mountains. Our evidence thus suggests that *viridis* monarchs keep to the central United States and *humistrata* monarchs keep to the eastern United States when migrating north. Perhaps the natural geographical barrier of the Appalachians causes this regional migration.

Even though we exerted substantially more field effort to collect monarchs in the southeast and northeast, we caught 6 times more monarchs in the more western samples from the south than in the southeast and 49 times more monarchs in the more western samples from the north than in the northeast. Hence, we suggest that our sample sizes reflect the regionality of monarch spring migration, from which we conclude that most monarchs are migrating from Mexico to Texas, where they breed extensively in early spring on the abundant *A. viridis* (see Cockrell et al., this volume; Lynch and Martin, this volume). Then their offspring migrate in very large numbers to the Great Lakes area. In contrast, considerably fewer monarchs remigrate from Mexico to Florida and correspondingly small numbers of their offspring appear to migrate north up the eastern coastal plain to the northeast. It is also possible that some of the spring butterflies in Florida may have overwintered in Florida along the

Table 7. Comparison of the wing conditions in southern spring (South) and northern spring (North) migrant monarchs. Samples are compared by a Wilcoxon two-sample test.

Monarch sample	N	Wing condition		Test statistic	
		Mean	SD	*z*	*p*
South	133	3.28	0.83	10.98	<0.0001
North	646	2.43	0.78		

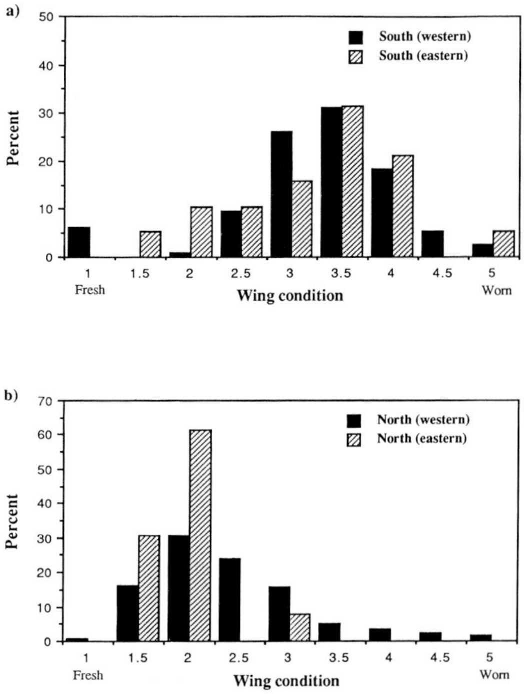

Figure 8. Wing wear distributions of (**a**) western and eastern monarch samples from the south and (**b**) western and eastern monarch samples from the north of the United States (1 = fresh wings in pristine condition, in 0.5 intervals to 5 = extremely worn and tattered wings). The samples are compared statistically in Table 8.

coast of the Gulf of Mexico (Fig. 1), where small and mobile overwintering colonies occur (Brower, 1985).

As monarchs remigrate into Texas each spring they encounter the widely distributed and abundant milkweeds *A. viridis* and *A. asperula capricornu* (Fig. 1; Lynch and Martin, this volume). These two

Table 8. Regional comparisons of wing conditions in spring migrant monarchs from the western and eastern subsamples of monarchs from the south and north of the United States, east of the Rocky Mountains. Samples are compared by Wilcoxon two-sample tests with significance at 5%.

Monarch sample	N	Wing condition		Test statistics	
		Mean	SD	*z*	*p*
South					
Western	114	3.29	0.82	0.29	0.78 NS
Eastern	19	3.24	0.86		
North					
Western	633	2.44	0.79	2.70	0.007
Eastern	13	1.92	0.40		

plants are chemically and morphologically similar and are known to hybridize in their limited region of sympatry from central Texas to Oklahoma (Lynch and Martin, 1987, this volume; Martin and Lynch, 1988). Since no other milkweed species approaches the abundance of these two, it is curious that we have failed to identify *A. asperula*-derived monarchs in our field-collected samples. Two explanations may account for this discrepancy. Either the majority of our northern samples (from Minnesota, Wisconsin, and Michigan) were collected too far east to include *A. asperula* monarchs (which may migrate due north to the Dakotas), or we have failed to distinguish between the *A. viridis* and *A. asperula* fingerprints. Although the cardenolide fingerprints of these two species appear very distinct in the analyses of Lynch and Martin (1987) and Martin and Lynch (1988), the authors claim strong similarity in the two patterns. Furthermore, our *A. viridis* fingerprint from Florida (Fig. 3) is similar to that of *A. asperula* from Texas (Martin and Lynch, 1988, fig. 4). At present we cannot distinguish between these two alternative explanations and further work is needed for their resolution.

In addition to showing that the host-plant recolonization strategy of the monarch butterfly is predominantly one of successive brood migration, our data contradict prior generalizations about the physiology of insect migration and reproduction. Thus, previous authors (Johnson, 1960, 1969; Kennedy, 1961, 1985) have suggested that the reproductive physiology of migratory insects is repressed during migration, much like the reproductive physiology of nonmigratory insects that diapause and overwinter *in situ* (Rogers, 1983). Such reproductive repression in migratory insects is commonly referred to as the "oogenesis-flight syndrome" (Johnson, 1969; Dingle, 1980; Rankin et al., 1986) and reproduction occurs only when migratory behavior has ended with the arrival of a migrant at an appropriate resource site. The monarch butterfly clearly violates this syndrome since both overwintered butterflies and the subsequent first-generation butterflies lay eggs on available milkweeds as they migrate from Mexico to southern Canada. Furthermore, monarchs begin to mate before they leave the Mexican overwintering sites (Brower, 1985; Van Hook, this volume), and females almost certainly can lay eggs at any time or location along their northward migration. Although the oogenesis-flight syndrome does appear to exist in autumn migrant monarchs, on their way south to overwintering sites, Rankin (1985) and Rankin et al. (1986) found that it was much weaker in field-collected monarchs, especially in spring, than in laboratory-reared monarchs (see also Herman, 1985, this volume). This suggests that the oogenesis-flight syndrome is a dynamic phenomenon in monarchs that may or may not exist according to the interplay between the cues used to switch the syndrome on or off. The syndrome may be stronger in the autumn when monarchs migrate south to nonreproductive overwintering sites, and weaker, or absent in the spring when the same monarchs migrate north, especially since the spring remigrants need to lay eggs on the southern milkweeds they encounter. Thus, rather than definable in physiological terms (Kennedy, 1985), monarch migration is primarily an ecological phenomenon in which evolution has struck a dynamic balance between the benefits for monarchs in finding the best food resources for their larvae and the costs of carrying eggs and finding mates. The next generation faces the same problems, but the abiotic daylength and temperature cues in May (see Cockrell et al., this volume), to which an oogenesis-flight syndrome might respond, are weak and so we should expect a correspondingly weak or nonexistent syndrome. Nevertheless, the first generation of spring monarchs must be migratory in order to colonize the superabundant milkweed resource of *A. syriaca* in the north. These monarchs may well minimize the costs of carrying heavy, mature eggs and spermatophores by laying eggs as fast as they develop while migrating northward.

In conclusion, the differences between the cardenolide concentrations, fingerprints, and wing wear of southern and northern monarchs are definitive evidence for successive brood migration. These differences are also interesting in terms of the use of cardenolides by monarchs in their aposematic defense against avian predation (Brower, 1984). Southern, overwintered monarchs with low cardenolide concentrations will be poorly protected against naive bird predators, which could result in high mortality of these spring remigrants. However, these monarchs arrive before the major fledging period of immature and naive bird predators and so they are likely to encounter experienced, mature birds that may ignore them as aposematic prey. In contrast, the northern monarch arrivals, with high cardenolide contents, will be well protected against experienced and also naive birds, of which the first broods will be fledging at the time of arrival of these toxic monarchs in late May and early June (cf. Waldbauer and Sheldon, 1971; Waldbauer, 1988).

In comparison with other danaid species, a large investment in chemical defense is highly characteristic of the monarch (Ackery and Vane-Wright, 1984; Brower, 1984; Ackery, this volume; Schneider, this volume) and so predation is likely to be a strong selective force in shaping and maintaining the monarch's life history. In fact, the remarkable distinction between the cardenolide contents of southern and northern spring migrants may contribute strongly to the maintenance of selection for monarch migration (Malcolm and Brower, 1989). Since monarchs can tolerate the moderate freezing temperatures of their overwintering sites (Anderson and Brower, this volume) it is curious that they have not evolved sufficient freeze tolerance to diapause within the distributions of abundant milkweeds like *A. viridis* and *A. syriaca* and eliminate the costs of migration (Baker, 1978). However, monarchs are also forced out of the southern states each summer by the onset of lethally high temperatures after completing, at most, two spring generations (Malcolm et al., 1987; Cockrell et al., this volume). By migrating farther north to *A. syriaca* (Fig. 1), monarchs can complete an additional three summer generations on this host plant (Malcolm et al., 1987) and maximize the number of autumn migrants that journey to Mexican overwintering sites. Thus, we suggest that an annual cycle of well-defended, first-generation spring migrants arriving in the northern United States, followed by less well-defended, breeding nonmigrants and then similarly defended autumn migrants, would allow monarchs to exploit any of their *Asclepias* host plants, in both space (from northern Mexico to southern Canada) and time (from late March to September). The weakest link in this cycle is the apparent loss of cardenolides during migration (Malcolm and Brower, 1989) resulting in the relatively poor defense of aggregated, overwintering monarchs in Mexico that suffer considerable predation from both birds and mice (Calvert et al., 1979; Fink and Brower, 1981; Fink et al., 1983; Brower and Calvert, 1985; Brower and Fink, 1985; Brower et al., 1985; Glendinning et al., 1988; Glendinning, this volume; Arellano G. et al., this volume).

ACKNOWLEDGMENTS

We are grateful to the following for major efforts with helping us collect spring migrant monarchs: Tonya Van Hook (FL), Alan Masters, Julie Gregg, Andrew Bunyan, and Paul Davis (TX), Steve Lynch, Ron Martin, and Tom Riley (LA), Andrew Brower (PA-ND), Susan Borkin (WI), Mark Scriber with J. Thorne, J. Sibenhorn, V. Ware, and M. Evans (WI), and Theodore Sargent (MA), and to William Calvert for help in collecting overwintering monarchs in Mexico. Tonya Van Hook, Julia Frey, Susan Glazier, Lee Hedrick, and Margaret Shepard assisted extensively with laboratory analyses. The research was funded by NSF grant BSR-8500416.

LITERATURE CITED

Ackery, P.R., and R.I. Vane-Wright. 1984. *Milkweed butterflies: Their cladistics and biology.* London: British Museum (Natural History); and Ithaca, NY: Cornell University Press, 425 pp.

Baker, R.R. 1978. *The evolutionary ecology of animal migration.* London: Hodder and Stoughton, 1012 pp.

Brower, L.P. 1961. Studies on the migration of the monarch butterfly I. Breeding populations of *Danaus plexippus* and *D. gilippus berenice* in south central Florida. *Ecology* 42:76–83.

Brower, L.P. 1984. Chemical defense in butterflies. In *The biology of butterflies,* eds. R.I. Vane-Wright and P.R. Ackery, *Symposia of the Royal Entomological Society of London* 11:109–134, London and New York: Academic Press (1989, Princeton, NJ: Princeton University Press).

Brower, L.P. 1985. New perspectives on the migration biology of the monarch butterfly, *Danaus plexippus* L. In *Migration: Mechanisms and adaptive significance,* ed. M.A. Rankin, 748–785. Austin: University of Texas, *Contributions in Marine Science,* Supplement, 27.

Brower, L.P., and W.H. Calvert. 1985. Foraging dynamics of bird predators on overwintering monarch butterflies in Mexico. *Evolution* 39:852–868.

Brower, L.P., and L.S. Fink. 1985. A natural toxic defense system: Cardenolides in butterflies versus birds. *Annals of the New York Academy of Sciences* 443: 171–188.

Brower, L.P., M. Edmunds, and C.M. Moffitt. 1975. Cardenolide content and palatability of a population of *Danaus chrysippus* butterflies from West Africa. *Journal of Entomology (A)* 49:183–196.

Brower, L.P., J.N. Seiber, C.J. Nelson, S.P. Lynch, and P.M. Tuskes. 1982. Plant-determined variation in the cardenolide content, thin-layer chromatography profiles, and emetic potency of monarch butterflies, *Danaus plexippus* reared on the milkweed, *Asclepias eriocarpa* in California. *Journal of Chemical Ecology* 8:579–633.

Brower, L.P., J.N. Seiber, C.J. Nelson, S.P. Lynch, and M.M. Holland. 1984a. Plant-determined variation in the cardenolide content, thin-layer chromatography profiles, and emetic potency of monarch butterflies, *Danaus plexippus* L. reared on milkweed plants in California: 2. *Asclepias speciosa. Journal of Chemical Ecology* 10:601–639.

Brower, L.P., J.N. Seiber, C.J. Nelson, S.P. Lynch, M.P. Hoggard, and J.A. Cohen. 1984b. Plant-determined variation in cardenolide content and thin-layer chromatography profiles of monarch butterflies, *Danaus plexippus* reared on milkweed plants in California: 3. *Asclepias californica. Journal of Chemical Ecology* 10:1823–1857.

Brower, L.P., B.E. Horner, M.A. Marty, C.M. Moffitt, and B. Villa-R. 1985. Mice (*Peromyscus maniculatus, P. spicilegus,* and *Microtus mexicanus*) as predators of overwintering monarch butterflies (*Danaus plexippus*) in Mexico. *Biotropica* 17:89–99.

Calvert, W.H, and L.P. Brower. 1986. The location of monarch butterfly (*Danaus plexippus* L.) overwintering colonies in Mexico in relation to topography and climate. *Journal of the Lepidopterists' Society* 40:164–187.

Calvert, W.H., L.E. Hedrick, and L.P. Brower. 1979. Mortality of the monarch butterfly (*Danaus plexippus* L.): Avian predation at five overwintering sites in Mexico. *Science* 204:847–851.

Cohen, J.A. 1985. Differences and similarities in cardenolide contents of queen and monarch butterflies in Florida and their ecological and evolutionary implications. *Journal of Chemical Ecology* 11:85–103.

Dingle, H. 1980. Ecology and evolution of migration. In *Animal migraton, orientation, and navigation,* ed. S.A. Gauthreaux, Jr., 1–101. New York: Academic Press.

Fink, L.S., and L.P. Brower. 1981. Birds can overcome the cardenolide defense of monarch butterflies in Mexico. *Nature* 291:67–70.

Fink, L.S., L.P., Brower, R.B. Waide, and P.R. Spitzer. 1983. Overwintering monarch butterflies as food for insectivorous birds in Mexico. *Biotropica* 15: 151–153.

Glendinning, J.I., A. Alonso Mejía, and L.P. Brower. 1988. Behavioral and ecological interactions of foraging mice (*Peromyscus melanotis*) with overwintering monarch butterflies (*Danaus plexippus*) in Mexico. *Oecologia* 75:222–227.

Herman, W.S. 1985. Hormonally mediated events in adult monarch butterflies. In *Migration: Mechanisms and adaptive significance,* ed. M.A. Rankin, 799–815. Austin: University of Texas, *Contributions in Marine Science,* Supplement, 27.

Johnson, C.G. 1960. A basis for a general system of insect migration and dispersal by flight. *Nature* 186: 348–350.

Johnson, C.G. 1969. *Migration and dispersal of insects by flight.* London: Methuen, 763 pp.

Kennedy, J.S. 1961. A turning point in the study of insect migration. *Nature* 189:785–791.

Kennedy, J.S. 1985. Migration, behavioral and ecological. In *Migration: Mechanisms and adaptive significance,* ed. M.A. Rankin, 5–26. Austin: University of Texas, *Contributions in Marine Science,* Supplement, 27.

Lynch, S.P., and R.A. Martin. 1987. Cardenolide content and thin-layer chromatography profiles of monarch butterflies, *Danaus plexippus* L., and their larval host-plant milkweed, *Asclepias viridis* Walt., in northwestern Louisiana. *Journal of Chemical Ecology* 13:47–69.

Malcolm, S.B. 1987. Monarch butterfly migration in North America: Controversy and conservation. *Trends in Ecology and Evolution* 2:135–138.

Malcolm, S.B., and L.P. Brower. 1986. Selective oviposition by monarch butterflies (*Danaus plexippus* L.) in a mixed stand of *Asclepias curassavica* L. and *A. incarnata* L. in south Florida. *Journal of the Lepidopterists' Society* 40:255–263.

Malcolm, S.B., and L.P. Brower. 1989. Evolutionary and ecological implications of cardenolide sequestration in the monarch butterfly. *Experientia* 45:284–295.

Malcolm, S.B., B.J. Cockrell, and L.P. Brower. 1987. Monarch butterfly voltinism: Effects of temperature constraints at different latitudes. *Oikos* 49:77–82.

Malcolm, S.B., B.J. Cockrell, and L.P. Brower. 1989. The cardenolide fingerprint of monarch butterflies reared on the common milkweed, *Asclepias syriaca. Journal of Chemical Ecology.* 15:819–853.

Martin, R.A., and S.P. Lynch. 1988. Cardenolide content and thin-layer chromatography profiles of monarch butterflies, *Danaus plexippus* L., and their larval host-plant milkweed, *Asclepias asperula* subsp. *capricornu* (Woods.) Woods., in north central Texas. *Journal of Chemical Ecology* 14:295–318.

Rankin, M.A. 1985. Endocrine influence on flight behavior. In *Migration: Mechanisms and adaptive significance,* ed. M.A. Rankin, 817–841. Austin: University of Texas, *Contributions in Marine Science,* Supplement, 27.

Rankin, M.A., M.L. McAnelly, and J.E. Bodenhamer. 1986. The oogenesis-flight syndrome revisited. In *Insect flight: Dispersal and migration,* ed. W. Danthanarayana, 27–48. Heidelberg and Berlin: Springer-Verlag.

Roeske, C.N., J.N. Seiber, L.P. Brower, and C.M. Moffitt. 1976. Milkweed cardenolides and their comparative processing by monarch butterflies (*Danaus plexippus* L.). *Recent Advances in Phytochemistry* 10: 93–167.

Rogers, D. 1983. Pattern and process in large-scale animal movement. In *The ecology of animal movement,* ed. I.A. Swingland and P.J. Greenwood, 160–180. Oxford: Clarendon Press.

Seiber, J.N., L.P. Brower, S.M. Lee, M.M. McChesney, H.T.A. Cheung, C.J. Nelson, and T.R. Watson. 1986. Cardenolide connection between overwintering monarch butterflies from Mexico and their larval food plant, *Asclepias syriaca. Journal of Chemical Ecology* 12:1157–1170.

Urquhart, F.A. 1960. *The monarch butterfly.* Toronto: University of Toronto Press, 361 pp.

Urquhart, F.A. 1966. A study of the migrations of the Gulf Coast population of the monarch butterfly (*Danaus plexippus* L.) in North America. *Annales Zoologici Fennici* 3:82–87.

Urquhart, F.A. 1976. Found at last: The monarch's winter home. *National Geographic* 150:160–173.

Urquhart, F.A. 1987. *The monarch butterfly: International traveler.* Chicago: Nelson-Hall, 232 pp.

Urquhart, F.A., and N.R. Urquhart. 1976a. Migration of butterflies along the Gulf Coast of northern Florida. *Journal of the Lepidopterists' Society* 30:59–61.

Urquhart, F.A., and N.R. Urquhart. 1976b. A study of the peninsular Florida populations of the monarch butterfly (*Danaus p. plexippus:* Danaidae). *Journal of the Lepidopterists' Society* 30:73–87.

Urquhart, F.A., and N.R. Urquhart. 1976c. The overwintering site of the eastern population of the monarch butterfly (*Danaus p. plexippus:* Danaidae) in southern Mexico. *Journal of the Lepidopterists' Society* 30:153–158.

Urquhart, F.A., and N.R. Urquhart. 1977. Overwintering areas and migratory routes of the eastern population of the monarch butterfly (*Danaus p. plexippus,* Lepidoptera: Danaidae) in North America with special reference to the western population. *Canadian Entomologist* 109:1583–1589.

Urquhart, F.A., and N.R. Urquhart. 1978. Autumnal migration routes of the eastern population of the monarch butterfly (*Danaus p. plexippus* L.; Danaidae; Lepidoptera) in North America to the overwintering site in the neovolcanic plateau of Mexico. *Canadian Journal of Zoology* 56:1759–1764.

Urquhart, F.A., and N.R. Urquhart. 1979. Vernal migration of the monarch butterfly (*Danaus p. plexippus,* Lepidoptera: Danaidae) in North America from the overwintering site in the neo-volcanic plateau of Mexico. *Canadian Entomologist* 111:15–18.

Waldbauer, G.P. 1988. Asynchrony between Batesian mimics and their models. In *Mimicry and the evo-*

lutionary process, ed. L.P. Brower, S103–S121. Chicago: University of Chicago Press, *The American Naturalist*, Supplement to Vol. 131.

Waldbauer, G.P., and J.K. Sheldon. 1971. Phenological relationships of some aculeate hymenoptera, their dipteran mimics and insectivorous birds. *Evolution* 25:371–382.

Williams, C.B. 1958. *Insect migration*. London: Collins, 235 pp.

Woodson, R.E., Jr. 1954. The North American species of *Asclepias* L. *Annals of the Missouri Botanical Garden* 41:1–211.

SPRING MIGRATION AND OVIPOSITION OF THE MONARCH BUTTERFLY IN LOUISIANA[1]

Thomas J. Riley

ABSTRACT. Occurrence of remigrant monarch butterflies and their oviposition on two species of *Asclepias* were documented in southern and central Louisiana from 1983 to 1988. Butterflies first appeared during the third week of March and were most abundant during early April. Oviposition on *Asclepias longifolia* and *Asclepias amplexicaulis* was greatest during the second week of April; it then rapidly declined and ceased by 20 April. The movement of remigrant monarchs through Louisiana was temporally consistent from year to year and was closely synchronized with the availability of suitable milkweed host species.

INTRODUCTION

The annual fall migration of the eastern United States population of the monarch butterfly, *Danaus plexippus* (L.), is a well-known phenomenon (Urquhart and Urquhart, 1978). Beginning in September, the butterflies leave temperate North America, escaping senescing larval food plants and freezing winter temperatures, and migrate to overwintering sites in the oyamel fir forests of the transvolcanic mountain range of central Mexico (Urquhart and Urquhart, 1976a). Here the interaction of climate, topography, and vegetation maintains the insects in a state of reduced reproductive and physical activity (Brower et al., 1977; Calvert and Brower, 1981, 1986; Calvert and Cohen, 1983; Calvert et al., 1983; Brower, 1985; Anderson and Brower, this volume; Calvert and Lawton, this volume; Van Hook, this volume). By late February and early March, increasing temperatures and day lengths are thought to trigger renewed activity in the colonies, culminating with the dispersal of the butterflies as they begin their remigration north to the United States (Brower, 1985).

Little is known of the fate of the monarchs once they leave their overwintering areas in Mexico and begin flying northward each spring. An early observation of this remigration was reported by Heitzman (1962). He sighted many monarchs moving northeast in southern Tamaulipas, Mexico, and extreme southern Texas on 21–27 March 1962. Urquhart and Urquhart (1976b, 1979) reported that monarchs begin to appear in Texas and Florida in early April, and Cohen and Brower (1982) and Malcolm et al. (1987) reported monarch oviposition on *Asclepias humistrata* Walt. in northern Florida during April (see also Cockrell et al., this volume; Lynch and Martin, this volume; Malcolm et al., this volume).

Recent research suggests that changing environmental conditions in February and March stimulate activity and oogenesis in the Mexican overwintering colonies (Brower, 1985; Calvert and Brower, 1986; Herman, 1985, this volume; Van Hook, this volume). This, along with the earlier reports of monarch movement, suggests that the timing of the spring remigration of overwintered monarchs into eastern North America is temporally consistent from year to year.

In this paper, I report on the spring occurrence of remigrant monarchs in Louisiana over a 6-yr period and their oviposition on two species of *Asclepias* during 1984 and 1985.

METHODS

Adult monarchs were recorded through capture and visual sightings. The latter were made during the months of March–May of 1983–1986. The first sightings of monarchs in Baton Rouge in 1987 and 1988 are also included. Sightings were made in the vicinity of Baton Rouge and

1. Approved for publication by the director of the Louisiana Agricultural Experiment Station as manuscript number 88-17-2437.

Table 1. **Date of first spring sightings of adult monarchs in East Baton Rouge Parish, Louisiana, 1983–1987.**

Year	Date
1983	27 March
1984	30 March
1985	26 March
1986	24 March
1987	21 March
1988	13 March

during 10-hr periods per week spent traveling by car from Baton Rouge to Alexandria and from Baton Rouge to Monroe. When a butterfly was sighted, its physical condition was estimated based on its coloration. Faded, light-colored, and worn butterflies were assumed to be remigrants, whereas fresh butterflies, with deep bright colors, were considered to be newly emerged, in agreement with Malcolm et al. (this volume).

Egg and larval counts were made from 7 April to 7 May 1984 and from 30 March to 25 May 1985 in the Red Dirt Wildlife Area, Kisatchie National Forest, Natchitoches Parish, Louisiana (31°30′N, 93°30′W). In 1984, counts were made only on *Asclepias longifolia* Michx., whereas the 1985 counts were made on *A. longifolia* and *Asclepias amplexicaulis* Sm. Plants were marked with numbered wooden stakes. Oviposition and larval development were determined by counting eggs and larvae on each marked plant at approximately 5-day intervals.

Asclepias obovata Ell., *Asclepias tuberosa* L., *Asclepias verticillata* L., and *Asclepias viridis* Walt. also occur in the study area. These species were not located until 15 April through mid-May 1985 and thus, were not monitored simultaneously with *A. longifolia* and *A. amplexicaulis.* Because they were not observed regularly, and individuals of each species were not monitored over an equal period of time, only the presence or absence of monarch eggs or larvae was noted for these four species.

RESULTS AND DISCUSSION

The first monarch sightings were consistently recorded during March in East Baton Rouge Parish (Table 1). The number of monarchs sighted was greatest during the last week of March through mid-April (Table 2). All monarchs observed in March and April were moderately to severely worn, and their colors were faded. Thus, their physical

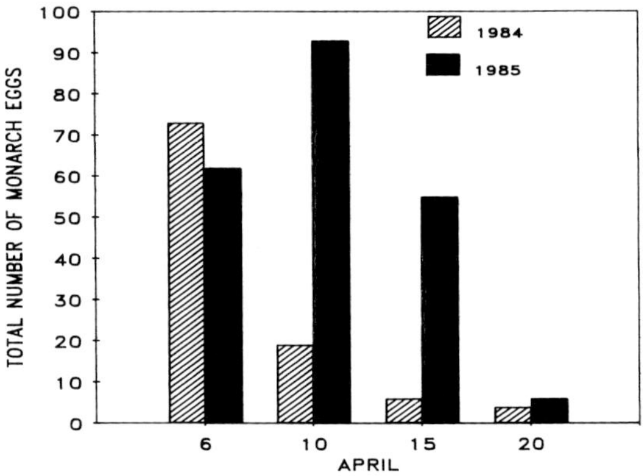

Figure 1. Total number of monarch eggs by date on 26 *A. longifolia* in 1984 and on 36 *A. longifolia* and 15 *A. amplexicaulis* in 1985. Actual dates of the first three observations in 1984 were 7, 12, and 14 April.

condition and date of occurrence in Louisiana, when combined with the timing of the breakup of the overwintering colonies in Mexico, strongly suggests that these monarchs were remigrants from the Mexican colonies.

Stems of *A. longifolia* and *A. amplexicaulis* most likely emerge through the soil in mid-March. Marked plants of both species examined on 25 February 1986 and 1987 had no emergent stems, whereas on 6 April 1985, heights ranged from 6.35 to 40.64 cm and 2.79 to 22.86 cm for 36 *A. longifolia* and 15 *A. amplexicaulis,* respectively.

In 1984, the greatest number of monarch eggs on *A. longifolia* were recorded on the earliest sample date of 7 April (Fig. 1). Sampling was started a week earlier in 1985 but no eggs were found on either *A. longifolia* or *A. amplexicaulis* on 30 March, although actively flushing plants were available. However, oviposition had occurred on these two species by 6 April and the greatest number of eggs were found on 10 April (Fig. 1). In both years, no eggs were found on any species of milkweed in the study area after 20 April.

Thus, the spring migration of monarchs through Louisiana appears to coincide with the availability of these two milkweed species for oviposition.

Asclepias longifolia was the most frequently ob-

Table 2. **Summary of adult monarch sightings in Louisiana, 20 March–25 May, 1984–1986.**

	Date					
Year	20–31 March	1–10 April	11–20 April	21–30 April	1–10 May	11–25 May
1984	5	31	25	0	1	
1985	8	23	9	11 + 1[a]	1	1[a]
1986	18	15	10	1	1	5[a]
Totals	31	69	44	12 + 1[a]	3	6[a]

[a] Freshly emerged butterflies.

Table 3. Summary of monarch eggs and larvae collected on 26 *A. longifolia* in the Kisatchie National Forest, Natchitoches Parish, Louisiana, April 1984.

Monarch life stage	Date			
	7 April	12 April	14 April	20 April
Eggs	73	19	6	4
First instar		17	8	
Second instar			6	

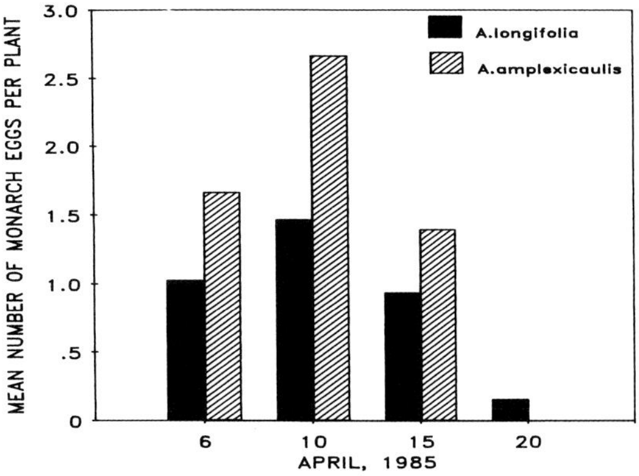

Figure 2. Number of eggs per plant by date on *A. longifolia* and *A. amplexicaulis* in 1985.

served milkweed species in the study area and accounted for the majority of oviposition and larval survival data (Tables 3 and 4). However, the number of eggs per plant (Fig. 2) and number of larvae surviving to the fifth instar (Table 4) were greatest for *A. amplexicaulis*. This may indicate a preference for *A. amplexicaulis* due to its cardenolide concentration, or an artifact of plant size or location relative to other *Asclepias* spp. in the area. However, Zalucki and Kitching (1982) and Malcolm and Brower (1986) have shown that monarchs select milkweed host plants with high cardenolide content. Considering the very low cardenolide levels in *A. amplexicaulis* (Roeske et al., 1976), cardenolides are unlikely to be implicated in monarch exploitation of *A. amplexicaulis*.

Asclepias amplexicaulis has large ovoid leaves and a large inflorescence relative to *A. longifolia*, which, in the study area, is characterized by long and very narrow leaves, usually <7 mm wide. The larger leaves and flowers of *A. amplexicaulis* are possibly more attractive for monarch oviposition and more likely to sustain larvae through the fifth instar. A similar relationship among plant size, monarch oviposition, and fifth-instar survival is reported by Cohen and Brower (1982) on *A. humistrata* in Florida.

Asclepias obovata and *A. viridis* were the only other milkweed species in the study area on which monarch life stages were found. Eggs and first instars were found on *A. obovata* and eggs through second instars were found on *A. viridus* on 15 April 1985. No monarch eggs, larvae, or signs of larval feeding were found on 20 *A. tuberosa* and 7 *A. verticillata*.

Although the relative abundance of any *Asclepias* species in the study area is unknown, the number of individuals of each species observed is presented in Table 5. *Asclepias amplexicaulis*, *A. obovata*, and *A. tuberosa* were scattered throughout the forest, occurring mostly as individual plants along roadsides. *Asclepias longifolia* was also widespread and occurred as individuals and in groups of 15 or fewer plants along 5–30 m of roadside. The 7 *A. verticillata* were growing together and were separated by <3 m, and the 18 *A. viridis* were located along a 150-m section of roadside. Despite their apparent low density, *A. obovata* and *A. viridis* may also serve as host plants for the larvae of remigrating monarchs in the Kisatchie National Forest.

The scattered distribution of milkweeds in the Kisatchie National Forest habitat contrasts to the study sites of Malcolm et al. (1987), Lynch and Martin (1987), and Cohen and Brower (1982), where plants were growing in abundance in open pastures. In northwestern Louisiana, where *A. viridis* is abundant in open habitats it is utilized as a major host plant by remigrant monarchs moving through that area (Lynch and Martin, 1987). The widely dispersed nature of milkweed plants in the Kisatchie National Forest would appear to be a poor resource

Table 4. Summary of monarch eggs and larvae collected on 15 *A. amplexicaulis* (amp.) and 36 *A. longifolia* (long.) in the Kisatchie National Forest, Natchitoches Parish, Louisiana, March and April 1985.

Monarch life stage	Date											
	30 March		6 April		10 April		15 April		20 April		29 April	
	long.	amp.	long.	amp.	long.	amp.	long.	amp.	long.	amp.	long.	amp.
Eggs			37	25	53	40	34	21	6			
First instar					7	11	17	12	5	3		
Second instar							5	8	3	4		
Third instar									2	10		
Fourth instar									4		1	
Fifth instar										5	1	5

Table 5. Summary of *Asclepias* species and monarch life stages observed in the Kisatchie National Forest, Natchitoches Parish, Louisiana, 1984 and 1985.

Asclepias species	No. of plants observed 1984	1985	Monarch life stages observed	Dates plants observed
longifolia	26	36	egg–fifth instar	6 April–22 May
amplexicaulis	0	15	egg–fifth instar	6 April–22 May
obovata	0	10	egg–first instar	15 April–22 May
tuberosa	0	20	0	11 May and 22 May
viridis	0	18	egg–second instar	15 April–22 May
verticillata	0	7	0	11 May and 22 May

for monarchs compared to more concentrated plant communities. This could explain the relatively poor utilization of *A. viridis* by monarchs in the Kisatchie habitat.

Mature fifth-instar larvae were found on large plants of *A. longifolia* and *A. amplexicaulis* on 20 and 29 April 1985 (Table 4). Successful completion of development by these larvae would suggest that a new generation of monarchs, offspring of the overwintered generation, can develop on these two milkweed species in Louisiana. Further evidence of a new generation of butterflies is provided by the sighting of fresh adults in late April and early May (Table 2). Although it is not visually possible to know where and on what species of *Asclepias* these fresh butterflies developed as larvae, the date of their sighting suggests that they are very likely among the first generation of adults from the remigrants (see Cockrell et al., this volume).

Also in 1985, 18 additional fifth instars were collected from *A. longifolia* and *A. amplexicaulis* that were not among the plants used in the study. When reared in the laboratory, all pupated between 30 April and 7 May, and adult emergence was between 9 and 17 May. Had these insects developed in the field, their dates of adult eclosion would correlate well with the sightings of fresh monarchs mentioned above. These first-generation adults are probably also migratory, because monarchs are not part of the summer butterfly fauna in Louisiana. If this is the case, it would support the suggestion of Malcolm et al. (1987)—that first-generation adults continue migrating north to exploit cooler northern temperatures, more favorable for larval development.

The observation of monarchs and their eggs and larvae during March, April, and May in Louisiana supports the hypothesis of Brower (1985) that overwintered monarchs utilize milkweed species encountered in Gulf Coast states during their recolonization of eastern North America (see also Malcolm et al., 1987, this volume; Cockrell et al., this volume; Lynch and Martin, this volume). The sudden abundance of adults, eggs, and larvae, followed by a rapid decline, suggests that the butterflies move through Louisiana, ovipositing on available *Asclepias* species as they go. Dates of adult arrival and commencement of oviposition in Louisiana suggest that northward movement of remigrant monarchs into the southern Mississippi valley is synchronized with the availability of suitable host plants.

LITERATURE CITED

Brower, L.P. 1985. New perspectives in the migration biology of the monarch butterfly, *Danaus plexippus* L. In *Migration: Mechanisms and adaptive significance,* ed. M.A. Rankin, 748–785. Austin: University of Texas, *Contributions in Marine Science,* Supplement, 27.

Brower, L.P., W.H. Calvert, L. E. Hedrick, and J. Christian. 1977. Biological observations on an overwintering colony of monarch butterflies (*Danaus plexippus* L.) in Mexico. *Journal of the Lepidopterists' Society* 31:232–242.

Calvert, W.H., and L.P. Brower. 1981. The importance of forest cover for the survival of overwintering monarch butterflies (*Danaus plexippus,* Danaidae). *Journal of the Lepidopterists' Society* 35:216–225.

Calvert, W.H., and L.P. Brower. 1986. The location of monarch butterfly (*Danaus plexippus* L.) overwintering colonies in Mexico in relation to topography and climate. *Journal of the Lepidopterists' Society* 40:164–187.

Calvert, W.H., and J.A. Cohen. 1983. The adaptive significance of crawling up onto foliage for the survival of grounded overwintering monarch butterflies (*Danaus plexippus*) in Mexico. *Ecological Entomology* 8:471–474.

Calvert W.H., W. Zuchowski, and L.P. Brower. 1983. The effect of rain, snow and freezing temperatures on overwintering monarch butterflies in Mexico. *Biotropica* 15:42–47.

Cohen, J.A., and L.P. Brower. 1982. Oviposition and larval success of wild monarch butterflies (Lepidoptera: Danaidae) in relation to host plant size and cardenolide concentration. *Journal of the Kansas Entomological Society* 55:343–348.

Heitzman, R. 1962. Butterfly migrations in March in northern Mexico. *Journal of the Lepidopterists' Society* 16:249–250.

Herman, W.S. 1985. Hormonally mediated events in adult monarch butterflies. In *Migration: Mechanisms and adaptive significance,* ed. M.A. Rankin,

799–815. Austin: University of Texas, *Contributions in Marine Science*, Supplement, 27.

Lynch, S.P., and R.A. Martin. 1987. Cardenolide content and thin-layer chromatography profiles of monarch butterflies, *Danaus plexippus* L., and their larval host-plant milkweed, *Asclepias viridis* Walt., in northwestern Louisiana. *Journal of Chemical Ecology* 13:47–70.

Malcolm, S.B., and L.P. Brower. 1986. Selective oviposition by monarch butterflies (*Danaus plexippus* L.) in a mixed stand of *Asclepias currasavica* L. and *A. incarnata* L. in south Florida. *Journal of the Lepidopterists' Society* 40:255–263.

Malcolm, S.B., B.J. Cockrell, and L.P. Brower. 1987. Monarch butterfly voltinism: Effects of temperature constraints at different latitudes. *Oikos* 49:77–82.

Roeske, C.N., J.N. Seiber, L.P. Brower, and C.M. Moffitt. 1976. Milkweed cardenolides and their comparative processing by monarch butterflies (*Danaus plexippus* L.). *Recent Advances in Phytochemistry* 10: 93–167.

Urquhart, F.A., and N.R. Urquhart. 1976a. The overwintering site of the eastern population of the monarch butterfly (*Danaus p. plexippus:* Danaidae) in southern Mexico. *Journal of the Lepidopterists' Society* 30:153–158.

Urquhart, F.A., and N.R. Urquhart. 1976b. A study of the peninsular Florida populations of the monarch butterfly (*Danaus p. plexippus;* Danaidae). *Journal of the Lepidopterists' Society* 30:73–87.

Urquhart, F.A., and N.R. Urquhart. 1978. Autumnal migration routes of the eastern population of the monarch butterfly (*Danaus p. plexippus* L.; Danaidae: Lepidoptera) in North America to the overwintering site in the neovolcanic plateau of Mexico. *Canadian Journal of Zoology* 56:1754–1764.

Urquhart, F.A., and N.R. Urquhart. 1979. Vernal migration of the monarch butterfly (*Danaus p. plexippus,* Lepidoptera: Danaidae) in North America from the overwintering site in the neo-volcanic plateau of Mexico. *Canadian Entomologist* 111:15–18.

Zalucki, M.P., and R.L. Kitching. 1982. Dynamics of oviposition in *Danaus plexippus* (Insecta: Lepidoptera) on milkweed *Asclepias* spp. *Journal of Zoology* 198:103–116.

ORIENTATION OF AUTUMN MIGRATION IN THE MONARCH BUTTERFLY[1]

Klaus Schmidt-Koenig

ABSTRACT. Vanishing bearings of 1,599 individual, autumn-migrating monarch butterflies (*Danaus plexippus* (L.)) were recorded at various inland locations of the eastern United States and at coastlines parallel and nonparallel to the direction of migration. Due to the high degree of directionality of individuals, sample mean vector lengths were mostly near 1.0. The intersample variability was largely due to direct wind action from drift in crosswinds and to indirect wind action with monarchs flying along the lee side of obstacles (dunes, dams, wood margins, etc.), even where this led them to deviate from the exact direction of migration. At the coast in unfavorable winds, including strong tailwinds, monarchs avoided heading out over the ocean. Butterflies took off repeatedly but returned after crossing the shoreline. Inland, across the Piedmont Plain as well as in the Appalachian Mountains, monarchs even migrated in strong tailwinds, flying very rapidly at altitudes near, or beyond, the range of powerful binoculars. Monarchs can probably keep track of and compensate for displacements by wind. Overcast conditions had no significant effect on headings or scatter. The mean directions summarized at two representative multisample inland locations agreed with the directions predicted by Kiepenheuer's (1984) model of the magnetoclinic compass.

INTRODUCTION

Insect migration is a widespread, sometimes spectacular phenomenon (Williams, 1958; Johnson, 1969; Baker, 1978; Dingle, 1978). However, despite much research and several books on the subject, even the phenology of insect migration is, on the whole, less well understood than that of other migrants (Schmidt-Koenig, 1975). With the exception of very large insects and economic pests such as locusts, most insect migrations are inconspicuous. Butterfly migration is often spectacular, but even so the phenology of migration, origin, destination, and mode of orientation of migrating butterflies are as yet poorly understood (Walker, 1985). Among butterflies, the monarch, *Danaus plexippus* (L.), ranks as the best-studied long-distance migrant inasmuch as its winter quarters have been located in California (Downes in Williams et al., 1942) and Mexico (Urquhart, 1976; Brower, 1977; Urquhart and Urquhart, 1978). How the butterflies find their routes and destinations, and how they navigate, is, however, as yet entirely obscure.

In a number of publications, various aspects of orientation by migrant monarchs have been discussed (e.g., Urquhart, 1960, 1976; Urquhart and Urquhart, 1978; Kanz, 1977; Gibo and Pallett, 1979). Furthermore, in a short communication (Schmidt-Koenig, 1979) I reported an initial series of observations on the orientation of monarch migration in the eastern United States. I was able to extend such observations considerably in the autumn migration periods of 1978, 1980, and 1981 (Schmidt-Koenig, 1985). In addition to continuing to record data of inland migration, I emphasized directional observations of migration under special circumstances, such as at coastlines parallel and nonparallel to the general direction of migration, at tips of islands, at mountain ridges, and under different meteorological conditions.

METHODS

Monarch migration in the eastern United States is highly directional, each individual flying along a seemingly straight line. Vanishing bearings of migrants were, therefore, easily measured with a magnetic compass, allowing for the local declination. Monarchs migrating at low altitude within

1. Based on "Migration strategies of monarch butterflies" by Schmidt-Koenig (1985) and republished with permission from the University of Texas at Austin.

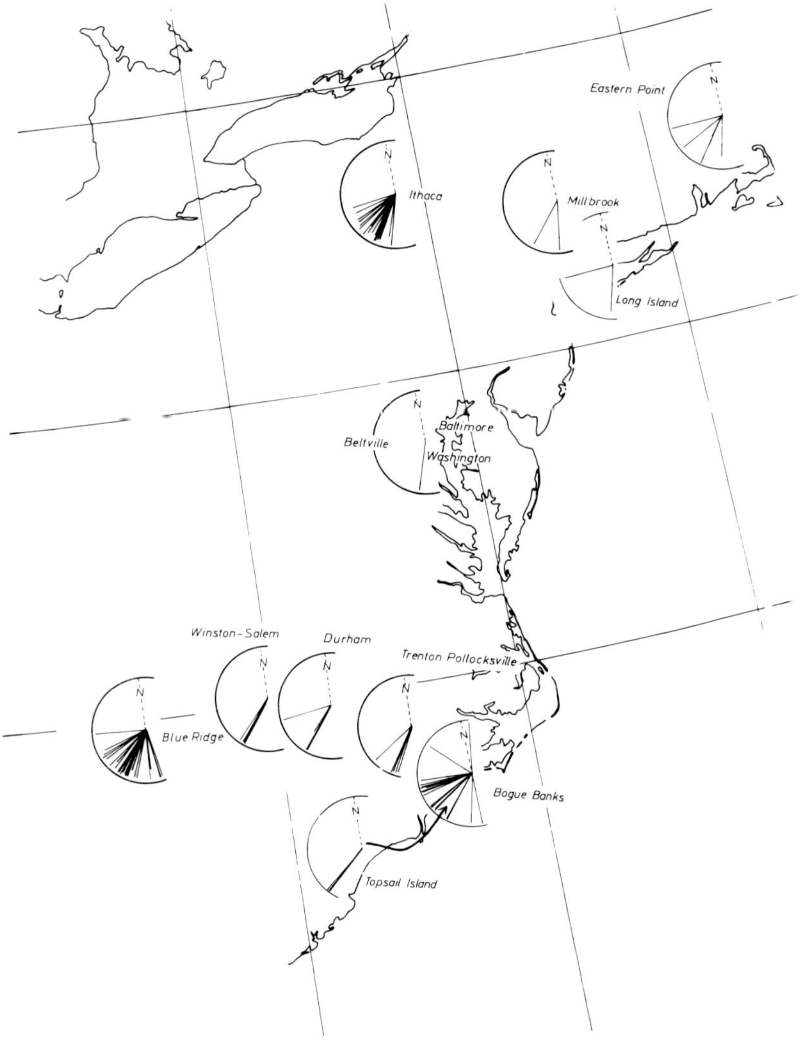

Figure 1. Eighty-eight mean vectors of monarch vanishing bearings (n ≤ 20), each obtained and summarized at the location indicated in the eastern United States during four autumn migration periods. For more details see text, especially for the high proportion of westerly vectors at Bogue Banks; the north pointing vector is explained in Figure 3.

range of the naked eye (usually in headwinds or crosswinds) were tracked by eye until vanishing 50–100 m away, the distance depending upon color and brightness of the background. Monarchs migrating overhead beyond that range were tracked by 10 × 40 binoculars until they vanished. During heavy migration, or high-altitude migration requiring tracking with binoculars, I stayed in one location and tracked only those monarchs that passed overhead. During smaller-scale migrations, I positioned myself at a line roughly perpendicular to the direction of migration such as a road, a fence, or the shoreline, waited for a monarch to cross that line, and walked to the crossing point to track the insect to the vanishing point. This eliminated parallax. Standard sample size was ≤20. Cloud cover, atmospheric visibility, and visibility of the sun were noted. Surface wind direction was measured by tracking light objects thrown up into the air. Whenever possible, wind directions at higher altitudes were measured by tracking floating objects such as spider webs. Wind direction is given as is customary in meteorology, aviation, and navigation; i.e., a 90° wind blows from due East. Wind strength was estimated on the basis of extensive prior personal experience with a hand-held wind meter. Samples were processed using the standard methods and statistical tests for circular data (Batschelet, 1981).

RESULTS

The mean vectors of 88 samples comprising 1,599 individual vanishing bearings at 11 locations in the eastern United States, including those already given in Schmidt-Koenig (1979, 1985), are summarized in Figure 1. Each sample represents a snapshot out of a stream of migrating monarchs. Although the mean

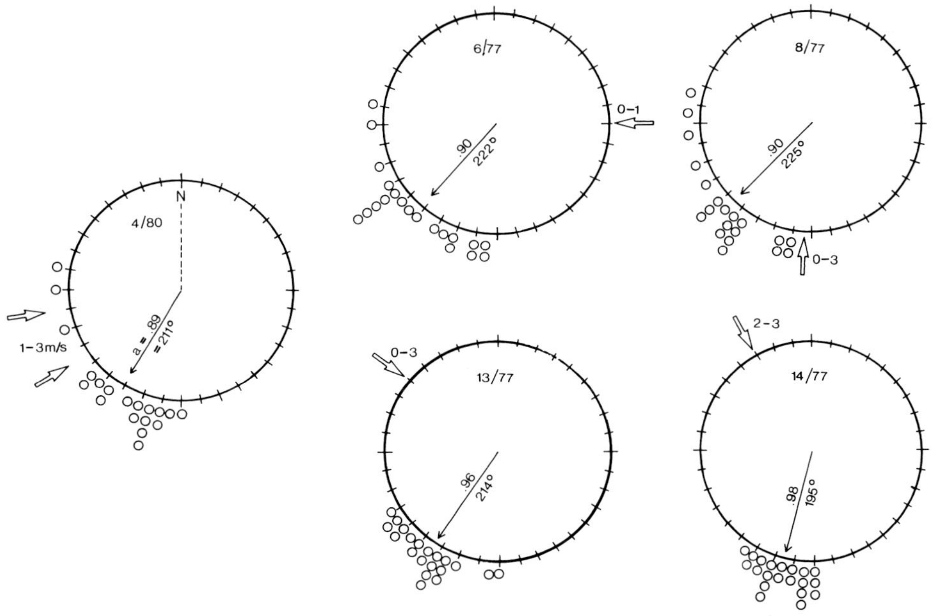

Figure 2. Five samples of 20 vanishing bearings, each collected at Dodge Road near Ithaca (New York) under different conditions of light wind. The centripetal arrow at the periphery of each diagram indicates wind direction. Wind strength is given in m·sec⁻¹. Each open symbol at the periphery indicates an individual vanishing bearing. Mean vectors with direction and length and sample identification numbers are given. Geographical north is given in sample 4/80.

vectors are bunched at each location, even samples with mean vectors 10° apart may be significantly different from each other due to the narrow range of individual monarch vanishing directions and correspondingly near unit-length vectors. Intersample variability is larger in the Blue Ridge Mountains (North Carolina) in a highly structured area than at Ithaca (New York) in much less-structured terrain. This intersample variability is largely due to direct or indirect wind action. Direct wind action means that the butterflies are drifted by crosswinds. Indirect action means that in strong headwind or crosswinds the butterflies fly under the lee of suitable structures such as dams, wood margins, dunes, etc., even if they deviate from the general direction of migration.

Five samples were collected at the same location near Ithaca under a variety of light wind conditions. Some shift according to wind direction and strength is obvious in every sample (Fig. 2), although the butterflies always visibly advance their headings to offset drift. The stronger the headwind or crosswinds, the lower the butterflies fly, and they fly under the lee of obstacles, if necessary putting up with directions diverging from the general direction of migration rather than allowing themselves to be drifted (e.g., out to sea). On Bogue Banks (North Carolina), a narrow island 40 km long, running roughly 80–260°, monarchs migrated along the northern shore in southerly winds; in northerly winds, they migrated along the southern shore or along inland wood margins. This behavior is the reason for the exceptionally large number of mean vectors pointing west at Bogue Banks in Figure 1. These data were collected during two autumn migration periods and in wind directions ranging clockwise from 190 to 55°.

Monarchs do fly out over the ocean and cross major bodies of water under suitable wind conditions. I saw monarchs leave the southern shores or southern or western tips of Eastern Point (Massachusetts), Bogue Banks, and Topsail Island (North Carolina). I watched monarchs from boats over open waters, or arriving at the northern shore of Long Island (New York) (Fig. 1), or at the northern tip of Topsail Island.

Monarchs can apparently assess the risk of heading out over the sea if the wind is unmanageable. This applies not only to headwinds or crosswinds but also to strong tailwinds. For example, in early October 1980 at Bogue Banks, large numbers of butterflies accumulated for days at the western tip of the island. Even on days with seemingly ideal tailwinds of 10–50° and speeds of 3–5 m·sec⁻¹ under clear skies, the butterflies appeared to leave the island at the western end of the dunes of Bogue Banks or at the shoreline but in fact could be seen returning close to the surface. I tracked monarchs that had taken off from stands of flowering plants such as goldenrod (*Solidago* sp.) or sea-myrtle (*Baccharis halimifolia*) (both Asteraceae) behind the dunes, heading in southerly directions, seem-

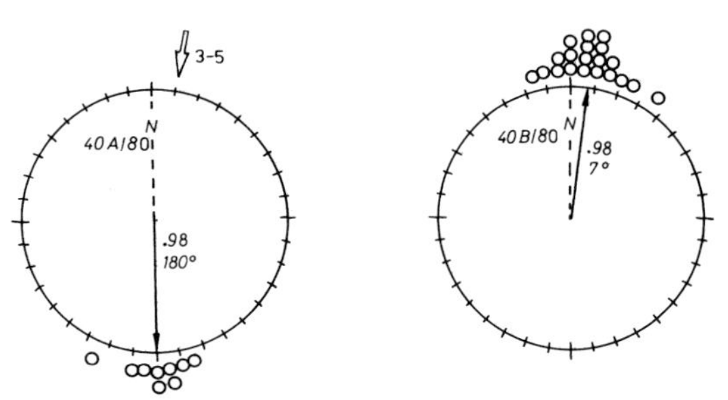

Figure 3. Two samples taken near the shoreline south of the western end of the dunes of Bogue Banks (North Carolina) on 6 October 1980 during a strong northerly wind. Looking up into the sky (40A/80) I recorded monarchs that had taken off from behind the dunes. Looking down to the ground (40B/80) I recorded monarchs that had turned around and dropped to the surface beating the wind back to the dunes. Symbols as in Figure 2.

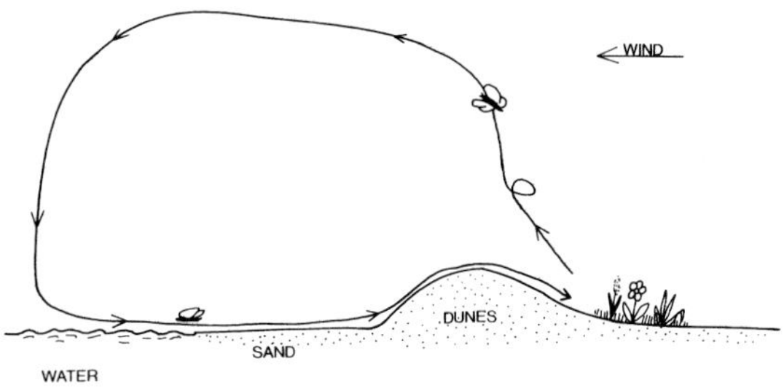

Figure 4. Diagrammatic sketch (not drawn to scale) of the observations made from 5 to 7 October 1980 at the western tip of Bogue Banks (North Carolina). For more details see text.

ingly out to sea. However, I also tracked monarchs that had turned around, dropped to the surface, and were beating the wind back to the dunes as low as possible over the surface of the water or sand. Samples 40A/80 and 40B/80 (Fig. 3) were taken simultaneously, looking up and down, respectively. A continuous, cyclic process of monarchs taking off downwind but coming back low against the wind (Fig. 4) went on from 5 to 7 October 1980, with winds 10–55° and 3–5 $m \cdot sec^{-1}$. A large proportion of monarchs turned around after having crossed the shoreline on their way out to the sea. Some proportion was certainly blown out to sea. On 7 October 1980, I witnessed a calming of the 40° wind from around 5 $m \cdot sec^{-1}$ to 1–2 $m \cdot sec^{-1}$. Monarchs continued to take off toward the south–southwest but no longer dropped down to the surface after having crossed the shoreline to head back to the dunes. Within a short time, the population of several thousand monarchs that had accumulated over several days had left the island. Such sequences of meteorological events at certain locations qualify as "migration traps" and may explain how the sometimes spectacular "clouds" of butterflies originate. The phenomenon could be called "meteorological synchronizaton."

On 9 September 1980, under clear skies, monarchs arrived at Eastern Point with southwesterly tracks low over sheltered land in southerly winds 1–3 $m \cdot sec^{-1}$ (Fig. 5). When encountering the shoreline, they did not fly out to sea but changed headings by about 60° and flew under the shelter of a massive breakwater. At the end, however, they headed out over the water despite the prevailing crosswind of up to 3 $m \cdot sec^{-1}$, which noticeably shifted the bearings clockwise to almost 270° from the standard south–southwest heading. In strong headwinds (205–220°, 6–11 $m \cdot sec^{-1}$) a day later, monarchs accumulated at Eastern Point—another example of "meteorological synchronization." They were able to approach the buildings at the tip of Eastern Point under lee shelter and then try over and over again to move on, but they were blown back inevitably when leaving the wind shadow.

The threshold at which monarchs treat wind conditions as unmanageable must be a function of wind direction, wind strength, local conditions, and possibly other variables. Many more data need to be collected before an assessment of that threshold can be made.

In Figure 1, 11 of the 13 samples from North Carolina at Trenton-Pollocksville, Durham (two at Durham are from 1977), and Winston-Salem were collected during an unusual meteorological condition between 9 and 13 October 1981. During this period, fairly strong north–northeasterly winds (up to 7 $m \cdot sec^{-1}$) were blowing. While very few single monarchs could be seen near the ground, intense migration was occurring at altitudes of at least 300–400 m above the ground ("high altitude" migration), at which monarchs could be detected and tracked only by field glasses and often only against white clouds. To estimate migration across a large area, I collected samples along a line from the coast to a point just northwest of Winston-Salem ("N.C. transect") from 9 to 13 October 1981. Although high-altitude migration in following winds can occur beyond the power of field glasses, Figure 6 shows samples of high-altitude migration observed near Grandfather Mountain (North Carolina) above about 1,600 m elevation. Samples 27/80 and 28/80 were recorded overhead with field glasses. In contrast, sample 31/80 was recorded at the same location without field glasses a day later when opposing winds forced the monarchs to fly within 5 m of the ground and to exploit the lee area of obstacles. Furthermore, I have also watched several monarchs migrating >3,600-m altitude in the Colorado Rocky Mountains and the California White Mountains, again high up in tailwinds or calm conditions, but close to the ground in headwinds or crosswinds.

DISCUSSION

Monarchs are not powerful flyers (Gibo and Pallett, 1979; Gibo, 1986) and are subject to considerable

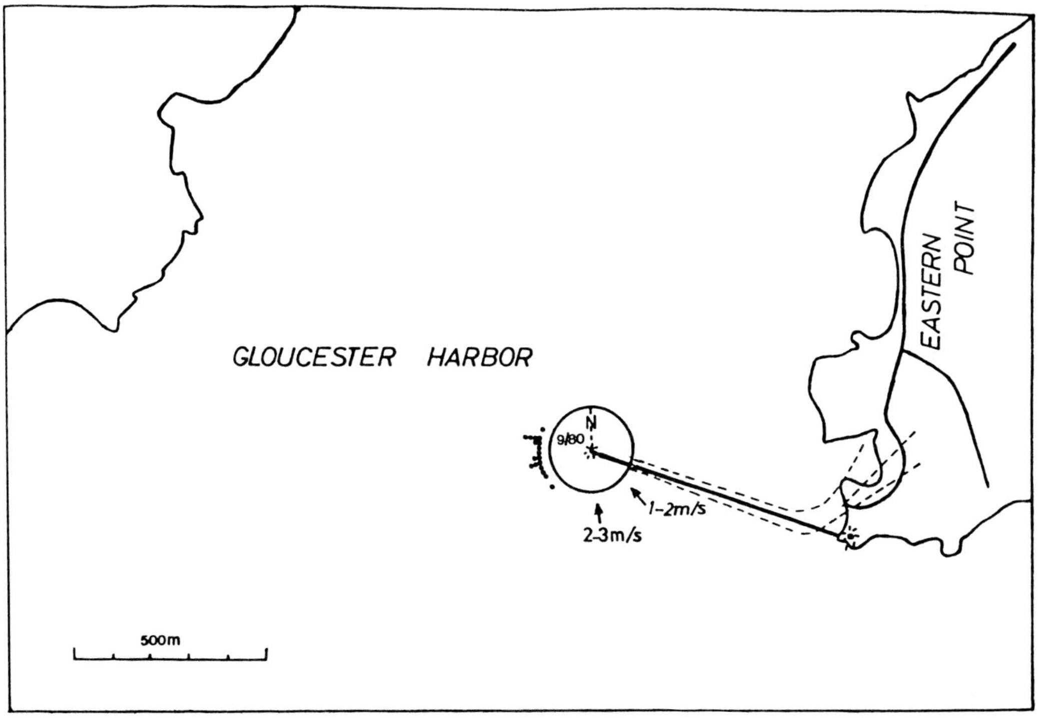

Figure 5. Migration at Eastern Point (Massachusetts) on 9 September 1980 in southeasterly winds and clear skies. Monarchs (dashed lines) approached the shoreline near the ground and then followed the breakwater (solid straight line pointing northwest). Eighteen vanishing bearings (black dots not drawn to scale) of butterflies heading out over the water were recorded at the lighthouse at the tip of the breakwater. Geographical north and sample identification number are given inside the circle. The centripetal arrows at the periphery indicate wind direction. Wind strength is given in $m \cdot sec^{-1}$.

drift. Large-scale cross-country drifts must occur. Accumulations of butterflies at the Atlantic coast are probably the result of westerly winds. My data are as yet insufficient to confirm corresponding observations made in 1980 and 1981, when considerable numbers of monarchs showed up at the coast only during and after days with westerly winds. Likewise, I cannot as yet quantify the degree of deviation tolerated by migrants from the exact migration direction either by passive drift or as the result of active exploitation of lee sides of shelters. The small angular range of mean vectors in Figure 1 suggests a fairly low tolerance limit, the monarchs apparently preferring windward deviation to downwind drift. To pinpoint successfully their wintering grounds in Mexico, the butterflies must be able to keep track of and to compensate for displacements forced upon them by wind. I can present one "snapshot" out of a stream of migrating monarchs to support this notion. The samples shown in Figure 7 were taken on 7 September 1980 in fairly strong northwesterly winds at Tompkins County Airport, Ithaca. Monarchs flying low in lee areas were more westerly directed after having crossed the airfield (6/80) and more southerly directed before having crossed it (7/80). They were certainly exposed to southeast drift while crossing the unsheltered airfield (see also Gibo, 1986).

Two explanations may account for the observed orientations of migratory monarchs. They may use either a sun compass (Kanz, 1977) or a magnetic compass (Jones and MacFadden, 1982; Jungreis, 1982).

A sun compass has been demonstrated in every animal, vertebrate or invertebrate, that has been tested. Standard experimental tests to demonstrate the use of the sun for compass orientation are those in which predictable shifts of direction can be obtained by using mirrors (e.g., Santschi, 1911; Kramer, 1951; von Frisch, 1967) or upon clock-shifts (e.g., Schmidt-Koenig, 1960). In using mirrors, the deviation accomplished corresponds to the angle subtended by the mirror. In clockshift experiments, the deviation accomplished corresponds to the number of hours of time-shift to which the animal and its internal clock, respectively, are subjected. The angular speed of the sun's azimuth being roughly 15°/hr, a 6-hr time-shift produces roughly 90° of shift in direction. The experiments by Kanz (1977), therefore, do not disprove the use of a sun compass by monarchs. Kanz (1977) recorded the position of monarchs in opaque or transparent orientation cag-

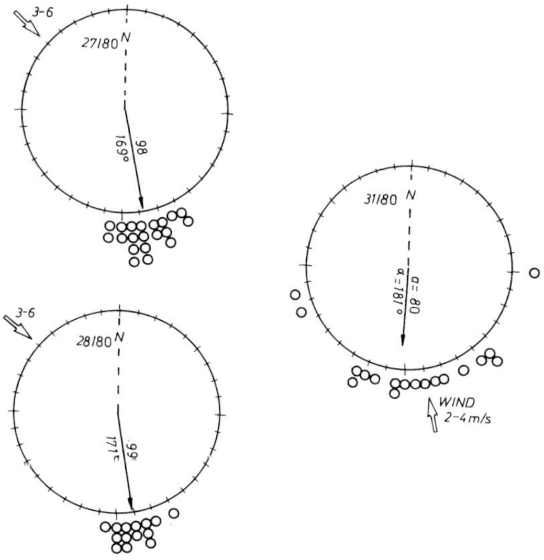

Figure 6. Three samples recorded near Grandfather Mountain (North Carolina) above 1,600 m elevation on 26 and 27 September 1980. Symbols as in Figure 2. For more details see text.

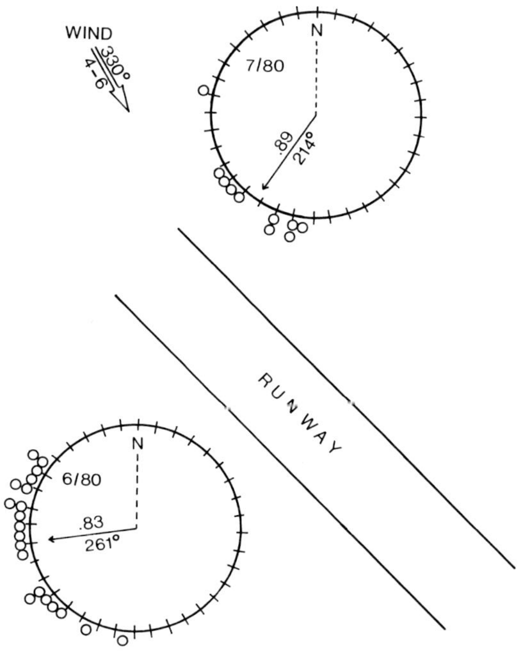

Figure 7. Two samples possibly demonstrating immediate compensation of drift. Symbols as in Figure 2. For more details see text.

es under various sun conditions, unfortunately without experimental manipulations that might have given more conclusive results.

If we extrapolate from findings on other insects (e.g., von Frisch, 1967), as a first approximation we can assume that monarchs have a sun compass. Overcast skies, however, do not disorient migrating monarchs (see also Gibo, 1986). Figure 8 shows six samples recorded under changing cloud conditions so that for one part of the sample the sun was visible (white symbols) but not for the other (black symbols). There was no obvious effect on the migrants. Likewise, five samples taken under completely overcast conditions (Fig. 9) give no indication of disorientation or other differences from samples recorded under sun. Thus, if they have a sun compass, monarchs must have other means for orientation, possibly one based on the Earth's magnetic field. Indeed, if one calculates (Table 1) the total mean vector for the two inland locations that may be considered fairly representative (because of large sample sizes and the absence of coastlines), the mean directions agree strikingly with the magnetoclinic headings calculated according to Kiepenheuer's (1984) model of the magnetic compass under the assumption of a broad-front monarch migration parallel to that documented by Urquhart and Urquhart (1978) for the central-eastern United States.

According to Kiepenheuer's (1984) model, magnetoclinic courses imply that the animal does not follow a specific course with respect to magnetic north (α_m) but continuously maintains the same "subjective" angle of magnetic dip. This subjective angle of dip (γ') is the projection of the true dip (γ) on a plane vertical to the direction of the animal's long axis (direction of heading). Changing angles of true dip (e.g., when moving to southern latitudes) therefore results in changing course directions with respect to magnetic north. In the present case of the monarch butterfly, the mean direction of migration in the Ithaca area (γ = 71°30′, δ = −8°) and in the Blue Ridge Mountain area (γ = 67°00′, δ = −4°) changes by about 10°30′ to the south, but the same subjective angle dip (γ = 75°12′) at both sites may be supposed to control this course. A course parallel to this with approximately the same true dip may be calculated from the data given by Urquhart and Urquhart (1978) for monarchs tagged in the Toronto area and in the Mexican winter quarters, respectively.

Based on the demonstration of induced magnetism in monarchs by Jones and MacFadden (1982) and Jungreis (1982), I conclude tentatively that monarchs use a magnetoclinic system similar to that suggested for birds by Kiepenheuer (1984). My earlier interpretation of the initial series of observations (Schmidt-Koenig, 1979)—that monarchs follow the great circle route (see also Gibo, 1986)—is no longer supported by the much more extensive data analyzed in the present paper.

My observations confirm and support those made by Gibo and Pallett (1979) and Gibo (1986) on flying strategies during migration. These authors report observations similar to mine as to the effect of wind direction and strength on initiation or discontinuation of migration, on altitude of flight, the use of

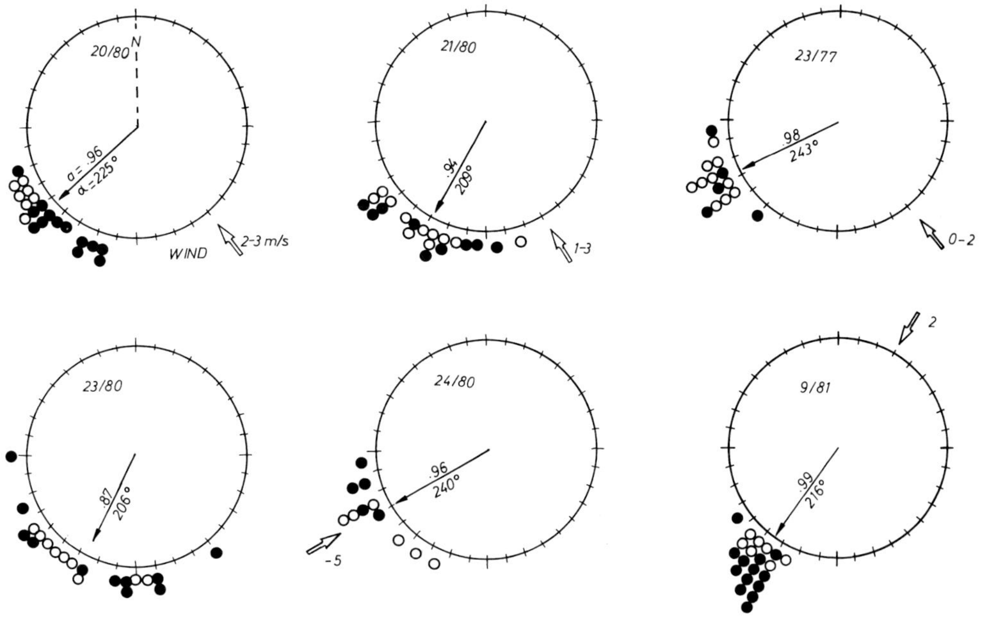

Figure 8. Six samples recorded partly under the sun (white symbols) and partly under overcast conditions (black symbols; otherwise symbols as in Fig. 2) without simultaneous shift of wind direction. In sample 20/80, directions under sun and without sun are significantly different ($p < 0.01$; Watson U^2 test). There is no significant difference ($p > 0.05$; same test) in any of the other samples nor in the summary of all six samples. Sample 24/80 was recorded at Black Mountain Gap, 1,450 m elevation, near Mt. Mitchell, North Carolina, partly in light rain and clouds drifting across the gap in the opposite direction to monarch migration.

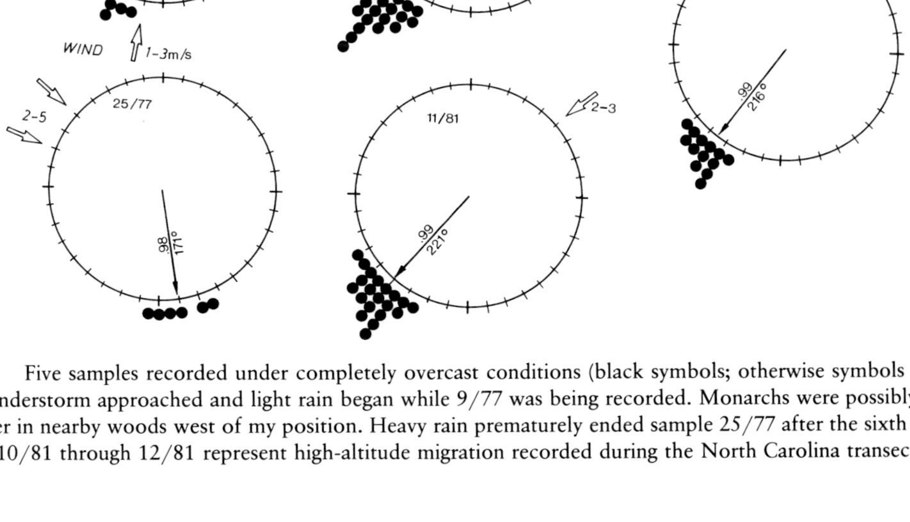

Figure 9. Five samples recorded under completely overcast conditions (black symbols; otherwise symbols as in Fig. 2). A thunderstorm approached and light rain began while 9/77 was being recorded. Monarchs were possibly heading for shelter in nearby woods west of my position. Heavy rain prematurely ended sample 25/77 after the sixth butterfly. Samples 10/81 through 12/81 represent high-altitude migration recorded during the North Carolina transect.

Table 1. Geographical and statistical data of the two multisample inland summaries for Ithaca and Blue Ridge (Fig. 1). N = total number of individual vanishing bearings, $\bar{\alpha}_N$ = total mean vector of N vanishing bearings, $C_{99\%}$ = 99% confidence interval of $\bar{\alpha}_N$, α_m = magnetoclinic heading according to Kiepenheuer (1984).

Location	N	$\bar{\alpha}_N$	$C_{99\%}$	α_m
Ithaca (42°30′N, 76°30′W)	348	224°03′	±4°40′	224°
Blue Ridge (36°00′N, 82°00′W)	486	214°42′	±3°51′	214°

wind shadow, and the flying techniques used in crossing obstacles. In particular, I can confirm observations on the use of thermals. Many times I have observed monarchs soar past at high speed in tailwinds and then immediately begin to circle when they encountered thermal lift such as that originating from black surfaces of parking lots. In calm conditions or light winds with a tail component, I observed frequently, as did Gibo and Pallett (1979) and Gibo (1986), monarchs circling out of sight in thermals overhead, sometimes together with hawks such as *Buteo platypterus.* I agree with these authors that monarch butterflies are highly adapted to exploit meteorological conditions to save energy. During migration monarchs forage for flower nectar to fuel their flight to wintering grounds in Mexico. Without the adaptive use of meteorological conditions, monarchs would have to both spend much more time with active migration and expend considerable time and energy in foraging, possibly at the expense of one reproductive generation in late summer. Most amazing of all is, however, that the monarch's highly flexible orientation and flight tactics are accomplished each autumn by a population of entirely naive animals.

Our observations (Gibo and Pallett, 1979; Schmidt-Koenig, 1979, 1985, and herein; Gibo, 1986) are relevant to the questions about butterfly orientation during migration, raised by Walker and Riordan (1981) and Walker (1985) in response to Williams (1958), Johnson (1969), and Baker (1978). Instead of passive or indirect migration on wind systems above or below the boundary layer, monarch butterflies can determine their direction of migration directly, possibly using the geomagnetic field according to Kiepenheuer's (1984) model. However, they also rely on favorable wind speeds and directions as a major means of transport even at high altitudes, at high velocities, and presumably for long distances.

ACKNOWLEDGMENTS

I am grateful to J. Kiepenheuer, Tübingen, for calculating the magnetoclinic heading, to him and to P.H. Klopfer, Durham (North Carolina), for critically reading and improving the manuscript.

LITERATURE CITED

Baker, R.R. 1978. *The evolutionary ecology of animal migration.* London: Hodder and Stoughton, 1012 pp.

Batschelet, E. 1981. *Circular statistics in biology.* London: Academic Press, 371 pp.

Brower, L.P. 1977. Monarch migration. *Natural History* 86:40–52.

Dingle, H. (ed.). 1978. *Evolution of insect migration and diapause.* New York, Heidelberg, and Berlin: Springer-Verlag, 284 pp.

Gibo, D.L. 1986. Flight strategies of migrating monarch butterflies (*Danaus plexippus* L.) in southern Ontario. In *Insect flight: Dispersal and migration,* ed. W. Danthanarayana, 172–184. Berlin and Heidelberg: Springer-Verlag.

Gibo, D.L., and M.J. Pallett. 1979. Soaring flight of monarch butterflies, *Danaus plexippus* (Lepidoptera: Danaidae), during the late summer migration in southern Ontario. *Canadian Journal of Zoology* 57:1393–1401.

Johnson, C.G. 1969. *Migration and dispersal of insects by flight.* London: Methuen, 763 pp.

Jones, D.S., and B.J. MacFadden. 1982. Induced magnetization in the monarch butterfly, *Danaus plexippus* (Insecta, Lepidoptera). *Journal of Experimental Biology* 96:1–9.

Jungreis, S.A. 1982. Biomagnetism: A possible orientation mechanism in migrating and non-migrating insects. M.Sc. thesis. Gainesville: University of Florida, 63 pp.

Kanz, J.E. 1977. The orientation of migrant and nonmigrant monarch butterflies, *Danaus plexippus* (L.). *Psyche* 84:120–141.

Kiepenheuer, J. 1984. The magnetic compass mechanism of birds and its possible association with the shifting course direction of migrants. *Behavioral Ecology and Sociobiology* 14:87--99.

Kramer, G. 1951. Eine neue Methode zur Erfassung der Zugorientierung und die bisher erzielten Ergebnisse. *Proceedings of the 10th International Ornithological Congress, Uppsala* 1950:269–280.

Santschi, F. 1911. Observations et remarques critiques sur le mécanisme de l'orientation chez les fourmis. *Review Swisse Zoologie* 19:303–338.

Schmidt-Koenig, K. 1960. Internal clocks and homing. *Cold Spring Harbor Symposia on Quantitative Biology* 25:389–393.

Schmidt-Koenig, K. 1975. *Migration and homing in animals.* Series Zoophysiology and Ecology 6. Berlin, Heidelberg, and New York: Springer-Verlag, 99 pp.

Schmidt-Koenig, K. 1979. Directions of migrating monarch butterflies (*Danaus plexippus;* Danaidae; Lepidoptera) in some parts of the eastern United States. *Behavioural Processes* 4:73–78.

Schmidt-Koenig, K. 1985. Migration strategies of monarch butterflies. In *Migration: Mechanisms and adaptive significance,* ed. M.A. Rankin, 786–798. Austin: University of Texas, *Contributions in Marine Science,* Supplement, 27.

Urquhart, F.A. 1960. *The monarch butterfly.* Toronto: University of Toronto Press, 361 pp.

Urquhart, F.A. 1976. Found at last: The monarch's winter home. *National Geographic* 150:161–173.

Urquhart, F.A., and N.R. Urquhart. 1978. Autumnal migration routes of the eastern population of the monarch butterfly (*Danaus p. plexippus* L.; Danaidae; Lepidoptera) in North America to the overwintering site in the neovolcanic plateau of Mexico. *Canadian Journal of Zoology* 56:1759–1764.

von Frisch, K. 1967. *The Dance language and orientation of bees.* Harvard: Belknap Press of Harvard University Press, 566 pp.

Walker, T.J. 1985. Butterfly migration in the boundary layer. In *Migration: Mechanisms and adaptive significance,* ed. M.A. Rankin, 704–723. Austin: University of Texas, *Contributions in Marine Science,* Supplement, 27.

Walker, T.J., and A.D. Riordan. 1981. Butterfly migration: Are synoptic-scale wind systems important? *Ecological Entomology* 6:433–440.

Williams, C.B. 1958. *Insect migration.* London: Collins, 235 pp.

Williams, C.B., G.F. Cockbill, M.E. Gibbs, and J.A. Downes. 1942. Studies in the migration of Lepidoptera. *Transactions of the Royal Entomological Society of London* 92:101–283.

8

OVERWINTERING BIOLOGY

Black-eared mouse (*Peromyscus melanotis*) eating a monarch butterfly. Of the five mouse species at the Mexican overwintering sites, this is the only one that feeds heavily on the butterflies. The other species are deterred by the monarch's chemical defenses. Photograph by John I. Glendinning and W. Perry Conway.

VEGETATION TYPES OF MONARCH BUTTERFLY OVERWINTERING HABITAT IN MEXICO

José Carmen Soto Núñez and Leonila Vázquez García

ABSTRACT. The structure and diversity of plant communities in which monarch butterflies overwinter were surveyed. The study sites are in northeastern Michoacán, near the state of México border at altitudes between 2,700 and 3,400 m, in the sierras Chincua and El Campanario and Cerro El Chivatí. This region is physiographically included in the Transverse Neovolcanic Belt, whose origins are Miocenic. Geologically, this area is comprised of cinder cones, lava beds, and basaltic pyroclasts. Soils are of the andosol type. The prevailing climate is temperate—humid, with a summer rainy season. The mean minimum and maximum temperatures of the coldest month are −3 and 18°C, respectively.

The area as a whole is about 65 km² and comprises three sections that correspond approximately with the main overwintering centers known for the monarch. Sites were defined on the basis of observations of the dynamics of the butterfly colonies made between 1979 and 1984. Plant samples were collected at each site.

Vegetational types formed two climax and four secondary communities. The former include *Abies* forests and upland cloud forest. The latter include mixed *Abies–Pinus* forest, *Pinus* forest, *Potentilla* meadows, and *Juniperus* shrub patches. The most important of these communities for monarch overwintering is the climax *Abies* forest, followed by secondary mixed *Abies–Pinus* forest.

The original *Abies* forests showed noticeable alteration in the El Rosario and El Chivatí areas, principally due to four factors: (1) clearing of the forest to open new land for agriculture; (2) uncontrolled timber exploitation; (3) forest fires, generally caused through negligence; and (4) livestock grazing and other range activities.

RESUMEN. Se presenta la estructura y la diversidad de las comunidades vegetales en donde la mariposa monarca hiberna. El área estudiada se localiza en el noreste del estado de Michoacán en el límite con el estado de México entre los 2,700 y 3,400 m, en las sierras de Chincua y El Campanario y el cerro El Chivatí. Fisiográficamente esta área corresponde al Eje Volcánico Transversal, de origen Miocénico. Geológicamente, esta región está formada por conos, lava y piroclastos basálticos. Los suelos se clasifican como andosoles. El clima predominante es el templado-húmedo con lluvias en verano. El promedio de las temperaturas mínima y máxima del mes más frío son −3 y 18°C respectivamente.

El área total es de aproximadamente 65 km², y comprende tres áreas que corresponden en general con los principales centros de hibernación de la monarca. Los sitios fueron delimitados en base a las observaciones realizadas sobre la dinámica que tuvieron las colonias de mariposas entre 1979 y 1984. Se colectó material botánico en cada sitio.

Los tipos de vegetación fueron agrupados en dos comunidades clímax y cuatro comunidades secundarias. Las primeras incluyen el bosque de *Abies* y el bosque mesófilo de montaña. Las segundas incluyen el bosque mixto de *Abies–Pinus*, el bosque de *Pinus*, praderas de *Potentilla* y los matorrales de *Juniperus*. La comunidad más importante para la hibernación de la monarca es el bosque climax de *Abies*, seguido por el bosque secundario mixto de *Abies–Pinus*.

El bosque original de *Abies* tuvo un mayor grado de alteración en los sitios de El Rosario y Chivatí, principalmente debido a cuatro causas: (1) el aclareo del bosque para ampliar zonas de cultivo; (2) la tala inmoderada de los bosques con fines de explotación; (3) los incendios, generalmente provocados por imprudencia; y (4) las actividades ganaderas.

INTRODUCTION

Since the Instituto de Biología of the Universidad Nacional Autónoma de México started research in relation to the "Programa General de Estudios y Protección de la Mariposa Monarca," one of the

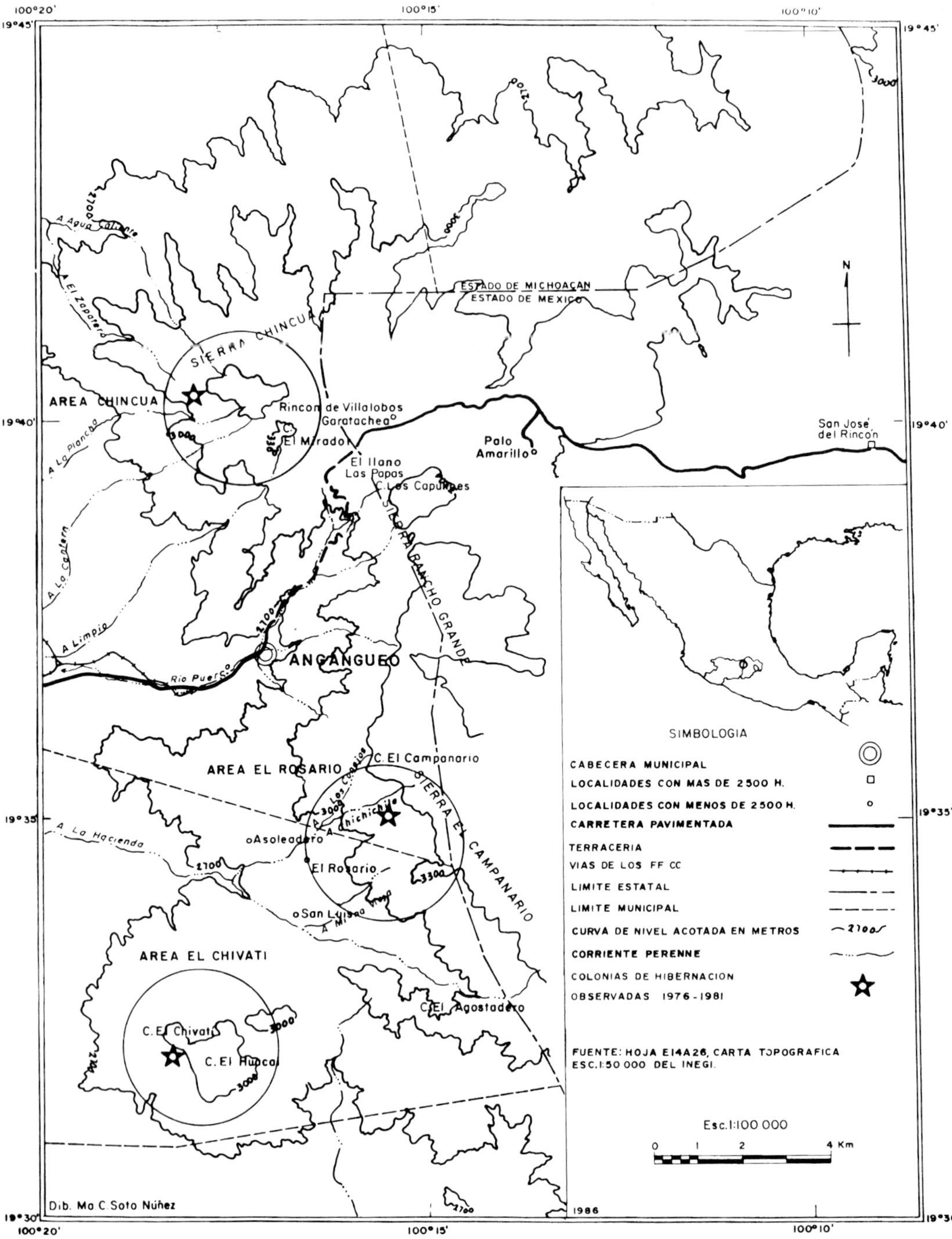

Figure 1. Map of the general study area; the overwintering sites of monarch butterflies are located within the circled areas (based on Anonymous, 1981).

needs most strongly felt was for basic research to understand the ecological interactions between the monarch butterfly and its natural environment at both overwintering sites and the general overwintering area. Such information would allow Mexican government agencies to design actions directed toward an adequate management of the winter roosting sites of this insect migrant.

Because vegetation is one of the key factors maintaining the conditions required by the butterfly, we examined the structure and diversity of the plant communities in the general overwintering area and at specific overwintering sites. Simultaneously, we assessed the degree of disturbance of the original vegetation in relation to monarch overwintering biology.

DESCRIPTION OF THE STUDY SITE

The monarch overwintering area considered here covers approximately 65 km^2 at altitudes between 2,700 and 3,400 m, in the sierras Chincua and El Campanario and Cerro El Chivatí, in eastern Michoacán (municipalities of Angangueo and Ocampo, near the western edge of the state of México (Fig. 1).

These mountains belong to the Transverse Neovolcanic Belt, an area of intense volcanism, of late Miocenic age (ca. 36 my), traversing Mexico in an east–west direction between 18 and 20°N (De Cserna, 1974).

This region is shaped by cineritic cones, lava beds, and basaltic pyroclasts (Guzmán and De Cserna, 1963). According to the FAO soil classification (Dudal, 1968, in Flores, 1974) and Correa et al. (1974), soils of the monarch overwintering sites are andosolic, with 6–13% of organic matter, pH 5.4–6.4, and gross to clay-spongy texture with a high content of active aluminum.

Local hydrology includes mostly year-round flowing creeks such as El Zapatero, Agua Caliente, La Cantera, La Plancha, Arroyo Limpio, and Arroyo Puerco (in the Chincua site); Los Conejos, Chichichile, Las Tinajas, La Mina Vieja, and La Hacienda (El Campanario site); and Borbollón (Cerro Chivatí site). These creeks are confluent at the south-facing slope and form the Río Zitácuaro, which ultimately discharges to the Río Tuzuntla or Cutzamala, well into the Balsas River Basin.

Climatic conditions are those of the C (w) group in the system proposed by Köppen (1948), characterized as temperate–humid with summer rains, and mean minimum and maximum temperatures of the coldest month are −3 and 18°C, respectively. However, as García (1981) has noted, altitudinal variations create climatic changes within this mountainous region, related to alterations of the precipitation : temperature ratio. Thus, using García's classification, the main climatic types would be C (w,/w_z) b (i′) g and C (w,) (w) bi, corresponding to the meteorological stations of Zitácuaro (Michoacán) and San Felipe del Progreso (state of México), respectively; humidity is intermediate, with a precipitation : temperature ratio from 43.2 to 55.0. The subtype C (w_2) (w) b (i) g has been recorded at the Villa de Allende Station (state of México); it is the moistest of the subhumid–temperate types, with a precipitation : temperature ratio >55.0.

THREE OVERWINTERING SITES STUDIED

The three overwintering sites considered here occur within the general area described above. They were identified by direct observation and from a report (Anonymous, 1983) and are characterized as follows.

The **Chincua site** is part of the privately owned land known as Jesús de Nazareno (municipality of Angangueo, Michoacán) with an approximate area of 25 km^2, centered at 19°40′20″N, 100°18′00″W, at an altitude of 3,150 m. This is the site where the first colony was reported to the scientific world (Urquhart, 1976). The colony establishes itself in the area known as El Rincón de Villalobos.

The **El Rosario site** is located on the foothills of the Sierra El Campanario, near the locality of El Rosario (municipality of Ocampo, Michoacán), centered near 19°35′00″N, 100°15′30″W, at 2,900 m, covering approximately 17 km^2.

The **Chivati site** is located between the slopes of the El Chivatí and El Huacal mountains, approximately centered at 19°32′00″N, 100°17′20″W, and at 3,000 m, with an area of ca. 24 km^2.

METHODS

Each year between 1979 and 1984, two field trips were made to the Angangueo region (Fig. 1), one at the onset of the overwintering season and the other at its end. This was done to locate the settling areas used by the butterfly colonies, as well as to determine the movements of the insects during each overwintering season, so as to establish the areas, and area sizes, for our vegetational studies. The study area was mapped with the topographic map of Angangueo (Anonymous, 1981).

To determine the different plant communities, we collected plant samples from several altitudinal transects, including the various access trails to the general area. We also collected within the specific areas of the colonies, to assess plant species diversity in the overwintering sites. Plant collecting was also extended through several seasons of the year, to assess phenological changes in the plant species, and thus obtain as complete a floristic inventory as possible. Herbarium specimens were identified and deposited in the National Herbarium of the Universidad Nacional Autónoma de México (MEXU).

RESULTS

Within the general study area, we recognized a total of six vegetational communities, which were grouped into climax and secondary successional associations. The climax group includes *Abies* forest and upland cloud forest. The secondary group includes mixed *Abies-Pinus* forest, *Pinus* forest, *Potentilla* meadow, and patches of *Juniperus* shrubland.

The main components of these communities are perennial species. Most of the tree species renew their leaves during the driest season of the year in a continuous sequence, resulting in a perennially green aspect to the forest.

THE *ABIES* FOREST

The climax *Abies* forest community is dominated by the oyamel fir, *Abies religiosa* (H.B.K.) Cham. & Schl., and is well-defined structurally and ecologically (Madrigal, 1967; Manzanilla, 1974; Rzedowski, 1978). This is the most important plant community for monarch overwintering within the study area, because of both its ample distribution and its role as substrate for aggregated monarchs.

Leopold (1950) considered the *Abies* forests of the country as boreal, on the basis of floristic, fau-

Table 1. Plant species of the *Abies* forest community, ranked in order of abundance.

a) Understory shrub species

1. *Senecio angulifolius* D.C. (Asteraceae)
2. *Satureja macrostemma* (Benth.) Briq. (Lamiaceae)
3. *Arracacia atropurpurea* (Lehm.) Benth. & Hook. (Apiaceae)
4. *Eupatorium mairetianum* D.C. (Asteraceae)
5. *Symphoricarpus microphyllus* H.B.K. (Caprifoliaceae)
6. *Senecio barba-johannis* D.C. (Asteraceae)
7. *Arctostaphylos longifolia* Benth. (Ericaceae)

b) Understory herb species

1. *Alchemilla procumbens* Rose. (Rosaceae)
2. *Stellaria cuspidata* Willd. (Caryophyllaceae)
3. *Cerastium nutans* Raf. (Caryophyllaceae)
4. *Acaena elongata* L. (Rosaceae)
5. *Senecio callosus* Sch. Bip. (Asteraceae)
6. *S. tolucanus* D.C. (Asteraceae)
7. *Salvia helianthemifolia* Benth. (Lamiaceae)
8. *Penstemon campanulatus* Willd. (Scrophulariaceae)
9. *Piqueria laxiflora* B.L. Rob. & Seat. (Asteraceae)
10. *Stevia monardifolia* H.B.K. (Asteraceae)
11. *Salvia fulgens* Cav. (Lamiaceae)
12. *S. gracilis* Benth. (Lamiaceae)
13. *S. sapinea* Epling (Lamiaceae)
14. *S. elegans* Vahl (Lamiaceae)
15. *Cirsium ehrenbergii* Sch. Bip. (Asteraceae)
16. *Bidens triplinervia* H.B.K. (Asteraceae)
17. *Sigesbeckia jorullensis* H.B.K. (Asteraceae)
18. *Geranium lilacinum* Kunth (Geraniaceae)
19. *Sibthorpia pichichensis* H.B.K. (Scrophulariaceae)
20. *Brachypodium mexicanum* Link. (Poaceae)
21. *Stipa virescens* H.B.K. (Poaceae)
22. *Archibaccharis hieracioides* (Heering.) Blake (Asteraceae)
23. *Senecio sanguisorbae* D.C. (Asteraceae)
24. *S. sinuatus* H.B.K. (Asteraceae)
25. *S. peltiformis* Hemsl. (Asteraceae)
26. *Myosotis scorpioides* L. (Boraginaceae)

nistic, structural, and ecological similarities, in comparison to the large forests covering the northern portions of North America and Eurasia (Rzedowski, 1978). However, Rzedowski (1978) concludes that true boreal and Mexican forests diverge noticeably in phenological patterns. During winter in the true boreal forests, the plant associations remain dormant due to low temperature and snow, while in Mexican *Abies* forests photosynthetic activities are not interrupted because of the milder climate. The more hospitable climate of the Mexican *Abies* forests, and their lack of extreme seasonal changes, could account for the selection of these communities in central Mexico as overwintering sites by the tens of millions of monarch butterflies that migrate each autumn from northeast Canada and the United States.

The *Abies* forest within the study area generally forms dense masses of considerable extension, located approximately between 2,800 and 3,400 m of altitude, on the high and exposed mountain tops of the sierras Chincua and El Campanario as well as Cerro El Chivatí. Some slopes and ravines at lower elevations are also covered by *Abies* forest. The height of the tree canopy varies between 25 and 30 m. The understory shrub layer is 2–5 m in height, and its species diversity is relatively high (Table 1a), increasing with increased tree and canopy disturbance. The understory herb layer contains most species (Table 1b), and the number of species increases with increasing disturbance of the community.

The least disturbed of the fir forests in the three sites was that at the Sierra Chincua, followed by that at the Chivatí area. In contrast, the El Rosario site was subject to marked disturbance, and the oyamel trees formed mixed fir–pine communities, with *Pinus pseudostrobus* Lindl.; the typical fir forest being restricted to the highest peaks of the Sierra El Campanario.

THE UPLAND CLOUD FOREST

Although the type of vegetation of the upland cloud forest occupies very restricted areas within the overwintering locality of the monarch butterfly, some deep ravines include this community and they are utilized by some monarch colonies. Examples are Barranca Honda (Chincua site) and Arroyo El Banquito (Chivatí area).

The upland cloud forest is a well-defined community, irregularly distributed along ravines and deep canyons, where high air moisture is prevalent. The cloud forest is usually restricted to a narrow altitudinal range between 2,500 and 2,750 m. Structurally, this type of vegetation is formed by two tree layers (Table 2), plus one or two shrub layers. The herbaceous understory is relatively impoverished (Table 2), except under conditions of disturbance to the original forest.

Ibarra (1983) reports a similar floristic composition (see Table 2) of the cloud forest at Cerro El Cacique (municipality of Zitácuaro, Michoacán), where colonies of monarch butterflies are known to roost during the winter (Anonymous, 1983).

THE MIXED *ABIES–PINUS* FOREST

The mixed *Abies–Pinus* forest community is of secondary origin and represents a pronounced stage of disturbance of the *Abies* forest. It is located approximately between 2,800 and 3,000 m within the altitudinal range characteristic of the climax fir forest, with which it is in contact and to which it shows many similarities in floristic composition of the shrub and herb layer (Table 3a and 3b).

This community extends over a vast area of El Chivatí and almost all of El Rosario; its distribution, however, is the most restricted in the Sierra de Chincua. Compared to the fir forest the mixed tree

Table 2. Plant species of the upland cloud forest community, ranked in order of abundance.

a) Commonest tree species

1. *Abies religiosa* (H.B.K.) Cham. & Schl. (Pinaceae)
2. *Pinus pseudostrobus* Lindl. (Pinaceae)
3. *Quercus laurina* Humb. & Bonpl. (Fagaceae)
4. *Cleyera integrifolia* (Benth. Choisy) (=*Eurya mexicana* (Turcz.) Syzsz.) (Theaceae)
5. *Quercus rugosa* Nee (Fagaceae)
6. *Cornus disciflora* D.C. (Cornaceae)
7. *Symplocos prionophylla* Hemsl. (Symplocaceae)
8. *Arbutus xalapensis* H.B.K. (Ericaceae)
9. *Cupressus benthamii* var. *lindleyi* Klotsch. (Cupressaceae)
10. *Alnus acuminata* ssp. *arguta* (Schlecht.) Furlow (Betulaceae)
11. *Ternstroemia lineata* D.C. ssp. *lineata* (Theaceae)
12. *Styrax ramirezii* Greenm. (Styracaceae)
13. *Ilex tolucana* Hemsl. (Aquifoliaceae)

b) Understory shrub and herb species

1. *Solanum torvum* Swartz. (Solanaceae)
2. *Eupatorium mairetianum* D.C. (Asteraceae)
3. *Solanum cervantesii* Lag. (Solanaceae)
4. *Salvia fulgens* Cav. (Lamiaceae)
5. *S. gesneraeflora* Lindl. & Paxton (Lamiaceae)
6. *Cestrum benthami* Miers. (Solanaceae)
7. *Moninna xalapensis* H.B.K. (Polygalaceae)

Table 3. Plant species of the mixed *Abies-Pinus* forest community, ranked in order of abundance.

a) Understory shrub species

1. *Senecio angulifolius* D.C. (Asteraceae)
2. *Arracacia atropurpurea* (Lehm. Benth. & Hook. (Apiaceae)
3. *Salvia fulgens* Cav. (Lamiaceae)
4. *Fuchsia microphylla* H.B.K. (Onagraceae)
5. *Arctostaphylos longifolia* Benth. (Ericaceae)
6. *Sambucus mexicana* Presl. (Caprifoliaceae)

b) Understory herb species

1. *Alchemilla procumbens* Rose (Rosaceae)
2. *Acaena elongata* L. (Rosaceae)
3. *Piqueria laxiflora* B.L. Rob. & Seat. (Asteraceae)
4. *Salvia helianthemifolia* Benth. (Lamiaceae)
5. *Senecio callosus* Sch. Bip. (Asteraceae)
6. *Brachypodium mexicanum* Link. (Poaceae)
7. *Senecio tolucanus* D.C. (Asteraceae)
8. *Sigesbeckia jorullensis* H.B.K. (Asteraceae)
9. *Salvia fulgens* Cav. (Lamiaceae)
10. *S. elegans* Vahl (Lamiaceae)
11. *Senecio sanguisorbae* D.C. (Asteraceae)
12. *Lupinus uncinatus* Schldl. (Fabaceae)
13. *Bidens triplinervia* H.B.K. (Asteraceae)
14. *Salvia gracilis* Benth. (Lamiaceae)
15. *Cirsium ehrenbergii* Sch. Bip. (Asteraceae)
16. *Stachys lindenii* Benth. (Lamiaceae)

layer is sparser, although there is no noticeable difference between the heights of two communities.

The mixed forest is dominated by variable proportions of *Abies religiosa* and *Pinus pseudostrobus* in the higher canopy, although *A. religiosa* is generally dominant. A second, lower level of the canopy is composed of a limited number of trees of species such as *Ternstroemia lineata* D.C. ssp. *lineata* and *Cleyera integrifolia* (Benth.) Choisy (=*Eurya mexicana* (Turcz.) Syzsz.), Theaceae, *Cornus disciflora* D.C., Cornaceae, *Arbutus xalapensis* H.B.K., Ericaceae, and *Symplocos prionophylla* Hemsl., Symplocaceae.

Due to the lower tree density in this community, the shrub and herb layers increase in diversity (Table 3a and 3b) and in the area covered, reaching 100% coverage in some cases. Some climbing plants also appear in this community, largely owing to a high degree of disturbance; among them are *Vicia pulchella* H.B.K. and *Astragalus guatemalensis* Hemsl., Fabaceae, and *Smilax moranensis* Mart. & Gal., Liliaceae.

THE *PINUS* FOREST

The pine forest, either pure, or more frequently combined with broadleaf species, mainly *Quercus*, comprises most of the forested portions of the high regions of the Transverse Neovolcanic Belt. In our study area, these forests thrive at altitudes between 2,700 and 3,000 m, but they are usually restricted to below the fir forest level.

Pine forest is the secondary plant community usually found in areas strongly altered by man (Ibarra, 1983). Within our general study area, various pine associations were located; in the Sierra Chincua (north and northeast sections), in Michoacán and México, between approximately 2,900 and 3,000 m, these forests grow in strongly to moderately sloping terrain and are formed by *P. pseudostrobus* and *Pinus montezumae* Lamb., with frequent addition of *A. religiosa.*

At lower altitudes, in the state of México near Palo Amarillo (municipality of San Felipe del Progreso), along the highway between 2,800 and 2,900 m, we found both *Pinus rudis* Endl. and *P. montezumae* growing on gently sloping ground. Both species were severely disturbed. Small patches of *Pinus hartwegii* Lindl. were also found among dense stands of *A. religiosa,* particularly associated with mountain peaks and rough slopes where coarse rocky outcrops were evident. This is especially true for altitudes >3,200 m such as, for example, the top of Cerro El Mirador (within the Chincua area). *Pinus hartwegii* was also observed in similar habitat on some rocky slopes near El Llano de las Papas, just on the México–Michoacán borderline.

THE *POTENTILLA* MEADOW

According to Cruz (1969, in Rzedowski, 1978) the *Potentilla* meadow is characteristic of clearings in coniferous forests of the high mountains in the Transverse Neovolcanic Belt, between 3,000 and

Table 4. Plant species of the *Potentilla* meadow community, collected primarily during the rainy season and ranked in order of abundance. Adjacent to the meadows are patches of matorral, *Juniperus monticola* Martínez (Cupressaceae).

1. *Potentilla candicans* H. & B. (Rosaceae)
2. *Sisyrinchium scabrus* Schldl. & Cham (Iridaceae)
3. *S. angustifolium* Mill. (Iridaceae)
4. *Eryngium carlinae* Delar. (Apiaceae)
5. *Prunella vulgaris* L. (Lamiaceae)
6. *Trifolium mucronatum* Willd. (Fabaceae)
7. *Senecio helodes* Benth. (Asteraceae)
8. *Halenia brevicornus* (H.B.K.) G. Don (Gentianaceae)
9. *Helenium scorsoneraefolia* (D.C.) A. Gray (Asteraceae)
10. *Plantago australis* Lam. (Plantaginaceae)
11. *Solanum tuberosum* L. (Solanaceae)
12. *Castilleja schaffneri* Hemsl. (Scrophulariaceae)
13. *Conyza schiedeana* (Less) Cronq. (Asteraceae)
14. *Stachys kerlii* Benth. (Lamiaceae)
15. *Trisetum* sp. (Poaceae)
16. *Muhlenbergia* sp. (Poaceae)
17. *Cirsium* sp. (Asteraceae)

3,500 m. In our study area, small meadows were common in the fir-forest areas between 3,000 and 3,300 m, on poorly drained soils and usually on flat or slightly inclined ground; they are locally called "llanos." Examples include Llano de las Papas, Llano del Rincón de Villalobos, Llano de los Toros, and Llano Redondo (all within the Chincua area) and Llano del Chivatí and Llano de la Media Luna (both at the top of Cerro Chivatí).

These meadows are dominated in the early months of the year by *Potentilla candicans* H. & B., Rosaceae, and later, by the onset of the rainy season, there is considerable growth of two grass genera, which gives them an appearance of dense, low grassland, locally referred to as "zacatal" (Rzedowski, 1978). Other species characteristic of this community are listed in Table 4.

THE *JUNIPERUS* SHRUBLAND

Adjacent to the *Potentilla* meadows, and along the banks of small watercourses, or on adjacent gently sloping areas, we found some patches or strips of *Juniperus monticola f. compacta* Martínez (Matorral). Most important are those of Llano de las Papas and Llano del Rincón de Villalobos, near the Chincua site, and some others near El Rosario.

THE PRESENT CONDITION OF THE LOCAL VEGETATION

The original vegetation is more extensively disturbed in the areas of El Rosario and El Chivatí than at Sierra Chincua because most of the areas involved are dominated by secondary communities of *A. religiosa.*

The main causes of disturbance are related to four kinds of human activities: (1) clearing of the forests to increase the area of agricultural land, (2) uncontrolled timber exploitation and poor timber removal techniques, (3) forest fires, usually induced through negligence by visitors or local sheep farmers, and (4) livestock grazing and inadequate management of pasturelands.

CONCLUSIONS

Of the six vegetational types considered in the study area, the forests of *A. religiosa* surpass all others in importance for three reasons: (1) their ample coverage of the monarch overwintering area, either in climax condition (Chincua site) or in secondary succession (e.g., the mixed *Abies–Pinus* forests of El Rosario and El Chivatí); (2) they are the best-defined plant communities in relation to structure and species diversity; and (3) they include the main overwintering centers known for the monarch butterfly, especially at the Chincua area.

ACKNOWLEDGMENTS

The authors wish to recognize the help and facilities provided by the Instituto de Biología of the Universidad Nacional Autónoma de México and by the Colegio de Ciencias y Humanidades of the same university.

LITERATURE CITED

Anonymous. 1981. Carta topográfica Angangueo E14A26, excala 1:50,000. México DETENAL-SPP.

Anonymous. 1983. Plan de manejo para la zona de hibernación de la mariposa Monarca. Unidad de Apoyo Técnico de la Subsecretaria Forestal y de la Fauna. México: SARH.

Correa, P.G., et al. 1974. *Geografía del Estado de Michoacán,* Tomo I. Gobierno del Estado. Morelia, Mich., 454 pp.

De Cserna, Z. 1974. La evolución geológica del panorama fisiográfico actual de México. In *El escenario geográfico. Serie México: Panorama histórico y cultural,* Tomo I, 19–56. México: Instituto Nacional de Antropología e Historia.

Flores, D.A. 1974. Los suelos de la República Mexicana. In *El escenario geográfico. Serie México: Panorama histórico y cultural,* Tomo II, 7–108. México: Instituto Nacional de Antropología e Historia.

García, E. 1981. *Modificaciones al sistema de clasificación climática de Köppen,* 3a ed. Enriqueta García de Miranda, Indianapolis 30, México, D.F., 252 pp.

Guzmán, E.J., and Z. De Cserna. 1963. Tectonic history of Mexico. In *The backbone of the Americas—Tectonic history from pole to pole,* 113–129. Symposium. Memoir No. 2. Assoc. Geol.

Ibarra, C.G. 1983. Comunidades vegetales del cerro El Cacique, ubicado en el Eje Neovolcánico; Zitácuaro, Michoacán. Tesis de licenciatura en Biología. México: Facultad de Ciencias, UNAM, 91 pp.

Köppen, W. 1948. *Climatología.* México: Fondo de Cultura Económica, 478 pp.

Leopold, A.S. 1950. Vegetation zones of Mexico. *Ecology* 31:507–518.
Madrigal, S.X. 1967. Contribución al conocimiento de la ecología de los bosques de oyamel (*Abies religiosa* (HBK.) Schlecht. & Cham.) en el Valle de México. *Instituto Nacional de Investigaciones Forestales. Boletín Técnico 18. México, D.F.*, 94 pp.
Manzanilla, H. 1974. *Investigaciones epidométricas y silvícolas en bosques mexicanos de Abies religiosa.* Dirección General de Información y Relaciones Públicas, México, D.F., Secretaría de Agricultura y Ganadería, 165 pp.
Rzedowski, J. 1978. *Vegetación de México.* México, D.F.: Limusa, 432 pp.
Urquhart, F.A. 1976. Found at last: The monarch's winter home. *National Geographic* 150(2):160–173.

INVESTIGATIONS OF POSSIBLE MONARCH BUTTERFLY OVERWINTERING AREAS IN CENTRAL AND SOUTHEASTERN MEXICO

Javier de la Maza E. and
William H. Calvert

ABSTRACT. Since the discovery of the wintering colony of monarch butterflies, *Danaus plexippus* (L.), in the state of Michoacán, Mexico, by Fred Urquhart in 1976, a number of researchers have devoted time and effort to searching for other monarch colonies in Mexico. Between 1977 and 1982 approximately 30 colonies at nine sites were found in a small area near the boundary of the states of Michoacán and México in the Eje Neovolcánico Transversal (Transvolcanic Belt) of central Mexico. Also during this period, a few reports of colonies in other areas were received. These could not be verified due to the lack of precise information. In an attempt to locate additional overwintering areas, the Mexican conservation group Monarca, A.C. (see Ogarrio, this volume), launched a cartographic survey of the Eje Neovolcánico Transversal. This study was designed to identify areas within the range having suitable conditions to host wintering colonies of monarch butterflies.

For some of the areas identified as suitable in the cartographic survey, reliable information on the presence of monarchs was obtained. Also, data on migration movements were gathered for the country's southeastern region, mainly from the southern portion of the Sierra Madre Oriental, the Sierra Madre del Sur, and the highlands (Los Altos) of the state of Chiapas.

Considerations of possible temporary colonies north and south of the known wintering region, as well as their importance in the overall protection of this unique biological phenomenon, are discussed.

INTRODUCTION

In 1976, Fred Urquhart reported the discovery in Mexico of the largest wintering area of the eastern North American population of the monarch butterfly, *Danaus plexippus* (L.). This area is located in the mountains of the state of Michoacán in the Eje Neovolcánico Transversal (Transvolcanic Belt; Urquhart, 1976; Urquhart and Urquhart, 1976a). Between 1976 and 1982, nine overwintering sites were located in the mountains of the Mexican states of Michoacán and México, and 30 colonies have been described by Calvert and Brower (1986). The news of the discovery of the overwintering areas of the monarch in Mexico stimulated others to report information about additional areas where monarch butterflies have been seen migrating or where colonies settle.

When the senior author worked for the Mexican Wildlife Service (Dirección General de Fauna Silvestre, SARH) in 1980, his agency received numerous reports of the presence of monarchs on the volcanoes Popocatépetl and Ajusco and on mountain ranges to the south such as the Sierra de Oaxaca and the Sierra de Juárez. On several occasions we, or our colleagues, traveled to these areas but were unable to locate the reported aggregations. Several other sightings of mass movements of monarchs were made by us or were reported in the literature. For example, a large movement of monarchs was observed nearly 100 years ago by the British author Rider Haggard, who saw them flying over the town of Orizaba in the state of Veracruz in the autumn of 1890 (in Williams, 1958). In 1977, while collecting butterflies in the Sierra Madre del Sur south of the city of Oaxaca, the senior author observed a large group of monarchs flying at a high altitude in a southeasterly direction. This group had left the coastal mountains behind and was flying toward the Pacific Ocean (de la Maza E. et al., 1977). During the autumn of 1980, the junior author observed a group of monarchs flying to the southeast along the escarpment above Orizaba in the direction of the Sierra de Juárez in the state of Oaxaca. These observations raised the possibility of finding monarch colonies south of the Eje Neovolcánico Transversal.

SEARCHING IN SOUTH AND CENTRAL MEXICO

In February 1982 we searched the mountains of the states of Guerrero, Oaxaca, and Chiapas for overwintering monarchs. Although we did obtain reports of the presence of monarch aggregations, we were unable to confirm the presence of permanent overwintering groups of any size in this area. In 1984, Monarca, A.C., initiated a cartographic study of the Eje Neovolcánico Transversal, the volcanic range that crosses Mexico from Colima near the Pacific Ocean to Veracruz on the Gulf of Mexico. More than 60 potential sites were identified. Nearly all sites with appropriate floral characteristics, namely pine (*Pinus* spp.) and oyamel fir (*Abies religiosa*) associations, were located in areas where intense commercial lumbering has resulted in the destruction of much of the original vegetation and in reduction of their value as overwintering habitats of the monarch butterfly. Nonetheless, some of these areas have been the subject of recent reports of monarch aggregations.

One of the most reliable reports of monarch colony occurrence is that for Cerro Las Capillas in the Sierra de Manantlán of Jalisco (Rafael Guzmán, Universidad de Guadalajara, Manantlán Biosphere Reserve, pers. comm.). Another report comes from a forester who worked in a sawmill in the mountains south of the city of Oaxaca for 10 years. He informed us that in the zone known as El Tlacuache, near the town Sola de Vega, he has seen monarch butterfly colonies for several winters (Sergio Contreras, pers. comm.).

The field chief of the Corporación de Fomento Forestal of the state of Chiapas also informed us that trees covered with orange-colored butterflies are occasionally found near the town of San Vicente, east of San Cristóbal de la Casas.

In April 1986, at Lagos de Montebello, near the border of Guatemala, a large population of monarchs was seen in erratic flight by the senior author. In spite of a large random component to their flight, most appeared to be flying to the northwest. South of the Lagos de Montebello in Guatemala are the Cuchumatanes, a large 3,000-m mountain range with extensive fir forest. Monarchs have been observed in the northern portion of this range. The presence of migrating monarchs in April this far south may suggest either very late departure from overwintering sites, characteristic of monarchs that migrate north to the United States (see Calvert and Lawton, this volume; Van Hook, this volume), or seasonal migrants, characteristic of monarchs in Central America (see Haber, this volume).

The director of the Sianka'an Biosphere Reserve in the state of Quintana Roo reported an abundance of monarchs in November 1985 flying south along the coast. This supports a possible autumn migration route from Florida, across the Caribbean, to Yucatán (Urquhart and Urquhart, 1976b). Large numbers of monarchs have also been observed flying ashore south of Veracruz in the vicinity of Alvarado (N. Smith, pers. comm.; H. Perez, pers. comm.).

DISCUSSION

Based on current information on monarch aggregations, and the movements of butterfly groups within Mexico, we conclude the following: Permanent overwintering areas of the monarch butterfly are all located within the Eje Neovolcánico Transversal, between altitudes of 2,700 and 3,400 m. The major colonies occur in a small rectangular area of about 800 km^2 in the states of México and Michoacán between latitudes 19°20′N and 19°45′N and longitudes 100°10′W and 100°20′W. Other colonies also form each year, but most of these are unstable and disintegrate rapidly. It is likely that these temporary colonies move to reform at the principal colonies.

The migratory groups observed in Oaxaca and Guerrero, much to the south of the principal overwintering areas in the Eje Neovolcánico Transversal, may be groups "lost" from the principal migration. These groups probably form small temporary aggregations in different places each year, and warmer temperatures may stimulate movement and possible reaggregation later at the principal overwintering areas, or an early end to diapause and local breeding.

The colonies reported from Chiapas are more difficult to interpret. We think that these are not groups that have overflown the major colonies of the Eje Neovolcánico Transversal, but rather they may be groups that migrated across the Gulf of Mexico and the Caribbean from Florida. The aggregations reported along the Guatemalan border and in the mountains of Guatemala may also have traveled from Florida.

In summary we suggest that migration and overwintering of monarchs in Mexico follows this pattern. (1) Permanent overwintering colonies are located in a reduced area in the states of México and Michoacán. This is the principal overwintering zone. (2) Temporary colonies are located to the north and the periphery of the principal, central overwintering areas. These colonies are unstable, disperse early, and may gather at the principal areas. (3) Groups that have overflown the principal overwintering areas may also form colonies that disperse early and eventually join the major colonies. (4) Butterflies that migrate through Florida and across the Gulf of Mexico may overwinter in Chiapas and Guatemala.

ACKNOWLEDGMENTS

We wish to thank Enrique Fuentes and the hundreds of informants that we encountered in our travels through Mexico. This research was supported by the World Wildlife Fund–US, grant number 1968, and by the Ministry of Urban Development and Ecology of the Mexican Government.

LITERATURE CITED

Calvert, W.H., and L.P. Brower. 1986. The location of monarch overwintering colonies in relation to topography and climate. *Journal of the Lepidopterists' Society* 40(3):164–187.

de la Maza E., R.G., J. de la Maza E., and A. Díaz Francés. 1977. Movimientos migratorios de "Monarcas" en el estado de Oaxaca, Mexico. *Boletín Informativo, Sociedad Mexicana de Lepidopterología* 3(5):12–13.

Urquhart, F.A. 1976. Found at last: The monarch's winter home. *National Geographic* 150:160–173.

Urquhart, F.A., and N.R. Urquhart. 1976a. The overwintering site of the eastern population of the monarch butterfly (*Danaus p. plexippus:* Danaidae) in Southern Mexico. *Journal of the Lepidopterists' Society* 30:153–158.

Urquhart, F.A., and N.R. Urquhart. 1976b. A study of the peninsular Florida populations of the monarch butterfly (*Danaus p. plexippus;* Danaidae). *Journal of the Lepidopterists' Society* 30:73–87.

Williams, C.B. 1958. *Insect migration.* London: Collins, 235 pp.

COMPARATIVE PHENOLOGY OF VARIATION IN SIZE, WEIGHT, AND WATER CONTENT OF EASTERN NORTH AMERICAN MONARCH BUTTERFLIES AT FIVE OVERWINTERING SITES IN MEXICO

William H. Calvert and Robert O. Lawton

ABSTRACT. Monarch butterflies overwinter each year in aggregated colonies of millions of reproductively diapausing individuals within the high-altitude fir forests of central Mexico. Colonies that vary in topographical, climatic, and habitat characteristics are located on peaks within Mexico's Transvolcanic Mountain Range. To determine whether or not variation among colony locations can influence monarchs, forewing lengths, wet and dry weights, and water content of monarch butterflies were monitored at the five colony locations during the 1985–1986 overwintering season. Three, near Angangueo, Michoacán, were considered core colonies because they were larger and more stable than two outlying colonies. The two outlying colonies were located near the volcano Nevado de Toluca.

Forewing lengths remained the same at all colonies until mid-February but declined significantly in late February and March. Colony movement intensified after mid-February; mating activity and apparent butterfly exodus from the colonies increased in early to mid-March. The reduction in butterfly forewing length was probably caused by larger butterflies leaving the colonies before smaller ones.

All weight components declined during the overwintering season in both sexes. Male wet weights were significantly heavier than female wet weights at all times during the season. Among colonies there were differences in the rate of wet weight decline, but only for females. Core-colony females declined at a greater rate than those in outlying colonies. Dry weight also declined in both sexes. The rates of dry weight decline and the elevations of the regressions were the same for both sexes. There were differences in rates of dry weight decline among colonies. Rates of decline in core colonies were greater than in outlying colonies. Water content declined in both sexes. Males had higher water content than females. No differences in the rate of water content decline were observed among colonies. Percent moisture declined during the season for males but not for females.

Differences in dry weight decline among colonies suggest that the colony environments impose different metabolic demands on overwintering monarchs. Colonies differed in forest structure, slope characteristics, distances to open sources of water, and probably in distances to, and quantities of, available nectar. Contrary to expectation, however, the colonies closest to water did not have the lowest rates of dry weight decline. Monarchs may get water with little energy expenditure by obtaining it in or near their roosts and by soaring extensively in their flights to water. The amount of nectar available near colonies is potentially a major influence on the lipid economy of overwintering monarchs. This and the relationship of weight decline to forest structure and slope characteristics are currently under study.

INTRODUCTION

Each autumn monarch butterflies in eastern North America migrate from breeding grounds at the latitudes of the Great Lakes to overwintering areas in the high-altitude (ca. 3,000 m) fir forests of the Transvolcanic Mountain Belt in Mexico (Urquhart and Urquhart, 1978). Monarchs arrive at the onset of the dry season in late October and early November (Fig. 1) and form small aggregations that move around and coalesce during November into large colonies averaging 0.85 ha in area (Calvert and Brower, 1986). Most colonies occur along an arc stretching from the western slopes of the volcano Nevado de Toluca in the state of México, northwest to the town of Zitácuaro in the state of Michoacán, and then north, past Angangueo to Contepec, Michoacán (Fig. 2; Calvert and Brower, 1986). The total length of the arc from the most easterly

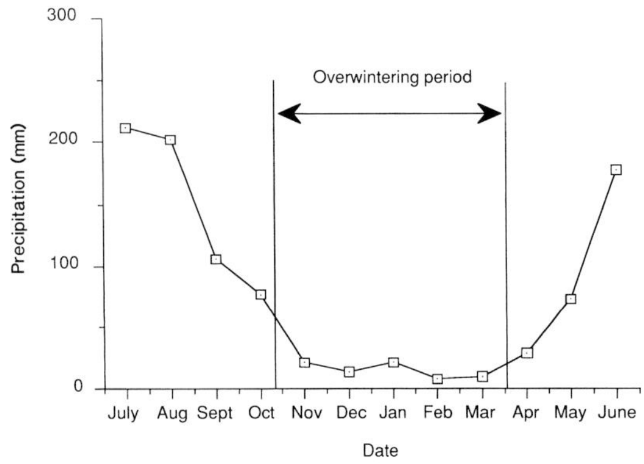

Figure 1. Average rainfall at eight high-altitude (>2,500 m) meteorological stations in or near monarch butterfly overwintering areas in Mexico's Transvolcanic Belt. The number of years each station has been in operation averaged 22 yr.

colony, Palomas, to the most northerly colony, Altamirano, is approximately 143 km. Chincua, Campanario, and Picacho are located in west-draining canyons on a single massif centered around Angangueo, Michoacán. Approximately 26 km to the south lies another massif that includes the 3,500-m peak El Pelón. Colonies located here tend to have a more southerly exposure than those of the region around Angangueo. The colony at Herrada, located approximately 40 km to the southeast of Cerro Pelón, normally forms on the southwest face, but in 1985–1986 and 1986–1987, it formed on the northeast face. Palomas is located another 18 km to the southeast on an old side vent of the volcano Nevado de Toluca. The most northerly colony, Altamirano, is located approximately 45 km to the northeast of Angangueo on a small massif above the town of Contepec. Colonies do not form here every year. When they do, they usually face south (Calvert and Brower, 1986; Calvert, unpubl. data).

The position of the colonies is relatively stable during the middle of the overwintering period from December through mid-February, but as the habitats dry out, they move toward a water source. Due to seasonal warming, butterfly activity increases in late February and March. Daily flights out of the colony to water and nectar, and returns to lower positions on the colony periphery, increase and result in more rapid change in colony position. In mid-March mating activity greatly increases and the butterflies begin the remigration northward (Brower, 1985; Calvert and Brower, 1986; Van Hook, this volume).

Autumnal migrants feed on floral nectar to store a mean fat content of 73% of their lean dry weight on arrival at the overwintering sites. Fat weights then decline during the winter until at the end of March mean monarch fat weight is only 37% of their lean dry weight (Brower, 1985). A similar phenology of weight loss occurs in monarchs that overwinter in California (Tuskes and Brower, 1978) and in Australia (James, 1984).

Overwintering survival is threatened by too rapid lipid depletion before metabolic reserves can be replenished in the spring, as well as by predation, freezing, and desiccation (Calvert et al., 1979; Glendinning et al., 1988, this volume; Masters et al., 1988; Alonso et al., this volume; Anderson and Brower, this volume). All of these factors are related in a complicated manner. For example, high temperatures within the forests where monarchs overwinter result in high metabolism, rapid weight loss, and possible starvation before nectar becomes plentiful in the spring (Calvert and Brower, 1986; Masters et al., 1988). Low temperatures result in low monarch body temperatures and low metabolic rates (Chaplin and Wells, 1982; Masters et al., 1988). Consequently water loss is reduced and lipids are conserved. Low temperatures, however, also render monarchs inactive and increase the time they are vulnerable to bird predation (Brower and Calvert, 1985; but see Alonso et al., this volume). Furthermore, overwintering at high altitude exposes monarchs to occasional bouts of severely cold weather and the physical abuse of storms that can dislodge them from relatively safe perches in the trees (Calvert et al., 1983).

Monarch overwintering sites differ in many attributes including location, slope aspect, slope declination, forest structure, and availability of water (Fig. 2 and Table 1). These factors, which affect monarch survival, can be monitored indirectly by following trends in monarch weight change. We asked: Do monarchs lose water and lipids at different rates in different colonies? If so, which environmental parameters are important in determining these differences? To ensure that weight loss was not simply the by-product of size-related mortality or the emigration of larger monarchs, we also considered wing length changes of the monarch population during the overwintering period.

METHODS

SAMPLE COLLECTION

From November through March 1985–1986 we made repeated collections of roosting monarchs at five overwintering colonies (Fig. 2 and Table 1). In the central core of the overwintering zone near Angangueo, Michoacán, collections averaged once every 8 days at Chincua, once every 27 days at Picacho, and once every 50 days at Chivatí/Huacal. Collections averaged once every 24 days at Herrada and once every 29 days at Palomas, both located east of the core area near the volcano Nevado de Toluca. By March 1986 the colonies at Herrada and Picacho had dispersed. No collections were made at these colonies after mid-February. [The naming of colonies and their location is discussed in Calvert and Brower (1986). Names refer either to the location of a specific colony that formed during the winter of 1985–1986 or to the butterflies roosting there.]

During each visit to a colony, samples of 30 females

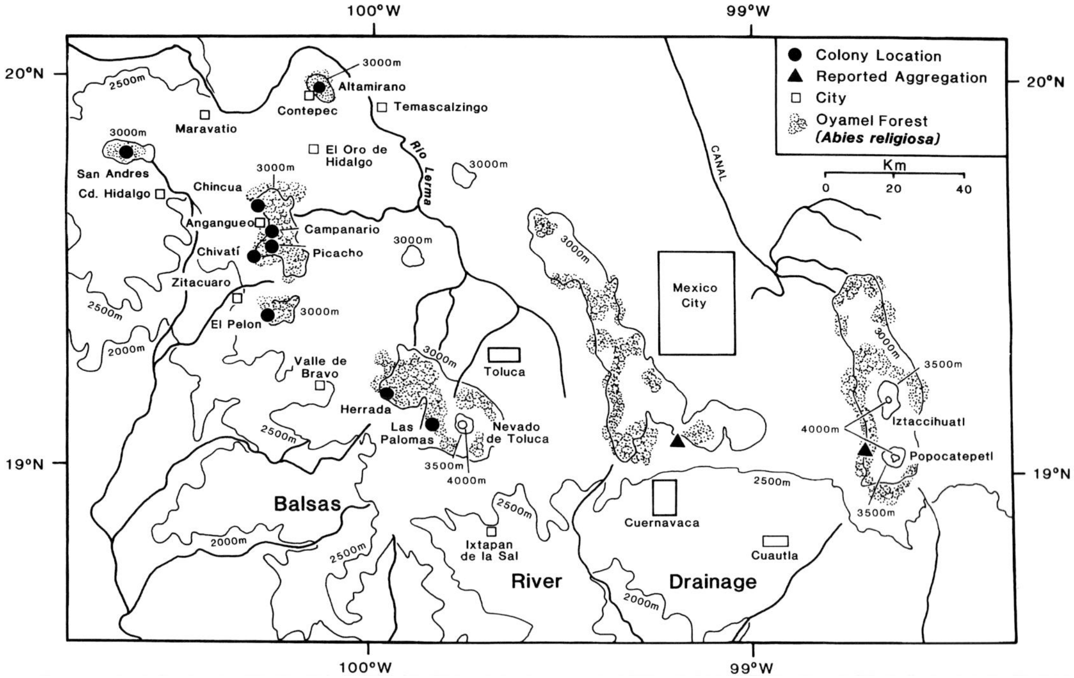

Figure 2. Location of monarch butterfly overwintering areas in relation to topography, drainage patterns, and oyamel (*Abies religiosa*) forest of Mexico's Transvolcanic Belt. Most sites are located along an arc stretching from the western slopes of the volcano Nevado de Toluca, in the state of México, northwest to the town of Zitácuaro, in the state of Michoacán, and then north, past Angangueo to Contepec, Michoacán. The total length of the arc from the most easterly colony, Palomas, to the most northerly colony, Altamirano, is approximately 143 km. Colonies are indicated by solid black dots; cities by open squares. Contour interval = 500 m (from Calvert and Brower, 1986).

and 30 males were taken from shaded bough clusters near the center of the colony. These animals were grouped by sex and weighed immediately on a 50-g Pesola scale. After weighing they were placed on ice until they could be frozen—a period that ranged from several hours to 5 days. At the end of the overwintering season, the forewing length of each butterfly was measured. Samples were recombined into their original groups by sex, dried in a forced-air oven at 50°C for 20–24 hr, and reweighed. To assess the amount of water in each sample, dry weights were subtracted from wet weights.

SAMPLE ANALYSIS

Student's t test was used to determine sex differences in forewing length, rates of wet, dry, and water weight decline, and forewing length differences between stable and unstable phases of the overwintering period (see Fig. 3). Temporal trends in all components of butterfly weight and percent moisture were examined by linear regression. Since butterfly weight did not decrease at a linear rate during the 5-mo overwintering period, the date (expressed as days since the first collection day, 20 November) was logarithmically transformed prior to computation of regressions. Temporal trends in butterfly size (forewing length) were examined by linear regression until mid-February. Analyses of covariance and Tukey's multiple-range tests were used to compare regression slopes and elevations among colonies (Zar, 1984).

Fat extractions and lean weight determinations were not made on these samples. To calculate percent moisture (weight of water divided by lean weight) for the Chincua colony, lean weight data for the 1979–1980 season were used (Brower et al., this volume).

RESULTS

SIZE

For samples taken between 20 November 1985 and 17 March 1986, male forewing lengths ($\bar{X} = 5.36$ cm, SE = 0.007, n = 947, range = 4.30–5.95 cm) were slightly, but significantly, longer than female forewing lengths ($\bar{X} = 5.32$, SE = 0.006, n = 987, range = 4.40–5.85 cm; $t = -5.01$, $p < 0.0001$, n = 1934; Fig. 3). All subsequent forewing length analyses were performed separately by sex.

Analysis of variance showed no significant differences among the five monarch colonies in either male or female wing lengths ($F = 0.87$, $p < 0.48$, n = 986, and $F = 2.17$, $p < 0.07$, n = 947 for females and males, respectively).

Forewing lengths in both male and female monarch samples did not change significantly during the

Table 1. Physical characteristics of the major 1985–1986 overwintering colonies, during the middle of the overwintering period.

Colony	Distance to next colony (km)[a]	Size (ha)	Altitude (m)	Facing azimuth	Forest density (trees/ha)	Basal area (m^2/ha)
Core colonies						
Chincua	9	0.53	3,265	172°	620	43.8
Campanario	5	2.06	3,290	241°	254	25.3
Picacho	6	0.27	3,430	265°	433	39.5
Chivatí/Huacal	18	0.18	3,095	248°	315	37.8
El Pelón	41	0.49	—	—	90	12.1
Outlying colonies						
Herrada	18	0.07	3,280	170°	797	85.2
Palomas	—	0.15	3,315	255°	467	42.2

[a] Distances are along the arc of colonies to the next colony to the south (Fig. 2).

immigration and stable phases of the overwintering period between November and mid-February ($F = 0.002$, $p < 0.96$, n = 27, and $F = 1.27$, $p < 0.27$, n = 27, for females and males, respectively; Fig. 3). Forewing lengths measured during the period 27 February to 18 March, when the colony positions were changing rapidly and monarchs were remigrating northward, were significantly smaller than the immigration and stable period from November to mid-February ($t = 6.48$, $p < 0.0001$, n = 31, and $t = 5.43$, $p < 0.0001$, n = 31, for females and males, respectively).

WET WEIGHT LOSS AMONG COLONIES

Both sexes lost wet weight during the overwintering season ($F = 27.21$, df = 30, $p = 0.0001$, and F 62.58, df = 30, $p = 0.0001$, for females and males, respectively), and both sexes lost wet weight at the same rate ($t = 1.00$, df = 58, $p > 0.2$). Males were consistently heavier in wet weight than females ($t = -3.17$, df = 60, $p < 0.0024$). Subsequent analyses of rates of wet weight decline among colonies were performed on pooled male and female data.

Although pooled sexes lost wet weight at the same rate among colonies ($F = 1.44$, df = 60, $p > 0.2$), a test of overall differences in the elevation of regressions of wet weight versus time did reveal significant differences among colonies, but only for females. Females in the three core area sites, Chincua, Picacho, and Huacal, were heavier in wet weight than females in the two outlying colonies, Herrada and Palomas (Table 2).

DRY WEIGHT

The decline of dry weight was significant for both sexes during the overwintering period ($F = 45.80$, df = 30, $p = 0.0001$, and $F = 77.48$, df = 30, $p = 0.0001$, for females and males, respectively). An analysis of covariance comparing regressions of monarch dry weight versus time in the overwintering season showed no differences between the sexes, in either the rate of dry weight decline (the regression slope, $t = 0.98$, df = 58, $p > 0.2$) or the dry weight at a given time (the regression elevation, $t = 1.38$, df = 58, $p > 0.2$). There were, however, differences among colonies for both sexes combined ($F = 6.92$, df = 60, $p < 0.0005$; Table 3). Pairwise comparisons using Tukey's test showed that the rate of dry weight decline for each colony was the same as that of its nearest neighbor(s) but differed from all others (Table 3 and Fig. 4). The lowest rates of dry weight decline were found in the outlying colonies, Palomas and Herrada, and the highest rates of decline were found in the two core colonies, Picacho and Huacal, while the core colony of Chincua was intermediate.

WATER CONTENT

Water content decline during the overwintering season was significant for both females and males ($F = 6.69$, df = 30, $p = 0.02$, and $F = 20.40$, df = 30, $p = 0.0001$, for females and males, respectively). Since males have higher wet weights than, but the same dry weights as females, it follows that males have a higher water content. The elevation of male water loss during the season was significantly higher than female water loss ($t = -3.92$, df = 58, $p < 0.0005$; Fig. 5). The rates of water decline were the same for the two sexes ($t = 0.73$, df = 58, $p > 0.2$). Subsequent analysis of differences among colonies were performed on data with the sexes pooled. A portion of these data are graphed for the Chincua colony, where rainfall records were kept (Fig. 6). The sharp rise in water content for both sexes on 6 February is undoubtedly related to the 14.5 mm rain that fell in that area between 4 and 6 February.

Although the differences among colonies in the rate of dry weight decline along with the lack of observed differences in wet weight decline suggest that there should be differences among colonies in water content decline, neither slope nor elevation differences were detectable by analysis of covari-

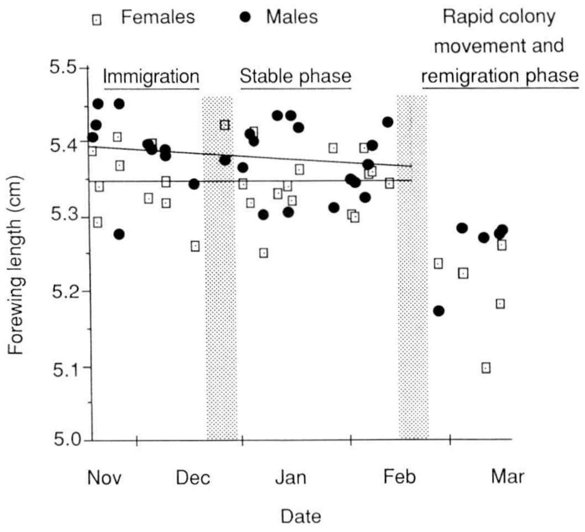

Figure 3. Average forewing lengths of roosting female and male monarch butterflies overwintering in colonies in Mexico's Transvolcanic Belt. Each point represents the average of 30 butterflies. After mid-February the average forewing length for both sexes significantly declines.

ance ($F = 1.03$, df = 4,60, $p > 0.5$, and $F = 1.76$, df = 4,60, $p > 0.2$, for slopes and elevations, respectively).

Percent moisture was calculated as water content divided by lean weight for the 1985–1986 Chincua colony, using water content for 1985–1986 butterflies and lean weights from the 1979–1980 overwintering season (Brower, 1985). Percent moisture declined during the overwintering season for males ($F = 7.56$, df = 14, $p = 0.02$) but not for females ($F = 2.12$, df = 14, $p = 0.17$). Males contained significantly greater moisture ($\bar{X} = 178.8$, SE = 3.03, n = 15) than did females ($\bar{X} = 162.4$, SE = 3.59, n = 15; $t = -3.49$, df = 28, $p = 0.002$).

WEIGHT LOSS RELATIVE TO THE FOREWING LENGTH

The general decline of weight during the overwintering season and the decrease of forewing length at the end of the season suggested that weight decline might be related to the decrease in size. Wet and dry weights were divided by forewing length^2 to neutralize the effect of size. For male and female data combined, polynomial regressions of wing-loading (= weight ÷ wing length^2) versus time in the overwintering season show that both variables declined from late November until early February and then rose slightly until 17 March ($F = 40.83$, df = 61, $p = 0.0004$, and $F = 88.61$, df = 61, $p = 0.0004$, for wet weight/forewing length^2 and dry weight ÷ forewing length^2 regressions, respectively).

DISCUSSION

TRENDS IN BUTTERFLY SIZE

Butterfly sizes, as measured by forewing length, did not change in samples of either males or females during the initial formation and consolidation phases of November or during the stable phases of December to mid-February (Fig. 3). However, samples

Table 2. Decline of wet weights in overwintering monarch butterflies at five colony locations in Mexico's Transvolcanic Belt. Butterfly colony samples with different letters are significantly different at 5% with Tukey's multiple-range test.

Colony		N	Mean wet weight (mg)	SD	Range	Slope	Y	r
Females								
Core colonies								
Chincua	A	15	513.9	45.6	615–445	−81.8	644.4	0.79
Picacho	A C	3	510.0	32.5	547–486	−105.4	686.7	0.91
Huacal	A	3	537.3	58.9	600–483	−112.7	583.7	0.98
Outlying colonies								
Herrada	B C	5	501.2	43.9	560–453	−63.0	583.7	0.98
Palomas	B	5	478.0	45.2	550–440	−50.7	553.0	0.94
Males								
Core colonies								
Chincua	A	15	546.7	46.1	645–490	−92.5	694.1	0.89
Picacho	A	3	525.0	56.0	582–470	−196.1	851.4	0.97
Huacal	A	3	55.47	47.6	609–520	−104.3	734.2	0.97
Outlying colonies								
Herrada	A	5	565.0	45.1	625–526	−64.0	648.8	0.97
Palomas	A	5	529.2	61.3	617–473	−59.7	617.5	0.82

Table 3. Decline of dry weights in combined male and female overwintering monarch butterflies at five colony locations in Mexico's Transvolcanic Belt. Butterfly colony samples with different letters are significantly different at 5% with Tukey's multiple-range test.

Colony		N	Mean dry weight (mg)	SD	Range	Slope	Y	r
Core colonies								
Chincua	A B D	30	252.8	25.2	308–207	−52.9	337.2	0.92
Picacho	D E	6	242.7	31.8	283–207	−125.2	451.2	0.98
Huacal	A D E	6	254.7	32.8	302–223	−79.2	391.0	0.91
Outlying colonies								
Herrada	A B C	10	247.7	29.6	302–207	−43.7	304.9	0.95
Palomas	B C	10	235.9	33.3	303–207	−33.9	291.6	0.90

taken during the more active phase of late February and March show butterfly forewing length to be significantly smaller than that of the earlier period. This later period is characterized by a general seasonal warming of the environment and rapid changes in colony position. The colonies usually accelerate downslope movement nearer to water sources. During the middle and latter portions of this active period, the frequency of mating increases, and outside the overwintering areas butterflies are seen flying northward.

The most likely explanation for the decrease in butterfly forewing length after mid-February is that larger butterflies leave the colonies first. This is suggested by changes in colony size during this period. By mid-February, Chincua, Picacho, and Palomas colonies had moved downslope to sites as far as 1 km from their November locations and were smaller than they had been in February. It is likely that many butterflies had left these colonies to begin the northward migration. By March, the colonies at Herrada and Huacal had disappeared from the areas they had occupied since November. These butterflies had either begun the remigration northward or dispersed to other colonies.

Colonies may be gradually vacated in a piecemeal fashion. In March, forays to nectar and water and mating activity become increasingly common. Butterflies may fly several kilometers downslope from their colony to take nectar in large flower fields around agricultural plots. Mating may take place anywhere along the route to nectaring sites or while nectaring. During the afternoon or early evening, these butterflies often do not return to the original colony site. Instead they form nighttime roosting clusters near the flower fields. Butterflies returning north may exit from these transient roosts while others come down from the main colony to take nectar, mate, and take their place.

Our indirect evidence of an exodus of large butterflies from the colonies in early to mid-March is particularly interesting when sexual selection by reproductively active monarchs at the overwintering sites is considered (see Van Hook, this volume). Contrary to expectations of a species with a long migratory life-history phase, females do not mate in and near the colony with large males in good condition; instead, they mate with smaller males in average to poor condition (Van Hook, this volume). While it is the rule in Lepidoptera, no data are available on last male sperm precedence in monarchs (Drummond, 1984). If it does occur in this species, the female behavior of avoiding egg laying immediately after mating would ensure that later matings were the ones that counted. If larger monarchs are more likely to survive the rigors of migratory flight, last male sperm precedence and the early departure of large males followed by late matings would together have a selective advantage, relative to the simpler strategy of panmictic mating occurring totally within the colony.

Other possible reasons for the decline in forewing length at the end of the season include differential predation by mice and birds on larger, and presumably fatter, butterflies in preference to smaller ones. Adverse weather conditions might also cause differential mortality of larger butterflies relative to smaller ones. However, intact (nonpreyed upon) butterflies dead without apparent cause collected from colony floors are smaller and have lower dry weights than monarchs collected from roosting clusters at the same time (Calvert, unpubl. data). These preliminary findings suggest that small rather than large butterflies are more likely to die from causes other than predation, such as those related to climate. Further research is needed to establish whether or not birds and mice select larger-than-average butterflies as prey.

The decrease in wingloading from November to early February, despite a decline in average monarch wing length, demonstrates a real metabolic impact on the overwintering butterflies in early and mid-season. In February the decline in wingloading stops and begins to rise and is probably a direct reflection of increased opportunities for feeding. At this time the abundant spring composites in the genera *Senecio* and *Eupatorium* begin to flower, and tens of thousands of monarchs fly from the

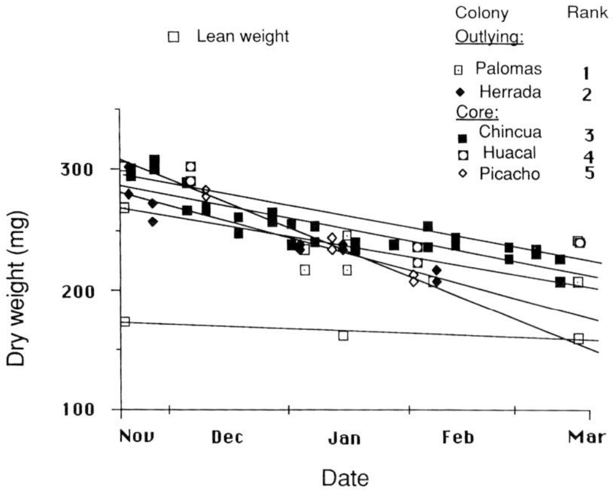

Figure 4. Decline in dry weight at five overwintering colonies from 20 November to 18 March 1985–1986. The colonies are ranked according to their rates of dry weight change with highest ranks for the lowest rates (see Table 3). Thus, the core colonies, Chincua, Huacal, and Picacho, show greater rates of weight decline than do the outlying colonies, Palomas and Herrada.

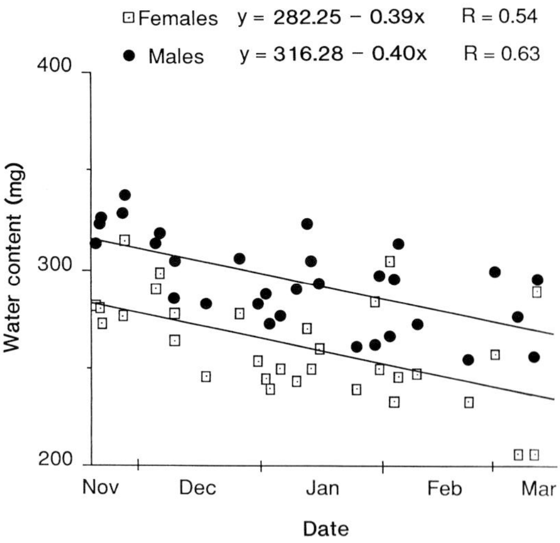

Figure 5. Decline in average water content in female and male monarch butterflies at all colonies in Mexico's Transvolcanic Belt during the 1985–1986 season. Males contain significantly more water than females.

colonies to flowers located principally in disturbed areas in the canyons and around fields located near the forest edge (Calvert and Brower, 1986).

Eastern North American monarchs are larger than those in western North America and Australia (Tuskes and Brower, 1978; James, 1984), and the maximum distance that butterflies from these populations travel to their overwintering areas are approximately 5,000, 2,000, and 400 km, respectively. Wing length, dry weight, and wing loading appear positively proportional to the distance of migration. This relationship may reflect a greater degree of provisioning for the longer journey.

TRENDS IN BUTTERFLY WEIGHT

Since monarch dry weight is highly correlated with the quantity of stored lipids (Brower, 1985), the depletion of lipid reserves during the winter should be directly related to dry weight decline. Differences in the rate of dry weight decline among the colonies suggests that metabolic demands, from whatever source, differ among the colonies. Three potential sources for variation among colonies in these metabolic demands are (1) differences in resting metabolism, as might be the case if the sites differed in slope exposure or forest density so that the amount of radiation entering the forest was affected, (2) differences in the availability of nectar sources and the costs of foraging for nectar, and (3) differences in the costs of maintaining a suitable water balance.

The dry season in Mexico's Transvolcanic Belt usually begins in October and persists until April or May (Fig. 1). Drying during the period when butterflies are overwintering is severe, especially in February and March. Over a 22-yr period at eight meteorological stations within or close to the overwintering areas, rainfall between November and March averaged only 14.4 mm, <7% of their summer maximum (Fig. 1; Anonymous, 1976).

The distance from colonies to open sources of water varies greatly from colony to colony, from year to year, and within seasons (Calvert and Brower, 1986). During the last 10 yr Chincua and Picacho have always formed within 200 m of an open source of water, while Palomas and Huacal are usually located >1 km from water. Late in the season colonies are occasionally located directly above streams, but in some areas (San Andrés and Altamirano; Fig. 2) open streams or reservoirs are >2 km away.

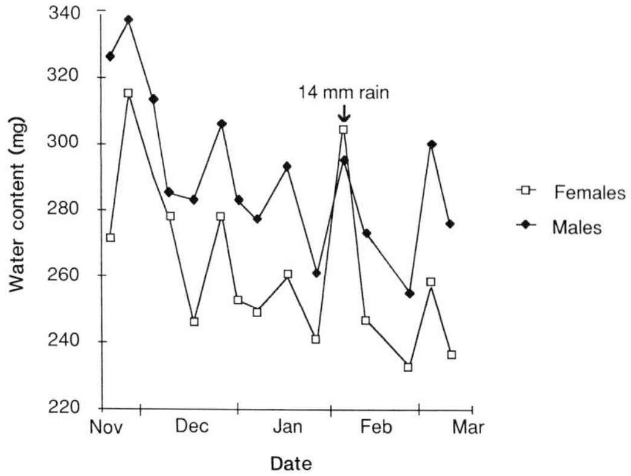

Figure 6. Decline in water content of female and male monarch butterflies at the Chincua colony from 20 November to 18 March 1985–1986. Light rains fell on several days in November and 14.5 mm of rain fell between 4 and 6 February.

Table 4. Decline of water content in combined male and female overwintering monarch butterflies at five colony locations in Mexico's Transvolcanic Belt. Butterfly colony samples with the same letter are significantly different at 5% with Tukey's multiple-range test.

Colony		N	Mean water content (mg)	SD	Range	Slope	Y	r
Core colonies								
Chincua	A	30	277.5	27.1	337–233	−34.2	332.0	0.55
Picacho	A	6	274.8	22.3	305–243	−26.0	317.8	0.29
Huacal	A	6	291.3	22.7	319–250	−29.1	341.8	0.49
Outlying colonies								
Herrada	A	10	285.4	31.8	329–244	−19.9	311.4	0.40
Palomas	A	10	267.7	31.0	324–233	−17.5	293.0	0.45

If the maintenance of water balance is costly in terms of lipid expenditure, colonies far from water should have greater rates of dry weight decline. Rates of dry weight loss, however, do not correlate well with distance to open water. Picacho, one of the closest colonies to water, had the greatest rate of dry weight decline while Chincua's, also close to water, was intermediate. In contrast, Huacal and Palomas, which were the greatest distance from water, had the least rate of weight decline (Fig. 4). This inverse relationship between lipid expenditure and distance to water suggests that maintenance of water balance is more complicated than previously believed (see also Masters et al., 1988).

At Chincua, the significant decline of percent moisture (water content ÷ lean weight) in males and lack of a decline in females suggest that females maintain water balance while males do not. Males have a higher water content than females (Fig. 5). The extra water in males possibly allows them to lose more water than females do without disrupting physiological functions.

Perhaps water balance can be cheaply maintained: i.e. the metabolic cost of acquiring water could be low for several reasons. First, relative humidity increases sharply at night within the forests where the monarchs roost (Calvert and Lawton, unpubl. data). Within their roosts butterflies may passively take up water directly from the air or actively ingest it from moisture that has condensed on their exoskeletons or on nearby plant surfaces. The increase in wet weight of butterflies after the rainy period from 4 to 6 February (Fig. 6) suggests that such absorption or ingestion may occur. James (1984) also found that moisture content of Australian monarchs varied with ambient humidity and the phenomenon of water absorption from a high humidity atmosphere is apparently common in plants (Breazeale et al., 1950). The proximity of dew in areas where nighttime radiational cooling is most intense such as in clearings and on tree canopies may save monarchs metabolically costly flight to open water throughout most of the season (Calvert and Brower, 1981). However, dew is likely to be less available as a water source as the dry season intensifies in February and March.

Flight itself may not be as energetically costly as often believed. The tens of thousands of butterflies that circle the colonies on warm sunny days expertly use winds rising against the mountains to soar. Soaring flight is used on forays for water and nectar as well. Masters et al. (1988) report that monarchs "flying" in the vicinity of a Chincua colony spent on average only 1.9 min out of each 10-min period in powered flight and the rest gliding.

The sharpest differences in weight decline appear to be between core and outlying colonies. Environmental differences between these two colony groups, besides distance to water, include the availability of nectar and openness of the habitats to incident radiation. The low rates of dry weight decline and the high average basal area forest (Table 1) of outlying colonies suggest that shielding of incident radiation by high basal area forests may play a role in determining rates of weight decline. However, there were insufficient data to show a correlation between these factors. The availability of nectar and moisture and factors affecting the entrance of solar radiation into the forest such as forest structure and slope aspect are currently being investigated.

ACKNOWLEDGMENTS

We are indebted to a great number of people who provided help, inspiration, and guidance during the course of our research in Mexico. The collaboration of a team of biologists from the Secretariat of Urban Development and Ecology, Morelia, Michoacán, was of special importance. Participants included Elizabet Verduzco, María de la Paz Espitia, Agustín Espinosa, María Macaria Mejía, and María Concepción Huerta. María de la Luz Ocampo and Juan Cristóbal Palatto of the University of San Nicolas, Morelia, also helped collect the data. We are greatly indebted to members of Monarca, A.C., the Mexican conservation organization, who provided logistical support while we were in the field. Especially helpful were Rodolfo Ogarrio, Javier de la Maza, Marlena Castro, and Fernando Ortíz Monasterio. Carlos Gottfried and family provided friendship and material support during out stay in Mexico; we owe them special thanks. Elizabeth O'Berry, Tonya Van Hook, Lincoln Brower, Myron Zalucki, David James, and especially Steve Malcolm provided many useful comments, which served to improve the manuscript. Jim Trostle and Lynn Morgan generously provided the use of their word processor in Costa Rica.

LITERATURE CITED

Anonymous. 1976. *Normales climatológicas.* Dirección General de Geografía y Meterologia. México, D.F: Servicio Meterológico Nacional.

Breazeale, E.L., W.T. McGeorge, and J.F. Breazeale. 1950. Moisture absorption by plants from an atmosphere of high humidity. *Plant Physiology* 25: 413–419.

Brower, L.P. 1985. New perspectives on the migration biology of the monarch butterfly, *Danaus plexippus* L. In *Migration: Mechanisms and adaptive significance,* ed. M.A. Rankin, 748–785. Austin: University of Texas, *Contributions in Marine Science,* Supplement, 27.

Brower, L.P., and W.H. Calvert. 1985. Foraging dynamics of bird predators on overwintering monarchs in Mexico. *Evolution* 39: 852–868.

Calvert, W. H., and L.P. Brower. 1981. The importance of forest cover for the survival of overwintering monarch butterflies (*Danaus plexippus,* Danaidae). *Journal of the Lepidopterists' Society* 35:216–225.

Calvert, W.H., and L.P. Brower. 1986. The location of monarch butterfly (*Danaus plexippus* L.) overwintering colonies in Mexico in relation to topography and climate. *Journal of the Lepidopterists' Society* 40:164–187.

Calvert, W.H., L.E. Hedrick, and L.P. Brower. 1979. Mortality of the monarch butterfly (*Danaus plexippus* L.): Avian predation at five overwintering sites in Mexico. *Science* 204:847–851.

Calvert, W.H., W. Zuchowski, and L.P. Brower. 1983. The effect of rain, snow and freezing temperatures on overwintering monarch butterfly colonies in Mexico. *Biotropica* 15:42–47.

Chaplin, S.B., and P.H. Wells. 1982. Energy reserves and metabolic expenditures of monarch butterflies overwintering in southern California. *Ecological Entomology* 7:249–256.

Drummond, B.A. 1984. Multiple mating and sperm competition in the Lepidoptera. In *Sperm competition and the evolution of animal mating systems,* ed. R.L. Smith, 291–370. London: Academic Press.

Glendinning, J.I., A. Alonso M., and L.P. Brower. 1988. Behavioral and ecological interactions of foraging mice (*Peromyscus melanotis*) with overwintering monarch butterflies (*Danaus plexippus*) in Mexico. *Oecologia* 75:222–227.

James, D.G. 1984. Phenology of weight, moisture and energy reserves of Australian monarch butterflies, *Danaus plexippus. Ecological Entomology* 9:421–428.

Masters, A.R., S.B. Malcolm, and L.P. Brower. 1988. Monarch butterfly (*Danaus plexippus*) thermoregulatory behavior and adaptations for overwintering in Mexico. *Ecology* 69:458–467.

Tuskes, P.M., and L.P. Brower. 1978. Overwintering ecology of the monarch butterfly, *Danaus plexippus* L., in California. *Ecological Entomology* 3:141–153.

Urquhart, F.A., and H.R. Urquhart. 1978. Autumnal migration routes of the eastern population of the monarch butterfly (*Danaus p. plexippus* L.; Danaidae: Lepidoptera) in North America to the overwintering site in the neovolcanic plateau of Mexico. *Canadian Journal of Zoology* 56:1754–1764.

Zar, J.H. 1984. *Biostatistical analysis.* Englewood, NJ: Prentice-Hall.

THE INFLUENCE OF TEMPERATURE ON CRAWLING, SHIVERING, AND FLYING IN OVERWINTERING MONARCH BUTTERFLIES IN MEXICO

Alfonso Alonso-M., John I. Glendinning, and Lincoln P. Brower

ABSTRACT. Tens of thousands of monarch butterflies land on the ground each day in overwintering colonies in Mexico. If they remain there through the night, they may be killed by freezing temperatures or mice. Because of the low temperatures, monarchs often cannot fly back to tree-borne clusters. Through a controlled field experiment, we studied their behavioral mechanisms for moving off the ground when flight was impossible, including crawling and shivering. We determined that (1) crawling alone did not elevate thoracic temperature (T_{th}) above the ambient temperature (T_a), but crawling and shivering simultaneously did; (2) as a monarch's T_{th} increased, so did its degree of wing openness and shivering rate; (3) wing position and shivering rate were closely interrelated; (4) crawling rate increased significantly with increasing T_{th}; (5) the crawling threshold (±95% confidence interval) was 4.6 ± 1.4°C, the shivering threshold 8.0 ± 1.4°C, and flying threshold 13.0 ± 3.1°C; and (6) T_{th} is the primary determinant of the type of locomotory behavior monarchs exhibit. Crawling and shivering thus enable monarchs to warm-up and to get off the ground when they are too cold to fly.

RESUMEN. En las colonias de hibernación de la mariposa monarca en México, miles de mariposas caen diariamente al suelo. Debido a las bajas temperaturas características de la zona muchas no pueden regresar volando a las agrupaciones de los árboles. Si pasan la noche en el suelo, las mariposas tienen un mayor riesgo de mortandad por congelación ó por depredación por ratones. Con un experimento de campo se estudiaron los mecanismos conductuales de la monarca para alejarse del suelo [i.e. las conductas de caminar y de elevar su temperatura torácica (T_{th}) al sacudir las alas]. Se encontró que: (1) aquellas que solo caminaron no elevaron su T_{th} por arriba de la temperatura ambiente (T_a), mientras que las que caminaron y sacudieron las alas al mismo tiempo, elevaron su T_{th} muy por arriba de la T_a; (2) conforme la T_{th} de las mariposas aumenta, así también el ángulo con el que abren las alas y la fuerza con la que las sacuden; (3) la posición de las alas y la frecuencia de aleteo están altamente correlacionadas; (4) la velocidad de ascenso está altamente relacionada con la T_{th}; (5) la temperatura mínima en la que el 50% de las mariposas caminan (±intervalo de confianza al 95%) es de 4.6 ± 1.4°C, para sacudir las alas es de 8.0 ± 1.4°C, y para volar es de 13.0 ± 3.1°C; y (6) la T_{th} es el factor primordial que determina el tipo de conducta locomotora que las mariposas presentan. Por tanto, los conductas de caminar y aletear permiten a las mariposas incrementar su T_{th} y alejarse del suelo cuando no pueden volar.

INTRODUCTION

During the 5-mo overwintering period of monarch butterflies, *Danaus plexippus* (L.), in the high-altitude fir forests of central Mexico, an average of 60,000 individuals per hectare per day land on the ground, either naturally or because birds dislodge them from tree-borne clusters (Brower and Calvert, 1985). Substantially greater numbers fall to the ground during storms. Because grounded monarchs are subject to extensive nocturnal mortality by freezing (Calvert et al., 1983; Calvert and Cohen, 1983) and mouse predation (Brower et al., 1985; Glendinning et al., 1988; Glendinning, this volume), it is important for them to reach above-ground supports in either trees or understory vegetation before nightfall. Owing to the low ambient temperatures (T_a's) characteristic of the overwintering period, monarchs that fall into shaded areas often cannot warm sufficiently to fly back into the clusters (Masters et al., 1988). However, they can crawl up onto understory vegetation and tree trunks (Fig. 1A and 1B).

Crawling monarchs often shiver (Kammer, 1970) their wings back and forth, which is a thermogenic behavior well documented in moths (Bartholomew and Heinrich, 1973; Heinrich and Mommsen, 1985).

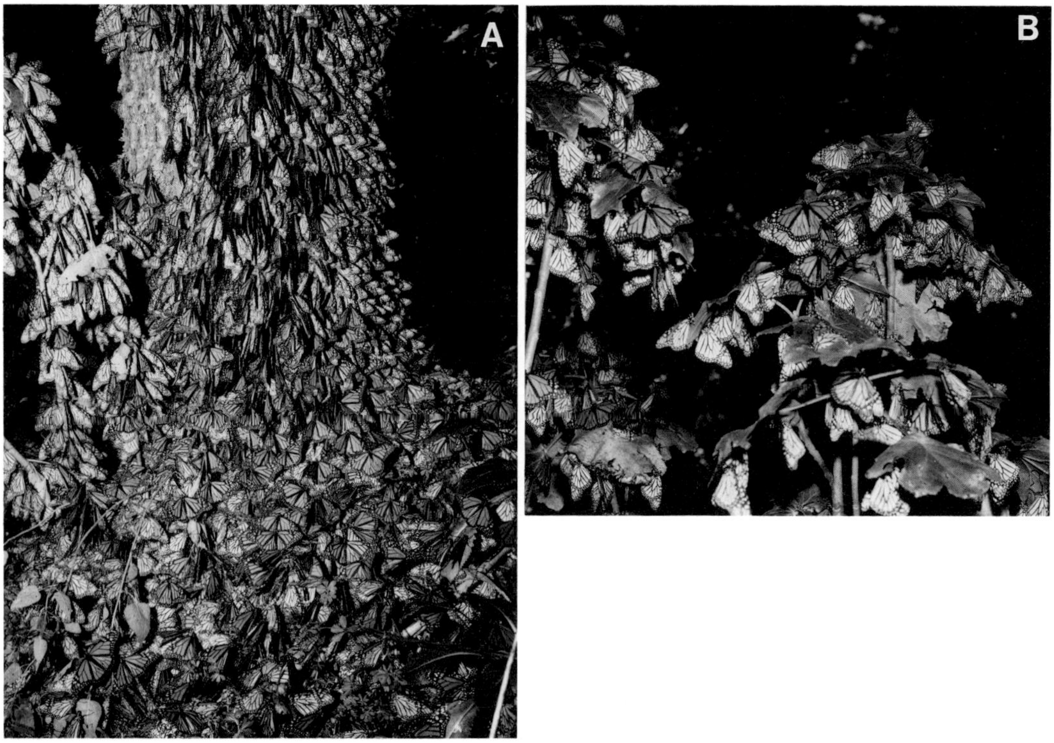

Figure 1. (**A**) Monarchs recently grounded by a severe winter storm crawling up a tree trunk and (**B**) roosting in understory vegetation. Photographs by L.P. Brower.

We hypothesized that shivering facilitated crawling when ambient temperatures were near freezing and flight was impossible. Shivering in insects has not been related to critical activities other than flight, except for the observation that certain orthopterans shiver to warm sufficiently to sing (Heath and Heath, 1982).

In this paper we (1) describe crawling and shivering behaviors of grounded monarchs, (2) explore the interrelationships of crawling, shivering, and flying, (3) determine the thermal thresholds of the three behaviors, and (4) discuss them as an integrated series of responses to low T_a's.

METHODS

THE STUDY AREA

We conducted the study from 28 January to 4 March 1986 on the slope of Arroyo La Plancha of the Sierra Chincua mountain massif, in northeastern Michoacán, Mexico, at an altitude of 3,200 m (latitude 19°41′N, longitude 100°18′W; Calvert and Brower, 1986). The habitat is a montane, boreal-like coniferous forest dominated by the oyamel fir *Abies religiosa* H.B.K. (see Soto-Núñez and Vázquez García, this volume; Snook, this volume).

EXPERIMENTAL DESIGN

We cut a 160 × 2.5 cm stem of the common *Senecio angulifolius* (Asteraceae) and marked centimeter increments with an indelible pen along its length. We then vertically implanted the stem in the path of crawling monarchs. Approximately 65% of the monarchs ascended the stem. We measured the thoracic temperature (T_{th}) of each monarch within 5 sec after it reached the top or stopped crawling by inserting a thermal probe ventrally into the thorax at the junction of the thorax and abdomen, as in Masters et al. (1988). To prevent our body heat from affecting the measurement, we held the forewings with forceps at least 4 cm from the thorax. We used the same probe to measure T_a's within 2 cm of each monarch just before we measured T_{th}. All measurements were made in late afternoon to early evening (1600–1900 hours), either beneath the *A. religiosa* canopy or at the edge of a clearing in a naturally shaded area.

Monarchs shiver at different rates and hold their wings in different positions while crawling. To quantify these behaviors, we recorded none, slow, and fast shivering, and, following Rawlins (1980), we recorded closed, half-open, or fully open wing positions. We used a stopwatch to record the time each monarch spent ascending the stake.

STATISTICAL ANALYSES

Statistical tests followed Zar (1984). For the regression analyses, we used either the untransformed data or ln-transformed data, depending on which form of data yielded the highest r^2 value. To explore for sexual differences, we regressed $\ln(T_{th})$ of crawling males and females on

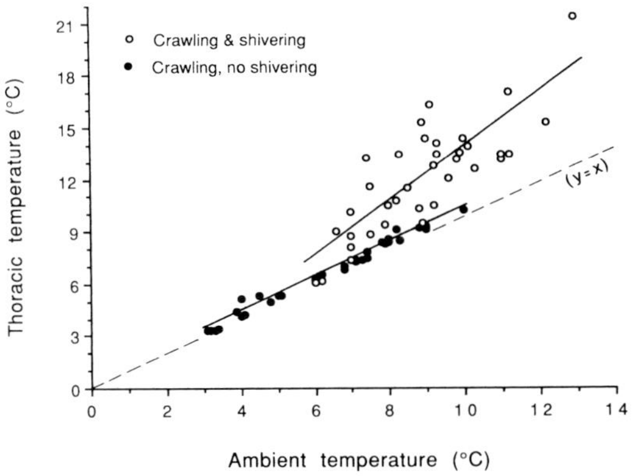

Figure 2. Relationship of T_{th} to T_a in crawling monarchs that were either crawling and shivering (open circles) or crawling but not shivering (closed circles). Thus, monarchs elevate their T_{th} above T_a by shivering but remain at T_a if they do not shiver. The dashed line $y = x$ is isothermal.

$\ln(T_a)$ separately and compared the slopes with a *t* test and the *y*-intercepts with ANCOVA. To determine whether crawling and/or shivering elevated the monarch's T_{th} above T_a, we regressed T_{th} on T_a separately for monarchs that were either crawling or crawling and shivering, and then compared the slopes of each regression line to the line of isothermality ($y = x$) with a one-sample (one-tailed) *t* test.

We explored the relationships between wing position and T_{th}, and between shivering rate and T_{th}, with Kruskal-Wallis one-way analyses of variance. We then compared the T_{th} of monarchs exhibiting different wing positions or shivering rates with the Tukey-type nonparametric multiple comparison. We correlated (Spearman rank test) wing position and shivering rate for each butterfly to determine the interrelatedness of these two behaviors.

We regressed crawling rate on T_{th} to explore the interrelationships among crawling rate, shivering rate, and wing position. We distinguished three categories of crawling butterflies: (1) wings closed and no shivering, (2) wings half open and slow shivering, and (3) wings fully open and fast shivering. Those 11 butterflies that did not fall into these categories (e.g. wings closed but shivering slowly) were classified as exceptions.

Finally, we estimated T_{th} thresholds for crawling and shivering by calculating the percentage of monarchs exhibiting either behavior during 1°C temperature intervals from 3 to 24°C. For example, of the 10 monarchs with T_{th} between 7 and 7.9°C, four were shivering, thus yielding a value of 40% shivering in this temperature interval. These percentages and binomial estimates of the standard errors were plotted as a function of T_{th}. We then estimated the T_{th} at which 50% of the monarchs shivered or crawled (henceforth, shivering and crawling thresholds). We provide 95% confidence limits for these estimations.

We modified this technique slightly to estimate the flight threshold. We knew each monarch's crawling rate just as it took flight (i.e., sustained itself in the air with strong wingbeats; Masters et al., 1988). Thus, by using reverse interpolation (Zar, 1984) with the regression of crawling rate on T_{th}, we estimated the T_{th} at its moment of flight. These T_{th} values were then used to determine flight threshold as described for crawling and shivering.

RESULTS

The regressions of T_{th} against T_a for naturally grounded monarchs of both sexes were logarithmic, and each was highly significant after ln-transformation (for 39 males, $\ln(T_{th}) = 1.33(\ln T_a) - 1.02$, $r^2 = 0.85$, $p < 0.001$; and for 39 females, $\ln(T_{th}) = 1.66(\ln T_a) - 2.93$, $r^2 = 0.86$, $p < 0.0001$). Since neither the slopes nor *y*-intercepts differed significantly from one another (for slope, $t = -0.42$, df = 74, $p > 0.50$; and for *y*-intercept, $F = 2.33$, df = 1,75, $p = 0.127$), we pooled data for both sexes in all subsequent analyses.

When we divided the data into two groups, based on locomotory behavior, the logarithmic relationship between T_{th} and T_a disappeared (Fig. 2). The slope of the regression line for monarchs that crawled and shivered ($y = 1.55x - 1.73$; $r^2 = 0.68$) was significantly higher than the isothermal line ($y = x$; $t = 4.75$; df = 36; $p < 0.001$), whereas that for monarchs that crawled without shivering ($y = 1.02x + 0.17$; $r^2 = 0.99$) was not ($t = 1.10$; df =

Table 1. Relationship of wing position and shivering rate to T_{th} (mean ± SE) in 78 male and female monarchs. Subscripts a, b, and c indicate significant differences among the wing positions or shivering rates ($p < 0.001$; Tukey-type nonparametric multiple comparison).

Behavior	n	T_{th} (°C)	Kruskal-Wallis statistic	df	*p* value
Wing position					
Closed	47	$6.6 (\pm 0.3)_a$			
Half open	15	$10.9 (\pm 0.6)_b$	47.36	2	<0.0001
Fully open	16	$14.3 (\pm 0.6)_c$			
Shivering rate					
None	41	$6.2 (\pm 0.3)_a$			
Slow	14	$10.4 (\pm 0.8)_b$	47.83	2	<0.0001
Fast	23	$13.0 (\pm 0.6)_c$			

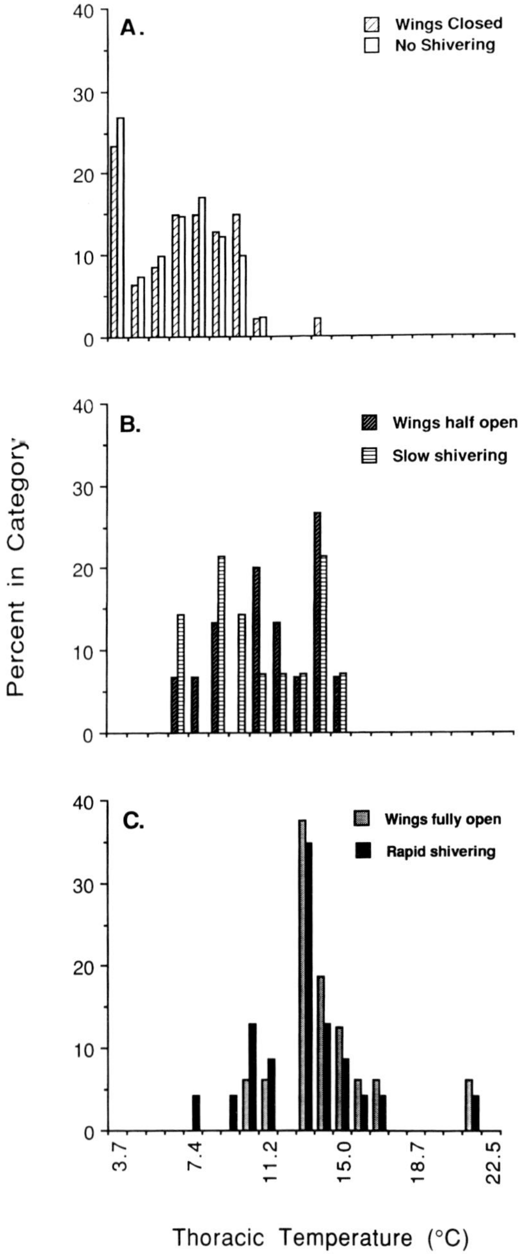

Figure 3. Histograms showing the interrelatedness of wing position with shivering rate in monarchs over a range of T_{th}'s. We present alongside one another the frequency distributions of those monarchs with (**A**) wings closed (n = 47) versus those not shivering (n = 41); (**B**) wings half open (n = 15) versus those shivering slowly (n = 14); and (C) wings fully open (n = 16) versus those shivering rapidly (n = 23).

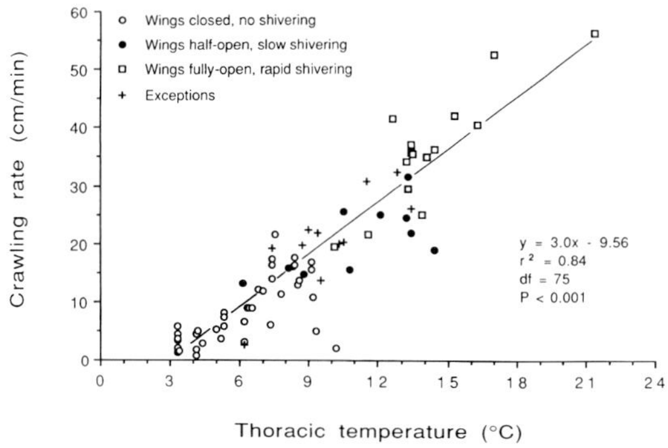

Figure 4. Regression of crawling rate in cm/min (=ascending velocity) to T_{th} in 78 monarchs. We distinguished the same categories of wing closure and shivering rate as in Figure 3. "Exceptions" are explained in text.

40; $p > 0.10$). Thus crawling monarchs can elevate their T_{th} above T_a by shivering. This capacity is accentuated at higher temperatures.

Thoracic temperature also increased significantly with both the degree of wing openness and shivering rate (Table 1). Moreover, there were strong correlations between these two behaviors over a range of T_{th}'s from 3.7 to 22.5°C ($r_s = 0.86$; df = 76; $p < 0.0001$). The interrelatedness of wing position and shivering rate is shown in Figure 3A–3C by the overlapping frequency distributions. Thus, butterflies with wings closed do not shiver (3A), those with wings half open shiver slowly (3B), and those with wings fully open shiver (3C).

The relationship between crawling rate and T_{th} was linear over the range of wing positions and shivering rates measured (Fig. 4). There was an increase in crawling rate from nonshivering monarchs with their wings closed to shivering monarchs with their wings fully open.

We show the relationship of T_{th} to the percentages of monarchs that either crawled, crawled and shivered, or crawled, shivered, and eventually took flight (Fig. 5). As these relationships were all linear, we fitted regression lines to them: crawling % = $19.7T_{th} - 41.6$, n = 41, $r^2 = 0.95$; shivering % = $13.9T_{th} - 57.6$, n = 94, $r^2 = 0.94$; and flying % = $4.9T_{th} - 14.1$, n = 70, $r^2 = 0.94$.

There was little overlap of the T_{th} ranges for each behavior. Once 100% of the monarchs were warm enough to crawl (top of the first line; Fig. 5), 42% were warm enough to shiver (bottom of second line); then, once 100% of the monarchs were warm enough to shiver (top of second line), 41% were warm enough to fly (bottom of third line).

DISCUSSION

Our results demonstrate that the crawling and shivering behaviors extend the thermal range over which monarchs can move when it is impossible for them to fly. Continuous hygrothermograph recordings taken in the late afternoon (1600–1900 hours) during February of 1979 (see Methods in Brower and Calvert, 1985) and of 1986 in a colony on Sierra Chincua are summarized in Figure 6. They reveal that the minimum T_a rarely dropped below the crawling threshold, whereas it was frequently below the shivering threshold. Late afternoon is when

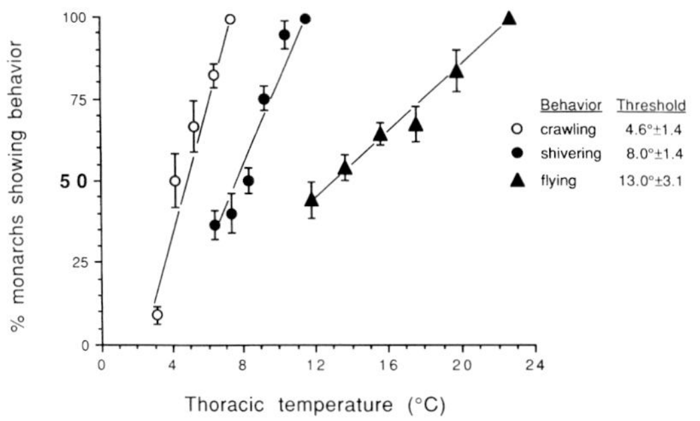

Figure 5. Relationships of T_{th} to the percentage of monarchs (means ± SE) that either crawled, crawled and shivered, or took flight. The three regressions (solid lines) were used to estimate the thresholds (±95% confidence interval) for crawling, shivering, and flying (i.e., the T_{th}'s at which 50% of the butterflies showed each behavior).

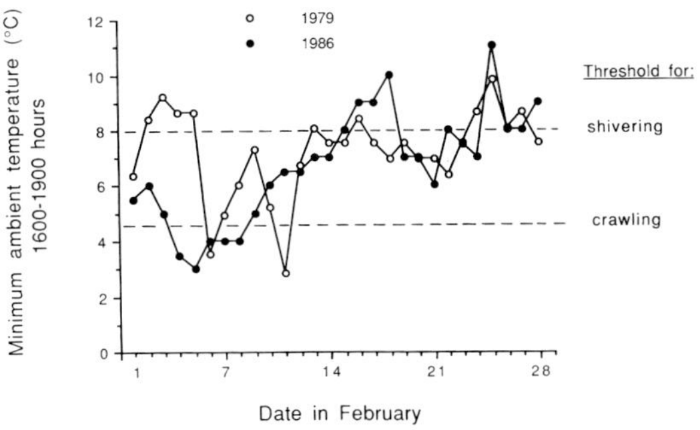

Figure 6. Minimum hygrothermograph recordings between 1600 and 1900 hours during February 1979 and 1986 in monarch colonies on Sierra Chincua. The crawling and shivering thresholds are indicated. Note that T_a rarely drops below the crawling threshold.

monarchs are most likely to become thermally stranded on the ground because T_a's fall rapidly and bird predators feeding at this time dislodge numerous monarchs from clusters. Because of the large mortality risk posed by freezing ground temperatures and mouse predation, it is important for monarchs to get off the ground before nightfall. Further work is needed to determine the actual proportion of butterflies grounded during the day that succeed in climbing up vertical supports.

Monarchs appeared to crawl quite rapidly at the cool overwintering sites. For example, at 4.6°C they crawled at approximately 4.8 cm/min, or 2.9 m/hr, and at 8.0°C they crawled about three times as fast. There are several instances in which rapid crawling would reduce a monarch's risk of becoming stranded on the ground. First, when a monarch falls to the ground in late afternoon, rapid movement may allow it to crawl to nearby vegetation before dark. Second, when storms dislodge large numbers of butterflies, there is often insufficient space on understory vegetation and tree trunks for all individuals, and the ones that crawl the quickest may be more successful at securing an overnight roost (Fig. 1A).

Monarchs that crawled and shivered simultaneously elevated their T_{th} significantly above T_a, which in turn resulted in a faster crawling rate. This thermogenic effect of shivering becomes more pronounced with increasing T_a. Therefore, shivering not only facilitates crawling, but also enables monarchs to reach the flight threshold. These benefits of shivering most likely outweigh the large metabolic costs, which are about 25 times those of resting monarchs at 16°C (Kammer, 1970), although the costs of shivering (or crawling) over a range of temperatures are not known.

As monarchs crawled up the experimental stake, they adopted different shivering rates and wing positions, depending on their T_{th}. In general, as T_{th} increased, so did the rates of shivering and crawling, as well as the degree of wing openness. We suggest that at higher T_{th}'s (i.e. >10.9°C), a positive, warming feedback loop develops among shivering, T_{th}, and wing position such that crawling rate is enhanced. Wing position appears to influence T_{th} indirectly by making shivering more efficient. The thoracic ganglia may monitor T_{th} (Hanegan and Heath, 1970) and orchestrate these complex, temperature-dependent behaviors.

The regression line for the percentage of monarchs flying as a function of T_{th} in Figure 5 was shallow compared to that of crawling and shivering monarchs. This suggests that many monarchs did not take flight upon reaching the flight threshold. Given the dangers associated with remaining on the ground, one might expect monarchs to crawl up vegetation reflexively when it is too cold to fly. However, once monarchs reach the flight threshold, they can maintain a high T_{th} by utilizing exogenous (basking) and endogenous (shivering) heat sources, together with limiting heat loss via the insulating hairlike scales located dorsally on their thorax (Church, 1960). Flight would then be possible in the event of bird attack, when clouds move over the colony and the temperature plummets, or when they overheat.

The flight threshold as reported here (13.0°C ± 3.1) and independently by Masters et al. (1988; 12.7–16.0°C) is substantially lower than those reported for papilionids and satyrids (28–33°C; Rawlins, 1980; Shelly and Ludwig, 1985) and for several moths (33–39°C; Bartholomew and Heinrich, 1973; Heinrich and Casey, 1973). Although Heinrich and Mommsen (1985) reported some moths have flight thresholds near 0°C, these moths remain isothermic with T_a. Thus, for an insect of tropical origin (Ackery and Vane-Wright, 1984), the monarch is well suited for movement at the cold, high-altitude overwintering sites.

ACKNOWLEDGMENTS

We thank Carlos Cordero, Luis Eguiarte, Steve Malcolm, Irene Sánchez, Ward Watt, and Myron Zalucki for comments on the manuscript; Carlos Martínez del Rio for statistical advice; Jorge Soberón and Monarca, A.C., for logistical support; and James B. Anderson, Alfredo Arellano G., and Tonya Van Hook for field assistance. Financial support was provided by U.S. National Science Foundation grant BSR-8500416 to L.P.B. and by CONACyT grant 56486 to A.A.M. We are grateful for a scientific collecting permit in Mexico from the Dirección General de Flora y Fauna Silvestre of SEDUE.

LITERATURE CITED

Ackery, P.R., and R.I. Vane-Wright. 1984. *Milkweed butterflies: Their cladistics and biology.* London: British Museum (Natural History); and Ithaca, NY: Cornell University Press, 425 pp.

Bartholomew, G.A., and B. Heinrich. 1973. A field study of flight temperatures in moths in relation to body weight and wing loading. *Journal of Experimental Biology* 58:123–135.

Brower, L.P., and W.H. Calvert. 1985. Foraging dynamics of bird predators on overwintering monarch butterflies in Mexico. *Evolution* 39:852–868.

Brower, L.P., B.E. Horner, M.A. Marty, C.M. Moffitt, and B. Villa R. 1985. Mice (*Peromyscus maniculatus, P. spicilegus,* and *Microtus mexicanus*) as predators of overwintering monarch butterflies (*Danaus plexippus*) in Mexico. *Biotropica* 17:89–99.

Calvert, W.H., and L.P. Brower. 1986. The location of monarch butterfly (*Danaus plexippus* L.) overwintering colonies in Mexico in relation to topography and climate. *Journal of the Lepidopterists' Society* 40:164–187.

Calvert, W.H., and J.A. Cohen. 1983. The adaptive significance of crawling up onto foliage for the survival of grounded overwintering monarch butterflies (*Danaus plexippus*) in Mexico. *Ecological Entomology* 8:471–474.

Calvert, W.H., W. Zuchowski, and L.P. Brower. 1983. The effect of rain, snow and freezing temperatures on overwintering monarch butterflies in Mexico. *Biotropica* 15:42–47.

Church, H.K. 1960. Heat loss and the body temperature of flying insects. II. Heat conduction within the body and its loss by radiation and conduction. *Journal of Comparative Physiology* 37:187–212.

Glendinning, J., A. Alonso-M., and L.P. Brower. 1988. Behavioral and ecological interactions of foraging mice (*Peromyscus melanotis*) and overwintering monarch butterflies (*Danaus plexippus*) in Mexico. *Oecologia* 75:222–227.

Hanegan, J.L., and J.E. Heath. 1970. Temperature dependence of the neural control of the moth flight system. *Journal of Experimental Biology* 53:629–639.

Heath, J.E., and M.S. Heath. 1982. Energetics of locomotion in endothermic insects. *Annual Review of Physiology* 44:133–143.

Heinrich, B., and T.M. Casey. 1973. Metabolic rate and endothermy in sphinx moths. *Journal of Comparative Physiology* 82:195–206.

Heinrich, B., and T.P. Mommsen. 1985. Flight of winter moths near 0°C. *Science* 228:177–179.

Kammer, A.E. 1970. Thoracic temperature, shivering, and flight in the monarch butterfly, *Danaus plexippus* (L.). *Zeitschrift für Vergleichende Physiologie* 68:334–344.

Masters, A.R., S.B. Malcolm, and L.P. Brower. 1988. Thermoregulatory behavior and adaptations for overwintering in the monarch butterfly (*Danaus plexippus*) in Mexico. *Ecology* 69:458–467.

Rawlins, J.E. 1980. Thermoregulation by the black swallowtail butterfly *Papilio polyxenes* (Lepidoptera: Papilionidae). *Ecology* 61:345–357.

Shelly, T.E., and D. Ludwig. 1985. Thermoregulatory behavior of the butterfly *Calisto nubila* (Satyridae) in a Puerto Rican forest. *Oikos* 44:229–233.

Zar, J.H. 1984. *Biostatistical analysis.* Englewood Cliffs, NJ: Prentice-Hall, 718 pp.

INTERSPECIFIC COMPARISONS OF THE FORAGING DYNAMICS OF BLACK-BACKED ORIOLES AND BLACK-HEADED GROSBEAKS ON OVERWINTERING MONARCH BUTTERFLIES IN MEXICO

Alfredo Arellano G., John I. Glendinning, James B. Anderson, and Lincoln P. Brower

ABSTRACT. We compared the foraging dynamics of black-backed orioles and black-headed grosbeaks on overwintering colonies of monarch butterflies during two successive seasons in Mexico. By suspending nets within a colony, we intercepted bird-damaged butterflies as they fell to the ground. Based on damage patterns, we found a strong correlation between the two birds' predation rates over each season, which indicates that they feed synchronously in mixed flocks. In 1985, grosbeaks killed more monarchs than did orioles, whereas in 1986, we found the opposite, suggesting a reversal in the proportion of the two bird species in the mixed-species flocks. Grosbeaks killed more males than females, while orioles killed both sexes equally. Daily variation in predation rate was large: 11.4-fold for orioles and 15.8-fold for grosbeaks. This variation, 14 and 25%, respectively, could be attributed to temperature: More butterflies were eaten on cold days. A 4-day feeding cycle in the 1986 oriole data was incorporated into the regression equation and significantly increased the r^2 from 14 to 26%. We conclude that mortality caused by both birds as well as the period of the orioles' feeding cycle are strongly influenced by temperature.

INTRODUCTION

Of the 37 species of omnivorous and insectivorous birds present at the Sierra Chincua overwintering site in Mexico (Table 1), only black-headed grosbeaks, *Pheucticus melanocephalus* Swainson, and black-backed orioles, *Icterus galbula abeillei* Lesson, are known to feed extensively on monarch butterflies, *Danaus plexippus* (L.) (Fink et al., 1983; Brower and Calvert 1985). These two species also appear to be the primary avian predators of monarchs at other overwintering sites in Mexico (Calvert et al., 1979). In one 2.25-ha colony on Sierra Chincua (density = 10 million monarchs/ha), they were estimated to have killed 2.03 million butterflies, or about 9% of the colony, over the entire overwintering period (Brower and Calvert, 1985). In smaller colonies, total bird-caused mortality may reach levels as high as 44% (Brower and Calvert, 1985). Other studies of bird predation on relatively discrete patches of insects have found far less total mortality. For instance, during outbreaks (>5.1 million larvae/ha) of the larch sawfly, *Pristiphora erichsonii* (Htg.), and the spruce budworm, *Choristoneura fumiferana* (Clem.), in boreal Canadian forests, <1% of the larvae were taken by birds (Morris et al., 1958; Buckner and Turnock, 1965).

Whereas grosbeaks can ingest large amounts of the toxic cardenolides stored in monarchs without vomiting, orioles vomit after eating only small amounts (Fink and Brower, 1981). Orioles reduce intake of cardenolides after they have killed monarchs by the way in which they eat them (Fink and Brower, 1981; Brower et al., 1988). Thus, while both species appear to be able to circumvent the monarch's cardenolide defense, Brower and Calvert (1985) detected a 7.9-day feeding cycle in their combined daily predation rates. To explain this cycle, they formulated the cardenolide toxicity hypothesis: Grosbeaks and orioles fed intensively on monarchs until they began to sicken and then decreased their consumption (presumably shifting to alternate foods) for several days until they recovered, after which they again fed intensively on monarchs. Although this cycle is supported generally by pharmacokinetic theory (Bowman and Rand, 1980, chapter 40), and was independent of temperature, Brower and Calvert's model assumed that both species were feeding synchronously and they also did not address the basis for the specific 7.9-day period of the cycle.

Here we explore interspecific similarities and dif-

Table 1. Thirty-seven known and potential bird predators of monarch butterflies at the Sierra Chincua, Michoacán, Mexico, overwintering site. Species names are from American Ornithologists' Union (1983) and are grouped according to feeding guild (Nocedal, 1984; Martin et al., 1951; L. Eguiarte, unpubl. obs.) and whether or not they have been observed feeding upon monarchs: extensively (+++), occasionally (++), rarely (+), or never (−). The estimates of predation by the different bird species are derived from (a) Fink et al. (1983) and (b) qualitative observations by the authors.

Species name	Common name	Known to eat monarchs in field	Source
	OMNIVORES		
Pheucticus melanocephalus	Black-headed grosbeak	+++	a
Icterus galbula abeillei	Black-backed oriole	+++	a
Icterus parisorum	Scott's oriole	++	a
Cyanocitta stelleri	Steller's jay	++	a
Sialia mexicana	Western bluebird	+	a
Catharus occidentalis	Russet nightingale thrush	+	a
Catharus guttatus	Hermit thrush	−	a
Toxostoma curvirostre	Curve-billed thrasher	−	b
Myadestes obscurus	Brown-backed solitaire	−	b
Atlapetes virenticeps	Green-striped brush-finch	−	a
Turdus assimilis	White-throated robin	−	a
Pipilo erythrophthalmus	Rufous-sided towhee	−	a
	INSECTIVORES		
	Sallying		
Myiarchus tuberculifer	Dusky-capped flycatcher	−	b
Myiarchus tyrannulus	Brown-crested flycatcher	+	a
Empidonax affinis	Pine flycatcher	+	a
Empidonax fulvifrons	Buff-breasted flycatcher	−	b
Ptilogonys cinereus	Gray silky-flycatcher	−	a
Mitrephanes phaeocercus	Tufted flycatcher	−	a
Myioborus miniatus	Slate-throated redstart	−	b
Nyctidromus albicollis	Common pauraque	−	b
Contopus pertinax	Greater pewee	−	b
	Foliage-gleaning		
Regulus satrapa	Golden-crowned kinglet	−	b
Regulus calendula	Ruby-crowned kinglet	+	a
Parus sclateri	Mexican chickadee	+	a
Vermivora celata	Orange-crowned warbler	−	a
Dendroica townsendi	Townsend's warbler	+	a
Dendroica coronata	Yellow-rumped ("Audubon's") warbler	−	b
Dendroica nigrescens	Black-throated gray warbler	−	b
Dendroica occidentalis	Hermit warbler	−	b
Ergaticus ruber	Red warbler	−	a
Troglodytes aedon	House ("brown-throated") wren	−	b
Campylorhynchus megalopterus	Gray-barred wren	+	a
	Bark-gleaning		
Piculus auricularis	Gray-crowned woodpecker	−	b
Picoides stricklandi	Strickland's ("brown-barred") woodpecker	−	b
Picoides villosus	Hairy woodpecker	−	a
Lepidocolaptes leucogaster	White-striped woodcreeper	−	b
Certhia americana	Brown creeper	−	b

ferences in the rates of grosbeak and oriole predation on monarchs over two consecutive seasons. This was possible because grosbeaks usually detach and eat the whole abdomen, whereas orioles commonly strip out the abdominal contents and/or thoracic muscles and reject the cuticle (Fink, 1980; Fink and Brower, 1981). Fink and Brower never observed grosbeaks stripping abdomens or thoraces or orioles detaching whole abdomens. We compared five aspects of each species' foraging patterns: (1) Do both birds feed synchronously in mixed flocks? (2) Do both species exhibit cyclical feeding patterns? (3) Do daily fluctuations in temperature influence predation rates in both species equally? (4) Does one species consistently kill more monarchs? (5) Do both species feed selectively on one sex of monarch?

METHODS AND MATERIALS

THE STUDY AREA

This study was conducted during two overwintering seasons, 1985 and 1986, on the slope of the Arroyo La Plancha of the Sierra Chincua mountain massif in northeastern Michoacán, Mexico, at approximately 20°N latitude. This massif, referred to as Site Alpha in previous papers, is one of the principal overwintering areas of the eastern North American population of the monarch butterfly (Calvert and Brower, 1986). The vegetation is montane, boreal, coniferous forest dominated by the oyamel fir (*Abies religiosa* H.B.K.). The altitude of the colonies ranged between 3,000 and 3,200 m.

SAMPLING PROCEDURES: 1985 OVERWINTERING SEASON

A grid that sampled 0.96 ha (8 × 12 stations with 10-m spacing between intersection points) was set beneath a 0.76-ha monarch colony in January 1985. Because the grid's sampling area was larger than the colony, it accommodated the minor shifts in the colony's position. Brower and Calvert (1985) showed that forest gaps influence where birds feed. To control for this, we first designated (using subjective estimations) all quadrats with <20% canopy coverage as open areas and those with >80% canopy coverage as closed areas. Then, we used a stratified random sampling scheme (Snedecor and Cochran, 1980) to select 21 open and 21 closed areas.

To monitor bird predation, we used nets (1 m^2, diameter 1.13 m, depth 45 cm) nailed to three 1.5-m tall stakes. One net was placed in the center of each of the open and closed areas, for a total of 42 nets. These nets intercepted falling monarchs that had been attacked by birds in trees; thus, monarchs attacked on the forest floor or low-lying vegetation by small mammals (Brower et al., 1985; Glendinning et al., 1988; Glendinning, this volume) could not be confused with those attacked by birds.

We cleared the nets daily between 9:00 and 11:00 A.M. from 28 January through 24 February (28 days) and tallied only that subset of monarch remains clearly damaged to various degrees by grosbeaks (i.e., those with their entire abdomen removed) and by orioles (i.e., those with a longitudinal slit running the length of their abdominal cuticle). We did not record the degree to which each monarch was eaten nor the number of disembodied abdomens or thoraces found. Consequently, in comparison to Brower and Calvert's (1985) study, our results underestimate total mortality and, thus, can only address the relative rates of predation by the two bird species. The sex of these monarchs was determined by wing and abdominal morphology.

SAMPLING PROCEDURES: 1986 OVERWINTERING SEASON

The 1985 data collection regime was used in this season with the following differences. Four parallel transects were set in a 0.6-ha colony in January 1986. A total of 26 nets were strung along these transects on two parallel nylon lines run between trees approximately 1.5 m above the ground at 3-m intervals. Transects were positioned in areas with high densities of monarchs; half of the nets were in open areas and the others in closed areas. We tallied all items in each net daily between 7:00 and 8:30 A.M., from 16 January through 28 February (n = 44 days), at which point the monarch colony moved downslope out of our sampling area. Daily temperature fluctuations were recorded with Taylor min–max thermometers. One thermometer was in each of the four transects inside one of the nets located in the closed area. Because the closed areas were covered by a dense canopy, exposure of the thermometers to insolation and radiational cooling to the open sky was limited. Temperature data were collected while tallying monarch damage and provided the min–max temperatures over the previous 24 hr. For the data analysis, we used the mean minimum temperatures of the four thermometers.

ESTIMATION OF PREDATION RATES

We tallied the number of male and female butterflies killed by each bird species each day. That number was then divided by the number of nets in the sampling regime (42 in 1985 and 26 in 1986) to give an estimate of the daily predation rate per m^2 for each species. We henceforth refer to this estimate as the predation rate.

PREDATION RATES BY BOTH BIRDS AND THEIR TEMPORAL COINCIDENCE

We compared predation rates of orioles and grosbeaks using a paired (two-tail) *t* test. Each season was analyzed independently because the different sampling regimes may have influenced the estimated predation rates. Synchronous feeding within each year's data set was tested by correlating, with the Pearson product moment correlation, the two birds' feeding rates.

PREDATION RATES IN RELATION TO BUTTERFLY SEX

Based on Fink (1980) and Brower and Calvert (1985), we expected one or both species to eat a smaller percentage of females. To test whether or not this trend was consistent during both years, we used a two-way ANOVA. Because the percentage of females collected in the net samples is unlikely to have been influenced by our different sampling procedures, we felt justified using this test.

Determining whether either bird species attacked male or female butterflies disproportionately is complicated by the observations that (1) trunk and bough clusters had slightly, but not significantly, different sex ratios during

Table 2. Mean percentage of female monarchs (±SE) in samples collected randomly from bough and trunk clusters in monarch colonies between 20 January and 28 February 1985 and 1986. Data are from Tonya Van Hook (unpubl. data). The means of the trunk and bough samples are compared with the Mann-Whitney *U* test (with normal approximation).

Location of clusters	Mean % females	Total no. of butterflies in all samples	No. of samples	*Z* value	*p*
Trunk	51.5 (±2.8)	4,310	11	1.10	0.2726
Bough	55.5 (±1.8)	2,657	14		

1985 and 1986 (Table 2), and (2) grosbeaks may feed more commonly at trunk clusters, whereas orioles may feed equally at both trunk and bough clusters (Bond, 1979; Fink, 1980). In an attempt to accommodate these caveats, we compared (with nonpaired *t* tests) the mean proportion of females in bird-predated samples with that of females in random samples taken from bough and trunk clusters (see Table 2).

CYCLIC PREDATION AND THE EFFECT OF DAILY TEMPERATURE

Based on Brower and Calvert (1985), we expected (1) an inverse correlation between temperature and daily predation rate, and (2) each bird species to exhibit a similar cyclical pattern of predation. To test these expectations, we subjected the predation rates in 1986 (independently for each species) to spectral analysis (Priestly, 1981; SPECTRA: SAS, 1982).

TESTS FOR NORMALITY AND DATA TRANSFORMATIONS

All data were normally distributed (Kolmogorov *D* statistic, $p > 0.05$) except for the sex ratio data, which were transformed (arcsine $\sqrt{x}$); nontransformed data are presented in the tables.

RESULTS

SPECIES-SPECIFIC PREDATION RATES

We identified a total of 1,347 grosbeak-predated and 475 oriole-predated monarchs in 1985, and 1,131 grosbeak-predated and 2,291 oriole-predated monarchs in 1986. Grosbeaks attacked about three times as many monarchs as did the orioles in 1985, whereas this relationship was reversed in 1986, with orioles killing twice as many monarchs as the grosbeaks (Table 3). During all sampling days in 1985, grosbeaks killed more butterflies, and during 93% of the sampling days in 1986, orioles killed more butterflies, irrespective of daily fluctuations in predation rate as great as 15.8-fold for grosbeaks and 11.4-fold for orioles (Fig. 1A and B).

Highly significant correlations were found between the predation rates of both species in both years (in 1985, $r^2 = 0.56$, n = 28 days, $p < 0.001$; in 1986, $r^2 = 0.55$, n = 44 days, $p < 0.001$). These data support the prior hypothesis of a high degree of feeding synchrony by the two birds (Fig. 2).

SEX RATIO OF THE PREDATION SAMPLES

During both years, grosbeaks killed a significantly lower ($p < 0.001$) percentage of females, as compared with orioles. There were no significant differences between the 2 yr or the interaction between bird species and year (Table 4). We determined that temperature did not affect the sex ratio of predated butterflies by regressing the percentage of females in the daily predation samples on daily temperature (for orioles, $r^2 = 0.005$, $p = 0.11$; for grosbeaks, $r^2 = 0.002$, $p = 0.41$) during 1986 (the only year for which we had temperature records).

Grosbeak-damaged samples had lower proportions of females ($p < 0.05$ in all comparisons) than both random bough and cluster samples (Tables 2 and 4). In contrast, neither oriole-damaged sample differed ($p > 0.20$ in all comparisons) from the bough and cluster samples.

Table 3. Mean (±SE) predation rates (mean no. of butterflies killed/net/day) of orioles (ori.) and grosbeaks (gros.). Forty-two nets were set each day between 28 January and 24 February 1985, and 26 nets between 16 January and 28 February 1986. Each species' respective mean predation rates are compared within each season.

Year	No. of sampling days	Bird species	Mean predation rate	Paired *t* value	*p*	Predation rate of ori./gros.
1985	28	Oriole	0.40 (±0.06)	−6.63	<0.0001	0.35
		Grosbeak	1.15 (±0.14)			
1986	44	Oriole	2.00 (±0.14)	10.18	<0.0001	2.02
		Grosbeak	0.99 (±0.09)			

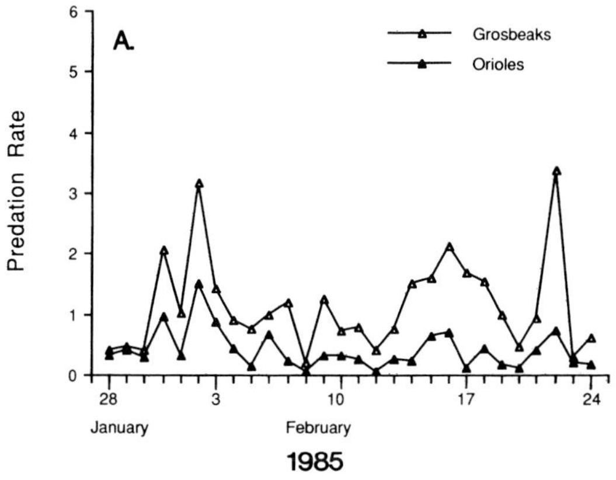

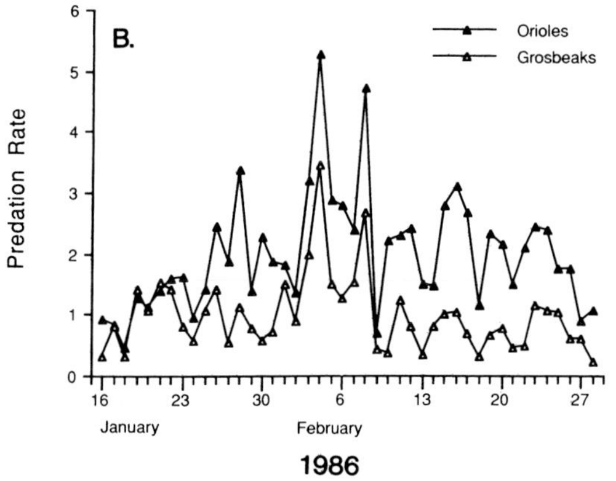

Figure 1. Predation rates (mean no. of monarch butterflies killed/net/day) of grosbeaks and orioles in the Sierra Chincua colony during (**A**) 28 January–24 February 1985 and (**B**) 16 January–28 February 1986.

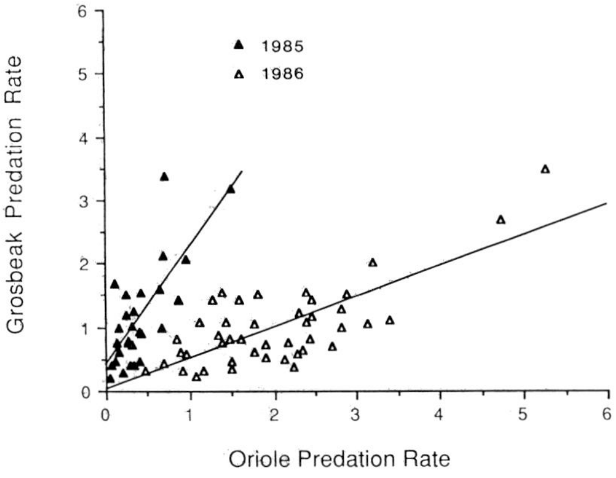

Figure 2. Correlations of the predation rates (mean no. of butterflies killed/net/day) between grosbeaks and orioles during 28 January–24 February 1985 and 16 January–28 February 1986.

CYCLICAL PREDATION AND THE EFFECT OF DAILY TEMPERATURE

A regression of the 1986 data indicated a significant effect of temperature on each bird's respective predation rate (for grosbeaks, $r^2 = 0.25$, $p = 0.0006$; for orioles, $r^2 = 0.14$, $p = 0.012$; Fig. 3). The regression slopes were negative, demonstrating that both bird species killed more butterflies on colder days.

We next subjected the same data to a two-step time-series analysis to investigate whether or not a feeding cycle existed independently of the temperature effect. This analysis removes the temperature effect from each species' regression model and compares the pattern of residual variation over time against a random pattern. For grosbeaks, no significant periodicity could be detected (Fisher's Kappa = 3.13, $p = 0.653$), suggesting that the cyclical trend for grosbeaks apparent in Figure 2 is due to temperature fluctuations and as-yet unknown factors. In contrast, for orioles, a statistically significant period of approximately 4.0 days was determined (Fisher's Kappa = 5.98, $p = 0.026$). We therefore reanalyzed the oriole data to estimate simultaneously the effect of the feeding cycle and temperature on their predation rate. The following model was derived:

$$\text{predation rate} = 2.80 - 0.28x_t + 0.434\cos(1.571t) - 0.201\sin(1.571t) + e_t$$

where 2.80 is the y-intercept; x_t denotes the minimum temperature over the night prior to net collection for day t; the value 1.571 in the sinusoidal function [a cos(1.571t) − b sin(1.571t)] denotes the first harmonic of the cyclic trend (with the period $= 2\pi/1.5708 = 4.0$ days) for day t; and the parameter e_t represents white noise with a variance of 0.757 for day t. To illustrate the 4-day cycle, we overlaid the oriole predation rate adjusted to the temperature effect [i.e., predation rate − $(2.80 - 0.28x_t)$] on the sinusoidal function (Fig. 4). By adding the sinusoidal function to the temperature model, r^2 increased significantly from 0.14 to 0.26 ($p < 0.05$). Thus, the temperature effect and the cyclic trend are both important in explaining daily variation in oriole predation.

DISCUSSION

RELATIVE PREDATION RATES AND SYNCHRONOUS FEEDING BY GROSBEAKS AND ORIOLES

This study provides the first quantitative evidence that, within each season, grosbeaks and orioles feed synchronously, thereby corroborating the observations of Calvert et al. (1979), Fink (1980), Fink and Brower (1981), and Brower and Calvert (1985). Grosbeaks killed a higher proportion of monarchs in 1985. These data agree with those of Brower and Calvert (1985), who deduced that grosbeaks killed more butterflies than did orioles in 1979. However, orioles killed a higher proportion in 1986. This reversal was unexpected and may be due to

Table 4. Mean percentage of female monarchs (±SE) killed per day by orioles and grosbeaks. Percentage data were transformed (arcsine $\sqrt{x}$) prior to the ANOVA. Data were collected between 28 January and 24 February 1985, and 16 January and 28 February 1986.

Year	No. of sampling days	Bird species	Mean % females per day
1985	28	Oriole	57.9 (±3.0)
		Grosbeak	35.6 (±1.9)
1986	44	Oriole	54.7 (±1.9)
		Grosbeak	37.6 (±1.7)

Two-way ANOVA of the % of females per day in the nets

Source of variation	df	SS	MS	F	p
Bird species	1	4,927.0	4,927.0	77.7	<0.001
Year	1	12.6	12.6	0.2	0.66
Species × year	1	111.4	111.4	1.8	0.2
Error	140	8,876.2	63.4		

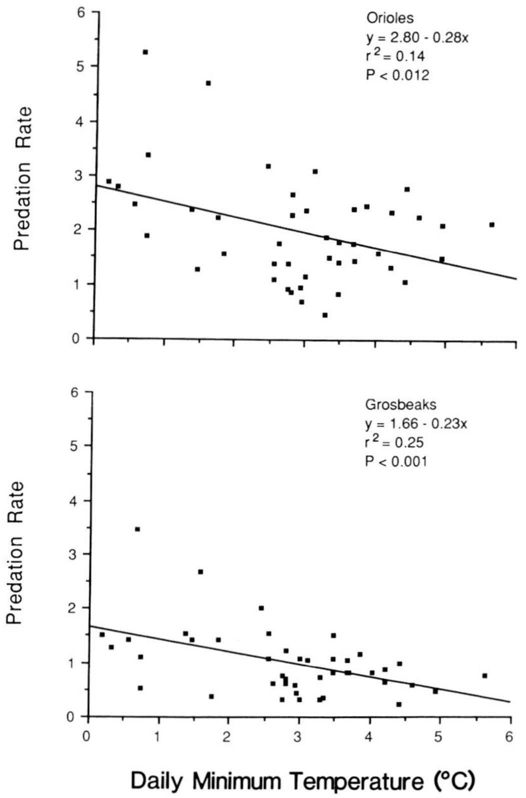

Figure 3. Regressions of predation rate (mean no. of butterflies killed/net/day) on daily minimum temperature for orioles and grosbeaks in the Sierra Chincua colony during 16 January–28 February 1986.

changes in the number of individuals of each species in the mixed-species flocks. Further study is needed.

Even though other bird species also prey upon monarchs occasionally (Table 1; especially Steller's jays and Scott's orioles), they do so irregularly and in groups of one to about five birds (Fink, 1980; the authors, unpubl. obs.). In contrast, orioles and grosbeaks visit colonies daily, both in the morning and afternoon, and they do so in flocks of 5 to at least 60 birds (Fink, 1980). Because grosbeaks and orioles attack many more monarchs than do the other birds, the predation rates we report herein are not likely to have been influenced by their feeding activities.

PREFERENCE OF THE BIRDS FOR FEMALE OR MALE BUTTERFLIES

We found that grosbeaks attacked a significantly lower percentage of female monarchs than did the orioles, which is consistent with Fink's (1980) findings. In addition, grosbeaks killed male monarchs disproportionately, whereas orioles killed both sexes in proportion to their abundance within monarch clusters. This pattern persisted over both seasons even though the relative numbers of monarchs killed by each species reversed between seasons. Because these results are derived from proportion data, they are not influenced by flock size and species composition and strongly indicate a pattern that is consistent among both colonies and years. Brower and Calvert (1985) concluded that both grosbeaks and orioles avoided females; however, they may have been incorrect because they did not separate their data according to species. If we had lumped our data, we would have drawn conclusions similar to Brower and Calvert's (mean % females in 1985 = 46.8, in 1986 = 46.2, and during January and February 1979 = 39.7, when Brower and Calvert conducted their study).

It is unclear why grosbeaks ate a lower proportion of female monarchs. Although female monarchs on average have 30% higher concentrations of cardenolides than do males (Brower and Calvert, 1985), grosbeaks appear to kill and eat monarchs randomly with respect to the concentration of these substances (Fink and Brower, 1981). Female monarchs also appear to have on average a 50% higher pyrrolizidine alkaloid (PA) concentration (Kelley et al., 1987). However, the unpalatability and toxic effects of monarch PAs to grosbeaks is unknown. Alternatively, there are potential sex-dependent nutritional benefits: Female monarchs have on average 26% more lipids, whereas males have on average 9% greater lean weight (Brower and Calvert, 1985). The relative importance of toxin avoidance versus nutritional benefits in determining feeding decisions of grosbeaks (and orioles) is unknown.

It should be stressed that our tallying procedure included only those butterflies that had been partially eaten by birds. Thus, those dead butterflies

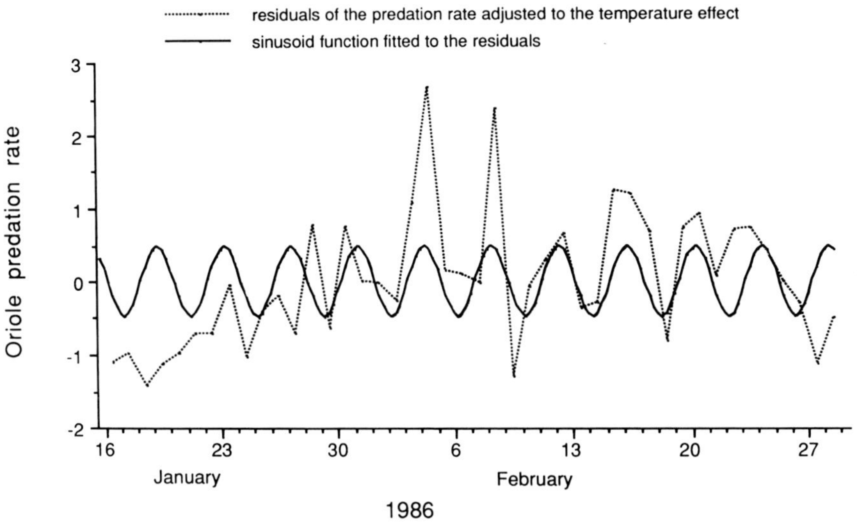

Figure 4. Overlay of the oriole predation rate adjusted to the temperature effect [i.e., predation rate − (2.80 − 0.28x_t)] on the sinusoidal function [0.434 cos(1.571t) − 0.201 sin(1.571t)], which has a 4-day period. Predation rate values above and below zero indicate days when the predation rate was greater and less, respectively, than that predicted by the temperature effect.

found in nets without damage patterns clearly attributable to one of the bird species would not have been recorded. Therefore, grosbeaks may have killed an equal number of male and female monarchs, but they ended up eating more males. Feeding experiments with caged birds would help resolve this issue.

THE INFLUENCE OF TEMPERATURE ON BIRD PREDATION

On colder days, both birds killed more butterflies. Further experimentation is needed to determine whether this was because on cold days (1) individual birds eat more butterflies (as has been found for other birds; Schildmacher, 1929; Thiollay, 1976), (2) more birds join the flock(s), and/or (3) butterflies are less active and, thus, easier to capture.

To explore possible effects of temperature on predation in different years, we compared the daily minimum temperatures from the period of 18 January to 19 February in 1986 (this study) with the same period in 1979 (Brower and Calvert, 1985, table 1). Paired *t* test comparisons between days revealed that 1986 was consistently and significantly colder than 1979 ($t = 4.01$, mean difference = 1.3°C, df = 39, $p = 0.0003$). Thus, Brower and Calvert's (1985) estimate of 9% mortality of the entire colony might have been substantially higher had the 1979 season been colder.

THE FEEDING CYCLE AND DAILY TEMPERATURE

Orioles, but not grosbeaks, fed cyclically on monarchs during 1986. Brower and Calvert (1985) also detected a feeding cycle during 1979. However, because they did not distinguish predation by both species, it is unclear whether the observed cyclicity was due to the orioles alone or to both species. The fact that a feeding cycle has been detected in both 1979 and 1986 suggests that it is a general phenomenon. Moreover, if only orioles feed cyclically, the cycle may also illustrate a pharmacological limitation in their dealing with long-term ingestion of cardenolides. Such speculation seems plausible given the divergent pharmacological sensitivities of these birds to short-term exposure to cardenolides (i.e., Fink and Brower, 1981). The effects of long-term cardenolide ingestion on these birds needs to be studied. Finally, if we assume that orioles ingest more cardenolides on the colder days, and also that the feeding cycle detected by Brower and Calvert (1985) was due to the orioles, we can offer a hypothesis to explain why the 7.9-day feeding cycle in 1979 was 3.9 days longer than that in 1986: The daily temperatures were significantly lower during 1986, and orioles may thus have eaten more butterflies each day. This would result in them building up toxic levels of cardenolides in their blood serum more quickly, thus causing a shorter feeding cycle.

ACKNOWLEDGMENTS

We thank William Calvert, Linda Fink, and Diana Schulmann for editorial comments; Mark Yang for the time series analyses; Tonya Van Hook for the sex ratio data in Table 2; Alfonso Alonso-M. for field assistance; Luis Eguiarte for help in grouping the birds in Table 1 according to feeding guild; María Ester M. and María Macaria M. for help in identifying the birds on Sierra Chincua; and Jorge Soberón and Monarca, A.C., for logistical support. Financial support was provided by U.S. National Science Foundation grant BSR-8500416 to L.P.B. We are grateful to the Dirección General de Flora y Fauna Silvestre of SEDUE for permission to conduct this research in the Sierra Chincua.

LITERATURE CITED

American Ornithologists' Union. 1983. *Check-list of North American birds*, 6th ed. Lawrence, KS: American Ornithologists' Union, xxxix × 877 pp.

Bond, C., Jr., 1979. Implications of avian predation on monarch butterflies in Mexico. Honors thesis. Amherst, MA: Department of Biology, Amherst College.

Bowman, W.C., and M.J. Rand. 1980. *Textbook of pharmacology*. Oxford: Blackwell.

Brower, L.P., and W.H. Calvert. 1985. Foraging dynamics of bird predators on overwintering monarch butterflies in Mexico. *Evolution* 39(4):852–868.

Brower, L.P., B.E. Horner, M.A. Marty, C.M. Moffit, and B. Villa R. 1985. Mice (*Peromyscus maniculatus, P. spicilegus,* and *Microtus mexicanus*) as predators of overwintering monarch butterflies (*Danaus plexippus*) in Mexico. *Biotropica* 17(2):89–99.

Brower, L.P., C.A. Nelson, J.N. Seiber, L.S. Fink, and C. Bond. 1988. Exaptation as an alternative to coevolution in the cardenolide-based chemical defense of monarch butterflies (*Danaus plexippus* L.) against avian predators. In *Chemical mediation of coevolution,* ed. K.C. Spencer, 447–476. San Diego: Academic Press.

Buckner, C.H., and W.J. Turnock. 1965. Avian predation of the larch sawfly, *Pristiphora erichsonii* (Htg.), (Hymenoptera: Tenthredinidae). *Ecology* 46(3):223–236.

Calvert, W.H., and L.P. Brower. 1986. The location of monarch butterfly (*Danaus plexippus* L.) overwintering colonies in Mexico in relation to topography and climate. *Journal of the Lepidopterists' Society* 40(3):167–187.

Calvert, W.H., L.E. Hedrick, and L.P. Brower. 1979. Mortality of the monarch butterfly (*Danaus plexippus* L.): Avian predation at five overwintering sites in Mexico. *Science* 204:847–851.

Fink, L.S. 1980. Bird predation on overwintering monarch butterflies. Honors thesis. Amherst, MA: Department of Biology, Amherst College.

Fink L.S., and L.P. Brower. 1981. Birds can overcome the cardenolide defence of monarch butterflies in Mexico. *Nature* 291:67–70.

Fink L.S., L.P. Brower, R.B. Waide, and P.R. Spitzer. 1983. Overwintering monarch butterflies as food for insectivorous birds in Mexico. *Biotropica* 15(2): 151–153.

Glendinning, J.I., A. Alonso, and L.P. Brower. 1988. Behavioral and ecological interactions of foraging mice (*Peromyscus melanotis*) with overwintering monarch butterflies (*Danaus plexippus*) in Mexico. *Oecologia* 77:222–227.

Kelley, R.B., J.N. Seiber, A.D. Jones, H.J. Segall, and L.P. Brower. 1987. Pyrrolizidine alkaloids in overwintering monarch butterflies (*Danaus plexippus*) from Mexico. *Experientia* 43:943–946.

Martin, A.C., H.S. Zim, and A.L. Nelson. 1951. American wildlife and plants: A guide to wildlife food habits. New York: Dover Publications.

Morris, R.F., W.F. Cheshire, C.A. Miller, and D.G. Mott. 1958. The numerical response of avian and mammalian predators during a gradation of the spruce budworm. *Ecology* 39(3):487–494.

Nocedal, J. 1984. Estructura y utilización del follaje de las comunidades de pájaros en bosques templados del Valle de México. *Acta Zoológica Mexicana* 6: 1–45.

Priestly, M.B. 1981. *Spectral analysis and time series, Vol. 1: Univariate series.* London: Academic Press.

SAS. 1982. *SAS user's guide.* Cary, NC: SAS Institute.

Schildmacher, H. 1929. Über den Wärmehaushalt kleiner Körnerfresser. *Ornithologische Monatsberichte* 37:102–106.

Snedecor, G.W., and W.G. Cochran. 1980. *Statistical methods,* 7th ed. Ames: Iowa State University Press.

Thiollay, J.M. 1976. Quantitative food requirements of some tropical birds. *Terre et la Vie* 30:229–245.

Zar, J.H. 1984. *Biostatistical analysis.* Englewood Cliffs, NJ: Prentice-Hall.

COMPARATIVE FEEDING RESPONSES OF THE MICE *PEROMYSCUS MELANOTIS, P. AZTECUS, REITHRODONTOMYS SUMICHRASTI*, AND *MICROTUS MEXICANUS* TO OVERWINTERING MONARCH BUTTERFLIES IN MEXICO

John I. Glendinning

ABSTRACT. Of four potential mouse predators (*Peromyscus melanotis, Peromyscus aztecus hylocetes, Reithrodontomys sumichrasti*, and *Microtus mexicanus salvus*), only *P. melanotis* feeds extensively on monarch butterflies at their overwintering sites in Mexico. Since overwintering monarchs contain cardiac glycosides (CGs) derived from their larval host plants, I tested the hypothesis that only *P. melanotis* tolerates the taste and/or toxic effects of monarch CGs. I offered active monarchs and mouse chow for two nights to all mouse species except *M. m. salvus*. *Microtus m. salvus* received inactive monarchs and the same artificial diet because it would not approach active monarchs. *Peromyscus melanotis* and *R. sumichrasti* fed heavily on abdomens, commonly discarding the cuticular material, which is known to contain high concentrations of CGs. In contrast, *M. m. salvus* and *P. a. hylocetes* did not discriminate between the cuticle and contents, and they ate far less monarch tissue. Individual *P. melanotis* were also offered a choice between equal numbers of male and female monarchs. They ate fewer female than male monarchs and showed a stronger aversion to their cuticular material, presumably because the females have higher concentrations of CGs than males. I then offered all four species a pure diet of monarchs for six consecutive nights. While both *P. melanotis* and *R. sumichrasti* ate similar mass-specific amounts of monarch tissue, *P. melanotis* gained and *R. sumichrasti* lost weight. Both *M. m. salvus* and *P. a. hylocetes* ate only small mass-specific amounts of monarch tissue and both lost weight, particularly the former species. I conclude that *M. m. salvus* avoids eating monarchs primarily because of an aversion to eating insects in general, whereas *P. a. hylocetes* and *R. sumichrasti* do so primarily because of an aversion to the monarch's CGs. In contrast, *P. melanotis* not only eats monarchs readily, but is also much less sensitive to CGs than are the other species.

INTRODUCTION

Aggregations of overwintering monarch butterflies, *Danaus plexippus* (L.), in Michoacán, Mexico, are vast and lipid- and protein-rich potential food sources for foraging predators (Brower and Calvert, 1985). However, few predators utilize these colonies. Of the 37 species of omnivorous and insectivorous birds present in the Sierra Chincua overwintering area (Arellano G. et al., this volume), only black-headed grosbeaks and black-backed orioles feed extensively on the monarchs (Calvert et al., 1979; Fink and Brower, 1981; Fink et al., 1983; Brower and Calvert, 1985; Arellano G. et al., this volume). Likewise, of the four most common species of mouse in the same overwintering area, *Peromyscus melanotis* J.A. Allen and Chapman, *Peromyscus aztecus hylocetes* Merriam, *Reithrodontomys sumichrasti* Saussure, and *Microtus mexicanus salvus* Saussure, only *P. melanotis* immigrates extensively to the monarch colonies (Brower et al., 1985, 1988a; Glendinning et al., 1988; Glendinning and Brower, 1990). There it feeds upon live, moribund, and recently dead monarchs on the forest floor and initiates high levels of reproduction. No other mouse species establish residency inside the colonies. Outside the monarch colonies, all four species show comparatively low levels of reproduction during the overwintering period.

I hypothesize that the four mouse species respond differently to this food resource because they have differing sensitivities to the cardiac glycosides (CGs) present in overwintering monarchs (Seiber et al., 1986; Malcolm and Brower, 1989; Malcolm et al., this volume). CGs are emetic, toxic, and bitter tasting to many vertebrates (Brower, 1984). Differential sensitivities to the emetic and toxic effects of

Table 1. Categories of monarch damage by mice.

1. No visible damage
2. Sampled: <25% of abdomen and/or thorax eaten
3. 25–50% (cuticle and contents) of thorax eaten
4. 51–100% (cuticle and contents) of thorax eaten
5. Abdomen eaten nonselectively (≥25% of cuticle and contents)
6. Abdomen eaten selectively (≥25% of contents; cuticle rejected)

monarch CGs appear to be major factors determining which bird species utilize monarchs at the Mexican overwintering sites (Fink and Brower, 1981; Fink et al., 1983). However, rodents do not vomit and are relatively insensitive to oral dosages of CGs (Barnes and Eltherington, 1973; Marty, 1983; Glendinning et al., 1990). Thus, it may be the bitter taste, rather than toxicity per se, of the CGs that determines whether a mouse species eats monarchs.

To compare the feeding responses of the four species to monarchs, I conducted three experiments with caged, individual mice. I measured how many monarchs each species ate and the degree to which they avoided cuticular material. Monarchs store higher concentrations of CGs in their cuticle compared to their body contents (Brower et al., 1988b). Second, I further explored CG avoidance in *P. melanotis,* which feeds naturally on monarchs, by offering them a choice between male and female monarchs. Male monarchs have on average 30% lower CG concentrations than females (Brower and Calvert, 1985). I compared how many abdomens of each sex were eaten and the extent to which each sex's cuticle was avoided. Third, I examined how well all four species could maintain weight on a diet consisting solely of monarchs.

METHODS AND MATERIALS

THE STUDY AREA

I conducted experiments from 20 January to 28 February 1986 on the slope of the Arroyo La Plancha of the Sierra Chincua mountain massif in northeastern Michoacán, Mexico, at approximately 20°N latitude. This site is one of the principal overwintering areas of the eastern population of the monarch butterfly (Calvert and Brower, 1986). At an altitude of approximately 3,200 m, the community is a montane, boreal, coniferous forest dominated by the oyamel fir (*Abies religiosa* H.B.K.) (Soto Núñez and Vázquez García, this volume).

TRAPPING AND MAINTENANCE OF MICE

I trapped mice in Sherman live traps (7.6 × 8.9 × 22.9 cm) at least 800 m from the monarch aggregation to reduce the possibility that they had prior experience eating monarchs. Because of a high variance in food consumption in reproductive mice, particularly females (Sadleir et al., 1973; Stebbins, 1977; Millar, 1979), only reproductively inactive adults (nonlactating or nonpregnant females and nonscrotal males) were used in the experiments. Each mouse was used only once.

The mice were housed and tested individually in wire mesh cages (ca. 30 cm high × 25 cm in diameter) set on a tarpaulin-covered table in the shade. Dacron batting was added for nesting material and pieces of cardboard were placed between the cages to isolate the mice visually. For six consecutive nights prior to the feeding experiments, each mouse was maintained on Purina laboratory chow no. 5001 (henceforth, mouse chow) and water. All of them fed and drank regularly and either maintained or gained weight.

MONARCH COLLECTION

Monarchs were collected with butterfly nets from accessible clusters on fir branches within a monarch aggregation. For experiments 1 and 3, monarchs were offered to mice without reference to their sex. Branch clusters are ca. 58% females during January and February (n = 7 samples, for a total of 1,095 butterflies; Tonya Van Hook, unpubl. data). For Experiment 2, equal numbers of each sex were offered to mice.

EXPERIMENT 1: PATTERNS OF FEEDING BY FOUR MOUSE SPECIES

In this experiment I compared the feeding responses of the four species to monarchs. Each mouse was weighed at the beginning of the experiment. Then four males and four females of each species were each offered 40 monarchs and mouse chow and water *ad libitum* for two consecutive nights. In this and the next experiment, all mice received live (or active) monarchs, except for *M. m. salvus,* which received inactive monarchs (i.e., ones whose thorax had been squeezed firmly) because they would not approach active ones. At 0900 hours all mouse chow was removed from each mouse's cage. At 1900 hours, the monarchs and fresh mouse chow were given to each mouse. At 0900 hours the following morning, each mouse was removed from its cage to tally patterns of feeding damage to dead monarchs and then replaced and deprived of food until 1900 hours that day, at which time the feeding trial was repeated.

I determined from preliminary feeding trials that mice consumed variable portions of thoraces but always ate the cuticle together with the contents. In contrast, when feeding upon abdomens, they characteristically ate either 100% of the cuticle and contents (i.e., fed nonselectively) or 100% of the contents only and discarded the cuticle (i.e., fed selectivity). Based on these observations, I chose *a priori* six categories of monarch damage (Table 1). I assumed that sampling a monarch represents active rejection of it based on taste (cf. Glendinning, 1990).

Determination of the amount of monarch tissue, cuticle, and CGs eaten by each mouse in experiments 1 and 3 involved several steps. First, I estimated the amount of tissue, cuticle, and CGs in the body parts of overwintering monarchs (see Table 2). For these estimations, I modified the data from Brower et al. (1988a) because they were derived from freshly eclosed, *Asclepias syriaca*-reared monarchs, which have higher CG contents and substan-

Table 2. Mean dry weights and CG contents of the abdominal cuticle and tissue, thoracic cuticle and tissue, and wings of freshly eclosed monarchs reared on *A. syriaca* (from Brower et al., 1988a, table 3). These data, along with the mean dry weights and CG contents (all body parts combined) of monarchs overwintering in Mexico (Malcolm and Brower, 1989) were used to estimate the dry weights and CG contents of the different body parts in overwintering monarchs. Data from *A. syriaca*-reared monarchs are appropriate for these estimations because the evidence suggests that 85–92% of the overwintering monarchs in Mexico have fed as larvae on this milkweed (Seiber et al., 1986; Malcolm et al., this volume). I assume that (1) the dry weights of the cuticle, wings, and thoracic tissue were the same in both types of monarchs, and that only the abdominal tissue weights differed, owing to increased amounts of fat associated with diapause; and (2) the distribution of CGs in the different body parts was the same in both types of monarchs, but that the CG contents were proportionally lower in each body part in overwintering monarchs.

	Freshly eclosed monarchs				Estimated values for overwintering monarchs				
	Dry weight[a]		CG content[a]		Dry weight		CG content		CG concentration (μg/0.1 g dry weight)
Body part	mg	%	μg	%	mg	%	μg	%	
Abdominal cuticle	11	6	91	26	11	5	24	26	218
Abdominal tissue	70	37	37	11	123	51	10	11	8
Thorax cuticle	22	11	41	12	22	9	11	12	50
Thorax tissue	41	22	37	11	41	17	10	11	24
Wings	45	24	145	41	45	19	39	41	87
Abdominal + thoracic tissue	111	59	74	21	165	68	20	21	12
Wings + cuticle	78	41	277	79	78	32	74	79	95
Sum of all parts	189	100	351	100	242[b]	100	94[b]	100	39

[a] Values are rounded to the nearest whole value.

[b] These are means derived from 563 overwintering monarch butterflies at Sierra Chincua (Malcolm and Brower, 1989).

tially lower amounts of fat in their abdomens than monarchs reared on the same food plant that have migrated to the Mexican overwintering grounds (Brower, 1985; Malcolm and Brower, 1989). Second, I assumed that whenever a mouse ate (1) an abdomen nonselectively, it ate 100% of the cuticle and contents; (2) an abdomen selectively, it ate 100% of the contents; (3) 25–50% of a thorax, it ate 38% of the cuticle and contents; and (4) 51–100% of a thorax, it ate 75% of the cuticle and contents. These assumptions agreed generally with my preliminary observations. Third, I calculated the total quantity of tissue, cuticle, and CGs eaten by a given mouse over a given time period with the following equations, which incorporate data from Table 2:

tissue eaten
= [(no. abdomens eaten selectively and nonselectively)(123 mg)]
+ [(no. thoraces eaten 25–50%)(0.38)(41 mg)]
+ [(no. thoraces eaten 51–100%)(0.75)(41 mg)];

cuticle eaten
= [(no. abdomens eaten nonselectively)(11 mg)]
+ [(no. thoraces eaten 25–50%)(0.38)(22 mg)]
+ [(no. thoraces eaten 51–100%)(0.75)(22 mg)];

and

CG eaten
= [(no. abdomens eaten selectively)(10 μg)]
+ [(no. abdomens eaten nonselectively)(10 + 24 μg)]
+ [(no. thoraces eaten 25–50%)(0.38)(10 + 11 μg)]
+ [(no. thoraces eaten 51–100%)(0.75)(10 + 11 μg)].

Fourth, I added together the predation records from both nights in Experiment 1, whereas for Experiment 3, I analyzed each night separately so as to monitor changes in consumption of monarch tissue, cuticle, and CGs over the six nights.

Because the four mouse species differed greatly in size (Table 3), I could not compare directly the amounts of monarch tissue, cuticle, or CG eaten. Therefore, I standardized consumption by computing the ratio of the weight of tissue, cuticle, or CG eaten to mouse weight$^{0.7}$. The power function, weight$^{0.7}$, is derived from a regression of consumption rate on weight for a wide range of mammals (Farlow, 1976; Peters, 1983). Mass-specific consumption values were calculated separately for each mouse, using their weight at the beginning of the experiment.

EXPERIMENT 2: PREDATION BY *P. MELANOTIS* IN RELATION TO BUTTERFLY SEX

The previous experiment examined whether either of the species discriminated between abdominal cuticle and contents. This experiment addressed a more complex question: When given a choice between monarchs of differing CG concentrations (in this case males vs. females), will *P. melanotis* eat more low CG ones and feed selectively on more of the high CG ones? Male monarchs have on average 30% lower CG concentrations than females (Brower and Calvert, 1985). I could not estimate separately for both sexes the amounts of tissue, cuticle, and CGs eaten because the data were not available; those in Table 2 are from pooled samples of male and female monarchs.

Seven female and 8 male *P. melanotis* were each offered 25 female and 25 male monarchs and mouse chow and water *ad libitum* per night for 2 consecutive nights.

Table 3. Initial weights (mean ± SE), weight changes and % mortality in *P. melanotis, R. sumichrasti, P. a. hylocetes,* and *M. m. salvus* given 55 monarchs per mouse per night for 6 consecutive nights, without alternative food.

Mouse species	N	Initial weight (g)	Weight change (g)	% weight change	% of mice that lost weight	% mortality
P. melanotis	6	19.6 (±0.4)	+0.83 (±0.2)	+4.2	0	0
R. sumichrasti	5	13.4 (±0.3)	−0.38 (±0.2)	−2.8	60	0
P. a. hylocetes	4	36.8 (±0.6)	−1.70 (±0.7)	−4.6	75	0
M. m. salvus	5	30.0 (±0.6)	−4.30 (±0.7)	−14.3	100	100[a]

[a] All died within 4 days.

Five hours before each feeding trial, I squeezed the thoraces of the female and male monarchs and then placed a 1-mm-diameter, red or gold spot of nail polish on each butterfly abdomen, thorax, head, and wings so as to enable tallying of the parts by sex after the mice dismembered them. The assignment of gold or red for males or females was determined randomly for each mouse. Because sexual differences in wing length are negligible and wet weights vary unpredictability (Brower and Calvert, 1985), I considered size and weight unlikely bases for killing an excess of one sex.

EXPERIMENT 3: EFFECTS OF LONG-TERM CONSUMPTION OF MONARCHS

In this experiment, I explored whether all four species could maintain or gain weight on a pure diet of monarchs. Each mouse (six *P. melanotis*, five *R. sumichrasti*, four *P. a. hylocetes*, and five *M. m. salvus*; sex ratios were approximately equal) was pre-exposed to 40 monarchs and mouse chow *ad libitum* per night for 2 nights. Then for the next 6 nights, each mouse received 55 monarchs per night and water *ad libitum*, without mouse chow. Preliminary feeding trials indicated that all species ate fewer than this number of monarchs per night. Each mouse's weight was taken at the beginning and end of the experiment, 8 days later. All mice were tested between 7 and 21 February.

STATISTICAL ANALYSES

All statistical tests followed Zar (1984). In Experiment 1, paired (two-tailed) *t* tests were run separately on each species to compare the number of abdomens versus thoraces eaten, and the number of abdomens eaten selectively versus nonselectively. One-way ANOVAs and Scheffé *F* tests were used to compare the species in terms of the mass-specific amounts of monarch tissue, cuticle, or CGs eaten, as well as the ratios of the amounts of CGs to tissues eaten. I was justified in using the mass-specific consumption values in the ANOVAs because the coefficient of variation of the scaling variable (i.e. mouse weight) was consistently less than that of the dependent variable (i.e. amount of tissue, cuticle, or CG consumed) (Anderson and Lydic, 1977; Packard and Boardman, 1988). The unpaired *t* test was used to compare the percentage of abdomens eaten selectively by *P. melanotis* and *R. sumichrasti.* Data were transformed (arcsin $\sqrt{x}$) for this and all subsequent statistical comparisons of percentage data.

Paired *t* test comparisons were used in Experiment 2 to compare the total male and female monarch abdomens eaten (selectively and nonselectively), as well as the percentage of those abdomens that were eaten selectively.

Two-factor ANOVAs (repeated on time) were used in Experiment 3 to determine the effects of species (*P. melanotis, P. a. hylocetes,* and *R. sumichrasti*) and time on tissue, cuticle, and CG consumption. The same test was used again to determine the effect of species and time on the percentage of abdomens eaten selectively by *P. melanotis* and *R. sumichrasti. Microtus m. salvus* was excluded from all statistical analyses in this experiment.

RESULTS

EXPERIMENT 1: PATTERNS OF FEEDING BY ALL FOUR MOUSE SPECIES

Peromyscus melanotis, R. sumichrasti, and *P. a. hylocetes* each ate >52% of the monarchs offered

Table 4. Comparison of the feeding patterns of all four mouse species upon monarch butterflies. Eight individuals per species were each offered 40 monarchs and mouse chow *ad libitum* per night for 2 consecutive nights. For each species, the mean number of monarchs eaten, sampled, or left undamaged (±SE) by each species are presented. Because mice ate both the abdomen and thorax of some monarchs, the row totals are all >80.

	Mean number of monarchs ± SE in each feeding category						
			Thoraces eaten		Abdomens eaten		
Species	No visible damage	Sampled	25–50%	51–100%	Non-selectively	Selectively	No. offered/ mouse
P. melanotis	14.3 ± 3	15.4 ± 4	5.1 ± 2	0.0 ± 0	18.8 ± 1	31.3 ± 2	80
R. sumichrasti	26.0 ± 3	10.9 ± 2	4.5 ± 1	0.9 ± 0.4	19.3 ± 2	23.3 ± 1	80
P. a. hylocetes	20.5 ± 1	17.8 ± 2	8.5 ± 1	4.8 ± 1	39.4 ± 4	2.8 ± 1	80
M. m. salvus	57.6 ± 5	11.9 ± 5	4.1 ± 2	3.8 ± 1	10.1 ± 1	0.1 ± 0.1	80

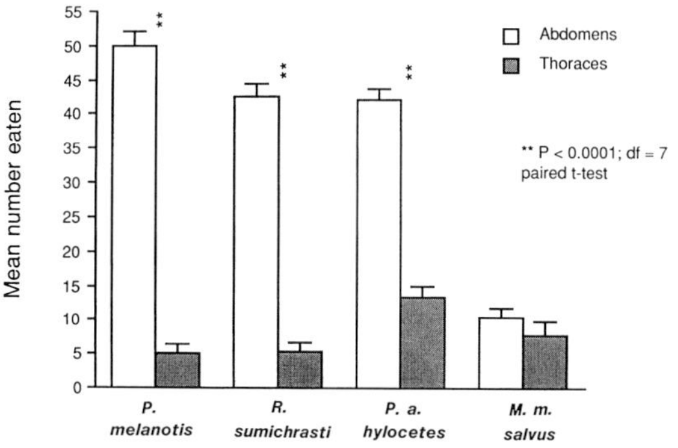

Figure 1. Mean number of monarch abdomens and thoraces (±SE) eaten by eight individuals of each of four mouse species. Each mouse received 40 monarchs and mouse chow *ad libitum* per night for 2 consecutive nights. The significant within-species comparisons between the number of abdomens and thoraces eaten are indicated by asterisks.

to them (Table 4) and significantly more abdomens than thoraces (Fig. 1). In contrast, *M. m. salvus* ate <13% of the monarchs offered to them and virtually identical numbers of abdomens and thoraces. Moreover, it approached only immobilized monarchs. None of the mice consumed the wings, head, or legs.

The standardized consumption values are shown in Table 5. *Peromyscus melanotis* and *R. sumichrasti* consumed significantly more monarch tissue than did *P. a. hylocetes* and *M. m. salvus*. However, *P. a. hylocetes* ate significantly more cuticle and both *R. sumichrasti* and *P. a. hylocetes* ate more CGs than did the other two species. To help interpret these patterns of consumption, I computed for each species the ratio of the amounts of CGs to tissues eaten. The ratios for *P. melanotis* and *R. sumichrasti* were significantly lower than those for the other two species (Table 5).

The large interspecific differences in these ratios can be explained by the way each mouse species consumed monarch abdomens. *Peromyscus a. hylocetes* and *M. m. salvus* rarely ate abdomens selectively, whereas *P. melanotis* and *R. sumichrasti* both ate significantly more abdomens selectively than nonselectively (Fig. 2). *Peromyscus melanotis* fed selectively on a significantly greater percentage of abdomens than did *R. sumichrasti* (means = 62.2 and 55.1, respectively; unpaired t value = 2.89; df = 14; $p < 0.013$). Thus, by feeding selectively, *P. melanotis* and *R. sumichrasti* reduced substantially the CG concentration of the monarch tissue they ingested.

By examining mouse-damaged monarchs and watching mice eat them, I discovered that all four species did not use the same technique for extracting abdominal contents. *Peromyscus melanotis* and *R. sumichrasti* commonly made longitudinal slits down the full length of the abdomen and then sucked and/or licked out the abdominal contents. In contrast, *P. a. hylocetes* and *M. m. salvus* usually bit off the end of the abdomen and pulled the contents out with their teeth.

EXPERIMENT 2: PREDATION BY *P. MELANOTIS* IN RELATION TO BUTTERFLY SEX

Peromyscus melanotis killed (feeding categories 2–6; Table 1) equal numbers of male and female monarchs, which suggests no initial difference in the risk of male and female monarchs to mouse attack (Table 6). However, adding together the number of abdomens eaten selectively and nonselectively reveals that *P. melanotis* ate a larger number of male than female abdomens (means = 28.4 and 22.7, respectively; paired t value = 3.2, df = 14; p = 0.006). Of those abdomens eaten, *P. melanotis* fed selectively on a greater percentage of female abdomens (female and male means = 68.1 and 50.0, respectively; paired t value = 10.63; df = 14; p =

Table 5. Interspecific comparison of the standardized amounts of monarch tissue, cuticle, and CGs eaten. The mean ratio of the amounts of CGs to tissue ingested (±SE) is also compared among species. Eight mice from each species were each offered 40 monarchs and mouse chow *ad libitum* per night for 2 consecutive nights. Between-species comparisons are made with one-way ANOVA. Different subscripts (a, b, and c) indicate significant differences among species within each column ($p < 0.05$; Scheffé F test).

Mouse species	Standardized amounts eaten (mean ± SE): Tissue ($g/kg^{0.7}$ mouse)	Cuticle ($g/kg^{0.7}$ mouse)	CGs ($mg/kg^{0.7}$ mouse)	CGs/tissue ratio*
P. melanotis	$676.9 \pm 25.9_a$	$32.0 \pm 1.8_a$	$96.4 \pm 4.2_a$	$0.14 \pm 0.01_a$
R. sumichrasti	$703.4 \pm 22.6_a$	$40.9 \pm 2.6_{a,b}$	$120.0 \pm 7.0_b$	$0.17 \pm 0.01_a$
P. a. hylocetes	$378.4 \pm 13.2_b$	$46.1 \pm 1.8_b$	$126.4 \pm 4.3_b$	$0.33 \pm 0.01_b$
M. m. salvus	$120.4 \pm 15.4_c$	$18.8 \pm 2.7_c$	$46.1 \pm 5.8_c$	$0.39 \pm 0.01_c$
F value	190.4	27.1	44.7	212.3
df	3	3	3	3
p	0.0001	0.0001	0.0001	0.0001

* Determined from individual ratios.

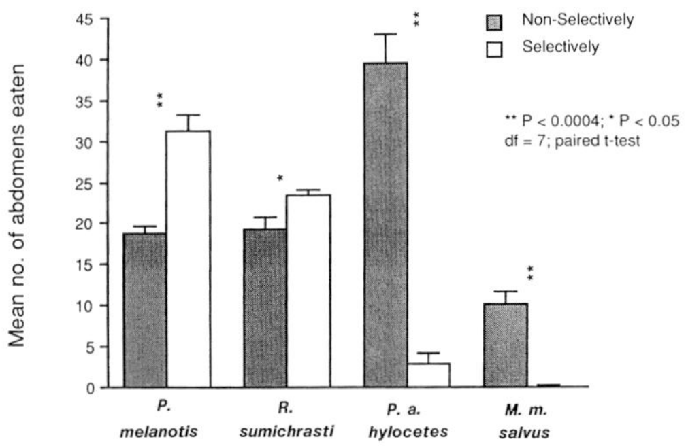

Figure 2. Mean number of monarch abdomens (±SE) eaten selectively and nonselectively by four mouse species (n = 8/species). Each mouse received 40 monarchs and mouse chow *ad libitum* per night for 2 consecutive nights.

0.0001). These results support the hypothesis that both the quantity of tissue eaten and the tendency to feed selectively are influenced by CG concentration.

EXPERIMENT 3: EFFECTS OF LONG-TERM CONSUMPTION OF MONARCHS

All *P. melanotis* gained weight over the 6 nights, whereas 60–100% of the individuals of each of the other species lost weight (Table 3). *Microtus m. salvus* were excluded from the statistical analyses because they ate very few monarchs and all died within 4 days.

The mass-specific amounts of monarch tissue, cuticle, and CGs eaten nightly by *P. melanotis, P. a. hylocetes,* and *R. sumichrasti* are plotted in Figure 3a–3c. The two-factor ANOVAs performed separately on tissue, cuticle, and CGs all indicated significant effects of species (for tissue, $F = 25.5$, df = 2, $p < 0.0001$; for cuticle, $F = 6.41$, df = 2, $p = 0.01$; and for CGs, $F = 7.4$, df = 2, $p = 0.008$) and of time for CGs ($F = 2.9$; df = 5, $p = 0.020$). There were not significant effects of time for tissue ($F = 1.0$, df = 5, $p = 0.445$) nor cuticle ($F = 2.0$, df = 5, $p = 0.09$). The interaction of species and time was significant for tissue ($F = 3.0$, df = 10, $p = 0.004$), cuticle ($F = 3.5$, df = 10, $p = 0.001$), and CGs ($F = 2.5$, df = 10; $p = 0.014$). These results demonstrate that (1) *P. melanotis* and *R. sumichrasti* consistently ate more tissue than did *P. a. hylocetes*; (2) *P. a. hylocetes* consistently ate more cuticle and CGs than did the other two mouse species; and (3) nightly consumption of cuticle and CGs by *R. sumichrasti* and *P. a. hylocetes* remained roughly constant over the 6 nights, whereas that of *P. melanotis* tended to diminish.

As in Experiment 1, *P. melanotis* and *R. sumichrasti* often fed upon abdomens selectively (Fig. 4). The two-factor ANOVA indicated a significant effect of time ($F = 3.3$, df = 5, $p = 0.014$), but not of species ($F = 2.0$, df = 1, $p = 0.19$) or of the interaction ($F = 1.8$, df = 5, $p = 0.14$), on the nightly rate of selective feeding by these two species. Because *P. a. hylocetes* rarely fed selectively on monarch abdomens (mean = 4.3% ± 2.1 of total abdomens eaten), it was excluded from Figure 4 and the analysis. Figure 4 suggests that only *P. melanotis* increased its nightly rate of selective feeding with time. This apparent interspecific difference may not have been detected by the ANOVA because of the large amount of intraspecific variation and small sample sizes.

The ratio of CGs to tissue ingested remained nearly constant over the 6 nights for all three species (Fig. 5). The two-factor ANOVA indicated a significant effect of species ($F = 21.3$, df = 2, $p < 0.0001$), but not of time ($F = 1.4$, df = 5, $p = 0.242$) or of the interaction ($F = 1.1$, df = 10, $p = 0.39$). Thus, by feeding selectively on abdomens, *P. melanotis* and *R. sumichrasti* reduced substantially the ratios of CGs to tissue in their diet over the 6 nights.

DISCUSSION

PATTERNS OF FEEDING BY MICE

The two species that fed selectively on monarch abdomens, *P. melanotis* and *R. sumichrasti,* also ate the most monarch tissue on a weight-specific basis. By discarding the abdominal cuticle, they substantially reduced the amount of CGs they ingested. *Peromyscus a. hylocetes* and *M. m. salvus,* on the other hand, rarely discarded cuticular material and thus ate significantly more cuticle and CGs.

There are several possible reasons why *M. m.*

Table 6. Comparative feeding patterns by *P. melanotis* upon male and female monarch butterflies. Fifteen mice were each offered 25 male and 25 female butterflies together with mouse chow *ad libitum* per night for 2 consecutive nights. For each monarch sex, the mean number eaten, sampled, or left undamaged (±SE) by the 15 mice are presented. Because mice ate both the abdomen and thorax of some monarchs, the row totals are all >50.

	Mean number of monarchs ± SE in each feeding category						
			Thoraces eaten		Abdomens eaten		
Monarch sex	No visible damage	Sampled	25–50%	51–100%	Non-selectively	Selectively	No. offered/ mouse
Male	8.5 ± 1	12.8 ± 1	1.8 ± 0.4	3.6 ± 1	14.1 ± 1	14.3 ± 1	50
Female	8.7 ± 2	17.8 ± 2	1.4 ± 0.4	1.3 ± 0.3	7.3 ± 1	15.4 ± 1	50

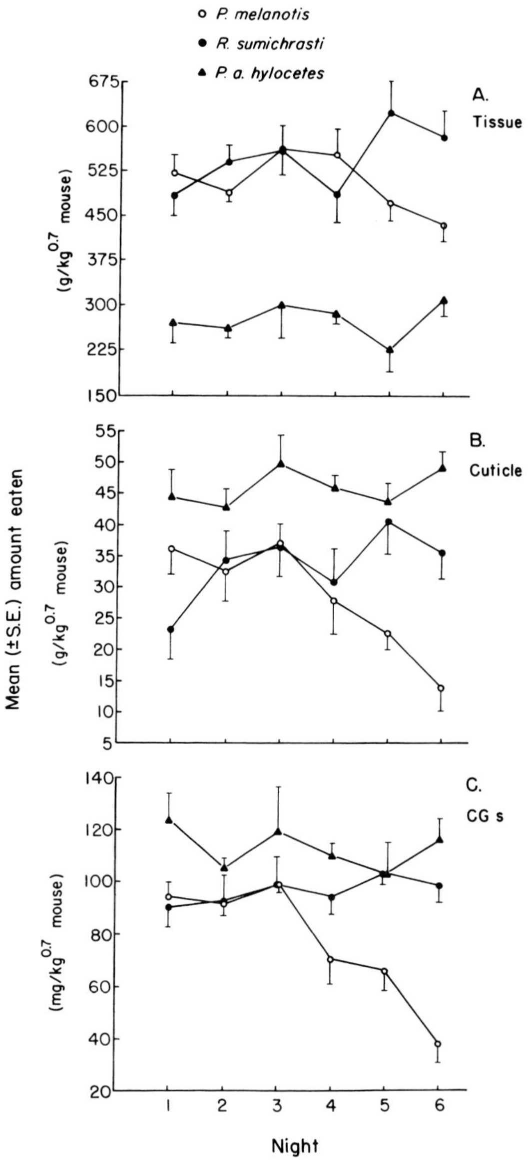

Figure 3. Standardized amounts of monarch tissue, cuticle, and CGs eaten by six *P. melanotis,* five *R. sumichrasti,* and four *P. a. hylocetes.* Each mouse was given 55 monarchs per night for 6 consecutive nights, after a 2-night pre-exposure to monarchs and mouse chow.

salvus and *P. a. hylocetes* ate so little monarch material as compared to the other two species. Even though *Microtus* species are known to eat insects to varying degrees (Batzli, 1985; Hanski and Parviainen, 1985), they tend to be strongly herbivorous and thus may be less motivated to attack and eat insects, particularly moving ones, than are more omnivorous species (Cyr, 1972). In fact, *M. m. salvus* was the only species that would not attack active monarchs. However, this explanation seems

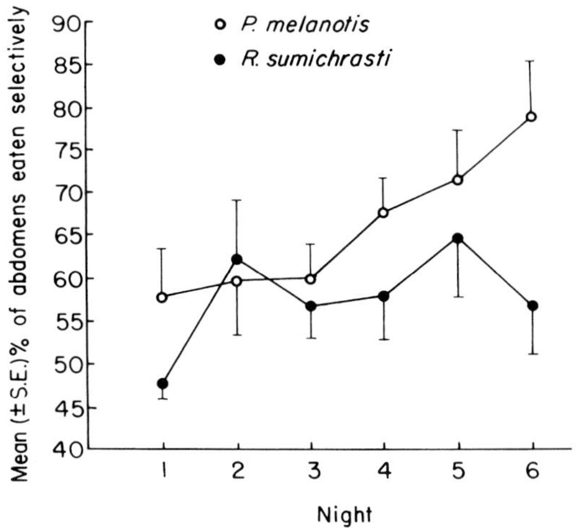

Figure 4. Mean percentage of abdomens (±SE) eaten selectively by six *P. melanotis* and five *R. sumichrasti.* Each mouse was given 55 monarchs per night for 6 consecutive nights, after a 2-night pre-exposure to monarchs and mouse chow.

unlikely in the case of *P. a. hylocetes,* given the strongly omnivorous feeding habits of its genus (Landry, 1970; Whittaker, 1966; Wolff et al., 1985) and the large numbers of monarchs that individuals ate (Fig. 1 and Table 4). Rather, it appears more likely that it was deterred from eating monarchs primarily by the CGs. In comparison to *P. melanotis, P. a. hylocetes* is highly sensitive to the taste of CGs (Glendinning et al., 1990; Glendinning, 1990).

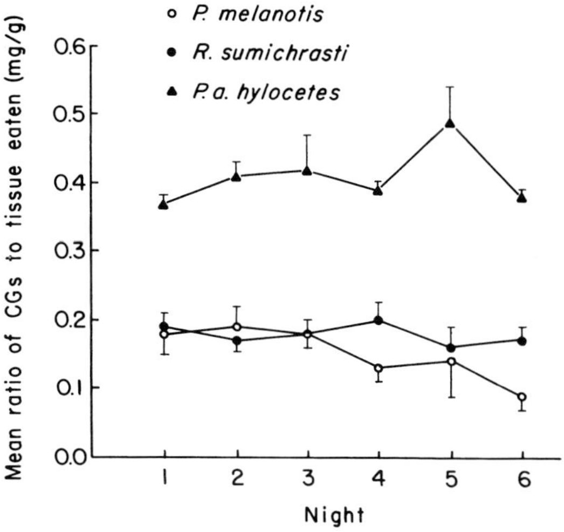

Figure 5. Ratios (mean ± SE) of monarch CGs (mg) to tissues (g) eaten by six *P. melanotis,* five *R. sumichrasti,* and four *P. a. hylocetes* over 6 consecutive nights. Each mouse received 50 monarchs per night (and no mouse chow), after a 2-night pre-exposure to monarchs and mouse chow.

While *R. sumichrasti* is also strongly deterred by these compounds, it limited ingestion of them through its selective feeding behavior.

It is unclear why only *P. melanotis* and *R. sumichrasti* commonly showed the selective feeding behavior. They may have modified normal fruit- or insect-eating behaviors, as has been suggested for black-backed orioles, which also feed selectively on monarch abdomens (Brower and Fink, 1985). However, why would *P. a. hylocetes* and *M. m. salvus* not have modified feeding behaviors of their own? Moreover, the selective feeding behavior is not unique to *P. melanotis* and *R. sumichrasti*. Red tree mice (*Phenacomys longicaudus*) peel off and discard the externally located resin ducts of Douglas fir [*Pseudotsuga taxifolia* (now *Pseudotsuga menziesii*)], needles before eating them (Benson and Borell, 1931). Second, chisel-toothed kangaroo rats (*Dipodomys microps*) shave off the hypersaline peripheral tissue of the leaves of saltbush (*Atriplex confertifolia*) so that the internal tissue can be ingested (Kenagy, 1972). Third, when grasshopper mice (*Onychomys torridus*) are offered house crickets (*Acheta domesticus*) whose heads have been treated with a highly bitter solution (quinine-hydrochloride and ethanol), they first chew off and discard the heads before eating the bodies (Cyr, 1972). Further study of insectivory and food-handling behavior in the four mouse species associated with the monarch colonies will be valuable in explaining the different ways they fed upon monarchs.

In addition to CG avoidance, two other reasons may explain why mice feed selectively on monarchs. Because small mammals (<100 g) appear incapable of digesting cuticle to any significant extent (Griffith and Gates, 1985; M. Allen, unpubl. data; but see Jeuniaux, 1961), they could increase their net energy yield of foraging by avoiding cuticle, and several foraging models predict animals should avoid indigestible foods (e.g. Krebs, 1978; Belovsky, 1978). Cuticle represents 10% of an overwintering monarch's abdominal biomass and 27% of its thoracic biomass (Table 2). Second, mice may have found the cuticle too tough and leathery. However, the results from Experiment 2 fail to support these two interpretations. *Peromyscus melanotis* ate more male abdomens and fed selectively on a greater percentage of female ones. Moreover, in an analogous experiment, where seven *P. melanotis* were given a choice between monarchs with high or low CG concentrations (monarchs reared on *A. curassavica* or *A. syriaca*, respectively), they preferred low-CG monarchs and fed selectively on more high-CG monarchs (Glendinning, 1990). If *P. melanotis* had avoided cuticle because of its indigestibility and/or leathery quality, they should have done so irrespective of CG concentration. These results support the hypothesis that *P. melanotis* and *R. sumichrasti* reject cuticular material because of an aversion to the high CG concentrations found there.

These results are in contrast to Brower et al. (1985), who rarely observed selective feeding by *P. melanotis* (erroneously referred to as *P. maniculatus* in their paper; Brower et al., 1988a). This discrepancy is possibly explained by them having given each mouse fewer than 30 monarchs without chow. In comparison, I gave either 40 monarchs with chow or 55 without chow. In fact, *P. melanotis* did not commonly feed selectively on abdomens when offered fewer than 30 monarchs, particularly when no mouse chow was available (Glendinning, 1989).

ABSENCE OF CONDITIONED FEEDING AVERSIONS TO MONARCH CGs

Peromyscus melanotis, R. sumichrasti, and *P. a. hylocetes* all consumed extremely high nightly dosages of monarch CGs in experiments 1 and 2 (means > 20 and 30 mg/kg mouse, respectively) without developing any signs of CG toxicity (e.g., ataxia, trembling, rolling convulsions, hyperexcitability; Marty, 1983). The LD_{50} for an oral dosage of digitoxin in laboratory mice (*Mus domesticus*), rats (*Rattus norvegicus*), and cats (*Felis domestica*) is 32.7, 23.8, and 0.30, respectively (Barnes and Eltherington, 1973). The mice in this study may have been able to tolerate such high dosages because they absorbed only a small percentage of the monarch CGs they ingested. CG absorption may be limited for other reasons. First, the gastrointestinal tracts of two species of mice closely related to the species studied herein are highly impermeable to monarch CGs (*P. maniculatus* and *M. domesticus*; Marty, 1983). Second, overwintering monarchs in Mexico have a high proportion of high polarity CGs (Seiber et al., 1986; Malcolm et al., this volume), which are absorbed by *M. domesticus* less readily than are low polarity ones (Lauterbach, 1981). Third, some of the ingested monarch CGs may have remained bound to the cuticle during digestion and were thus unavailable for absorption (see Brower et al., 1988b). Further study of the toxicological sensitivity of these three species to CGs, as well as the bioavailability of CGs bound in cuticle, will help determine whether or not either species could develop cardenolide toxicity and, hence, conditioned feeding aversions, following prolonged consumption of monarchs.

EFFECTS OF LONG-TERM CONSUMPTION OF MONARCHS

All species of mice ate nearly constant amounts of monarch tissue over the 6 nights, except for *M. m. salvus*. What differed among the other species of mice was the standardized amount of tissue eaten per night. *Peromyscus a. hylocetes* ate consistently less than did *P. melanotis* and *R. sumichrasti*.

These results suggest that three-fourths of the *P. a. hylocetes* lost weight because they ate insufficient amounts of food. However, it seems paradoxical that three-fifths of the *R. sumichrasti* also lost weight considering they ate similar standardized amounts of tissue as did the *P. melanotis*. One possible explanation is that because *R. sumichrasti* is highly averse to the taste of CGs, the monarch diet may have caused stress (sensu Selye, 1983), a condition known to increase heart rate, blood pressure, oxygen consumption (Porges, 1985), and possibly search activity; all of these factors elevate energetic demands. Stress also leads to weight loss in a variety of mammals, apparently owing to reduced food intake and/or a decrease in digestive efficiency (Klasing, 1985). Under such stress, *R. sumichrasti* would have lost weight even with the same mass-specific energy intake as *P. melanotis.*

EFFECTIVENESS OF PYRROLIZIDINE ALKALOIDS AGAINST MICE

Adult monarchs also contain pyrrolizidine alkaloids (PAs; Edgar et al., 1976; Kelley et al., 1987). These compounds are bitter tasting to humans (Brower, 1984) and lethal to several species of rodents, birds, and ruminants following chronic oral dosages (Mclean, 1970; Goeger et al., 1983; Cheeke and Pierson-Goeger, 1983). However, it has not been demonstrated that monarch PAs deter vertebrate predators, although this is widely assumed (e.g. Eisner, 1980; Boppré, 1986; Schneider, this volume). Recent feeding experiments suggest that consumption of monarchs by *P. melanotis, P. a. hylocetes,* and *R. sumichrasti* is probably not influenced strongly by PAs (Glendinning et al., 1990). Neither of these species avoided diets treated with free amine or N-oxide monocrotaline (a PA) until the concentration was equal to or well above the mean for overwintering monarchs.

WHY IS *P. MELANOTIS* THE ONLY SPECIES THAT FEEDS UPON OVERWINTERING MONARCHS?

Two ecological explanations might account for why the three other mouse species do not eat monarchs: (1) *P. melanotis* competitively excludes them, or (2) the microhabitat characteristics (e.g., understory vegetation) of overwintering areas may only suit *P. melanotis.* However, elsewhere I rule these out (Glendinning and Brower, 1990), and the explanation most consistent with my results is that *P. melanotis* is less sensitive to the taste of CGs in monarchs.

My findings raise several issues regarding the ability of vertebrates to discriminate between harmless and harmful bitter substances. Even though many naturally occurring bitter foods are toxic to vertebrates, there are also many such foods that are not (Bate-Smith, 1972; Garcia and Hankins, 1975; Brower, 1984). Do some species avoid all bitter foods, irrespective of their potential toxicity? If so, then such species would be vulnerable to Batesian taste mimicry (e.g., the Batesian mimic could be bitter and harmless, and the model bitter and harmful; Brower, 1984). Alternatively, are some species generally less sensitive to the bitter taste and toxicity of a variety of poisons? A comparative study of the relative sensitivities of *P. melanotis, P. a. hylocetes,* and *R. sumichrasti* to naturally occurring bitter substances that vary widely in chemical structure and differ in toxicity should begin to unravel these questions.

ACKNOWLEDGMENTS

I thank Lincoln Brower, Louis Guillette, Linda Fink, Steve Malcolm, Diana Glendinning, and Myron Zalucki for valuable editorial comments. I also thank James Anderson, Alfonso Alonso-M., Alfredo Arellano G., and Tonya Van Hook for stimulating fireside discussions; Tonya Van Hook and Mary Allen for the unpublished data; Bernardo Villa R., Jorge Soberón, and Monarca, A.C., for logistical support. Financial support was provided by U.S. National Science Foundation grant BSR-8500416 to L. Brower and by grants from the Sigma Xi Society, the Explorers Club, and the Department of Zoology, University of Florida. I am grateful for a scientific collecting permit in Mexico from the Dirección General Conservación Ecológia de los Recursos Naturales.

LITERATURE CITED

Anderson, D.E., and R. Lydic. 1977. On the effect of using ratios in the analysis of variance. *Biobehavioral reviews* 1:225–229.

Barnes, C.D., and L.G. Eltherington. 1973. *Drug dosages in laboratory animals: A handbook.* Berkeley: University of California Press.

Bate-Smith, E.C. 1972. Attractants and repellents in higher animals. In *Phytochemical ecology: Proceedings of the phytochemical society symposium,* ed. J.B. Harborne, 45–56. London: Academic Press.

Batzli, G.O. 1985. Nutrition. In *Biology of New World Microtus,* ed. R.H. Tamarin, 779–806. Special Publication No. 8, American Society of Mammalogy.

Belovsky, G.E. 1978. Diet optimization in a generalist herbivore: The moose. *Theoretical Population Biology* 14:105–134.

Benson, S.B., and A.E. Borell. 1931. Notes on the life history of the red tree mouse, *Phenacomys longicaudus. Journal of Mammalogy* 12:226–233.

Boppré, M. 1986. Insects pharmacophagously utilizing defensive plant compounds (pyrrolizidine alkaloids). *Naturwissenschaften* 73:17–26.

Brower, L. P. 1984. Chemical defence in butterflies. In *The biology of butterflies,* eds. R.I. Vane-Wright and P.R. Ackery, *Symposia of the Royal Entomological Society of London* 11:109–134. London and New York: Academic Press (1989, Princeton, NJ: Princeton University Press).

Brower, L.P. 1985. New perspectives on the migration biology of the monarch butterfly, *Danaus plexippus* L. In *Migration: Mechanisms and adaptive significance,* ed. M.A. Rankin, 748–785. Austin: University of Texas, *Contributions in Marine Science,* Supplement, 27.

Brower, L.P., and W.H. Calvert. 1985. Foraging dynamics of bird predators on overwintering monarch butterflies in Mexico. *Evolution* 39(4):852–868.

Brower, L.P., and L.S. Fink. 1985. A natural toxic defense system: Cardenolides in butterflies versus birds. *Annals of the New York Academy of Science* 443: 171–186.

Brower, L.P., B.E. Horner, M.A. Marty, C.M. Moffit, and B. Villa R. 1985. Mice (*Peromyscus maniculatus, P. spicilegus, Microtus mexicanus*) as predators of overwintering monarch butterflies (*Danaus plexippus*) in Mexico. *Biotropica* 17(2):89–99.

Brower, L.P., B.E. Horner, M.A. Marty, C.M. Moffit, and B. Villa R. 1988a. Errata. *Biotropica* 20:135.

Brower, L.P., C.A. Nelson, J.N. Seiber, L.S. Fink, and C. Bond. 1988b. Exaptation as an alternative to coevolution in the cardenolide-based chemical defense of monarch butterflies (*Danaus plexippus* L.) against avian predators. In *Chemical mediation of coevolution*, ed. K.C. Spencer, 447–476. Academic Press: New York.

Calvert, W.H., and L.P. Brower. 1986. The location of monarch butterfly (*Danaus plexippus*) overwintering colonies in Mexico in relation to topography and climate. *Journal of the Lepidopterists' Society* 40(3):167–187.

Calvert, W.H., L. E. Hedrick, and L.P. Brower. 1979. Mortality of the monarch butterfly (*Danaus plexippus* L.): Avian predation at five overwintering sites in Mexico. *Science* 204:847–851.

Cheeke, P.R., and M.L. Pierson-Goeger. 1983. Toxicity of *Senecio jacobaea* and pyrrolizidine alkaloids in various laboratory animals and avian species. *Toxicology Letters* 18:343–349.

Cyr, M.A. 1972. Predatory behavior of the grasshopper mouse (*Onychomys*). Ph.D. dissertation. Los Angeles: University of California.

Edgar, J.A., P.A. Cockrum, and J.L. Frahn. 1976. Pyrrolizidine alkaloids in *Danaus plexippus* L. and *Danaus chrysippus* L. *Experientia* 32:1535–1536.

Eisner, T. 1980. Chemistry, defense, and survival: Case studies and selected topics. In *Insect biology in the future*, eds. M. Locke and D.S. Smith, 847–878. New York: Academic Press.

Farlow, J.O. 1976. A consideration of the trophic dynamics of a late Cretaceous large dinosaur community (Oldman Formation). *Ecology* 54:841–857.

Fink, L.S., and L.P. Brower. 1981. Birds can overcome the cardenolide defence of monarch butterflies in Mexico. *Nature* 291:67–70.

Fink, L.S., L.P. Brower, R.B. Waide, and P.R. Spitzer. 1983. Overwintering monarch butterflies as food for insectivorous birds in Mexico. *Biotropica* 15: 151–153.

Garcia, J., and W.G. Hankins. 1975. The evolution of bitter and the acquisition of toxiphobia. In *Olfaction and taste. V. Proceedings of the 5th international symposium in Melbourne, Australia*, eds. D.A. Denton and J.P. Coghlan, 39–45. New York: Academic Press.

Glendinning, J.I. 1989. Comparative responses of five sympatric species of mice to overwintering colonies of monarch butterflies. Ph.D. dissertation. Gainesville: University of Florida.

Glendinning, J.I. 1990. Responses of three mouse species to deterrent chemicals in the monarch butterfly. II. Taste tests using intact monarchs. *Chemoecology* 1:124–130.

Glendinning, J.I., and L. P. Brower. 1990. Feeding and breeding responses of five mice species to overwintering aggregations of the monarch butterfly. *Journal of Animal Ecology* 59:1091–1112.

Glendinning, J.I., A. Alonso-M., and L.P. Brower. 1988. Behavioral and ecological interactions of foraging mice (*Peromyscus melanotis*) with overwintering monarch butterflies (*Danaus plexippus*) in Mexico. *Oecologia (Berlin)* 75:222–227.

Glendinning, J.I., L.P. Brower, and C.A. Montgomery. 1990. Responses of three mouse species to deterrent chemicals in the monarch butterfly. I. Taste and toxicity tests using artificial diets laced with digitoxin or monocrotaline. *Chemoecology* 1:114–123.

Goeger, D.E., P.R. Cheeke, H.S. Ramsdell, S.S. Nicholson, and D.R. Buhler. 1983. Comparison of the toxicities of *Senecio jacobaea, Senecio vulgaris* and *Senecio glabellus* in rats. *Toxicology Letters* 15:19–23.

Griffith, L.A., and J.E. Gates. 1985. Food habits of cave-dwelling bats in the central Appalachians. *Journal of Mammalogy* 66:451–460.

Hanski, I., and P. Parviainen. 1985. Cocoon predation by small mammals, and pine sawfly population dynamics. *Oikis* 45:125–136.

Jeuniaux, C. 1961. Chitinase: An addition to the list of hydrolases in the digestive tract of vertebrates. *Nature* 192(4798):135–136.

Kelley, R.B., J.N. Seiber, A.D. Jones, H.J. Segall, and L.P. Brower. 1987. Pyrrolizidine alkaloids in overwintering monarch butterflies (*Danaus plexippus*) from Mexico. *Experientia* 43:943–946.

Kenagy, G.J. 1972. Saltbush leaves: Excision of hypersaline tissue by a kangaroo rat. *Science* 178:1094–1096.

Klasing, K.C. 1985. Influence of stress on protein metabolism. In *Animal stress*, ed. G.P. Moberg, 269–280. Bethesda, MD: American Physiological Society.

Krebs, J.R. 1978. Optimal foraging: Decision rules for predators. In *Behavioral ecology: An evolutionary approach*, eds. J.R. Krebs and N.B. Davies, 23–63. Sunderland, MA: Sinauer Associates.

Landry, S.O., Jr. 1970. The Rodentia as omnivores. *Quarterly Review of Biology* 45:351–372.

Lauterbach, F. 1981. Intestinal absorption and secretion of cardiac glycosides. In *Cardiac glycosides, Part II: Pharmacokinetics and clinical pharmacology*, ed. K. Greeff, 105–139. New York: Springer-Verlag.

Malcolm, S.B., and L.P. Brower. 1989. Evolutionary and ecological implications of cardenolide sequestration in the monarch butterfly. *Experientia* 45:284–294.

Marty, M. 1983. Metabolism and disposition of milkweed (Asclepiadaceae) cardenolides in monarch butterfly larvae, *Danaus plexippus*, and in a predator, *Peromyscus maniculatus*. Ph.D. dissertation. University of California, Davis.

Mclean, E.K. 1970. The toxic actions of pyrrolizidine (*Senecio*) alkaloids. *Pharmacology Review* 22(4):429–483.

Millar, J.S. 1979. Energetics of lactation in *Peromyscus*

maniculatus. Canadian Journal of Zoology 57:1015–1019.

Packard, G.C., and T.C. Boardman. 1988. The misuse of ratios, indices and percentages in ecophysiological research. *Physiological Zoology* 61:1–9.

Peters, R.H. 1983. *The ecological implications of body size.* Cambridge, MA: Cambridge University Press.

Porges, S.W. 1985. Spontaneous oscillations in heart rate: Potential index of stress. In *Animal stress,* ed. G.P. Moberg, 281–294. Bethesda, MD: American Physiological Society.

Sadleir, R.M.F.S., K.D. Casperson, and J. Harling. 1973. Intake and requirements of energy for breeding of wild deermice, *Peromyscus maniculatus. Journal of Reproductive Fertility* (supplement) 19:237–252.

Seiber, J.N., L.P. Brower, S.M. Lee, M.M. McChesney, H.T.A. Cheung, C.J. Nelson, and T.R. Watson. 1986. Cardenolide connection between overwintering monarch butterflies from Mexico and their larval food plant. *Journal of Chemical Ecology* 12: 1157–1170.

Selye, H. 1983. The stress concept: Past, present and future. In *Stress research,* ed. C.L. Cooper, 1–20. New York: John Wiley and Sons.

Stebbins, L.L. 1977. Energy requirements during reproduction of *Peromyscus maniculatus. Canadian Journal of Zoology* 55:1701–1704.

Whittaker, J.O. 1966. Food of *Mus musculus, Peromyscus maniculatus bairdi* and *Peromyscus leucopus* in Vigo County, Indiana. *Journal of Mammalogy* 47:473–486.

Wolff, J.O., R.D. Dueser, and K.S. Berry. 1985. Food habits of sympatric *Peromyscus leucopus* and *Peromyscus maniculatus. Journal of Mammalogy* 66: 795–798.

Zar, J.H. 1984. *Biostatistical analysis.* Englewood Cliffs, NJ: Prentice-Hall, 718 pp.

OVERWINTERING MONARCH BUTTERFLIES IN CALIFORNIA: PAST AND PRESENT

John Lane

ABSTRACT. The overwintering biology of monarch butterflies in California is reviewed and the characteristics of overwintering sites are described. Three aspects of overwintering are considered in more detail: (1) the current location and range of monarch overwintering in California and both native and introduced tree species used by overwintering monarchs, (2) the historical location and nature of overwintering sites, and (3) the earliest published accounts of monarchs and monarch overwintering in California.

INTRODUCTION

Events in California have moved quickly since the Symposium on the Biology and Conservation of the Monarch Butterfly at Cocoyoc, Mexico, in 1981—now dubbed "Moncon-1." My contribution (Lane, 1984) stressed the need for a modern survey of the state's coastline to locate the overwintering sites of monarchs as a first step toward their conservation. In 1984, *The Monarch Project* was initiated by the Xerces Society (see Allen and Snow, this volume), and an overwintering site survey became its first priority. A first edition of the survey was produced with a grant from World Wildlife Fund–US and incorporated the data collected through the 1984–1985 overwintering season (Nagano and Lane, 1985). Reference to most of these sites, and a few additional ones, can be found scattered throughout Urquhart (1960) although Urquhart (1987) inexplicably omits mention of a number of localities found in his earlier work. The purpose of the present paper is thus to revise Lane (1984) and address the following three questions. (1) Where do monarchs overwinter today and which native tree species do they use? (2) Where might they have overwintered in the historical past? (3) What do we know of the events surrounding the original discovery of monarchs overwintering along the coast of California? Furthermore, in light of Vane-Wright's "Columbus Hypothesis" (this volume), I address the question: How early can monarchs (the species as distinct from overwintering masses) be documented in California?

DISTRIBUTION OF COASTAL OVERWINTERING COLONIES

The early works on butterflies of the state only briefly mention the locations of monarch aggregations. Wright (1906) states only that "I have noted these gatherings in spring and in the autumn; on the shore of the Pacific, near Los Angeles, and in Texas . . . ." Comstock (1927) says of "butterfly trees," "We have on the west coast a number of localities in which such groves of trees are located. A particularly noted 'resting station' is located at Monterey, where a grove of Eucalyptus is visited annually by the southward moving swarms. A grove of similar character is located at Santa Monica [near Los Angeles]." In addition, a photograph of monarchs clustered on *Eucalyptus* is shown with a caption stating "Courtesy of the San Diego Natural History Museum," likely taken on a museum field trip to a butterfly grove in San Diego in the 1920s.

Prior to Lane (1984) and Nagano and Lane (1985) only one list of monarch aggregation sites in California has been published. Downes and Williams (in Williams et al., 1942) listed about two dozen sites along the coast from San Diego to Marin County. Subsequently, a survey was produced for the California Department of Fish and Game (Sakai et al., 1989) and this data entered into the California Natural Areas Diversity Data Base in Sacramento. While continuing the survey through the years since 1981 (when Lane (1984) was written) significant new information has become available and has modified my understanding of historical overwintering in the state.

Nagano and Lane (1985) and Sakai et al. (1989) record slightly over 200 reports of monarch aggregations and actual overwintering sites along most of the California coastline. The northernmost reported overwintering location is in northern Mendocino County (Nagano and Sakai, 1988; Sakai et al., 1989), and the southernmost is San Diego, by the Mexican border. This is a span of about 620 miles (998 km), extending from 32°45′ to 39°45′ north latitude and encompassing about 80% of the state's coastline. Thus, most of the state's coastline is climatically suitable to support overwintering. (Along the Pacific Coast outside of California, monarch aggregations are known only a short distance below the border, fom a single report at Ensenada, Baja California, Mexico, located about 100 km south of San Diego.)

COLONY FORMATION AND PERSISTENCE

A bewildering array of terms appears in the literature to describe monarch aggregations. Urquhart et al. (1965) define as "transient hibernal roosting colonies" those colonies "that establish roosting sites for a portion of the winter," and "hibernal roosting colonies" as those "that are of a more permanent nature." I prefer the simpler terms "temporary" and "permanent," distinguishing the two kinds of colonies by whether or not the butterflies remain past the winter solstice, i.e., long enough to allow mating activity to take place. In some conservation circles, the term "temporary" is feared to imply lesser importance, and they suggest "autumnal" in its place; for symmetry, then, "hibernal" can be used for aggregations persisting through the winter. "Bivouac," a term used widely (and loosely) for a long time by Urquhart and others, might best be restricted to those ephemeral aggregations of monarchs formed along the migratory route well away from their winter destinations, such as are commonly reported in late summer at Point Pelce, Ontario, Cape May, New Jersey, and many overnight aggregations observed at various other locations. "Aggregation" is a useful neutral term specifying neither autumnal ("temporary") nor hibernal ("permanent") status. The terms "colony," "roost," "site," and "location" are all general and interchangeable.

Monarchs arrive at the coast and begin forming aggregations in trees in late September. Many temporary colonies break up in October to early December, as their monarchs disperse to other permanent sites where they will spend the winter. The number of these temporary autumn aggregation sites is far larger than the number of permanent overwintering sites where mating occurs in January and February (and sometimes March) just prior to colony breakup (Tuskes and Brower, 1978; Nagano et al., this volume). Typically, many of the temporary autumn aggregation sites are made up of smaller numbers of butterflies while the permanent overwintering aggregation sites continue to grow in size through the autumn and often form the largest colonies (Lane 1984, 1985; Ackery and Vane-Wright, 1984).

Colony formation and persistence is dynamic. A colony that is permanent one year may be temporary the next. Whether the site is usually temporary or permanent, it may form most years but not form all years—thus, site "fidelity" can vary. Generally, sites having larger aggregations of butterflies tend to form more predictably (i.e., have higher fidelity) and with more permanence (i.e., persisting after the solstice) than sites with smaller numbers of butterflies. Perhaps several factors bear on this dynamic variability, such as site quality, local seasonal environmental variability, and cyclic variation in the overall abundance of individuals from year to year (see Tuskes and Brower, 1978).

The date of colony breakup varies with seasonal weather. In warm, dry years, sexual activity commences earlier and colonies may break up in late January. If cold (and usually wet) weather persists, colony breakup may be delayed into March. Sexual activity may begin in early January in particularly warm winters, but more typically probably peaks in the latter half of February.

CHARACTERISTICS OF COASTAL AGGREGATION AND OVERWINTERING SITES

Virtually all of California's monarch aggregations form at sites sharing a set of common characteristics (Lane, 1984, 1985; Nagano and Lane, 1985). Such sites occur close to the coast, at low elevations, and in protected locations. Most sites occur within half a mile of the water (ocean or bay). Many sites are essentially at sea level, with most <200-ft and probably none >300-ft elevation. Shelter is usually afforded by a combination of the cover provided by the canopy of surrounding taller vegetation and local topography. Most sites occur in shallow canyons, gullies, or on the lee side of hills. Sites frequently occur on points, peninsulas, and where the coastline runs generally in an east–west direction offering protection from the predominating winds out of the northwest. However, where tree cover is especially dense, sites may occur on flat, exposed terrain. Thus, surrounding cover of the tree canopy and topography provide wind protection, and the underlying shrub and herb layers are also likely to contribute to the specific, narrow microclimatic requirements of the butterflies, as has been shown in Mexico (Calvert and Brower, 1982; Calvert et al., 1984). Obviously, for conservation purposes, it is an entire habitat that requires protection—not just specific "butterfly trees" but the surrounding forest as well (Lane, 1985).

Most California overwintering sites occur along natural watercourses. However, fog and winter rains likely provide adequate moisture for overwintering colonies to persist without having a permanent source of running water. Large numbers of monarchs are rarely observed coming to watercourses in California, unlike the mass downhill flights to moisture noted some years in Mexico (Brower, 1985; Calvert and Brower, 1986; Calvert and Lawton, this volume).

Abundant nectar resources may also play an important role in the formation of temporary autumn aggregations. A number of such aggregations form each year in locations having an abundant source of nectar. Whether or not temporary colony dispersal is sometimes triggered by diminishing nectar resources deserves future study.

INTERIOR AGGREGATION OR OVERWINTERING SITES

Occasionally, monarchs form persistent aggregations some distance from the coast. Such aggregations remain longer than the transient aggregations of migratory butterflies that may rest for the night ("bivouacs"), or for a few days, to wait out a short period of inclement weather such as (Urquhart 1960, 1987) reports for autumn migrant monarchs in the eastern United States.

The most surprising records of inland aggregations are for Saline Valley in Inyo County (Nagano and Lane, 1985), located in true desert east of Mount Whitney and north of Death Valley (Fig. 1). Small numbers of monarchs (up to low hundreds) have been observed here in late November for several years by Derham Giuliani of Big Pine, California. Several sites occur along the eastern base of the Inyo Mountains along permanent watercourses. It is not clear that these monarchs actually persist through the winter, but the mere fact that they aggregate and persist here is of interest and may indicate conditions suitable for overwintering could occur in unexpected, highly specialized inland locations.

Davenport (1983) reported monarchs every month of the year at Bakersfield in Kern County, near the extreme southern end of California's Central Valley (Fig. 1). He has noticed very small numbers flying around several areas each winter for over 15 yr. The particular locations are regional warm areas frequently lacking winter ground fog typical of the Central Valley (K. Davenport, pers. comm., 1988). It is possible that other local microclimatic patches suitable for monarch overwintering may exist along the margin of the Central Valley bordering the Sierran foothills.

A small inland overwintering colony (numbering up to the low hundreds) has also been observed by me for several years at Salinas, Monterey County. However, this site is only about 10 mi inland from Monterey Bay and is located on a broad agricultural plain subject to a constant flow of marine air. Thus, although located unusually far inland, its environment is essentially coastal, and the occurrence of overwintering monarchs here is not exceptional. Similar occurrences are known from San Luis Obispo.

Figure 1. Map of California showing the coastal counties, islands, and two inland counties mentioned in the text. Overwintering colonies are documented from San Diego County near the Mexican border north to northern Mendocino County. A single record of aggregations (possibly not overwintering) comes from central Humboldt County. Records of aggregations from Del Norte County are lacking at present. Locations of aggregations from Kern and Inyo counties are indicated by black dots.

BUTTERFLY TREES: INTRODUCED AND NATIVE

The great majority of overwintering sites along the coast today are associated with *Eucalyptus* trees, primarily blue gum, *Eucalyptus globulus* (Tuskes and Brower, 1978; Lane, 1984; Nagano and Lane, 1985; Sakai et al., 1989). These Australian trees clearly were not available for monarchs to use prior to their introduction in the 1850s. With their arrival and widespread planting for windbreaks, fuel lots, and landscaping, there was a dramatic increase in potential monarch overwintering habitat throughout the state. At the same time, the extent of dense groves of native species for monarchs to use along the coast declined due to cutting for lumber and firewood and to land clearing for agricultural and urban development (Lane, 1984).

THE *EUCALYPTUS* DILEMMA

Eucalyptus is an aggressive exotic whose allelopathic chemicals inhibit germination of the native vegetation and generally reduce biotic diversity. State and Federal officials responsible for managing public lands frequently express the desire to eradicate such exotic vegetation and return the land to its native condition. Thus, ironically, monarch conservation efforts may be viewed by some as frustrating the otherwise laudible goal of promoting

California's unique native vegetation through *Eucalyptus* removal.

It is important to note that while the introduction and subsequent success of *Eucalyptus* has no doubt increased the amount of dense, tree-covered overwintering habitat for monarchs along the coast, much of California's native tree-cover along the coast has been lost. Therefore, it is desirable that those areas where *Eucalyptus* provides monarchs with suitable overwintering habitat be protected. Fortunately, only a tiny fraction of all the *Eucalyptus* growing on public lands along the coast is actually utilized by monarchs, and concern for the butterflies should not impede most *Eucalyptus* removal efforts. Where significant numbers of butterflies do use *Eucalyptus*, it is hoped that the trees can remain, or their removal can be phased and more suitable trees be planted so that viable monarch habitat will be continuously conserved.

CALIFORNIA'S NATIVE BUTTERFLY TREES

Prior to European contact and the introduction of exotic tree species such as *Eucalyptus*, monarchs were presumably restricted to using those native trees that grew along the coast. Yet, much of California's native coastal landscape consisted of coastal prairie grassland or low, dense chaparral scrub. Dense groves of trees were essentially absent from most of the immediate coast, especially south of Sonoma County (Munz and Keck, 1959; Sudworth, 1967; Griffin and Critchfield, 1972; Barbour and Major, 1977; Mayfield, 1980). Therefore, the sites where monarchs could have overwintered along the coast originally must have been restricted to those relatively few areas where dense stands of native trees grew close to the shoreline.

Finding monarchs using native trees growing in remnant stands along the coast today suggests that they used native stands of those trees along the coast in historical times. The number of California native tree species (growing in native stands) known to support overwintering monarchs today is considerably greater than the number known when I wrote the first "status" report for Moncon-1 in 1981 (Lane, 1984). At that time, only two native species were thought to support overwintering, and so only those species were thought to have been available to monarchs in historical times. This led to the suggestion that prior to European contact, overwintering monarchs would have only been found in three small districts of the California coast (see the sections on Monterey pine and Monterey cypress, below). Field investigations of overwintering activity along the coast through the intervening years have greatly increased the number of native tree species known to support overwintering today. As a result, our view of where monarchs may have overwintered historically has undergone significant change. A variety of native trees probably grew in sufficiently dense groves close to the coast to support overwintering. These groves could have been more or less widely scattered along nearly the entire state coastline.

The data upon which this historical view is based are presented below. Monarch utilization patterns are discussed for each recorded native roost tree species, beginning with Monterey pine and Monterey cypress. Subsequent tree species are presented in an order generally representing their distribution from north to south. Native tree distribution data come from Sudworth (1967), Griffin and Critchfield (1972), Barbour and Major (1977), and other sources as cited below.

Monterey Pine
(*Pinus radiata* D. Don.)

Monterey pine has only three widely separated, localized native stands on the mainland of California: (1) the Monterey Peninsula (including Point Lobos and Carmel Highlands a few kilometers to the south) in Monterey County, (2) in the vicinity of Cambria in San Luis Obispo County, and (3) in the vicinity of Año Nuevo Point along the Santa Cruz–San Mateo county line. The famous Pacific Grove sites are on the Monterey Peninsula where the forest has been extensively reduced and thinned by development. Overwintering persists in the Cambria district although some sites appear to have been all but lost to development. A butterfly cluster was observed (once) on the Coastways Ranch in the Año Nuevo district (Mr. and Mrs. Hank Bradley, Sr., pers. comm., 1985), but additional clusters have not been observed despite repeated visits. It is possible that an overwintering colony still remains to be discovered, but it is more likely that clearing the forest for agriculture about 70 yr ago claimed the best of the potential overwintering habitat.

Monterey Cypress
(*Cupressus macrocarpa* Hartw.)

Monterey cypress is restricted in its native range to a narrow coastal strip along the southern Monterey Peninsula near Carmel and at Point Lobos a few kilometers to the south. Although the peninsular stand is but a few miles from the famous Pacific Grove butterfly sites (associated with Monterey pine), no monarch aggregations are known from native cypress stands, nor from the Monterey pine forest growing immediately adjacent to the cypress. Much potential habitat has been lost in this area with clearing of the forest for homes and several famous golf courses. However, this cypress has been planted extensively along the coast and is used by monarchs in a number of locations outside of its highly restricted native range.

Bishop Pine
(*Pinus muricata* D. Don.)

Bishop pine grows near the coast at a number of widely separated locations from Humboldt to Santa

Barbara counties (Fig. 1), but it is known to be used by monarchs only along the Sonoma and southern Mendocino county coast. At Anchor Bay, this pine provides the northernmost record for overwintering in a native forest setting. Other colonies may yet be found associated with this pine where it occurs even farther north. (The northernmost overwintering site reported for the state is associated with *Eucalyptus* about 65 mi farther north at Rockport, Mendocino County.)

Bishop pine occurs at some places near the coast at elevations considered too high for monarch overwintering, such as on Inverness Ridge in Marin County and where it occurs on the mainland in San Luis Obispo and Santa Barbara counties. On Santa Cruz Island, Santa Barbara County, monarchs overwinter in small numbers on *Eucalyptus* growing close to sea level but are not reported from the Bishop pine stands growing at higher elevations.

An intriguing report of clustered butterflies on the sand spit bordering Humboldt Bay at Eureka, Humboldt County, places monarch aggregations yet another 70 mi north (Fig. 1). Whether this record properly refers to overwintering or just a temporary autumn aggregation is unclear. Monarchs migrating down the coast in autumn could be expected to form overnight "bivouacs" or even temporary autumn aggregations, even if they didn't spend the winter this far north. The dense native stands of shore pine (*Pinus contorta* Dougl. ex Loud.) in the area had been extensively cut when I visited the site in November 1986. Perhaps other cluster sites will be found associated with stands of shore pine or Sitka spruce (*Picea sitchensis* (Bong.) Carr.) occurring nearby within the Lanphere–Christensen Dunes Preserve owned by the Nature Conservancy.

Grand Fir
(*Abies grandis* (Dougl. ex D. Don) Lindl.)

Grand fir supports autumn aggregations and likely overwintering in northernmost Sonoma County on "The Sea Ranch." (True overwintering does occur nearby in association with other native and nonnative tree species.) The species occurs along the coastal strip from Mendocino County north, scattered within north coastal coniferous and redwood forests (Munz and Keck, 1959).

Coast Redwood
(*Sequoia sempervirens* (D. Don) Endl.)

Native stands of coast redwood occur in the California coast ranges from the Oregon border south to the Big Sur coastline of southern Monterey County. The species does not tolerate salt well and is typically absent from the immediate coast except in unusually well-protected, local situations. The species has been extensively cut for timber throughout its range.

Monarchs form temporary autumn aggregations and perhaps overwinter on redwood in northernmost Sonoma County, very close to sites where they overwinter on other native trees. Near Big Sur south of Monterey, monarchs likely overwinter on second-growth redwood growing in a shallow arroyo, downcut into the coastal terrace just behind the shoreline. Nearby, deeper canyons are typical of a number of large canyons on the long "Big Sur coastline" (essentially the length of Monterey County) in having extensive redwood stands but apparently lacking monarchs. Perhaps these sites are too cold, being largely shaded in the day and "drowned" in a river of frigid air flowing down these steep, narrow canyons each night from the high slopes of the Santa Lucia Mountains. It is possible that even in these large, deep canyons, trees once grew close enough to the water so that the moderating influence of the ocean might have provided a suitable microclimate for monarchs. However, redwoods were extensively logged from this region, with the most accessible trees growing closest to the coast being taken first (Fink, 1982).

Monarchs are recorded overwintering today in association with *Eucalyptus* or other introduced trees at several sites where there is evidence that redwood once grew. At Palo Colorado Canyon on the Big Sur coast in Monterey County, old redwood stumps remain within the *Eucalyptus* grove where the butterflies overwinter. On Hot Springs Creek on the Big Sur coast, monarchs overwinter each year on introduced trees growing just below a dense native stand of redwood. A single redwood still grows within the butterfly area, and it seems likely that additional redwoods were removed for the construction of a highway bridge that separates the intact redwood forest from the butterfly trees. On the bank of Soquel Creek in the city of Capitola in Santa Cruz County, where a few native redwoods still grow adjacent to introduced *Eucalyptus* and pine where monarchs aggregate each autumn and in some years overwinter in small numbers. A similar situation is found above Harvey West Park in the city of Santa Cruz.

Douglas Fir
(*Pseudotsuga menziesii* (Mirb.) Franco)

Douglas fir is one of the dominant coastal forest trees from Sonoma County north, but it is much less common to the south and rare south of Santa Cruz County. In northernmost Sonoma County, monarchs aggregate in the autumn and overwinter (at least in some years) in association with this species (together with redwood and grand fir). Near Año Neuvo Point (the San Mateo–Santa Cruz county line), remnant Douglas firs growing along riparian corridors support autumn aggregations of monarchs. It seems likely that true overwintering could have occurred in this district before much of the Douglas fir and Monterey pine forest was removed for agriculture and timber. Douglas fir has been and remains a principal timber species in California, and, like redwood, many of the coastal stands were the first to be cut.

Coast Live Oak
(*Quercus agrifolia* Née)

Coast live oak is common along parts of California's coast from Santa Cruz County south but appears not to provide suitable monarch habitat. In its native stands, it does not appear to provide sufficient wind protection because in open woodland each tree stands alone, or in shallow coastal canyons it grows as a dense thicket of uniform modest height without additional taller trees to provide good wind protection. The few sites in the state where monarchs have been reported on this oak (from Santa Cruz to Santa Barbara counties) apparently represent either very small temporary autumn aggregations or (more commonly) one or a few oaks being used only marginally because they grow within or adjacent to a larger stand of some other species of non-native tree which grows taller, forms denser groves, and which provides more suitable monarch habitat.

Sycamore
(*Platanus racemosa* Nutt.)

Sycamore is perhaps the only native tree species used in southern California (south of San Luis Obispo County). This was once a common riverine species, but virtually none of the few streams or rivers running to the coast in the southern part of the state retain their original floodplain forest vegetation. Overwintering monarchs cluster on sycamore in several canyons close to the ocean at the foot of the Santa Monica Mountains between Ventura and Los Angeles, sometimes clinging to bare twigs after all the leaves have fallen.

Torrey Pine
(*Pinus torreyana* Parry ex Carr.)

The native stand of this species is confined on the mainland to a tiny district by La Jolla and Del Mar in San Diego County. Those reports of monarchs utilizing this species in this general area come from planted groves of pines growing outside its highly restricted native range. In its native setting, Torrey pine grows in open woodland amid open chaparral scrub and does not appear to form sufficiently dense stands to support either autumn aggregations or true overwintering. The native stands of this pine growing on Santa Rosa Island (Santa Barbara County) may possibly be suitable for overwintering by monarchs, but I am unaware of records of monarchs from this island (see Wenner and Harris, this volume).

POSSIBLE OVERWINTERING HABITATS NOW LOST

At the time of the first European contact with California, most of the coastline lacked significant tree cover, and where dense groves of trees did grow along the coast, they were frequently logged early for timber or firewood. Much of the tree cover that did exist grew along the watercourses running to the sea. Some early historical accounts from the coast of central California, where today monarchs overwinter in large numbers, suggest that the vegetative cover along these riparian corridors might well have been extensive enough to support monarch overwintering.

The Pajaro River forms the boundary between Santa Cruz and Monterey counties where it flows into Monterey Bay, about midway between the monarch overwintering colonies at Pacific Grove and Santa Cruz. The river was described by the first European land party to explore the region at a point not far inland from the coast, near the present town of Watsonville. On 8 October 1769, Fr. Crespi wrote in his diary: "... We halted on the bank of the river ... which was near its very verdant and pleasant plain, full of cottonwoods, alders, tall oaks, live oaks, and other species not known to us" (Bolton, 1966: 197). On 10 October they encountered redwood for the first time: "... traveled ... over plains and low hills, well forested with very high trees of a red color, not known to us. ... In this region there is a great abundance of these trees and because none of the expedition recognizes them, they are named redwood from their color" (Bolton, 1966: 198). Then, on 17 October, they crossed the San Lorenzo River, at the present site of Santa Cruz. Fr. Crespi wrote: "Not far from the sea we came to a large river .... In its bed there is a thick growth of cottonwoods and alders .... We made camp on the other side of the river, the descent and ascent of which cost some trouble to clear and make a passage .... Besides the growth along the river there are many redwoods" (Bolton, 1966: 202). Virtually nothing is left of these once extensive riverine forests. Nearby, close to some few remaining redwoods growing in a steep draw, monarchs still aggregate in autumn and overwinter in some years on *Eucalyptus* in Santa Cruz.

In southern California, it seems possible that monarchs might have overwintered in sycamore trees and perhaps other riverine species that once grew along the creeks and rivers of the Los Angeles basin, and at similar sites south to the Mexican border.

THE DISCOVERY OF OVERWINTERING MONARCH MASSES IN CALIFORNIA

Mystery still surrounds the circumstances of the original discovery of overwintering masses of monarch butterflies in California. Somewhere, one expects, there must be a written account of extreme surprise and delighted fascination with the discovery of trees covered with countless multitudes of live butterflies. The further delight in discovering this "lost" account still awaits us.

The earliest account describing overwintering masses of monarch butterflies that I am aware of was published by the *Monterey Weekly Herald* on 30 May 1874. The author (anonymous) describes a walk taken in the autumn of 1873 to the Point of Pines (Point Pinos) near the present-day location of overwintering monarchs in Pacific Grove. "On my ramble I passed through the favorite haunt of butterflies, a large, brown species, spotted with black, the only kind known to perch in flocks. Millions [*sic*] were fluttering around; while overhead stout branches of firs dropped with their weight; and all twinkling in shade and sunlight, seemed the personification of happiness in their leafy home. The beach was soon in sight again . . . ." The "firs" are almost certainly Monterey pines, rather than Douglas fir (Griffen and Critchfield, 1972).

It is important to note that this account suggests the locals already knew something of the uniqueness of the phenomenon. While Monterey was at this time only a very small, "sleepy" American port, it had been the capital of both Spanish and then Mexican Alta California, and one of only a few ports available for replenishing provisions and making repairs on the entire central California coast (Fink, 1982). Many ships had put in over the years, including a number attached to major expeditions with ship's naturalists and instructions to explore the new and exotic places visited and collect natural history specimens. Botanists especially knew the Monterey Peninsula for its highly endemic flora. But exactly how, when, and by whom were the masses of butterflies discovered? And how did the locals get a perspective on the uniqueness of the phenomenon? At this time, I believe the answer is we don't know. It remains a mystery.

Ackery and Vane-Wright (1984) and Vane-Wright (this volume) report the first records in the scientific literature as Thaxter (1880) and Bush (1881), and the latter is quoted here in its entirety:

> Trees Attractive to Butterflies—Mrs. A. E. Bush, of San Jose, Cal., writes: "I have been to Monterey, and was fortunate enough to see the 'butterfly tree', or trees, as there are three of them. These trees are the Monterey Pine (*Pinus insignis* Dougl.) [*Pinus radiata* today] and are probably over one and a half feet in diameter, and completely covered with live butterflies. To say that there were as many butterflies as leaves upon the trees would not be a very great exaggeration. I saw them in the morning when it was cool and they could not fly very well, and picked up a dozen from the grass in a few seconds. A lady resident informed me that for the twelve years she has lived there the appearance has been the same."

The identity of the "lady resident" is unknown, but her statement dates her earliest observation of the butterflies there back to about 1869. An informative booklet on the "Butterfly Trees" of Pacific Grove (Shepardson, 1914) states that the butterflies were first noticed by locals about 50 years earlier. This suggests the butterflies were first noticed in about 1864, although the ambiguity of the statement may not allow a firm distinction from the 1869 date (Lane, 1984; Ackery and Vane-Wright, 1984; Vane-Wright, this volume). It is clear from Shepardson's book, however, that she is not the "lady resident" noted in the 1881 report. Shepardson also states that there was no notice taken of of the butterflies in Monterey's earlier Spanish/Mexican era.

PACIFIC GROVE IN LATER YEARS

It appears that little note was taken of the butterflies at Pacific Grove until the early 1930s, when, during the Great Depression, the local Chamber of Commerce hit upon the idea of making their town "Butterfly Town, U.S.A." It was at this time that the City Council enacted the famous butterfly protection ordinance, whereby anyone found molesting the butterflies could be subject to a $500 fine. In actuality, I am not aware of anyone ever having been fined for harming the butterflies, and history has shown that the town's ordinance quite missed the point. Over the years, much butterfly overwintering habitat has been lost to urban development and thinning of the forest (Shepardson, 1914; Lane, 1984). Fink (1982) states of Pacific Grove: "Much of the pine forest has been cut, resulting in a colder climate." Nothing resembling the spectacle described in 1881 survives today in Pacific Grove (or anywhere else in California).

EARLY RECORDS OF OVERWINTERING MONARCH MASSES IN CALIFORNIA

Although I have not attempted anything like an exhaustive search for early written accounts of overwintering monarchs in the state (as might be found in local newspapers, magazines, etc.), it appears that little was written about them in the last century and early in this century. Thus, the two following early accounts appear notable.

During a visit to Santa Cruz on 30 December 1888, Theophilous d'Estrella observed that monarchs "fly in the hundreds and rest in the thousands" on cedar [Monterey cypress] and gum [*Eucalyptus*] trees at Point Santa Cruz (Anonymous, 1889). (A few cypress and *Eucalyptus* trees remain at this site and monarchs still gather there in small numbers, persisting through the winter in some years. The land, known today as Lighthouse Field, is an undeveloped state park and is an excellent candidate for selective planting of additional trees to restore a robust overwintering colony to this public site.)

Four years after Shepardson's excellent booklet on Pacific Grove's butterfly trees appeared (Shepardson, 1914), a similar but less informative booklet also appeared about the butterfly trees in Bolinas (Marin County) (Barber, 1918).

THE EARLIEST RECORDS OF MONARCHS IN CALIFORNIA

A few years before monarchs were first reported massing at Monterey (about 1864), we have evidence of the species occurring in the San Francisco area. While discussing "*Danais*" worldwide, Behr

Figure 2. Plate 7 from volume 3 of the report of von Kotzebue's expedition (Kotzebue, 1821) illustrating a monarch butterfly taken in the vicinity of San Francisco in October 1816 (photo courtesy J.P. Donahue).

(1862) reported: "There is a great gathering of this *Danais* on the outer Telegraph Hill, and a considerable number is constantly seen sea-faring themselves out on the Pacific." Behr read this paper at a meeting of the California Academy of Sciences on 17 March (perhaps late for overwintering aggregations). Whether he refers to a winter aggregation or perhaps only large numbers nectaring, either in late autumn or early spring, is unclear. However, it is interesting to note that some years later (Behr, 1885), in an ongoing series on California butterflies, he states that the monarch is frequent and local, has more than one brood, feeds on Asclepiadaceae, and (very significantly) hibernates as an egg. It would appear that he was not aware (even in 1885) that this species overwinters in large numbers as adults. Thus, I take his 1862 report as evidence of the species in the state, but not evidence for overwintering masses in San Francisco, several years before such masses were noticed on the Monterey Peninsula.

The earliest record of a monarch butterfly occurring in California appears to be a specimen taken around San Francisco in 1816 (J. Donahue, pers. comm.). A monarch (identified as *Idea plexippus*) was illustrated in the original scientific account (Kotzebue, 1821) of the von Kotzebue expedition (Fig. 2). This Russian expedition made only one landing in California, stopping at San Francisco for most of October 1816 (Mahr, 1932). Monarchs would be expected to begin appearing on the coast by late September, making the capture entirely reasonable.

At this point, it is worth considering just how far back in time it might be reasonable to expect to find any written evidence for monarchs in California. Beginning with Cabrillo in 1542–1543, a few explorers sailed along California's coast but hardly came onto land at all. Vizcaíno, discoverer of Monterey Bay, came onto the Monterey Peninsula in the winter of 1602–1603 and his accounts include notes of the native people and the natural setting, but nothing is mentioned pertaining to monarchs (Fink, 1982). The first land party to explore the coastal regions of the state immediately began establishing missions, the first at San Diego in 1769 followed by a military garrison at Monterey and a mission at nearby Carmel the next year. Foreign vessels were not welcome in the ports of early Spanish California and only began stopping at Monterey in the late 1780s. This was only 30 years before von Kotzebue sailed into San Francisco Bay. It seems unlikely that any future finds will significantly change the date of our earliest record of monarchs in the state, unless some dramatic evidence comes from studies of the native people of California.

It is disappointing that so little is known of the native people who lived along much of California's coast before they were brought into the Spanish missions (Heizer, 1978). Each tribe knew their (generally small) home territories intimately. They were attentive and thorough observers of their natural setting. They had to be. Could they have possibly failed to notice masses of colorful butterflies assembling at the same locations every winter? I think not. It is my suspicion that even if the padres had cared to inquire about local natural wonders, or what the natives believed about their natural world (which they did not; see Geiger, 1976), the native people might well not have shared such knowledge with them lightly. At any rate, I have been unable to find any evidence that any of the several native tribes who lived along the California coast (such as the Pomo, Coast Miwok, Costonoan or Ohlone, or Chumash peoples) knew of masses of overwintering butterflies, from either their (few) preserved legends, ethnographic studies, or inquiries of their descendants.

ACKNOWLEDGMENTS

I gratefully acknowledge my indebtedness and offer my thanks to Sandy Lydon at Cabrillo College, for the reference to early monarch overwintering at Monterey from an early Monterey newspaper; Frank Perry, Santa Cruz City Museum, for the reference to early monarch overwintering at Lighthouse Point, Santa Cruz; Derham Giuliani, Big Pine, California, for his observations on monarchs in Saline Valley and field assistance in visiting them; and most especially to Julian Donahue, Natural History Museum of Los Angeles County, for his generous spirit of cooperation and gracious sharing of the von Kotzebue monarch record and providing the illustration.

LITERATURE CITED

Ackery, P.R., and R.I. Vane-Wright. 1984. *Milkweed butterflies: Their cladistics and biology.* London: British Museum (Natural History) and Ithaca, NY: Cornell University Press, 425 pp.

Anonymous. 1874. *Monterey Weekly Herald,* Monterey, May 30, 1874.

Anonymous. 1889. *The Weekly News* (Newspaper of the California School for the Deaf), Berkeley, January 5, 1881.

Barber, M.D. 1918. *Winter butterflies in Bolinas.* San Francisco: Paul Elder and Co., 18 pp.

Barbour, M.G., and J. Major. 1977. *Terrestrial vegetation of California.* New York: John Wiley and Sons, 1002 pp.

Behr, H. 1862. Butterflies of California. *Proceedings of the California Academy of Natural Sciences* 2:167–169.

Behr, H.H. 1885. *Biological synopsis of California Lepidoptera.* Bulletin #3, California Academy of Sciences, pp. 63–65.

Bolton, H.E. (ed.). 1966. *Historical memoirs of New California,* Vol. 2, by Fray Francisco Palou, O.F.M. New York: Russel and Russel, 390 pp.

Brower, L.P. 1985. New perspectives on the migration biology of the monarch butterfly, *Danaus plexoppus* L. In *Migrations: Mechanisms and adaptive significance,* ed. M.A. Rankin, 748–785. Austin: University of Texas, *Contributions in Marine Science,* Supplement, 27.

Bush, A.E. 1881. Trees attractive to butterflies. *The American Naturalist* 15:572.

Calvert, W.H., and L.P. Brower. 1982. The importance of forest cover for the survival of overwintering monarch butterflies (*Danaus plexippus,* Danaidae). *Journal of the Lepidopterists' Society* 35:216–225.

Calvert, W.H., and L.P. Brower. 1986. The location of monarch butterfly (*Danaus plexippus* L.) overwintering colonies in Mexico in relation to topography and climate. *Journal of the Lepidopterists' Society* 40:164–187.

Calvert, W.H., W. Zuchowski, and L.P. Brower. 1984. Monarch butterfly conservation: Interactions of cold weather, forest thinning and storms on the survival of overwintering monarch butterflies (*Danaus plexippus* L.) in Mexico. *Atala* 9:2–6.

Comstock, J.A. 1927. *Butterflies of California.* Los Angeles: J.A. Comstock, 334 pp.

Davenport, K. 1983. Geographic distribution and checklist of the butterflies of Kern County, California. *Journal of the Lepidopterists' Society* 37:46–69.

Fink, A. 1982. *Monterey County: The dramatic story of its past.* Santa Cruz, CA: Western Tanager Press, 254 pp.

Geiger, M. 1976. *As the padres saw them: California Indian life and customs as reported by the Franciscan missionaries 1813–1815.* Santa Barbara, CA: Santa Barbara Mission Archive Library, 170 pp.

Griffin, J.R., and W.B. Critchfield. 1972. *The distribution of forest trees in California.* USDA Forest Service Research Paper PSW-82/1972. Berkeley, CA: Pacific Southwest Forest and Range Experiment Station, 118 pp.

Heizer, R.F. 1978. California. In *Handbook of North American Indians,* Vol. 8, ed. W.C. Sturtevant. Washington, D.C.: Smithsonian Institution, 800 pp.

Kotzebue, O. von. 1821. *Entdeckungsreise in die Südsee und nach der Beringsstrasse zur Entdeckung einer nordöstlichen Durchfahrt, unternommen in den Jahren 1815–18 auf Kosten Sr. Erlaucht des Herrn Reichskanzlers Grafen Romanzoff auf dem Schiffe Rurik,* 3 vols. Weimar.

Lane, J. 1984. The status of monarch butterfly overwintering sites in Alta California. *Atala* 9:17–20.

Lane, J. 1985. California's monarch butterfly trees. *Pacific Discovery* 38:13–15.

Mahr, A.C. 1932. The visit of the "Rurik" to San Francisco in 1816. *Stanford University Publications, University Series: History, Economics, and Political Science* 2:125–187.

Mayfield, D.W. 1980. Ecology of a discovered land. *Pacific Discovery* 33:12–20.

Munz, P.A., and D.D. Keck. 1959. *A California flora.* Berkeley and Los Angeles: University of California Press, 1681 pp.

Nagano, C.D., and J. Lane. 1985. *A survey of the location of monarch butterfly (Danaus plexippus (L.)) overwintering roosts in the state of California, U.S.A.: First year 1984/1985.* Portland, OR: The Monarch Project, 30 pp. (+ 58 tables and 19 maps).

Nagano, C.D., and W. Sakai. 1988. Making the world safe for monarchs. *Outdoor California* 49:5–9.

Sakai, W.H., J. Lane, A.V. Evans, J. Schrumpf, and M. Monroe. 1989. The wintering colonies of the monarch butterfly [*Danaus plexippus* (L.): Nymphalidae: Lepidoptera] in the state of California, U.S.A. *California Department of Fish and Game Contract Report FG7551.* Sacramento, California, 38 pp.

Shepardson, L. 1914. *The butterfly trees.* Monterey, CA: Herald Printers, 32 pp.

Sudworth, G.B. 1967. *Forest trees of the Pacific slope.* New York: Dover Publications, 455 pp.

Thaxter, R. 1880. Swarming of *archippus. Canadian Entomologist* 12:38–39.

Tuskes, P.M., and L.P. Brower. 1978. Overwintering ecology of the monarch butterfly, *Danaus plexippus* L., in California. *Ecological Entomology* 3:141–153.

Urquhart, F.A. 1960. *The monarch butterfly.* Toronto: University of Toronto Press, 361 pp.

Urquhart, F.A. 1987. *The monarch butterfly: International traveler.* Toronto: University of Toronto Press, 232 pp.

Urquhart, F.A., P. Beard, and R. Brownlee. 1965. A population study of a hibernal roosting colony of the monarch butterfly (*D. plexippus*) in northern California. *Journal of Research on the Lepidoptera* 4:221–226.

Williams, C.B., G.F. Cockbill, M.E. Gibbs, and J.A. Downes. 1942. Studies in the migration of lepidoptera. *Transactions of the Royal Entomological Society of London* 92:155–165.

Wright, W.G. 1906. *The butterflies of the west coast of the United States,* 2nd ed. San Bernadino, CA: W.G. Wright, 257 pp.

OVERWINTERING OF THREE AUSTRALIAN DANAINES: *TIRUMALA HAMATA HAMATA*, *EUPLOEA TULLIOLUS TULLIOLUS*, AND *E. CORE CORINNA*

Elly Scheermeyer

ABSTRACT. The behavior and reproductive state of three danaine species, *Euploea core, Euploea tulliolus,* and *Tirumala hamata,* which cluster together along the east coast of Australia, were studied at two overwintering sites. Using a capture–mark–recapture program in 1982, data on condition of the butterflies, sex ratio, longevity, and movement of the adults were obtained. The condition of all butterflies deteriorated slowly over autumn and winter and rapidly in spring. Fresh individuals of *T. hamata* and *E. tulliolus* were absent in spring, but not of *E. core.* Some *E. core* continued to breed locally throughout the winter. Long-term recaptures of several months duration were made at the overwintering sites of all three species. At least 10% of marked *E. core* and *E. tulliolus* were recaptured, but only 3% of *T. hamata,* which appeared much more mobile than the other two species. *Euploea core* and *T. hamata* had predominantly male-biased sex ratios at the overwintering sites, but not *E. tulliolus.* The sex ratio of *E. tulliolus* fluctuated from male-biased in autumn (April) to female-biased in spring (September).

The reproductive state of the females was assessed by dissections. Most females were reproductively dormant in autumn and winter. In spring most *E. core* had eggs, but *T. hamata* and *E. tulliolus* had at most only partially developed ovaries. Throughout the overwintering period most females of *E. tulliolus* were mated. In *E. core* and *T. hamata* most females dissected in winter were virgin. Immatures collected from natural and introduced host plants substantiated the results obtained by dissections.

There were important differences in the behavior and reproductive state of the species at the overwintering sites, and these are discussed in the context of environmental variables and host-plant distribution and availability. Similarities and differences in the behavior of these species and *Danaus plexippus* when facing an adverse period for breeding are pointed out, so as to aid in designing future studies on the overwintering ecology of danaines.

INTRODUCTION

The adoption by adult *Danaus plexippus plexippus* (L.) of a migration and overwintering strategy to survive seasonal adversity has been well documented in the past two decades (e.g., Urquhart, 1960, 1976; Brower, 1985). Other danaines may behave in similar ways during adverse periods for breeding. In general, when adult butterflies delay reproduction, the maintenance of condition and vigor are necessary to ensure survival. Each species must enhance its chances of surviving unfavorable climate, predation, and metabolic depletion. How each species copes with these factors may depend on the nature of the adverse period (cold, hot, dry, or wet) and on when and where its host plants will again be suitable for oviposition and larval development.

Several species of Danainae, including *Tirumala hamata hamata* (W.S. Macleay), several species of *Euploea,* and occasionally *Danaus affinis affinis* (Fab.), form clusters during the winter, or dry season, along the north and east coast of tropical Australia (Bailey and Richards, 1975; Monteith, 1982). These authors suggest that reproduction is suspended during that period, but data on the reproductive state of these species and their general behavior are lacking. Recently, four of these species that cluster along the tropical east coast of Australia, *T. h. hamata, Euploea tulliolus tulliolus* (Fab.), *Euploea core corinna* (W.S. Macleay), and *Euploea sylvester sylvester* (Fab.) have been studied at various overwintering sites near Townsville and inland (Scheermeyer, 1987). One of these, *E. core,* has also been studied in subtropical Brisbane (Kitching and Zalucki, 1981; Scheermeyer, 1987). Here I will focus on the first three species, which are common at the Townsville sites, and discuss their reproductive state and general behavior while drawing heavily upon results obtained in the study of Scheermeyer (1987). *Euploea sylvester* was not common at any of the study sites and few data could be collected.

Various general trends are summarized here using data obtained between March and November, the period in which the butterflies usually cluster and referred to as the overwintering period. Similar trends were seen in other years (Scheermeyer, 1987).

MATERIALS, METHODS, AND STUDY SITES

In 1980 (a pilot study) and 1982, the behavior of *T. hamata, E. core,* and *E. tulliolus* was studied in detail at two overwintering sites in the Townsville area: Pallarenda on the mainland (19°11′S, 146°46′E) and Horseshoe Bay on nearby Magnetic Island (19°07′S, 146°52′E). The study sites cover approximately 3 km^2 each and consist of open woodland intersected by some densely forested gullies, creek beds, and swampy areas in which the butterflies clustered.

The host plant of *E. tulliolus* larvae, *Malaisia scandens* (Lour.) Planch. (Moraceae), is a strangling woody vine found along the edge of coastal rainforest and moist gullies and occurs at both sites. *Euploea core* uses a wide range of host plants, including many native species and introduced ornamentals. *Cryptostegia grandiflora* R. Br. (Asclepiadaceae) appears to be the main host plant in the Townsville area and is especially abundant in Pallarenda. The primary host plant of *T. hamata* in Australia is *Secamone elliptica* R. Br. (Asclepiadaceae), a corky vine found in dry and littoral rainforest. It has not been recorded in the Townsville area. A secondary host plant is *Cynanchum carnosum* R. Br. (Asclepiadaceae), a slender vine that grows in marshes. It occurs in the vicinity of Townsville but is rarely used by *T. hamata* (pers. obs.). (For host-plant records see Common and Waterhouse, 1981; Scheermeyer and Zalucki, 1985, and references therein). Most host plants have no fresh foliage in winter, except for *C. carnosum,* which remains in a reasonable condition until mid-winter.

The reproductive status of the butterflies was examined on a fortnightly basis by dissecting a sample of females of each species from both study sites. I recorded whether the ovaries had no development, partial development (oocytes filled with yolk for less than half their volume), or full development (oocytes filled with yolk for more than half their volume; usually some chorionated eggs were present). Furthermore, I recorded the number and condition (fresh, intermediate, or depleted) of spermatophores in the bursa copulatrix. Recently mated females were individuals that contained a fresh spermatophore. The condition of the butterflies was assessed as fresh, slightly worn, or worn, depending on fading of wing color and loss of scales.

To obtain data on age structure (as inferred from wing condition) and mobility of the overwintering butterflies, an extensive mark and recapture program was carried out for 6 days each fortnight at a large number of capturing localities within the two main study areas. Butterflies were captured with a net while in flight. Rarely, and only on cold days in mid-winter, could butterflies be stalked close enough to capture them just as they were taking flight. Each captured butterfly was individually marked with a paper tag on the left forewing and date and site of capture or recapture were recorded. Butterflies were assessed for sex and condition at capture and reassessed for condition at subsequent captures. To obtain any recaptures at all, previously marked butterflies, when discovered, were captured selectively. This allowed results on longevity and movement of the butterflies to be obtained but prevented population size estimates.

RESULTS AND DISCUSSION

DISTRIBUTION AND BEHAVIOR IN SUMMER AND AUTUMN

The distribution of *T. hamata, E. core,* and *E. tulliolus* is mainly coastal from tropical to warm temperature regions (Common and Waterhouse, 1981; Scheermeyer, 1985). *Euploea core* is abundant throughout the year in Townsville, but *E. tulliolus* is rare in summer. Both species breed locally. *Tirumala hamata* is virtually absent in mid-summer (late December to mid-February) and, except for an odd individual, does not breed in the Townsville area (Scheermeyer, 1987).

In March, small numbers of butterflies start to cluster at the numerous overwintering sites in the Townsville area and by May several hundred individuals of each species were present at the overwintering sites I studied. In mid-winter several thousand individuals of each species cluster at Magnetic Island, but at Pallarenda the number fluctuates around several hundred (Scheermeyer, 1987). Butterflies cluster loosely over areas from 20 m up to 1 km wide. Individuals are generally spaced from each other by a typical wing-flapping behavior on settling. Individuals of different species may settle below or opposite each other.

REPRODUCTIVE STATUS OF FEMALES DURING THE OVERWINTERING PERIOD

In all three species, most females (65–98%) at the overwintering sites had no eggs in autumn and winter (Fig. 1). Furthermore, except for a small proportion of *E. tulliolus* in autumn, females of *T. hamata* and *E. tulliolus* rarely had eggs throughout the study period. However, a proportion of females of *E. core* had eggs in autumn and winter, and the majority in spring. Many females were mated (25–100%; Fig. 2), especially when they had eggs.

Immatures of *T. hamata* and *E. tulliolus* were observed only in autumn (a few *T. hamata* in April on *C. carnosum,* and some *E. tulliolus* in March and April on *M. scandens*). Immatures of *E. core* were found throughout the study period but were uncommon in winter (Scheermeyer, 1987).

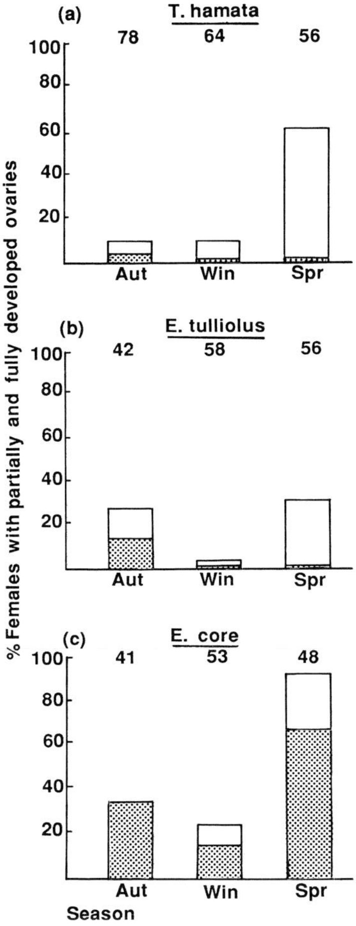

Figure 1. The percentage of females with partially and fully developed ovaries in samples of (**a**) *T. hamata,* (**b**) *E. tulliolus,* and (**c**) *E. core* in autumn, winter, and spring 1982. Samples of both study sites have been combined. Numbers indicate sample sizes; dark shading, fully developed ovaries; no shading, partially developed ovaries. The remainder had no ovarian development at all.

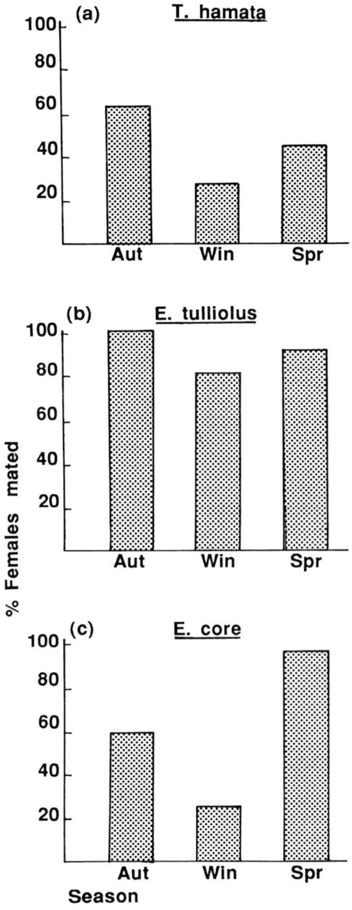

Figure 2. The percentage of females mated in the samples of (**a**) *T. hamata,* (**b**) *E. tulliolus,* and (**c**) *E. core* in autumn, winter, and spring 1982. Sample sizes as in Figure 1.

Reproduction of the three species in Townsville is limited in autumn and winter, and in spring remains suppressed in *T. hamata* and *E. tulliolus* only. This is in accordance with the absence of a primary host plant of *T. hamata* in the Townsville area and the usual lack of fresh foliage on local *M. scandens,* the host plant of *E. tulliolus,* until the monsoonal rainstorms in summer.

CONDITION AND SURVIVAL OF THE OVERWINTERING BUTTERFLIES

Figure 3 presents the condition of the overwintering butterflies. The following trends were the same for all three species: (1) the proportion of fresh individuals decreased during autumn; (2) the proportion of slightly worn individuals changed little during autumn and early winter, but decreased at the end of winter and rapidly in early spring; and (3) the proportion of worn individuals increased during autumn, winter, and early spring.

These trends suggest that recruitment of fresh individuals is limited in autumn and winter in all three species. Furthermore, the number of fresh individuals of *T. hamata* and *E. tulliolus* declined over winter and reached zero in spring. This supports the suggestion that in winter and spring reproduction of *T. hamata* and *E. tulliolus* does not occur in Townsville. The fresh individuals in the population of *E. core* in winter and spring supports the observation that some *E. core* continue to breed in winter and many breed in spring.

The increase in slightly worn individuals of *T. hamata* and *E. tulliolus* at the end of spring may indicate that there is an influx of slightly worn individuals, but it may also indicate that worn individuals are disappearing (by death or dispersal) at a faster rate than slightly worn individuals. This is

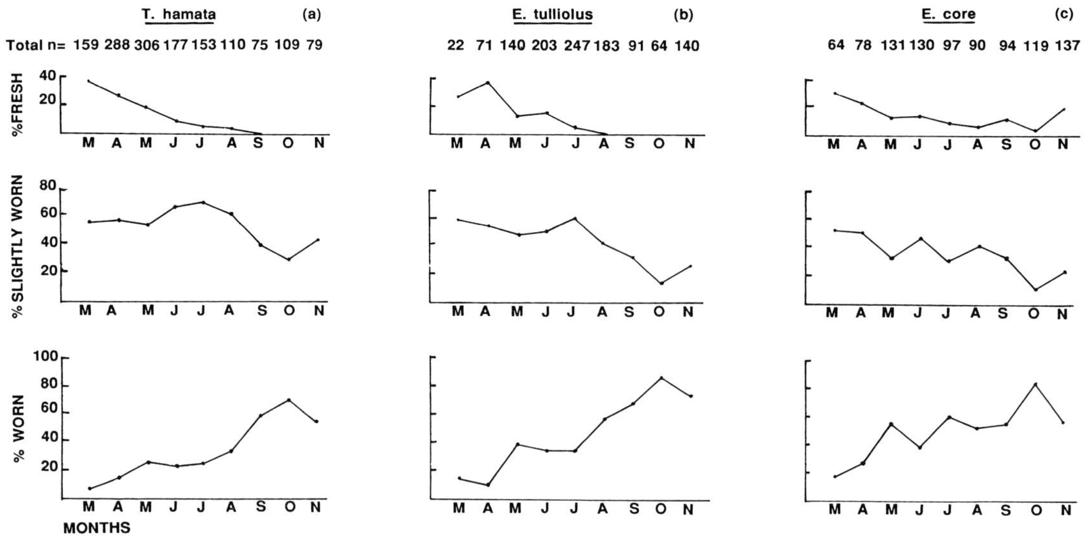

Figure 3. Condition of (**a**) *T. hamata,* (**b**) *E. tulliolus,* and (**c**) *E. core* from March to November 1982. Fortnightly samples were combined for each month as no significant differences were present. Graphs present the proportion of individuals in fresh, slightly worn, and worn condition. Numbers indicate sample sizes.

consistent with the observed rapid population decline in late spring.

The condition of the butterflies during the overwintering period suggests that a large proportion survives until spring, because fresh individuals are almost absent in winter and early spring. Overwintering butterflies also deteriorate slowly in winter, but rapidly in spring, so that most butterflies were worn by October.

Survival was further assessed from the results of recaptures (Table 1). The number of individuals recaptured of each species was significantly different ($\chi^2 = 123.4$, df $= 2$, $p < 0.001$). Few recaptures were made of *T. hamata* (only 3%), although 1,753 butterflies were marked, which is almost twice the number of the other species. Fourteen percent of the marked individuals of *E. tulliolus* and 10% of *E. core* were recaptured at least once.

In all species some long-term recaptures were made (Table 1). Two recaptures of individuals of *T. hamata* and *E. tulliolus* span almost 6 months between marking and recapture. One was a fresh male of *T. hamata* marked in April and recaptured in slightly worn condition in September; the other was a slightly worn female of *E. tulliolus* marked in May and recaptured in worn condition in November. Other recaptures included some after 3 and 4 months at the overwintering sites (Table 1).

The long-term recaptures support the suggestion that individuals of all three species are capable of surviving until well into spring. The low percentage of recaptures, especially of *T. hamata,* suggests either high mortality or considerable movement of the butterflies during the overwintering period. There was very little direct evidence of butterfly mortality. However, a small flock of six migratory white-breasted woodswallows, *Artamus leucorhynchus* (L.), was observed feeding on *T. hamata* at the overwintering site for several days before the birds left (pers. obs.).

MOVEMENT DURING THE OVERWINTERING PERIOD

Winter temperatures in the Townsville area are high (mean daily maximum is 25°C; Australia Bureau of Meteorology, 1975), and the butterflies are active to a certain extent. Although most of the day is spent roosting, some movement in and near the cluster site does occur throughout the day. When disturbed all roosting butterflies fly up rapidly, either dispersing in the surrounding areas or taking several minutes, or more, before they settle down again.

The minimal distance flown by butterflies recaptured after 14 days or more is summarized in Table 2. Although the majority of butterflies were recaptured at the site of marking or last release, many were also recaptured somewhere else. Since all regular sampling was confined to the two study areas, Pallarenda on the mainland and Horseshoe Bay on Magnetic Island, recaptures made in other bays on the island were either returned by the public or obtained during my casual visits to those locations. Other marked butterflies were seen at different locations but could not be caught. Movement occurred in all three species and is likely to be much more extensive than indicated by my sampling.

Since each study area comprised several sampling locations, movement could generally be distin-

Table 1. Recapture results of *T. hamata, E. tulliolus,* and *E. core.* All butterflies were marked in the Townsville area between March and October of 1980 and 1982.

Species	Number marked	Number recaptured	Total number of recaptures made in weeks since marking						
			0–3	4–7	8–11	12–15	16–19	20–23	24–27
T. hamata	1,753	52	41	9	2	1	0	1	0
E. tulliolus	1,079	154	118	32	10	4	2	0	1
E. core	1,005	98	82	21	9	3	4	0	0

guished as being either to another cluster site or to a source of food or chemical attractant, such as nectar, mud puddles, and plants containing pyrrolizidine alkaloids (PAs). PAs are used in the synthesis of the male courtship pheromones (Edgar et al., 1973; Boppré et al., 1978; Boppré, 1986). Feeding on PA plants was observed only in autumn and spring. This was in accordance with the mating activity seen in these months.

Movement during the winter seemed to be associated with (1) obtaining energy for metabolic maintenance and (2) seeking another shelter. The latter may occur when cluster sites lose their sheltering qualities (e.g., when the trees start to defoliate; the dominant wind direction changes).

SEX RATIO OF THE CLUSTERING BUTTERFLIES

Overall, the sex ratio of *E. tulliolus* did not differ from a 1:1 ratio, but the sex ratios of *T. hamata* and *E. core* were skewed toward males (*T. hamata,* 30% females, n = 1,456, $\chi^2_C = 230.2$, df = 1, $p < 0.001$; *E. tulliolus,* 48% females, n = 1,161, $\chi^2_C = 2.5$, df = 1, $p > 0.05$; *E. core,* 21% females, n = 940, $\chi^2_C = 304.5$, df = 1, $p < 0.001$).

From laboratory rearing of field-collected immatures of *E. core* between March and November 1982 there was no significant difference in the sex ratio at emergence (44% females, n = 150, $\chi^2_C = 1.9$, df = 1, $p > 0.05$). The strongly skewed sex ratio in the field may have been caused by differing survival or differing behavior of males and females. Survival of field-caught overwintering *E. core* in a flight cage was not significantly different between males and females (Scheermeyer, 1987), hence differing behavior of the sexes may be responsible for the skewed sex ratios.

Figure 4 shows the sex ratio in each month of the study as there was little fluctuation on a fortnightly basis. The sex ratio of *T. hamata* and *E. core* were male-biased throughout the study period, expect for *T. hamata* in mid-winter (July, $\chi^2_C = 3.8$, df = 1, $p > 0.05$). The sex ratio of *E. tulliolus* fluctuated and was male-biased in April ($\chi^2_C = 8.1$, df = 1, $p < 0.005$) and female-biased in September ($\chi^2_C = 8.6$, df = 1, $p < 0.005$).

The sex ratio of *E. core* in autumn varied considerably from 13% females in March to 35% in April (March–May: $\chi^2_C = 11.4$, df = 2, $p < 0.005$). In April, several *Melaleuca* trees were flowering near the overwintering sites and most butterflies were captured while feeding or roosting near these trees, suggesting that both sexes may participate equally in this prewinter feeding surge. The overall male bias in the sex ratio of *E. core* at the overwintering sites may be caused by the absence of reproductive females, which are searching for host plants. As noted above, a small proportion of *E. core* females had eggs even in winter and eggs and larvae were found throughout the overwintering period.

The male-biased sex ratio of *T. hamata* in autumn could be explained similarly and is not necessarily the result of a difference in emergence or differential survival. Since *T. hamata* appears to migrate to the Townsville area to overwinter, ovipositing females may remain longer in the breeding

Table 2. Change in locality and distance moved (in km) of individuals of *T. hamata, E. tulliolus,* and *E. core* since their last release at either Horseshoe Bay (HB) on Magnetic Island or at Pallarenda on the mainland. Values shown are the total number of recaptures made at least 2 wk after release. Data are combined results of recaptures made between March and November of 1980 and 1982. D = distance between localities.

Species	Number of butterflies recaptured		
	Same location (D = 0)	New location within study area (1 km < D < 3 km)	New location outside study area (2 km < D < 20 km)
T. hamata	12	5	2 from HB to Pallarenda 2 from HB to other bay
E. tulliolus	67	20	1 from HB to Pallarenda 2 from HB to other bay
E. core	53	12	2 from HB to Pallarenda 3 from HB to other bay

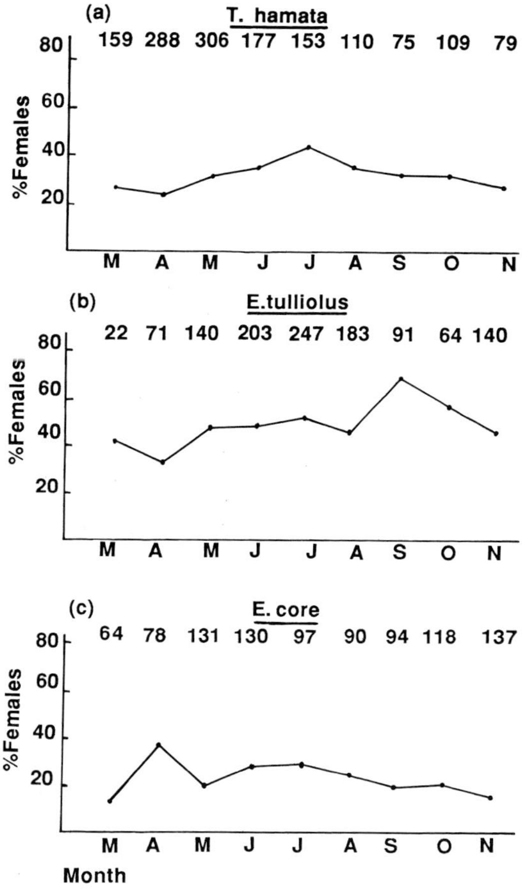

Figure 4. The percentage of females of (**a**) *T. hamata,* (**b**) *E. tulliolus,* and (**c**) *E. core* from March to November 1982. Fortnightly samples were combined for each month as no significant differences were present. Numbers indicate sample sizes.

areas when migration begins in autumn.

It was noted that the biased sex ratios of *T. hamata* and *E. core* were caused in particular by the samples taken from Pallarenda (Fig. 5). In Pallarenda, PA sources were frequently available, and in addition, abundant host plants of *E. core,* but not of *T. hamata,* were present. The Pallarenda site appeared to function as a courting site, as well as an overwintering site. It seemed that in addition to overwintering males, sexually active males of *E. core* and *T. hamata* aggregated at this site throughout the overwintering period (Scheermeyer, 1987). Courtship pursuits by these sexually active males may have caused overwintering females to leave this site and settle in other cluster sites where such males occurred less frequently. Male–female interactions leading to emigration of females has also been noted in other butterflies (Shapiro, 1970; Zalucki and Kitching, 1984).

The opposite tendency was noted in *E. tulliolus* (Fig. 5b). Throughout the overwintering period and spring females aggregated in Pallarenda, where the host plant *M. scandens* was abundant.

DISPERSAL

At the end of winter, in August and September, the total number of individuals of each species decreased dramatically to several hundred in Horseshoe Bay. During spring, September and October, the number fluctuated but generally decreased to very few in November at both study sites. The abundance of *E. core* in Pallarenda, however, showed little variance in spring from that recorded in winter (for actual counts, captures, and visual estimates per unit of time, see Scheermeyer, 1987). It appeared that large numbers of *T. hamata* and *E. tulliolus* left the overwintering sites on several occasions in spring, although migrations in spring have rarely been observed and only of *T. hamata* (Smithers, 1983; Scheermeyer, 1987).

REPRODUCTIVE STATE OF THE BUTTERFLIES IN SPRING

As noted above, dispersal was not related to the presence of eggs in *T. hamata* and *E. tulliolus,* but in spring a considerable number of females had partially developed ovaries (Fig. 1). Increased reproductive activity during spring was also indicated by an increase in number of mated females (Fig. 2), yet most females of *T. hamata* were still virgin. Mass mating was not recorded for any of the species, although on some days several mating couples of one or more danaine species (including *Danaus affinis*) were observed at the study sites.

In spring, the male-biased sex ratio of *T. hamata* (Fig. 4), the decreasing number present at the study sites, and the absence of breeding individuals in the Townsville area suggest that initially more females than males leave the overwintering sites. The small proportion of females mated or with eggs in spring in Townsville suggests that females of *T. hamata* leave before they have mated, or soon after they mate, while their ovaries are only partially developed (Figs. 1 and 2). Males may accompany them, but some may linger at the overwintering sites until receptive females become scarce.

Data from females captured near Brisbane during a southward spring migration of *T. hamata* (recorded along several hundred kilometers of the subtropical east coast of Australia on 17 October 1983, followed by a smaller migration on 27 October) support this hypothesis; 31% of captured females ($n = 16$) that I dissected had recently mated and the rest were virgin. Furthermore, 81% of the migrating females had partially developed ovaries, and 19% had eggs, although a few females with eggs had not mated. The sex ratio of captured individuals on the first day of the migration was 67% females ($n = 27$), which contrasted with the overall male bias (70% males) at the northern overwintering sites ($\chi^2_C = 15.6$, df $= 1$, $p < 0.01$). Ten days later the sex ratio of the migrating butterflies was 75%

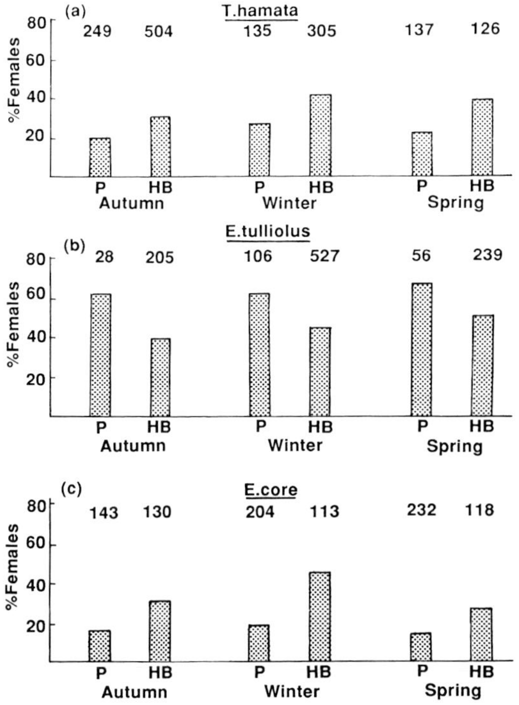

Figure 5. The percentage of females of (**a**) *T. hamata,* (**b**) *E. tulliolus,* and (**c**) *E. core* for each study area in autumn, winter, and spring 1982. The sex ratio of each species was significantly different between the sites in each season [Pallarenda (P) vs. Horseshoe Bay (HB) from autumn to spring: *T. hamata,* $\chi^2_C = 9.2, 8.9, 7.5$, df = 1, all $p < 0.005$; *E. tulliolus,* $\chi^2_C = 3.9, 9.5, 4.0$, df = 1, all $p < 0.05$; *E. core,* $\chi^2_C = 6.8, 22.8, 8.4$, df = 1, all $p < 0.01$]. Numbers indicate sample sizes.

males (n = 12). As expected, the condition of the migrating butterflies (8% fresh, 51% slightly worn, and 41% worn; n = 39) suggests that they had overwintered.

Euploea core breeds locally in Townsville and, in contrast to *T. hamata,* almost all females at the study sites had mated in spring. Examination of the condition of the spermatophores showed that 58% of the dissected females (n = 48) had mated recently. Mated females may leave the overwintering sites in search of host plants, of which many are locally available. They may return to the overwintering sites for shelter, to remate, and to oviposit when host plants are available. Females with eggs are therefore occasionally found at the overwintering sites in spring, but the small proportion and absence of recaptures of females in spring suggest that they only stay for a short period.

Euploea tulliolus appears to adopt a different strategy. First, the female-biased sex ratio in early spring, September and October, suggests that males disperse before females (sex ratio: September to November; $\chi^2_C = 10.5$, df = 2, $p < 0.01$) (Fig. 4). Second, although most females of *E. tulliolus* had already mated before spring, usually in autumn (Fig. 2), few females (only 20%) were recently mated in spring. This suggests that most males leave the overwintering sites in spring before they start mating with females in Townsville. Third, females tended to congregate at the site where PA sources and host plants were abundant (Fig. 5), but the host plant *M. scandens* lacked fresh foliage in spring and both sexes had dispersed before fresh foliage was present in summer. It is also not known whether or not females collect PA. Because *M. scandens* is the primary, and probably the sole, host plant of *E. tulliolus* in Australia, and because it has a limited and mainly clumped distribution along the east coast of Australia, it is plausible that males of *E. tulliolus* mate preferentially near patches of host plants suitable for oviposition, which were absent in Townsville. Accordingly, males may disperse in spring to seek these host patches. Females may accompany them or follow shortly after.

CONCLUSIONS

The behavior of the three butterfly species overwintering in Townsville differed in several aspects. Most of the differences could be explained in terms of host-plant presence and availability. A primary host plant of *T. hamata* was not present in the Townsville area and gravid females were rarely observed. *Euploea core,* as a local breeder on many different species of native and ornamental host plants, had more females with eggs at the overwintering sites than the other two species. Not only did some females breed throughout the winter but it was also the only species with fresh individuals in spring. The local host plant of *E. tulliolus,* although present, was in 1982 only suitable for oviposition and larval feeding in autumn, which coincided with the occurrence of a small number of reproducing females, while most females were reproductively dormant. In some aspects, however, *E. tulliolus* behaved quite differently from *T. hamata* and *E. core.* The sex ratio of *E. tulliolus* was overall equal but female-biased near host plants and PA sources. Most females were mated at any time of the year and males seemed to initiate dispersal in spring. Most of these differences may also be related to the particular distribution and availability of its single host plant. The data also suggest that females of *E. tulliolus* may "solicit" matings, but this needs further investigation.

The reproductive state of several Australian species of Danainae in winter is presented in Table 3. In all species reproductive dormancy seems to occur, but not in *D. plexippus* and *D. affinis* in the subtropics. In some species, although the majority of the population is reproductively dormant, a small number of individuals may continue to breed. *Danaus affinis* and *D. chrysippus* may have been reproductively dormant in the tropics during midwinter in 1982. Both species joined the clusters for a short time in June and July of that year and no

Table 3. Reproductive state of several species of Danainae, and their host-plant condition during winter in Australia. In some species, both reproductively dormant (RD) and reproductive (R) individuals occur; the smaller proportion is indicated by parentheses. No. recorded is the number of species recorded as host plants by Common and Waterhouse (1981); temp = temperate (Sydney); subt = subtropical (Brisbane); trop = tropical (Townsville); + = in good condition; − = wilted or absent.

Species	Region	Reproductive state	Host plants: Condition	Host plants: No. recorded	Reference*
D. plexippus	temp	RD/(R)	−/+	6	a
	subt	R	+	6	b
D. affinis	subt	R	+	1	c
	trop	R/(RD)	+/(−)	1	d
D. chrysippus	trop	R/(RD)	+/(−)	10	d
T. hamata	trop	RD	−	2	d
E. core	subt	RD/(R)	−/(+)	22	e,d
	trop	RD/(R)	−/(+)	22	d
E. tulliolus	trop	RD	−	1	d
E. sylvester	trop	RD	−	1	d

* a = James (1979), b = Zalucki and Kitching (1984), c = A.G. Orr (pers. comm.), d = Scheermeyer (1987), e = Kitching and Zalucki (1981).

matings were observed in that period. The few dissections showed no ovarian development and very few immatures were found in the field. Generally, where suitable host plants are present, breeding continues, to a certain extent, throughout the year.

The results of this study and a number of other studies (Smithers, 1973, 1977; Tuskes and Brower, 1978; Kitching and Zalucki, 1981; James, 1982, 1984; Brower, 1985) indicate that danaines may have some features in common, whereas they differ widely in others when facing adverse periods for breeding. These features include the following.

1. Reproductively dormant danaines are not necessarily virgin: Although most *T. hamata*, *E. core*, and *D. plexippus* are usually virgin, most *E. tulliolus* are in fact mated (see also Van Hook, this volume).

2. Not all individuals within a species are necessarily reproductively dormant during an adverse period for breeding (e.g. in *E. core* and *D. plexippus* breeding and nonbreeding individuals may occur in the same area).

3. Mating before dispersal from the overwintering sites seems common in *E. core* and *D. plexippus*, but seems much less frequent in *T. hamata* and *E. tulliolus* (see also Van Hook, this volume).

4. Longevity of danaines is considerable; overwintering individuals of *Tirumala* and *Euploea* may live 6–9 months in the field.

5. The sex ratio observed at an overwintering site may be strongly influenced by the presence of host plants, PA sources, and the absence of reproducing females.

6. Dispersal in spring may be initiated by either sex and this sex may vary among the species. In Australia, females seem to initiate dispersal in *T. hamata*, *E. core*, and *D. plexippus*, but males initiate dispersal in *E. tulliolus*.

7. Mobility of overwintering individuals varies among species and may be considerable in tropical locations.

Although many danaine species can enter reproductive dormancy during adverse environmental conditions, there appears to be considerable variability in reproductive behavior among species, and within species at different locations (Scheermeyer, 1987), which depends to a great extent on host-plant distribution and availability.

ACKNOWLEDGMENTS

My special thanks to Professor R.E. Jones for her stimulating support throughout this study and her criticism of the manuscript and to Dr. P.C. Bailey for his advice on the final draft. Financial support for this study was provided by a Griffith University Postgraduate Research Award.

LITERATURE CITED

Australia Bureau of Meteorology. 1975. *Climatic averages Australia*, metric ed. Canberra: A.G.P.S.

Bailey, W.J., and K.T. Richards. 1975. A report on the insect fauna of the Prince Regent River Reserve, north-west Kimberley, Western Australia. In *A biological survey of the Prince Regent River Reserve, north-west Kimberley, Western Australia in August, 1974*, eds. J.M. Miles and A.A. Burbidge, 101–112. *Western Australian Wildlife Research Bulletin No. 3.*

Boppré, M. 1986. Insects pharmacophagously utilizing defensive plant chemicals (pyrrolizidine alkaloids). *Naturwissenschaften* 73:17–26.

Boppré, M., R.L. Petty, D. Schneider, and J. Meinwald. 1978. Behaviorally mediated contacts between scent organs: Another prerequisite for pheromone production in *Danaus chrysippus* males (Lepidoptera). *Journal of Comparative Physiology* 126:97–103.

Brower, L.P. 1985. New perspectives on the migration biology of the monarch butterfly, *Danaus plexippus* L. In *Migration: Mechanisms and adaptive significance,* ed. M.A. Rankin, 748–785, Austin: University of Texas, *Contributions in Marine Science,* Supplement, 27.

Common, I.F.B., and D.F. Waterhouse. 1981. *Butterflies of Australia.* Sydney: Angus and Robertson, 682 pp.

Edgar, J.A., C.C.J. Culvenor, and G.S. Robinson. 1973. Hairpencil dihydropyrrolizines of Danainae from the New Hebrides. *Journal of the Australian entomological Society* 12:144–150.

James, D.G. 1979. Observations on two overwintering clusters of *Danaus plexippus* (L.) (Lepidoptera: Nymphalidae) in the Sydney area during 1978. *Australian entomological Magazine* 5:81–85.

James, D.G. 1982. A transient non-breeding population of *Danaus plexippus* (L.) (Lepidoptera: Nymphalidae) near Picton, New South Wales. *Australian entomological Magazine* 8:96–99.

James, D.G. 1984. Population and general biology of non-reproductive colonies of the monarch butterfly, *Danaus plexippus* (L.) (Lepidoptera: Danaidae) in New South Wales. *Australian Journal of Zoology* 32:663–670.

Kitching, R.L., and M.P. Zalucki. 1981. Observations on the ecology of *Euploea core corinna* (Nymphalidae) with special reference to an overwintering population. *Journal of the Lepidopterists' Society* 35: 106–119.

Monteith, G.B. 1982. Dry season aggregations of insects in Australian monsoon forests. *Memoirs of the Queensland Museum* 20:533–543.

Scheermeyer, E. 1985. Some factors affecting the distribution of *Euploea core corinna* (W. S. Macleay) (Lepidoptera: Danainae). *Australian Journal of Zoology* 33:339–348.

Scheermeyer, E. 1987. Seasonality or opportunism in reproduction of Australian danaine butterflies: *Euploea core, E. tulliolus* and *Tirumala hamata* (Lepidoptera). Unpublished Ph.D. thesis. Brisbane: Griffith University.

Scheermeyer, E., and M.P. Zalucki. 1985. Food plant records of *Euploea core corinna* (W. S. Macleay) with some notes on larval colouration. *Australian entomological Magazine* 11:87–90.

Shapiro, A.M. 1970. The role of sexual behavior in density-related dispersal of pierid butterflies. *The American Naturalist* 104:367–372.

Smithers, C.N. 1973. A note on length of adult life of some Australian butterflies. *Australian entomological Magazine* 1:62–66.

Smithers, C.N. 1977. Seasonal distribution and breeding status of *Danaus plexippus* (L.) (Lepidoptera: Nymphalidae) in Australia. *Journal of the Australian entomological Society* 16:175–184.

Smithers, C.N. 1983. Migration records in Australia. 3. Danainae and Acraeinae (Lepidoptera: Nymphalidae). *Australian entomological Magazine* 10:21–27.

Tuskes, P.M., and L.P. Brower. 1978. Overwintering ecology of the monarch butterfly, *Danaus plexippus* L., in California. *Ecological Entomology* 3:141–153.

Urquhart, F.A. 1960. *The monarch butterfly.* Toronto: University of Toronto Press, 361 pp.

Urquhart, F.A. 1976. Found at last: The monarch's winter home. *National Geographic* 150:161–173.

Zalucki, M.P., and R.L. Kitching. 1984. The dynamics of adult *Danaus plexippus* L. (Danaidae) within patches of its food plant, *Asclepias* spp. *Journal of the Lepidopterists' Society* 38:209–219.

9

CONSERVATION

Sign along the interpretive trail in the protected El Campanario monarch overwintering site near Angangueo, Michoacán, Mexico. Photograph by Julian P. Donahue.

CONSERVATION OF MONARCH BUTTERFLY MIGRATION IN NORTH AMERICA: AN ENDANGERED PHENOMENON

Stephen B. Malcolm

Passenger pigeons (*Ectopistes migratorius*) declined in numbers from some 2 billion individuals in the late 1870s to complete extinction of the species less than 40 years later (Ziswiler, 1967; Ehrenfeld, 1970; Schorger, 1973). These birds ranged over much of North America east of the Rocky Mountains and were intensely gregarious when nesting in vast colonies that could extend over several hundred square kilometers of woodland. They were also equally gregarious when forming enormous roosts during the nonbreeding season and during migration between breeding and nonbreeding areas. Such intense gregariousness made the pigeons extremely vulnerable to exploitation as food for livestock and humans, and billions were killed up until the late 1890s. Vast flocks of hundreds of millions of individuals led observers to believe that the species was indestructible. Then suddenly, around 1890, passenger pigeons vanished. Only a small number of nestlings could be found and the species clung to survival until the last individual died in the Cincinnati Zoo in 1914. The dismay caused by the demise of the passenger pigeon has perhaps been expressed most elegantly by Leopold (1970, p. 118):

> The pigeon was a biological storm. He was the lightning that played between two opposing potentials of intolerable intensity: the fat of the land and the oxygen of the air. Yearly the feathered tempest roared up, down, and across the continent, sucking up the laden fruits of forest and prairie, burning them in a traveling blast of life. Like any other chain reaction, the pigeon could survive no diminution of his own furious intensity. When the pigeoners subtracted from his numbers, and the pioneers chopped gaps in the continuity of his fuel, his flame guttered out with hardly a sputter or even a wisp of smoke.

Such an incredibly dramatic extinction almost defies comprehension and leaves us dazed and bewildered by our avaricious stupidity. Ziswiler (1967) suggested five reasons why the passenger pigeon was vulnerable to extinction: (1) the clearing of the American forests removed food and nesting resources, (2) intense gregariousness may have been necessary to ensure successful breeding, (3) small groups were below some critical density to establish feeding and nesting locations, (4) the intense extermination wrought by man was exacerbated by natural enemies, and (5) mate location became a limiting factor at low population density.

Yet more than seven decades after the last passenger pigeon vanished, we are little closer to reconciling man's destructively exploitative existence with the rest of the biotic diversity of our planet (Wilson, 1985). To be fair to the pioneers of the comparatively new discipline of conservation biology (Soulé, 1985, 1986), our inability to conserve biotic diversity is not so much a problem of a lack of understanding of the ecological processes involved (although these processes are by no means well understood). Rather, the primary problems are ones of conservation initiatives constantly thwarted by the anthropocentric considerations of sociology, politics, and economics. We are finding it very difficult to integrate humanity within natural environments without complete dominance and manipulation. It would seem to be a comparatively simple task for mankind to develop an ethic that is adequate for preserving species and biotic diversity, but despite occasional successes our overall response has been a dismal failure (Rolston, 1985). Thus our inability to conserve species and viable habitats centers much more on problems of the implementation, both philosophically and practically, of ecologically sound conservation initiatives rather than on our ability to comprehend the nature and likelihood of extinction.

The passenger pigeon illustrated clearly the conservation concept of an "endangered phenomenon," which Brower and Malcolm (1990) define as "a spectacular aspect of the life history of an animal or plant species involving a large number of individuals that is threatened with impoverishment or demise; the species per se need not be in peril, rather, the phenomenon it exhibits is at stake." Wells et al. (1983) used similar criteria to define their "red data book status category" of a "threatened phenomenon" as "aggregates or populations of organisms that together constitute major biological phenomena, endangered as phenomena but not as taxa."

Of course, phenomena that involve a high proportion of species numbers are likely to be sensitive to rapid and large changes of scale, and populations are often likely to crash and become extinct. In the case of the passenger pigeon, gregariousness must have been critical to the species' survival and so both the phenomenon and the species were extinguished—perhaps because gregariousness was on such a massive scale that "population" became synonymous with "species." A lack of fragmentation of the species into populations or metapopulations would have made the species vulnerable to sudden changes in population size (Quinn and Hastings, 1987; Hanski, 1989).

The monarch butterfly, *Danaus plexippus*, has many similarities to the passenger pigeon. Although the passenger pigeon was a prime candidate for the status of an endangered phenomenon, it was not until 1983 that the migration and overwintering phenomenon of the monarch became the first such aspect of an organism's life history to be designated the "red data" category of a threatened phenomenon (Wells et al., 1983; Pyle, 1984) by the International Union for Conservation of Nature and Natural Resources (IUCN) and the World Wide Fund for Nature (WWF).

Despite this designation, it is not thought that the monarch is threatened with extinction as a species (Brower and Malcolm, 1991), even though, like the passenger pigeon of the mid-19th century, many millions breed each year in North America and almost all of them migrate and coalesce into enormous roosts during the nonbreeding winter months. Ironically, the same clearing of the forests across the eastern United States that destroyed the passenger pigeon's food and breeding resources has been suggested to have benefitted monarchs by increasing the abundance of milkweed larval food resources in open farmland (Vane-Wright, 1986, this volume). Thus, perhaps as a consequence of human activities, the monarch is now a familiar and much loved insect throughout North America and appears likely to be adopted as the "national insect" of the United States.

Nevertheless, despite the wide distribution of the monarch and its conspicuous annual migration to and from breeding areas, the phenomenon of migration and overwintering remains threatened simply because entire populations coalesce from almost continent-wide distributions to a few tiny overwintering areas in Mexico and California. It is these overwintering areas that are threatened by our sociological, political, and economic inabilities to solve ecologically simple conservation problems.

Efforts to conserve the overwintering habitats of the monarch butterfly are a good example of much needed attempts to integrate human activities within the well-defined winter range of this remarkable butterfly. These efforts also illustrate very nicely the difficulties of achieving biologically obvious conservation goals. Although breeding habitat is threatened to some extent by the use of herbicides and agricultural practices that destroy the milkweed host plants of their larvae, it is the overwintering phase of their life cycle that is the most vulnerable and poses the greatest threat to the phenomenon of monarch migration.

Each autumn, hundreds of millions of butterflies east of the Rocky Mountains and Great Plains coalesce into 10 sites near the center of Mexico (Calvert and Brower, 1986; Calvert et al., 1989). These range in area from 0.03 to 3.34 ha, with 11–12 million butterflies per hectare, and are located in the Transvolcanic Mountains west of Mexico City (Brower and Calvert, 1985; Calvert and Brower, 1986; Calvert et al., 1989). Thus, the largest sites contain tens of millions of monarchs tightly clustered on the trunks, leaves, and branches of oyamel fir trees (*Abies religiosa*) in high-altitude forest at approximately 3,000-m elevation.

In contrast, west of the Rockies, monarchs coalesce into as many as 129 colonies at sites that span almost 8° of latitude along the coast of California, from north of San Francisco in Mendocino County (39°40′N), south to Ensenada (31°53′N) in Baja California, northern Mexico (Nagano and Lane, 1985; Sakai et al., 1989). The trees of most overwintering sites are dominated by blue gum (*Eucalyptus globulus*): the aggressively invasive eucalypt species that was introduced from Australia in the last century. Overwintering sites in California are much smaller than those in Mexico and support from dozens to a recorded maximum estimate of approximately 171,000 monarchs (Nagano and Lane, 1985). Tuskes and Brower (1978) estimated a maximum of 95,000 butterflies at the largest of their three overwintering colonies in California. Almost the entire North American fauna of the monarch butterfly spends the winter at these overwintering locations in California and Mexico. Thus, any loss or perturbation of their overwintering habitat is potentially catastrophic.

Despite these differences in the location, size, and habitat of overwintering colonies in Mexico and California the problems for conservation of the migratory phenomenon are ecologically similar. In both cases the butterflies spend up to 5 mo in the densely packed aggregations. Here, they are extremely vulnerable to destruction of their wooded overwintering habitat and to any kind of disturbance that might force them to move excessively and use up the stored fat they carry (Masters et al., 1988; Calvert et al., 1989).

Inevitably, short-term economic pressures pose the greatest threat to overwintering habitat. In Mexico, the oyamel fir is a valuable timber resource (Snook, this volume), whereas coastal land in California is prime territory for the development of resort complexes. The different political, economic, and sociological structures of Mexico and California also pose additional and sometimes unique problems to be solved before overwintering habitat is safe (Allen and Snow, this volume; Camus de Castro, this volume; Gottfried Joy, this volume; Ortíz Monasterio, this volume).

At first sight, the Mexican overwintering sites appear to be safe under the protection of an October 1986 Presidential decree (Diario Oficial, 1986) that clearly defines the extent and level of protection for five important locations (the other five known sites are not protected). Such protection is almost unprecedented in Mexico and owes its existence to the remarkably effective efforts of the Mexican nonprofit conservation organization Monarca, A.C., which focuses its efforts to protect the monarch overwintering sites in Mexico (Ogarrio, 1984, this volume). In addition to federal protection, Monarca, A.C., has also worked very hard to include members of the agrarian collectives (*ejidos*), which legally own the land on which monarchs aggregate each winter, as part of their conservation efforts (Rojas González de Castilla, this volume). Thus, many of the local people, the ejido members, or *ejidatarios,* have become interpretive guides to the two sites that are open to the public. In addition, Monarca, A.C., has opened food and trading concessions and designed souvenirs to be made locally and sold to tourists. Tourists are now flooding in each winter to see the extraordinary colony of millions of aggregated monarchs at Rosario (also called Sierra El Campanario), which is accessible from the picturesque towns of Angangueo and Ocampo in Michoacán. An additional colony is open to the public at Macheros near Donato Guerra in the state of México. The economic boost that these tourists bring is helping to turn the local subsistence economy that exploits the forest into a cash economy that will, one hopes, protect the forest.

Although the five protected sites in Mexico appear to be safe, only 40,000–50,000 ha of the high-altitude oyamel fir forests, used exclusively by monarchs, are now left in all of Mexico (Manzanilla, 1974). These forests have traditionally been exploited by a system of selectively removing only the largest trees. Unfortunately, the traditional method of selective cutting—the "*método mexicano de ordenacíon forestal*"—has led to overexploitation of overwintering forests. Thus six out of seven measured overwintering forests have tree basal areas (m^2/ha) less than those of the forests that Manzanilla (1974) categorized as having been exploited commercially by selective cutting (Calvert et al., 1989). All five of the supposedly completely protected core areas of the overwintering sites covered by the presidential decree are included among these overexploited forests! Thus, although impressive initiatives exist for monarch protection in Mexico, there is little doubt that they are not being implemented and that protection decreed for buffer and core areas of overwintering forests is being ignored by ejidatarios and foresters alike.

Furthermore, Calvert et al. (1989) suggest that Mexican foresters could be seduced by the economic rewards that accrue from European methods of clear-cutting blocks of forest. After cutting and timber removal, cleared blocks are replanted immediately, with little opportunity for successional diversity and nutrient fixation. This produces single-age stands of timber that facilitate future exploitation (Mueller-Dombois, 1988). Unfortunately, the oyamel forests are all on very steep slopes and are extremely unlikely to sustain even one cycle of exploitation before the forests are destroyed by nutrient loss, soil erosion, and the destruction of watersheds. Even in Europe, where the forests are on either flat or gently sloping ground, the decline of the forests after several cycles of exploitation is now readily apparent.

If such exploitation occurs throughout the oyamel forests of Mexico, not only will the overwintering habitat of all monarchs east of the Rocky Mountains be destroyed, but so will the high-altitude forests and with them the water supplies for millions of Mexicans. Moreover, although the motivation to protect monarchs is evident in Mexico, it is worrying that much of the implementation necessary to ensure protection is being ignored. Thus even in the protected overwintering sites, trees are still being cut and at least one of the five protected sites is dangerously close to complete destruction (Calvert et al., 1989).

It is easy to patronize Mexicans with pompous platitudes about overexploitation since we tacitly assume that endangered species and their habitats are well conserved in first-world countries like the United States. Unfortunately, the economics of land development on the coast of California is also destroying many monarch overwintering sites and threatening almost every site in the state. Sakai et al. (1989) have recorded the destruction or loss of 21 overwintering colonies in California, with a further 7 having been severely damaged. Sites have been, and are being, destroyed to expand and build motels, hotels, and resort complexes, to clear agricultural land, to build roads and new houses, and to clear picturesque views of the Pacific Ocean for those sufficiently wealthy to afford coastal real estate. Of course, many landowners are delighted to have the spectacular occurrence of a monarch overwintering colony on their land and they do everything they can to protect them through legal easements and registrations of the site. Like Monarca, A.C., in Mexico, *The Monarch Project* of the Xerces Society is also a nonprofit organization that works exclusively to conserve monarchs and it has negotiated many such easements and registrations throughout the state (Allen and Snow, this volume). However, where monarch sites clash with multimillion dollar developments monarchs usually lose. Some coastal counties have ordinances that protect overwintering trees and such protection is proving partially effective in counties like Santa Barbara. Further north, the city of Pacific Grove has long been known as "Butterfly Town USA" because monarchs have been known to overwinter on eucalypts and native pines within the city since at least 1864 (Lane, 1984, 1985, this volume; Vane-Wright, this volume). Sadly, for many years the city neglected to conserve overwintering habitat and ironically, the very trees that monarchs used, winter after winter, were cut down to expand the motels that housed the visitors that came to see the butterflies. Happily, the situation is slowly being reversed and now the city of Pacific Grove sees more value to integrating monarchs and people in the same environment, and their Washington Park is being protected, researched, and slowly restored to pristine overwintering habitat.

Still other overwintering sites in California are threatened by another problem that is unlikely to rear its controversial head in Mexico. Introduced eucalypts are considered by many to be a pest species in California. Thus in some state parks, attempts to eradicate eucalypts are in progress. The enthusiasm for eradicating these exotics has even extended to cutting trees in the middle of one of the largest monarch overwintering colonies in California, at Montana de Oro State Park, near San Luis Obispo. Since it is extremely unlikely that such successful tree colonizers could be eradicated from California, this kind of thoughtless habitat destruction is especially lamentable.

Like Mexico, the most valuable protection for monarchs in California has been granted by state legislation. In 1988, the State of California set aside $2 million from a bond issue of $776 million for acquisition of land on which monarchs overwinter along the Pacific coastline (Proposition 70, the "California Wildlife, Coastal, and Park Land Conservation Act"). Such constructive legislation is important, but California also needs to develop a rational policy for the conservation of overwintering monarchs that defines each overwintering habitat and restricts any negative and avoidable impact on overwintering monarchs. We now think that all permanent overwintering sites, no matter how small or large, are important to the success of monarch overwintering and the survival of the migratory phenomenon.

The destruction of habitat crucial to endangered phenomena associated with species like the monarch butterfly and the passenger pigeon, and our inability to implement measures to conserve such species, are in many ways products of Hardin's (1968) "tragedy of the commons." Until recently, the Mexican ejidos in which monarch overwintering colonies occur have exemplified Hardin's contention that common land is only justifiable when it is exploited at low intensity by low densities of human populations with little environmental impact. Once human populations become more affluent and increase in size, as in central Mexico and the west coast of California today, or the late 19th century eastern United States, Hardin contends that access to common resources, like those used by monarchs and passenger pigeons, leads to gross overexploitation and environmental disaster. This influential idea has been challenged recently by Reader (1988), who points out that access to common resources has never been as free and unlimited as Hardin assumed. Access is usually regulated so that resources are sustainable. Although Reader's qualification is undoubtedly true for many stable human communities worldwide, the destruction of passenger pigeons and monarch overwintering habitat is closer to Hardin's original concept of the commons because few controls on exploitation were, or are being implemented. Passenger pigeons were destroyed in their nests and roosts by unlimited access and almost unparalleled systematic destruction (Schorger, 1973). Similarly, monarch overwintering habitat is being destroyed in Mexico by both ejidatarios and foresters with almost unlimited access to the forests, despite the presence of armed guards putatively posted to protect monarchs (Calvert et al., 1989). These two examples illustrate a lack of restraint to exploitation that is reaching similar proportions in California. The most obvious distinction between monarch overwintering land in California and Mexico is that Mexican ejido land is exploited by ejidatarios for their basic survival and by the federal government for profit, whereas in California most coastal land is privately owned and is exploited for profit alone.

It is rare for monarchs to overwinter on common land in California because high population densities and inflated land prices have resulted in division of originally common land into countless, privately owned parcels. This is perhaps the ultimate extension of "the tragedy of the commons," where control has become subject to the whims of single owners. The almost infinite diversity of land uses available to individual land owners poses a nearly insurmountable problem for conservationists bent on protecting endangered species. Confronted with irresistible recreational and economic rewards it takes great enlightenment and environmental awareness for a California landowner in, for example, Marin County or Santa Barbara County to agree to an easement that protects monarch overwintering habitat. Nevertheless, such easements have been negotiated and many organizations in California, from private local interest groups to state and federal bodies, are motivated to integrate the full gamut of the ecological, sociological, political, and economic facets of successful conservation.

Both Mexico and California should realize that motivation and implementation of conservation goals must go hand-in-hand before we can protect monarchs and their migration. Unless these custodians of an entire continental fauna can reverse the present destruction of monarch overwintering habitat, it is going to be both expensive and distressing to have to travel to Australia or Central America to see North America's most loved and admired insect. The passenger pigeon never had this option but it too was admired for its elegance and vitality. It is an extraordinary privilege to sit among 100 million overwintering monarchs high in the mountains of central Mexico. It must have been equally incredible to sit on the heights over Hartford, Kentucky, in 1847 and watch a flock of an estimated 1 billion pigeons cloud the sky, blot out the sun (Schorger, 1970), and dwarf our arrogance.

LITERATURE CITED

Brower, L.P., and W.H. Calvert. 1985. Foraging dynamics of bird predators on overwintering monarch butterflies in Mexico. *Evolution* 39(4):852–868.

Brower, L.P., and S.B. Malcolm. 1991. Animal migrations: Endangered phenomena. *American Zoologist* 31:265–276.

Calvert, W.H., and L.P. Brower. 1986. The location of monarch butterfly (*Danaus plexippus* L.) overwintering colonies in Mexico in relation to topography and climate. *Journal of the Lepidopterists' Society* 40(3):164–187.

Calvert, W.H., S.B. Malcolm, J.I. Glendinning, L.P. Brower, M.P. Zalucki, T. Van Hook, J.B. Anderson, and L.C. Snook. 1989. Conservation biology of monarch butterfly overwintering sites in Mexico. *Vida Silvestre Neotropical* 2(1):38–48.

Diario Oficial. 1986. Organo del Gobierno Constitucional de los Estados Unidos Mexicanos. México, D.F., Jueves 9 de Octubre de 1986, 398(27):33–41.

Ehrenfeld, D.W. 1970. *Biological conservation.* New York: Holt, Rhinehart and Winston, 226 pp.

Hanski, I. 1989. Metapopulation dynamics: Does it help to have more of the same? *Trends in Ecology and Evolution* 4(4):113–114.

Hardin, G. 1968. The tragedy of the commons. *Science* 162:1243–1248.

Lane, J. 1984. The status of monarch butterfly overwintering sites in alta California. *Atala* 9(1&2):17–20.

Lane, J. 1985. California's monarch butterfly trees. *Pacific Discovery* 38(1):13–15.

Leopold, A. 1970. *A Sand County almanac: With essays on conservation from Round River.* New York: Sierra Club/Ballantine Books, 295 pp.

Manzanilla, H. 1974. *Investigaciones epidométricas y silvícolas en bosques Mexicanos de Abies religiosa.* Ciudad de México: Dirección General de Información y Relaciones Públicas, Secretaría de Agricultura y Ganadería, 165 pp.

Masters, A.R., S.B. Malcolm, and L.P. Brower. 1988. Monarch butterfly (*Danaus plexippus*) thermoregulatory behavior and adaptations for overwintering in Mexico. *Ecology* 69(2):458–467.

Mueller-Dombois, D. 1988. Forest decline and dieback—A global ecological problem. *Trends in Ecology and Evolution* 3(11):310–312.

Nagano, C.D., and J. Lane. 1985. A survey of the location of monarch butterfly (*Danaus plexippus* [L.]) overwintering roosts in the state of California, U.S.A.: First year 1984/85. Portland, OR: *The Monarch Project,* 106 pp.

Ogarrio, R. 1984. Development of the civic group, Pro Monarca, A.C., for the protection of the monarch butterfly wintering grounds in the Republic of Mexico. *Atala* 9(1&2):11–13.

Pyle, R.M. 1984. International efforts for monarch conservation, and conclusion. *Atala* 9(1&2):21–22.

Quinn, J.F., and A. Hastings. 1987. Extinction in subdivided habitats. *Conservation Biology* 1(3):198–208.

Reader, J. 1988. Human ecology: How land shapes society. *New Scientist* 119(1629):51–55.

Rolston, H., III. 1985. Duties to endangered species. *BioScience* 35(11):718–726.

Sakai, W.H., C.D. Nagano, A.V. Evans, J. Schrumpf, J. Lane, and M. Monroe. 1989. The wintering colonies of the monarch butterfly (*Danaus plexippus* (L.): Nymphalidae: Lepidoptera) in the state of California, U.S.A. *Report for the California Department of Fish and Game,* 26 pp.

Schorger, A.W. 1973. *The passenger pigeon: Its natural history and extinction.* Norman: University of Oklahoma Press, 424 pp.

Soulé, M.E. 1985. What is conservation biology? *BioScience* 35(11):727–734.

Soulé, M.E. (ed.). 1986. *Conservation biology: The science of scarcity and diversity.* Sunderland, MA: Sinauer, 584 pp.

Tuskes, P.M., and L.P. Brower. 1978. Overwintering ecology of the monarch butterfly, *Danaus plexippus* L., in California. *Ecological Entomology* 3:141–153.

Vane-Wright, R.I. 1986. White monarchs. *Antenna* 10: 117–118.

Wells, S.M., R.M. Pyle, and N.M. Collins. 1983. *The IUCN invertebrate red data book.* Gland, Switzerland: International Union for Conservation of Nature and Natural Resources, 632 pp.

Wilson, E.O. 1985. The biological diversity crisis. *BioScience* 35(11):700–706.

Ziswiler, V. 1967. *Extinct and vanishing animals: A biology of extinction and survival.* New York: Springer-Verlag, 133 pp.

CONSERVATION OF THE MONARCH BUTTERFLY RESERVES IN MEXICO: FOCUS ON THE FOREST

Laura C. Snook

ABSTRACT. The oyamel fir forests (*Abies religiosa* (H.B.K.) Schl. & Cham.), in which monarch butterflies, *Danaus plexippus* (L.), overwinter in Mexico, are subject to severe degradation. Exogenous factors, including extraction of trees for industrial and domestic uses, grazing, fires, and conversion of land to agriculture, must be controlled for the forest to survive. This requires not only the redirection of current fir forest exploitation to the neighboring pine–oak (*Pinus–Quercus*) forests and plantations, but also a socioeconomic development strategy to provide economic alternatives for subsistence agriculturalists living in and around the reserves. Endogenous factors such as dwarf mistletoe (*Arceuthobium abietis-religiosae*), bark beetles (probably *Scolytus mundus*), and defoliators (*Evita hyalinaria blandaria*) also contribute to forest mortality and impede regeneration, but to a lesser extent than exogenous factors. Practices used in commercial forestry can be applied to reduce the impact of these pests, if necessary.

RESUMEN. Los bosques de oyamel (*Abies religiosa* (H.B.K.) Schl. & Cham.) donde las mariposas Monarca, *Danaus plexippus* (L.), pasan el invierno en México, se encuentran amenazados. Factores exógenos, incluyendo la extracción de madera para finalidades domésticas e industriales, el pastoreo, los incendios, y la conversión de áreas forestales a campos agrícolas, requieren un control adecuado para que estos bosques se conserven. Es necesario proveer los bienes materiales actualmente derivados de los bosques de oyamel, de bosques aledaños de pino–encino (*Pinus–Quercus*) o de plantaciones, así como implementar una estrategia de desarrollo socioeconómico en la zona, con el fin de dar alternativas económicas a los campesinos que viven a nivel de subsistencia alrededor de las reservas. Factores endógenas, incluyendo el muérdago (*Arceuthobium abietis-religiosae*), escarabajos descortezadores (probablemente *Scolytus mundus*), y defoliadores (*Evita hyalinaria blandaria*) también contribuyen a la mortandad de oyameles, y reducen la capacidad regenerativa del bosque. Al ser necesario, estos se pueden controlar con prácticas aplicadas en el manejo forestal comercial.

INTRODUCTION

The migration of the eastern North American populations of monarch butterflies, *Danaus plexippus* (L.), to Mexico every winter is one of the outstanding examples of long-distance animal migration (Brower, 1985), as other papers in this volume attest. Since 1975, when the monarch's overwintering habitat, the high-altitude oyamel fir (*Abies religiosa* (H.B.K.) Schl. & Cham.) forest, was found and publicized (Urquhart, 1976), biologists and conservationists from Canada, the United States, Mexico, and elsewhere have expended considerable effort to support and encourage institutional and government actions to protect these Mexican forests.

The most recent accomplishment in this decade-long process was the signing of a decree by Mexican President Miguel de la Madrid, in September 1986, which defined as protected natural areas five forested mountaintops in the states of Michoacán (Chincua, Campanario, Chivatí-Huacal, and Cerro Altamirano) and México (Cerro Pelón), where the monarch butterflies overwinter (Diario Oficial, 1986) (Fig. 1). Eighty hectares of these forests were expropriated by the government of Michoacán, and 700 more were bought by the Secretariat of Urban Development and Ecology (SEDUE) from a private property owner (SEDUE, pers. comm.). Almost all of the remaining area (nearly 15,000 ha) belongs to more than 30 ejidos, land grants made through the Agrarian Reform to groups of "campesinos," or peasants who live off the land. Ejido lands cannot be bought, sold, or transferred.

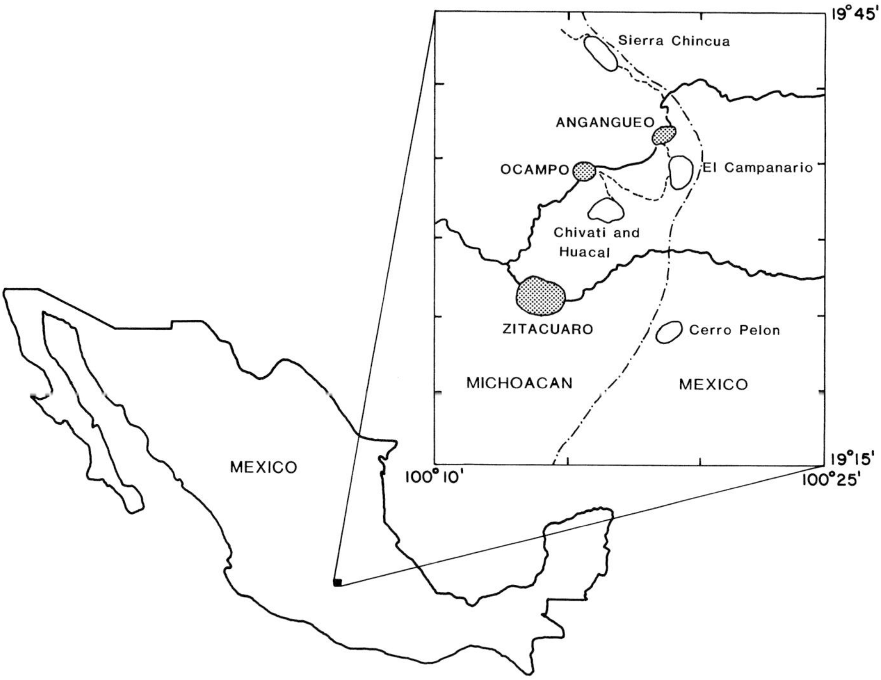

Figure 1. Location of the four areas first proposed as monarch butterfly reserves in Mexico (from Espejo et al., 1985). Cerro Altamirano, which was added later, is located slightly north and east of the area included in the inset map (see also Calvert and Lawton, this volume; Soto Núñez and Vázquez García, this volume).

To translate this legislative protection into reality, a forest conservation plan must be designed and implemented. Such a plan must integrate knowledge not only of the importance of these forests to butterfly survival, but also of the characteristics and status of the forests, the forests in relation to the local populace, the factors that are causing forest degradation and destruction, and the potential of different strategies and tactics to protect and restore them (for accounts of plans, see the papers by Camus de Castro, Gottfried Jay, Ogarrio, Ortíz Monasterio, and Rojas González de Castilla, in this volume).

THE FOREST AS MONARCH BUTTERFLY HABITAT

The importance of the Mexican oyamel fir forests for the survival of overwintering monarch butterflies has been well documented by Calvert et al. (1979, 1982), Calvert and Brower (1981), Brower and Calvert (1985), and Brower (1985). In his 1985 paper, Brower states, "Survival of the monarchs depends on a balance of climatic factors which requires that the weather be (1) cold enough to maintain the butterflies in a state of reproductive torpor, but not so cold as to kill them, (2) warm enough to allow them to maintain the integrity of their clusters, but not so warm as to result in excessive activity and (3) wet enough to prevent desiccation and forest fires, but not so wet and cold as to cause excessive inactivity." (p. 759) . . . "the high altitude fir forests in this narrow zone between 19° and 20°N latitude of Mexico's Transvolcanic Belt provide a cool climate, stable minimum nightly temperatures and a balance of sufficient moisture and sunshine which seem ideally suited for the survival of these densely packed monarch overwintering colonies" (p. 761).

The modification of this forest structure by mortality or extraction of trees, has been found to affect negatively the survival of the overwintering monarchs. "Even high quality forestry practice [low-intensity selective extraction], degrades the Mexican overwintering habitat both by causing more butterflies to freeze (Calvert and Brower, 1981; Calvert et al., 1982) and by allowing easier access to avian predators which kill up to 34,300 butterflies per day (Brower and Calvert, 1985). If forest degradation results in smaller numbers of overwintering monarchs at these sites, mortality may increase because the intensity of bird predation is inversely proportional to colony size (Calvert et al., 1979). Selective tree removal also lets more sunlight into the colony, butterfly activity increases and critical lipid reserves must diminish" (p. 770).

Therefore, the cornerstone of any strategy to conserve the "endangered phenomenon" of the monarch's annual winter migration to Mexico (Wells et al., 1983) must be the protection and enhancement of the fir forests where the monarchs overwinter. The design of such a strategy requires that the forest be understood not only as butterfly habitat, but also as part of a larger ecosystem, and as a source of goods and services for local people.

CHARACTERISTICS AND STATUS OF THE FIR FORESTS

Today's fir forests in Mexico are relicts of extensive boreal forests that advanced southward as the cold climates descended to tropical latitudes during periods of glaciation (Rzedowski, 1978). In the 10,000 yr since the glaciers receded, these forests have been displaced by temperate and tropical floras adapted to the warmer climatic conditions of today. Now only 40,000–50,000 ha of fir forests remain in Mexico (Manzanilla, 1974), distributed as isolated islands at elevations between 2,400 and 3,600 m (Rzedowski, 1978), where the cold climate excludes most other genera and permits the firs to dominate. This reduced area and patchy distribution pattern make the fir forest perhaps the most vulnerable to deforestation pressures of any forest type in Mexico.

It is interesting to note that the warming trends, which reduced the extension of the fir forests to <2% of Mexico's area (Flores, 1961, in Rzedowski, 1978), may have contributed to the monarch butterflies' parallel expansion to northern latitudes, where they exploit the spring and summer milkweed flora (Young, 1982; Brower, 1985), thus resulting in the migration phenomenon that is the focus of our interest.

The composition and relative dominance of species in the Mexican fir forests respond to an elevational gradient. Above 3,000 m, the forests are dominated by fir, with occasional pure stands of pine (*Pinus* spp.) (Rzedowski, 1978; Espejo et al., 1985). This light-demanding genus typically plays a successional role in these forests, temporarily replacing shade-tolerant firs subsequent to disturbances that produce large clearings (Rzedowski et al., 1977; author's obs.). Individuals of madrone (*Arbutus glandulosa*), willow (*Salix paradoxa*), oak (*Quercus* spp.), and alder (*Alnus firmifolia*) occur in the understory (Rzedowski, 1978; Espejo et al., 1985; see also Soto Núñez and Vázquez García, this volume).

The forests occur primarily on slopes. In flat areas with closed drainage, probably subject to soil freezing, trees do not grow, and attempts to plant firs in these areas have failed. These "llanos," dominated by grasses and forbs (notably *Potentilla candicans*), are bordered by the bush-sized *Juniperus monticola* (Rzedowski, 1978). As a result of grazing pressures, many of these areas have been invaded by unpalatable bushes whose barbed seeds are dispersed by sheep.

Wide bands of *Baccharis conferta* occur between the forest and the llanos in many places. This shrub, an indicator of disturbance, is known as "la madrina del pino" (the godmother of pines), because it favors the reestablishment of both pines and firs, probably by protecting the seedlings from freezing temperatures and grazing pressure.

Below 2,900 m, oak dominance increases over that of fir and pine, and tree species characteristic of the lower elevation hardwood forests (notably *Clethra mexicana*) appear as well. Under certain ecological conditions, stands of *Cupressus lindleii* Klozsch are found (Rzedowski, 1978; Espejo et al., 1985).

Firs are not observed below 2,700 m, where other species increase in frequency and dominance to integrate the most extensive forest in Mexico, the pine–oak type (Rzedowski, 1978). This elevation represents the lower limit used to define the extent of monarch overwintering habitat (Reyes and de la Maza, 1982; Monarca, A.C., 1984).

The oyamel fir forests are dynamic communities. The processes of forest regeneration subsequent to disturbance are revealed by patches of *B. conferta* with pine and fir saplings beginning to emerge above them, pure stands of pine within the fir zone, and bands of dying *J. monticola* completely overtopped by young stands of fir, which originally regenerated under their cover.

Despite this potential for regeneration after disturbance, the most obvious characteristic of the fir forests of the monarch reserves is their degraded state and shrinking area. Large old stumps remain from past logging operations and lumber roads, many severely eroded, crisscross the area (Fig. 2). Scars of past fires are apparent on stumps, and some areas have burned recently. Recently felled trees (Fig. 3), piles of stacked cordwood (Fig. 4), and fresh stumps (Fig. 5) are widespread, and many trees are scarred with "window cuts" used to check the straightness of the grain for splitting into roof shingles (Figs. 3 and 6). Within the limits of the reserves, areas have been cleared and plowed for cultivation as the agricultural frontier expands up the mountain (Fig. 7). Neighboring hilltops are completely deforested (Fig. 8).

FACTORS OF FOREST DEGRADATION AND DESTRUCTION: THEIR EFFECTS

Ranked in order of their apparent and potential importance, the factors of forest degradation and destruction in the monarch reserve forests are (1) extraction of trees, (2) fire, (3) biotic pathogens and pests (bark beetles, dwarf mistletoe, and defoliators), (4) livestock grazing, and (5) tourism. Each one has different effects on the forest (Table 1). The conversion of forested land to agriculture is in a category by itself as the ultimate and usually terminal factor of forest destruction.

The extraction of trees is one of the principal threats to the monarch's winter habitat. Tree density in some young stands within the reserves has been reduced from 1,000 to 150/ha. This thinning and the associated creation of clearings, modifies the microclimatic characteristics of the forest and

Figure 2. Old stumps and logging roads left by commercial logging operations in the Chincua overwintering area.

Figure 3. A recently felled tree in one of the reserves is hewn with an axe into a beam.

thus threatens butterfly survival (Calvert et al., 1982). In the absence of other stresses, such openings could also be expected to stimulate forest regeneration processes, favoring seed production and seedling establishment and growth. However, in these forests not only are mature trees being lost, but regeneration and growth are inhibited by fire, grazing, and infestations of dwarf mistletoe (*Arceuthobium abietis-religiosae*) (Rzedowski, 1978; Rodríguez, 1983).

Fires, almost always associated with the clearing of neighboring agricultural land or used to stimulate the growth of grasses for grazing animals, destroy existing regeneration and can damage or kill even mature trees over large areas under certain conditions. Livestock grazing impedes establishment of regeneration and modifies species composition in the herb layer. If fire and grazing were excluded, forest regeneration in some areas could be expected to increase significantly.

Dwarf mistletoe has infested large numbers of trees in some areas. This fir parasite deforms and reduces growth of mature trees, reduces seed production and viability (Boyce, 1961; Smith, 1970), and kills seedlings and saplings (Horst, 1979). Thus, while dwarf mistletoe does not generate openings in the canopy, it does severely affect stand vigor and regeneration capacity and probably prevents establishment of artificial (planted), as well as natural, regeneration. Mistletoe infestations also weaken trees, predisposing them to attack by other organisms that do cause mortality. Principal among these are bark beetles. Hernández and Cibrián (1981) indicate that the following species of bark beetles (Scolytidae) have been found to attack firs in Mexico: *Scolytus ventralis, Scolytus mundus, Scolytus hermosus, Pseudohylesinus variegatus,* and *Pityophthorus blackmanii.* While standing dead trees in the reserve areas may indicate bark beetle attack,

Figure 4. Cordwood stacked in the Chivatí-Huacal overwintering area for sale to neighboring conglomerate board of papermaking industries.

Figure 5. High, fresh stumps of small trees indicate domestic cutting in the Cerro Pelón overwintering area.

Figure 6. Many trees, such as these in the El Campanario area, are scarred with "window cuts," made to check the straightness of their grain before felling for splitting into roof shingles (see also Fig. 3).

Figure 7. Agricultural clearing occurs even within the boundaries of the reserve areas protected for the monarchs. This new field is within the Campanario reserve area.

it is not known which species are present in the fir forests of the monarch reserves, or to what extent.

Adult bark beetles usually attack trees weakened by stress (high density, overmaturity, drought, fire, parasitism, defoliation, pathogen attack or damage), drill through the bark, and excavate galleries in the phloem, where they reproduce. The larvae continue to feed on the phloem, almost always girdling the tree and killing it. When they emerge as adults, the new generation of bark beetles searches out another stressed target tree, where they mate and reproduce. Thus, an infested tree represents a threat to neighboring trees. Some species of bark beetle experience periodic population outbreaks, attacking even healthy trees, and can destroy thousands of hectares of forest, greatly increasing the risk of fire (Graham and Knight, 1965).

The oyamel fir forests have also suffered periodic outbreaks of a geometrid moth (*Evita hyalinaria blandaria* Dyar), whose larvae feed on fir needles and are capable of completely defoliating a tree. Outbreaks affecting 500–7,000 ha were reported in the central Mexican fir forests in 1957, 1958, 1968, 1977, and 1978. The most recent outbreak, in 1986, occurred in the state of México, bordering on the Chincua monarch reserve. Within an affected area of 4,613 ha, 23% of the trees were 51–75% defoliated, and 24% of the trees were 76–100% defoliated. Damage was found to be more severe in pure stands of fir than in mixed stands, where fir grew together with *Cupressus, Pinus,* and *Quercus.* The larvae feed preferentially on tender needles, attacking juvenile trees more frequently than adults. Juvenile trees are attacked from the top down, and adult trees from the bottom up (Carbajal and López, 1987). Older trees that have been defoliated are likely to die as a result. Trees that have been partially defoliated may be attacked by bark beetles (Sangri, 1962; Carbajal and López, 1987).

Bark beetles, mistletoe, and insect defoliators are endogenous factors of forest stress, factors that have evolved as part of the forest ecosystem. Under natural conditions and over the long term, they do not represent a threat of forest destruction but, rather, act as mechanisms for the replacement of overmature forest stands by younger, more vigorous ones. However, where forest extent is limited, as in the case of the monarch reserve "islands," such endogenous factors may require monitoring and control.

Tourism is probably the least damaging factor of forest degradation, even though 12,000–15,000 people visited the forested reserves during the 1983–1984 butterfly season (SEDUE, pers. comm.) and 40,000–50,000 visited during the 1985–1986 season (R. Ogarrio, pers. comm.; also see Camus de Castro, this volume; Rojas González de Castilla, this volume). Heavy use of access roads by trucks carrying tourists stimulates erosion. Tacos sold to visitors to the reserves are cooked over firewood cut from the forest. If the shortage of local hotel facilities were to encourage tourists to camp, other kinds of problems could arise: the cutting of wood for

Figure 8. The agricultural frontier moves up the hillsides around El Campanario, replacing the forests with fields.

Table 1. Factors that contribute to the degradation and destruction of high-altitude fir forests in Mexico.

Factors	Effects
Forest degradation factors	
Extraction of trees	Opens gaps and clearings; modifies microclimate
Fire	Destroys regeneration; may increase soil erosion
Bark beetles[a]	Kill trees; can destroy large areas of forest
Defoliators[a]	Kill trees; can affect large areas and predispose trees to beetle attack
Mistletoe[a]	Reduces regeneration capacity; predisposes trees to bark beetle attack
Livestock grazing	Reduces regeneration
Tourism	Soil erosion (access); could stimulate wood cutting
Forest destruction factors	
Conversion of land to agriculture	Replaces system; destroys regeneration capacity
(Fire)	Kills trees and understory; favors pine regeneration
(Bark beetles)[a]	Kill trees; may stimulate regeneration

[a] Endogenous factors.

campfires, for example, and possible contamination of the streams where the monarchs drink. Although many of the potential threats of tourism to the monarch forest reserves remain speculative and unsubstantiated, it is notable that conservation efforts have emphasized the control of tourists rather than other factors that affect the forest in much more immediate and significant ways.

The conversion of forest land to agricultural use in the Mexican central highlands is the terminal factor of forest destruction. Driving from Mexico City to the monarch reserves, one sees how agriculture has displaced previous forests. When agricultural production drops off, the land is abandoned, and "zacatón" (*Muhlenbergia* sp.), a large bunch grass, soon colonizes the site. This is excavated for its roots, which are used to make brushes. Extraction of zacatón roots destroys soil stability and encourages erosion. The final step in this process is the conversion of the soil into adobe bricks used in local construction, leaving the land completely mined and unproductive.

FACTORS OF FOREST DEGRADATION AND DESTRUCTION: THEIR CAUSES

While mistletoe, bark beetles, and defoliators are endogenous to the forest system and tourism is a spectator response to the phenomenon of monarch butterfly migration, the principal exogenous factors of forest degradation and destruction are more complex. To design adequate responses to them, it is necessary to differentiate between symptoms (tree cutting, fires, or land clearing) and the various levels of causes that drive them.

Extraction of trees is of two types: commercial/industrial and domestic. Commercial forest exploitation in the monarch reserve forests provides raw materials for local sawmills and neighboring conglomerate board and papermaking industries. Cutting is carried out according to Mexican Forest Service guidelines and controlled through marking of the trees to be cut and the issuing of permits for extraction and transport of authorized volumes. Data in Monarca, A.C. (1984), indicate that authorized cuts in the area of three proposed reserve sites in Michoacán in 1981 totalled 13,400–15,200 m^3 of pine and 13,100–14,300 m^3 of fir. The authorized cut in this area in 1982 was 14,900–16,700 m^3 of pine and 820–2,020 m^3 of fir (Monarca, A.C., 1984).

The driving forces behind these extractions are both the need for raw materials on the part of local forest industries and the need for income on the part of the local forest landowners, principally ejidos (Mexican agricultural communes) (Table 2). Ejido members (ejidatarios) are paid for cutting the wood, while the ejido community as a whole receives forest-use payments known as "derecho de monte."

In addition, at least one permit was granted to cut 3,000 m^3 in the nucleus area of the Chincua reserve, as a sanitation measure to extract trees infected with mistletoe (Monarca, A.C., 1984). This action indicated either a lack of familiarity with the biology of this parasite and the means to control it (see subsequent section) or other reasons for granting the permit.

Timber cutting carried out by individuals for domestic purposes is probably responsible for tree extractions two or three times greater than the volumes authorized for commercial harvesting in 1981 and 1982. Such domestic extractions have traditionally taken place outside the realm of forest service control. This occurs because they fall into a legal category different from commercial extractions (Ley y Reglamento Forestal, 1960, 1986); because domestic wood is transported by burro or mule on trails, rather than by trucks on roads, which require permits; and because professional foresters are reluctant to penalize poor rural people for taking from the forest products that they have traditionally used and currently need.

Table 2. The use by local people of oyamel fir forests in Mexico as a source of materials and goods for consumption and income.

Forest resource and product	For consumption	For income
Trees		
Fuelwood	X	X
Charcoal		X
Shingles, beams	X	X
Bolt wood		X
Saw timber		X
Christmas trees		X
Land		
Grazing	X	X
Agriculture	X	X
Tourism		X

Table 3. Three primary exogenous factors responsible for forest degradation and destruction and some of their causes and possible solutions.

Exogenous factors of forest degradation and destruction
1. Extraction of trees
2. Livestock grazing and associated fire risk
3. Conversion of forest land to agriculture and associated fire risk

Immediate causes (human needs)
- Fuelwood
- Construction materials
- Food
- Income

Short-/medium-term solutions
- Alternate sources of materials
- Alternative materials
- Alternative land management systems to increase productivity
- Alternative sources of food
- Alternative sources of income (e.g., change from subsistence to market economy)

Root causes
- Poverty
- Demographic increase

Medium-/long-term solutions
- Economic development
- Reduction in population growth
- Emigration

Fifteen thousand people live in the rural areas in and around the 5,600 ha of the Chincua, Chivatí-Huacal, and Cerro Campanario reserves in Michoacán (Fig. 1), and the situation is similar in the other overwintering areas. Most of the population lives at a subsistence level, deriving their food from agricultural production and grazing, their fuel and construction materials from the forest, and their income from the sale of agricultural goods (if and when production exceeds consumption) and of wood products (Fundación Mexicana para el Desarrollo Rural, 1984) (Table 2).

In the fir zone, local people preferentially heat their houses and cook with fir wood. In the fir forest zone at similar elevations in the state of Veracruz, wood consumption was found to be about 15 tons (35 m^3) per house per year (Snook and Hammond, unpubl. obs.). Assuming seven people per house and similar levels of consumption, total domestic fuelwood use by the households in the areas of Chincua, Chivatí-Huacal, and Cerro Campanario is estimated at approximately 75,000 m^3/yr, about 40,000 fir trees of average size (at 1–3 m^3/tree). Charcoal is another important fuel in the area, and mounds for charcoal production were observed within the nucleus of the Campanario reserve in 1984.

Most of the houses in local communities are constructed of wooden beams and boards, with roofs of fir shingles (called "tejamanil") split by hand. In addition to the enormous number of trees cut for these purposes for use by rural households, significant quantities of construction materials, fuelwood, and charcoal are sold to nearby urban centers and represent important sources of income for rural people (Table 2).

Forest degradation from livestock grazing, forest fires, and the conversion of forest land to agriculture derive from Mexico's rural subsistence economy, compounded by a rapid rate of demographic increase (Table 3). In a similar mountain region of Veracruz, an analysis of census data between 1970 and 1980 revealed a population growth of 4%/yr (Snook and Hammond, unpubl. obs.). At this rate, a population doubles in 18 yr, and with it, the demand for fuel, food, and income. While the rate of emigration from the Michoacán and México highlands to Mexico City and elsewhere may result in a lower rate of population growth in this region, local demographic tendencies have not yet been analyzed.

There are several levels of causes for the anthropogenic pressures on the fir forest. Immediate causes are the needs of local people for fuel, construction materials, food, and income (Table 2). On a more fundamental level, however, these demands are symptoms of poverty which does not permit the purchase of goods currently obtained "for free" from the forest and its land, reflecting both a lack of local employment alternatives and the rate of population increase (Table 3).

The range of alternative responses to these pressures on the forests of the monarch butterfly reserves is limited by the predominant communal system of land tenure, which does not permit the transfer of land ownership. Difficulties in designing and carrying out a cohesive and integrated forest conservation plan are further compounded by multiple jurisdictions over the land on the part of nine municipalities, two states, and three federal secretariats [SEDUE, the Secretariat of Agriculture and Water Resources (SARH), and the Secretariat of Agrarian Reform (SRA)].

EXOGENOUS FACTORS OF FOREST DEGRADATION AND DESTRUCTION: POTENTIAL SOLUTIONS

Extraction of trees for industrial purposes can be controlled by the Forest Service. To date, however, they seem simply to have prohibited cutting during the 5 months of the monarch butterfly's overwintering period. This probably reflects their position as part of the SARH, whose objective is to stimulate and support primary production, as well as pressure by industrialists and ejidos who need the raw materials and the income derived from the sale and cutting of trees.

Yet it would probably be feasible to supply these industrial needs from the pine–oak forests below 2,700 m and outside the reserve areas. Forests in this region, like virtually all of Mexico's forests, are currently managed according to a system known as the "Método Mexicano de Ordenación de Montes" (the Mexican Method for Forest Control) (Musalem, 1979). The Mexican Method consists of selective, low-intensity extractions that take out 35–40% of the volume of desired trees (firs or pines). In mixed-species forests, this translates into only 15–20% of the total standing volume at the time of harvest (Snook and Negreros, 1986).

This selective system impedes regeneration of shade-intolerant conifers (Negreros and Snook, 1984; Snook and Negreros, 1986) and results in lower growth rates than those that result from other silvicultural systems. If cutting intensity were increased in pine–oak forests, sustained yields could probably be maintained at up to three times the current level, barring regeneration problems resulting from subsequent agricultural invasion or erosion (Snook-Cosandey, 1984; Snook and Negreros, 1986). Thus, with appropriate silviculture, it would be feasible to substitute the industrial raw materials supplied from the fir forest with pine extracted from the pine–oak forests outside the reserves.

This same silvicultural approach (and the same pine–oak forests, if their extent were sufficient) could potentially provide for the material needs of local people, currently met through exploitation of the fir forest. Pine, like fir, can be used for both fuelwood and construction materials, including roof shingles. Oak makes even better fuelwood and charcoal than fir and is rarely in demand by industry. Other species (e.g., alder and madrone) are currently unmanaged and underexploited and, although limited in extent and volume, have potential for use in making furniture and small craft items. To evaluate the potential for changing current patterns of forest use, however, a number of additional factors would have to be investigated. Among these are the boundaries of individual ejidos with regard to different forest types, the nearness of different forests to users, and user preferences for certain kinds of woods for particular uses.

If the productive potential of existing pine–oak and hardwood forests were insufficient to supply all these regional needs, plantations could conceivably be established to supplement them. However, the constantly increasing demand for agricultural and grazing land represents a severe limitation to tree planting, and plantations are so costly that they would have to be heavily subsidized.

Fuels and building materials derived from sources other than the forest could entirely supplant this material dependence on the forest. However, any product that must be bought rather than collected as a "free good" or produced through one's own labor requires a corresponding buying power that local people currently lack. There are other problems as well. For example, gas stoves are not designed to provide an adequate surface for cooking tortillas, the staple food.

Increased transformation of current forest products, such as raw pine resin, could stimulate economic development and employment in the area. Local sawmills provide raw materials for a few small furniture-making industries, and this level of transformation yields much more economic gain per m^3 of harvested timber than selling fuelwood, charcoal, shingles, or beams. Christmas tree plantations could yield high returns and simultaneously protect fir germplasm. An experimental Christmas tree plantation is currently being established at the site of an old sawmill in the Chincua area (Ogarrio, this volume).

Just as intensified silviculture could provide more wood from the existing forests outside the reserves, agricultural production could be intensified outside the fir forest zone to reduce the need for the continual opening of new fields. Current yields average only 500–1,000 kg/ha corn (Fundación Mexicana para el Desarrollo Rural, 1984), one-sixteenth the production of an Iowa cornfield. Similarly, extensive livestock grazing in the reserve areas could perhaps be supplanted by intensive livestock production methods combined with intensive forage production. By removing agricultural activities and grazing from the reserve areas, the threat of fires spreading into the forest from burns used to clear agricultural lands or renew pastures would be lessened.

However, while agricultural and silvicultural yields can be increased arithmetically through good management and technology, the population depending on these products is increasing exponentially. This means that local production will have to be supplemented by food produced elsewhere, implying a possible need for government subsidies and a cash economy.

Tourism development, although seasonal, could provide an alternative source of employment and economic development in the region, if it were

Table 4. Methods and limitations for the control of endogenous mistletoe, bark beetle, and defoliator infestations that contribute to the degradation of oyamel fir forests.

Alternatives for control	Limitations
Mistletoe	
Clearcut and replacement	Habitat destruction
Pruning of affected branches	Expensive, continuous habitat modification
Use of herbicides	Uncertain effectiveness, expensive, potential toxicity
Bark beetles	
Maintenance of forest vigor to prevent attack	Requires control of mistletoe and fires; not possible in infested areas
Extraction of affected trees	Habitat modification; requires knowledge of pest biology, competitors, and predators
Use of toxic substances to kill broods	Potentially toxic; expensive
Use of pheromones to trap or repel	Expensive to develop; complex to use
Defoliator	
Spraying with B.T.	Potentially toxic to monarchs
Encouragement of mixed-species stands	Long term; might require canopy opening, modifying habitat conditions
Introduction of hyperparasites	Requires considerable research; might affect monarchs

carefully designed for that purpose. Already, local people are being paid to keep tourists from disturbing the butterflies, and local truck drivers charge to bring visitors to the reserve areas from nearby towns. Scientific researchers studying the monarchs have also employed local people. By defining the most appropriate locations and forms of access and use and incorporating adequate infrastructure, planned development of tourism could also reduce the impact of visitors and associated services on the reserves' forests.

ENDOGENOUS FACTORS OF FOREST DEGRADATION AND DESTRUCTION: POTENTIAL SOLUTIONS

Although endogenous factors represent a lesser threat to the oyamel forests than exogenous factors such as cutting, clearing, and fires, dwarf mistletoe, bark beetles, and insect defoliators may significantly influence mortality and regeneration of the oyamel fir forests. Control of the mistletoe, *A. abietis-religiosae,* may or may not be a top priority for forest conservation and enhancement, depending on the extent and impact of the infestations. Management strategies based on the biology of the parasite have been developed to control mistletoe in commercial forests (Tables 1 and 4).

The dwarf mistletoe of oyamel fir flowers in March and April (Rodríguez, 1983). The female plants are wind-, insect-, or bird-pollinated (Smith, 1970; Rodríguez, 1983), and fruits mature in October or November. The mature fruit explodes, sending sticky seeds from 4 to 30 m (Rodríguez, 1983), where they adhere to neighboring branches, germinate, and begin to penetrate the cambium. There are no easily visible signs of infestation until years later, when the branches produce "witches brooms," or the mistletoe produces fruiting structures (Boyce, 1961; Smith, 1970). Mistletoe inoculated on firs in California did not fruit until 8 yr later (Scharpf and Parmeter, 1982). During this interval, infected trees cannot be recognized.

Mistletoe is most commonly controlled by clearcutting to eradicate seed sources and replanting with uninfected stock (Boyce, 1961). This does not seem an appropriate option for the monarch reserves. The pruning of infested branches to reduce the seed source is a less drastic alternative for management, but an expensive one, because pruning must be repeated at intervals in response to the long period of invisible development of the mistletoe (Boyce, 1961; Smith, 1970; Heidmann, 1983). It is possible that the resulting reduction of crown density could affect butterfly behavior and survival.

The spraying of chemical herbicides on aerial shoots of the mistletoe has also been proposed (Livingston, 1983). The effectiveness of these substances and their potential effects on butterfly populations would require analysis. Research is required to determine both the extent and the implications for forest mortality and regeneration (both natural and artificial) of the mistletoe infestations in the monarch reserves, to determine whether or not an applied control strategy is necessary. Then the relative costs and benefits of each control option, for forest health and butterfly survival, must be evaluated (Table 4).

The defoliator, *E. h. blandaria,* is a geometrid moth that lays its eggs on the underside of fir needles or understory plants. Larvae emerge and feed on fir needles from June to April, when they become

Table 5. ***In situ* and *ex situ* forestry activities for the conservation of monarch butterfly overwintering habitat in the high-altitude oyamel fir forests of Mexico.**

In situ
Control of cutting
Protection and stimulation of natural regeneration
Artificial regeneration (planting)
Tourism management
Management of dwarf mistletoe
Management of bark beetles
Management of defoliators
Ex situ[a]
Plantations for fuelwood
Plantations for income (Christmas trees)
Intensified management of neighboring pine and pine–oak forests
Establishment/improvement of forest-based transformation industries
Tourism development
Prevention of spread of pests into reserve forests

[a] Other activities, outside the realm of forestry, would have to be undertaken as well.

chrysalids. Swinging from branch to branch on silken threads and carried by the wind to neighboring trees, they defoliate an expanding patch of forest. Apparently the larvae become dormant during the cold winter months. Adult moths emerge in June and July. One generation occurs per year, and outbreak populations occur approximately every 10 yr (Sangri, 1962; Carbajal and López, 1987). When larval populations were sampled during the 1986 outbreak, the highest density was found on 1,131 ha of relatively monospecific fir forest bordering the Chincua monarch reserves. To prevent their spreading into the monarch overwintering area, the SARH decided to spray this area.

While DDT and Sevin had been used to control previous outbreaks, the risks associated with these toxins led the pest control team to choose Dipel, a commercial preparation of *Bacillus thuringiensis* (B.T.), a bacterial pathogen of Lepidoptera used to control many agricultural pests and the so-called spruce budworm, an important forest pest in the United States. During October, November, and December of 1986, a helicopter was used to spray 750 g/ha of Dipel powder, mixed with an adherent substance, from 10 m above the treetops, over 2,195 ha of oyamel fir forest. Twenty-one days after spraying, the number of larvae had dropped by 75%, although it is not known how much of this was due to the spraying and how much was due to hailstorms and cold weather (López and Carbajal, 1987). Laboratory experiments on the effectiveness of B.T. in reducing defoliation of oyamel fir by *E. h. blandaria* larvae revealed that 50% of sample trees dipped in 10,000 ppm of B.T. were 50% defoliated. Only 30% of the trees dipped in concentrations of 1,000 ppm were >50% defoliated. All of the untreated control trees were >50% defoliated. Forty-eight hours after treatment, 91% of the larvae subject to the highest concentration were dead, as compared with 81% at 1,000 ppm. Only 4% of the control larvae had died. B.T. was most effective in reducing defoliation when applied to larvae during the third instar (Alatorre, 1987).

The decision to spray a lepidopteran pathogen so close to the monarch's overwintering area aroused an outcry on the part of scientists and conservationists concerned with the monarch butterfly. As pointed out by Brower (1986), the effects of Dipel on adult butterflies was unknown; carriers and contaminants in the Dipel liquid could include toxins; sugars in the spray could stimulate overwintering monarchs to excessive drinking, leading to ingestion of high quantities of B.T. spores, which could be spread in monarch feces during their return migration, contaminating the milkweed food source of the next generation's larvae; and B.T. spores would remain present and active in the sprayed forests for months. Data are not yet available on the impact of the spraying on overwintering monarchs. However, since future outbreaks of *E. h. blandaria* populations are certain to occur, research on alternative control methods including silviculture (favoring mixed over pure fir stands), and introduction of parasitoids of the moth (work initiated by Sangri, 1962) might be worthwhile.

The most important approach to the control of bark beetles is silvicultural: the prevention of conditions of stress to reduce the susceptibility of trees to attack (Table 4). This is achieved by cutting to control stand density and assure vigor and by preventing fire and disease infestations, including parasites and defoliators (Graham and Knight, 1965). Cutting is usually oriented toward trees determined to be susceptible to attack, according to a risk-rating system (Ferrell, 1973). The fir forests in question are not dense and appear to be made up of relatively young trees. Thus, forest thinning, which would negatively affect butterfly habitat quality, would seem unnecessary. However, control of mistletoe, defoliators, and fires might be important to reduce the risk of bark beetle outbreaks.

If bark beetle infestations were located, they could be controlled on a tree-by-tree basis to ensure that the larval broods are killed or removed from the site before their emergence as adults. This requires knowledge of the biological cycle of the beetles, which varies among species, among hosts, and among climatic conditions (Graham and Knight, 1965). *Scolytus mundus* produces two generations a year in firs in Hidalgo, with adults emerging May–July and October–January (Hernández and Cibrián, 1981), and although this is likely to be the species that attacks oyamel fir in the states of Michoacán and México, there has been no research done on its biology in the monarch reserves.

Bark beetle broods can be killed by extraction and processing of affected trees, by felling and burning, by bark peeling and sunkill, or by application of toxic substances to infested logs (Anderson, 1960; Table 4). Endosulfan and ethylene dibromide are effective in killing *S. ventralis* broods, but they contaminate water and are toxic to fish, bees, and other wildlife, so they would hardly be appropriate in the context of conserving monarch overwintering habitat (Hamel, 1981). In addition, toxic substances reduce beetle enemy and predator populations as well, thus militating against natural population control mechanisms (Swezey and Dahlsten, 1983).

Other bark beetle control methods (used principally against *Dendroctonus* species) include "trapping" of beetles in trees baited with attractant pheromones, and then removed (Furniss et al., 1972), or in traps baited with such substances (Table 4). Other substances can be applied to trees to repel beetles (Rudkinsky, 1968; Furniss et al., 1972; Payne and Richardson, 1979; Nijholt and McMullen, 1980). In Mexico, copper sulfate and sodium carbonate injections in healthy pines have been used with mixed success, combined with felling and covering of infested trees, to prevent the spread of *Dendroctonus* infestations (González, 1981; Perusquia, 1981). Pheromones can also be used to contain infestations (Richerson et al., 1980) or to reduce attack (St. Clair et al., 1977).

While potentially valuable pest management methods have been applied in other forests to manage dwarf mistletoe, defoliators, and bark beetles, pest control must be both species- and site-specific and compatible with management objectives. In the case of the monarch butterfly reserves, neither the bark beetle species and their biological cycles nor the extent of infestation is known, nor is the relationship of infestation to fir mortality. Observations elsewhere in Mexico indicate that *S. mundus* does not always kill infested trees, although trees weakened by this beetle are often killed by secondary infestations of *P. variegatus* (Hernández and Cibrián, 1981).

FORESTRY ACTIVITIES FOR THE CONSERVATION OF MONARCH OVERWINTERING HABITAT

Forest management for the conservation of monarch overwintering habitat in Mexico requires simultaneous *in situ* endeavors within the reserves and *ex situ* undertakings outside the reserve areas (Table 5). In combination, these activities should accomplish two integrated objectives: (1) the maintenance and improvement of forest structure to protect and enhance butterfly habitat characteristics, and (2) the reduction of anthropogenic pressures through economic development projects designed to provide, from other sources, the material goods and income currently obtained from the fir forest.

To design specific tactics to accomplish these two objectives, considerable research is required to quantify and evaluate the spectrum of threats to the forests where the monarch butterflies overwinter and their range of solutions. Such research should be carried out as part of a series of actions integrating the coordinated participation of the three secretariats responsible for land-use management in the area (SARH, SEDUE, and SRA), the state governments of Michoacán and México, nine county governments (Municipios), more than 36 ejidos, and private property owners. Strategies and tactics should incorporate adequate incentives for local residents and decisionmakers to modify the current situation.

CONCLUSIONS

The monarch butterfly has adapted to climatic change and modifications in the distribution of vegetation over time by expanding its range. Over the same period, without recourse to seasonal mobility, the tropical fir forest has been in retreat. While monarchs have evolved an elaborate life history characterized by chemical defenses and long-distance migration (see Cockrell et al., Malcolm et al., Lynch and Martin, and Nelson, this volume), which contributes to their success as a species, the high-altitude fir forests of Mexico have become increasingly fragmented, isolated, stressed, and vulnerable to natural and human depredations. Consequently, the entire eastern population of the monarch butterfly, numbering hundreds of millions of individuals, is now threatened because the Mexican fir forest is endangered, a victim of structural and systemic features of Mexican forest land use. Outstanding among these is the subsistence economy of Mexico's rural population.

The degradation and destruction of Mexico's forests have inspired concern among a conscious few for decades, and yet the trend of deforestation has accelerated with time. Perhaps international recognition of the value of the monarch migration as an outstanding feature of the biological heritage of the planet will provide an impetus for the design and implementation of a plan to counter this pattern. To the extent that this international concern can be translated into an effective strategy permitting local people and institutions to profit from conservation, the monarch butterfly will have contributed greatly to the development of an alternative to deforestation in Mexico.

ACKNOWLEDGMENTS

This paper is based on fieldwork carried out as part of a consultancy with Dr. J. Kathleen Parker, to World Wildlife Fund–US and Monarca, A.C. Information about the *Evita hyalinaria* outbreak was made available through the kindness of Biol. Ignacio Carbajal V. and Biol. Jorge Macías, of the Dirección General de Sanidad y Protección Agropecuaria y Forestal, SARH.

LITERATURE CITED

Alatorre, R. 1987. Susceptibilidad del defoliador del oyamel *Evita hyalinaria blandaria* (Dyar) (Lepidoptera: Geometridae) al bioinsecticida a base de *Bacillus thuringiensis*. *Memoria, IV Simposio sobre Parasitología Forestal, Durango, Durango,* 28–30 Octubre 1987.

Anderson, R.F. 1960. *Forest and shade tree entomology*. New York: John Wiley, 428 pp.

Boyce, J.S. 1961. *Forest pathology*. New York: McGraw-Hill.

Brower, L.P. 1985. New perspectives on the migration biology of the monarch butterfly, *Danaus plexippus* L. In *Migration: Mechanisms and adaptive significance*, ed. M.A. Rankin, 748–785. Austin: University of Texas, *Contributions in Marine Science*, Supplement, 27.

Brower, L.P. 1986. Commentary: The potential impact of DIPEL spraying on the monarch butterfly overwintering phenomenon. *Atala* 14(1):17–19.

Brower, L.P., and W.H. Calvert. 1985. Foraging dynamics of bird predators on overwintering monarch butterflies in Mexico. *Evolution* 39(4):852–868.

Calvert, W.H., and L.P. Brower. 1981. The importance of forest cover for the survival of overwintering monarch butterflies (*Danaus plexippus*, Danaidae). *Journal of the Lepidopterists' Society* 35:216–225.

Calvert, W.H., W.H. Hedrick, and L.P. Brower. 1979. Mortality of the monarch butterfly (*Danaus plexippus* L.): Avian predation at five overwintering sites in Mexico. *Science* 204:847–851.

Calvert, W.H., W. Zuchowski, and L.P. Brower. 1982. The impact of forest thinning on microclimate in monarch butterfly (*Danaus plexippus* L.) overwintering sites in Mexico. *Boletín de la Sociedad Botánica de México* 42:11–18.

Carbajal, I., and R. López. 1987. Evaluación del daño causado por el defoliador del oyamel *Evita hyalinaria blandaria* (Dyar) (Lep: Geometridae) en el Municipio de San Felipe del Progreso, Edo. de México. *Memoria, IV Simposio sobre Parasitología Forestal, Durango, Durango,* 28–30 Octubre 1987.

Diario Oficial. 1986. Organo del Gibierno Constitucional de los Estados Unidos Mexicanos. México, D.F., Jueves 9 de Octubre de 1986 398(27):33–41.

Espejo, A., J.L. Brunhuber, G. Segura, and J. Ibarra. 1985. Estudio de la vegetación de la zona de hibernación de la mariposa Monarca en la Sierra Chincua. *Contribución voluntaria, IX Congreso Forestal Mundial, México, D.F.,* 1–11 Julio 1985.

Ferrell, G.T. 1973. *Stand and tree characteristics influencing density of fir engraver beetle attack scars in white fir*. USDA Forest Service Research Paper No. PSW-97, 8 pp.

Fundación Mexicana para el Desarrollo Rural. 1984. Reporte Socioeconómico sobre los Municipios de Angangueo y Ocampo, Michoacán (unpubl.).

Furniss, J., L.N. Kline, R.F. Schmitz, and J.A. Rudinsky. 1972. Tests of three pheromones to induce or disrupt aggregation of Douglas fir beetles (Coleoptera: Scolytidae) on live trees. *Annals of the Entomological Society of America* 65:1227–1232.

González, C.E. 1981. Algunas experiencias sobre el combate de descortezadores de los pinos. *Memoria, Primera Reunión sobre Plagas y Enfermedades Forestales Mexicanos,* 12–14 Septiembre 1979. Publicación Especial, Instituto Nacional de Investigaciones Forestales 32:45–48.

Graham, S.A., and F.B. Knight. 1965. *Principles of forest entomology*. New York: McGraw-Hill, 417 pp.

Hamel, D.R. 1981. *Forest management chemicals*. USDA Forest Service Agricultural Handbook #585, p. 145.

Heidmann, L.J. 1983. *Silvicultural control of dwarf mistletoe in southwestern ponderosa pine*. Research Note No. RM-433, USDA Forest Service, 4 pp.

Hernández, E., and D. Cibrián. 1981. Observaciones sobre la biología de *Scolytus mundus* Wood, un descortezador de *Abies religiosa* (H.B.K.) Schl. et Cham. en el Parque Nacional "El Chico", Hidalgo. *Chapingo, Nueva Epoca* 31–32:14–20.

Horst, R.K. (reviser). 1979. *Westcott's plant disease handbook*, 4th ed. New York: Van Nostrand Reinhold, 803 pp.

Livingston, W.H. 1983. Ethephon stimulates abscission of eastern dwarf mistletoe aerial shoots on black spruce. *Plant Disease* 67(8):909–910.

López, R., and I. Carbajal. 1987. Aplicación aerea de *Bacillus thuringiensis* para el control del defoliador del oyamel *Evita hyalinaria blandaria* (Dyar) (Lep. Geometridae) en el municipio de San Felipe del Progreso, Edo. de México. *Memoria, IV Simposio sobre Parasitología Forestal, Durango, Durango,* 28–30 Octubre 1987.

Manzanilla, H. 1974. *Investigaciones Epidométricas y Silvícolas en Bosques Mexicanos de Abies religiosa*. Dirección General de Información y Relaciones Públicas, Secretaría de Agricultura y Ganadería, México, D.F.

Monarca, A.C. 1984. *Propuesta del Plan Maestro para la Protección de la Mariposa Monarca en México,* Anexo 1, Antecedentes. R. Brañes, J.L. Cortés, X. Chiappa, D.F. de Castro, F. Ortíz, V. Sánchez.

Musalem, F.J. 1979. *Las Bases y Primeras Acciones del Programa Nacional de Mejoramiento Silvícola, en Bosques de Coníferas*. Productos Forestales de Michoacán (PROFORMICH), Morelia, Michoacán, 40 pp.

Negreros, P., and L.C. Snook. 1984. Análisis del effecto de la intensidad de corta sobre la regeneración de pinos en un bosque de pino-encino. *Ciencia Forestal* 47:48–61.

Nijholt, W.W., and L.H. McMullen. 1980. Pine oil prevents mountain pine beetle attack on living lodgepole pine trees. *Bimonthly Research Notes* 36(1):1–2 [Pacific Forest Research Centre, Victoria, British Columbia, Canada. (Forestry Abstracts 1980 #6982).]

Payne, R.L., and J.V. Richardson. 1979. Management implications of inhibitors for *Dendroctonus frontalis*. *Proceedings, IUFRO Conference, Zurich and Zuoz, Switzerland,* 4–9 September, 1978, pp. 323–331. (Forestry Abstracts 1980 #2552.)

Perusquia, O. 1981. Principales plagas forestales. *Memoria, Primera Reunión sobre plagas y enfermedades forestales, México, D.F.,* 12–14 Septiembre 1979. (Publicación Especial, Instituto Nacional de Investigaciones Forestales 32:15–18.)

Reyes, J.J., and J. de la Maza. 1982. *Plan de manejo*

para las áreas de hibernación de la mariposa Monarca (Danaus plexippus) en el estado de Michoacán, México. SARH/Subsecretaría Forestal, y de la Fauna, Dirección de Fauna Silvestres, México, D.F.

Richerson, J.V., F.A. McCarthy, and T.L. Payne. 1980. Disruption of southern pine beetle infestations with Frontalure. *Environmental Entomology* 9(1):90–93.

Rodríguez, A. 1983. Muérdago enano sobre *Abies, Pinus,* y *Pseudotsuga* en México. *Ciencia Forestal* 8(45):7–45.

Rudkinsky, J.A. 1968. Pheromone mask by the female *Dendroctonus pseudotsugae,* an attraction regulator. *Pan-Pacific Entomologist* 44:248–250.

Rzedowski, J. 1978. *Vegetación de México.* México, D.F.: Editorial Limusa, 432 pp.

Rzedowski, J., L. Vela, and X. Madrigal. 1977. Algunas consideraciones acerca de la dinámica de los bosques de coníferas en México. *Ciencia Forestal* 2(5):15–35.

Sangri, L. 1962. Observaciones y combate de la plaga defoliadora de oyamel (*Evita hyalinaria blandaria*). México y sus Bosques, Epoca II, No. 2, Abril–Mayo–Junio, pp. 23–27.

Scharpf, R.F., and J.R. Parmeter, Jr. 1982. *Population dynamics of dwarf mistletoe on young true firs in the central Sierra Nevada, California.* USDA Forest Service Research Paper No. PSW-161, 9 pp.

Smith, W.H. 1970. *Tree pathology: A short introduction.* New York: Academic Press.

Snook, L.C., and P. Negreros. 1986. Effects of Mexico's selective cutting system on pine regeneration and growth in a mixed pine–oak (*Pinus–Quercus*) forest. *Current Topics in Forest Research: Emphasis on Contributions by Women Scientists.* USDA Forest Service General Technical Report SE-46, pp. 27–31.

Snook-Cosandey, L. 1984. El papel del desarrollo tecnológico en la resolución de algunos problemas silvícolas. *Alternativas Tecnológicas VI,* Consejo Nacional de Ciencia y Tecnología/Academia Mexicana de Ingeniería, México, D.F., pp. 219–224.

St. Clair, A., R.L. Goulding, and J.A. Rudinsky. 1977. Controlled-release dispensing system for 3,2-MCH, antiaggregative pheromone of the Douglas fir beetle (Col.: Scolyt.) *Zeitschrift für Angewandte Entomologie* 83(3):297–300.

Swezey, S.L., and D.L. Dahlsten. 1983. Effects of remedial application of lindane on emergence of natural enemies of western white pine beetle *Dendroctonus brevicomis* (Coleoptera: Scolytidae). *Environmental Entomology* 23(1):210–214.

Urquhart, F.A. 1976. Found at last: The monarch's winter home. *National Geographic* 150:160–173.

Wells, S.M., R.M. Pyle, and N.M. Collins. 1983. *The IUCN invertebrate red data book.* Gland, Switzerland: IUCN, 632 pp.

Young, A.M. 1982. An evolutionary-ecological model of the evolution of migratory behavior in the monarch butterfly and its absence in the queen butterfly. *Acta Biotheoretica* 31:219–237.

CONSERVATION ACTIONS TAKEN BY MONARCA, A.C., TO PROTECT THE OVERWINTERING SITES OF THE MONARCH BUTTERFLY IN MEXICO

Rodolfo Ogarrio

The migratory and overwintering phenomenon of the monarch butterfly population, which migrates from southern Canada and the United States, east of the Rocky Mountains, to the Transvolcanic Belt in central Mexico, is unique in nature. The phenomenon is also an especially complex problem for conservationists in Mexico.

Each year approximately 100 million monarch butterflies form overwintering colonies from November through March in two neighboring states: Michoacán and México. These states have different administrative structures to deal with state parks, reserves, and natural resources. At the federal level, the Ministry of Urban Development and Ecology and the Ministry of Agriculture and Hydraulic Resources share jurisdiction and responsibilities for the protection of the high-altitude forests in which the butterflies overwinter. These overwintering colonies are also located in areas with different land tenure systems. A small part of the overwintering area was privately owned, and so it had few restrictions on property rights. Therefore, this property of 700 ha was bought by a trust established by Monarca, A.C., and both state and federal governments. However, the largest portion of the overwintering area is held by the local inhabitants under a community property system called "ejido." Ejidatarios, or beneficiaries of the system, can work the land, but they cannot sell or rent the land. Conditions for ejidatarios are extremely difficult. Increasing population numbers, lack of education, meager incomes, and low agricultural productivity are among the major problems facing them. This situation is generating great pressure upon the forests where the butterflies overwinter, since the ejidatarios are clearing the forests to obtain wood for fuel and construction and to enlarge the agricultural frontier.

Since its inception, Monarca, A.C., realized that the most effective means to preserve nature was through the socioeconomic improvement of those communities that exploit the natural forest resources. This promotes local involvement and cooperation in an environmentally sound conservation process, including economic development. Based on this premise Monarca, A.C., began to implement its conservation objectives.

To achieve the objectives, the association encouraged the social interaction and participation of all sectors related to protection of the monarch butterfly's overwintering sites. Thus, we involved the government, the local communities and private sector, the general public, and the international conservationist community in our conservation efforts.

Monarca, A.C., has searched for practical solutions to conservation problems, based on Mexican realities such as Mexico's political situation, existing economic constraints, and the cultural background of those communities involved. In September 1984 a trust was established by Monarca, A.C., with the participation of the federal government, through the Ministry of Urban Development and Ecology, the Ministry of Agriculture and Hydraulic Resources, and the state governments of Michoacán and México, to preserve the monarch's habitat in Mexico. The trust purchased the aforementioned 700 ha on 4 September 1986, at Sierra Chincua, in the state of Michoacán, where the butterflies overwinter. On 9 October 1986, 16,000 ha were declared ecological reserves by a presidential decree, which gives legal protection to the areas. This was achieved through the joint participation of Monarca and the government.

The type of management proposed by the association at the butterfly sanctuaries has been chosen by the government as an example to be followed on other protected areas in the country. This confidence led the Ministry of Urban Development and Ecology to entrust Monarca with the management of part of the sanctuary at Sierra Chincua. Among the specific projects that Monarca, A.C., has promoted are the following (see Camus de Castro, this volume).

1. In El Rosario, one of the sanctuaries open to the public, Monarca, A.C., has worked with the ejidatarios to manage tourism in the sanctuary. The association has also helped ejidatarios to organize a community shop and to sell articles related to the monarch butterfly phenomenon, such as postcards, books, hand-painted tiles, and other handcrafts. These are sold at a profit for the community. Profits go in their entirety to the local people, who thus obtain an alternative source of income to improve their living standards (see Rojas Gonzales de Castilla, this volume).

2. The local community at El Rosario has been organized to manage tourism, and visitors are provided with parking lots, restrooms, guides, and a snack section. More importantly, an interpretative trail was established and a visitors' center erected, with information related to the monarch butterfly and its migratory phenomenon, as well as its importance and the necessity to protect it.

3. Through the media, Monarca has fostered a broad public awareness of the monarch phenomenon, and an educational kit was designed to promote environmental education among children (see Camus de Castro, this volume; Ortíz Monasterio, this volume).

4. Educational programs are being implemented, directed at schoolchildren, teachers, and the general public, as well as training courses for guards and guides.

5. A tree nursery was created in the Sierra Chincua sanctuary, an area devoted to scientific research. The nursery will provide trees for both reforestation of denuded areas in or around the sanctuaries and to sell as Christmas trees for the benefit of local ejidatarios.

The association has also financed scientific research on the butterfly and its migratory phenomenon, through grants from international agencies such as the World Wildlife Fund. We have also received the cooperation of the Canadian Government, through the Canadian Embassy in Mexico. Monarca, A.C., is participating with Mexican and American scientists to investigate the interaction between the monarch and its forested overwintering habitat. For example, we are researching the relationship between mortality and habitat structure at several overwintering colonies; the capacity of the forest to regenerate; dendroecology and dendrochronology; and the importance of low vegetation for the survival of the monarch butterfly.

Since its inception in 1977, the private conservation association Monarca, A.C., has worked closely with the Mexican government, local inhabitants in conservation areas, and the international community to try to demonstrate that working together and taking into account the real needs of the local inhabitants is the only way to advance toward a better balance between man and his environment.

MONARCH CONSERVATION IN MEXICO: THE CHALLENGE OF MEMBERSHIP AND FUND-RAISING FOR MONARCA, A.C.

Carlos F. Gottfried Joy

ABSTRACT. On 19 September 1985 the Mexican conservation group Monarca, A.C., dedicated to the conservation of monarch butterflies, underwent a self-reorganization in which several new programs were initiated, one of which was to open the association to public membership. The main objectives of these new programs are to create both a source of volunteers and a broad base of public support, to ensure the successful realization of the association's ambitious conservation goals. This paper deals with the major obstacles before the membership program could be implemented, the activities that have taken place, the present status, and future plans.

INTRODUCTION

On 19 September 1985 Monarca, A.C., organized a 4-day meeting to plan for self-reorganization and the development of new strategies that would further ensure conservation of the monarch butterfly overwintering phenomenon. On that day, Mexico, and especially Mexico City, experienced the strongest and most devastating earthquake in its known history. Despite the devastation and chaos the group was sufficiently determined to proceed with the meeting and convened 80 km from Mexico City. That meeting was the beginning of a new strategy to develop a membership and fund-raising program.

Until the meeting Monarca, A.C.'s role had been principally advisory, giving assistance and information to the Mexican government, to local inhabitants of some of the overwintering sites, to interested conservation organizations, and to interested individuals. Monarca, A.C., also undertook social, economic, and scientific studies related to the monarch butterfly overwintering phenomenon. Monarca, A.C., decided that more action was required and opened the organization to public membership so that more effective conservation of the overwintering phenomenon could be achieved.

Going public in any organization is difficult, but in Mexico matters are complicated by many complex factors such as fiscal laws, governmental and public acceptance, and current economic status.

The first challenge for the foundation, however, was not external to the organization but internal: We had to define what we stood for, what we wanted to accomplish, and how we were going to do it. In other words, what was going to be the role and image of Monarca, A.C., both internally and publicly? The following account was a result of team work by Monarca, A.C., members and officers.

OBJECTIVES AND ACTIVITIES OF MONARCA, A.C.

Our overall mission is to guarantee the conditions that will permit the continuance of the migration and overwintering monarch butterfly phenomenon on the North American continent.

In order to achieve this mission we consider that the following values, responsibilities, objectives, performance indicators and authority are important.

VALUES

We encourage team work, dedication, perseverance, joint efforts, open-mindedness, moderation, decision implementation, commitment, and conviction in our mission and responsibility for the outcome.

Furthermore, we consider that it is important to (1) respect the integrity of the ecosystem of which the monarch butterfly is a part; (2) recognize and respect the political and administrative authorities who are legally responsible for protecting the migrating and overwintering areas; (3) recognize and respect the basic needs of the local inhabitants that share the forest resources with the monarch butterflies; (4) recognize the obligation that we have to protect a unique international phenomenon on our planet which occurs across at least three countries and of which we are only its trustees; (5) recognize that the role of the foundation consists of ensuring the effective protection of this unique phenomenon, while acknowledging and cooperating with other conservation efforts; and (6) acknowledge that human existence depends on the maintenance of the integrity of natural ecosystems and that wildlife species, both plant and animal, have a right to existence, without the need to justify that existence in terms of its usefulness to mankind. Man is also a part of, and not external to, natural ecosystems, and human existence must consequently be guided by ecological principles.

RESPONSIBILITIES

Monarca, A.C., is assuming responsibility for providing rational, feasible, and economic alternative life-styles for the local inhabitants affected by the phenomenon, so that the migrating and overwintering forest resources may be conserved as an adequate habitat for monarch butterflies. To do this we are attempting to obtain the necessary human, financial, and material resources for an effective conservation operation. We are also interacting with the political and administrative authorities who are legally responsible for protection of the overwintering areas.

These responsibilities are being evaluated in relation to their effectiveness in fulfilling our objectives. We take responsibility for setting an example of sound conservation principles and practices in our individual, everyday activities and for communicating our objectives, values, acquired knowledge, and results. We also consider it a prime responsibility to develop local, national, and international representation.

OBJECTIVES

Within our overall mission our primary objectives are fivefold: (1) to strengthen the structure, membership, and participation in the foundation to assure its effective growth and development; (2) to promote and support the protection of the overwintering habitat through promotion of concrete and tangible actions; (3) to encourage conservation of the migrating and overwintering areas through local action and participation; (4) to promote respect, love, and social consciousness toward nature in the public, through the implementation of educational and training programs related to the monarch butterfly migrating and overwintering phenomenon; and (5) to promote and support basic, applied, and scientific research that will permit wider knowledge and conservation of the migrating and overwintering phenomenon.

PERFORMANCE INDICATORS

We are assessing the success of our operations by reviewing the levels of contributions and collaboration from the sympathizers and members of the foundation and the public image of the foundation. Our research activities also allow us to review regularly the status of the monarch butterfly migrating and overwintering phenomenon.

AUTHORITY

Our authority to implement monarch conservation extends to doing anything necessary that the law does not forbid and only what the bylaws of the foundation authorize it to do. We also consider it important to implement only those actions allowed according to community consensus.

CONSTRAINTS

External challenges to the activities and organization of Monarca, A.C., begin with the Mexican fiscal laws. There are basically two government permits required to operate successfully as a nonprofit organization. The first permit allows for operation and filing as a nonprofit organization. The second allows for the tax deduction of donations and membership fees by the donor or member. We obtained the first permit, but we also required the second permit since failure to obtain this status is the main reason most private nonprofit organizations fail, by preventing them from operating in a realistic manner. Without donations and membership fees being tax-deductible, the possibilities of obtaining substantial funds to operate are greatly reduced, since most organizations and individuals can only afford substantial donations if they are deductible from their taxes. To complicate matters, Monarca, A.C., had previously applied for and been denied this permit. The law specifically states that only those donations going to previously approved organizations, which are principally used for education, may be entitled to the tax-deductible status. There are exceptions, but these are extremely rare, with approval only having been given at ministerial or presidential level. Knowing that our chances were very small, we nevertheless proceeded to develop the framework of our application for this special permit.

The second of the external challenges is that of the traditional implementation of conservation in Mexico. By virtue of cultural and legal traditions, it is broadly accepted by the public and the government that conservation must be implemented by the government. This has left little room for the active participation of private organizations, and as a result, private civic action with a volunteer force had been indirectly discouraged, by both the law and tradition. Volunteerism as a way of life, and as a way of conserving and improving the environment, has not been developed in Mexico. However, we believe that there is a largely frustrated and caring public that wants to participate actively in conservation of the monarch butterfly, but until now has not found the vehicle for doing so. We see the membership program in Monarca, A.C., as the vehicle for positively joining the concern and efforts of many individuals into a coordinated volunteer force.

The third of the external challenges is the critical economic status of the country and its people. With a progressively weakening economy, the available private and public resources that can be given to conservation of the monarch butterfly overwintering phenomenon are being reduced. Rallying support for conservation of this phenomenon when other very pressing needs are unsatisfied is difficult.

SOLUTIONS

Obtaining the fiscal permit that allows donations and memberships to be tax-deductible was the most important and pressing solution to the financial constraints on our activities; without it we could not operate our fund-raising and membership program. In November 1985, we convened with the Minister of Treasury, Lic. Jesús Silva Herzog. Not only were we well received, but the Minister showed a personal interest in the butterflies. On 20 August 1986 all permits and necessary documents were issued to Monarca, A.C., enabling the membership and fund-raising campaign to continue.

On 15 June 1986 the Minister of Urban Development and Ecology called upon Monarca, A.C., to develop jointly the policies and laws regulating the Mexican overwintering sites, thus beginning a new era of cooperation between the Government and private conservation organizations. This new governmental attitude was instigated by the President of Mexico, in a speech in which he acknowledged the limitations of the Government and the national economy and encouraged private civic action to assist and participate where possible in conservation efforts. This attitude will dramatically increase the cooperation between private, nonprofit organizations and Government agencies and will provide the opportunity to use volunteer resources in important joint programs. It will be up to Monarca, A.C., to motivate current members, attract new members, and organize these programs.

Since the declaration of governmental support for private conservation efforts in Mexico, Monarca, A.C., has gratefully received grants of financial support from the following national and international organizations: The World Wildlife Fund, The McArthur Foundation, International Union for the Conservation of Nature, International Fund for Animal Welfare, The Nature Conservancy, The Mans Foundation, The Los Angeles County Museum of Natural History, Office of Technology Assessment, and the U.S. Congress. We also propose to raise funds from our membership as well as other conservation-conscious individuals, organizations, and corporations. Membership of Monarca, A.C., has been grouped into individual and corporate memberships to encourage fund-raising. Individual memberships include student, single, family, contributor, donor, benefactor, sponsor, and patron categories, while corporate memberships include benefactor, sponsor, and patron categories.

Fund-raising will also be helped by the production and sale of monarch-related artifacts such as postcards, ceramic hand-crafts, tiles, calendars, books, guides, posters, T-shirts, and hats (see Rojas González de Castilla, this volume). We will also sell to private and public corporations the use of the monarch trademark, audio-visual presentations, and both movie and video films on the monarch (artistic, scientific, and documentary) along with other written or filmed materials (see Ortíz Monasterio, this volume).

Within the overwintering habitat we are also organizing guided tours to monarch aggregations from November to February. Tours are not available in March because this is the critical remigration period when monarchs leave their overwintering sites in Mexico.

PLANS FOR THE FUTURE

The membership program began in October 1986, with a mass-mailing campaign. Membership from all parts of the world will be gladly welcomed. This will be done from a previously compiled list of: previous interested supporters (ca. 500), recipients of the "MONARCAS" book (ca. 3,000), "MONARCAS CALENDAR" (ca. 2,000), recorded visitors with addresses from 10 yr of records from over 20 different countries (ca. 10,000). The mass-mailing campaign will be followed by a publicity campaign in newspapers, magazines, periodicals and journals, television interviews, and radio interviews.

A permanent Monarca, A.C., exhibition and orientation lodge will be set up at the Cerro Campanario (El Rosario) overwintering site, where visitor membership will be encouraged.

CONCLUSION

This is the first time in Mexico that a private conservation organization has attempted a campaign of habitat and species conservation. It is a vulnerable, highly exposed campaign that will be observed critically by the government, the general public, and all other interested parties. Thus, Monarca, A.C., has set an important precedent for future success or failure in the conservation of Mexico's natural environment.

EDUCATION TO PROMOTE PROTECTION OF THE MONARCH BUTTERFLY IN MEXICO: A DISCUSSION OF THE CONSERVATION EDUCATION PROGRAM INITIATED BY MONARCA, A.C.

Fernando Ortíz Monasterio

The objective of the educational program of Monarca, A.C., is to "promote a change of attitudes in the population, both locally and nationally, making the butterfly the focal point of awareness towards a more balanced relationship between society and nature." This statement is to us a paradigm of integrated conservation and has been the key to a number of educational policies, programs, and actions undertaken by Monarca, A.C.

The declaration of Tblissí (UNESCO/PNUMA, 1977) seeks to promote among member countries of the United Nations environmental education as part of educational curricula. At Monarca, A.C., we consider that the monarch butterfly is an ideal model for environmental education. The monarch is now a symbol of the need to protect nature, as well as a symbol of optimism for successful environmental conservation. Just as soil must be prepared for the seed to grow, human groups require an atmosphere of confidence and credibility before any change of attitudes can take place (Sánchez, 1981). In the years I have worked on nature conservation, one of my richest experiences has been witnessing the power of the monarch as a model system, which illustrates the complex interactions between society and nature and the many opportunities for learning from these interactions (Ortíz Monasterio P. and Ortíz Monasterio Garza, 1984).

Monarca, A.C., has been contributing to the environmental education of authorities and local populations in areas where monarchs overwinter so that the importance of conservation is understood. We are also establishing that alternative sources of income can be generated by protecting the monarch (see Rojas González de Castilla, this volume). Through education, local people, or ejidatarios, have developed an appreciation for long- and short-term benefits that they can obtain by conserving monarchs.

As the foundation of our educational program we have sought to understand the historical basis of current land-use and socioeconomic patterns so that we can steer communities toward greater environmental awareness. Although it is not known for how long monarchs have migrated to and from Mexico, the United States, and Canada, we are able to communicate our current understanding of this migratory phenomenon at different levels through lectures, books, films, television, and radio. Thus education has been promoted by Monarca, A.C., at international, federal, state, municipal, and ejido levels of organization. At each of these levels, the content, language, style, and visual presentations are aimed specifically to promote environmental awareness and responsibility. One of the most stimulating successes of Monarca, A.C., is that it has been able to promote environmental responsibility at all levels and obtain both practical and legislative support (Ogarrio, this volume) (see also the other papers on Mexican conservation in this volume).

A decade ago it was considered essential that the location of monarch overwintering sites in Mexico should be kept secret to protect the monarch. Time has shown that most environmental destruction to overwintering habitat is produced by random and commercial logging activities. These activities were performed in ignorance of the forest habitat requirements for monarch overwintering and so Monarca, A.C., promoted education so that forest exploitation could be made compatible with conservation. Now we consider that it is important to communicate as much as possible about the location and biology of monarch overwintering. In this way we will promote enlightened appreciation for the monarch and environmental conservation in general. For example, during an interview with an ejidatario some years ago at the El Rosario overwintering site in Michoacán, he stated that he was cutting down the trees for his children. The same man, a decade later, was building a wooden fence to establish the limit of the protected core of the monarch sanctuary, and when asked why he was building it, he answered again, "for my children." Thus what was a major problem of habitat destruction has become a major opportunity of resource renewal for local people, as a result of education.

To implement our educational program Monarca, A.C., has developed an "educational kit" comprised of a map, poster, child's coloring book, informative text, and graphics for local guides. Other items include decals, seals to print monarchs, mobiles, postcards, and a game in which players represent the social elements that meet in the field of conservation.

The style of this educational and entertainment kit is simple, yet informative, positive, and realistic. The kit is available for sale, among other places, at the El Rosario overwintering site, to which the public has access.

Finally, it must also be stressed that a key element in protecting the sanctuaries has been scientific information about the biology of monarch migration and overwintering. Excellence in research can never be underestimated. Knowledge of the behavior of the species, patterns of migration, the forest ecosystem, local socioeconomics, and other aspects of monarch overwintering in Mexico provide a sound basis for the implementation of conservation tactics. Practical results of research, such as that of Calvert et al. (1984), which shows that forest thinning increases overwintering monarch mortality, are vital to understanding the phenomenon of monarch overwintering and to implementing constructive programs of education and conservation.

In summary, the underlying rationale for our educational program emphasizes the following five points.

1. The beauty and complexity of the monarch migratory phenomenon provide an excellent example of interactions between nature and society.

2. Scientific research can, and should, be linked directly to protection and to education. The contemporary world in which we live makes mass media, especially television, a most important educational tool.

3. If informed, local people and local authorities ought to deal with the everyday, direct management of the protected areas. External support is also needed. An efficient, collaborative approach has been the key to constructive conservation efforts.

4. International support through Mexican organizations has been valuable in collecting information for dissemination within the education program.

5. Monarch-related publications and arts and crafts have helped the effectiveness of our education program.

Thus, in essence, the educational process has been guided by the goal of making conservation a self-sustaining activity.

LITERATURE CITED

Calvert, W.H., W. Zuchowski, and L.P. Brower. 1984. Monarch butterfly conservation: Interactions of cold weather, forest thinning and storms on the survival of overwintering monarch butterflies (*Danaus plexippus* L.) in Mexico. *Atala* 9(1&2):2–6.

Ortíz Monasterio P., F., and V. Ortíz Monasterio Garza. 1984. *Mariposa Monarca: Vuelo de Papel.* México: CIDCLI, S.C., 62 pp.

Sánchez, V. 1981. Papel de la educación en la intervención entre estado de desarrollo y medio ambiente. In *Estilos de Desarrollo y Medio Ambiente en América Latina,* eds. O. Sunkel and N. Gligo. México: Fondo de Cultura Económica.

UNESCO/PNUMA. 1977. Conferencia Integralmente sobre Educación Ambiental, Documento 7, Tblissí, URSS, 14–26 Octubre de 1977.

OPERATIVE PROGRAMS IN THE MONARCA, A.C., PROJECT

María Elena Camus de Castro

ABSTRACT. As a result of a number of coordinated actions taken by Monarca, A.C., since its inception, an *Action Plan* has been elaborated for both short- and medium-term conservation. This plan is based on methodology developed by Kenton Miller for planning ecological reserves in Latin America and is in accordance with guidelines from the Ministry of Programming Budget.

Taking into consideration the complex nature of monarch conservation, this *Action Plan* consists of 10 subprograms: Land Use, Administration, Protection, Scientific Research, Forest Management, Socio-economic Development, Training and Education, Construction, Tourism, and Coordination.

Based upon the Plan's objectives and strategies of implementation, Monarca, A.C., has established activity priorities to optimize its limited resources.

INTRODUCTION

This paper outlines the development of operative programs initiated by Monarca, A.C., for the conservation of monarch butterflies, *Danaus plexippus* (L.), and their overwintering habitat in Mexico.

Development of an *Action Plan* to protect the monarch migratory phenomenon has taken time and great effort because political jurisdiction in the overwintering areas has complicated the implementation of conservation plans. Before October 1986, monarch overwintering habitat was under the jurisdiction of two federal ministries, two state governments, and both municipal and ejido authorities. Another factor has been the lack of regulated land use in the ecological reserves established for monarch conservation until September 1988. At present, this regulation is being developed and it will be published in the official diary in Mexico.

The cultural and historical background, poverty, and the lack of industrial development by the local population have led to intensive exploitation of timber resources within the forested monarch butterfly overwintering habitat. Consequently this has changed tree densities, temperatures, and humidities in the high-altitude oyamel fir forests of the Transvolcanic Belt across central Mexico, in which monarchs overwinter (Calvert and Brower, 1981, 1986; Calvert et al., 1982, 1986, 1988; Calvert and Lawton, this volume).

Since Monarca, A.C., was legally constituted in 1980 it has been collaborating with federal, state, and municipal authorities as well as with "ejidatarios," the local residents in overwintering areas, to achieve its conservation goals.

OPERATIVE PROGRAMS: OBJECTIVES AND STRATEGIES

Based on a workshop held in January 1985, and Miller's (1980) method for planning ecological reserves, Monarca, A.C., elaborated an *Action Plan.* The plan was developed with the collaboration of the Ministry of Urban Development and Ecology and the assistance of social ecologist Kathy Parker. This *Action Plan* is divided into 10 subprograms and has two objectives. First, to plan the direction and development of specific resources in specific locations and, second, to provide a basic guide model for the future organization of all national parks in Mexico.

The 10 subprograms of our *Action Plan* have specific objectives and strategies for implementation and are outlined as follows.

1. *Land Use*

To determine the most adequate land use, in accordance with the physical and legal characteristics of the overwintering area to achieve effective protection.

The strategies for this subprogram are to delimit and investigate the regime of land tenure in each area, and to estimate optimum use within the constraints imposed by core and buffer protective forest zones and the necessary legal actions to achieve the above.

2. *Administration*

Administer reserves so that human, material, and financial resources are maximized, as well as achieving efficient protection of monarch butterfly colonies and their habitat. Integration of the socioeconomic development of the region through a permanent planning process is also considered of paramount importance.

Planning, programming, and executing the necessary actions for the adequate administration of the reserve is to be implemented through the Technical Committee.

3. *Protection*

Avoiding damage to the monarch butterfly and its overwintering habitat is the objective of protection to assure monarch survival and the continuation of the migratory phenomenon.

This subprogram will be executed by the employment of people to prevent damage to overwintering reserves and to give security to visitors.

4. *Scientific Research*

To increase knowledge of the natural and cultural resources of the area in order to protect the monarch butterfly, its habitat, and the socioeconomic development of the region.

Basic and applied research in biology, ecology, forestry, and socioeconomics by different Mexican and foreign institutions is to be coordinated by the Technical Committee.

5. *Forestry Management*

To achieve a balance between monarch butterfly protection and the socioeconomic development of the local population through the implementation of adequate technologies and methods for viable forest exploitation.

A management plan that controls forest exploitation so that forest resource use, socioeconomic development, and monarch protection are maximized is being developed.

6. *Socioeconomic Development*

To promote a better quality of life for the communities within the reserve areas so that monarch conservation has a positive impact on the local population.

The strategy is to understand these rural communities through the analysis of demography, income, occupation, public services, and cultural preferences so that social and economic development can be implemented effectively.

7. *Education*

To provide information about the existing relationships among man, butterfly, and nature to promote a more responsible attitude toward habitat and species conservation. The strategy is to design, produce, communicate, and distribute educational material.

8. *Construction*

To develop and implement construction regulations for the reserves, by comparison of the construction needs and the zoning rules of each area.

9. *Tourism*

To provide tourist services and general information to facilitate the educational and recreational aspects of tourism. Implementation strategies will use interpretation and visitors' centers to provide guided tours. The characteristics and numbers of people who visit the conservation areas will also be determined. The impact of visitors on the ecology of reserves will be monitored.

10. *Coordination*

The last subprogram will require an efficient interaction among the official agencies and entities, institutions, organizations, and communities involved in the protection and management of monarch reserves. These interactions will be coordinated by the Technical Committee.

OPERATIVE PROGRAMS: ACHIEVEMENTS

Based on the recommendations of Monarca, A.C., six overwintering areas were declared as ecological reserves on 22 August 1986 by the Minister of Urban Development and Ecology representing the former President of Mexico, Lic. Miguel de la Madrid. The declaration took place at the Sierra Chincua overwintering site above Angangueo, Michoacán, with the attendance of the President, the Secretary, and members of the Technical Committee of the trust established by Monarca, A.C., as well as ejidatarios and municipal authorities, with representatives of several conservation organizations from Canada, the United States, and Mexico.

In our Education Subprogram we have given courses on the biology of monarchs, their overwintering habitat, and the need for protection to reserve guards and other local inhabitants so they can lead tour groups through the overwintering areas as knowledgeable guides and protect monarchs and their habitat from damage. An interpretative path has already been designed. A 1-week educational and recreational course has also been designed for primary school children.

A bibliography of literature and information with respect to the monarch butterfly and oyamel (*Abies religiosa* H.B.K.) ecosystem is being compiled by Monarca, A.C. This information will be used for management purposes, education, and academic studies. This data base will facilitate the detection of those studies that need to be performed in the future to aid monarch preservation.

With reference to the Construction and Tourism subprograms, Monarca, A.C., with the support of the World Wildlife Fund, has constructed a visitors' center at the overwintering site "El Rosario." Here visitors can obtain general recreational and educational information.

In the 1985–1986 overwintering season, an estimated 50,000 people visited the two monarch overwintering areas open to the public. For this reason Monarca, A.C., has contacted travel agencies that take tourists to the sites to maintain tourism at a scale that will not adversely affect the success of monarch overwintering.

Another center has been constructed by the Ministry of Urban Development and Ecology, which contains a store run by "ejidatarios" to fulfill one of our goals in the Socioeconomic Development Subprogram. In the 1985–1986 season Monarca, A.C., designed and produced several articles to be sold by "ejidatarios" for profit. Monarca, A.C., delivers these articles to the "ejidatarios" at cost price so that profits accrue to the community. The diversity and number of these artifacts will be increased.

It is important to mention that "ejidatarios" have used the profit from the 1985–1986 season to connect their village school to the national educational TV system for high-school students.

During 1986 Monarca, A.C., collaborated with TV and radio stations to bring to the public more information on habitat and species conservation, using the monarch to illustrate the activities of Monarca, A.C. This has provoked increased public awareness toward both the monarch reserves and our natural world.

In conjunction with this short to midterm *Action Plan,* studies of the monarch and its habitat continued with collection of data related to the socioeconomics of the region and a detailed topographic and floristic data base. This information will be used to design a long-term management plan.

PUBLIC DOCUMENTS RELATING TO MONARCH PROTECTION IN MEXICO

Fideicomiso para la Protección de la Mariposa Monarca. 1984. Public instrument No. 42,423. Notary Public No. 129 Mexico City. Mexico.

Fundación Mexicana para el Desarrollo Rural. 1984. Estudio socioeconómico descriptivo de Ocampo y Angangueo, Michoacán. Mexico.

Mariposa Monarca, A.C. Statues. Public instrument No. 11,603. Notary Public No. 147 Mexico City. Mexico.

SEDUE-MONARCA, A.C. 1985. Presupuesto por Programas. Mexico.

LITERATURE CITED

Calvert, W.H., and L.P. Brower. 1981. The importance of forest cover for the survival of overwintering monarch butterflies (*Danaus plexippus,* Danaidae). *Journal of the Lepidopterists' Society* 35:216–225.

Calvert, W.H., and L.P. Brower. 1986. The location of monarch butterfly (*Danaus plexippus* L.) overwintering colonies in Mexico in relation to topography and climate. *Journal of the Lepidopterists' Society* 40(3):164–187.

Calvert, W.H., W. Zuchowski, and L.P. Brower. 1982. The impact of forest thinning on microclimate in monarch butterfly (*Danaus plexippus* L.) overwintering areas of Mexico. *Boletin de la Sociedad Botánica de México* 42:11–18.

Calvert, W.H., N.P Mendoza V., and R.O. Lawton. 1986. Relationship between mortality and habitat structure at several Mexican monarch butterfly overwintering colonies. Field report to Monarca, A.C., and The World Wildlife Fund.

Calvert, W.H., M. Hyatt, and N.P. Mendoza. 1988. The effects of understory vegetation on the survival of overwintering monarch butterflies (*Danaus plexippus* L.) in Mexico. *Acta Zoologica Mexicana* 18:1–17.

Miller, K.R. 1980. *Planificación de Parques Nacionales para el ecodesarrollo en Latino-América.* Madrid, España. Fundación para la Ecología y la Protección del Medio Ambiente (FEPMA).

THE IMPORTANCE OF ALTERNATIVE SOURCES OF INCOME TO "EJIDATARIOS" (LOCAL RESIDENTS) FOR CONSERVATION OF OVERWINTERING AREAS OF THE MONARCH BUTTERFLY

Susana Rojas González de Castilla

ABSTRACT. Monarch butterflies overwinter in high-altitude oyamel fir forests in central Mexico. Most local residents, or ejidatarios, within the overwintering areas of the monarch butterfly live by exploiting the forest for wood trade and opening areas for agriculture. Because of this exploitation the conservation organization Monarca, A.C., has taken special interest in developing other sources of income for ejidatarios, so that forest clearing can be minimized or avoided. Past clearing and thinning of the forest for various human needs has greatly reduced the number of suitable habitats available to overwintering monarch butterflies. Thus we believe that it is necessary to redirect the income of ejidatarios to other sources so that the forest will be subject to less exploitation. Some of the alternative sources of income proposed by Monarca, A.C., to the ejidatarios include profits obtained from tourist services and souvenirs.

Another short-term strategy for habitat protection includes rental of specific trees in the core areas from the ejidatarios, thereby competing with the lumber companies who buy trees with concessions. At the very least, these projects offer ejidatarios immediate and tangible monetary incentives for leaving trees standing and conserving monarch overwintering habitat. These incentives will also hopefully give ejidatarios cause to believe that the trees are an economic resource that must be managed and not rapidly expended.

INTRODUCTION

Year after year, millions of monarch butterflies overwinter in the Transvolcanic Mountain Belt of central Mexico in the states of México and Michoacán. The overwintering sites occur in high-altitude coniferous forests dominated by the oyamel fir *Abies religiosa*. These forests are critical to the conservation of the long-distance, international migration of monarch butterflies to, and from, Canada, the United States, and Mexico. Like many other forests around the world, these Mexican forests are endangered by severe, local exploitation.

The forests in which monarchs overwinter are owned by the ejidatarios; the lands were given to them by the Mexican government under the community system of property ownership established in Mexico as the "ejido." Such community property was given legally to peasants as a result of Agrarian Reform legislation earlier this century.

At the time of the land-tenure reform in 1917 it was not a priority to manage natural resources so that exploitation could be sustained at a renewable level. Local populations expanded in size very rapidly and destroyed areas of forest for agricultural land use. Additionally, lumber companies began to work in the area, resulting in still faster clearing and thinning of the forest, thus reducing suitable overwintering habitat available to the monarchs.

One of the primary objectives of the Mexican conservation organization Monarca, A.C., is to address the socioeconomic problems of local residents in the monarch overwintering areas.

We believe that alternative sources of income for local inhabitants, or ejidatarios, will prevent further loss of the forest. Reduction of forest exploitation will protect local ecosystems and conserve renewable resources such as aquifers, nutrients, and soil as well as minimize soil erosion.

The first steps to pursue to conserve forest resources is to provide ejidatarios with information on the long-term importance of the forest to their environment and also to show them environmentally and economically viable ways to use their natural resources.

BACKGROUND INFORMATION

We face two kinds of problems related to forest exploitation: one from local forest exploitation and the other from large-scale commercial forestry activity. Wood from the forest is used by ejidatarios for fuel and construction purposes, and large areas are also cleared for agricultural use. Agriculture has become a major source of forest habitat destruction since cultivation needs have increased at the same rate as local population growth. Thus, the agriculture frontier now extends to the hilltops in many areas. The use of poor agricultural techniques has exacerbated the situation since new land has to be cleared to replace wasted, nutrient-poor, and eroded land. Hence, it is essential that the ejidatarios understand the appropriate use of their natural resources, so that they are sustainable indefinitely without additional exploitation.

Commercial forestry activities have had an enormous impact on the structure and extent of the oyamel fir forests in the overwintering area. Where commercial forestry activities are permitted, the implementation of plans to manage the forest resource is inadequate and long-term forest improvement and conservation plans have yet to be implemented. The forest economy has been inadequately administered since most of the profits derived from the exploitation go to the lumber companies, leaving the local residents without either most of the benefits of this activity or the forest resource.

Both problems have a great impact on the monarch population because the butterflies avoid clearings and areas radically thinned by logging, thus reducing the moderating effect of the canopy on temperature extremes. As shown by other papers in this volume, the monarch requires a complex array of environmental factors, found in a well-preserved forest, to ensure overwintering survival. These include microclimatic characteristics such as humidity, temperature, certain types of flowers as nectar sources, and many other factors that influence monarch overwintering (see Anderson and Brower, this volume; Calvert and Lawton, this volume). Furthermore, in the event of partial or complete overwintering habitat destruction it is not known whether or not the physiology and behavior of overwintering monarchs is sufficiently flexible to allow shifts of overwintering locations.

MEASURES TAKEN BY MONARCA, A.C.

Monarca, A.C., is trying to find the most adequate way to preserve the overwintering habitat of the monarch butterfly by developing a variety of viable alternative sources of income. These alternatives should assist in the protection of forests in the overwintering area and benefit the local population. Implementation of a conservation education program for local residents will help them to justify conservation efforts and appreciate the potential for higher standards of living as well as better ways to preserve the overwintering forests of monarch butterflies.

This is not a simple task. To implement our objectives we have set short-, medium-, and long-term goals. Our short-term achievements include the use of tourism as a source of income and the collection of data on forest management in the area.

On medium- and long-term bases we will continue with tourism and educational activities and implement a forest management plan compatible with the results of research in the overwintering area.

In the sanctuary management plan that Monarca, A.C., has been using, we have recognized four principal overwintering sites, each one having a different management treatment. Sierra Chincua and Chivatí-Huacal (see Calvert and Lawton, this volume) are destined only for research and conservation purposes; no tourism is allowed in this area. The Cerro Pelón and Cerro El Campanario (El Rosario) areas are also used for research and conservation activities, but general admission is permitted under strict control. El Rosario is most influenced by tourism since it is easily accessible. The Mexican government and Monarca, A.C., have also agreed on a Plan of Action for the overwintering season, including the designation of El Rosario as open to the public and equipping the site with facilities financed in large part by Monarca, A.C., supported by the World Wildlife Fund–US. El Rosario was visited by approximately 20,000 persons in the 1985–1986 season.

Monarca, A.C., has organized local residents to charge an access fee to the public visiting the sanctuary. Due to the large number of visitors the community has been able to benefit from a significant income to build new school classrooms and purchase many of the items needed by small communities.

Visitor access to overwintering monarchs is strictly monitored and controlled by guides and registration at the entrance post.

Monarca, A.C., has also created a "Tienda Ejidal" or Ejido's Store, which is run by the ejidatarios and supplied and supervised by us. Among the articles sold in the store are stickers, postcards, and ceramic tiles with monarch themes on them. These articles are produced by Monarca, A.C., and sold to the ejidatarios at cost price. The selling price is then set by mutual arrangement and profits are used by the community to improve the quality of life in the ejido. By doing this we now have the trust of these people since they realize that tourism can be a profitable activity. With their approval we are increasing the number of different items sold in the store each winter.

In addition, Monarca, A.C., in coordination with the National Handcraft Fund (FONART) and the Ministry of Urban Development and Ecology (SEDUE), has started a nationwide monarch-related handcraft contest. The most appealing articles from this contest will in turn be made by the ejidatarios and their families and sold in the store.

Because tourism is now an important source of alternative income to the local residents, we decided to provide the public with more facilities to visit the sanctuary. Thus we have constructed a Visitor's Center at El Rosario that will contain a photographic exhibit describing the life history of the monarch butterflies and the nature of the forest they rely on. The center's main objective is to educate the general public about the importance of the phenomenon and the means of preserving it. Among other activities that we have initiated to facilitate public visits, in addition to the Visitor's Center, is the construction of interpretative trails, a parking lot, restrooms, and eating facilities and the provision of knowledgeable guides to inform visitors to the sanctuary. Eating facilities, guides, and construction activities also provide employment for some families. Furthermore, since guides are given education courses on the monarch overwintering phenomenon they are able to justify monarch conservation within their community.

We believe that we have a very positive and constructive program of monarch conservation integrated within the local community. Now tourists can enjoy additional benefits and facilities. The local inhabitants are benefiting. Their community is improving with the increased organization of public access to the sanctuary and most importantly, the monarch butterfly will continue to visit its oyamel fir overwintering forest each winter, for the foreseeable future.

THE MONARCH PROJECT: A PROGRAM OF PRACTICAL CONSERVATION IN CALIFORNIA

Melody Mackey Allen and Katrin B. Snow

The Monarch Project came to life through the efforts of Lincoln P. Brower, Robert M. Pyle, and Melody Mackey Allen on a trip to the monarch butterfly overwintering sites in Mexico in 1983. At Brower's invitation, the three gathered to assess the conservation potential. Brower believed an opportunity existed for a private conservation organization to help protect the monarch migration phenomenon in both Mexico and California. Monarca, A.C., organizers were then initiating Mexican conservation efforts (see Ogarrio, this volume). After meeting with Monarca, A.C., and with state and federal officials in Mexico, Brower, Pyle, and Allen returned to the United States and drafted a biological protection plan and a conservation strategy for both Mexico and California.

The Xerces Society, a nonprofit conservation group dedicated to the preservation of invertebrate diversity, was the logical organization to adopt a project of monarch conservation. Within the broad frame of preserving invertebrate diversity, the conservation of an endangered phenomenon such as the monarch butterfly's three-nation migration is an obvious target: It is the conservation of a unique life history. Allen began to seek funding for *The Monarch Project* master plan and initial protection of the California monarch sites, and *The Monarch Project* became a formal undertaking of the Xerces Society on 1 July 1984. A Xerces board member, Christopher Nagano, raised funds from World Wildlife Fund–US for a California survey of monarch sites (see Lane, this volume). By September of 1984, initial funding for habitat preservation had been secured from the David and Lucile Packard Foundation, and the task of negotiating protective easements and landowner registrations began.

In coastal California, conservation easements and the International Register of Monarch Sites work effectively to provide the legal and biological management framework for cost-effective, long-term preservation of numerous overwintering sites. Landowners who donate conservation easements protect their monarch site legally and permanently, while retaining ownership of their land. Landowners who voluntarily register their site in the International Register can protect their site without deeding property rights and become actively involved in conservation themselves.

Habitat preservation through conservation easements and landowner registry will remain *The Monarch Project*'s top priority as long as real-estate values along the coast continue to skyrocket and development pressure remains pitched. *The Monarch Project* accomplishes preservation by working in a broad network of state and local government officials, conservationists, consultants, scientists, land trusts, park districts, and public agencies. Working collaboratively, *The Monarch Project* grounds its position in scientific research, negotiating protection for monarch sites threatened with development and developing adequate long-term management for protected sites. For example, the first conservation easement in Santa Cruz was the result of a collaboration among the developer, *The Monarch Project,* and the city planning department. Other easements continue to be negotiated under similar circumstances.

In June 1988, Californians passed a Parks and Wildlife bond act, Proposition 70, which allocated \$2 million for monarch habitat site acquisitions. The bond act, coming soon after state legislation requiring a survey of all monarch sites, is the closest thing to a state mandate for preservation that exists for the monarch butterfly. The monarch has lacked legal protection because it is not an endangered species.

The available recourse is to secure legal protections for monarch habitats through land-use planning and local zoning ordinances. *The Monarch Project* has submitted language protecting monarch habitat for consideration in the land-use plans of Pacific Grove and Santa Cruz, two critical areas for monarch overwintering in California.

For *The Monarch Project,* the passage of Proposition 70 laid the groundwork for completing the preservation of critical California monarch habitat. *The Project* worked with monarch scientists and state and local officials to establish priorities for potential acquisition of surveyed sites and to rank the sites accordingly. The first site has been purchased, and the state continues to negotiate with landowners on future acquistions.

The corollary to land preservation is land management. Looking to the future, *The Monarch Project* has begun a unique habitat revegetation project to enhance monarch habitat in a native Monterey pine forest. The plan is a unique attempt to revegetate a monarch habitat and will serve as a model for other cities or counties experiencing similar problems. While development projects near monarch habitat typically include mitigations such as tree planting, no one, to our knowledge, has developed a scientifically based management plan that aims to remedy the chronic, ecologically intricate problems that threaten the health of a forested overwintering habitat. To preserve monarch habitat for the long-term future, it is essential to understand forest succession and natural processes of change and degradation.

Since education is an integral part of community involvement and support for monarch habitat conservation, *The Monarch Project* sponsors slide presentations and lectures. Right now, two volunteer teachers are helping develop curriculum and class projects for grades preschool through sixth.

At the national level, the Entomological Society of America is spearheading legislation to make the monarch butterfly the national insect. The Xerces Society is a sponsoring organization, along with the National Audubon Society, the National Wildlife Federation, The Nature Conservancy, and others. *The Monarch Project* is distributing brochures describing the legislation and contacting West Coast congressional representatives in a search for sponsors of the legislation.

The Monarch Project efforts to preserve this extraordinary, birdlike migration over the long term require understanding and protection of its three principal components: overwintering habitats, migratory behavior, and breeding habitats. Hence, *The Project* and other California scientists launched scientific research programs so that the precise habitat requirements and biology of overwintering monarchs can be fully understood.

The research programs generated new data on monarch biology throughout the annual cycle of this remarkable insect in California. As new research on the California monarch overwintering sites proceeds, new, published data will be an invaluable aid to policymakers whose decisions affect monarch habitat.

10

CONCLUSIONS

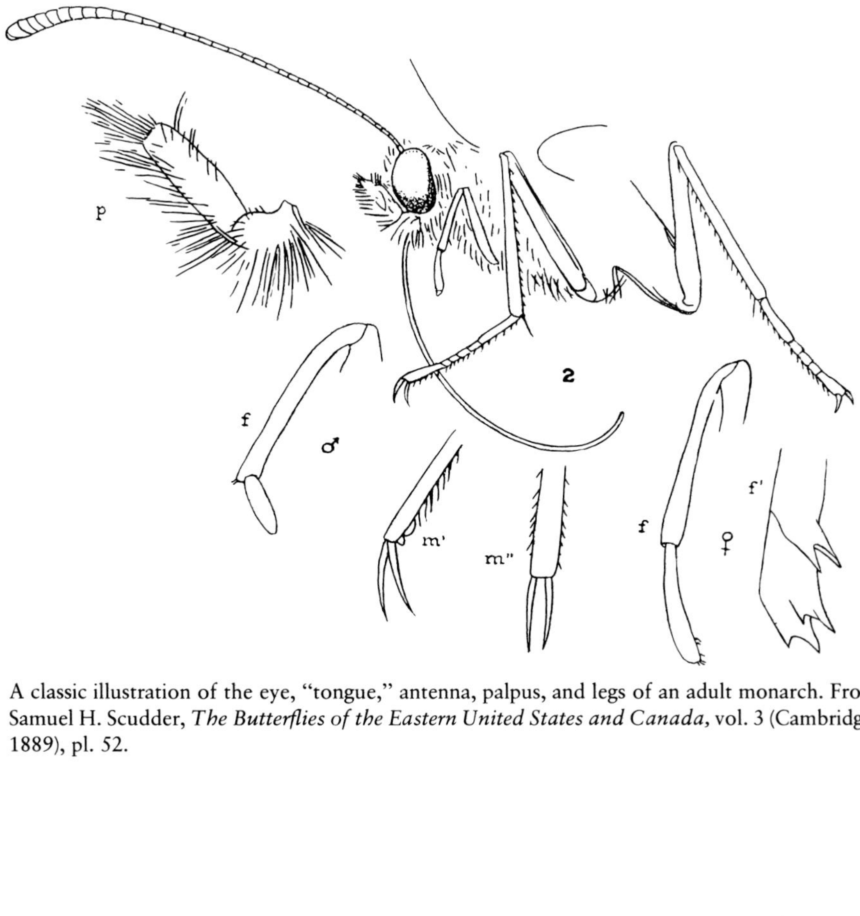

A classic illustration of the eye, "tongue," antenna, palpus, and legs of an adult monarch. From Samuel H. Scudder, *The Butterflies of the Eastern United States and Canada*, vol. 3 (Cambridge, 1889), pl. 52.

BIOLOGY AND CONSERVATION OF THE MONARCH BUTTERFLY: CONCLUDING REMARKS

Stephen B. Malcolm and Myron P. Zalucki

BIOLOGY

Four themes dominate the considerations of monarch biology in this volume: host-plant exploitation, migration, overwintering, and chemical defense against natural enemies. Each contribution is focused on, or at least alludes to, one or more of these themes. Such domination is not surprising given the use of the monarch as a model to further our conceptual understanding of each of these areas of biology, and the early interest in the monarch as a prime example of both aposematic coloration (Poulton, 1902; Brower, 1969; Malcolm, 1990; Ritland and Brower, this volume) and long-range insect migration (Williams, 1930, 1958; Urquhart, 1960; Johnson, 1969; Brower, 1977, 1985; Malcolm, 1987). We now understand in considerable detail how, why, and when monarchs exploit particular host plants, migrate, and overwinter.

Curiously, despite much effort to understand the monarch's defenses against natural enemies, a realistic appraisal of the interaction between monarch defenses and natural enemy foraging still eludes us. Although the laboratory experimentation of J.V.Z. Brower (1958), L.P. Brower et al. (1968), Brower (1969), Brower and Glazier (1975) and the field observations of Calvert et al. (1979), Fink and Brower (1981), Fink et al. (1983), Glendinning et al. (1988), Arellano G. et al. (this volume), and Glendinning (this volume) reflect a vigorous and instructive implementation of chemical ecology, we still know very little about monarch–natural enemy interactions, especially during the periods when this conspicuous butterfly migrates and breeds.

Since migration between narrow overwintering distributions and wide breeding distributions is a dominant feature of monarch biology, 14 of the 27 papers in chapters 5, 6, 7, and 8 that consider the four dominant themes also consider migration either directly or indirectly. Inevitably, the four papers of chapter 6 that consider various aspects of the monarch's annual cycle also use each of the main themes as a frame to support considerations of endocrinology (Herman), thermoregulation (Masters), cold-hardiness (Anderson and Brower), and comparative biology of Australian danaines (Kitching and Scheermeyer). These four papers, in conjunction with the three contributions by James, Cockrell et al., and Schmidt-Koenig in chapter 7, offer valuable perspectives on the proximate control of monarch migration. Thus, temperature and daylength are the predictable cues to which monarchs fine-tune their physiology so that they can exploit the continent-wide distribution of larval food plants. Although it is yet to be shown whether the seasonal directionality of monarch migration is a response to magnetism, light polarization, or rates of daylength change, we do know from the papers by Cockrell et al., James, Lynch and Martin, Malcolm et al., and Riley (this volume) that the timing of monarch migration is closely tuned to temperature and daylength changes in both North America and Australia.

Such predictable spatial and temporal patterns of host-plant use provide perhaps the best clues to discovering the ultimate reasons for monarch migration. Monarchs probably do not have to migrate to escape freezing winter temperatures since Anderson and Brower (this volume) show that they exhibit considerable inducible freeze-tolerance, and it is not difficult to believe that selection for increased levels of freeze-tolerance, like that of other nonmigratory insects (Tauber et al., 1986), is a simple evolutionary step. Instead, Malcolm et al. (this volume) consider that both milkweed host-plant distribution and defense against predation provide simultaneous selection for a migratory pattern of host-plant selection and use. Migration thus facilitates exploitation of almost any milkweed, at almost any time, across the entire North American continent. Since the monarch's spatial and temporal pattern of milkweed exploitation also appears to influence its chemical defense against predators, Malcolm and Brower (1989) have suggested that monarchs and milkweeds show multispecies coevolution (Gould, 1988), and that coevolution is mediated via predation. Despite the conclusion of Brower et al. (1988), that interactions among overwintering monarchs and avian predators are examples of exaptations and not coevolution, we believe that this monarch-centered, three trophic level interaction provides a useful and testable model for the kind of chemically mediated, coevolved interaction discussed in Spencer (1988).

Although monarchs have long been known to migrate each year across North America (Williams, 1930, 1958; Urquhart, 1960; Johnson, 1969; Vane-Wright, this volume), surprisingly little is known about monarch migration in California, especially in view of the fact that monarch overwintering sites have been documented in California for at least 100 years longer than the much larger sites in Mexico (Lane, this volume; Vane-Wright, this volume).

Since so few data on monarch migration in California exist, it is understandable that Wenner and Harris (this volume) question the accepted dogma that western monarchs behave in the same way as eastern monarchs. However, their range expansion model of monarch movement does not agree with the directed migrations observed by Nagano et al. (this volume), who found that populations of spring migrating monarchs in California moved bimodally over distances >10 km. The discrepancies between these two interpretations of monarch movement in California may perhaps be explained by the timing and distance of monarch movements among overwintering sites and away from overwintering sites to larval food resources. Short-distance, early spring movements of monarchs are likely to be more in accordance with the observations of Wenner and Harris and we look forward to the publication by Nagano and Sakai of their short-distance mark–release–recapture data for monarch movements of <10 km.

Malcolm (1987) also suggested that the range expansion hypothesis of Wenner and Harris was untenable because the onset of lethally cold autumn temperatures would select against range expansion and select for increased freeze tolerance. But no mechanism would exist to return monarchs to their well-documented overwintering sites. As soon as monarchs are acknowledged to return from breeding sites to coastal overwintering sites then migration contracts their range. Our own feeling is that any movement by monarchs between spatially separated overwintering resources and larval food resources, however short, constitutes migration (sensu Kennedy, 1975, 1985; Southwood, 1977; Denno, 1983; Rogers, 1983). However, as both Wenner and Harris and Nagano et al. point out, there is no doubt that interpretation and description of monarch migration in California is confounded by the distribution of Coastal and Transverse Mountain ranges that influence the spatial and temporal distributions of their milkweed food resources.

All of the papers in this volume that mention the distribution of monarchs in North America assume that monarchs are separated by the Rocky Mountains into two populations—eastern and western. However, no evidence exists to show that the populations are in fact separated and that there is not a one- or even two-way mixing of butterflies during spring and autumn migrations. Instead, Funk (1968) describes overwintering by monarch butterflies in southwestern Arizona, which suggests that monarchs may cross the Rocky Mountains. Indirect evidence also indicates that monarchs may occur high in the Rocky Mountains during the autumn migration since an accurate, sex-limited mimic of the monarch (females of the pierid butterfly *Neophasia terlooti*) is active in the Chiricahua Mountains of southeastern Arizona in November (Poulton, 1914; Scott, 1986). Moreover, the seasonal migration of monarchs across Costa Rican mountains (described by Haber, this volume) and the eastward movement of monarchs from overwintering sites in California (Nagano et al., this volume) suggest that monarchs are able to traverse considerable, high-altitude barriers. It is even possible that monarchs in California could be a sink population, continuously maintained by eastern monarchs as a source population, much like the scenario modeled by Pulliam (1988). A large-scale, simultaneous, mark–release–recapture effort at both Mexican and Californian overwintering sites would help to show whether or not monarchs can cross the Rocky Mountains and Mexican Sierras during spring migration. The question of population mixing is also ripe for a molecular biology solution since description of the mitochondrial DNA of eastern and western monarchs might illustrate gene flow between the two populations (see Brower and Boyce, 1991).

An additional, indirect impact on monarch migration is considered by Vane-Wright (this volume), who argues that the postsettlement destruction of North America's forests in the last century dramatically increased both milkweed availability and the numbers of historically migrating monarchs (also Vane-Wright, 1986, 1987). This, he suggests, led to large numbers of chance movements that resulted in the range expansion of monarchs across the Pacific Ocean to Australia. In response to this suggestion, Malcolm and Brower (1987) argued that the prior distribution of milkweeds by man to Australia and the Pacific was a more parsimonious explanation for the present-day global distribution of monarchs. Indeed, prior to the westward movement of North American settlers into the Great Lakes region during the 19th century, much of the area covered by the present-day distribution of *Asclepias syriaca* (undoubtedly the monarch's most abundant larval food resource) was either open prairie grassland (McManis, 1964; Brewer, 1970) or open oak woodland with trees sparsely scattered through grassland (Peters, 1970). Thus, forest removal may not have influenced the biomass of the current milkweed resource. Instead, in the absence of objective information, one could argue that milkweed availability is now much reduced through the widespread agricultural use of herbicides, in comparison with presettlement habitat. Either way, this debate hinges on the need for information on the distribution, biomass, and utilization of the different milkweed species exploited by monarchs. Perhaps the use of remote sensing techniques, like the work of Zalucki et al. (1981), will facilitate the collection of such information and lead to better understanding of the general dynamics of resource exploitation by a highly mobile herbivore.

Closer examinations of host-plant exploitation by monarchs, in the papers by Zalucki, Ackery, Nelson, Lynch and Martin, and Borkin, in this volume (see also Cockrell et al., Malcolm et al., and Riley, this volume), illustrate much of the complexity of the interactions between monarchs and milkweeds. Like many danaid butterflies, monarchs feed almost exclusively on milkweeds, especially in the genus *Asclepias* (Malcolm and Brower, 1986) and the closely allied African genus, *Gomphocarpus,* growing in Australia (=*Asclepias* in Zalucki and Kitching, 1982). Borkin's re-examination of whether or not *Apocynum* species can constitute food for monarch larvae illustrates the need for caution in the acceptance of published records of larval host plants, such as the highly dubious records of monarch larvae feeding on *Gossypium, Euphorbia,* or *Citrus* (Ackery and Vane-Wright, 1984; Ackery, 1988).

Like Malcolm and Brower (1989), the two papers by Nelson in this volume show that monarchs have a highly specific interaction with the chemistry of their milkweed host plants (see also Lynch and Martin, and Malcolm et al., this volume). Monarchs are able to increase their sequestered cardenolide concentrations from plants with low cardenolide titers, and conversely, the butterflies can reduce the available cardenolide concentrations when larvae feed on plants with very high cardenolide concentrations. Thus the butterflies show a logarithmic pattern of cardenolide storage, with an asymptote at approximately 350 μg of cardenolide/0.1 g dry butterfly. Interestingly, Oyeyele and Zalucki (1990) and Zalucki et al. (1990) have shown recently that ovipositing female monarchs, in both the southeastern United States and Australia, select milkweed host plants with dry leaf cardenolide concentrations close to this butterfly asymptote. Both low and high cardenolide plants were rejected by free-flying females postalighting.

Despite such effective host-plant selection and cardenolide sequestration, Malcolm and Brower (1989) showed that adult monarchs lose considerable amounts of their stored cardenolide during migration. This unexplained loss of cardenolides is perhaps the simplest explanation for the extensive predation by birds and mice on overwintering monarchs, after their autumn migration, as described by Arellano G. et al. and Glendinning (this volume). The limited distributions, locations, and vegetation of monarch overwintering habitats, with their tightly aggregated butterflies, in Mexico (Soto Núñez and Vázquez García, de la Maza E. and Calvert, Calvert and Lawton, this volume), California (Lane, this volume), and Australia (Scheermeyer, this volume) make overwintering monarchs with reduced chemical defenses vulnerable to predation.

In addition to the four major themes of host-plant exploitation, migration, overwintering, and defense that dominate this volume, a fifth theme of reproduction percolates through many of the papers and is considered in detail in chapters 3 and 4. This fifth theme is interesting in that it provides a further link between defense and host-plant use. In this case, the link is provided by pyrrolizidine alkaloids (PAs). In their strong advocacy of the monarch as an instructive model of chemical ecology, both Schneider and Boppré (this volume) argue that although PAs are clearly implicated in chemical communication between males and females of most danaid butterflies during courtship and mating (but apparently not in the monarch), they are also a likely component of the monarch's chemical defense against predators. Although monarch larvae appear able to sequester PAs from host plants (Rothschild and Edgar, 1978), PAs are not known to occur in the milkweed host plants of monarch larvae. Instead, monarchs and other danaids sequester PAs as adults from the nectar of PA-rich flowers. Schneider and Boppré stress that, in comparison with other danaids, the monarch is unusual both for its ability to sequester cardenolides effectively and for the apparent degeneration of the general danaid dependence on PAs as communication chemicals used in courtship (e.g., Kittmann and Schneider, this volume).

Instead, the monarch stands almost alone among danaids in the utilization of highly aggressive, visual male selection of females, with little courtship. Thus, Van Hook (this volume) shows that, as monarchs leave an overwintering site in Mexico at the onset of spring migration, small males in poor condition select and mate with large females in excellent condition. Such non-random mate selection implies that small males attempt to maximize their fitness at overwintering sites because they are unlikely to survive the migration north and mate with successfully migrated females at northern breeding sites (see Cockrell et al., this volume). However, the significance of non-random mating at overwintering sites is difficult to interpret. To assess the significance of this mating it would be interesting to know how much of the sperm from these males is used to fertilize eggs, rather than simply offer nutrients for female use (Wells et al., this volume). It would also be interesting to know where and when most of the mating occurs that produces the first spring generation of monarchs in the southern United States (Cockrell et al. and Malcolm et al., this volume), particularly in view of the importance of milkweed patches to monarch reproduction (Zalucki, this volume).

Finally, the systematic perspective of Kitching et al. (this volume) strongly suggests that the monarch is the most evolutionarily derived danaid butterfly in the world today. Although *Danaus* (*Danaus*) *plexippus* and its two closely related sister species, *Danaus* (*D.*) *cleophile* and *Danaus* (*D.*) *erippus* are remarkably similar, *D. plexippus* remains exceptional for its global distribution (Vane-Wright, this volume) in comparison with *D. cleophile,* restricted to the Greater Antilles, and the doubtfully distinct *D. erippus* south of the Amazon in South America (Ackery and Vane-Wright, 1984).

As shown by this volume, such distinct differences provide excellent opportunities for comparative research. Moreover, the opportunities for research into biological concepts using the monarch as a model seem almost limitless. Thus, via the monarch, we are learning more about chemical and physiological ecology, mate selection, host-plant use, herbivory, pharmacology, interactions between prey defense and predator foraging, biogeography, metapopulation dynamics, tritrophic level interactions, migration, and overwintering, and how all these processes come together to shape life-history evolution in complex and variable environments.

CONSERVATION

Apart from the conceptual value inherent in understanding monarch biology, part of the value to such understanding lies in its use for habitat and species conservation. As we have seen, these habitats are spread across all three countries of North America—Mexico, the United States, and Canada. As such, the monarch offers an excellent opportunity for international cooperation in conservation. Calvert et al. (1989) pinpoint monarch overwintering sites as the keystone necessary to conserve the endangered monarch migratory phenomenon in North America (Wells et al., 1983; Brower and Malcolm, 1991). Although monarchs form a metapopulation that waxes and wanes seasonally across the continent, the continent-wide coalescence into a handful of overwintering sites in Mexico and California make the North American monarch extraordinarily vulnerable to overwintering habitat perturbation and destruction.

In the introduction to the conservation section of this volume, Malcolm compares the vulnerability of the monarch to that of the extinct passenger pigeon. The dynamic population fluctuations in space and time of both species have many similarities, characterized by periods of intense aggregation. A cynical view of conservation might suggest that, unlike the passenger pigeon, the only reason that monarchs exist today is that we cannot eat or sell them. We hope that the Mexican efforts described in this volume to use monarchs to bring tourist money to a poor rural economy will be successful. Much of the research described in this volume has helped to justify the efforts of the Mexican conservation organization Monarca, A.C. In particular, objective information on the high-altitude overwintering forest structure in Mexico (Snook, this volume) and its influence on overwintering monarchs (Calvert et al., 1989; Calvert and Lawton, this volume) has helped to protect the fir forests through federal and state legislation and through local education and employment (Ogarrio, Gottfried Joy, Ortíz Monasterio, Camus de Castro, Rojas Gonzâlez de Castilla, this volume).

Similarly, research on overwintering monarchs in California has facilitated the provision of state funds for conservation, the implementation of voluntary land easements, and a small but discernible swing toward more responsible and environmentally conscious land development (Allen and Snow, this volume; Malcolm, this volume).

Nevertheless, despite valiant efforts to conserve monarch overwintering habitat in Mexico and California, it is apparent that conservation guidelines and legislation to protect monarch overwintering habitat are being circumvented or even ignored. Two reasons account for much of the ineffectiveness of current conservation measures: (1) local ignorance of conservation legislation to protect forested monarch overwintering habitat, and (2) deliberate exploitation of overwintering habitat in the face of ineffective implementation of conservation legislation.

The new challenge that faces conservation biologists is to integrate and communicate objective research, to influence the politics, sociology, and economics of responsible land use, and to ensure that our current conservation goals will be relevant and realistic in the face of both global environmental change and global economic constraints and dynamics. Let biologists continue to provide the information necessary to conserve and appreciate the biology of this model species.

LITERATURE CITED

Ackery, P.R. 1988. Hostplants and classification: A review of nymphalid butterflies. *Biological Journal of the Linnean Society* 33:95–203.

Ackery, P.R., and R.I. Vane-Wright. 1984. *Milkweed butterflies: Their cladistics and biology.* London: British Museum (Natural History); and Ithaca, NY: Cornell University Press, 425 pp.

Brewer, R. 1970. Death by the plow. *Natural History* 79(7):28–35, 110.

Brower, A.V.Z., and T.M. Boyce. 1991. Mitochondrial DNA variation in monarch butterflies. *Evolution* 4–5:1281–1286.

Brower, J.V.Z. 1958. Experimental studies of mimicry in some North American butterflies. Part I. The monarch, *Danaus plexippus,* and viceroy, *Limenitis archippus archippus. Evolution* 12(1):32–47.

Brower, L.P. 1969. Ecological chemistry. *Scientific American* 220(2):22–29.

Brower, L.P. 1977. Monarch migration. *Natural History* 86:40–53.

Brower, L.P. 1985. New perspectives on the migration biology of the monarch butterfly, *Danaus plexippus* L. In *Migration: Mechanisms and adaptive significance,* ed. M.A. Rankin, 748–785. Austin: University of Texas, *Contributions in Marine Science,* Supplement, 27.

Brower, L.P., and S.C. Glazier. 1975. Localization of heart poisons in the monarch butterfly. *Science* 188: 19–25.

Brower, L.P., and S.B. Malcolm. 1991. Animal migrations: Endangered phenomena. *American Zoologist* 31:265–276.

Brower, L.P., W.N. Ryerson, L.L. Coppinger, and S.C. Glazier. 1968. Ecological chemistry and the palatability spectrum. *Science* 161:1349–1351.

Brower, L.P., C.J. Nelson, J.N. Seiber, L.S. Fink, and C. Bond. 1988. Exaptation as an alternative to coevolution in the cardenolide-based chemical defense of monarch butterflies (*Danaus plexippus* L.) against avian predators. In *Chemical mediation of coevolution,* ed. K.C. Spencer, 447–475. San Diego: Academic Press.

Calvert, W.H., L.E. Hedrick, and L.P. Brower. 1979. Mortality of the monarch butterfly (*Danaus plexippus* L.): Avian predation at five overwintering sites in Mexico. *Science* 204:847–851.

Calvert, W.H., S.B. Malcolm, J.I. Glendinning, L.P. Brower, M.P. Zalucki, T. Van Hook, J.B. Anderson, and L.C. Snook. 1989. Conservation biology of monarch butterfly overwintering sites in Mexico. *Vida Silvestre Neotropical* 2(1):38–48.

Denno, R.F. 1983. Tracking variable host plants in space and time. In *Variable plants and herbivores in natural and managed systems,* eds. R.F. Denno and M.S. McClure, 291–341. New York: Academic Press.

Fink, L.S., and L.P. Brower. 1981. Birds can overcome the cardenolide defence of monarch butterflies in Mexico. *Nature* 291:67–70.

Fink, L.S., L.P. Brower, R.B. Waide, and P.R. Spitzer. 1983. Overwintering monarch butterflies as food for insectivorous birds in Mexico. *Biotropica* 15: 151–153.

Funk, R.S. 1968. Overwintering of monarch butterflies as a breeding colony in southwestern Arizona. *Journal of the Lepidopterists' Society* 22:63–64.

Glendinning, J.I., A. Alonso Mejía, and L.P. Brower. 1988. Behavioral and ecological interactions of foraging mice (*Peromyscus melanotis*) with overwintering monarch butterflies (*Danaus plexippus*) in Mexico. *Oecologia* 75:222–227.

Gould, F. 1988. Genetics of pairwise and multispecies plant-herbivore coevolution. In *Chemical mediation of coevolution,* ed. K.C. Spencer, 13–55. San Diego: Academic Press.

Johnson, C.G. 1969. *Migration and dispersal of insects by flight.* London: Methuen, 763 pp.

Kennedy, J.S. 1975. Insect dispersal. In *Insects, science, & society,* ed. D. Pimentel, 103–119. New York: Academic Press.

Kennedy, J.S. 1985. Migration, behavioral and ecological. In *Migration: Mechanisms and adaptive significance,* ed. M.A. Rankin, 5–26. Austin: University of Texas, *Contributions in Marine Science,* Supplement, 27.

Malcolm, S.B. 1987. Monarch butterfly migration in North America: Controversy and conservation. *Trends in Ecology and Evolution* 2:135–138.

Malcolm, S.B. 1990. Mimicry: Status of a classical evolutionary paradigm. *Trends in Ecology and Evolution* 5(2):57–62.

Malcolm, S.B., and L.P. Brower. 1986. Selective oviposition by monarch butterflies (*Danaus plexippus* L.) in a mixed stand of *Asclepias curassavica* L. and *A. incarnata* L. in south Florida. *Journal of the Lepidopterists' Society* 40(4):255–263.

Malcolm, S.B., and L.P. Brower. 1987. White monarchs. *Antenna* 11(1):2–3.

Malcolm, S.B., and L.P. Brower. 1989. Evolutionary and ecological implications of cardenolide sequestration in the monarch butterfly. *Experientia* 45:284–295.

McManis, D.R. 1964. The initial evaluation and utilization of the Illinois prairies, 1815–1840. Chicago: The University of Chicago, 109 pp.

Oyeyele, S.O., and M.P. Zalucki. 1990. Cardiac glycosides and oviposition by *Danaus plexippus* on *Asclepias fruticosa* in south-east Queensland (Australia), with notes on the effect of plant nitrogen content. *Ecological Entomology* 15:177–185.

Peters, B.C. 1970. Pioneer evaluation of the Kalamazoo County landscape. *The Michigan Academician* Fall 1970:15–25.

Poulton, E.B. 1902. Mimicry and natural selection. *Proceedings of the V International Zoological Congress, Berlin* 1901:171–179.

Poulton, E.B. 1914. Mimicry in North American butterflies: A reply. *Proceedings of the Academy of Natural Sciences of Philadelphia* January 1914:161–195.

Pulliam, H.R. 1988. Sources, sinks, and population regulation. *The American Naturalist* 132:652–661.

Rogers, D. 1983. Pattern and process in large-scale animal movement. In *The ecology of animal movement,* eds. I.A. Swingland and P.J. Greenwood, 160–180. Oxford: Clarendon Press.

Rothschild, M., and J.A. Edgar. 1978. Pyrrolizidine alkaloids from *Senecio vulgaris* sequestered and stored by *Danaus plexippus. Journal of Zoology, London* 186:347–349.

Scott, J.A. 1986. *Butterflies of North America.* Stanford, CA: Stanford University Press, 583 pp.

Southwood, T.R.E. 1977. Habitat, the templet for ecological strategies? *Journal of Animal Ecology* 46: 337–365.

Spencer, K.C. 1988. Editor, *Chemical mediation of coevolution.* San Diego: Academic Press, 609 pp.

Tauber, M.J., C.A. Tauber, and S. Masaki. 1986. *Sea-*

sonal adaptations of insects. New York and Oxford: Oxford University Press, 411 pp.
Urquhart, F.A. 1960. *The monarch butterfly*. Toronto: University of Toronto Press, 361 pp.
Vane-Wright, R.I. 1986. White monarchs. *Antenna* 10(3): 117–118.
Vane-Wright, R.I. 1987. Dick Vane-Wright replies .... *Antenna* 11(1):3.
Wells, S.M., R.M. Pyle, and N.M. Collins. 1983. *The IUCN invertebrate red data book*. Gland, Switzerland: International Union for Conservation of Nature and Natural Resources, 632 pp.
Williams, C.B. 1930. *The migration of butterflies*. Edinburgh: Oliver and Boyd, 473 pp.
Williams, C.B. 1958. *Insect migration*. London: Collins, 235 pp.
Zalucki, M.P., and R.L. Kitching. 1982. Dynamics of oviposition in *Danaus plexippus* (Insecta: Lepidoptera) on milkweed, *Asclepias* spp. *Journal of Zoology, London* 198:103–116.
Zalucki, M.P., A. Chandica, and R.L. Kitching. 1981. Quantifying the distribution and abundance of an animals resource using aerial photography. *Oecologia* 50:176–183.
Zalucki, M.P., L.P. Brower, and S.B. Malcolm. 1990. Oviposition by *Danaus plexippus* in relation to cardenolide content of three *Asclepias* species in the southeastern U.S.A. *Ecological Entomology* 15:231–240.

CONTRIBUTORS

Phillip R. Ackery, Department of Entomology, The Natural History Museum, Cromwell Road, London SW7 5BD, U.K.

Melody Mackey Allen, *The Monarch Project,* Xerces Society, 10 Southwest Ash Street, Portland, Oregon 97204, U.S.

Alfonso Alonso-M., Centro de Ecología, Universidad Nacional Autónoma de México, Apartado Postal 70-275, Coyoacán, México, D.F. 04510, Mexico. Present address: Department of Zoology, University of Florida, Gainesville, Florida 32611, U.S.

James B. Anderson, Department of Zoology, University of Florida, Gainesville, Florida 32611, U.S.

Alfredo Arellano G., Centro de Ecología, Universidad Nacional Autónoma de México, Apartado Postal 70-275, Coyoacán, México, D.F. 04510, Mexico. Present address: Secretaría de Ecología del Gobierno del Estado de Yucatán, Calle 64 #437X53 Y47-A, C.P. 97000, Mérida, Yucatán, Mexico.

Michael Boppré, Forstzoologisches Institut der Universität Freiburg, Fohrenbühl 27, D-7801 Stegen-Wittental, Federal Republic of Germany

Susan Sullivan Borkin, Invertebrate Zoology Section, Milwaukee Public Museum, Milwaukee, Wisconsin 53233, U.S.

Lincoln P. Brower, Department of Zoology, University of Florida, Gainesville, Florida 32611, U.S.

William H. Calvert, Department of Zoology, University of Florida, Gainesville, Florida 32611, U.S.

María Elena Camus de Castro, Monarca, A.C., Avenida Constituyentes 345-806, México, D.F. 11830, Mexico

Barbara J. Cockrell, Department of Biological Sciences, Western Michigan University, Kalamazoo, Michigan 49008, U.S.

Javier de la Maza E., Monarca, A.C., Avenida Constituyentes 345-806, México, D.F. 11830, Mexico

Julian P. Donahue, Entomology Section, Natural History Museum of Los Angeles County, 900 Exposition Boulevard, Los Angeles, California 90007, U.S.

John I. Glendinning, Department of Zoology, University of Florida, Gainesville, Florida 32611, U.S. Present address: Department of Entomology, University of Arizona, Tucson, Arizona 85721.

Carlos F. Gottfried Joy, Monarca, A.C., Avenida Constituyentes 345-806, México, D.F. 11830, Mexico

William A. Haber, Missouri Botanical Garden, P.O. Box 299, St. Louis, Missouri 63166, U.S.

Ann M. Harris, Department of Biological Sciences, University of California, Santa Barbara, California 93106, U.S.

William S. Herman, Department of Genetics and Cell Biology, University of Minnesota, St. Paul, Minnesota 55108, U.S.

David G. James, Yanco Agricultural Institute, New South Wales Department of Agriculture, Yanco, New South Wales 2703, Australia

Ian J. Kitching, Department of Entomology, The Natural History Museum, Cromwell Road, London SW7 5BD, U.K.

Roger L. Kitching, Department of Ecosystem Management, University of New England, Armidale, New South Wales 2351, Australia

Rolf Kittmann, Fakultät für Biologie, Universität Konstanz, D-7530 Konstanz, Federal Republic of Germany; and Max-Planck-Institut für Verhaltensphysiologie, D-8131 Seewiesen, Federal Republic of Germany. Present address: Institut für Zoologie II, Friedrich-Alexander Universität, Staudtstrasse 5, D-8520 Erlangen, Federal Republic of Germany.

John Lane, Santa Cruz City Museum, 1305 East Cliff Drive, Santa Cruz, California 95062, U.S.

Robert O. Lawton, Department of Biological Sciences, University of Alabama in Huntsville, Huntsville, Alabama 35899, U.S.

Steven P. Lynch, Department of Biological Sciences, Louisiana State University, Shreveport, Louisiana 71115, U.S.

Stephen B. Malcolm, Department of Biological Sciences, Western Michigan University, Kalamazoo, Michigan 49008, U.S.

Ronald A. Martin, Department of Chemistry, Louisiana State University, Shreveport, Louisiana 71115, U.S.

Alan R. Masters, Department of Zoology, University of Florida, Gainesville, Florida 32611, U.S.

Christopher D. Nagano, Sacramento Endangered Species Office, U.S. Fish and Wildlife Service, 2800 Cottage Way, Room E-1823, Sacramento, California 95825, U.S.

Carolyn J. Nelson, Department of Pharmacy, University of Sydney, New South Wales 2006, Australia. Present address: Department of Cardiovascular Medicine, The Prince Henry Hospital, Anzac Parade, Little Bay, New South Wales 2036, Australia.

Rodolfo Ogarrio, Monarca, A.C., Avenida Constituyentes 345-806, México, D.F. 11830, Mexico

Fernando Ortíz Monasterio, Monarca, A.C., Avenida Constituyentes 345-806, México, D.F. 11830, Mexico

Thomas J. Riley, Department of Entomology, Louisiana State University, Baton Rouge, Louisiana 70803, U.S.

David B. Ritland, Department of Zoology, University of Florida, Gainesville, Florida 32611, U.S.

Steffen H. Rogers, Department of Biological Science, University of Tulsa, Tulsa, Oklahoma 74104, U.S.

Susana Rojas González de Castilla, Monarca, A.C., Avenida Constituyentes 345-806, México, D.F. 11830, Mexico

Walter H. Sakai, Life Science Department, Santa Monica College, 1900 Pico Boulevard, Santa Monica, California 90405, U.S.

Elly Scheermeyer, School of Australian Environmental Studies, Griffith University, Nathan, Queensland 4111, Australia

Klaus Schmidt-Koenig, Department of Zoology, Duke University, Durham, North Carolina 27706, U.S.; and Abteilung Verhaltensphysiologie, Universität Tübingen, D-7400 Tübingen, Federal Republic of Germany

Dietrich Schneider, Max-Planck-Institut für Verhaltensphysiologie, D-8131 Seewiesen, Federal Republic of Germany

Laura C. Snook, Instituto Nacional de Investigaciones sobre Recursos Bióticos (INIREB), Apartado Postal 63, Xalapa, Veracruz, Mexico; and Yale School of Forestry and Environmental Studies, 205 Prospect Street, New Haven, Connecticut 06511, U.S.

Katrin B. Snow, *The Monarch Project,* Xerces Society, 10 Southwest Ash Street, Portland, Oregon 97204, U.S.

José C. Soto Núñez, Departamento de Botánica, Instituto de Biología, Universidad Nacional Autónoma de México, Apartado Postal 70-153, Coyoacán, México, D.F. 04510, Mexico

Richard I. Vane-Wright, Department of Entomology, The Natural History Museum, Cromwell Road, London SW7 5BD, U.K.

Tonya Van Hook, Department of Entomology and Nematology, University of Florida, Gainesville, Florida 32611, U.S.

Leonila Vázquez García, Departamento de Zoología, Instituto de Biología, Universidad Nacional Autónoma de México, Apartado Postal 70-153, Coyoacán, México, D.F. 04510, Mexico

Harrington Wells, Department of Biological Science, University of Tulsa, Tulsa, Oklahoma 74104, U.S.

Patrick H. Wells, Department of Biology, Occidental College, Los Angeles, California 90041, U.S.

Adrian M. Wenner, Department of Biological Sciences, University of California, Santa Barbara, California 93106, U.S.

Myron P. Zalucki, Department of Entomology, University of Queensland, St Lucia, Queensland 4067, Australia

INDEX

References to figures and tables are printed in boldface type.

Abies (Pinaceae),
grandis, 339
religiosa, 157, 163, 269, 287, 289–292, 296, **301,** 310, 317, 324, 358, 359, 363–373, 386, 389. *See also* overwintering (forest/habitat/Mexico/vegetation)
abiotic conditions. *See* life history (constraints), temperature, water content, weather
Acheta domesticus, 330
Ackery and Vane-Wright, 11–25, 31, 38, 81, 127, 131, 147, 158, 165, 167, 169, 181, 184, 313, 399
Acraeinae, 79–81, **202, 203**
Acridipsas illidgei, 173
Actinote (Acraeinae). *See Altinote*
advertisement. *See* communication signals
Aedes aegyptii, 98
Africa, 19, 21, 22, 179–181. *See also* South Africa
Agelaius phoeniceus. See red-winged blackbird
aggregations. *See* overwintering
agriculture, 390,
herbicides, 358
Agrotis infusa, 173
alar patches (glands). *See* androconia
Aleutian Islands, 11, 15
allelochemicals, 19–25, 39, 337. *See also* allomones, cardenolides, phenylglycosides, pyrazines, pyrrolizidine alkaloids, secondary metabolites, sterols, toxins
allomones, 25
allozymes, 11, 13–15, 181,
aspartate amino-transferase (AAT), 14
glucose-6-phosphate dehydrogenase (G6PD), 14
malate dehydrogenase (MDH), 14
nonspecific dehydrogenase-1 (NDH-1), 14
phosphoglucomutase (PGM), 14
phosphohexose isomerase (PHI), 14
Alnus firmifolia, 365
Altinote (Acraeinae),
leucomelas, **203,** 204–206
Amaurina. *See* Nymphalidae
Amauris (Nymphalidae), 12–25, **12,**
ochlea, 22
America, 171,
Central, 15, 179–181
North, 5, 12, 38, 49, 107, 118, 125, 129, 147, 149, 154, 157, 179, 181, 183–185, 190, 193–198, 233–250, 253–265, **254,** 269, 290, 295, 299, 304
Great Lakes, 247, 263
South, 5, 12, 23, 179, 400
tropical, 183
androconia, 19, 21–25, 29–39, **30–33,**
Amauris, 30
Danaus, 30
Danaus chrysippus, **30–32**
Danaus cleophile, 31
Danaus erippus, **30**
Danaus gilippus, **30,** 43
Danaus plexippus, **30–33**
glandular cells, 21, 30, **32, 33**
hairbrushes. *See* hairpencils
androconia *continued*
hairpencils, 11, **17,** 19–22, 29–39, **30, 31,** 34, 43–45
morphology, 29–39
Tirumala, 30
trichogen cells, 21
volatiles, **24,** 31, **35, 36,** 43–45
wing pockets (glands), 19, 22, 29, **30, 31**
Anesia, 11, **12**
Anosia. See Danaus
antennae, 22, 32, 44,
odor-sensitive cells, 19, 21, 22, 43–45. *See also* pheromones (receptor)
sensilla. *See* odor-sensitive cells
Anteos (Pieridae), 201,
clorinde, 204
maerula, 204
Aphelocoma coerulescens, **130, 131**
Aphrissa (Pieridae), 205,
statira, **203**
aphrodisiac. *See* pyrrolizidine alkaloids (danaidone)
Apocynaceae, 11, 20, 80, 125, 127, 169, **170, 171.** *See also Apocynum*
Carissa ovata, **170**
Ichnocarpus, **170**
frutescens, **170**
Mandevilla sp., 169, **170**
Nerium oleander, 127, 165, **170**
Parsonsia
straminea, **170**
eucalyptophylla, **170**

Apocynaceae *continued*
Prestonia, 127
Trachelospermum, 169, **170**
Apocynum, 125–127, 399,
androsaemifolium, 125–127, **126**
sibiricum, 125–127, **126**
apomorphy. *See* evolution
aposematism. *See* warning signals
Appalachian Mountains, 260, 263
apparency, 137. *See also* herbivory, predation
Arbutus glandulosa, 365
Arceuthobium abietis-religiosae, 366–373
Arctiidae, 21, 22, 25
Argentina, 179, 184
Argyreus hyperbius, 173
Aristolochia, 79
Artamus leucorynchus, 348
Asclepiadaceae, 4, 11, 20, 21, 37, 38, 64, **66,** 69, **77,** 91, 129, 147, 152, 169, **170, 171,** 180, 183, 184, 189, 196, 197. *See also Asclepias,*
Araujia hortorum, 169, **170**
Calotropis
gigantea, **170,** 171
procera, 94, 95, 98, **170,** 171, 189
Cryptostegia grandiflora, 169, **170,** 346
Cynanchum, 131
carnosum, **170,** 346
floribundum, **170**
laeve, 108, 119
Gymnanthera nitida, **170**
Hoya australis, **170**
introduced species, 169–171, **170, 171,** 179–181
Leichardtia australis, **170**
Marsdenia leichardtiana, **170**
Pentatropis
atropurpurea, **170**
quinquepartita, **170**
Pergularia extensa, 95, 98
Sarcostemma, 131
australe, **170**
clausum, 132
crispum, 108
cynanchoides, 108
Secamone elliptica, **170,** 346
Stephanotis, **170**
Asclepias, 21, 107, 122, 126, 127, 165, 169, **170,** 171, 180, 181, 209–217, 234, 253, 255, 256, **257,** 269, 272, 399,
amplexicaulis, 107–109, **110,** 119, **238,** 269–272, **270–272**
asperula, 85, 94, 237, **238,** 245, 250, **257,** 264
californica, 84, 85, **86, 87,** 95, 119, 120, **210,** 211, **212,** 213
capricornu, 107–122, **110, 111, 114, 115, 254,** 255, 256, 263
Asclepias continued
cordifolia, 84, 85, **86, 87,** 95, 98
cryptoceras, 108
curassavica, 25, 69, 70, 86, 87, **88, 89,** 94, 95, 97–101, 119, 121, 159, **170,** 189, 204, 206, **210,** 211, 256, 257, 330
distribution. *See* distribution (host plant)
engelmanniana, 108
eriocarpa, 85, **86–88,** 94, 97, 119, 120, **210,** 211, 212, **213**
erosa, 85, **86, 87,** 94, **210,** 211
exaltata, 120, 256, **258**
fascicularis, 98, **210,** 211, 212, **213, 214**
fruticosa (=*Gomphocarpus fruticosus*), 69, 70, 91–102, **170,** 189, **210,** 211, 216
habitat, 211
hirtella, **238**
humistrata, 86, 87, **88, 89,** 94, **131,** 136, 237, **238,** 245, 250, **254,** 255, 256, **257, 258,** 259, 260, **261, 262,** 263, 269, 271
hybridization, 120
incarnata, 25, 126, 132
latifolia, 108
longifolia, 119, **238,** 269–272, **270–272**
nivea, 257
obovata, 270, **272**
perennis, 108
physocarpa (=*Gomphocarpus physocarpus*), 86, **89,** 94, 98, 100, 189
purpurascens, 120
quadrifolia, **238**
rotundifolia, **170**
semilunata, **170**
solanoana, 108
speciosa, **77,** 84, 85, **86–88,** 99, 118–121, **238,** 256
subulata, 95
syriaca, 85, 87, **88,** 99, 118, 120, 121, 125, 126, 237, **238,** 246, 250, **254,** 255, 256, **257, 258,** 260, **261, 262,** 264, 265, **325,** 330, 399
tuberosa, 131, 135, 270–272, **272**
variegata, **238**
verticillata, 126, 270, **272**
vestita, 85, **86–88,** 94, 95, 97–99, 101, **210,** 211
viridiflora, 107–109, **110, 111**
viridis, 107–122, **110, 111, 114, 115,** 237, **238,** 245, 250, **254,** 255, 256, **257, 258,** 259, 260, **261, 262,** 263, 264, 270–272, **272**
Asclepiodora, 120
Asia, 12, 19, 38, 165, **166–168,** 179, 290
assortative mating. *See* behavior (mating—mate choice)
Asteraceae, 277, 310
Atlantic Ocean, 180, 181, 184, 279
Atriplex confertifolia, 330
Australia, 5, 14, 15, 22, 69, 144, 165–174, **166–168, 172,** 181, 184, 189–198, **190,** 216, 300, 304, 345–353, 358, 361, 399,
Cape York, 167, 180
central, 171
east coast, 171, 189
New South Wales, 62, 91, 165, 173, 180, 189–198
Northern Territory, 189
northwestern, 171
Queensland, 69, 171, 173, 180, 181, 189, 345–347, 350
South, 172, 189
Tasmania, 171, 180, 189
Victoria, 165, 193
Western, 171, 180, 189
automimicry. *See* mimicry
autoradiography, 61, 63
Azores, 180

Baccharis,
conferta, 365
halimifolia, 277
Bacillus thuringiensis, 372
Baja California, 336, 358
bark beetles. *See Scolytus*
basking. *See* behavior (basking), temperature (thermoregulation)
Bates, H.W., 79
behavior, 4, 154, 309–313, 345–353,
basking, **141,** 235, 247, 313. *See also* temperature (thermoregulation)
courtship. *See* behavior (mating)
crawing. *See* temperature (overwintering)
feeding, 20, 21, 38, 125, 272, 282, 300, 306. *See also* herbivory, predation, warning signals (taste)
flight. *See* migration (flight behavior)
mating, 21, 22, 24, 29–39, **37,** 43, **47,** 49–59, 61–67, 69–75, 150, 179, 197, 198, 202, 211, 219, 249, 296, 299, 304, 336, 345, 349, 350, 353, 399
age, 50, 304. *See also* wing condition
costs, 261
evolutionary shift, 38
male aggression, 29, 33, 34, 49–59, 69–75

behavior *continued*
mating *continued*
mate choice, 34, **47,** 49–59, **51–57,** 69–75, 304, 399, 400
multiple, 49–59, 61–67, 75, 304
non-random. *See* behavior (mate choice)
pre-migration, 49–59, 197, 215, 304
proportion, **205**
size, 49–59, **52–57,** 304. *See also* weight
tandem flight, 50
timing, 49–59, 62, 64, 229, 269, 304
movement. *See* migration, temperature (overwintering—crawling/shivering)
overwintering. *See* overwintering, temperature
oviposition, 25, 49, 57, 58, 66, 69–75, 206, 209–217, **212, 213,** 233–250, **236,** 269, 270, 304, 351, 399
costs, 261
host plants, 69–75, **212, 213,** 233–250, 262, 264, 270–272. *See also* Asclepiadaceae, herbivory
choice, 126, 127, 129, 271, 272
patch size, 69–75, **70,** 271–272
migratory, 233–250, **236,** 261, 264. *See also* oogenesis-flight syndrome
sites, 57, 58, 64
timing, 233–250, **236, 238–245**
predation. *See* predation
reproductive. *See* behavior (mating)
sexual. *See* behavior (mating)
thermoregulation. *See* temperature (thermoregulation)
wing flapping, 346
Bermuda, 180
bile salts, **98,** 99. *See also* cholic acid, cardenolides (absorption), fat content
bioassay, 99. *See also* cardenolides (emetic response)
biological models, 3, 5
biosynthesis, 19, 20, 32, 44, 135
biotic diversity. *See* evolution (diversity)
bird predators. *See* predators (birds)
bishop pine. *See Pinus muricata*
bitterness. *See* warning signals (taste)
bivouac. *See* migration
black-backed oriole. *See Icterus galbula abeillei*
black-eared mouse. *See Peromyscus melanotis*
black-headed grosbeak. *See Pheucticus melanocephalus*
blue gum. *See Eucalyptus globulus*
blue jay. *See Cyanocitta cristata*
bogong moth. *See Agrotis infusa*
Bombyx mori, 98
Boppré, Michael, 21, 25, 38, 135
Borneo, 180
Brazil, 15
breeding. *See* distribution
Brower, Jane V.Z., 4, 19–21, 81, 129, **130, 131,** 134, 136, 179, 397
Brower, Lincoln P., 1, 4, 19–21, 81, 126, 134, 179, 182, 393, 397
Brown, Keith S. Jr., 21, 36, 80
brown-crested flycatcher. *See Myiarchus tyrannulus*
Brugger, Ken and Cathy, 4
bursa copulatrix, 50, 61, 62, 63, **65,** 67, 257, 346. *See also* spermatophores
Buteo platypterus, 282
butterfly town. *See* California (Pacific Grove)

California, 5, 34, 36, 61, 108, 121, 144, 147, 149, 153, 179, 181, 183–185, 197, 209–217, 219–231, 245, **254,** 275, 278, 300, 335–344, **337,** 358, 359, 393, 394. *See also* overwintering (California), U.S.A.,
Cascade Mountains, 209, 210
Central Valley, 210
Death Valley, 337
Indians, 343
Los Angeles, **221,** 223, **224,** 229, 335, 340, 381
Marin County, 335, 339, 360
Mojave Desert, 219, 228, 229
Monterey, 335, 337, 338, 340, 341, 343
Pacific coast, 335
Pacific Grove, 341, **360**
Saline Valley, 337
San Bernardino Mountains, 228, 229
San Diego, 335, 340
San Gabriel Mountains, 228, 229
San Luis Obispo County, **220,** 224, 337–339, 340
Morro Bay, **220, 222,** 224
Pismo Beach, **220, 222**
San Rafael Mountains, 209, 210
Santa Barbara, 209–217
County, **220, 222,** 224, 338–340, 360
Santa Cruz Island, 211, 212, **214,** 216, 217, 338–341, 393
Santa Rosa Island, 212, 217
Santa Ynez Mountains, 213
Sierra Nevada Mountains, 209, 210, 219, 229
Ventura County, **220, 222,** 224
Calotropis. See Asclepiadaceae
Campylorhynchus megalopterus, **316**
Canada, 4, 209, 249, 253, **254,** 265, 315, 378, 383, 389. *See also* America (North),
Ontario, 121
southeastern, 118, 249, 377
Toronto, 219, 280
western, 209
Canary Islands, 180, 181, 185
cardenolides, 4, 5, 19–21, 25, 58, **77,** 83–89, 91–102, 107–122, 125, 126, 129, **131,** 152, 179, 235, 253–265, 321, 324, 399,
abdomen, 317, 323–333, **325, 329**
absorption, 91–102, 219, 330. *See also* cardenolides (sequestration), fat content (lipid absorption)
aglycones (genins), 83, 84, 91–102
afrogenin, 95
calotropagenin, 87, **88,** 93, **94,** 95, 97–99, 102, 120
digitoxigenin, 83, **84, 85,** 87, 92, **93,** 99, 109, 117, 120, **258, 262**
gomphogenin, 93, **94,** 95, 97–99, 102
uzarigenin, 83, **84, 85,** 87, 92, **93, 95,** 99, 120
avoidance, 330
concentration. *See* cardenolides (quantitative variation)
eggs, 79
emetic response, 38, 99, 100–102, 119, 121, 131, 135, 136, 315, 324. *See also* bioassay, blue jay
epoxy, 86, 87, **88,** 120, 121
excretion, 97, 100, **101**
extraction, 92
fingerprint, 37, 87, 101, 107, 121, 122, 216, 217, 250, 253–265, **258, 261, 262.** *See also* cardenolides (qualitative variation)
geographic variation, 107–122, 136, 253–265, **258–261,** 399. *See also* cardenolides (qualitative variation/quantitative variation/temporal variation)
Mexico, **259**
southern U.S.A., **259,** 399
northern U.S.A., **259**
glucosides, 99
glycosides, 83–89, 91–102
afroside, 92, **93–96,** 97, 99, 102
asclepin, **88,** 92, **93, 95, 96,** 97
aspecioside, 87, **88,** 99, 121, 256, 258
calactin, 86, 87, **88,** 92, **93–96,** 97–99, 102, 120
calotoxin, 87, **88, 94,** 98, 99, 102

cardenolides *continued*
glycosides *continued*
calotropin, 83, **84, 85,** 86, 87, **88,** 92, **93–95,** 97–99, 102, 120
desglucosyrioside, 86, **88,** 97, 99, 120
digitoxin, 92, 109, **113–115,** 118, **131,** 256, **258, 262**
frugoside, **94,** 98, 99, 102
gofruside, 93, **94,** 98, 99, 102
gomphoside, 92, **93, 95, 96,** 97, 102
humistratin, **88**
labriformidin, 86, **88,** 97, 120
labriformin, 86, **88,** 97, 120
ouabain, 119
syriobioside, 87, **88,** 99, 258
syrioside, 121, 258
uscharidin, 83, **84, 85,** 86, 92, **93, 95, 96,** 97–99, 120
uscharin, 92, **93, 95, 96,** 97, 102, 120
voruscharin, 120
head, 323
intrinsic, 38, 79
metabolism and transformation, 86–89, 91–102, **95, 96,** 107, 121
migration, 101, 121, 253–265, **259–262**
Na+, K+-ATPase inhibition, 119
overwintering. *See* overwintering (Mexico—cardenolides)
polarity, 86–89, 91–102, **101,** 107, 117, 118, 120, 121, 256, 258, 330
pre-ingestinal cue. *See* warning signals (taste)
qualitative variation, 37, 86–89, 91–102, 107, 117–122, **117,** 253–265, **258, 261, 262.** *See also* cardenolides (fingerprint/geographic variation/temporal variation)
quantitative variation, 83–86, 91–102, 107–122, **113–116,** 253–265, **257, 259–261, 325, 329.** *See also* cardenolides (geographic variation/temporal variation)
sensitivity, 323
sequestration, 20–21, 37, 83–89, **84–89, 93–96, 101,** 91–102, 107–122, 129, 253–265, 399
amphiphiles, 91–102
body parts, 100, 101
cost, 119
dose-dependence, 83, 321
efficiency, 84
emulsification, 91–102
larvae, 91–102
logarithmic, 83, 107, 114, **115,** 119

cardenolides *continued*
sequestration *continued*
mechanism, 91–102, 119
permeability, 96–102
physiology, 96–102
saturation, 96, 107
selective, 87–89, 91–102, **93,** 117
sexual difference, 84, 86, 101, 102, 107, 115, **116,** 119, 324
xenobiotic, 102
spatial variation. *See* cardenolides (geographic variation)
storage, 96–102, **101,** 219. *See also* cardenolides (sequestration)
structure, 86–89, **84, 88,** 93, **94,** 95, **96,** 97–102. *See also* cardenolides (polarity)
temporal variation, 120–122, 136, 253–265
thorax, 317, 323–325
toxicity, 38, 315, 323–333. *See also* cardenolides (emetic response)
wings, 323–325
cardiac-active steroids. *See* cardenolides
cardiac glycosides (CGs). *See* cardenolides
Caribbean islands, 5, 12, 179–181, 257, 296
Caroline Islands, 180
cat. *See Felis domesticus*
Catharus occidentalis, **316**
Celestrina (Lycaenidae) *pseudargiolus,* **152**
chauvinism. *See* behavior (mating)
Chelone, 79
chemical communication. *See* communication signals (chemical)
chemical defense, 1, 19–21, 24, 25, 34, 37–39, **77,** 91, 121, 129–137, 179, 253–265, **285,** 359, 386, 390, 397–400. *See also* cardenolides, pyrazines, pyrrolizidine alkaloids, sequestration, toxins, warning signals,
de novo synthesis, 135
herbivore prey, 19, 129–137
palatability spectrum, 129, 131, 264. *See also* warning signals (taste)
host plants, 1
chemical ecology, 1, 4, 19–25, 37, 135, 397, 400
chemoecology. *See* chemical ecology
chemoprints. *See* cardenolides (fingerprint)
China, 166, 180
chisel-toothed rat. *See Dipodomys microps*
cholesterol. *See* sterols
cholic acid, **98,** 99
Choristoneura fumiferana, 315

chromatography,
column (CC), 92
gas (GC), 35
high performance liquid (HPLC), 92, 97
thin layer (TLC), 87, 92–95, 98, 102, 107, 109, **117,** 117–122, 131, 256–265, **258, 261, 262**
Chrysomelidae, 135
Cincinnati Zoo, 357
Cissia (Satyrinae) *hermes,* 202
citrus, 399
classification. *See* evolution (origin, phylogenetic, systematics), Nymphalidae
Clethra mexicana, 365
climate. *See* weather
coastal plain willow. *See Salix caroliniana*
coast live oak. *See Quercus agrifolia*
coast redwood. *See Sequoia sempervirens*
Cocoyoc. *See* Mexico
Colias, 148. *See also* Pieridae,
eurytheme, **152**
meadii, 149
philodice, 149
colonization. *See* migration (range expansion)
coloration. *See* warning signals
Columbus hypothesis, 179–185, 335, 398. *See also* habitat (deforestation)
commercial forestry, 390
common crow butterfly. *See Euploea core*
communication signals. *See also* territoriality,
chemical (odor), 17, 19–25, 29–39, 399. *See also* pheromones, pyrazines, pyrrolizidine alkaloids, volatiles
visual, 21. *See also* behavior (mating)
competition, 234,
interspecific, 107, 111, 112, 131, 171, 181
intraspecific, 58, 121. *See also* behavior (mating)
conservation, ix, 5, 173, 287, 292, 295, 296, 357–361, 363–373, 377–384, 389–391, 393, 394, 400,
California, 360, 393–394. *See also* The Monarch Project
common land, 360. *See also* ejidatarios, ejidos
economic pressure, 358, 359, 365, 400
education, 378, 383, 384, 386, 387, 390
Mexico, 363–373, 377–381, 383–

conservation *continued*
Mexico *continued*
387, 389–391. *See also* Monarca, A.C.
politics, 357, 359, 400
red data, 358
threatened phenomenon, 5, 184, 358, 364, 400
tourism, 365–373
conspicuousness. *See* warning signals
copulation. *See* behavior (mating)
Costa Rica, 5, 15, 184, 201–206
cremaster, 14
Crotalaria, 21, 34, 44, 135
crypsis, 80
Cryptostegia. See Asclepiadaceae
Ctenuchidae, 21, 25
Culex pipiens, 98
Cupressus, 367,
lindleii, 365
macrocarpa, 338
cuticle. *See* cardenolides (abdomen/head/thorax/wings)
Cyanocitta,
cristata, **1,** 99, 121, 130
stelleri, **316,** 320
cyanogenic glycosides, 80
cycad, 79
Cynanchum. See Asclepiadaceae

Danaidae. *See* Nymphalidae
Danaina. *See* Nymphalidae
Danainae. *See* Nymphalidae
Danaini. *See* Nymphalidae
Danaus (*Anosia*), 11–25, 38, 165, 166, **167,**
chrysippus, 4, 12, **13,** 19, 22, 23, 165, 166, 169, **170, 171, 173,** 181, **352**
c. petilia, 91–102
eresimus, 12, **13,** 159, **160**
gilippus, **1,** 12, **13,** 19–25, 32, 36, 203, 206
g. berenice, 129–137, **131, 133, 134,** 159, **160**
plexaure, 12, **13**
Danaus (*Danaus*), 11–25, 38, 165, 166, **167,**
cleophile, 12, **13,** 22, 38, 400
erippus, **9,** 12, **13,** 23, 38, 179, 184, 400
plexippus,
p. nigrippus, **9**
p. plexippus, **1, 9, 13, 77, 152,** 166, **168,** 169, **171–173,** 174, 203, **204, 205,** 219, 315–321, 335–341, 345, 358, **352,** 386, **395,** 400 (throughout volume)
'fumosus', 181
'leucogyne', 181
'nivosus', 181, 184, 185
Danaus continued
plexippus continued
p. portoricensis, **9**
Danaus (*Salatura*), 11–25, 165, 166, **167,**
affinis, 12, **13,** 166, 169, **170, 171, 173,** 174, 350, **352**
genutia, 12, **13,** 166, 167, **168**
ismare, 12, **13,** 181
melanippus, 12, **13,** 181
day-degrees, 233–250, **242–244,**
development time, 233–250, **242–247**
day length. *See* physiology (day length)
death. *See* mortality
defense. *See* chemical defense, crypsis, warning signals
Dendroctonus, 373
Dendroica townsendi, **316**
Denmark, 180
Dermestes maculatus, 97, 98
desiccation. *See* overwintering (water), physiology (water loss)
d'Estrella, Theophilous, 341
development time. *See* day-degrees (development time)
diapause. *See* physiology (diapause)
diet breadth. *See* herbivory
digitized image, 63
Dipodomys microps, 330
dispersal. *See* migration
distastefulness. *See* warning signals (taste)
distribution, ix, 3, 4, 165, 169, 171, **172,** 179–185,
autumn, 122, 189, 346
breeding. *See* distribution (spring/summer), life history
evolutionary shift, 4, 38
historical, 171, **172,** 179–185, 335. *See also* migration (range expansion)
host plant, 165, 181, 211, 233–250, 253–265. *See also* Asclepiadaceae, *Asclepias,* behavior (oviposition—host plants), food supply, herbivory (host plant), milkweed
latitudinal, 233–250, **236, 238, 240, 244, 245**
milkweed, 38, 94, 161, 169, 233–250, 253–265. *See also* Asclepiadaceae, *Asclepias,* behavior (oviposition—host plants), food supply, herbivory (host plant)
range. *See* distribution (spatial), migration (range expansion)
spatial, 4, 189, 209–217, 233–250, 253–265, 335. *See also* migration
spring, 107, 121, 122, 147, 233–250,
distribution *continued*
spring *continued*
253–265, **259,** 269. *See also* life history (spring breeding), migration
summer, 121, 122, 147, 157, 159–162, **160, 161,** 165, 233–250, 253–265, **254,** 346. *See also* life history (summer breeding), temperature
temperate, 3, 4, 38, 118, 147, 233–250
temporal, 4, 107, 233–250, 265. *See also* migration
tropical, 3, 4, 38, 107, 118, 147, 201–206, 215, 313
winter. *See* overwintering
worldwide, 165, 179–185
diversity. *See* evolution (diversity)
DNA. *See* mitochondrial DNA
dogbane. *See Apocynum*
Donahue, Julian P., ix, 141
Douglas fir. *See Pseudotsuga menziesii*
drought, 189, 197
dwarf mistletoe. *See Arceuthobium*

eclosion, 144
ecological chemistry. *See* chemical ecology
Ectopistes migratorius, ix, 357, 360, 361
Edgar, John, 21, 22, 80, 81, 127, 135
eggs, 21, 25, 211. *See also* larvae, life history,
egg production, 49, 50, 64. *See also* behavior (oviposition)
Ehrlich and Raven, 19, 20
Eisner, Thomas, 21
ejidatarios, 359, 360, 377, 387, 389–391. *See also* conservation, ejidos, Mexico, overwintering
ejidos, 359, 360, 363, 377, 390. *See also* ejidatarios, Mexico, overwintering
electroantennogram, 25, 43–45. *See also* antennae
electrophoresis, 13–15
electrophysiology, 43–45
emetic response. *See* cardenolides
Empidonax affinis, **316**
endangered phenomenon. *See* conservation (threatened phenomenon)
endocrine system, 143–144, 397. *See also* fat content, physiology (reproductive),
corpora allata, 143
corpora cardiaca, 143
diapause. *See* physiology (diapause)
hormone precursors, 96
hormones, 58
adipokinetic (AKH), 143–144

endocrine system *continued*
hormones *continued*
diuretic (DH), 143–144
juvenile (JH), 143–144, 153, 161, 162, 189, 193–195, 235
neuropeptides, 143
red pigment concentrating (RPCH), 143–144
migration. *See* migration (physiology)
reproductive tract, 143–144, 189. *See also* physiology (reproduction—gonad development)
energetics. *See* fat content, overwintering (energetics)
Entomological Society of America, 394
Ericaceae, 291
Eucalyptus, 61, 335, 337, 340, 360. *See also* overwintering (Australia/California),
globulus, 337, 358
Eumaeus, 79
Eupatorium, 23, 34, 44, 135
Euphorbia, 399
Euphydryas, 79,
aurinia, 148
Euploea, 11–25, **12**, 165, 166, **167**,
alcathoe (=*eichhorni*), 169, **170, 171, 173,** 174
algea, 167, **170, 171, 173**
batesii, 167, 169, **170, 171, 173**
climena, 167
core, **17, 23, 24,** 79, 127, 165, 167, **168,** 169, **170, 171,** 172, **173,** 345–353, **352**
darchia, 167, **168, 170, 171, 173,** 174
klugii, **23, 24**
sylvester, **23, 24,** 169, **170, 171, 173,** 345, 346, **352**
tulliolus, **168,** 169, **170, 171,** 172, **173,** 174, 345–353, **352**
usipetes, 167, **168**
Euploeina. *See* Nymphalidae
Euploeini. *See* Nymphalidae
Euptychia (Satyrinae) *jesia*, 202
Eurema (Pieridae), 204, 205,
daira, 204, **205**
gratiosa, **203**
lisa, **152**
Europe, **184,** 185, 290
Eurycotis floridana, 97
Euschemon rafflesia, 173, 174
Evita hyalinaria blandaria, 367, 371, 372
evolution, 152, 264, 398. *See also* fitness, life history,
adaptation, 62, 79, 147, 154, 198, 313
age. *See* origin
ancestral food plant, 127
evolution *continued*
apomorphy, 14, 185
coevolution, 398
plant-herbivore, 398
prey-predator, 135
diversity, 165, 357
genetic variation, 181, 185
geographic variation, 129. *See also* cardenolides (geographic variation)
host plant range, 4, 127, 179–181. *See also* distribution (host plant/milkweed), migration (range expansion)
hybridization, 181, 264
migration. *See* migration (evolution)
origin, 11–15, 137, 180
phylogenetic, 11–15, 31, 80, 136
plesiomorphy, 14, 185
pre-adaptation. *See* herbivory (pre-adaptation)
reinforcing selection, 179, 184
sympatry, 120
systematics, 11–25, 400
extinction, 357

fat content, 61–67, **62,** 189, **192,** 193–195, 202, 204, **205,** 210, 256, 299, 300, 326,
diglycerides, 143–144
lipid absorption, 96–100. *See also* cardenolides (absorption/sequestration)
migration, 101, 143–144, 189, 193–195, 300
overwintering, 101, 143, 147–154, 157–161, 193, 299, 300, 305. *See also* overwintering (energetics)
reproductive, 193
sexual difference, 101, 299
feeding. *See* behavior (feeding), herbivory
feeding experiments. *See* herbivory, predation
Felis domesticus, 330
Ficus. *See* Moraceae
Fiji, 167
filaments. *See* larvae
fingerprints. *See* cardenolides
fire, 211
fitness, 49, 58, 62. *See also* evolution,
reproductive success, 58, 61, 67
survivorship, 61, 62, 64, **66,** 72, 149, 234
food plant. *See* Asclepiadaceae, *Asclepias*, behavior (oviposition—host plants), food supply, herbivory (host plant), milkweed
food supply, 64, 66, 109, 165, 234, 248–250, 264, 269, 282, 299, 300
Fraenkel, Gottfried, 19, 20
freezing. *See* temperature
fund-raising, 379

generations. *See* life history (generations)
genetic variation, 72. *See also* allozymes, electrophoresis
glaciation, 180
glandular cells. *See* androconia
Gomphocarpus. *See Asclepias*
Gossypium, 390
grand fir. *See Abies grandis*
grasshopper mouse. *See Onychomys torridus*
gray-barred wren. *See Campylorhynchus megalopterus*
Great Britain, **184**
Greater Antilles, 400
Greater Sunda Islands, 180
grosbeak, 37
Guam, 180
Guatemala, 296
Gulf of Mexico coast. *See* U.S.A.

habitat, 183, 184, 201–206, 211, 212, 337, 338, 346,
breeding, 229
defoliation, 349
deforestation, 179, 184, 185, 206, 338, 377
exploitation, 359, 389
forest, 180, 201–206. *See also Abies*, overwintering (forest)
gaps, 317
management, 386
native, 338
prairie, 180
protection, 338, 393
reforestation, 359
seasonal, 201–206
Haggard, Rider, 295
hairbrushes. *See* androconia
hairpencils. *See* androconia
Hardin, G., 360
Hawaii, 180, 181
Heliconiinae, **202, 203,** 204
Heliconius, 79–81
Heliothis, 144
Heliotropium, 21, 22, 34, 44
Hemileuca oliviae, 148
herbivores,
foraging. *See* herbivory
oligophage, 79
specialist, 3
herbivory, 4, 19–21, 102, 400. *See also* behavior (feeding), physiology (nutrition),
benefits, 20
costs, 20
feeding experiments, 125–127
host plant, 219, 228, 343–345. *See*

herbivory *continued*
host plant *continued*
also Asclepiadaceae, *Asclepias,* behavior (oviposition—host plants), food supply, milkweed
cattle damage, 109
chemistry. *See* allelochemicals, cardenolides, phenylglycosides pyrazines, pyrrolizidine alkaloids, secondary metabolites
choice. *See* behavior (feeding/oviposition—host plants)
development, 109, 110, **111, 112,** 118, 211, 212
distribution, 69–75, 107, 108, 136, 179–181, 197, 204–206, 209–217, **210,** 233–250, 264, 265, 345, 397
livestock poisoning, 197
patch residence, **71.** *See also* behavior (mating)
patch size, 69–75
records, 125, 165, 169, **170, 171**
timing, 109, 209–217, 265
use, 69–75, 107, 108, 118, 121, 125–127, 165, 180, 196, 197, 204, 205, 209–217, 233–250, 264, 265, 400
larvae, 91, 107–111, **110,** 118, 121, **126,** 211, 351. *See also* larvae
distribution, 118
nutrition, 205
pre-adaptation, 79, 127
Hesperiidae, 165, **202,** 204
Hispaniola, 22–23
history,
presettlement, 335–344
Holocene, 180
homocourtship, 51, **56.** *See also* behavior (mating)
homoplasy, 13
Hong Kong, 180
hormones. *See* endocrine system
host plant. *See* Asclepiadaceae, *Asclepias,* behavior (oviposition—host plants), herbivory (host plant)
host plant evolution. *See* evolution
host plant succession, 136
host plant use. *See* herbivory
Hoya. See Asclepiadaceae
Hyles lineata, 148

Icterus,
galbula abeillei, 315–321, **316**
parisorum, **316,** 320, 330
Idea, 11, **12,** 166
Ideopsis, 11, **12,** 19, 20, 166
India, 179–181
Indonesia, 166
interactions,
interactions *continued*
plant-herbivore. *See* herbivory
prey-predator. *See* predation
interdisciplinary research, 25
isotope labeling. *See* autoradiography
Ithaca, 277, 279, 280, **282.** *See also* U.S.A. (New York)
Ithomiinae, 11, 36, 79–81, 201, **203,** 204
Ituna, 11
Itunina. *See* Nymphalidae
IUCN, 5

Jamaica, 15, 22
Java, 166, 180
Jones, Frank Morton, 4

Kanz, J.E., 279–280
Kiepenheuer, J., 275, 280

land use, 385
Lantana camara (Verbenaceae), 194, 196
larvae. *See also* herbivory (host plant—use/larvae), life history,
development, 233–250. *See also* day-degrees, life history
filaments, 11, 13
mobility, 127
mortality, 107, 235, 271, 315
numbers, 107, **110,** 118, 211, 233–250, 271, 272
predation, 107, 315
latex, 125, 126. *See also* Asclepiadaceae, Apocynaceae
lek. *See* behavior (mating)
Leopold, Aldo, 357
Lepidium (Brassicaceae) *virginicum,* 204
lesser wanderer. *See Danaus* (*Anosia*) *chrysippus*
Liabum (Asteraceae), 204
life history, 62, 118, 120–122, 125–127, 147, 165, 233–250, 397, 400. *See also* day-degrees, eggs, fitness, food supply, larvae, migration, mortality, reproduction,
annual cycle, 144, 149, 150, 189–198, 201–206, 209–217, 233–250, 265. *See also* distribution, migration, overwintering
Australia, 189–198, **191, 192**
autumn breeding, 193
bet-hedging, 198
California, 209–217
cohorts. *See* migration (cohorts)
constraints, 233–250. *See also* temperature (constraints, cues)
abiotic, 233–250, **244,** 264
biotic, 245. *See also* competition, food supply, herbivory, predation
life history *continued*
continuous breeding, 185, 190, 198, 209–217, 233, 237, 245
generations (voltinism), 64, 69, 118, 121, 149, 190, 209–217, 233–250, **244,** 253–265. *See also* day-degrees
mixed, 185, 201–206, 209–217
spring breeding, 108–113, **110, 112,** 118–121, 209–217, 233–250, 296, 304
summer breeding, 189, 190, 197, 209–217, 233–250, 253–265
Limenitis (*Basilarchia*)
archippus, 4, 19, 25, 129–137, **152,** 179, 300, **302–306, 313,**
a. archippus, 129, 130, **130, 131**
a. floridensis, 129–137, **130, 133, 134**
arthemis
a. astyanax, **131,** 132, **133, 134**
lipid reserves. *See* fat content
livestock poisoning. *See* herbivory (host plant)
Lycaenidae, 148, 165, **202,** 203, **204,** 205, 206
Lycorea, 11, **12,** 22
Lygaeus, 108, 111, **112,**
kalmii, 112

Magnetic influence, 275, 280, **282**
Malaisia. See Moraceae
Malaysia, 166, 180
Manataria (Satyrinae) *maculata,* 202
Mandevilla. See Apocynaceae
Manduca, 144
mark-recapture. *See* population size
Marpesia (Nymphalidae) *chiron,* 206
Marquesas, 180
Mascarene Islands, 179, 180
mate choice. *See* behavior (mating)
mating. *See* behavior (mating)
maturation time. *See* day-degrees (development time)
Mauritius, 181
Meinwald, Jerrold, 21, 22, 31
Melaleuca styphelioides, 196, 349. *See also* overwintering (habitat)
Melinaea, 80
metabolism, 61, 62, 64, 144, 219, 345, 349. *See also* cardenolides (metabolism), fat content, overwintering (energetics)
metamorphosis, 99
Mexican chickadee. *See Parus sclateri*
Mexico, 4, 14, 22, 34, 36–38, 49, 50, 147–154, **151,** 157, **160, 161,** 179–181, 184, 233–235, 241, 249, 253, 265, 269, 275, 279, 282, 287–293, **288,** 295–297, 299–306, 309, 358, 377, 389, 393. *See also* overwintering (Mexico),

Mexico *continued*
Cocoyoc, 335
Diario Oficial, 359, 363
Mexico City, 379, 389
Ministry of Urban Development & Ecology (SEDUE), 322, 363, 390
Microtus mexicanus, 323–331
migration, 107, 136, 147, 179, 180, 185, 189–198, 201–206, 209–217, 219–231, 269–273, 275, 299, 305, 345, 357, 361, 397, 398. *See also* cardenolides, distribution, fat content, population size (mark-recapture), overwintering, physiology, temperature,
annual, 14, 38, 61, 107, 136, 147, 157, 179–181, 183, 184, 201–206, 209–217, 233, 265, 269, 296, 299, 398
Australia, 172–174, **173,** 189–198, **191, 192,** 345–353
autumn, 49, 107, 143, 144, 149, 159–162, **160, 161,** 184, 193, 196, 210, 213–215, 219, 233, 234, 269, 275–283, 290, 299
benefits, 234
bivouac, 336, 337, 339
California, 209–217, 219–230, 398
cardenolides. *See* cardenolides (migration)
cohorts, 233–250, **242–249,** 264
first arrivals, 233–250, **236, 242–249**
main cohort, 233–250, **236, 242–249**
colonization. *See* migration (range expansion)
Costa Rica, 201–206
costs, 58, 234, 261, 264, 282. *See also* oogenesis-flight syndrome
day length. *See* physiology (day length)
direction, 179, 189–198, 201–206, 209, 217, 219–231, 237, 253, 275–283, **276–281,** 295, 296. *See also* migration (autumn/spring)
angular headings, **226, 227**
bimodal, 219–231
dispersal, 62, 64, 209, 213, 226, 269, 270, 296, 304, 350. *See also* overwintering (dispersal)
distance, 189, 193, 196, 198, 209, 214, 215, 226, **228,** 352
evolution, 181, 183, 234, 264
fall. *See* autumn
feeding, 194, 282. *See also* nectar
flight altitude, 278, 280, 282, 295
flight behavior, 58, 189–198, 204, 210, 211, 275–283, **278, 279,** 296, 299, 306
migration *continued*
Kiepenheuer's model, 275, 280, 281
navigation and orientation, 275–283, **276–281,** 397
sun compass, 280
northward. *See* migration (spring)
oogenesis-flight syndrome. *See* oogenesis-flight syndrome
oviposition, 121, 233–250, 264. *See also* behavior (oviposition)
physiology, 144, 147, 253. *See also* endocrine system, physiology
range expansion, 61, 171, **172,** 179, **182, 184,** 209, 217, 221, 295, 296, 336, 352, 398
Columbus hypothesis, 179–185, 335, 398
expansion-contraction hypothesis, 215–217
Scudder hypothesis, 180
Walker hypothesis, 181
recolonization. *See* migration (spring)
resource location, 69, 264, 272, 299, 300. *See also* herbivory (host plant)
seasonal. *See* migration (annual)
sex ratio. *See* sex ratio
single sweep hypothesis, 233–250, 253–265, **255**
source-sink, 398
South America, 179
speed (rate), 233–250, **250**
spring, 49, 57, 58, 107, 108, 118, 121, 147, 149, 159–162, **160, 161, 177,** 189–198, 209–217, 219–230, 233–250, 253–265, 269–272, 296, 299, 300, 304
successive brood hypothesis, 233–250, 253–265, **255**
temperature, 189–198, 272
timing, 229, 233–250
wind influence, 275–283
thermals, 282
wing wear. *See* wing condition
migratory monarch, special collection, 224
milkweed. *See also* Asclepiadaceae, *Asclepias,* behavior (oviposition—host plants), distribution (milkweed), herbivory (host plant),
common. *See Asclepias syriaca*
narrow-leaf. *See Asclepias fascicularis*
sandhill. *See Asclepias humistrata*
xeric, 228
milkweed bug,
large. *See Oncopeltus fasciatus*
small. *See Lygaeus kalmii*
milkweed butterflies. *See* Nymphalidae (Danainae)
mimicry, 1, 4, 19, 79, 129–137, 179, 398,
automimicry, 20, 39, 131, 136
Batesian, 15, 20, 25, 38, 129–137, 398
dynamics, 136–137
model, 20, 79, 129–137
Müllerian, 129–137
mitochondrial DNA, 398
mixed-function oxidases, 38
models,
abiotic, 235, **236,** 237, **244,** 245
biological, 3, 5
chemical, 83–89
magnetic, 275, 280, 281
mathematical, 61–67, 131
Moluccas, 180, 181
Monarca, A.C., 295, 296, 359, 368, 377–391, 393, 400
monarch (wanderer) butterfly. *See Danaus (Danaus) plexippus plexippus,* white monarch
Moncon-1, ix, 335, 338
Moncon-2, ix
Monterey cypress. *See Cupressus macrocarpa*
Monterey pine. *See Pinus radiata*
Moraceae, 11, 169, **170, 171,**
Ficus, **170**
benjamina, **170**
coronata, **170**
eugenioides, **170**
muelleri, **170**
obliqua, **170**
platypoda, **170**
racemosa, **170**
Malaisia
scandens, **170,** 346, 347, 350, 351
tortuosa, **170**
morphology. *See also* behavior (mating—size), temperature (morphology),
eggs, 346
oocytes, 346
ovaries, 346, 350
size, 193, 299–301, 304
wing length, 49–59, **52–57,** 299–305, **303**
mortality, 64, 121, 315. *See also* larvae,
adult, 315, 319, 326
dosage dependent, 331
mountains, 209, 210, 213. *See also* California, Mexico, Rocky Mountains
movement. *See* migration
Musca domestica, 97, 330
Myiarchus tyrannulus, **316**

Nathalis iole, 149
National Audubon Society, 394

National Handcraft Fund (FONART), 390
national insect, 358, 394
National Wildlife Federation, 394
natural enemies, 248, 397. *See also* parasitism, parasitoids, pathogens, predators
Nature Conservancy, 339
nectar, 23, 64, 67, 150, 194, 209, 210, 215, 282, 299, 300, 304–306, 339, 349
Nei's genetic distance, 14
Neophasia (Pieridae) *terlooti*, 398
Nerium. *See* Apocynaceae
neuropeptide. *See* endocrine system
New Caledonia, 167, 180
New Guinea, 166, 167, 169, 180
New World, 11, 12, 86, 166, 169
New Zealand, 5, 180, 181, 184
Nicaragua, 183
nutrient loss, 359
nutritional ecology. *See* herbivory
nutritional physiology. *See* physiology (nutrition)
Nymphalidae, 4, 11–25, 79, 148, **202**,
  Danaidae, 4, 127, 136
    Danainae, 11, 11–25, 29–39, 43–45, 79, 158, 165–174, **166**
      Euploeini, 11–25, 127
        Euploeina, 11. *See also Euploea, Idea, Protoploea*
        Itunina, 11–25. *See also Anetia, Ituna, Lycorea*
      Danaini, 11–25
        Amaurina, 11–25. *See also Amauris, Ideopsis, Parantica, Radena*
        Danaina, 12. *See also Danaus, Tiradelphe, Tirumala*
Nymphalis (Nymphalidae), 144,
  antiopa, 161

Old World, 12, 180, 181
oleander. *See* Apocynaceae (*Nerium oleander*)
Oncopeltus fasciatus, 99, 108, 119
Onychomys torridus, 330
oogenesis. *See* physiology (reproduction—gonad development)
oogenesis-flight syndrome, 253, 264. *See also* behavior (oviposition—migratory), physiology (oogenesis)
orientation. *See* migration (direction/flight behavior)
oriole, 37. *See also Icterus*
overwintering, 5, 36, 38, 165, 179, 184, 219, 269, 275, 287–293, 295–297, 299–306, 309–313, 341, 345, 357–361, 387, 389–391, 397, 400.
overwintering *continued*
  See also California, Mexico, physiology (diapause), temperature, weight,
  Arizona, 398
  Australia, 172, 173, 189–198, **191, 192,** 230, 335, 344, 345–352
    aggregations, **173,** 345–352
    coastal, 189, 196, 345–352
    numbers, 173, 196, 197
    reproduction, 190, **352**
  behavior, 50–59, **52,** 61–67, 108, 121, 147–154, **162,** 197, 211, 296, 300, 306, 309–313, 357, 358
  biomass, 330
  California, 108, 121, 144, 147, 149, 153, 183, 197, 209–217, **254,** 275, 300, 336, 398
    aggregations, 61, 62, 224, 229, 336
    history, 183, 185, 335–341
    location, 61–63, 183, 335–341
  clusters, 150, 153, 224, 296, 304, 317, **318,** 336, 339, 345, 346, 348
  colonies, 336
    temporary, 339
    permanent, 339
  dispersal, 189, 197, 209, 213, 270. *See also* migration (dispersal)
  distribution, 335
  energetics, 61–67, **62,** 143, 147–154, 210, 299, 300, 305, 306, 313, 359. *See also* fat content
  feeding, 194, 209, 299, 300, 304–306. *See also* nectar
  Florida, 183, 209, **254**
  forest, 253, 287–292, 363–373. *See also Abies religiosa*, conservation, *Eucalyptus*, habitat, *Melaleuca*, overwintering (habitat)
    agriculture, 365–373
    canopy structure, 157, 290–292, **291**
    defoliation, 367
    deforestation, 340, 341, 359, 363–373
    disturbance, 290, 292
    exploitation, 365–373
    fire, 292, 365–373
    grazing, 365–373
    pathogens, 365–373
    reforestation, 359
    tourism, 365–373
    understory, 289–291, **290, 291**
  gregariousness, 357, 358
  Guatemala, 295, 296
  habitat, 195, 196, 287–292, 295, 296, 299, **302,** 309, 313, 317, 337, 338, 359, 363–373
  mating. *See* behavior (mating)
overwintering *continued*
  metacolony, 229
  Mexico, 107, 108, 118, 122, 125, 147–154, **151,** 157–163, 181, 184, 209, 233, 235, 237, 246, 247, 253, **254,** 256–258, **260–262,** 265, 269, 270, 275, 279, 280, 282, 287, 292, 295–297, 299–306, 315–321, 323–333, 358, 377, 378
    aggregations, 49, 61, 157, 253, 289, 295, 296, 299, 300, 309
    Angangueo, **301, 355,** 359
    Baja California, 336, 358
    Campanario, 287, **288,** 289, 290, 300, **301, 302, 355,** 359, 363, **364, 367,** 369, 378, 381, 383, 386, 390, 391
    cardenolides, 87, 99, 121, 253–265, **259–262.** *See also* cardenolides
    Cerro Altamirano, **259,** 300, **301,** 363
    Cerro Pelón, **259, 262,** 300, **301, 302,** 363, **364, 366,** 390
    Chivatí, 287, **288,** 290, **301, 302,** 363, **364, 366,** 369
    colony movement, 299, 300, 302–305, **303**
    colony size, 49, 50, 287, 295, 296, 299–306, **302.** *See also* population size (mark-recapture)
    discovery (of overwintering colonies), 157, 162, 181, 289, 295
    duration, 49–51, 296, 302. *See also* overwintering (dispersal)
    elevation, 61, 157, 159, 253, 287, 290–292, 299, 300, **302,** 313
    Herrada, 300, **301–306**
    Huacal, **302–306,** 363, **364, 366**
    location, 49, 50, **259,** 287–293, 295, 296, 299, 300, **301,** 304
    Michoacán, 49, 107, 121, 159, **259,** 287, 289, 291, 295, 299, 300, **301,** 310, **316,** 323, 324, **355,** 377, 383, 389
    Nevado de Toluca, 299, 300, **301**
    Palomas, 300, **301–306**
    Picacho, 300, **301–306**
    Rosario. *See* overwintering (Mexico—Campanario)
    San Andres, **301**
    Sierra Chincua, 50, 159, 223, 224, 257, **259, 288,** 289–292, **301,** 310, 312, 315, **316, 319,** 322, 363, **364, 366,** 368, 369, 377
    Sierra Madre ranges, 295

overwintering *continued*
Sierra Madre ranges *continued*
site alpha. *See* overwintering (Mexico—Sierra Chincua)
Transvolcanic Mountains, 4, 49, 107, 157, 253, 269, 287, 289, 291, 295, 296, 299, **301, 305, 306**
vegetation, 287–292, 296, 310
Abies forest, 287–292, **290,** 365–373
Juniperus shrubland, 292, 365
mixed *Abies-Pinus* forest, 290–292, **291**
Pinus forest, 291
Potentilla meadow, 291, 292, **292,** 365
upland cloud forest, 289–290, **291**
weather, 157, 158, 287, 289, 290, **300,** 305, 306, 309, **310,** 313
mortality, 61, 304, 313, 348
movement, 348
numbers, 194, 196. *See also* overwintering (Mexico—colony size)
predation, 150, 197, 285, 300, 304, 309, 313, 315–321. *See also* predation, predators
reproductive state, 345–353, **347**
riparian corridor, 340
roosting, 348
transient roosts, 193–197, 295, 296, 304, 336
water, 61, 150, 157, 162, 209, 299, 302, 303, **305, 306.** *See also* physiology (water loss)
weather, 196, 275, 280, **281, 310**
weight, **325**
oviposition. *See* behavior (oviposition)
oxygen consumption. *See* metabolism
oyamel fir. *See Abies religiosa*

Pacific Grove. *See* California
Pacific Ocean, 5, 12, 180, 181, 184, 185, 198, **210**
palatability. *See* warning signals (taste)
palatability spectrum. *See* warning signals (taste)
palpation. *See* spermatophores
Panama, 15
panmixia, 14, 50. *See also* behavior (mating)
Papilio, **130, 133, 134,** 144, 148,
glaucus, **131,** 132
palamedes, 132
polyxenes asterius, 148
troilus, 132
Papilionidae, 79, **202,** 313
Papilionoidea, 11
Paralucia spinifera, 173
Parantica, 11, **12,** 20, 166, **167,**
aglea, **23, 24**
sita, 38
parasitism, 126, 197, 211. *See also* herbivory
parasitoids, 113, 171, 205,
hymenoptera, 113
tachinids, 113
Parsons, John, 4, 37
Parsonsia, 34, 80
Parus sclateri, **316**
passenger pigeon. *See Ectopistes migratorius*
Passifloraceae, 80
pathogens, 181
Peromyscus,
aztecus, 323–331
melanotis, 285, 323–331
Peru, 15
pharmacophagy, 21, 34, 38
pharmokinetic theory, 315
Phenacomys longicaudus, 330
phenylglycosides, 135, 136
pheromones, 4, 19–25, 135, 349. *See also* androconia, antennae, behavior (mating), communication, signals, pyrrolizidine alkaloids, physiology (reproductive),
odor bouquet, 24, 29, 31
receptor, 43
pheromone transfer particles (PTPs), 22, 24, 30, 32
Pheucticus melanocephalus, 315–321
Philippines, 166, 180
photoperiod. *See* physiology (day length)
phylogeny. *See* evolution
physiological ecology, 20
physiological time. *See* day-degrees
physiology, 4, 154. *See also* cardenolides, fat content, metabolism, pheromones, sequestration,
day length, 153, 190, 196, 198, 234, 235, **244,** 264, 269
desiccation. *See* physiology (water loss)
diapause, 143–144, 150, 153, 157, 161, 184, 189–198, **192,** 205, 215, 296, 299
hormones. *See* endocrine system
migration. *See* migration
nutrition, 97–102. *See also* fat content, herbivory (nutrition), sequestration, sterols
reproduction, 49, 56–58, 161, 189–198, **205,** 249, 261. *See also* behavior (mating), endocrine system, pheromones, spermatophores
gonad development, 57, 58, 62, 69, 75, 143–144, 150, 189, 193, 197, 198
physiology *continued*
reproduction *continued*
oogenesis, 189, 197, **205,** 235, 264, 269. *See also* oogenesis-flight syndrome
senescence, 143–144
temperature. *See* day-degrees, temperature
water loss, 143–144, 157, 300, 302, **305, 306**
Picea (Pinaceae) *sitchensis,* 339
Pieridae, 148, 166, 201, **202, 203,** 204
Pieris (Pieridae), 144,
brassicae, 73, 100
rapae, 70
pine flycatcher. *See Empidonax affinis*
Pinus (Pinaceae), 365, 367,
contorta, 339
hartwegii, 291
montezumae, 291
muricata, 338, 339
pseudostrobus, 290, 291
radiata, 338
rudis, 291
torreyana, 340
Pityophthorus blackmanii, 366
plant–animal interactions. *See* herbivory
plant-derived toxins. *See* cardenolides, pyrazines, pyrrolizidine alkaloids, sequestration
plant poisons. *See* toxins (plant), warning signals
Platanus racemosa, 340
Pleistocene, 181
plesiomorphy. *See* evolution
Pliocene, 180
poisons. *See* toxins
pollination, 20
Polygonia sp., 161
polygyny, 49, 58
population dynamics, **66, 67,** 179, 211, 358, 400. *See also* life history, fitness, population size, population structure
population size, 64, 194. *See also* overwintering (colony size), population dynamics,
mark-recapture, **71,** 74, 190, 193, 196, 197, 214, 219–231, **226, 229,** 234, 326, 345, 346, 349, 398
population structure, 69
Potentilla (Rosaceae), 287, 289, 291, 292
Poulton, E.B., 1, 4, 12–15
pre-adaptation. *See* evolution
precursors. *See* biosynthesis
predation, 1, 4, 102, 130, 285, 315–321, 323–333, 345. *See also* be-

predation *continued*
havior (predation), larvae (predation),
feeding experiments, 38, 39, 129–137, 316
foraging cycles, 315–321
dynamics, 323–333
rates, 319–321
temperature, 318, 319, **321**
gourmand-gourmet hypothesis, **1**, 134
immigration, 323
learned aversion, 131, 132, 135, 136
overwintering, 150, 197, 265, 285, 300
pattern, **326**
predator foraging, 4, 132, 136, 300, 304, 400
selective consumption, 323–333
by body parts, **324, 325,** 327
by sex, 317, 318, 320, 323, 325
predators, 4, 107, 121, 171, 197, 205, 234,
ants, 107, 108, 111, **112,** 118
birds, 19, 25, 37, 91, 121, 129–137, 150, 197, 264, 265, 300, 304, 313. *See also* by species
mammals, 197
mice, 150, 265, 285, 304, 309, 313. *See also Peromyscus*
spiders, 108, 111, **112,** 118
prey. *See* chemical defense (herbivore prey)
Pristiphora erichsonii, 315
protection. *See* chemical defense, warning signals
protein content, 193
Pseudohylesinus variegatus, 366, 373
Pseudotsuga menziesii, 330, 339
Protoploea, 11, 12, **12,** 166,
apatela, 12
Pterocomma smithia, 161
pupae, 14, 113,
gold markings, 14
pupation, 112, 113, 118
pyrazines, 4, 20, 21,
methoxy alkyl, 38
pyrrolizidine alkaloids (PAs), 4, 19–25, 34–39, 43–45, 58, 80, 135, 136, 331, 349, 351, 399,
carcinogenesis, 38
defensive use, 24–25, 36–39, 135
diesters, 22, 36
dihydropyrrolizines, 19, 31, 34, 36, 43–45
hydroxydanaidal, 19, 34, 43–45
danaidal, 19, 43–45
danaidone, 19, 32, 34, 43–45
methyl hydroxy-danaidoate, 22
monocrotaline, 331
monoesters, 22, 36
N-oxide, 331

pyrrolizidine alkaloids *continued*
sequestration, 19–25, 34–39, 135
toxicity, 38

queen butterfly. *See Danaus gilippus*
Quercus (Fagaceae), 291, 365, 367,
agrifolia, 340

Radena, 11, 12
range. *See* distribution
range expansion. *See* migration
rape. *See* behavior (mating)
rat. *See Rattus norvegicus*
Rattus norvegicus, 330
recolonization. *See* migration (spring)
red-spotted purple butterfly. *See Limenitis arthemis astyanax*
red tree mouse. *See Phenacomys longicaudus*
red-winged blackbird, 129–137
Regulus calendula, **316**
Reichstein, Tadeus, 4, 19, 37, 87
Reithrodontomys sumichrasti, 323–333
repellance, 21. *See also* behavior (feeding), warning signals
reproduction, 4, 49–59, 61–67, 189–198. *See also* behavior (mating), fitness, life history, physiology (reproduction), spermatophores
reproductive success. *See* fitness
resource use, 69, 300, 304. *See also* behavior (mating), herbivory, migration, nectar
Rhabdophaga sp., 161
Riodinidae, **202**
Rocky Mountains, 5, 157, 180, 219, 253, **254,** 256, **262, 263,** 278, 358, 359, 377
roosting. *See* overwintering (behavior)
Rothschild, Miriam, 4, 19–21, 29, 34, 81
ruby-crowned kinglet. *See Regulus calendula*
russet nightingale thrush. *See Catharus occidentalis*

Salatura. See Danaus
Salicaceae, 135, 136
salicin, 135. *See also* phenylglycosides
Salix,
babylonica, 130
caroliniana, 132
paradoxa, 365
Satyrinae, 148, 165, 202, 313
scent organs. *See* androconia
Schneider, Dietrich, 19–22, 135
Scolytus,
hermosus, 366
multistriatus, 161
mundus, 366, 372
ventralis, 366, 372, 373

Scott's oriole. *See Icterus parisorum*
scrub jay. *See Aphelocoma coerulescens*
Scudder hypothesis. *See* migration (range expansion)
seasonality. *See* migration, overwintering, distribution (temporal)
secondary metabolites, 4, 19, 39, 79, 99, 125, 126, 205. *See also* allelochemicals, cardenolides, cyanogenic glycosides, phenylglycosides, pyrazines, pyrrolizidine alkaloids, sequestration, warning signals
secondary plant substances. *See* secondary metabolites
Seiber, James N., 83, 86, 98
Seitz, 12
Senecio, 34, 135, 304, 310
sequestration, 1, 4, 19–25, 34–39, **77,** 79, 91–102, 135, 179, 399. *See also* cardenolides, herbivory, larvae, phenylglycosides, physiology, pyrrolizidine alkaloids, warning signals
Sequoia sempervirens, 339
sex. *See* behavior (mating)
sex ratio, 34, **52,** 69–75, **71–74,** 173, 189, 193, 196, 197, 201–203, 206, 211, 261, 324, 345, 349–352, **351,**
density dependence, 74
sexual selection, 49. *See also* behavior (mating), physiology (reproductive)
shipping, 180, 181
shore pine. *See Pinus contorta*
Sialia mexicana, **316**
Sierra Chincua. *See* overwintering (Mexico)
Sierra Madre ranges. *See* overwintering (Mexico)
signals. *See* communication signals, warning signals
single sweep migration hypothesis. *See* migration
Siproeta (Nymphalinae) *stelenes biplagiata,* 204, 206
sister-group, 11, 12
Sitka spruce. *See Picea sitchensis*
Sitobion avenae, 159
size. *See* morphology
snow, 209. *See also* overwintering, temperature, weather
sociobiology, 20
socioeconomic development, 386
soil erosion, 359, 389
Solanaceae, 80
Solidago, 277
Solomons, 166, 167, 180
South Africa, 94, 171
Spain, 179–181, **184**

spectra,
chemical ionization/mass (CI/MS), 92, 93
nuclear magnetic resonance (NMR), 92
spectrophotometry, 92, 256
spermatophores, 21, 24, 25, 36, 49–59, 61–67, **65,** 204, **205,** 256, 264, 346, 351, 399. *See also* behavior (mating), bursa copulatrix, physiology (reproductive),
number. *See* behavior (mating—multiple)
nutrient transfer, 49, 50, 57, 58, 61–67, **64, 65,** 69, 75, 399
sperm precedence, 58, 304
Sri Lanka, 24
starvation. *See* food supply
Steller's jay. *See Cyanocitta stelleri*
Stephanotis. See Asclepiadaceae
steroids. *See* cardenolides
sterols, 96–98. *See also* fat content,
b-sitosterol, 97, 98
campesterol, 97
cholesterol, 97, **98**
phytosterols, 97, 98
storage. *See* cardenolides (storage), sequestration
succession. *See* host plant succession
successive brood migration hypothesis. *See* migration
Sulawesi, 180
Sumatra, 166
survivorship. *See* fitness
swallowtail butterfly. *See Papilio*
sycamore. *See Platanus racemosa*
symbiotes, 96
systematics. *See* evolution (systematics)

tagging. *See* population size (mark-recapture)
Taiwan, 180
taste. *See* warning signals
Tellervinae, 11, 80
temperate. *See* distribution
temperature, 147–154, 157–163, 171, 183, 189–198, 209, 233–250, 269, 300, 309–313, 315–321, 345. *See also* day-degrees,
abdomen, 159, 310
altitude, 149, 300, 310
ambient, 147–154, 189–198, **195,** 211, 233–250, 309
body, 189–198
coloration, 148, 152
cooling, **153**
constraints, 147, **148,** 233–250, **244,** 300
crawling threshold, 312
cues, 189–198, 233–250, 264, 269, 300

temperature *continued*
ectothermy, 147, 148, 152, 153
endothermy, 147, 148, 150, 153, 195, 313
flight threshold, **153,** 300, 309, 310, 313
freeze protection, 157–163, 397
freezing point, 158, 310
supercooling, 157–163, **160, 161**
thermal hysteresis, 158–162, **160**
freezing, 62, 147–154, 157–163, 189, 209, **210,** 213, 215, 237, 264, 269, 300, 309. *See also* overwintering
heating, **153,** 309–313
insolation (solar radiation), 148, 151, 157
insulation, 148, 152
lethal, **149,** 153, 235, **244,** 250, 265, 309
migration. *See* migration (temperature)
morphology, 147–154. *See also* morphology
overwintering, 61, 62, **141,** 147–154, **151,** 157–163, **162,** 189–198, 287, 309–313. *See also* overwintering
clustering, **162**
crawling, 148, **149,** 150, 151, 309–313, **310–313**
drinking, **162,** 300
flying, 148, **149,** 150, **162,** 309, 312, 313
nectaring, **162,** 300
shade, **154**
shivering, 148, **149,** 150, 151, 153, **154,** 309–313, **311–313**
wing position, 310–312, **311, 312**
polyphenism, 149, 153
predator foraging. *See* predation (predator foraging)
spring, **161,** 233–250
summer breeding, 147–154, **151,** 159–162, **160, 161,** 233–250
thermoregulation, **141,** 147–154, 194, 195, 219, 309–313, 397
basking, 148, 152, 153, **154,** 235, 247. *See also* behavior (basking)
breeding, 147–154
migrating, 157–163
overwintering, 147–154, 309–313
thorax, 147–154, **153, 154,** 159, 194, 309, **312,** 313
territoriality, 23
Tetrahymena corlissi, 98
Thaxter, 341
The Monarch Project, 335, 359, 393, 394
The Nature Conservancy, 381, 394
thermals, 282

thermoregulation. *See* temperature (thermoregulation)
thoracic ganglia, 313
threatened phenomenon. *See* conservation
tiger swallowtail butterfly. *See Papilio glaucus*
Timor, 167
Tiradelphe, **12,** 166
Tirumala, **12,** 19, 22, 23, 165, 166, **167,** 169,
hamata, 165, 166, 169, **170, 171,** 172, **173,** 345–353, **352**
limniace, **23, 24**
petiverana, **17**
septentrionis, **23, 24**
TNDP (2,2′ 4,4′-tetranitrodiphenyl), 92, 256
Torrey pine. *See Pinus torreya*
tough-bodied. *See* warning coloration
tourism, 359, 378, 386, 390. *See also* conservation
Townsend's warbler. *See Dendroica townsendi*
toxins, 4, 19–21, 38, 80, 121, 125. *See also* cardenolides, pyrrolizidine alkaloids, warning signals,
emetic response. *See* cardenolides
insect. *See* sequestration
plant, 19–21
plant-derived. *See* sequestration
Trachelospermum. See Apocynaceae
Transvolcanic Mountains. *See* Mexico, overwintering (Mexico)
trichogen cells. *See* androconia
trophic levels, 4, 99–102, **101,** 398, 400
tropical. *See* distribution

UNESCO/PNUMA, 383
unpalatability. *See* warning signals (taste)
Urquhart, Fred A., ix, 4, 33, 34, 108, 209, 210, 214–216, 219, 234, 249, 253–255, 280, 295, 397
U.S.A., **177,** 185, **254,** 377, 389. *See also* America (North), California,
Arkansas, 180, **238, 240, 242,** 247
Blue Ridge Mountains, 277, 280, **282**
California. *See* California
coastal plain, 260, 263
Colorado, 278
Congress, 381
Connecticut, **240**
Delaware, **240, 243**
District of Columbia, 181
eastern, 107, 119, 129, 233–250, **236,** 253–265, **260, 261, 263,** 269, 275
Florida, 108, 118, 121, 129–137,

U.S.A. *continued*
Florida *continued*
150, 152, 153, 159, **160,** 179, 233, 235, 237, **238, 240, 242, 244,** 245, **254,** 256, 257, **258, 259,** 260, 263, 269, 271, 296
Georgia, **238, 240, 242,** 245
Great Plains, 118, 263, 358
Gulf of Mexico coast, 107, 150, 159, 180, 233, 245, 263, 272
Idaho, 209, 210
Illinois, 241
Indiana, 241
Iowa, 237, **238,** 241, **242,** 246, 247
Kansas, 108, 118, 120, 121, 237, **238,** 241, **242,** 261
Kentucky, 241, 261, 361
Louisiana, 107–122, 235, **238, 240, 242,** 256, 257, **259,** 262, 269–273
Maryland, 121
Massachusetts, 121, 126, 127, 235, **238, 240, 242, 243,** 256, **259,** 261, 262, 277, **279**
Michigan, 150, 153, 235, **238, 240, 243,** 256, **259,** 261, **262,** 263, 264
Minnesota, 144, 233, 235, **238, 240, 242, 243,** 246, 247, 256, **258, 259,** 261, 263, 264
Missouri, 181, 237, **238,** 241, **242,** 247
Montana, 209, 210
Nebraska, **238, 242**
New Jersey, **240**
New York, **240,** 246, 277, 279
North Carolina, 235, 237, **244,** 277, 278, 281
North Dakota, **240, 242,** 256, **258, 259,** 263, 264
northern, **177,** 253–265, **254, 257, 260–263**
Ohio, 235, 237, **238, 242–244,** 248, 256, **259,** 261, 263
Oklahoma, 108, 118, 120, 121, **238, 240, 242,** 262, 264
Oregon, 209, 339
Pennsylvania, 181, **238, 240, 243,** 256, **259,** 263
southern, 107, 112, 149, **177,** 253–265, **254, 257, 260–263,** 399
Texas, 107–122, **177,** 235, 237, **238, 240,** 241, **242,** 243, 245, 246,

U.S.A. *continued*
Texas *continued*
247, **250,** 256, 257, **259,** 260, **262,** 264, 269
Virginia, **240,** 241, **258**
Washington, 209
western, 108, 112, **254**
West Virginia, 241
Wisconsin, 125–127, **151,** 152, **177,** 235, 237, **238, 240,** 241, **242–244,** 246, 248, 256, **259,** 261, **262,** 263, 264
Utetheisa ornatrix, 36

Vanessa (Nymphalidae), 144,
atalanta rubria, 148
urticae, 161
virginiensis, 148
viceroy butterfly. *See Limenitis archippus*
vitellogenesis. *See* physiology (reproductive—gonad development)
Vizcaíno, 343
volatiles. *See* androconia
volcanic highlands. *See* Transvolcanic Mountains
voltinism. *See* life history (generations)
von Kotzebue, **342,** 343

Walker hypothesis. *See* migration (range expansion)
wanderer (monarch) butterfly. *See Danaus (Danaus) plexippus plexippus*
warning signals, 19,
aposematism, 19–21, 79–81, 129–137, 264
larvae, 81
coloration, 1, 11, 19–21, 129–137, 181
emesis. *See* cardenolides (emetic response)
odor, 4, 19–21, 38. *See also* sequestration
taste, 19–21, 36–39, 79–81, 121, 125, 129–137, 323, 324, 330, 331. *See also* sequestration
bitterness, 135, 331
palatability spectrum, 129–137. *See also* chemical defense (herbivore prey)
water content. *See* overwintering (water), physiology (water loss)
watershed, 359
weather, 193, 196, 201, 202, 204, 206, 209–217, **210,** 275, 280, **281,** 309, 349, 390. *See also* migration, overwintering, temperature
weeping willow. *See Salix babylonica*
weight, 49–59, **52–57,** 107, 193, 299–301. *See also* behavior (mating), morphology (size/wing length), overwintering (water/weight),
change, **326,** 328
dry, 193, 299, 302–306, **304, 305**
gain, **325,** 326
lean, 193, 300, 303
loss, 323, 326, 330, 331
wet, 193, 299, 302
western bluebird. *See Sialia mexicana*
white-breasted wood swallow. *See Artamus leucorynchus*
white monarch. *See Danaus (Danaus) plexippus plexippus* 'nivosus', Hawaii
Whittaker and Feeny. *See* allelochemicals
Williams, C.B., 209, 214, 215
willow. *See* Salicaceae
wind. *See also* migration (wind influence),
trade winds, 202. *See also* Costa Rica
wing condition, 49–59, **52–57, 177,** 196, 197, 202, 204, **205,** 206, 210, 214, 235, 237, 253–265, **263,** 270, 304, 346, 347. *See also* behavior (mating age)
wing glands. *See* androconia
wing length. *See* behavior (mating—size), morphology (size/wing length), weight
wing load, 304, 305
wing scales, 58. *See also* androconia
wing veins. *See* androconia
wing wear. *See* wing condition
winter. *See* overwintering, temperature
Worldwide Fund for Nature. *See* World Wildlife Fund
World Wildlife Fund (WWF), 335, 378, 383, 390, 393

Xerces Society. *See* The Monarch Project

Zeiraphera diniana, 161